DATA ANALYSIS METH
IN PHYSICAL OCEANOGR. .. ...

THIRD EDITION

DATA ANALYSIS METHODS IN PHYSICAL OCEANOGRAPHY

THIRD EDITION

RICHARD E. THOMSON
Fisheries and Oceans Canada
Institute of Ocean Sciences
Sidney, British Columbia
Canada

and

WILLIAM J. EMERY
University of Colorado
Aerospace Engineering Sciences Department
Boulder, CO
USA

ELSEVIER

AMSTERDAM • BOSTON • HEIDELBERG • LONDON • NEW YORK • OXFORD
PARIS • SAN DIEGO • SAN FRANCISCO • SINGAPORE • SYDNEY • TOKYO

Elsevier
225, Wyman Street, Waltham, MA 02451, USA
The Boulevard, Langford Lane, Kidlington, Oxford OX5 1GB, UK
Radarweg 29, PO Box 211, 1000 AE Amsterdam, The Netherlands

Notice
No responsibility is assumed by the publisher for any injury and/or damage to persons or property as
a matter of products liability, negligence or otherwise, or from any use or operation of any methods,
products, instructions or ideas contained in the material herein. Because of rapid advances in the medical
sciences, in particular, independent verification of diagnoses and drug dosages should be made.

British Library Cataloguing in Publication Data
A catalogue record for this book is available from the British Library

Library of Congress Cataloging-in-Publication Data
A catalog record for this book is available from the Library of Congress

ISBN: 978-0-12-387782-6

For information on all **Elsevier** publications visit our
web site at http://store.elsevier.com/

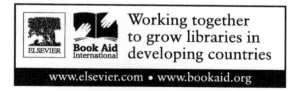

Working together
to grow libraries in
developing countries

www.elsevier.com • www.bookaid.org

- Coastal photo courtesy of Dr. Audrey Dallimore, School of Environment and Sustainability, Royal Roads
  University, British Columbia, Canada
- Satellite image from Thomson, R.E., and J.F.R. Gower. 1998. A basin-scale oceanic instability event in the
  Gulf of Alaska. J. Geophys. Res., 103, 3033-3040

Dedication

Richard Thomson dedicates this book to his wife Irma, daughters Justine and Karen, and grandchildren Brenden and Nicholas.

Bill Emery dedicates this book to his wife Dora Emery, his children Alysse, Eric, and Micah, and to his grandchildren Margot and Elliot.

Contents

Preface

There have been numerous books written on data analysis methods in the physical sciences over the past several decades. Most of these books are heavily directed toward the more theoretical aspects of data processing or narrowly focus on one particular topic. Few books span the range from basic data sampling and statistical analysis to more modern techniques such as wavelet analysis, rotary spectral decomposition, Kalman filtering, and self-organizing maps. Texts that also provide detailed information on the sensor and instruments that collect the data are even more rare. In writing this book we saw a clear need for a practical reference volume for earth and ocean sciences that brings established and modern processing techniques together under a single cover. The text is intended for students and established scientists alike. For the most part, graduate programs in oceanography have some form of methods course in which students learn about the measurement, calibration, processing, and interpretation of geophysical data. The classes are intended to give the students needed experience in both the logistics of data collection and the practical problems of data processing and analysis. Because the class material generally is based on the experience of the faculty members giving the course, each class emphasizes different aspects of data collection and analysis. Formalism and presentation can differ widely. While it is valuable to learn from the first-hand experiences of the class instructor, it seemed to us important to have available a central reference text that could be used to provide some uniformity in the material being covered within the oceanographic community. This 3rd Edition provides a much needed update on oceanographic instrumentation and data processing methods that have become more widely available over the past decade.

Many of the data analysis techniques most useful to oceanographers can be found in books and journals covering a wide variety of topics. Much of the technical information on these techniques is detailed in texts on numerical methods, time series analysis, and statistical methods. In this book, we attempt to bring together many of the key data processing methods found in the literature, as well as add new information on spatial and temporal data analysis techniques that were not readily available in older texts. Chapter 1 also provides a description of most of the instruments used in physical oceanography today. This is not a straightforward task given the rapidly changing technology for both remote and in situ oceanic sensors, and the ever-accelerating rate of data collection and transmission. Our hope is that this book will provide instructional material for students in the marine sciences and serve as a general reference volume for those directly involved with oceanographic and other branches of geophysical research.

The broad scope and rapidly evolving nature of oceanographic sciences has meant that it has not been possible for us to fully detail all existing instrumentation or emerging data analysis methods. However, we believe that many of the methods and procedures outlined in this book will provide a basic understanding of the kinds of options available to the user for interpretation of data sets. Our intention is to describe general statistical and analytical methods that will be sufficiently fundamental to maintain a high level of utility over the years.

Finally, we also believe that the analysis procedures discussed in this book apply to a wide readership in the geophysical sciences. As with oceanographers, this wider community of scientists would likely benefit from a central source of information that encompasses not only a description of the mathematical methods, but also considers some of the practical aspects of data analyses. It is this synthesis between theoretical insight and the logistical limitations of real data measurement that is a primary goal of this text.

Richard E. Thomson and William J. Emery
North Saanich, British Columbia
and Boulder, Colorado

Acknowledgments

Many people have contributed to the three editions of this book over the years. Dudley Chelton of Oregon State University and Alexander Rabinovich of Moscow State University and the Institute of Ocean Sciences have helped with several chapters. Dudley proved to be a most impressive reviewer and Sasha has provided figures that have significantly improved the book. For this edition, we also thank friends and colleagues who took time from their research to review sections of the text or to provide suggestions for new material. There were others, far too numerous to mention, whose comments and words of advice have added to the usefulness of the text. We thank Andrew Bennett of Oregon State University for reviewing the section on inverse methods in Chapter 4, Brenda Burd of Ecostat Research for reviewing the bootstrap method in Chapter 3, and Steve Mihály of Ocean Networks Canada for assisting with the new section on self-organizing maps in Chapter 4. Roy Hourston and Maxim Krassovski of the Institute of Ocean Sciences helped generously with various sections of the book, including new sections on regime shifts and wavelet analysis in Chapter 5. The contributions to Chapter 1 from Tamás Juhász and David Spear, two accomplished technicians at the Institute of Ocean Sciences, are gratefully acknowledged. Patricia Kimber of Tango Design helped draft many of the figures.

Expert contributions to the third edition—including reports of errors and omissions in the previous editions—were also provided by Michael Foreman, Robie Macdonald, Isaac Fine, Joseph Linguanti, Ron Lindsay, Germaine Gatien, Steve Romaine, and Lucius Perreault (Institute of Ocean Sciences, Fisheries and Oceans Canada), Philip Woodworth (Permanent Service for Mean Sea Level, United Kingdom), William (Bill) Woodward (President and CEO of CLS America, Inc., USA), Richard Lumpkin (NOAA/AOML, USA), Laurence Breaker (California State University, USA), Jo Suijlen (National Institute for Coastal and Marine Management, The Netherlands), Guohong Fang (First Institute of Oceanography, China), Vlado Malačič (National Institute of Biology, Slovenia), Ųyvind Knutsen (University of Bergen, Norway), Parker MacCready (University of Washington, USA), Andrew Slater (University of Colorado, USA), David Dixon (Plymouth, United Kingdom), Drew Lucas (Scripps Institute of Oceanography, USA), Wayne Martin (University of Washington, USA), David Ciochetto (Dalhousie University, Canada), Alan Plueddemann (Woods Hole Oceanographic Institution, USA), Fabien Durand (IRD/LEGOS, Toulouse, France), Jack Harlan (NOAA, Boulder Colorado, USA), Denis Gilbert (The Maurice Lamontagne Institute, Fisheries and Oceans Canada), Igor Yashayaev (Bedford Institute of Oceanography, Fisheries and Oceans Canada), Ben Hamlington (University of Colorado, USA), John Hunter (University of Tasmania, Australia), Irene Alonso (Instituto de Cienciias Mrinas de Andalucia, Spain), Yonggang Liu (University of South Florida, USA), Gary Borstad (ASL Environmental Sciences Ltd., Canada), Earl Davis and Bob Meldrum (Pacific Geosciences Centre, Canada), and Mohammand Bahmanpour (University of Western Australia, Australia).

1

Data Acquisition and Recording

1.1 INTRODUCTION

Physical oceanography is an ever-evolving science in which the instruments, types of observations, and methods of analysis undergo continuous advancement and refinement. The changes that have occurred since we completed the 2nd Edition of this book over a decade ago have been impressive. Recent progress in oceanographic theory, instrumentation, sensor platforms, and software development has led to significant advances in marine science and the way that the findings are presented. The advent of digital computers has revolutionized data collection procedures and the way that data are reduced and analyzed. No longer is the individual scientist personally familiar with each data point and its contribution to his or her study. Instrumentation and data collection are moving out of direct application by the scientist and into the hands of skilled technicians who are becoming increasingly more specialized in the operation and maintenance of equipment. New electronic instruments operate at data rates and storage capacity not possible with earlier mechanical devices and produce volumes of information that can only be handled by high-speed computers. Most modern data collection systems transmit sensor data directly to computer-based

data acquisition systems where they are stored in digital format on some type of electronic medium such as hard drives, flash cards, or optical disks. High-speed analog-to-digital converters and digital-signal-processors are now used to convert voltage or current signals from sensors to digital values. Increasing numbers of cabled observatories extending into the deep ocean through shore stations are now providing high-bandwidth data flow in near real time supported by previously impossible sustained levels of power and storage capacity. As funding for research vessels diminishes and existing fleets continue to age, open-ocean studies are gradually being assumed by satellites, gliders, pop-up drifters, and long-term moorings. The days of limited power supply, insufficient data storage space, and weeks at sea on ships collecting routine survey data may soon be a thing of the past. Ships will still be needed but their role will be more focused on process-related studies and the deployment, servicing, and recovery of oceanographic and meteorological equipment, including sensor packages incorporated in cabled observatory networks. All of these developments are moving physical oceanographers into analysts of what is becoming known as "big data". Faced with large volumes of information, the challenge to oceanographers is deciding how to approach these mega data and how to select the measurements and numerical simulations that are most relevant to the problems of interest. One of the goals of this book is to provide insight into the analyses of the ever-growing volume of oceanographic data in order to assist the practitioner in deciding where to invest his/her effort.

With the many technological advances taking place, it is important for marine scientists to be aware of both the capabilities and limitations of their sampling equipment. This requires a basic understanding of the sensors, the recording systems and the data processing tools. If these are known and the experiment carefully planned, many problems commonly encountered during the processing stage can be avoided. We cannot overemphasize the need for thoughtful experimental planning and proper calibration of all oceanographic sensors. If instruments are not in near-optimal locations or the researcher is unsure of the values coming out of the machines, then it will be difficult to believe the results gathered in the field. To be truly reliable, instruments should be calibrated on a regular basis at intervals determined by use and the susceptibility of the sensor to drift. More specifically, the output from all oceanic instruments such as thermometers, pressure sensors, dissolved oxygen probes, and fixed pathlength transmissometers drift with time and need to be calibrated before and after each field deployment. For example, the zero point for the Paroscientific Digiquartz (0–10,000 psi) pressure sensors used in the Hawaii Ocean Time-series at station "Aloha" 100 km north of Honolulu drifts about 4 dbar in three years. As a consequence, the sensors are calibrated about every six months against a Paroscientific laboratory standard, which is recalibrated periodically at special calibration facilities in the United States (Lukas, 1994). Even the most reliable platinum thermometers—the backbone of temperature measurement in marine sciences—can drift of order 0.001 °C over a year. Our shipboard experience also shows that

opportunistic over-the-side field calibrations during oceanic surveys can be highly valuable to others in the science community regardless of whether the work is specific to one's own research program. As we discuss in the following chapters, there are a number of fundamental requirements to be considered when planning the collection of field records, including such basic considerations as the sampling interval, sampling duration, and sampling location.

It is the purpose of this chapter to review many of the standard instruments and measurement techniques used in physical oceanography in order to provide the reader with a common understanding of both the utility and limitations of the resulting measurements. The discussion is not intended to serve as a detailed "user's manual" nor as an "observer's handbook". Rather, our purpose is to describe the fundamentals of the instruments in order to give some insight into the data they collect. An understanding of the basic observational concepts, and their limitations, is a prerequisite for the development of methods, techniques, and procedures used to analyze and interpret the data that are collected.

Rather than treat each measurement tool individually, we have attempted to group them into generic classes and to limit our discussion to common features of the particular instruments and associated techniques. Specific references to particular company's products and the quotation of manufacturer's engineering specifications have been avoided whenever possible. Instead, we refer to published material addressing the measurement systems or the data recorded by them. Those studies that compare measurements made by similar instruments are particularly valuable. On the other hand, there are companies whose products have become the "gold standard" against which other manufacturers are compared. Reliability and service are critical factors in the choice of any instrument. The emphasis of the instrument review section is to give the reader a background in the collection of data in physical oceanography. For those readers interested in more complete information regarding a specific instrument or measurement technique, we refer to the references at the end of the book where we list the sources of the material quoted. We realize that, in terms of specific measurement systems, and their review, this text will be quickly dated as new and better systems evolve. Still, we hope that the general outline we present for accuracy, precision, and data coverage will serve as a useful guide to the employment of newer instruments and methods.

1.2 BASIC SAMPLING REQUIREMENTS

A primary concern in most observational work is the accuracy of the measurement device, a common performance statistic for the instrument. Absolute accuracy requires frequent instrument calibration to detect and correct for any shifts in behavior. The inconvenience of frequent calibration often causes the scientist to substitute instrument precision as the measurement capability of an instrument. Unlike absolute accuracy, precision is a relative term and simply represents the ability of the instrument to repeat the observation without deviation. Absolute accuracy further requires that the observation be consistent in magnitude with some universally accepted reference standard. In most cases, the user must be satisfied with having good precision and repeatability of the measurement rather than having absolute measurement accuracy. Any instrument that fails to maintain its precision, fails to provide data that can be handled in any meaningful statistical fashion. The best instruments are those that provide both high precision and defensible absolute accuracy. It is sometimes advantageous to measure simultaneously the same variable with more than one reliable instrument. However, if the instruments have the same precision but not the same absolute accuracy, we are reminded

of the saying that "a man with two watches does not know the time".

Digital instrument resolution is measured in bits, where a resolution of N bits means that the full range of the sensor is partitioned into 2^N equal segments ($N = 1, 2...$). For example, eight-bit resolution means that the specified full-scale range of the sensor, say $V = 10$ V, is divided into $2^8 = 256$ increments, with a bit resolution of $V/256 = 0.039$ V. Whether the instrument can actually measure to a resolution or accuracy of $V/2^N$ units is another matter. The sensor range can always be divided into an increasing number of smaller increments but eventually one reaches a point where the value of each bit is buried in the noise level of the sensor and is no longer significant.

1.2.1 Sampling Interval

Assuming the instrument selected can produce reliable and useful data, the next highest priority sampling requirement is that the measurements be collected often enough in space and time to resolve the phenomena of interest. For example, in the days when oceanographers were only interested in the mean stratification of the world ocean, water property profiles from discrete-level hydrographic (bottle) casts were adequate to resolve the general vertical density structure. On the other hand, these same discrete-level profiles failed to resolve the detailed structure associated with interleaving and mixing processes, including those associated with thermohaline staircases (salt fingering and diffusive convection), that now are resolved by the rapid vertical sampling provided by modern conductivity-temperature-depth (CTD) probes. The need for higher resolution assumes that the oceanographer has some prior knowledge of the process of interest. Often this prior knowledge has been collected with instruments incapable of resolving the true variability and may, therefore, only be suggested by highly aliased (distorted) data collected using earlier techniques. In addition, laboratory and theoretical studies may provide information on the scales that must be resolved by the measurement system.

For discrete digital data $x(t_i)$ measured at times t_i, the choice of the sampling increment Δt (or Δx in the case of spatial measurements) is the quantity of importance. In essence, we want to sample often enough that we can pick out the highest frequency component of interest in the time series but not oversample so that we fill up the data storage file, use up all the battery power, or become swamped with unnecessary data. In the case of real-time cabled observatories, it is also possible to sample so rapidly (hundreds of times per second) that inserting the essential time stamps in the data string can disrupt the cadence of the record. We might also want to sample at irregular intervals to avoid built-in bias in our sampling scheme. If the sampling interval is too large to resolve higher frequency components, it becomes necessary to suppress these components during sampling using a sensor whose response is limited to frequencies equal to that of the sampling frequency. As we discuss in our section on processing satellite-tracked drifter data, these lessons are often learned too late—after the buoys have been cast adrift in the sea.

The important aspect to keep in mind is that, for a given sampling interval Δt, the highest frequency we can hope to resolve is the *Nyquist* (or *folding*) *frequency*, f_N, defined as

$$f_N = 1/2\Delta t \qquad (1.1)$$

We cannot resolve any higher frequencies than this. For example, if we sample every 10 h, the highest frequency we can hope to see in the data is $f_N = 0.05$ cph (cycles per hour). Equation (1.1) states the obvious—that it takes at least two sampling intervals (or three data points) to resolve a sinusoidal-type oscillation with period $1/f_N$ (Figure 1.1). In practice, we need to contend with noise and sampling errors so that it takes something like three or more sampling

increments (i.e., $\geq$ four data points) to accurately determine the highest observable frequency. Thus, f_N is an upper limit. The highest frequency we can resolve for a sampling of $\Delta t = 10$ h in Figure 1.1 is closer to $1/(3\Delta t) \approx 0.033$ cph. (Replacing Δt with Δx in the case of spatial sampling increments allows us to interpret these limitations in terms of the highest wavenumber (*Nyquist wavenumber*) the data are able to resolve.)

An important consequence of Eqn (1.1) is the problem *of aliasing*. In particular, if there is energy at frequencies $f > f_N$—which we obviously cannot resolve because of the Δt we picked—this energy gets folded back into the range of frequencies, $f < f_N$, which we are attempting to resolve (hence, the alternate name "folding frequency" for f_N). This unresolved energy does not disappear but gets redistributed within the frequency range of interest. To make matters worse, the folded-back energy is disguised (or aliased) within frequency components different from those of its origin. We cannot distinguish this folded-back energy from that which actually belongs to the lower frequencies. Thus, we end up with erroneous (aliased) estimates of the spectral energy variance over the resolvable range of frequencies. An example of highly aliased data would be current meter data collected using 13-h sampling in a region dominated by strong semidiurnal (12.42-h period) tidal currents. More will be said on this topic in Chapter 5.

As a general rule, one should plan a measurement program based on the frequencies and wavenumbers (estimated from the corresponding periods and wavelengths) of the parameters of interest over the study domain. This requirement may then dictate the selection of the measurement tool or technique. If the instrument cannot sample rapidly enough to resolve the frequencies of concern it should not be used. It should be emphasized that the Nyquist frequency concept applies to both time and space and the Nyquist wavenumber is a valid means

FIGURE 1.1 Plot of the function $F(n) = \sin(2\pi n/20 + \phi)$ where time is given by the integer $n = -1, 0, \ldots, 24$. The period $2\Delta t = 1/f_N$ is 20 units and ϕ is a random phase with a small magnitude in the range ± 0.1 radians. Open circles denote measured points and solid points the curve $F(n)$. Noise makes it necessary to use more than three data values to accurately define the oscillation period.

of determining the fundamental wavelength that must be sampled.

1.2.2 Sampling Duration

The next concern is that one samples long enough to establish a statistically significant determination of the process being studied. For time-series measurements, this amounts to a requirement that the data be collected over a period sufficiently long that repeated cycles of the phenomenon are observed. This also applies to spatial sampling where statistical considerations require a large enough sample to define multiple cycles of the process being studied. Again, the requirement places basic limitations on the instrument selected for use. If the equipment cannot continuously collect the data needed for the length of time required to resolve repeated cycles of the process, it is not well suited to the measurement required.

Consider the duration of the sampling at time step Δt. The longer we make the record the better we are to resolve different frequency components in the data. In the case of spatially separated data, Δx, resolution increases with

increased spatial coverage of the data. It is the total record length $T = N\Delta t$ obtained for N data samples that: (1) determines the lowest frequency *(the fundamental frequency)*

$$f_0 = 1/(N\Delta t) = 1/T \qquad (1.2)$$

that can be extracted from the time-series record; (2) determines the frequency resolution or minimum difference in frequency $\Delta f = |f_2 - f_1| = 1/(N\Delta t)$ that can be resolved between adjoining frequency components, f_1 and f_2 (Figure 1.2); and (3) determines the amount of band averaging (averaging of adjacent frequency bands) that can be applied to enhance the statistical significance of individual spectral estimates. In Figure 1.2, the two separate waveforms of equal amplitude but different frequency produce a single spectrum. The two frequencies are well resolved for $\Delta f = 2/(N\Delta t)$ and $3/(2N\Delta t)$, just resolved for $\Delta f = 1/(N\Delta t)$, and not resolved for $\Delta f = 1/(2N\Delta t)$.

In theory, we should be able to resolve all frequency components, f, in the frequency range $f_0 \le f \le f_N$, where f_N and f_0 are defined by Eqns (1.1) and (1.2), respectively. Herein lies a classic sampling problem. In order to resolve the

FIGURE 1.2 Spectral peaks of two separate waveforms of equal amplitude and frequencies f_1 and f_2 (dashed and thin line) together with the calculated spectrum (solid line). (a) and (b) are well-resolved spectra; (c) just resolved spectra; and (d) not resolved. Thick solid line is total spectrum for two underlying signals with slightly different peak frequencies.

frequencies of interest in a time series, we need to sample for a long time (T large) so that f_0 covers the low end of the frequency spectrum and Δf is small (frequency resolution is high). At the same time, we would like to sample sufficiently rapidly (Δt small) so that f_N extends beyond all frequency components with significant spectral energy. Unfortunately, the longer and more rapidly we want to sample, the more data we need to collect and store, the more power we need to provide, and the more time, effort, and money we need to put into the sensor design and sampling program.

Our ability to resolve frequency components follows from Rayleigh's criterion for the resolution of adjacent spectral peaks in light shone onto a diffraction grating. It states that two adjacent frequency components are just resolved when the peaks of the spectra are separated by frequency difference $\Delta f = f_0 = 1/(N\Delta t)$ (Figure 1.2). For example, to separate the spectral peak associated with the lunar–solar semidiurnal tidal component M_2 (frequency = 0.08051 cph) from that of the solar semidiurnal

tidal component S_2 (0.08333 cph), for which $\Delta f = 0.00282$ cph, it requires $N = 355$ data points at a sampling interval $\Delta t = 1$ h or $N = 71$ data points at $\Delta t = 5$ h. Similarly, a total of 328 data values at 1-h sampling are needed to separate the two main diurnal constituents K_1 and O_1 ($\Delta f = 0.00305$ cph). Note that since f_N is the highest frequency we can measure and f_0 is the limit of our frequency resolution, then

$$f_N/f_0 = (1/2\Delta t)/(1/N\Delta t) = N/2 \qquad (1.3)$$

is the maximum number of Fourier components we can hope to estimate in any analysis.

1.2.3 Sampling Accuracy

According to the two previous sections, we need to sample long enough and often enough if we hope to resolve the range of scales of interest in the variables we are measuring. It is intuitively obvious that we also need to sample as accurately as possible—with the degree of recording accuracy determined by the response characteristics of the sensors, the number of bits per data record (or parameter value) needed to raise measurement values above background noise, and the volume of data we can live with. There is no use attempting to sample the high end of the spectrum if the instrument cannot respond sufficiently rapidly or accurately to resolve changes in the parameter being measured. (A tell-tale sign that an instrument has reached its limit of resolution is a flattening of the high-frequency end of the power spectrum; the frequency at which the spectrum of a measured parameter begins to flatten out as a function of increasing frequency typically marks the point where the accuracy of the instrument measurements is beginning to fall below the noise threshold.) In addition, there are several approaches to this aspect of data sampling including the brute-force approach in which we measure as often as we can at the degree of accuracy available and then improve the statistical reliability of each data record through postsurvey averaging, smoothing, and other manipulation. This is the case for observations provided through shore-powered, fiber-optic, cabled observatories such as the ALOHA observatory located 100 km north of the island of Oahu (Hawaii), the Monterey Accelerated Research System (MARS) in Monterey Canyon, California, the Ocean Networks Canada cabled observatory systems Victoria Experimental Network Under the Sea (VENUS), and North-East Pacific Time Series Underwater Networked Experiments (NEPTUNE) extending from the Strait of Georgia to the continental margin and Cascadia Basin out to the Juan de Fuca Ridge off the west coast of British Columbia, the Ocean Observatories Initiative (OOI) Regional Scale Nodes off the coasts of Oregon and Washington in the Pacific Northwest of the United States (including Axial Seamount), and the Dense Oceanfloor Network System for Earthquakes and Tsunamis off the east coast of Japan. Data can be sampled as rapidly as possible and the data processing left to the postacquisition stage at the onshore data management facility.

1.2.4 Burst Sampling vs Continuous Sampling

Regularly spaced, digital time series can be obtained in two different ways. The most common approach is to use a *continuous sampling mode*, in which the data are sampled at equally spaced intervals $t_k = t_0 + k\Delta t$ from the start time t_0. Here, k is a positive integer. Regardless of whether the equally spaced data have undergone internal averaging or decimation using algorithms built into the machine, the output to the data storage file is a series of individual samples at times t_k. (Here, "decimation" is used in the loose sense of removing every nth data point, where n is any positive integer, and not in the sense of the ancient Roman technique of putting to death one in 10 soldiers in a legion guilty of mutiny or other crime.) Alternatively, we can use a *burst sampling mode*, in which rapid sampling is undertaken over a relatively short time interval Δt_B or "burst" embedded within each regularly spaced time interval, Δt. That is, the data are sampled at high frequency for a short duration starting (or ending) at times t_k for which the burst duration $\Delta t_B << \Delta t$. The instrument "rests" between bursts. There are advantages to the burst sampling scheme, especially in noisy (high frequency) environments where it may be necessary to average out the noise to get at the frequencies of interest. Burst sampling works especially well when there is a "spectral gap" between fluctuations at the high and low ends of the spectrum. As an example, there is typically a spectral gap between surface gravity waves in the open ocean (periods of 1−20 s) and the 12.4-hourly motions that characterize semidiurnal tidal currents. Thus, if we wanted to measure surface tidal currents using the burst-mode option for our current meter, we could set the sampling to a 2-min burst every hour; this option would smooth out the high-frequency wave effects but provide sufficient numbers of velocity measurements to resolve the tidal motions. Burst sampling enables us to filter out the high-frequency noise and obtain an improved estimate of the variability hidden underneath the high-frequency fluctuations. In addition, we can examine the high-frequency variability by scrutinizing the burst sampled data. If we were to sample rapidly enough, we could estimate the surface gravity wave energy spectrum. Many oceanographic instruments use (or have provision for) a burst sampling data collection mode.

A "duty cycle" has sometimes been used to collect positional data from Service Argos satellite-tracked drifters as a cost-saving form of burst sampling. In this case, all positional data within a 24-h period (about 10 satellite fixes) were collected only every third day. Tracking costs paid to Service Argos were reduced by a factor of three using the duty cycle. Unfortunately, problems arise when the length of each burst is too short to resolve energetic motions with periods comparable to the burst sample length. In the case of satellite-tracked drifters poleward of tropical latitudes, these problems are associated with highly energetic inertial motions whose periods $T = 1/(2\Omega\sin\theta)$ are comparable to the 24-h duration of the burst sample (here, $\Omega = 0.1161 \times 10^{-4}$ cycles per second is the earth's rate of rotation and $\theta \equiv$ latitude). Beginning in 1992, it became possible to improve resolution of high-frequency motions using a 1/3-duty cycle of 8 h "on" followed by 16 h "off". According to Bograd et al. (1999), even better resolution of high-frequency mid-latitude motions could be obtained using a duty cycle of 16 h "on" followed by 32 h "off".

A duty cycle is presently being used in the Deep-ocean Assessment and Reporting of Tsunamis (DART) buoys moored in several regions of the world ocean as part of enhanced tsunami warning systems. To save battery life and reduce data storage needs, the bottom pressure sensors (BPRs) in DART buoys report time-averaged pressure ($\cong$ water depth) every 15 min to orbiting satellites through an acoustic link to a nearby surface buoy. When the built-in algorithm

detects an anomalous change in bottom pressure due to the arrival of the leading wave of a tsunami, the instrument switches into "event mode". The instrument then transmits bottom pressure data every 15 s for several minutes followed by 1 min averaged data for the next 4 h (González et al., 1998; Bernard et al., 2001; González et al., 2005; Titov et al., 2005; Mungov, 2012). At present, DART buoys switch to event mode if there is a threshold change of 3 cm in equivalent water depth for earthquake magnitudes greater than 7.0 and epicenter distances of over 600 km. Problems arise when the leading wave form is too slowly varying to be detected by the algorithm or if large waves continue to arrive well after the 4-h cutoff. For example, several DART buoys in the northeast Pacific failed to capture the leading tsunami wave from the September 2009, $M_w = 8.1$ Samoa earthquake or to detect the slowly varying trough that formed the lead wave from the magnitude 8.8, February 2010 earthquake off the coast of Chile (Rabinovich et al., 2012). Vertical acceleration of the seafloor associated with seismic waves (mainly Rayleigh waves moving along the water—bottom interface) can also trigger false tsunami responses (Mofjeld et al., 2001). Because of the duty cycle, only those few buoys providing continuous 15-s internal recording can provide data for the duration of major tsunami events, which typically have frequency-dependent, e-folding decay timescales of around a day (Rabinovich et al., 2013).

1.2.5 Regularly vs Irregularly Sampled Data

In certain respects, an irregular sampling in time or nonequidistant placement of instruments can be more effective than a possibly more esthetically appealing than uniform sampling. For example, unequal spacing permits a more statistically reliable resolution of oceanic spatial variability by increasing the number of quasi-independent estimates of the dominant wavelengths (wavenumbers). Since oceanographers are almost always faced with having fewer instruments than they require to resolve oceanic features, irregular spacing can also be used to increase the overall spatial coverage (fundamental wavenumber) while maintaining the small-scale instrument separation for Nyquist wavenumber estimates. The main concern is the lack of redundancy should certain key instruments fail, as so often seems to happen. In this case, a quasi-regular spacing between locations is better. Prior knowledge of the scales of variability to expect is a definite plus in any experimental array design.

In a sense, the quasi-logarithmic vertical spacing adopted by oceanographers for bottle cast (hydrographic) sampling—specifically 0, 10, 20, 30, 50, 75, 100, 125, 150 m, etc.—represents a "spectral window" adaptation to the known physical–chemical structure of the ocean. Highest resolution is required near the surface where vertical changes in most oceanic variables are most rapid. Similarly, an uneven horizontal arrangement of observations increases the number of quasi-independent estimates of the horizontal wavenumber spectrum. Digital data are most often sampled (or subsampled) at regularly spaced time increments. Aside from the usual human propensity for order, the need for regularly spaced data derives from the fact that most analysis methods have been developed for regular-spaced, gap-free data series. Although digital data do not necessarily need to be sampled at regularly spaced time increments to give meaningful results, some form of interpolation between values may eventually be required. Since interpolation involves a methodology for estimating unknown values from known data, it can lead to its own sets of problems.

1.2.6 Independent Realizations

As we review the different instruments and methods, the reader should keep in mind the three basic concerns with respect to observations: accuracy/precision; resolution (spatial and temporal); and statistical significance (statistical sampling theory). A fundamental consideration in ensuring the statistical significance of a set of measurements is the need for independent realizations. If repeated measurements of a process are strongly correlated, they provide no new information and do not contribute to the statistical significance of the measurements. Often a subjective decision must be made on the question of statistical independence. While this concept has a formal definition, in practice it is often difficult to judge. A simple guide is that any suite of measurements that is highly correlated (in time or space) cannot be independent. At the same time, a group or sequence of measurements that is totally uncorrelated, must be independent. In the case of no correlation between each sample or realization, the number of "degrees of freedom" is defined by the total number of measurements; for the case of perfect correlation, the redundancy of the data values reduces the degrees of freedom to unity for a scalar quantity and to two for a vector quantity. The degree of correlation within the data set provides a way of estimating the number of degrees of freedom within a given suite of observations. While more precise methods will be presented later in this text, a simple linear relation between degrees of freedom and correlation often gives the practitioner a way to proceed without developing complex mathematical constructs.

As will be discussed in detail later, all of these sampling recommendations have statistical foundations and the guiding rules of probability and estimation can be carefully applied to determine the sampling requirements and dictate the appropriate measurement system. At the same time, these same statistical methods can be applied to existing data in order to better evaluate their ability to measure phenomena of interest. These comments are made to assist the reader in evaluating the potential of a particular instrument (or method) for the measurement of some desired variable.

1.3 TEMPERATURE

The measurement of temperature in the ocean uses conventional techniques except for deep observations, where hydrostatic pressures are high and there is a need to protect the sensing system from ambient depth/temperature changes higher in the water column as the sensor is returned to the ship. Temperature is the easiest ocean property to measure accurately. Some of the ways in which ocean temperature can be measured are:

1. Expansion of a liquid or a metal.
2. Differential expansion of two metals (bimetallic strip).
3. Vapor pressure of a liquid.
4. Thermocouples.
5. Change in electrical resistance.
6. Infrared radiation from the sea surface.

In most of these sensing techniques, the temperature effect is very small and some form of amplification is necessary to make the temperature measurement detectable. Usually, the response is nearly linear with temperature so that only the first-order term in the calibration expansion is needed when converting the sensor measurement to temperature. However, in order to achieve high precision over large temperature ranges, second, third and even fourth order terms must sometimes be used to convert the measured variable to temperature.

1.3.1 Mercury Thermometers

Of the above methods, (1), (5), and (6) have been the most widely used in physical oceanography. The most common type of the liquid expansion sensor is the mercury-in-glass thermometer. In their earliest oceanographic application, simple mercury thermometers were lowered into the ocean with hopes of measuring the temperature at great depths in the ocean. Two effects were soon noticed. First, thermometer housings with insufficient strength succumbed to the greater pressure in the ocean and were crushed. Second, the process of bringing an active thermometer through the oceanic vertical temperature gradient sufficiently altered the deeper readings that it was not possible to accurately measure the deeper temperatures. An early solution to this problem was the development of min–max thermometers that were capable of retaining the minimum and maximum temperatures encountered over the descent and ascent of the thermometer. This type of thermometer was widely used on the British Challenger expedition of 1873–76.

The real breakthrough in thermometry was the development of reversing thermometers, first introduced in London by Negretti and Zambra in 1874 (Sverdrup et al., 1942, p. 349). The reversing thermometer contains a mechanism such that, when the thermometer is inverted, the mercury in the thermometer stem separates from the bulb reservoir and captures the temperature at the time of inversion. Subsequent temperature changes experienced by the thermometer have limited effects on the amount of mercury in the thermometer stem and can be accounted for when the temperature is read on board the observing ship. This "break-off" mechanism is based on the fact that more energy is required to create a gas–mercury interface (i.e., to break the mercury) than is needed to expand an interface that already exists. Thus, within the "pigtail" section of the reversing thermometer is a narrow region called the "break-off point", located near appendix C in Figure 1.3, where the mercury will break when the thermometer is inverted.

The accuracy of the reversing thermometer depends on the precision with which this break occurs. In good reversing thermometers this precision is better than 0.01 °C. In standard mercury-in-glass thermometers, as well as in

Reservoir

C B

Enlarged section showing
pig-tail (A) appendix
dead arm (B) and break-off
point (C).

1-0 E D

Enlarged section showing
readings on main stem (D)
and auxiliary (E).

Bulb

Glass jacket

Protected   Unprotected

FIGURE 1.3 Details of a reversing mercury thermometer showing the "pigtail appendix".

Mercury expands in a near-linear manner with temperature. As a consequence, it has been the liquid used in most high precision, liquid-glass thermometers. Other liquids such as alcohol and toluene are used in precision thermometers only for very low temperature applications, where the higher viscosity of mercury is a limitation. Expansion linearity is critical in the construction of the thermometer scale, which would be difficult to engrave precisely if expansion were nonlinear.

In a mercury thermometer, the volume of the bulb is equivalent to about 6000 stem-degrees Celsius. This is known as the "degree volume" and usually is considered to comprise the bulb and the portion of the stem below the mark (a stem-degree is the temperature measured by any tube-like thermometer). If the thermometer is to retain its calibration, this volume must remain constant with a precision not commonly realized by the casual user. For a thermometer precision within $\pm 0.01\,°C$, the bulb volume must remain constant within one part in 600,000. Glass does not have ideal mechanical properties and it is known to exhibit some plastic behavior and deform under sustained stress. Repeated exposure to high pressures may produce permanent deformation and a consequent shift in bulb volume. Therefore, precision can only be maintained by frequent laboratory calibration. Such shifts in bulb volume can be detected and corrected by the determination of the "ice point" (a slurry of water and ice), which should be checked frequently if high accuracy is required. The procedure is more or less obvious but a few points should be considered. First the ice should be made from distilled water and the water–ice mixture should also be made from distilled water. The container should be insulated and at least 70% of the bath in contact with the thermometer should be of chopped ice. The thermometer should be immersed for five or more minutes during which the ice–water mixture should be stirred continuously. The control temperature of the bath can be taken by an

reversing thermometers, there are concerns other than the break point, which affect the precision of the temperature measurement. These are:

1. Linearity in the expansion coefficient of the liquid.
2. The constancy of the bulb volume.
3. The uniformity of the capillary bore.
4. The exposure of the thermometer stem to temperatures other than the bulb temperature.

accurate thermometer of known reliability. Comparison with the temperature of the reversing thermometer, after the known calibration characteristics have been accounted for, will give an estimate of any offsets inherent in the use of the reversing thermometer in question.

The uniformity of the capillary bore is critical to the accuracy of the mercury thermometer. In order to maintain the linearity of the temperature scale it is necessary to have a uniform capillary as well as a linear response liquid element. Small variations in the capillary can occur as a result of small differences in cooling during its construction or to inhomogeneities in the glass. Errors resulting from the variations in capillary bore can be corrected through calibration at known temperatures. The resulting corrections, including any effect of the change in bulb volume, are known as "index corrections". These remain constant relative to the ice point and, once determined, can be corrected for a shift in the ice point by addition or subtraction of a constant amount. With proper calibration and maintenance, most of the mechanical defects in the thermometer can be accounted for. Reversing thermometers are then capable of accuracies of $\pm 0.01\,°C$, as given earlier for the precision of the mercury break point. This accuracy, of course, depends on the resolution of the temperature scale etched on the thermometer. For high accuracy in the typically weak vertical temperature gradients of the deep ocean, thermometers are etched with scale intervals between 0.1 and 0.2 °C. Most reversing thermometers have scale intervals of 0.1 °C.

The reliability and calibrated absolute accuracy of reversing thermometers continue to provide standard temperature measurement against which all forms of electronic sensors are compared and evaluated. In this role as a calibration standard, reversing thermometers continue to be widely used. In addition, many oceanographers still believe that standard hydrographic stations made with sample bottles and reversing thermometers, provide the only reliable data. For these reasons, we briefly describe some of the fundamental problems that occur when using reversing thermometers. An understanding of these errors may also prove helpful in evaluating the accuracy of reversing thermometer data that are archived in the historical data file. The primary malfunction that occurs with a reversing thermometer is a failure of the mercury to break at the correct position. This failure is caused by the presence of gas (a bubble) somewhere within the mercury column. Normally all thermometers contain some gas within the mercury. As long as the gas bubble has sufficient mercury compressing it, the bubble volume is negligible, but if the bubble gets into the upper part of the capillary tube it expands and causes the mercury to break at the bubble rather than at the break-off point. The proper place for this resident gas is at the bulb end of the mercury; for this reason it is recommended that reversing thermometers always be stored and transported in the bulb-up (reservoir-down) position. Rough handling can be the cause of bubble formation higher up in the capillary tube. Bubbles lead to consistently offset temperatures and a record of the thermometer history can clearly indicate when such a malfunction has occurred. Again the practice of renewing, or at least checking, the thermometer calibration is essential to ensuring accurate temperature measurements. As with most oceanographic equipment, a thermometer with a detailed history is much more valuable than a new one without some prior use.

There are two basic types of reversing thermometers: (1) protected thermometers that are encased completely in a glass jacket and not exposed to the pressure of the water column; and (2) unprotected thermometers for which the glass jacket is open at one end so that the reservoir experiences the increase of pressure with ocean depth, leading to an apparent increase in the measured temperature. The increase in temperature with depth is due to the compression of the glass bulb, so that if the compressibility of the glass is known from the manufacturer, the pressure and hence the depth can be inferred from the temperature difference, $\Delta T = T_{\text{Unprotected}} - T_{\text{Protected}}$. The difference in thermometer readings, collected at the same depth, can be used to compute the depth of the temperature measurement to an accuracy of about ±1% of the depth. This subject will be treated more completely in the section on depth/pressure measurement. We note that the ±1% full-scale accuracy for reversing thermometers is better than the accuracy of ±2–3% normally expected from modern depth sounders but is much poorer than the ±0.01% full-scale pressure accuracy expected from strain gauges used in most modern CTD probes.

Unless collected for a specific observational program or taken as calibrations for electronic measurement systems, reversing thermometer data are most commonly found in historical data archives. In such cases, the user is often unfamiliar with the precise history of the temperature data and thus cannot reconstruct the conditions under which the data were collected and edited. Under these conditions one generally assumes that the errors are of two types; either they are large offsets (such as errors in reading the thermometer), which are readily identifiable by comparison with other regional historical data, or they are small random errors due to a variety of sources and difficult to identify or separate from real physical oceanic variability. Parallax errors, which are one of the main causes of reading errors, are greatly reduced through use of an eyepiece magnifier. Identification and editing of these errors depends on the problem being studied and will be discussed in a later section on data processing.

1.3.2 The Mechanical Bathythermograph

The mechanical bathythermograph (MBT) uses a liquid-in-metal thermometer to register temperature and a Bourdon tube sensor to measure pressure. The temperature-sensing element is a fine copper tube nearly 17 m long filled with toluene (Figure 1.4). Temperature readings are recorded by a mechanical stylus, which scratches a thin line on a coated glass slide. Although this instrument has largely been replaced by the expendable bathythermograph (XBT), the historical archives contain numerous temperature profiles collected using this device. It is, therefore, worthwhile to describe the instrument and the data it measures. Only the temperature measurement aspect of this device will be considered; the pressure/depth recording capability will be addressed in a later section.

There are numerous limitations to the MBT. To begin with, it is restricted to depths less than 300 m. While the MBT was intended to be used with the ship underway, it is only really possible to use it successfully when the ship is traveling at no more than a few knots. At higher speeds, it becomes impossible to retrieve the MBT without the risk of hitting the instrument against the ship. Higher speeds also make it difficult to properly estimate the depth of the probe from the amount of wire out. The temperature accuracy of the MBT is restricted by the inherent lower accuracy of the liquid-in-metal thermometer. Metal thermometers are also subject to permanent deformation. Since metal is more subject to changes at high temperatures than is glass it is possible to alter the performance of the MBT by continued exposure to higher temperatures (i.e., by leaving the probe out in the sun). The metal return spring of the temperature stylus is also a source of potential problems in that it is subject to hysteresis and creep. Hysteresis, in which the uptrace does not coincide with the downtrace, is especially prevalent when the temperature differences are small. Creep occurs when the metal is subjected to a constant loading for long periods. Thus, an MBT continuously used in the heat of the tropics may be found later to have a slight positive temperature error.

Most of the above errors can be detected and corrected by frequent calibration of the MBT. Even with regular calibration, it is doubtful that the stated precision of ±0.1 °F (±0.06 °C)

FIGURE 1.4 A bathythermograph showing its internal construction and sample bathythermograph slides.

can be attained. Here, the value is given in °F since most of the MBTs were produced with this temperature scale. When considering MBT data from the historical data files, it should be realized that these data were entered into the files by hand. The usual method was to produce an enlarged black-and-white photograph of the temperature trace using the nonlinear calibration grid unique to each instrument. Temperature values were then read off of these photographs and entered into the data file at the corresponding depths. The usual procedure was to record temperatures for a fixed depth interval (i.e., 5 or 10 m) rather than to select out inflection points that best described the temperature profile. The primary weakness of this procedure is the ease with which incorrect values can enter the data file through misreading the temperature trace or incorrectly entering the measured value. Usually these types of errors result in large differences with the neighboring values and can be easily identified. Care should be taken,

however, to remove such values before applying objective methods to search for smaller random errors. It is also possible that data entry errors can occur in the entry of date, time, and position of the temperature profile and tests should be made to detect these errors.

1.3.3 Resistance Thermometers (XBT)

Since the electrical resistance of metals, and other materials, changes with temperature, these materials can be used as temperature sensors. The resistance (R) of most metals depends on temperature (T) and can be expressed as a polynomial

$$R = R_0\left(1 + aT + bT^2 + cT^3 + \ldots\right) \qquad (1.4)$$

where a, b, and c are constants and R_0 is the resistance at $T = 0\,°C$. In practice, it is usually assumed that the response is linear over some limited temperature range and the proportionality can be given by the value of the coefficient a

(called the temperature resistance coefficient). The most commonly used metals are copper, platinum, and nickel, which have temperature coefficients, a, of 0.0043, 0.0039, and 0.0066/°C, respectively. Of these, copper has the most linear response but its resistance is low so that a thermal element would require many turns of fine wire and would consequently be expensive to produce. Nickel has a very high resistance but deviates sharply from linearity. Platinum having a relatively high resistance level is very stable and has a relatively linear behavior. For these reasons, platinum resistance thermometers have become a standard by which the international scale of temperature is defined. Platinum thermometers are also widely used as laboratory calibration standards and have accuracies of 0.001 °C.

The semiconductors form another class of resistive materials used for temperature measurements. These are mixtures of oxides of metals such as nickel, cobalt, and manganese, which are molded at high pressure followed by sintering (i.e. heating to incipient fusion). The types of semiconductors used for oceanographic measurements are commonly called thermistors. These thermistors have the advantages that: (1) the temperature resistance coefficient of −0.05/°C is about 10 times as great as that for copper; and (2) the thermistors may be made with high resistance for a very small physical size.

The temperature coefficient of thermistors is negative, which means that the resistance decreases as temperature increases. This temperature coefficient is not a constant except over very small temperature ranges; hence the change of resistance with temperature is not linear. Instead, the relationship between resistance and temperature is given by

$$R(T) = R_0 exp\left[\beta\left(T^{-1} - T_0^{-1}\right)\right] \qquad (1.5)$$

where $R_0 = R(T_0)$ is the conventional temperature coefficient of resistance, T and T_0 are absolute temperatures (K) with respective resistance values of $R(T)$ and R_0, and constant β is

determined by the energy required to generate and move the charge carriers responsible for electrical conduction. (As β increases, the material becomes more conducting.) Thus, we have a relationship whereby temperature T can be computed from the measurement of resistance $R(T)$.

One of the most common uses of thermistors in oceanography is in XBTs. The XBT was developed to provide an upper ocean temperature-profiling device that operated while the ship was underway. The crucial development was the concept of depth measurement using the elapsed time for the known fall rate of a "freely falling" probe. To achieve "free fall", independent of the ship's motion, the data transfer cable is constructed from fine copper wire with feed spools in both the sensor probe and in the launching canister (Figure 1.5). The details of the depth measurement capability of the XBT will be discussed and evaluated in the section on depth/pressure measurements.

The XBT probes employ a thermistor placed in the nose of the probe as the temperature-sensing element. According to the manufacturer (Sippican Corp.; Marion, Massachusetts, U.S.A.), the accuracy of this system is ±0.1 °C. This figure is determined from the characteristics of a batch of semiconductor material, which has known resistance—temperature ($R-T$) properties. To yield a given resistance at a standard temperature, the individual thermistors are precision-ground, with the XBT probe thermistors ground to yield 5000 Ω (here, Ω is the symbol for the unit of ohms) at 25 °C (Georgi et al., 1980). If the major source of XBT probe-to-probe variability can be attributed to imprecise grinding, then a single-point calibration should suffice to reduce this variability in the resultant temperatures. Such a calibration was carried out by Georgi et al. (1980) both at sea and in the laboratory.

To evaluate the effects of random errors on the calibration procedure, 12 probes were calibrated repeatedly. The mean differences between the measured and bath temperatures

FIGURE 1.5 Exploded view of a Sippican Oceanographic, Inc. XBT showing spool and canister. XBT, Expendable bathythermograph.

was ±0.045 °C with a standard deviation of 0.01 °C. For the overall calibration comparison, 18 cases of probes (12 probes per case) were examined. Six cases of T7s (good to 800 m and vessel speeds up to 30 knots) and two cases of T6s (good to 500 m and at less than 15 knots) were purchased newly from Sippican, while the remaining 10 cases of T4s (good to 500 m up to 30 knots) were acquired from a large pool of XBT probes manufactured in 1970 for the U.S. Navy. The overall average standard

deviation for the probes was 0.023 °C, which then reduces to 0.021 °C when consideration is made for the inherent variability of the calibration procedure.

A separate investigation was made of the $R-T$ relationship by studying the response characteristics for nine probes. The conclusion was that the $R-T$ differences ranged from +0.011 °C to −0.014 °C which then means that the measured relationships were within ±0.014 °C of the published relationship and that the calculation of new coefficients, following Steinhart and Hart (1968), is not warranted. Moreover the final conclusions of Georgi et al. (1980) suggest an overall accuracy for XBT thermistors of $\pm 0.06\,^{\circ}\text{C}$ at the 95% confidence level and that the consistency between thermistors is sufficiently high that individual probe calibration is not needed for this accuracy level.

Another method of evaluating the performance of the XBT system is to compare XBT temperature profiles with those taken at the same time with a higher accuracy profiler such as a CTD system. Such comparisons are discussed by Heinmiller et al. (1983) for data collected in both the Atlantic and the Pacific using calibrated CTD systems. In these comparisons, it is always a problem to achieve true synopticity in the data collection since the XBT probe falls much faster than the recommended drop rate of around 1 m/s for a CTD probe. Most of the earlier comparisons between XBT and CTD profiles (Flierl and Robinson, 1977; Seaver and Kuleshov, 1982) were carried out using XBT temperature profiles collected between CTD stations separated by 30 km. For the purposes of intercomparison, it is better for the XBT and CTD profiles to be collected as simultaneously as possible.

The primary error discussed by Heinmiller et al. (1983) is that in the measurement of depth rather than temperature. There were, however, significant differences between temperatures measured at depths where the vertical temperature gradient was small and the depth error should make little or no contribution. Here, the XBT temperatures were found to be systematically higher than those recorded by the CTD. Sample comparisons were divided by probe type and experiment. The T4 probes (as defined above) yielded a mean XBT–CTD difference of about $0.19\,^{\circ}\text{C}$ while the T7s (defined above) had a lower mean temperature difference of $0.13\,^{\circ}\text{C}$. Corresponding standard deviations of the temperature differences were $0.23\,^{\circ}\text{C}$, for the T4s, and $0.11\,^{\circ}\text{C}$ for the T7s. Taken together, these statistics suggest an XBT accuracy less than

the $\pm 0.1\,^{\circ}\text{C}$ given by the manufacturer and far less than the $0.06\,^{\circ}\text{C}$ reported by Georgi et al. (1980) from their calibrations.

From these divergent results, it is difficult to decide where the true XBT temperature accuracy lies. Since the Heinmiller et al. (1983) comparisons were made in situ, there are many sources of error that could contribute to the larger temperature differences. Even though most of the CTD casts were made with calibrated instruments, errors in operational procedures during collection and archival could add significant errors to the resultant data. Also, it is not easy to find segments of temperature profiles with no vertical temperature gradient and therefore it is difficult to ignore the effect of the depth measurement error on the temperature trace. It seems fair to conclude that the laboratory calibrations represent the ideal accuracy possible with the XBT system (i.e. better than $\pm 0.1\,^{\circ}\text{C}$). In the field, however, one must expect other influences that will reduce the accuracy of the XBT measurements and an overall accuracy slightly more than $\pm 0.1\,^{\circ}\text{C}$ is perhaps realistic. Some of the sources of these errors can be easily detected, such as an insulation failure in the copper wire, which results in single step offsets in the resulting temperature profile. Other possible temperature error sources are interference due to shipboard radio transmission (which shows up as high-frequency noise in the vertical temperature profile) or problems with the recording system. Hopefully, these problems are detected before the data are archived in historical data files.

In closing this section we comment that, until recently, most XBT data were digitized by hand. The disadvantage of this procedure is that chart paper recording does not fully realize the potential digital accuracy of the sensing system and that the opportunities for operator recording errors are considerable. Again, some care should be exercised in editing out these large errors, which usually result from the incorrect hand recording of temperature, date, time or position. It is becoming increasingly popular to use digital XBT recording systems, which improve the accuracy of the recording and eliminate the possibility of incorrectly entering the temperature trace. Such systems are described, for example, in Stegen et al. (1975) and Emery et al. (1986). Today, essentially all research XBT data are collected with digital systems, while the analog systems are predominantly used by various international navies.

1.3.4 Salinity/Conductivity-Temperature-Depth Profilers

Resistance thermometers are widely used on continuous profilers designed to replace the earlier hydrographic profiles collected using a series of sampling bottles. The in situ electronic instruments continuously sample the water temperature, providing much higher resolution information on the ocean's vertical and horizontal temperature structure. Since density also depends on salinity, electronic sensors had to be developed to measure salinity in situ and were incorporated into the profiling system. As discussed by Baker (1981), an early electronic profiling system for temperature and salinity was described by Jacobsen (1948). The system was limited to 400 m depth and used separate supporting and data transfer cables. Next, a system called the salinity-temperature-depth (STD) profiler was developed by Hamon and Brown in the mid-1950s (Hamon, 1955; Hamon and Brown, 1958). The evolution of conductivity measurement, the basic parameter for the derivation of salinity, will be discussed in the section on salinity. This evolution led to the introduction of the CTD profiling system (Brown, 1974). This name change identified improvements not only in the conductivity sensor but also in the temperature sensing system designed to overcome the mismatch in the response times between the temperature and conductivity sensors. This mismatch often resulted in erroneous salinity spikes in the earlier STD systems (Dantzler, 1974).

Most STD/CTD systems use a platinum resistance thermometer as one leg of an impedance bridge from which the temperature is determined. An important development was made by Hamon and Brown (1958) where the sensing elements were all connected to oscillators that converted the measured variables to audio frequencies that could then be sent to the surface via a single conducting element in the profiler support cable. The outer cable sheath acted as both mechanical support and the return conductor. This data transfer method has subsequently been used on most electronic profiling systems, despite the ever-present concern for possible ground-fault problems. The early STDs were designed to operate to 1000 m and had a temperature range of 0–30 °C with an accuracy of ±0.15 °C. Later STDs, such as the widely used Plessey Model 9040, had accuracies of 0.05 °C with temperature ranges of −2 to +18 °C or +15 to +35 °C (range was switched automatically during a cast). Modern CTDs, such as the Sea-Bird Electronics, Inc. (SBE) 25*plus* and 911*plus* (Figure 1.6), the General Oceanics Idronaut Ocean Sciences 316 (modified after the EG&G Mark V), the Valeport Midas CTD, the Falmouth Scientific, Inc. Integrated CTD profiler, the Applied Microsystems (AML) Plus v2 CTD, and the RBR (Branker) CTD typically have accuracies of ±0.001 °C over a range of roughly −2 to +35 °C. The glass-coated platinum alloy thermistor beads used in the SBE profilers are required to have a stability (drift) of less than 0.001 °C over a range of −5 to +35 °C during the six months prior to delivery. This compares with the drift of 0.001 °C/month from earlier platinum resistance thermometers (Brown and Morrison, 1978; Hendry, 1993).

1.3.5 Dynamic Response of Temperature Sensors

Before considering more closely the problem of sensor response time for STD/CTD systems, it is worthwhile to review the general dynamic characteristics of temperature measuring systems. For example, no temperature sensor responds

(a)

Lifting ring

Battery/Recorder housing

Par sensor

Guard cage 316 stainless

DO/pH/ORP sensor

Fluorometer

Transmissometer

Pump

Pressure sensor

Conductivity sensor

Thermometer

95 cm (38 in)

28 cm (11 in)

SHADED MODULES ARE INCLUDED IN BASIC SEALOGGER CTD

FIGURE 1.6 (a) Schematic of the Sea-Bird SBE 25 CTD and optional sensor modules *(courtesy, Doug Bennett, Sea-Bird Electronics, Inc)*. (b) schematic of General Oceanics MK3C/WOCE CTD and optional sensors; (c) schematic of electronics and sensors of General Oceanics MK3C/WOCE CTD. ORP, Oxidation-Reduction potential; DO, Dissolved oxygen. *(Courtesy, Dan Schaas and Mabel Gracia, General Oceanics.)*

FIGURE 1.6 *(continued)*.

(a) Input E_i

(b) Platinum thermometer response E_l

(c) Thermistor response E_t

(d) Servo-balanced bridge response E_h

(e) Combined output $E_l + E_h$

FIGURE 1.7 Combined output and response times of the resistance thermometer of a CTD (the different parts of the figure (a–e) are described to the right of each letter). CTD, Conductivity-Temperature-Depth profiler.

instantaneously to changes in the environment, which it is measuring. If the environment temperature is changing, the sensor element lags in its response. This can be seen in the response of the combined platinum thermometer and miniature thermistor probe in Figure 1.7 (from Brown and Morrison, 1978). In this case, the response of the thermistor probe was designed to match the 25 ms response time of the conductivity cell. A simpler example is a reversing thermometer which, when lowered through the water column, would at no time read the correct environment temperature until it had been stopped and allowed to equilibrate with the surrounding water for some time. The time (K) that it takes the thermometer to respond to the temperature of a new environment is known as the response time or "time constant" of the sensor.

The time constant K is best defined by writing the heat transfer equation for the temperature sensor as

$$-\frac{dT}{dt} = \frac{1}{K}(T - T_w) \qquad (1.6)$$

where T_w and T are the temperatures of the medium (water) and thermometer and t refers to the elapsed time. If we assume that the temperature change occurs rapidly as the sensor descends, the temperature response can be described by the integration of Eqn (1.6) from which:

$$(T - T_w)/(T_0 - T_w) = \Delta T/\Delta T_0 = e^{-t/K} \qquad (1.7)$$

In this solution, T_0 refers to the temperature of the sensor before the temperature change and K is defined so that the ratio $\Delta T/\Delta T_0$ becomes e^{-1} ($= 0.368$) when 63% of the temperature change, ΔT, has taken place. The time for the temperature sensor to reach 90% of the final temperature value can be calculated using $e^{-t/k} = 0.1$. A more complex case is when the temperature of the environment is changing at a constant rate; that is

$$T_w = T_1 + ct \qquad (1.8)$$

where T_1 and c are constants. The temperature sensor then follows the same temperature change but lags behind so that

$$T - T_w = -cK \qquad (1.9)$$

The response times, as defined above, are given in Table 1.1 for various temperature sensing systems. Values refer to the time in seconds for the sensor to reach the specified percentage of its final value.

TABLE 1.1 Response Times (in seconds) for Various Temperature Sensors

Device	$K_{63\%}$	$K_{90\%}$	$K_{99\%}$
Mechanical bathythermograph	0.13	0.30	0.60
STD	0.30	0.60	1.20
Thermistor	0.04	0.08	0.16
Reversing thermometer	17.40	40.00	80.00

STD, Salinity-Temperature-Depth profilers

The ability of the sensor to attain its response level depends strongly on the speed at which the sensor moves through the medium. An example of the application of these response times is an estimate for the period of time a reversing thermometer must be allowed to "soak" in order to register the appropriate temperature. Suppose we desired an accuracy of $\pm 0.01\,°C$ and that our reversing thermometer is initially $10\,°C$ warmer than the water. From Eqn (1.7), $0.01/10.0 = \exp(-t/K)$, so that for a 99% completed response, $t = 553$ s or 9.2 min. Thus, the standard recommended soak period of 5 min (for a hydrographic cast) is set by thermometer limitations rather than by the imperfect flushing of the water sample bottles.

Suppose that the thermistor listed in Table 1.1 is being used to profile in the thermocline, where the temperature change with depth is about $2\,°C/m$. To sense a change in temperature in the thermocline with a high-performance resolution of $0.001\,°C$ at the 99% response level for every meter of depth, the response equation requires that $\exp(-t/0.16) = 0.001/2.0$ from which we find $t = 1.22$ s. Thus, we have the usual recommendation for a CTD lowering rate of roughly 1 m/s.

1.3.6 Response times of CTD Systems

As with any thermometer, the temperature sensor on CTD profilers does not respond instantaneously to a change in temperature. Heat must first diffuse through the viscous boundary layer set up around the probe and through the protective coatings of the sensor (Lueck et al., 1977). In addition, the actual temperature head must respond before the temperature is recorded. These effects lead to the finite response time of the profiler and introduce noise into the observed temperature data (Home and Toole, 1980). A correction is needed to achieve accurate temperature, salinity, and density data. Fofonoff et al. (1974) discuss how the single-pole filter model Eqn (1.7) may be used to correct the temperature data. In this lag-correction procedure, the true temperature at a point is estimated from Eqn (1.6) by calculating the time rate of change of temperature from the measured record using a least-square linear estimation over several neighboring points.

Home and Toole (1980) argue that data corrected with this method may still be in error due to errors arising in the estimation of terms in the differential equation or the approximation of the equation to the actual response of the sensor. As an alternative, they suggest using the measured data to estimate a correction filter directly. This procedure assumes that the observed temperature data may be written as a convolution of the true temperature with the response function of the sensor such that

$$T(t) = H[T^*(t)] \qquad (1.10)$$

where T is the observed temperature at time t, T^* is the true temperature, and H is the transfer or response function of the sensor. The filter g is sought so that

$$g \cdot H = \delta(t) \qquad (1.11)$$

where δ is the Dirac delta function. The filter g can be found by fitting, in a least-squares sense, its Fourier transform to the known Fourier transform of the function H. This method is fully described in the appendix to Home and Toole (1980). The major advantage of this filter technique is only realized in the computation of salinity from conductivity and temperature.

In addition to the physical response time problem of the temperature sensor, there is the problem of the nonuniform descent of the CTD probe due to the effects of a ship's roll or pitch (Trump, 1983). From a study of profiles collected with a Neil-Brown CTD, the effects of ship's roll were clearly evident at the 5-s period when the data were treated as a time series and spectra were computed. High coherence between temperature and conductivity effects suggests that the mechanisms leading to these roll-induced features are not related to the sensors themselves but rather reflect an interaction between the environment and the sensor motion. Two likely candidates are: (1) the modification of the temperature by water carried along in the wake of the CTD probe from another depth; and (2) the effects of a turbulent wake overtaking the probe as it decelerates bringing with it a false temperature. Trump (1983) concludes by saying that, while some editing procedure may yet be discovered to remove roll-related temperature variations, none is presently available. He, therefore, recommends that CTD data taken from a rolling ship cannot be used to compute statistics on vertical fine structure and suggests that the best way to remove such contamination is to employ a roll-compensation support system for the CTD probe. Trump also recommends a series of editing procedures to remove roll effects from present CTD data and argues that of the 30,000 raw input data points in a 300 m cast that up to one-half will be removed by these procedures. A standard procedure is to remove any data for which there is a negative depth change between successive values on the way down and vice versa on the way up.

1.3.6.1 Limiting the Mismatch between Temperature and Conductivity Responses

As with other modern CTDs, the response time, $K_{63\%}$, of the thermistor sensor used in SBE CTDs (for which $K_{63\%} = 65 \pm 10$ ms) differs from the response time of the conductivity cell for commonly used CTD descent rates of 1 m/s. This mismatch in sensor response times can lead to major errors in salinity calculated using the conductivity, temperature, and pressure data. To minimize these salinity errors, selected makes of CTDs, such as the SBE 911*plus*, use pumps to maintain a constant flow rate across the sensors housed within a temperature-conductivity (T-C) duct (a small flow-through pipe). The specified pumped flow rate is that which makes the conductivity response time equal to that of the platinum thermistor bead of the instrument. Because the temperature

sensor is positioned just ahead of the conductivity sensor in the T-C duct, there is a time delay between the temperature and conductivity measurements for the same parcel of water. However, because the flow rate is constant, the time delay is constant. Hardware and software within the CTD can then be used to advance the conductivity measurement in time so it is coincident with the temperature measurement. Pumping also enables the user to factor in the bias of around $1-2 \times 10^{-3}\,°C$ in the temperature measurement caused by frictional heating as the water flows past the sensor. The idea that the T−C mismatch could be diminished by separating the thermometer from the conductivity cell (so that the spatial separation acts as a time delay as the unit falls through the water) was discussed by Topham and Perkins (1988).

1.3.6.2 Dealing with the Effects of Ship Motion

In addition to minimizing the response mismatch between temperature and conductivity sensors, pumps can be used to mitigate the effects of ship motion on CTD data by maintaining a steady rate of flow past the instrument sensors. Pumping also reduces the thermal mass effect due to the lagged exchange of heat between the sensor and the environment being measured (Lueck, 1990; Lueck and Picklo, 1990). It is during the processing stage that the analyst needs to remove data "spikes" arising from C−T mismatch generated when ship motion causes the CTD to rise abruptly through a sharp property gradient or into the air when the CTD is near the ocean surface. This can be accomplished by removing data collected during negative changes in depth during the downcast and positive changes in depth during the upcast, or by removing data for which the vertical gradient in density is negative (unless the inversions are very thin, in which case they may have arisen from salt fingering or diffusive convection; cf. Spear and Thomson, 2012 and references therein). A recent procedure discussed by

McTaggart et al. (2010) and Gatien et al. (pers. comm. 2012) consists of the following components for the *de-spiking, filtering, and removal of obviously bad data:* (1) Early in the processing, a SBI SeaSoft program is run that replaces single-point spikes with padded values. This is generally applied to pressure, temperature, and conductivity channels only; (2) SeaSoft program DELETE is run to remove data collected during the upcast. It is also used to remove data collected when the descent rate of the CTD is lower than a cutoff value, typically 0.3 m/s; this is not usually applied within 10 m of the top and bottom of the cast. Pressure is typically filtered at this stage, but this is done earlier for some CTDs; (3) if upcast data are required, files are reversed and then put through DELETE; (4) graphical editing is applied late in processing (but before bin averaging) to remove spikes and other obviously "erroneous" data. Such data include: instrumental spikes; data collected before the pumps were turned on; data collected near the surface before the sensors equilibrated; data corrupted by shedding from the wire or from ship wakes that were not removed by DELETE, especially in cases where the instrument actually reversed direction; (5) graphical editing is also used to clean salinity data by smoothing spikes that are likely due to small mismatches between temperature and conductivity arising from small variations in sensor alignment; (6) a median filter, width ~0.5 m, is usually applied to the fluorescence data; (7) the CLEAN routine is sometimes used to remove data if they appear to be erroneous in a particular range, for example removing all transmissivity values that are below a certain pressure (depth) or all fluorescence observations greater than some extreme value (e.g., $> 200\,\mu g/l = 200$ mg/ m^3, depending on region).

1.3.7 Temperature Calibration of CTD Profilers

Although the temperature sensors on modern CTD profilers are highly accurate and have essentially replaced the need for bottle casts with thermometers, there can be problems with electronic drift and other subtle instrument changes that often necessitate the need for in situ calibrations using one or more bottle samples. For this reason, and also to collect water samples for chemical and biological analyses, most CTDs are used in conjunction with Rosette (or Carousel) water samplers, which can be commanded to sample at desired depths. A Rosette sampler consists of an aluminum cage at the end of the CTD conducting cable to which are fixed six, 12, or more water bottles (e.g., General Oceanics Go-Flo type water samplers) that can be individually triggered electronically from the surface. Larger cages can accommodate larger volume bottles, typically up to 30 L. While such in situ calibrations are more important for conductivity measurements, it is good practice to compare temperatures from the reversing thermometers with the CTD values. Although much less of a requirement now compared to the early career days of the authors, this comparison must be done in waters with near-uniform temperature and salinity profiles so that the errors between the CTD values and water sample are minimized. One must pick the time of the CTD values that coincide exactly with the tripping of the bottle. As reported by Scarlet (1975), in situ calibration usually confirms the manufacturer's laboratory calibration of the profiling instrument. Generally, this in situ calibration consists of comparisons between four and six temperature profiles, collected with reversing thermometers. Taken together with the laboratory calibration data, these data are used to construct a "correction" curve for each temperature sensor as a function of pressure. Fofonoff et al. (1974) present a laboratory calibration curve obtained over an 18-month period for an early Neil-Brown CTD (the Neil-Brown CTD was the forerunner of the General Oceanics CTD systems). A comparison of 175 temperatures measured in situ with this profiler and those measured by reversing mercury

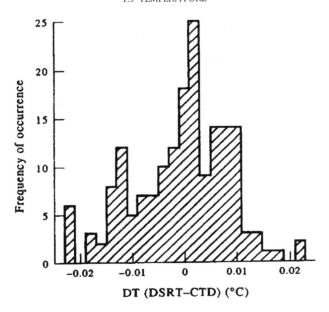

FIGURE 1.8 Histogram of temperature differences. Values used are the differences in temperature between a deep-sea reversing mercury thermometer (DSRT) and the temperature recorded by an early Neil-Brown CTD. DT, Difference in temperatures; CTD, Conductivity-Temperature-Depth profiler. *(From Fofonoff et al. (1974).)*

thermometers is presented in Figure 1.8. In the work reported by Scarlet (1975), these calibration curves were used in tabular, rather than functional, form and intermediate values were derived using linear interpolation. This procedure was likely adequate for the study region (Scarlet, 1975) but may not be generally applicable. Other calibration procedures fit a polynomial to the reference temperature data to define a calibration curve.

1.3.8 Sea Surface Temperature

Sea surface temperature (SST) was one of the first oceanographic variables to be measured in the open ocean. In the late 1760s, Benjamin Franklin led a study to map the position of the Gulf Stream using simple mercury-in-glass thermometers suspended from ships that were traveling between the U.S. Colonies and Europe. In his book "The Ocean: A general account of the

science of the Sea" first published in 1913, Sir John Murray (Naturalist on the H.M.S Challenger expedition of 1872—76) provided the first map of the annual range of the SST throughout the world ocean.

1.3.8.1 Ship Measurements

Due to its accessibility and importance to a broad range of oceanographic and meteorological disciplines, SST has become the most routine and widely observed oceanic parameter. As in the past, the measurement of SST continues to be part of the scheduled marine weather observations made by ships at sea. The introduction of routine underway SST measurements, in which SST is measured in a sample of surface water collected in a bucket, did away with the technique of suspending a thermometer from the ship. When only approximate SST measurements were required, this method was adequate. However, as the temperature sensors improved,

FIGURE 1.9 Frequency of occurrence of standard deviations associated with the 3.5-day mean sea surface temperature at Ocean Station "P" (50° N, 145° W). Difference between bucket temperature and ship-intake surface temperature. *(Modified after Tabata (1978a).)*

problems with the technique led to modifications of the bucket system. New buckets were built that contained the thermometer and captured only a small volume of near-surface water.

There are many possible sources of error with the bucket method including heating or cooling of the water sample temperature on the ship's deck, heat conduction through contact of the container with the thermometer, spillage, and the temperature change of the thermometer while it is being read (Tabata, 1978a). In order to avoid these difficulties, special sample buckets have been designed (Crawford, 1969), which shield both the container and the thermometer mounted in it from the heating/cooling effects of sun and wind. Comparisons between temperature samples collected with these special bucket samplers and reversing thermometers near the sea surface yielded temperature differences of ±0.1 °C (Tauber, 1969; Tabata, 1978a).

Seawater cooling for ship's engines makes it possible to measure SST from the temperature of the engine intake cooling system sensed by some type of thermometer imbedded in the cooling stream. Called "injection temperatures" these temperature values are reported by Saur (1963) to be on the average 0.7 ± 0.9 °C higher than corresponding bucket temperatures. For his study, Saur used SST data from 12 military ships transiting the North Pacific. Earlier similar studies by Brooks (1926) and Roll (1951) found smaller temperature differences, with the intake temperatures being only 0.1 °C higher than the bucket values. Brooks found, however, that the

engine-room crew usually recorded values that were 0.3 °C too high. More recent studies by Walden (1966), James and Fox (1972) and Collins et al. (1975) found temperature differences of 0.3, 0.3 ± 1.3, and 0.3 ± 0.7 °C, respectively. Tabata (1978b) compared three-day average SSTs from the Canadian weatherships at Station "P" in central northeast Pacific (which used a special bucket sampler) with ship-injection temperatures from merchant ships passing close by. He found an average difference of 0.2 ± 1.5 °C (Figure 1.9). Again, the mean differences were all positive suggesting the heating effect of the engine-room environment on the injection temperature.

The above comparisons between ship injection and ship bucket SSTs were made with carefully recorded values on ships that were collecting both types of measurements. Most routine ship-injection temperature reports are sent via radio or satellite by the ship's officers and have no corresponding bucket sample. As might be expected, the largest errors in these SST values are usually caused by errors in the data transmission or in the incorrect reporting or receiving of ship's position and/or temperature value (Miyakoda and Rosati, 1982). The resulting large deviations in SST can normally be detected by using a comparison with monthly climatological means and applying some range of allowable variation such as 5 °C.

This brings us to the problem of selecting the appropriate SST climatology—the characteristic SST temperature structure to be used for the global ocean. Until recently, there was little agreement as to which SST climatology was most appropriate. In an effort to establish a guide as to which climatology was the best, Reynolds (1983) compared the available SST climatologies with one he had produced (Reynolds, 1982). It was this work that led to the selection of Reynolds (1982) climatology for use in the Tropical Ocean Global Atmosphere (TOGA) research program. At present, the most widely used SST data are the gridded (1° latitude × 1° longitude) weekly and monthly mean Optimum Interpolation SST Version-2 data (Reynolds et al., 2002) provided by the US National Oceanic and Atmospheric Administration (NOAA). These data cover the entire world ocean and are a blend of ship, satellite, and meteorological buoy data (see Web site http://www.esrl.noaa.gov/psd/data/gridded/data.ncep.oisst.v2.html).

1.3.8.2 Satellite-Sensed SST (Radiation Temperature)

In contrast to ship and drifting buoy measurements that sample localized areas on a nonsynoptic basis, earth-orbiting satellites offer an opportunity to uniformly sample the surface of the globe on a nearly synoptic basis. This can be done by polar-orbiting satellites that sample the earth's surface at least four times per day per satellite (twice during the day and twice at night) depending on latitude (satellite swaths overlap at higher latitudes). In addition, geostationary satellites are capable of observing the earth's surface every 30 min in the thermal infrared, albeit at a somewhat lower resolution than the polar-orbiting spacecraft. Infrared sensors flown on satellites retrieve information that can be used to estimate SST, with certain limitations. Clouds are one of the major limitations in that they prevent long-wave, sea surface radiation from reaching the satellite sensor. Fortunately, clouds are rarely stationary and all areas of the earth's surface are eventually observed by the satellite. In general, researchers

wishing to produce "cloud-free" images of SST are required to construct composites over time using repeated satellite infrared data. These composites may require images spanning a couple of days to a week. A standard SST data product provided by NOAA consists of 8-day composites spanning an area of $9 \times 9\,km^2$. Even with the need for composite images, satellites provide almost global coverage on a fairly short timescale compared with a collection of ship, drifter, or glider SST observations. The main problem with satellite SST estimates is that their level of accuracy is a function of satellite sensor calibration and specified corrections for the effects of the intervening atmosphere, even under apparently cloud-free conditions (Figure 1.10).

A recent improvement has been the use of passive microwave satellite imagery to estimate SST. Unlike the thermal infrared, microwave emissions are transmitted through the cloudy atmospheres and hence can sense SST when atmospheric conditions preclude the use of the infrared signal for this purpose. The microwave sensing of SST has the drawback that the microwave images have a much lower spatial resolution and that microwave signals are quickly contaminated by land. Rainfall and moderate to

FIGURE 1.10 Sea Surface temperature (SST) differences, satellite minus ship, plotted as a function of the number of cloud-free pixels in a 50 × 50 pixels array. *(From Llewllyn-Fones et al. (1984).)*

strong winds can also corrupt passive microwave SST sensing. Still the ability to sense SST in regions of persistent clouds has promoted the use of passive microwave SST from a variety of sensors. Problems with the scanning multichannel microwave radiometer (SMMR) that flew on Nimbus-7 resulted in fairly unreliable SSTs. Later the success of the tropical rainfall mapping missions (TRMM) microwave imagery (TMI) made it possible to map SST on a microwave scale between ±40° latitude. Later, the advanced microwave scanning radiometer—Earth (AMSR-E) made it possible to accurately observe SST in an appropriate channel. Many products were developed to merge passive microwave and infrared SSTs to take advantage of the higher resolution infrared data with the cloud independent sensing of the passive microwave.

One of the first satellites capable of viewing the whole earth with an infrared sensor was ITOS 1, launched January 23, 1970. This satellite carried a scanning radiometer with an infrared channel in the 10.5—12.5 μm spectral band. The scanner viewed the earth in a swath with a nadir (subsatellite) spatial resolution of 7.4 km as the satellite traveled in its near-polar orbit. A method that uses these data to map global SST is described in Rao et al. (1972). This program uses the histogram method of Smith et al. (1970) to determine the mean SST over a number of pixels (picture elements), including those with some cloud. A polar-stereographic grid of 2.5° of latitude by 2.5° of longitude, encompassing about 1024 pixels per grid point per day, was selected. In order to evaluate the calculated SST retrievals from the infrared measurements, the calibrated temperature values were compared with SST maps made from ship radio reports. The resulting root-mean-square (RMS) difference for the northern hemisphere was 2.6 °C for three days in September 1970. When only the northern hemisphere ocean weathership SST observations were used, this RMS value dropped to 1.98 °C, a reflection of the improved ship SST observations. A comparison for all

ships from the southern hemisphere, for the same three days, resulted in an RMS difference of 2.45 °C. As has been discussed earlier in this chapter, one of the reasons for the magnitude of this difference is the uncertainty of the ship-sensed injection SST measurements.

The histogram method of Smith et al. (1970) was the basis for an operational SST product known as GOSSTCOMP (Brower et al., 1976). Barnett et al. (1979) examined the usefulness of these data in the tropical Pacific and found that the satellite SST values were negatively biased by 1—4 °C and so were, by themselves, not very useful. The authors concluded that the systematically cooler satellite temperatures were due to the effects of undetected cloud and atmospheric water vapor. As reported in Miyakoda and Rosati (1982), the satellite SST retrieval procedure was evolving at that time and retrieval techniques had improved (Strong and Pritchard, 1980). These changes included new methods to detect clouds and remove the effects of atmospheric water vapor contamination.

More recently, the advanced very high resolution radiometer (AVHRR)—a follow on to the very high resolution radiometer (VHRR) that was carried by ITOS—became the standard for infrared SST retrievals. In terms of SST the main benefits of the new sensor system were: (1) improved spatial resolution of about 1 km at nadir; and (2) the addition of other spectral channels, which improve the detection of water vapor by computing the differences between various channel radiometric properties. The AVHRR is a four- or five-channel radiometer with channels in the visible (0.6—0.7 μm), near-infrared (0.7—1.1 μm), and thermal infrared (3.5—3.9, 10.5—11.5, and 11.5—12.5 μm). The channel centered at 3.7 μm was intended as a nighttime cloud discriminator but was more useful when combined with the 11 and 12 μm channels to correct for the variable amounts of atmospheric water vapor (Bernstein, 1982). While there are many versions of this "two-channel" or "dual-channel" correction procedure (also called the "split-window" method), the most widely used was developed by McClain (1981). The final version of this radiometer (AVHRR-3) is designed to switch the mid-range infrared (3.7 μm) to 1.6 μm during daylight and back to 3.7 μm at night.

The above channel correction methods have been developed in an empirical manner using some sets of in situ SST measurements as a reference to which the AVHRR radiance data were adjusted to yield SST. This requires selecting a set of satellite and in situ SST data collected over coincident intervals of time and space. Bernstein (1982) chose intervals of several tens of kilometers and several days (Figure 1.11), while McClain et al. (1983) used a period of one day and a spatial grid of 25 km. In an evaluation of both of these methods, Lynn and Svejkovsky (1984) found that the Bernstein method yielded a mean bias of +0.5 °C, while the McClain equation a bias of −0.4 °C, relative

to in situ SST measurements. In each case, the difference from in situ values was smaller than the RMS errors suggested by the authors of the two methods. Bernstein (1982) compared mean maps made from 10 days of AVHRR retrievals with similar maps made from routine ship reports. He found the maps to agree within ±0.8 °C (one standard deviation) and concluded that this level of agreement was limited by the poor accuracy of the ship reports. He suggested that properly handled radiometer data can be used to study climate variations with an accuracy of 0.5−1.0 °C. This is consistent with the results of Lynn and Svejkovsky (1984) for a similar type of data analysis.

Another possible source of satellite SST estimates is the visible infrared spin scan radiometer (VISSR) carried by the geostationary orbiting Earth satellite (GOES). Unfortunately, the VISSR has a spatial resolution of about 8 km at the subsatellite point for the infrared channel. Another

FIGURE 1.11 Grid point by grid point cross plot of the mapped values of sea surface temperature from ship-based and AVHRR-based maps. AVHRR, advanced very high resolution radiometer. *(Bernstein (1982).)*

disadvantage of the VISSR is the lack of onboard infrared calibration similar to that which was available from AVHRR (Maul and Bravo, 1983). While VISSR does provide a hemispherical scan every half hour, its shortcomings have discouraged its application to the general estimation of SST. In some cases, VISSR data have been examined where there is a lack of suitable AVHRR data. In one such study, Maul and Bravo (1983) found that a regression between VISSR infrared and in situ SST data, using the radiative transfer equations, yielded satellite SST estimates that were no better than ±0.9 °C. The conclusion was that, in general, GOES VISSR SST estimates are accurate to within ±1.3 °C only. The primary problem with improving this accuracy is the presence of sub-pixel size clouds, which contaminate the SST regression.

Efforts have been made to improve the accuracy of retrievals from AVHRR through a better understanding of the onboard satellite calibration of the radiometer and by the development of regional and seasonal "dual-channel" atmospheric correction procedures. Evaluation of these correction procedures, compared with collections of atmospheric radiosonde measurements, has demonstrated the robust character of the "dual-channel" correction and improvements only require a better estimate of the local versus global effects in deriving the appropriate algorithm (Llewellyn-Jones et al., 1984). Thus, it appears safe to suggest that AVHRR SST estimates were made with accuracies of about ±0.5 °C, assuming that appropriate atmospheric corrections were performed.

Three workshops were held at the Jet Propulsion Laboratory (JPL) to compare the many different techniques of SST retrievals from existing satellite systems. The first workshop (January 27–28,1983) examined only the microwave data from the SMMR while the second workshop (June 22–24, 1983) considered SMMR, high-resolution infrared sounder (HIRS), and AVHRR for two time periods,

November 1979 and December, 1981. The third workshop (February 22–24, 1984) examined SST products derived from SMMR, HIRS, AVHRR, and VISSR atmospheric sounder (VAS, atmospheric sounder on the GOES) for an additional two months (March and July, 1982). A series of workshop reports is available from JPL and the results are summarized in journal articles in the November 1985 issue of the *Journal of Geophysical Research*.

In their review of third-workshop results, Hilland et al. (1985) reported that the overall RMS satellite SST errors range from 0.5 to 1.0 °C. In a discussion of the same workshop results, Bernstein and Chelton (1985) were more specific, reporting RMS differences between satellite and SST anomalies ranging from 0.58 to 1.37 °C. Mean differences for this same comparison ranged from −0.48 to 0.72 °C and varied substantially from month to month and season to season. They also reported that the SMMR SSTs had the largest RMS differences and time-dependent biases. Differences for the AVHRR- and HIRS-computed SSTs were smaller. When the monthly ship SST data were smoothed spatially to represent 600 km averages, the standard deviations of the monthly ship averages from climatology varied from 0.35 to 0.63 °C. Using these smoothed ship SST anomalies as a reference, the signal-to-noise variance ratios were 0.25 for SMMR and 1.0 for both the AVHRR and HIRS.

The workshop review by McClain et al. (1985) of the AVHRR-based multichannel SST (MCSST) retrieval method found biases of 0.3–0.4 °C (with MCSST lower than ship), standard deviations of 0.5–0.6 °C, and correlations of +0.3 to +0.7 (see also Bates and Diaz, 1991). They also discussed a refined MCSST technique being used with more recent NOAA-9 AVHRR data that yielded consistent biases of −0.1 °C and RMS differences (from ship SSTs) of 0.5 °C. In an application of AVHRR data to the study of warm Gulf Stream rings, Brown et al. (1985) discuss a calibration procedure, which provides

SST estimates accurate to ±0.2 °C. This calibration method was the result of thermal vacuum tests, which revealed instrument-specific changes in the relative emittance between internal (to the satellite) and external (deep space) calibration targets. By reviewing satellite pre-launch calibration data they found that there was an instrument-specific, nonlinear departure from a two-point linear calibration for higher temperatures. In addition, it was found that the calibration relationship between the reference platinum resistance thermocouples and the sensor systems changed in the thermal vacuum tests; hence a limited instrument retest, as part of the calibration cycle, was recommended as a way to improve AVHRR SST accuracy. Such higher accuracy absolute SST values are of importance for future climate studies where small, long-term temperature changes are significant.

The workshop results for the geostationary sounder unit (VAS) (Bates and Smith, 1985) revealed a warm bias of 0.5 °C with an RMS scatter of 0.8–1.0 °C. The positive bias was attributed to a diurnal sampling bias and a bias in the original set of empirical VAS/buoy matchups. Use of a second set of VAS/buoy matches reduced this warm bias making VAS SSTs more attractive due to the increased temporal coverage (every half hour) over that of the AVHRR (one to four images per day). All of these satellite SST intercomparisons were evaluated against either ship or buoy measurements of the near surface bulk temperature. As is often acknowledged in these evaluations, the bulk temperature is not generally equal to the sea surface skin temperature measured by the satellite. Studies directed at a comparison between skin and bulk temperatures by Grassl (1976), as well as Paulson and Simpson (1981), demonstrate marked (about 0.5 °C) differences between the surface skin and subsurface temperatures. In an effort to better evaluate the atmospheric attenuation of infrared radiance, Schlüessel et al. (1987) compare precision radiometric measurements

made from a ship with SST calculated using a variety of techniques from coincident NOAA-7 AVHRR imagery. In addition, subsurface temperatures were continuously monitored with thermistors in the upper 10 m for comparison with the ship and satellite radiometric SST estimates.

As part of this study, Schlüessel et al. (1987) examined the effects of radiometer scan angle on the AVHRR attenuation and concluded that differences in scan angle, resulting in different atmospheric paths, resulted in significant changes in the computed SST. To correct for atmospheric water vapor attenuation, HIRS radiances were used to correct the multichannel computation of SST from the AVHRR. The correspondence between the HIRS radiances and atmospheric water vapor content was found by numerical simulation of 182 different atmospheres. With the HIRS correction, AVHRR-derived SST was found to have a bias of +0.03 °C and an RMS error of 0.42 °C when compared with the ship radiometer measurements. Comparison between ship radiometric and in situ temperatures yielded a mean offset of 0.2 °C and a range of −0.5 to +0.9 °C about this value (Figure 1.12). According to Weinreb et al. (1990), even if the nonlinearity corrections for the sensor remained valid after satellite launch, the error in AVHRR data is still 0.55 °C of which 0.35 °C is traceable to calibration of the laboratory blackbody.

Originally called the "Global Ocean Data Assimilation Experiment (GODAE) High Resolution SST" group, GHRSST has now evolved into the Group for High Resolution Sea Surface Temperature (https://www.ghrsst.org/) with a funded project office and a science team. As stated on the Web site "The aim of the GHRSST is to provide the best quality sea surface temperature data for applications in short, medium, and decadal/climate time scales in the most cost-effective and efficient manner through international collaboration and scientific innovation." GHRSST is an international project with over

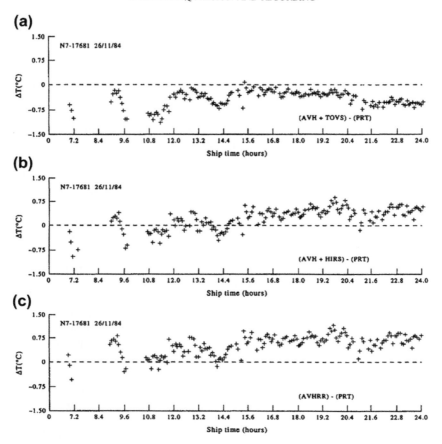

FIGURE 1.12 Difference between uncorrected (a) and corrected (b, c) satellite-sensed sea surface temperatures and ship radiometric in situ temperatures. HIRS, high-resolution infrared sounder; PRT, Platinum resistance thermocouples; AVHRR, advanced very high resolution radiometer.

$18 million US invested across all of its project activities. GHRSST generates data products for the wider science community while keeping a research front to further improve satellite SST products. Of particular interest are problems in diurnal variability, skin temperature deviations, and SST validation/verification. GHRSST data management is designed to make SST data fields more widely available for all science users.

In an effort to standardize the discussion of SST, GHRSST has introduced some definitions of SST (Minnett and Kaiser-Weiss, 2012). These definitions are best described using a figure, presented here as Figure 1.13.

Near-surface temperature gradients in the ocean are the result of three different processes: (1) absorption of solar insolation in the first few meters; (2) heat loss to the atmosphere by thermal emissions from the upper few millimeters of the surface; and (3) subsurface turbulent mixing. The temperature of the mean skin layer, which is typically only 10 μm thick, is derived as an average over length scales that are long compared to those of turbulent eddies that erode

FIGURE 1.13 Cartoon of near-surface temperatures with a highly nonlinear y-axis. Scales are for general guidance and are not to be used for derivations.

the skin layer. This is inherently true of satellite infrared SST measurements that are spatial averages over typically 1 km^2 or more. Even ship-based skin temperature radiometers, with typical m^2 spatial resolutions and a finite integration time, result in averages over areas of a few square meters.

The solar radiation that penetrates the air–sea interface is mostly absorbed (or scattered and then absorbed), subsequently heating the ocean to depths of a few meters or more (Figure 1.14). In most cases, the ocean is warmer than the overlying atmosphere and the net heat exchange is always from the ocean to the atmosphere.

The outgoing radiative emission from the sea surface maintains a gradient such that the temperature of the ocean surface in contact with the atmosphere is always cooler than the underlying water. As a result, the ocean supplies heat to the atmosphere both in terms of this radiant transfer and in terms of the sensible heat loss. This heat is then distributed within the atmosphere by turbulent eddies in the air. The very thin aqueous layer that supports this thermal

gradient is referred to as the "thermal skin layer" which, even though it is statically unstable (cold water overlying warmer water), it is hindered from achieving static stability through vertical convection by the viscosity of the water. The profiles in Figure 1.14 demonstrate that the emission measured in the thermal infrared originates in the thermal skin layer while the microwave emissions originate, at least in part, below the skin layer. There is no emission from the sea surface in the shorter-wavelength visible portion of the spectrum. However, it is these energetic shorter wavelengths that heat the upper ocean, which is then stored and transported by the ocean circulation.

Estimates of the average thickness of the thermal skin layer vary, ranging from a few millimeters (Katsaros et al., 1977) to a few tenths of a millimeters (Hanafin, 2002; Hanafin and Minnett, 2001). This variation is to be expected as the thermal skin layer is a dynamic feature being continually eroded by turbulence from below (e.g., Soloviev and Schlüssel, 1994). Nearly all of the infrared radiation and much of the microwave

Absorption and emission of radiation in the ocean skin layer

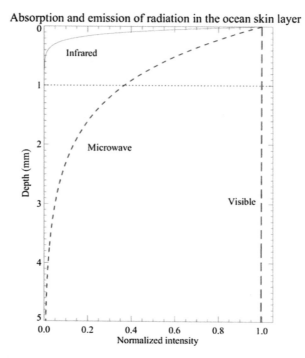

FIGURE 1.14 Profiles of normalized radiation intensity through the upper 5 mm of the ocean surface for visible, thermal infrared, and microwave radiation incident at the sea surface. The Beer–Lambert's Law absorption coefficients are 1/m in the visible, 10^{-4}/m in the infrared, and 10^{-3}/m in the microwave frequency bands. From Kirchoff's Law, the emission coefficient equals the absorption coefficient for local thermodynamic equilibrium.

energy from the sun and atmosphere that is incident at the sea surface is absorbed within the thermal skin layer. In contrast, about 99% of the incident energy in the visible band penetrates through the upper millimeter of the ocean.

Thus, at infrared wavelengths the electromagnetic skin layer is completely contained in the thermal skin layer of the ocean while at microwave frequencies the thermal and electromagnetic skin depths are comparable and microwave energy can extend into the subskin regime of the ocean. Infrared satellite radiometers only measure emissions at infrared wavelengths and, therefore, only sense the thermal skin temperature of the ocean and not the subsurface temperature (the temperature measured from buoys and ships). This subsurface temperature

measurement, which is frequently referred to as the "bulk SST", has the disadvantage that it can vary in depth from a few centimeters to 2 m below the surface.

The relationship between the skin and bulk SSTs also varies with external conditions while the relationship between the skin SST and the subskin SST is rather well behaved (Minnett et al., 2011). The difference between the skin and subskin SST is known to asymptote at −0.13 K at high wind speeds and can exceed −0.6 K at low winds, consistent with its existence at the radiative boundary layer of the ocean. The constant offset between the skin SST and the deeper bulk SST indicated in Figure 1.13 is the average difference during the night when there is no external heating; it also holds for winds stronger than 6 m/s (Donlon et al., 2002). Under low winds the relationship between skin SST and bulk SST is highly variable vertically, horizontally, and temporally (Minnett, 2003; Ward, 2006; Gentemann and Minnett, 2008). Under low wind conditions, heat in the upper ocean is not well mixed, causing thermal stratification. There is also a strong diurnal component to this stratification along with a dependence on cloud cover and wind speed. As a result, the variability of near-surface temperature gradients can be quite pronounced compared to the accuracy of the SST measurements needed for climate research (e.g., Ohring et al., 2005). Consequently, the effects of the thermal skin layer and diurnal heating must be taken into account when generating climate records of SST.

GHRSST continues to be an international community that works with data and provides services for a wide variety of research and operational agencies. Divided into focus groups, GHRSST science team members take on responsibilities regarding the generation and distribution of a wide variety of SST data products. GHRSST is led by elected international experts: the GHRSST Science Team.

1.3.9 The Modern Digital Thermometer

In many oceanographic institutions using sampling bottles for biogeochemistry measurements, the mercury thermometer has been replaced by a digital deep-sea reversing thermometer built by Sensoren-Instrumente-System (SIS) of Kiel that uses a highly stable platinum thermistor to measure temperature. The SIS RTM 4002 digital reversing thermometer has the outer dimensions of mercury instruments so that it fits into existing thermometer racks. Instead of the lighted magnifying glass needed to read most mercury thermometers, the user simply touches a small permanent magnet to the sensor "trigger" spot on the glass face of the thermometer to obtain a bright digital readout of the temperature to three decimal places. Since the instrument displays the actual temperature at the sample depth, there is no need to read an auxiliary thermometer to correct the main reading, as is the case for reversing mercury thermometers. This makes life much more pleasant on a rolling ship in the middle of the night. Because the response time of the platinum thermometer is rapid compared with the "soaking" time of several minutes required for mercury thermometers, less ship time is needed at oceanographic stations.

The SIS RTM 4002 has a range of -2 to $40\,°C$ and a stability of $0.00025\,°C$ per month. According to the manufacturer, the instrument has a resolution of $\pm0.001\,°C$ and an accuracy of $0.005\,°C$ over the temperature range -2 to $20\,°C$. Both resolution and accuracy are considerably lower for temperatures in the range $20-40\,°C$. A magnet is used to reset the instrument and to activate the light-emitting diode (LED). Sampling can be performed in the three sequential modes. The "Hold" mode displays the last temperature stored in memory; the "Cont" mode allows for continuous sampling for use in the laboratory; while the "Samp" mode is used for reversing thermometer applications. The instrument allows for a minimum of 2700 samples on two small lithium batteries.

A recent addition to stand-alone oceanographic temperature measurement is the SBE 35, a laboratory standards thermometer that is commonly used on carousel water samplers for separate validation of CTD temperature data. The SBE 35 can be used up to depths of 6800 m and, according to the manufacturer, is "unaffected by shock and vibration encountered in shipboard and industrial environments", making it ideal for use in calibration laboratories in the range of -5 to $+35\,°C$. The SBE 35 communicates via a standard RS-232 interface at 300 baud, 8 data bits, no parity. The instrument has a manufacturer's specified resolution of $0.000025\,°C$, an initial accuracy of $\pm0.001\,°C$, and a stability of $0.001\,°C$ per year.

1.3.10 Potential Temperature and Density

The deeper one goes into the ocean, the greater the heating of the water caused by the compressive effect of hydrostatic pressure. The ambient temperature for a parcel of water at depth is significantly higher than it would be in the absence of pressure effects. Potential temperature is the in situ temperature corrected for this internal heating caused by adiabatic compression as the parcel is transported to depth in the ocean. To a high degree of approximation, the potential temperature defined as $\theta(p)$, or T_θ, is given in terms of the measured in situ temperature $T(p)$ as $\theta(p) = T(p) - F(R)$, where $F(R) \approx 0.1\,°C/km$ is a function of the adiabatic temperature gradient R. The results can have important consequences for oceanographers studying water mass characteristics in the deep ocean. For example, measured ambient temperatures in the deeper waters of the 2750 km long, maximum 6669 m deep Middle America Trench

off the west coast of Central America gradually increase with depth, whereas the calculated potential temperature profiles reveal that the water column below about 4000 m depth is typically "isothermal" to within 0.001 °C.

The difference between the ambient temperature and θ increases slowly from zero at the ocean surface to about 0.5 °C at 5000 m depth. For depth differences of approximately 100 m and temperatures less than 5 °C, the difference between the two forms of temperature is roughly the absolute resolution ($\sim 0.01\,°C$) of most thermistors (Figure 1.15). Differences of this magnitude are significant in studies of deep ocean heating from hydrothermal venting or other heat sources where temperature anomalies of 10 millidegrees (0.010 °C) are considered large. In fact, if the observed temperatures are not converted to potential temperature, it is impossible to calculate the anomalies correctly.

The use of potential temperature in the calculation of density leads to the definition of potential density, $\rho_\theta = \rho(S, \theta, 0)$ in kg/m^3, as the value

FIGURE 1.15 Difference between in situ temperature (T) recorded by a CTD vs the calculated potential temperature (θ) for a deep station in the North Pacific Ocean (35° N, 152° W). Below about 500 m, this curve is applicable to any region of the world ocean. CTD, Conductivity-Temperature-Depth profiler. (*Data from Martin et al. (1987).*)

of ρ for a given salinity and potential temperature at surface pressure, $p = 0$. The corresponding counterpart to $\sigma_t = (10^3[\rho(S, T, 0) - 1000])$, is then called "sigma-theta", where $\sigma_\theta = 10^3[\rho(S, \theta, 0) - 1000]$. Since density surfaces (as well as isotherms) can be displaced vertically hundreds of meters by internal oscillations in the deep ocean, it is crucial that we compare temperatures correctly by taking into consideration the thermal compression effect. Readers familiar with the oceanographic literature will also note the use of σ_2, σ_4, and similar sigma expressions for density surfaces in the deep ocean. These expressions are used as reference levels for the calculation of density at depths where the effect of hydrostatic compression on density becomes important. For example, $\sigma_4 = 10^3[\rho(S, \theta, 4) - 1000]$ refers to density at the observed salinity and potential temperature referred to a pressure of 4000 dbar (40,000 kPa) or about 4000 m depth. Use of σ_θ in the deep Atlantic suggests a vertically unstable water mass below 4000 m whereas the profiles of σ_4 correctly increase toward the bottom (Pickard and Emery, 1992: Talley et al., 2011). As indicated by Table 1.2, the different sigma values differ significantly.

1.4 SALINITY

It is the salt in the ocean that separates physical oceanography from other branches of fluid dynamics. Most oceanographers are familiar with the term "salinity" but many are not aware of its precise definition. Physical oceanographers often forget, that salinity is a nonobservable quantity and was traditionally defined by its relationship to a measurable parameter, "chlorinity". For the first half of the 20th century, chlorinity was measured by the chemical titration of a seawater sample. In 1899, the International Council for the Exploration of the Sea (ICES) established a commission, presided over by Professor M. Knudsen, to study the problems of determining salinity and density from seawater samples. In its report (Forch et al., 1902), the commission recommended that salinity be defined as follows: "The total amount of solid material in grams contained in 1 kg of seawater when all the carbonate has been converted to oxide, all the bromine and iodine replaced by chlorine, and all the organic material oxidized."

Using this definition, and available measurements of salinity, chlorinity (Cl), and density

TABLE 1.2 Comparison of Different Forms of Sigma for the Western Pacific Ocean near Japan (39°41' N, 147°56' E)

Depth (m)	In situ T (°C)	Potential T (0) (°C)	Salinity (psu)	σ_θ	σ_2	σ_4
0	18.909	18.909	32.574	23.192	31.706	39.852
100	1.160	1.156	33.158	26.555	35.830	44.689
500	3.338	3.305	34.108	27.145	36.286	45.020
1000	2.697	2.632	34.410	27.447	36.619	45.382
2000	1.868	1.734	34.600	27.672	36.890	45.696
3000	1.528	1.311	34.661	27.752	36.993	45.820
4000	1.456	1.138	34.679	27.778	37.029	45.865
5000	1.503	1.069	34.686	27.788	37.043	45.883
5460	1.547	1.054	34.688	27.791	37.046	45.886

Columns 2 and 3 give the in situ and potential temperatures, respectively. Sigma units are kg/m^3. Salinity is now often written with no units attached to the values. (From Talley et al. (1988).)

for a relatively small number of samples (a few hundred), the commission produced the empirical relationship

$$S(0/00) = 1.805Cl(0/00) + 0.03 \qquad (1.12)$$

known as Knudsen's equation and a set of tables referred to as Knudsen's tables. The symbol ‰ indicates "parts per thousand" (ppt) in analogy to percent (%) which is parts per hundred. In the more modern Practical Salinity Scale, salinity is a unitless quantity but, because many oceanographers are uncomfortable with unitless values, salinity is sometimes written as "psu" for *practical salinity units*. (*Note*: Although we are aware that salinity is often written as a unitless variable, we are guilty of switching among the three formats in the text, especially, when discussing historical salinity measurements. We also find that including the unit *psu* helps eliminate ambiguity since it is clear that the variable under discussion is salinity, rather than some other nondimensional variable.) It is interesting to note that Knudsen himself considered using electrical conductivity (Knudsen, 1901) to measure salinity. However, due to the inadequacy of the apparatus available, or similar problems at the time, he decided that the chemical method was superior.

There are many different titration methods used to determine salinity but that most widely applied is the colorimetric titration of halides with silver nitrate ($AgNO_3$) using the visual end point provided by potassium chromate (K_2CrO_4), as described in Strickland and Parsons (1972). With a trained operator, this method is capable of an accuracy of ± 0.02‰ in salinity using the empirical Knudsen relationship. For precise laboratory work Cox (1963) reported on more sensitive techniques for determining the titration end point, which yield a precision of 0.002‰ in chlorinity. Cox also describes an even more complex technique, used by the Standard Sea-water Service, which is capable of a precision of about ± 0.0005‰ in chlorinity. It is fairly safe to say

that these levels of precision are not typically obtained by the traditional titration method and that preconductivity salinities are generally no better than ± 0.02‰ (± 0.02, or ± 0.02 psu).

1.4.1 Salinity and Electrical Conductivity

In the early 1950s, technical improvements in the measurement of the electrical conductivity of seawater turned attention to using conductivity as a measure of salinity rather than the titration of chlorinity. Seawater conductivity depends on the ion content of the water and is therefore directly proportional to the salt content. The primary reason for moving away from titration methods was the development of reliable methods of making routine, accurate measurements of conductivity. The saving in time and labor has been exceptional. As noted earlier, the potential for using seawater conductivity as a measurement of salinity was first recognized by Knudsen (1901). Later papers explored further the relationship between conductivity, chlorinity, and salinity. A paper by Wenner et al. (1930) suggested that electrical conductivity was a more accurate measure of total salt content than of chlorinity alone. The authors' conclusion was based on data from the first conductive salinometer developed for the International Ice Patrol. This instrument used a set of six conductivity cells, controlled the sample temperature thermostatically and was capable of measurements with a precision of better than ± 0.01‰. With an experienced operator the precision may be as high as ± 0.003‰ (Cox, 1963). The latter is a typical value for most modern conductive and inductive laboratory salinometers and is an order-of-magnitude improvement in the precision of salinity measurements over the older titration methods.

It is worth noting that the conductivity measured by either inductive or conductive laboratory salinometers, such as the widely used Guildline 8410A Portasal™ (portable)

Salinometer, are relative measurements that are standardized by comparison with "standard seawater". As an outgrowth of the ICES commission on salinity, the reference, or standard seawater, was referred to as "Copenhagen Water" due to its earliest production by a group in Denmark. This standard water is produced by diluting a large sample of seawater, until it has a precise salinity of 35‰ (Cox, 1963). Standard UNESCO seawater is now being produced by the "Standard Seawater Service" in England as well as at other locations in the U.S.A. (i.e., Woods Hole Oceanographic Institution).

Standard seawater is used as a comparison standard for each "run" of a set of salinity samples. To conserve standard water, it is customary to prepare a "secondary standard" with a constant salinity measured in reference to the standard seawater. A common procedure is to check the salinometer every 10−20 samples with the secondary standard and to use the primary standard every 50 or 100 samples. In all of these operations, it is essential to use proper procedures in "drawing the salinity sample" from the hydrographic water bottle into the sample bottle. Assuming that the hydrographic bottle remains well sealed on the upcast, two effects must be avoided: first, contamination by previous salinity samples (that have since evaporated leaving a salt residue that will increase salinity in the present bottle sample); and second, the possibility of evaporation of the present sample. The first problem is avoided by "rinsing" the salinity bottle and its cap two to three times with the sample water. Evaporation is avoided by using a screw cap with a gasket seal. A leaky bottle will give sample values that are distorted by upper ocean values. For example, if salinity increases and dissolved oxygen decreases with depth, deep samples drawn from a leaky bottle will have anomalously low salinities and high oxygens.

Salinity samples are usually allowed to come to room temperature before being run on a laboratory bench salinometer. In running the salinity samples one must be careful to avoid air bubbles and insure the proper flushing of the salinity sample through the conductivity cell. Some bench salinometers correct for the marked influence of ambient temperature on the conductivity of the sample by controlling the sample temperature while other salinometers merely measure the sample temperature in order to be able to compute the salinity from the conductivity and coincident temperature.

Another reason for the shift to conductivity measurements was the potential for in situ profiling of salinity. The development history of the STD/CTD profilers has been sketched out in Section 1.3.4 in terms of the development of continuous temperature profilers. The salinity sensing aspects of the instrument played an important role in the evolution of these profilers. The first STD (Hamon, 1955) used an electrode-type conductivity cell in which the resistance or conductivity of the seawater sample is measured and compared with that of a sample of standard seawater in the same cell. Fouling of the electrodes can be a problem with this type of sensor. Later designs (Hamon and Brown, 1958) used an inductive cell to sense conductivity. The inductive cell salinometer consists of two coaxial torodoial coils immersed in the seawater sample in a cell of fixed dimensions. An alternating current is passed through the primary coil, which then induces an electromagnetic force (EMF) and hence a current within the secondary coil. The EMF and current in the secondary coil are proportional to the conductivity (salinity) of the seawater sample. Again, the instrument is calibrated by measuring the conductivity of standard seawater in the same cell. The advantage of this type of cell is that there are no electrodes to become fouled. A widely used inductive type STD was the Plessey model 9040, which claimed a salinity accuracy of ±0.03. Precision was somewhat better, being between 0.01 and 0.02 depending on the resolution selected. Modern electrode-type cells measure the difference in voltage between conductivity elements at each end of the seawater passageway. With the conducting elements potted into the same material this type of salinity sensor is less prone to contamination by biological fouling. At the same time, the response time of the conductive cell is much greater than that of an inductive sensor leading to the problem of salinity spiking due to a mismatch with the temperature response.

The mismatch between the response times of the temperature and conductivity sensors is the primary problem with STD profilers. Spiking in the salinity record occurs because the salinity is computed from a temperature measured at a slightly different time than the conductivity measurement. Modern CTD systems record conductivity directly, rather than the salinity computed by the system's hardware, and have faster response thermal sensors.

In addition, most modern CTD systems use electrodes rather than inductive salinity sensors. As shown in Figure 1.16, this sensor has a set of four parallel conductive elements that constitute a bridge circuit for the measurement of the current passed by the connecting seawater in the glass tube containing the conductivity elements. The voltage difference is measured between the conducting elements in the bridge circuit of the conductivity cell. The primary advantage of the conductive sensor is its greater accuracy and faster time response. Moreover, as discussed in sections 1.3.6 and 1.4.2, the mismatch problem is further mitigated in CTD systems, which use pumps to maintain a constant flow of water past the conductivity sensors. The flow rate is designed to minimize the mismatch between the conductivity and temperature sensors. In their discussion of the predecessor of the modern CTD, Fofonoff et al. (1974) give an overall salinity accuracy for this instrument of $\pm 0.003\,\%_o$. This accuracy estimate was based on comparisons with in situ reference samples whose salinities were determined with a laboratory salinometer also accurate to this level (Figure 1.17). Accuracies of this level are the same as the standard deviation of duplicate

salinity samples run in the lab, demonstrating the high level of accuracy of CTD profilers.

1.4.1.1 A Comparison of Two CTDs

During sea trials in the North Atlantic, scientists at the Bedford Institute of Oceanography (Bedford, Nova Scotia) examined in situ temperature and salinity records from a EG&G Mark V CTD and a SBE 9 CTD (Hendry, 1993). The standards used for the comparisons were temperatures measured by SIS digital-reading reversing thermometers and salinity samples drawn from 10-l bottles on a Rosette sampler and analyzed using a Guildline Instruments Ltd Autosal 8400A salinometer standardized with IAPSO Standard Water. Here, IAPSO stands for the International Association for the Physical Sciences for the Oceans.

The Mark V was able to sample at 15.625 Hz and used two thermometers and a standard inductive salinity cell. The fast response (250 ms time constant) platinum thermometer was used to record the water temperature while the slower resistance thermometer, whose response time is more closely tuned to that of the conductivity cell, was used in the conversion of conductivity to salinity. Plots of the differences in temperature and salinity between the bottle samples and the CTD are presented in Figure 1.18. Using only the manufacturer's calibrations for all instruments, the Mark V CTD temperatures were lower than the reversing thermometer values by $0.0034 \pm 0.0023\,°C$ with no obvious dependence on depth (Figure 1.18(a)). In contrast, the Mark V salinity differences (Figure 1.18(c)) showed a significant trend with pressure, which may be related to the instrument used or a peculiarity of the cell. With pressure in decibar, regression of the data yields

Salinity Diff(bottle − CTD)

$$= 0.00483 + 6.259 10^{-4} \text{ Pressure (CTD)}$$

(1.13)

with a squared correlation coefficient $r^2 = 0.84$. Removal of the trend gives salinity values

FIGURE 1.16 Guildline conductivity (salinity) sensor showing the location of four parallel conductive elements inserted into the hallow glass tube. Conductivity is measured as the water flows through the glass tube. Cable plugs into the top of the CTD end plate on the pressure case. CTD, Conductivity-Temperature-Depth profiler.

accurate to about ±0.003 psu. Pressure errors of several decibars (several meters in depth) were noted.

The SBE 9 and SBE 25 sample at 24 Hz use a high-capacity pumping system and T-C duct to flush the conductivity cell at a known rate (e.g., 2.5 m/s pumping speed for a rate of 0.6–1.2/s). When on deck, the conductivity cell must be kept filled with distilled water. To allow for the proper alignment of the temperature and conductivity records (so that the computed salinity is related to the same parcel of water as the temperature), the instrument allows for a time shift of the conductivity channel relative to the temperature channel in the deck unit or in the system software Seasoft (module AlignCTD). In the Bedford

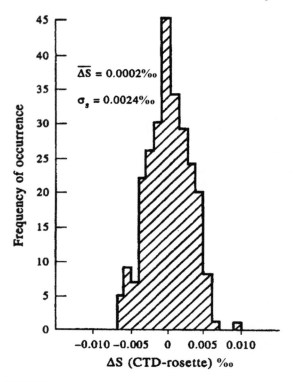

FIGURE 1.17 Histogram of salinity differences (in parts per thousand, $^0/_{00}$). Values used are the differences in salinity between salinity recorded by an early Neil-Brown CTD and deep-sea bottle samples taken from a Rosette sampler. $\overline{\Delta S}$ is the mean salinity differences and σ_s is the standard deviation. *(Modified after Fofonoff et al. (1974).)*

By comparison, the precision of a single bottle salinity measurement is ±0.0007 psu. Pressure errors were less than 1 dbar. Due to geometry changes and the slow degradation of the platinum black on the electrode surfaces, the thermometer calibration is expected to drift by 2 m°C/year and the electronic circuitry by 3 m°C/year.

Based on the Bedford report, modern CTDs are accurate to approximately ±0.002 °C in temperature, ±0.005 psu in salinity and <0.5% of full-scale pressure in depth. The report provides some additional interesting reading on oceanic technology. To begin with, the investigators had considerable difficulty with erroneous triggering (misfiring) of bottles on the Rosette. Those of us who have endured this notorious "grounding" problem appreciate the difficulty of trying to decide if the bottle did or did not misfire and if the misfire registered on the shipboard deck unit. If the operator triggers the unit again after a misfire, the question arises as whether the new pulse fired the correct bottle or the next bottle in the sequence. Several of these misfires can lead to confusing data, especially in well-mixed regions of the ocean. It is good policy to keep track of the misfires for sorting out the data later.

Another interesting observation was that variations in lowering speed had a noticeable influence on the temperature and conductivity measurements. Since most modern CTDs dissipate 5−10 W, the CTD slightly heats the water through which it passes. At a 1-m/s nominal fall rate, surface swell can cause the actual fall rate to oscillate over an approximate range of ±1 m/s with periodic reversals in fall direction at the swell period. (Heave compensation is needed to prevent the CTD from being pulled up and down.) As a result, the CTD sensors are momentarily yanked up through an approximately 1 m°C thermal wake that is shed from boundary layers of the package as it decelerates. Hendry (1993) claims that conditional editing based on package speed and acceleration is

study, the conductivity was shifted by 0.072 s earlier to align with temperature. (The deck unit was programed to shift conductivity by one integral scan of 0.042 s and the software the remaining 0.030 s.) Using the manufacturer's calibrations for all instruments, the SBE 9 CTD temperatures for the nine samples were higher than the reversing thermometer values by 0.0002 ± 0.0024 °C with only a moderate dependence on depth (Figure 1.18(b)). Salinity data from 30 samples collected over a 3000-db depth range (Figure 1.18(d)) gave CTD salinities that were lower (fresher) than Autosal salinities by 0.005 ± 0.002 psu, with no depth dependence.

FIGURE 1.18 CTD correction data for temperature (bottle—CTD) based on comparison of CTD data with in situ data from bottles attached to a Rosette sampler. (a) Temperature difference for the EG (b) same as (a) but for the Sea-Bird SBE 9 CTD; (c) salinity difference for the EG (d) same as (c) but for the SBE 9 CTD. Regression curves are given for each calibration in terms of the pressure, P, in decibar. r^2 is the squared correlation coefficient. (*Adapted from Hendry (1983).*)

reasonably successful in removing these arti-facts. Since turbulent drag varies with speed squared, mechanical turbulence was found to cause package vibration that affected the electri-cal connection from the platinum thermometer to the Mark V CTD. Mixing, entrainment, and thermal contamination caused differences in down- vs upcasts in both instruments. The correction for thermal inertia of the conductiv-ity cell in the SBE CTD resulted in salinity changes of 0.005 psu with negative downcast corrections when the cell was cooling and posi-tive upcast corrections when the cell was warming.

According to repeat deep CTD-Rosette casts at Ocean Station "P" in the central northeast Pacific (50° N, 145° W) carried out by investigators at the Institute of Ocean Sciences (Sidney, British Columbia), there is good repeatability in the pumped SBE 911*plus* CTD systems. For example, variations referenced to common density surfaces at around 1200-m depth during a June-2012 sur-vey found changes of 0.001 °C and 0.0002 psu over a period of 4 h and 0.002 °C and 0.0005 psu over a period of 2.5 days. Other years showed similarly small variations. No drift in the sensors was detected during postcruise calibrations, and some of the change was attributed to slight changes in salinity values obtained from the Rosette bottle samples (Germaine Gatien and Steve Romaine, pers, comm., 2012).

1.4.2 The Practical Salinity Scale

In using either chlorinity titration or the measurement of conductivity to compute salinity, one employs an empirical definition relating the observed variable to salinity. In light of the increased use of conductivity to measure salinity, and its more direct relationship to total salt content, a new definition of salinity has been developed. As a first step in establishing the relationship between conductivity and salinity, Cox et al. (1967) examined this relationship in a variety of water samples from various geographical regions. These results were used in formulating new salinity tables (UNESCO, 1966) from which Wooster et al. (1969) derived a polynomial fit giving a new formula for salinity in terms of the conductivity ratio (K) at 15 °C. The RMS deviation of this fit from the tabulated values was 0.002‰ in chlorinity for values greater than 15‰ and 0.005‰ for smaller values. It is worth noting that Cox et al. (1967) found that deep samples (>2000-m depth) had a mean salinity, computed from chlorinity, that was 0.003‰ lower than that for conductivity. This was not true for the surface samples.

As noted earlier, a new salinity definition has been adopted called the "practical salinity scale" or PSS 78 (Lewis, 1980). This scale has been accepted by major oceanographic organizations and has been recommended as the scale in which to report future salinity data (Lewis and Perkin, 1981). The primary objections to the earlier salinity definition of Wooster et al. (1969) were:

1. With salinity defined in terms of chlorinity it was independent of the different ionic ratios of seawater.
2. The mixtures of reference seawaters used to derive the relationship between chlorinity and conductivity ratio were nonreproducible.

3. The corresponding International Tables do not go below 10 °C which makes them unsuitable for many in situ salinity measurements.

In the practical salinity scale, it is suggested that standard seawater should be a conductivity standard corresponding to, and having the same ionic content as, Copenhagen Water. The salinity of all other waters will be defined in terms of the conductivity ratio (R_{15} or C_{15} in the nomenclature of Lewis and Perkins) derived from a study of dilutions of standard seawater. This becomes then a practical salinity scale as distinct from an absolute salinity defined in terms of the total mass of salts per kilogram of solution.

A major problem in applying this new salinity scale is its application to archived hydrographic data. As discussed by Lewis and Perkins (1981), the correction procedure for such data depends not only on the reduction formula used, but also on the calibration procedure used previously for the salinity instrument. Essentially, the correction procedure amounts to performing this calibration a second time using the differences, provided by Lewis and Perkins (1981) between the older salinity scale and PSS 78. Another alternative would be to return to the original raw data, if they have been saved, and to recompute the salinity according to PSS 78. From the discussion of Lewis and Perkins it is clear, however, that for salinities in the range of 33—37‰ differences of about ±0.01‰ can be anticipated between archived salinities and the corresponding values computed using PSS 78. This is about the same overall accuracy of modern CTD profilers.

It is interesting that the primary motivation for the development of PSS 78 came from people working in low salinity polar waters, where the UNESCO tables did not apply. In areas such as estuarine environments, which have very low salinities, and mid-ocean ridge regions with strong hydrothermal fluid venting, even PSS 78 is not adequate and there are still serious limitations to computing accurate salinities from conductivity measurements. There are several reasons for these limitations. First of all, the approach of relating specific conductance to salinity of total dissolved solids requires that the proportions of all the major ions in the natural electrolyte remain constant in time and space, and second, that the salinity expression represents all the dissolved solids in the fluid. Another factor to keep in mind is that the density calculated from the conductivity values is based on conducting ions in the fluid. If there are chemical components—or suspended particles—that contribute to the density of the fluid but not to the conductivity (i.e., nonconductive ions) then the density will be wrong. This is exactly the problem faced by McMannus et al. (1992) for deep CTD data from Crater Lake in Oregon. Here, silicic acid from hydrothermal venting in the south basin of the lake below 450-m depth did not contribute to the conductance but did

alter the density. Without accounting for silicic acid the water column was weakly stratified; after accounting for it, the bottom waters became stratified. The different combinations of ions in hydrothermal fluids from mid-ocean ridges can seriously alter the salinity structure observed at the source during a normal deep CTD measurement. However, because of rapid entrainment of ambient bottom water, the ion mix becomes similar to that of normal seawater a few meters above the vent orifice. Typical ratios of ambient to hydrothermal fluid volumes are 7000:1. The vertical structure in buoyant plumes, such as that in Figure 1.19 taken a few meters above venting fields on Juan de Fuca Ridge in the northeast Pacific, presumably results from unstable conditions arising from turbulent mixing in the rising plume, not from sensor response problems.

Suspended particles can also contribute to increases in water density without affecting the

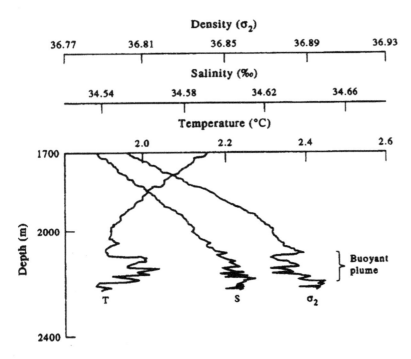

FIGURE 1.19 Vertical profiles of temperature, salinity (‰) and potential density collected from a CTD mounted on the submersible Alvin. Data collected during ascent away from the main hydrothermal vent field at Endeavour Ridge in the northeast Pacific (48° N, 129° W). Density is unstable over the depth range of the buoyant portion of the plume. *(From Lupton et al. (1985).)*

"salinity". In a recent study, Thomson et al. (2010) suggest that prolonged, along-axis episodic flows of 0.2 m/s and greater observed at depths of over 4000 m at the bottom of the Middle America Trench off Central America are likely rotationally modified, autosuspending turbidity currents initiated by tidal current resuspension of sediments above the 100-km long segment of the trench that shoals by 1000 m to the southeast of the mooring site. Suspended particles in the turbidity currents are estimated to be only about 0.0003—0.006% by volume. CTD measurements reveal that the potential density of the water is uniform in the lower 1000 m of the trench and therefore unlikely to account for such strong bottom currents. The Woods Hole submersible *Alvin* was sometimes forced to hang onto bottom-mounted equipment to stop the vehicle from drifting downtrench while working on borehole sites in the region (Earl Davis, pers. com., 2009). Results suggest that tidally induced turbidity currents may be common to steep, well-mixed regions of the deep ocean adjacent to sediment rich continental margins.

1.4.3 Nonconductive Methods

Efforts have been made to infer salinity directly from measurements of refractive index and density. Since the refractive index (n) varies with the temperature (T) and salinity (S) of a water sample (and with the wavelength of the illumination), measurements of n and T can be used to obtain in situ estimate of salinity. In order to achieve a salinity accuracy of ± 0.01 psu, it is necessary to measure n to within 20×10^{-7} and to control temperature to within $\pm 0.005\,°C$. Some refractometers are capable of measuring n to 100×10^{-7}, leading to a salinity precision of 0.06 psu. Handheld refractometers are simple and easy to use but yield salinity measurements no better than ± 0.2 psu. For higher sensitivity, interference methods can be used giving a precision in n of 5×10^{-7} corresponding to a salinity

precision of ± 0.003 psu. This is a comparative interference technique and requires a reference seawater sample. Since it is a comparative method, knowledge of the exact temperature is not critical as long as both samples are observed at the same temperature. Direct measurements of seawater density can yield a precision of ± 0.008 in sigma-t (Kremling, 1972), which can be used to calculate salinities to within ± 0.02 psu. Since the measurement of density is much more complicated than those of temperature and electrical conductivity, these latter quantities are usually observed and used to compute the in situ density.

1.4.4 Remote Sensing of Salinity

The measurement of sea surface salinity (SSS) from space was first attempted on Skylab (Lerner and Hollinger, 1977) using a 1.4 GHz microwave radiometer. Although many of the corrections needed to adjust for ambient atmospheric and oceanographic conditions were not well understood in the 1970s, the strong correlation between the sensor data (after correcting for other effects) and SSS were sufficiently encouraging to proceed with the technology. Spacecraft remote sensing of SSS using low-frequency microwave radiometry was first proposed by Swift and McIntosh (1983). The AMSR-E was placed in orbit on the Aqua satellite in 2002 but because of its limited ocean salinity measurement accuracy, it was not suitable for measuring the small salinity gradients of the open ocean. However, AMSR-E data taken over the Amazon River plume was originally used to demonstrate the feasibility of measuring ocean surface salinity with microwave radiometers from space (Reul et al., 2009; Klemas, 2011).

Two satellite missions are measuring (or have measured) SSS: the European Space Agency (ESA) Soil Moisture and Ocean Salinity (SMOS) mission launched in November 2009 and the NASA Aquarius mission launched in June 2011. Both missions were designed to last about three years. At present, the satellites are providing large-scale SSS measurements at accuracies of ± 0.5 and ± 0.2 psu, respectively, at temporal scales of weeks to months over spatial scales of 50–100 km. The SMOS mission is measuring soil moisture over continental surfaces and surface salinity over the ocean while the Aquarius mission is observing SSS only. SMOS uses an interferometric antenna system to avoid the need for a very large antenna to measure the small salinity signal. Aquarius uses a large reflector to sense the same low frequency microwave signals as SMOS.

The Microwave Imaging Radiometer with Aperture Synthesis on the SMOS satellite measures the passive microwave emission of earth's surface (the brightness temperature, T_b) at a frequency of 1.400–1.427 GHz in the L-band, which is the band best suited to determining SSS. In this case, radiation downwelling from space at ~ 1.4 GHz and impinges on the ocean surface and is partly absorbed within an ~ 1-cm thick upper layer and partially reflected toward the upper atmosphere. Both moisture and salinity decrease the microwave radiation emitted from the earth's surface. SMOS collects data at 6 AM and 18 PM (local time) each day and the surface of the earth is fully covered every three days. Aquarius has an active microwave system coupled to the passive system, yielding a sensitivity that is a least an order of magnitude better than SMOS but over a larger spatial footprint and a longer 8-day repeat cycle for full coverage of the earth.

The polarized radiometric brightness temperature (the microwave radiation from the sea surface) used to determine SSS is given by $T_b = e \times T_s$ where T_s is the actual SST, and e is the emissivity of the sea surface, which itself is a function of SSS and T_s. The effects on T_b from ocean roughness and Faraday rotation must also be taken into consideration. Once the apparent brightness temperature of a body of water is measured and where the thermodynamic SST is measured by other means, the salinity at the surface can be determined. For

typical oceanic ranges of surface salinity and T_s, T_b has a brightness range of about 4–6 K at L-band frequencies. The sensitivity of T_b to changes in SSS vs T_s is greatest in warm water (0.7 K/psu at 30 °C) and least in cold water (0.3 K/psu at 0 °C).

Salinity measurements from space are complicated by several factors, of which the most significant is the effect of sea surface roughness on microwave emission (Yueh et al., 2001). Sea surface roughness associated with wind waves, swell, and foam from breaking waves give rise to a change in the apparent brightness temperature of up to ~ 5 K. Because of the short timescales of ocean winds, accurate measurements of sea surface roughness must be simultaneous with the brightness measurements. These stringent requirements pose technical challenges for achieving the required radiometric accuracy and stability. Finally, the low frequency involved requires the use of very large antennas (or an innovative antenna solution such as was used in SMOS) to achieve a moderate spatial resolution on the ground. The brightness temperature at L-band for a space orbiting radiometer is also affected by galactic emissions ($\Delta T_b \approx 2$–8 K), atmospheric emissions ($\Delta T_b \approx 2.4$–2.8 K), and emission from water vapor and cloud liquid water (both of which have a small effect on ΔT_b).

In summary, calculation of SSS involves: (1) determination of T_b at the sea surface by correcting for ionospheric, atmospheric, and extraterrestrial radiation; (2) correcting for sea surface roughness and SST; and (3) calculation of SSS from T_b. Radiative transfer models allow correction for up- and downwelling emission from the atmosphere, for atmospheric and ionospheric attenuation, and for Faraday rotation of the polarized microwave emissions as the radiation passes through the ionosphere. Downwelling galactic emissions are taken into account using maps provided by radio astronomers. With these corrections, and knowledge of the SST and roughness, salinity can be calculated from T_b.

1.5 DEPTH OR PRESSURE

1.5.1 Hydrostatic Pressure

The depths of profiling instruments are mainly derived from measured hydrostatic pressure, p. This is possible because of the almost linear relationship between hydrostatic pressure, $p = p(z)$, and geometric depth, z. The relationship is such that the "pressure expressed in decibars is nearly the same as the numerical value of the depth expressed in meters" (Sverdrup et al., 1942). The validity of this approximation can be seen in Table 1.3 in which we have compared values of hydrostatic pressure and geometric depth for a standard ocean. At depths shallower than 4000 m, the difference is less than 2%. For many applications, this error is sufficiently small that it can be neglected and hydrostatic pressure values can be converted directly into geometric depth. The cause for the slight difference

between pressure in decibars and depth in meters is found in the familiar hydrostatic relation,

$$p(z) = -g \int_z^0 \rho(z)\mathrm{d}z$$

where $g = 9.81 \text{ m/s}^2$ is the acceleration due to gravity and $\rho \approx 1.025 \times 10^3 \text{ kg/m}^3$ is the mean water density. Units of p are (kg/m^3) (m/s^2) $(\text{m}) = (\text{kg/m})(1/\text{s}^2)$. Also, $p = \text{force/area}$ has units $(\text{N/m}^2) = (\text{Pa}) = (10^{-5} \text{ bar})$. One Newton $(\text{N}) = 1 \text{ kg m/s}^2$, so that $p \approx 1.025 \times 10^3$ $(9.81)z = 1.005525z$ (where depth z is expressed in meters). A different value of density gives a slightly different p vs z relation.

Certain techniques allow for continuous measurement of hydrostatic pressure while others can be carried out at discrete depths only. An example of the latter is the computation of "thermometric depth" using a combination of protected and unprotected reversing thermometers to sense the effects of pressure on the temperature reading. This is still considered one of the most accurate methods of determining hydrostatic pressure and is often used as an in situ calibration procedure for CTD profilers. Specifically, the pressure, p (in decibars, dbar), obtained from a CTD is related to the temperature difference, ΔT, between the protected and unprotected thermometers by $p \approx g\Delta T/k$, where the pressure constant for each individual thermometer is $k \approx 0.1 \text{ °C/(kg/cm}^2)$. The details of this procedure are well described in Sverdrup et al. (1942, p. 350) but with a significant printing error (a missing plus sign in the second bracket), which is not corrected in Defant (1961, Vol. 1, p. 35). When correctly applied (see LaFond, 1951; Keyte, 1965), the thermometer technique is capable of yielding pressure measurements accurate to ±0.5% (Sverdrup, 1947). Most modern CTD systems claim a similar accuracy using strain-gauge sensors to directly measure pressure. The accuracy of early CTD pressure sensors was a function of the depth (pressure) itself and varied from 1.5 dbar in the upper 1500 dbar to

TABLE 1.3   Comparison of Pressure (dbar) and Depth (m) at Standard Oceanographic Depths Using the UNESCO Algorithms

Pressure (dbar)	Depth (m)	Difference (%)
0	0	0
100	99	1
200	198	1
300	297	1
500	495	1
1000	990	1
1500	1483	1.1
2000	1975	1.3
3000	2956	1.5
4000	3932	1.7
5000	4904	1.9
6000	5872	2.1

Percent difference = (pressure−depth)/pressure × 100%

over 3.5 dbar below 3500 dbar (Brown and Morrison, 1978). A test of a SBE 9 CTD (Hendry, 1993) found pre- and postcruise pressure calibration offsets of less than 1 dbar. Nonlinearity and hysteresis were less than 0.5 dbar over the full range of the sensor.

The MBT introduced earlier in this chapter measures pressure with a Bourdon tube sensor. The problem with this sensor is that the response of the tube to volume change is nonlinear and any alteration in tube shape or diameter will lead to changes in the pressure response. As a result of the nonlinear scaling of the MBT, pressure readout required a special optical reader to read the scales; this read-out error added to the inaccuracies of the Bourdon tube, resulting in the limited accuracy of the MBT.

1.5.2 Free-fall Velocity

Unlike the MBT, the more commonly used XBT, does not measure depth directly but rather infers it from the elapsed time of a "freely falling" probe. While this is a key element that makes such an expendable system feasible it is also a possible source of error. In their study, Heinmiller et al. (1983) first corrected XBT profiles for systematic temperature errors and then compared the XBT profiles with corresponding CTD temperature profiles. In all cases, the XBT isotherm depths were less than the corresponding CTD isotherm depths for observations deeper than an intermediate depth (150 m for T4s and 400 m for T7s) with the largest differences at the bottom of each trace. Near the bottom of the XBT temperature profile, the difference errors exceeded the accepted limit of 2% error, with the deviation being far greater for the shallower T4 probes (Figure 1.20(a)). Added to this systematic error is an RMS depth error of approximately 10 m regardless of probe type (Figure 1.20(b)). Based on the data they analyzed, Heinmiller et al. (1983) provide a formula to correct for the systematic depth error. There are two primary sources of this depth error: first, the falling probe loses weight (and density) as the wire runs out of the probe supply spool, thus changing the fall rate; and second, frictional forces increase as the probe enters more dense waters. The increasing length of copper wire being paid out behind the probe must also add to the net drag on the probe.

The issue of XBT depth error, first reported by Flierl and Robinson (1977), has been extensively investigated by many groups (Georgi et al., 1980; Seaver and Kuleshov, 1982; Heinmiller et al., 1983; Green, 1984; Hanawa and Yoritaka, 1987; Roemmich and Cornuelle, 1987; Hanawa and Yoshikawa, 1991; Hanawa and Yasuda, 1991; Rual, 1991) with varying results There is general agreement that the XBT probes fall faster than specified by the manufacturer and that some corrections are needed. Most of these assessments have been performed as a comparison with nearly coincident CTD profiles. The concentration assessments in the western Pacific shows the interest in this problem in Japan, Australia, and Noumea, New Caledonia.

A sample comparison between XBT and CTD temperature profiles (Figure 1.21(a)) shows the differences between the XBT and CTD temperature profiles as a function of depth. These profiles have not been corrected using the standard depth equation. The sample in Figure 1.21(b) has been depth corrected using the formulation given by Hanawa and Yoritaka (1987). Note the substantial changes in the shape of the difference profile and how the depth correction eliminates apparent minima in the differences. The overall magnitude of the differences has also been sharply reduced, demonstrating that many of the apparent temperature errors are, in reality, depth errors.

These XBT depth errors are known to be functions of depth since they depend on an incorrect fall-rate equation. This is clearly demonstrated in Figure 1.22, which gives the mean depth difference of a collection of 126 simultaneous temperature profiles, along with the standard deviation (shown as bars that represent the standard

FIGURE 1.20 Vertical Profiles of XBT-CTD depth differences for T4 and T7 XBTs for different data sets. (a) Mean values, $\overline{\delta_z}$ (m); (b) Standard deviation $\overline{\sigma_{\delta_z}}$ (m).

deviation on either side of the mean line) at the various depths. Also shown are the ±2% or ±5 m limits, which are given as depth error bounds by the manufacturer. From this figure it is clear that there is a bias with the XBT falling faster than specified by the fall-rate equation, resulting in negative differences with the CTD profiles. The mean depth error of 26 m at 750-m depth translates into 3.5%,

which obviously exceeds the manufacturer's specification.

Various investigators reduced these comparisons to new fall-rate equations, for which the depth (z) is given in terms of elapsed time, t, by

$$z = at - bt^2 \qquad (1.14)$$

and found that the coefficients were not very different (Figure 1.23). Along with the

FIGURE 1.20 *(continued).*

FIGURE 1.21 Average temperature error profiles (TXBT-TCTD) for XBT/CTD comparisons on FR0487 using the SEAS II XBT systems; center line gives the mean value. (a) Depth uncorrected; (b) depth corrected using the formulation given by Hanawa and Yoritaka (1987). *(From IOC-888, Annex IV, p. 6.)*

coefficients, this figure also shows contours of maximum deviations in depth relative to the revised equation of Hanawa and Yoshikawa (1991) for these different combinations of constants a and b in the fall-rate Equation (1.14). Most of the errors in Figure 1.23 lie within the ±10-m envelope of depth deviations, suggesting that it might be possible to develop a new fall-rate equation (i.e., new coefficients) that represents a universal solution to the fall-rate problem for XBT probes. An effort was made to develop this universal equation be reanalyzing existing XBT-CTD comparison profiles. This revised equation is

$$z = 6.733t - 0.00254t^2 \qquad (1.15)$$

This revised fall-rate equation only applies to the T7 (roughly 700-m depth) XBT probes that were used in the comparisons. It was concluded that similar comparisons must be carried out for the other types of XBT probes.

1.5.3 Echo Sounding

Acoustic depth sounders are now standard equipment on all classes and sizes of vessels. Marketed under a variety of names including echo sounder, fish finder, depth sounder, or depth indicator, the instruments all work on the same basic principle: The time it takes for an acoustic signal to make the round trip from a source to an acoustic reflector, such as the

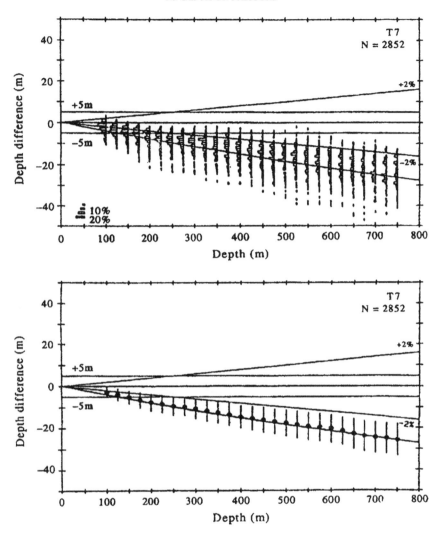

FIGURE 1.22 The mean depth difference of a collection of 126 simultaneous temperature profiles, along with the standard deviation (shown as bars that represent the standard deviation on either side of the mean line) at the various depths. *(From IOC/INF-888, p. 13.)*

seafloor, is directly proportional to the distance traveled. Water supports the propagation of acoustic pressure waves because it is an elastic medium. The acoustic waves radiate spherically and travel with a speed $c(E, \rho)$ which depends on the elasticity (E) and density [$\rho = \rho(S, T, P)$] of the water. If the speed of the sound is known at each time t along the sound path, then the distance d from the sound source to the seafloor is given in terms of the two-way travel time by

$$d = \frac{1}{2} \int_{t_t}^{t_r} c(t) \, dt \qquad (1.16)$$

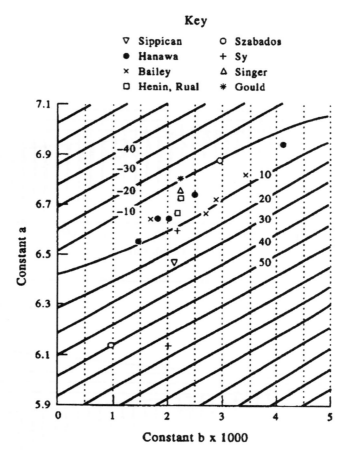

FIGURE 1.23 Fall-rate equation coefficients for different XBT studies listed in the key. *(From IOC/INF-888.)*

where

$$\Delta t = t_r - t_t = 2 \int_{t_t}^{t_r} [1/c(S, T, P; t)]dz(t) \quad (1.17)$$

is the time between transmission time (t_t) of the sound pulse and reception time (t_r) of the reflected pulse or echo. In practice, the values of c along the sound paths are not known and Eqn (1.16) must be approximated by

$$d = \frac{1}{2}\langle c \rangle \Delta t \quad (1.18)$$

where $\langle c \rangle$ is a mean sound speed over the pathlength, a value normally entered into the echo sounder during its calibration. The depth determined using the time delay is called a "sounding". In hydrography, a "reduced" sounding is one that is referenced to a particular datum. As noted by Watts and Rossby (1977), Eqn (1.17) is similar in form to the equation for dynamic height (geopotential anomaly), suggesting that travel time measurements from an inverted echo sounder (IES) can be used to measure geostrophic currents (cf. Section 1.6.3).

Since the bulk properties of water depend on the temperature (T), salinity (S), and pressure (P), sound speed also depends on these parameters through the relation

$$c = c_{0,35,0} + \Delta c_T + \Delta c_S + \Delta c_P + \Delta c_{S,T,P} \quad (1.19)$$

in which $c_{0,35,0} = 1449.22$ m/s ($=$ the speed of sound at $0\,°C$, 35 psu, and pressure $P = 0$, corresponding to depth $z = 0$). The remaining terms are the first-order Taylor expansion corrections for temperature, salinity, and hydrostatic pressure; the final term, $\Delta c_{S,T,P}$, is a nonlinear corrective term incorporating the simultaneous variation of all three properties. A well-known set of values for this equation, having a stated experimental standard deviation of 0.29 m/s, is attributed to W. Wilson (Hill, 1962; p. 478). To a close approximation (Calder, 1975; MacPhee, 1976)

$$c \text{ (m/s)} = 1449.2 + 4.6T - 0.055T^2 + 0.00029T^3$$
$$+ (1.34 - 0.010T)(S - 35) + 0.016z$$
$$(1.20)$$

or (Mackenzie, 1981)

$$c \text{ (m/s)} = 1448.96 + 4.591T - 5.304 \times 10^{-2}T^2$$
$$+ 2.374 \times 10^{-4}T^3 + 1.340(S - 35)$$
$$+ 1.630 \times 10^{-2}z + 1.675 \times 10^{-7}z^2$$
$$- 1.025 \times 10^{-2}T(S - 35) - 7.139$$
$$\times 10^{-13}Tz^3$$
$$(1.21)$$

where T is the temperature ($°C$) and S is the salinity (psu) measured at depth z (m). Accurate profiles of sound speed clearly require accurate measurement of temperature, which may not be available in advance. A commonly used oceanic approximation is the mean calibration speed $\langle c \rangle = 1490$ m/s that is generally applied to ship's sounders. Note that the speed of sound increases with increasing temperature, salinity, and pressure, with temperature having by far the greatest effect (Figure 1.24). For example,

c increases by 1.3 m/s per 1 *psu* in salinity (range 34–35 *psu*); increases by 4.5 m/s per 1.0 $°C$ for temperature (range 0–10 $°C$); and increases by 1.6 m/s per 100-m depth. The depth capability of any sounder is limited by the power output of the transducer transmitting the sound pulses, by the sensitivity of the receiver listening for the echo returns, and by the capability of the instrument electronics and software to resolve signal from noise. In modern sounders, pulse lengths typically range from 0.1 to 50 ms and a single transducer with a transmit/receive switching arrangement is used to both generate and receive the acoustic signals. The depth capability of an echo sounder is limited also by a number of important environmental factors. In general, sound waves are attenuated rapidly in water according to the relation:

$$\text{propagation loss (dB)} = 20 \log(r) + \alpha r/1000$$

where the propagation loss is measured in decibels (dB), r is the distance (or depth range in the case of depth sounders) in meters, and α (Figure 1.25) is the attenuation coefficient (dB/km) in seawater as a function of frequency, temperature, and salinity (Urick, 1967). The first term in the equation accounts for geometrical spreading of the transmitted and received signal while the second encompasses scattering and absorption. Diffraction and refraction arising from density gradients have a minor effect on the attenuation compared with these other factors. The higher the frequency of the source the greater the attenuation due to absorption and the more limited the depth range (Figure 1.26). It is for this reason that most deep-sea sounders operate in the 1.0–50 kHz range. Even though high-frequency sounders can provide more precise depth resolution through shorter wavelengths and narrower beam widths, they cannot penetrate deeply enough to be of use for general soundings. However, they do have other important applications, including bioacoustic studies of the distribution and biomass of zooplankton, fish, and marine

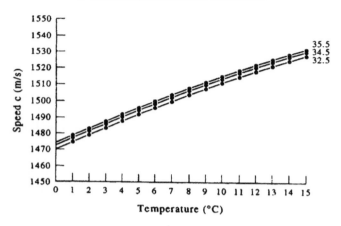

FIGURE 1.24 Speed of sound as a function of temperature for different mean salinities (32.5, 34.5, and 35.5 psu) and fixed depth, $z = 1500$ m.

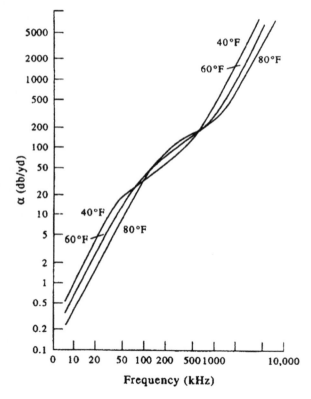

FIGURE 1.25 Absorption coefficient in seawater at salinity 35 psu as a function of frequency for three temperatures (40 °F = 4.4 °C; 60 °F = 15.6 °C; 80 °F = 26.7 °C). Conversion factor; 1 dB/km = 1.0936 dB/kyard; 1 kHz = 1000 cps; db ≡ dB. *(After Urick (1967).)*

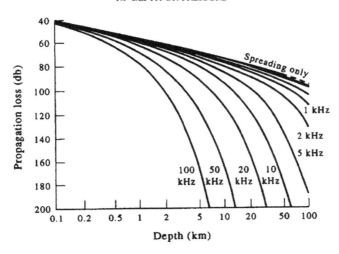

FIGURE 1.26 One-way sound attention (Propagation loss, PL) in water as a function of sounder frequency (1 khz = 1000 cps). Curves are derived using PL = 20 log r + αr/1000, where r is in kilometers and α is taken from Figure 1.25 using the conversion to db/km. Note that, in the figure, db denotes dB.

mammals. It was echo-sounder observations of tidally induced undular bores in the lee of a shallow sill in Knight Inlet (British Columbia) by Farmer and Smith (1980a,b) that helped rekindle interest in the observation and modeling of hydraulic jumps, internal waves, and soliton formation in a variety of oceanic settings (Apel et al., 1985; Farmer and Armi, 1999).

The output transducer converts electrical energy to sound energy and the receiving transducer converts sound vibrations to electrical energy. Loss of the acoustic signal through geometrical spreading is independent of frequency and results from the spherical spreading of wave fronts as a function of distance, while frequency-dependent absorption leads to the conversion of sound into heat through viscosity, thermal conductivity, and inframolecular processes. Signal loss due to geometric wave front spreading follows a $1/r^2$ inverse square law. Scattering is caused by suspended particles, density microstructure, and living organisms. In the upper 25 m of the water column, air bubbles from breaking waves and gas exchange processes are major acoustic scatterers. If I_0 is the

intensity (e.g., power in watts) of the transducer and I_a is some reference intensity (nominally the output intensity in watts measured at 1-m distance from the transducer head) then the measured backscatter I_r is given by

$$I_r/I_a = b \exp\left(-\gamma r\right)/r^2 + A_n \qquad (1.22)$$

where $b = I_0/I_a$ is the gain of the transducer, γ is the inverse scale length for absorption of sound in water, $1/r^2$ gives the effect of geometric spreading over the distance r from which the sound is being returned, and A_n is a noise level. A return signal intensity reduction by a factor of two corresponds to a loss in intensity of -3 dB [$=10$ log $(1/2)$] while a reduction by a factor of 100 corresponds to a loss of -20 dB [$=10$ log $(1/100)$]. Echo sounders are generally limited to depths of 10 km. The low (1–15 kHz) frequencies needed for these depths result in poor resolutions of only tens of meters. Since it takes roughly 13 s for a transducer "ping" to travel to 10 km and back, the recorded depth in deep water is not always an accurate measure of the depth beneath a moving ship. Better resolution is provided for depths less than 5 km using

FIGURE 1.27 Schematic of how an acoustic pinger and the ship's sounder in the receive mode can be used to accurately determine proximity of a probe to the bottom. c is the speed of sound in water and t is time.

frequencies of 20–50 kHz. High-resolution sounders operate in a few hundred meters of water using frequencies of 30–300 kHz. Transducer beam width, side-lobe contamination, and side echoes ultimately reduce the resolution of any sounder.

1.5.3.1 Height above the Bottom

In many oceanographic applications, the investigator is interested in real-time, highly accurate measurements of the altitude of his or her instrument package above the bottom. A real-time reporting echo sounder on the package will serve as an altimeter, enabling the operator to safely lower the sampling package, such as a CTD-Rosette system, to within 10 m of the bottom. Alternatively, the investigator can choose a cheaper route and attach a high-power omnidirectional "pinger" to the package. Rather than

trying to measure the total depth of water, one uses the pinger on the package together with ship's transducer (or hydrophone lowered over the side of the ship) to obtain the difference in depth (i.e., difference in time) between the signal, which has taken a direct route from the pinger and one which has been reflected from the bottom at an angle θ, where θ is roughly the angle the tow line makes with the bottom (Figure 1.27). The direct path PB takes a time t_d while the path reflected from the bottom (PAB) takes a total time t_r. The height, d, of the package above the bottom is

$$d = \tfrac{1}{2}c\Delta t \sin \theta = \tfrac{1}{2}c(t_r - t_d) \sin \theta \qquad (1.23)$$

where Δt is the time delay between the direct and reflected pings and c is the speed of sound in water. As the instrument package approaches the bottom, the two strongest analog traces on the depth recorder can be seen converging toward a "crossover" point. When the time taken to cover the direct and reflected paths are equal ($\Delta t = 0$), the package has hit the bottom. The novice operator will be confused by the number of "false" bottom crossovers or wrap-around points when working in water that is deeper than integer multiples of the chosen sounder range. To avoid high levels of stress as each crossover is approached, the operator must know in advance how many false crossovers or wrap-arounds to expect before the instrument is truly in proximity to the bottom. For example, in water of depth 3230 m, a recorder chart set for a full-scale depth range of 0–750 m will register false bottoms when the pinger reaches water depths of 230, 980, and 1730 m; i.e. $3230\,\text{m} - (n \times 750\,\text{m})$, where $n = 4$, 3, 2. As far as the depth recorder is concerned, $3230\,\text{m} - (4 \times 750\,\text{m})$ is the same as 230 m.

Analog sounder devices such as the PDR (Precision Depth Recorder) are the only real choice for this application since the analog trace provides a continuous visual record of how many crossovers have passed and how rapidly the final crossover is being approached. (The depth is first obtained from the ship's sounder, which can then be turned to "receive mode" only.) The two traces give a history of what has been happening so that it is easier to project in one's mind what to expect as the instrument nears the bottom. Problems arise if the package gets too far behind the vessel and the return echoes become lost in the ambient noise or if the bottom topography is very rugged and numerous spurious side echoes and shadows begin to appear. We recommend omnidirectional rather than strongly directional pingers so that if the package streams away from the ship or twists with the current and cable there is still some acoustic energy making its way to the ship's hull. To help avoid hitting the bottom, it is best to have the ship's sounder output turned off so that only receive mode is working and the background noise is reduced; the operator can check on the total depth every once in a while by reconnecting the "transmit pulse".

The ships echo sounder correctly measures the height above bottom since it is programed to divide any measured time delay by a factor of two to account for two-way travel times. The depth accuracy of the method improves as the package approaches the bottom because both the direct and reflected paths experience the same sound speed c. A value of c more closely tuned to deep water is applicable here since all that really counts is c near the seafloor in the region of study. With a little experience and a good clean signal, one can get accuracies of several meters above the bottom using 8–12 kHz sounders in several kilometers of water. Attaching both a pinger and an altimeter is clearly the preferable solution.

Note that the depth errors using the pinger method are negligible while the actual sounding depths from a depth sounder can be quite large. For example, if spatial differences in sound speed vary from 1470 to 1520 m/s over the sounding depth, then the maximum percentage depth errors are $\Delta c/c = 50/1500 = 0.033 = 3.3\%$. In 4000 m of water, this would amount to an error of $0.033 \times 4000\,\text{m} \approx 133\,\text{m}$.

1.5.4 Other Depth Sounding Methods

For sake of completeness, several remote sensing, depth-sounding methods are introduced in the following sections.

1.5.4.1 *Laser Induced Detection and Ranging*

Laser induced detection and ranging (LIDAR) is an active electro-optical (LASER) remote-sounding method using a pulsed laser system as a radiation source flown from an aircraft. An airborne sensor measures the distance to the surface of the ocean and to the seafloor along the appropriate light path by measuring the time interval between emission of the pulse and the reception of its reflection from the surface and the bottom. A typical LIDAR unit consists of a pulsed laser transmitter, a receiver, and a signal analyzer—recorder. The technique is good to depths of a few tens of meters in coastal waters where extinction coefficients are typically around 0.4—1.6/m. The rapid spatial sampling capability of this technique makes it highly useful to hydrographers wanting to map shoals, rocks and, other navigational hazards. LIDAR topographic data are also important for generating coastal inundation maps for storm surge and tsunami runup.

1.5.4.2 Synthetic Aperture Radar

One of the surprising aspects of synthetic aperture radar (SAR) is its ability to "see" shallow banks, ridges, and shoals in the coastal ocean. In this case, SAR does not measure the bottom topography directly but, instead, detects the distortion of the wave ripplet field over the feature caused by deflection and/or acceleration of the ocean currents. For a discussion of this effect, the reader is referred to Robinson (1985). SAR is also being used to detect surface wakes generated by submarines and ships, and to delineate ocean fronts arising from changes in current shear and marked thermal gradients such as the "Wall of the Gulf Stream" (Belkin and O'Reilly, 2009; Williams et al., 2013)

1.5.4.3 Satellite Altimetry

The suggestion that there is a close connection between oceanic gravity anomalies and water depth was postulated as early as 1859 (Pratt, 1859, 1871). The idea of using measured gravity anomalies to estimate water depth began with Siemens (1876) who designed a gravity meter that obviated the need to spend hours for a single sounding (Vogt and Jung, 1991). It was not until the launching of the SEASAT radar altimeter in 1978 that this concept could be used as alternative to large-scale sounding measurements. Indeed, the first SEASAT-derived gravity map of the world oceans (Haxby, 1985) closely resembles a bathymetric map with a horizontal resolution of about 50 km (Vogt and Jung, 1991). The idea of using satellite radar altimetry

as a "bathymeter" is based on the good correlation observed between gravity anomalies and bathymetry in the 25—150 km wavelength radar band. Satellite bathymetry is especially valuable for sparsely sounded regions of the world ocean such as the South Pacific, and in regions where depths are based on older soundings in which navigation errors are 10 km or more. As pointed out by Vogt and Jung, however, one-dimensional predictions cannot be accurately ground-truthed with a single ship survey track since the geoid measured along the track is partly a function of the off-track density distribution. (Geoid refers to a constant geopotential reference surface, such as long-term mean sea-level, which is everywhere normal to the earth's gravitational field.) A broad swath of shipborne data is needed to overlap the satellite track swath.

In addition to SEASAT, early satellites equipped with radar altimeters to map sea surface topography include GEOS-3, GEOSAT, ERS-1, and TOPEX/Poseidon. Modern satellites with high-resolution altimeters include Jason-1, Jason-2, and HY-2A (China). Of the earlier satellites, only data from the U.S. Navy's GEOSAT have been processed to the accuracy and density of coverage needed to clearly resolve tectonic features in the marine gravity field on a global basis (Marks et al., 1993; EOS). This mission (which has been superseded by the Gravity Recovery And Climate Experiment, GRACE) was designed to map the marine geoid to a spatial resolution of 15 km. Detailed maps of the seafloor topography south of 30 °S from the GEOSAT Geodetic Mission were declassified in 1992. These maps have a vertical RMS resolution of about 10—20 cm and, together with SEASAT data, have been used to delineate fracture zones, active and extinct mid-ocean ridges, and propagating rifts. Satellite altimetry from ERS-1 has been used to map the marine gravity field over the permanently ice-covered Arctic Ocean (Laxon and McAdoo, 1994). Future declassification of military status satellite data will lead to further analysis of the seafloor structure over the remaining portions of the world ocean.

The launch of TOPEX/Poseidon satellite in 1992 with accurate radar altimeters resulted in important data sets for tidal analysis of global sea levels (e.g., Schrama and Ray, 1994; Ray 1998, Cherniawsky et al., 2001). In 2002, the satellite was replaced in the same orbit by Jason-1 with a very accurate altimeter. A companion altimetry satellite, Jason-2, was launched in 2008. As we discuss later in more detail, satellite altimetry has proven invaluable for interpreting large-scale circulation previously deduced from dynamic height data. Satellite altimetry also is particularly well suited to examining the spatial structure, temporal variability, and propagation of mesoscale features in the ocean, including mesoscale eddies (Di Lorenzo et al., 2005), geostrophic surface currents, and Rossby waves (cf. Chelton and Schlax, 1996).

GRACE uses a highly accurate microwave ranging system to measure changes in the speed and distance between two identical spacecraft (nicknamed "Tom" and "Jerry") flying in a polar orbit about 220 km apart, 500 km above earth (Wikipedia, 2013). Launched on March 17, 2002, the ranging system on GRACE is able to detect separation changes as small as 10 μm (approximately one-tenth the width of a human hair) over the satellite separation distance. As they circle the globe 15 times a day, the satellites measure variations in earth's gravitational pull. When the first satellite passes over a region of slightly stronger gravity (corresponding to a positive gravity anomaly), it is pulled slightly ahead of the trailing satellite. This causes the distance between the satellites to increase. The first spacecraft then passes the anomaly, and slows down again while the following spacecraft accelerates, then decelerates over the same point. By measuring the constantly changing distance between the two satellites and combining that data with precise global positioning systems (GPS) positioning measurements, scientists can construct a detailed map of earth's gravity.

This information, in turn, can be used to study a variety of global processes, including the rate at which mass is being added to the oceans from the land (Peltier, 2009).

1.6 SEA-LEVEL MEASUREMENT

The measurement of sea-level is one of the oldest forms of oceanic observation. Pytheus of Marseilles, who is reported to have circumnavigated Britain around 320 BC, was one of the first to actually record the existence of tides and to note the close relationship between the time of high water and the transit of the moon. Nineteenth-century sea-level studies were related to vertical movements of the coastal boundaries in the belief that, averaged over time, the height of the mean sea level was related to movements of the land. More recent applications of sea-level measurements include the resolution of tidal constituents for coastal tide height predictions, assisting in the prediction of El Niño/La Niña events in the Pacific, and determining the effects of climate change. Tide gauge data are essential to studies of wind-generated storm surges, which can lead to devastating flooding of highly populated low-lying areas such as Bangladesh and the Eastern Seaboard of the United States. (In late October 2012, parts of the northeast coast of the US were inundated by a major storm surge during Hurricane Sandy. New York registered 10-m waves and a maximum 4.23-m storm surge.) Tide gauges located along the perimeter of the Pacific Ocean and on Pacific islands are integral to the Pacific Tsunami Warning Center (PTWC) headquartered in Honolulu (Hawaii) and Palmer (Alaska) that alerts coastal residents to possible seismically generated waves associated with major underwater earthquakes and crustal displacements. The changing emphasis of digital tide gauge observations on tsunami warning and research, rather than for tidal analysis, means that the gauges must record at intervals of

6 min or shorter (preferably 1 min) and to be directly accessible online at all times.

In addition to measuring the vertical movement of the coastal land mass, long-term sea-level observations reflect variations in large-scale ocean circulation, surface wind stress, and oceanic volume. Because they provide a global-scale integrated measure of oceanic variability, long-term (>50 years) sea-level records from the global tide-gauge network provide some of the best information available on global climate change. Long-term trends in mean sea level are called *secular* changes while changes in mean sea level that occur throughout the world ocean are known as *eustatic* changes. As described later in this section, changes in eustasy are associated with variations in land-based glaciation, variations in steric height arising from changes in global ocean temperature and salinity, fluctuations in the accumulation of oceanic sediments, and tectonic activity, such as the change in ocean volume and the shape of the ocean basins. Since coastal stations really measure the movement of the ocean relative to the land, land-based sea-level measurements are referred to as relative sea-level measurements. Mean sea level is the long-term average sea level taken over a periods of months or years. Datum levels used in hydrographic charts can be defined in many ways and generally differ from country to country (Thomson, 1981; Woodworth, 1991), but are generally referenced to some measure of low water, such as "lower low water, large tides". For geodetic purposes, mean sea-level needs to be measured over many years.

1.6.1 Specifics of Sea-Level Variability

As the previous discussion indicates, observed sea-level variations about some mean equilibrium, "datum" level can arise from four principal components: (1) *short-term* temporal fluctuations in the height of the sea surface including those associated with wind waves and oceanic tides forced by the changing alignment of the sun, moon, and earth. In addition, there are changes due to atmospheric pressure (the inverse barometer effect, corresponding to a ~1-cm *rise* in sea level for a 1-mb *drop* in atmospheric pressure), to wind-induced current setup/setdown along the coast (including Coriolis effects on the alongshore component of wind-generated currents), to changes in river runoff, and to changes in the large-scale ocean currents and gyres caused by fluctuations in the oceanic wind field and water mass redistribution; (2) *Long-term* eustatic changes resulting from changes in the mass of the ocean due to melting or accumulation of land-based ice in the major continental ice sheets that sit atop Antarctica and Greenland, changes in grounded polar ice caps, and changes in the smaller ice sheets and mountain glaciers. There are also major long-term variations in sea level due to *Steric effects*— slow variations in sea level arising from changes in ocean volume (i.e., density) without a change in ocean mass. At present, steric heights are generally increasing throughout much of the world ocean as a result of global warming (expansion) of the upper ocean (IPCC, 2007, 2013). Diminishing the salt concentration of the upper ocean due to enhanced rainfall or riverine input has the same steric effect as thermal heating; (3) *Coastal subsidence* involving the lowering of the land brought about by reduction in the thickness (compaction) of unconsolidated coastal sediments, erosion, sediment deposition and with the withdrawal of fluids (water, oil, etc.) from the sediments; (4) Large-scale vertical crustal land movements that produce sea-level change through *tectonic processes* (mountain building) and ongoing *glacio-isostatic adjustment* arising from the continued viscoelastic response and rebound of the earth to melting of glaciers during the last ice age. In addition to the change in land level due to the unloading of the crust with the removal of glacial ice, sea levels are affected by the gravitational effects of continental ice sheets on the adjacent sea water. This spatially varying "sea-level fingerprinting" occurs because, as the ice disappears, relative sea levels fall at decreasing rates from highly glaciated areas to lesser glaciated areas (Mitrovica et al., 2001; Riva et al., 2010; U.S. National Research Council, 2012).

The principal semidiurnal (M_2) and diurnal (K_1) tidal constituents, with respective periods of 12.42 and 23.93 h, can be accurately resolved using a 15.3-day tidal record of hourly values. Further resolution of the important *spring-neap* cycle of the tides (the 15-day fortnightly cycle associated with the alignment of the sun, moon, and earth) and the tropic (declinational) cycle (the roughly 15-day cycle arising from the tilt of earth's axis relative to the moon's orbital plane) require a record length of roughly 29 days (or 0.98 lunar months; one lunar day = 25 h). For most practical purposes, this is the minimum length of record that is acceptable for construction of local tide tables. In fact, many countries maintain primary tide-gauge stations as reference locations for secondary (short-term) tide-gauge stations. Differences in the tide heights and times of high/low water are tabulated relative to the primary location. Accurate resolution of all 56 principal tidal constituents requires a record length of 365 days while an accurate measure of all components for long-term tidal applications requires a record of 18.6 years. The 18.6-year "Metonic" cycle or nuation is linked to the 5-degree tilt of the plane of the moon's orbit with respect to the plane of the earth's orbit and is the time it takes the line of intersection of these two planes to make one complete revolution (Thomson, 1981). Other tidal constituents include: The centimeter-scale Pole Tide (Chandler Effect) with a period of 14.3 months that arises from the Chandler Wobble in the instantaneous axis of the earth's rotation; an 8.8-year cycle associated with alterations in the eccentricity of the moon's orbit about the earth; and a 20,940-year cycle due to a wobble in the earth's orbit about the sun (precession of the equinox). Meteorological tides are caused by local atmospheric forcing and include

large (~ 1 m) sea-level changes associated with storm surges that often flood low-lying areas (Murty, 1984). Their periodicities are related to changes in wind and atmospheric pressure.

In addition to the relatively small changes in sea level associated with the Metonic cycle and other orbital factors, there are a variety of major variations due to geological processes. Geological techniques such as coring of Greenland glaciers show a relatively rapid rise of sea-level from 18,000 years ago, when global sea levels were roughly 130 m lower than today. The rate of sea-level rise, which averaged 10 mm/year during the glacial—interglacial transition period, slowed dramatically about 8000 years ago when the levels were 15 m below those of today. Present levels were reached roughly 4000 years ago. Since that time, mean sea-level changes have consisted of oscillations of small amplitude (Barnett, 1983). However, there is now concern that global sea levels are rising at over 3 mm/year due to global warming through buildup of CO_2 in the atmosphere. Some studies (e.g. Pfeffer, 2008 and Grinsted et al., 2009) estimate that the mean rate during the twenty-first Century could be as high as 20 mm/year. Many long-term stations show definite long-term trends (Figure 1.28) that are most likely related to global climate change. There is presently a general increase of about 0.5—1 m per century for gauges located in geologically "stable" regions of the world. This could reach 2 m per century if some of the higher estimates are correct. It now generally accepted that the long-term increase in global sea level is linked to melting of water locked in the polar ice sheets and northern hemisphere glaciers due to global warming. Changes in the land-based ice cover alter both the ocean volume and the geodetic loading. However, it should be noted that the rate of rise varies considerably from location to location and can be strongly dependent on regional tectonic activity and the lingering effects of the continuing glacio-isostatic response (Peltier, 1990; Tushingham and Peltier, 1992;

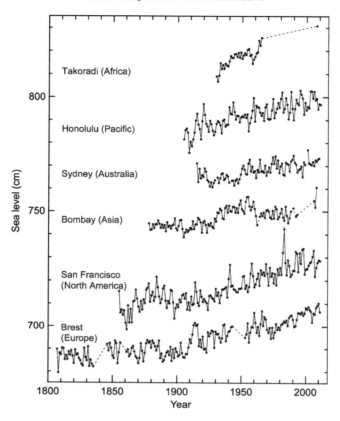

FIGURE 1.28 Annual mean sea-level values for the longest records for each continent. Data are Revised Local Reference (RLR) records from the Permanent Service for Mean Sea-level at the Bidston Observatory in Merseyside. Each record has been given an arbitrary offset for presentation purposes. The Takoradi record was truncated in 1965 when major problems with the gauge were reported. *(Courtesy, Philip Woodworth, PSMSL.)*

James et al., 2011). Investigators attempting to extract a possible climate change component from global sea-level records spend considerable effort generating spatially smoothed data sets consisting of a relatively small subset of the total tide gauge data available from the world archives.

Mean sea levels are usually computed from long series of hourly observations. Generally, a simple arithmetic average of hourly values is computed, but other methods, including the application of low-pass numerical filters to eliminate tides and storm surges, may be used before the means are computed. The average of all high

and low water levels is called the mean tide height; it is close to, but not identical with, mean sea level. Monthly and annual mean sea-level series from a global network of stations are collected and published by the Permanent Service for Mean Sea level (PSMSL) in England, together with details of gauge location and the definitions of the datums to which the measurements are referred. Data at PSMSL are held for 2167 "Metric" file stations of which 1384 have had their data adjusted to a tide gauge benchmark datum to form the Revised Local Reference (RLR) data set. This datum is approximately

7.000 m below mean sea level, with the arbitrary choice made to avoid negative numbers in the resulting RLR monthly and annual mean values. Of these stations, 133 have data from before 1900. Most of these stations are in the northern hemisphere so that careful analysis is necessary to avoid geographic bias in their interpretation. Amsterdam has the longest tide-gauge record in the world but the oldest data that satisfy the selection criteria of the PSMSL are from Brest, starting in 1807. For many stations, the PSMSL Website (www.psmsl.org) provides links to other sources of sea level and land level (GPS) information. The site also has a considerable amount of background information on making sea-level measurements and analyzing the data. Recent developments at PSMSL have been described by Holgate et al. (2013).

Since the 1990s, tide gauge records have been augmented by satellite altimetry measurements of global sea level. The long-term trend in the spatially averaged altimetry data (Figure 1.29) supports arguments for accelerated eustatic sea-level rise compared to the beginning of the twentieth century. According to satellite records, global sea levels are presently rising at a mean rate of about 3.2 mm/year.

1.6.2 Tide and Pressure Gauges

Although pressure and acoustic gauges are becoming increasingly more popular around the world, many sea-level measurements are still made using a float gauge, in which the float rises and falls with the water level (Figure 1.30). Modern recording systems replace the analog pen with a digital recording system that allows for unattended, real-time data acquisition, control, and communication. Data are recorded on flash memory with removable media support using SD cards, MMC cards, and USB thumb drives. Measurements can be event-driven or independently scheduled, with sample intervals ranging from 1 s to a day. Loggers can have a wide variety of built-in functions such as min/max and averaging over specified periods. Many of the digital recording systems have been equipped with telemetering systems that send sea-level heights and time via satellite transmitters (e.g., GOES, INMARSAT, METEOSAT, INSAT, IRIDIUM), modem (Cell, CDPD (Cellular Digital Packet Data), telephone landline), radio (UHF/VHF, spread spectrum), or other serial communication devices. Direct and reliable communication links are critical for tsunami warning systems.

FIGURE 1.29 Change in mean sea level between 1993 and 2005 from satellite altimetry including corrections for Glacial Isostatic Adjustment (GIA) and seasonal variations *(After Nerem (2005).)*

FIGURE 1.30 (a) A basic float stilling-well gauge used to measure water levels on the coast; (b) Schematic of tide-gauge station with the gauge, network of benchmarks and advanced geodetic link. TGBM, Tide gauge benchmark. *(After Woodworth (1991).)*

The most important aspect of this type of sea-level measurement is the installation of the float stilling well. The nature of this installation will determine the frequency response of the float system and will help damp out unwanted high-frequency oscillations due to surface gravity waves. Stilling wells can have their inlet at the bottom of the well or use a pipe inlet connected to the lower part of the well. Both designs damp out the high-frequency sea-level changes. Maintenance of the sea-level gauge insures that the water inlet orifice is kept clear of obstructions from silt, sand, or marine organisms. Also, in areas of strong stratification such as rivers or estuaries, the water in the stilling well can be of a different density than the water surrounding it. When installing such a measurement system, it is important to provide adequate protection from contamination of the stilling well and from damage to the recorder. A potential hazard in all harbor gauge installations is damage from ship traffic or contamination of the float response from ship wakes.

Proper installation of coastal sea-level gauges requires that they be surveyed into a legal benchmark so that measured changes will be known relative to a known land elevation (Figure 1.30). When properly tied to a benchmark height, changes in gauge height relative to land can be taken into account when computing the mean sea-level. This tie-in with the local benchmark datum is done by running the level back to the nearest available geodetic datum. Modern three-dimensional satellite-based GPS and long-baseline telemetry systems now make it possible to accurately determine the vertical movement of a tide gauge relative to the geoid. Satellite altimeter measurements initiated in the early 1990s provide global measurements of the relative sea surface, which can then be compared with the conventional sea-level measurements. The importance of the altimeter sea level records will continue to grow as the measurement accuracy and record length continue to improve. The usual test of a sea-level instrument (called the Van de Casteele test) involves operating the instrument over a full tidal cycle and comparing the results against simultaneous measurements made with a manual procedure. This procedure only shows if the recording device is operating properly so that a separate test of the stilling well is needed to accurately measure the response of the float. Other than mechanical problems and poorly documented repositioning errors, timing errors are one of the major sources of error in sea-level records. Sea-level gauges have either mechanical or electronic clocks, which must be periodically checked to insure that there is no significant drift in the timing of

the mechanism. When possible these checks should be made weekly. Depending on the specific instrument, well-maintained sea-level recorders are capable of measurements accurate to within a several millimeters.

Another type of sea-level measuring device is the pneumatic or bubbler gauge (Figure 1.31). This system links changes in the hydrostatic pressure at the outlet point of the bubbles to variations in sea level. Like other pressure sensing gauges, this gauge measures the combined sea-level height and atmospheric pressure. As a consequence, most bubbler gauges operate in the differential mode whereby the recorded value is the difference between the measured pressure and the atmospheric pressure. While these instruments are somewhat less accurate than float gauges, they are useful in installations where a float gauge would be subject to either damage from ship traffic or strongly influenced by wave motion. In a study in Tasmania in the 1970s, Australian technicians reported that the plastic pressure tubes leading from the electronics package to the ocean were constantly being destroyed by curious wombats.

Sea-level heights are recorded in a variety of formats. Graphical records must be digitized and care taken to record only values properly resolved by the instrument. Differences in recording scale will lead to variations in the resolution of the gauge thereby limiting the accuracy of the digitized data. Modern gauges eliminate this possible problem by recording digital data. (For most tsunami work, digital sampling intervals of less than 6 min—and preferably 1 min or less—are highly recommended.) During digitization of analog data, it is important to edit out any of the obvious errors due to pen-ink problems or mechanical failures in the advance mechanism. Also, when long-term sea-level variations are of interest, one must be careful to filter out high-frequency fluctuations due to waves and seiches. The choice of time reference is important in creating a sea-level time series. The usual convention is to use the

Basic pneumatic bubbling system

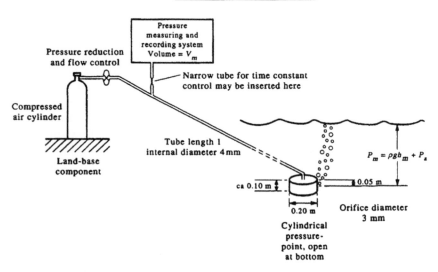

FIGURE 1.31 The pneumatic or bubble gauge. This system links changes in the hydrostatic pressure, P_m, at the outlet point of the bubbles to variations in sea-level, h_m, water density, ρ, and atmospheric pressure, P_a.

local time at the location of the tide gauge, which can then be referenced to Greenwich Mean Time (GMT, now called Universal Time Coordonné, UTC). The sea-level record should also contain some information about the reference height datum on which the sea-level heights are based. Digital recording systems are subject to clock errors and care should be taken to correct for these errors when the digital records are examined.

In recent years, sensitive and accurate pressure sensors have been developed for measurement of deep-sea tides and trans-oceanic tsunamis where fluctuations of the order of 1 mm need to be detected in depths of thousands of meters. At first, these sensors were largely based on the "Vibraton" built by United Control Corporation, which measured pressure by changes in the frequency of oscillation of a wire under tension. This frequency change was measured to an accuracy of 6×10^{-4} Hz and led to a sea-level accuracy of 0.8 mm (Snodgrass, 1968). To maintain this high level of accuracy, it was necessary to correct for temperature effects

to a resolution of 0.001 °C. When Vibraton sensors ceased to be commercially available they were replaced by resonating quartz crystal transducers (Wimbush, 1977), which are now the standard for measurement in both deep sea and coastal pressure gauge recorders. Manufactured by Paroscientific (Paros, 1976), these sensors have a depth sensitivity of 1×10^{-4} dbars (approximately 10^{-4} m or 0.1 mm) for both shallow and deep-sea measurements. Most modern pressure gauges used in coastal- and deep-sea tidal and tsunami measurements make use of these types of sensors. In addition, Aanderaa commonly used pressure sensors manufactured by the Finnish company Vaisala in their water level gauges; these sensors have a resolution of 0.01 dbars (0.01 hPa) over a range of 500–1100 hPa. (Pressure gauges used in coastal waters are often known as water level gauges.) Temperature correction is required to maintain accurate depth measurements. Wearn and Baker (1980) report measurements made by such quartz sensors from year-long moorings in the

Southern Ocean. Unfortunately, instabilities in the quartz sensors lead to sensor drifts, which limited the use of the sensors in long-term, deep pressure measurements. The use of dual pressure sensors helps to correct for drift since each pressure sensor will have somewhat different drift characteristics but will produce similar responses to higher frequency oceanic variability. A technical report by Paroscientific that compared 10-min data from a bottom pressure recorder (BPR) in 7000 m of water with a surface nano-barometer that was measuring ambient atmospheric pressure (an instrument that is 700 times more accurate) shows that the BPR-tracked changes in barometric pressure with a resolution of 0.25 Pa or about 0.025 mm of equivalent water depth (Schaad, 2009, Technical Note, Paroscientific, Inc.). Oceanographic instruments can now resolve pressure variations to a fraction of a millimeter at full ocean depth, corresponding to parts per billion in water depth. Tides, infragravity waves, tsunamis, meteotsunamis (tsunami-like oscillations generated by atmospheric pressure pulses), and seasonal variations in sea level are readily resolved at this precision. However, sensor drift of roughly 1 cm/year precludes long-term accuracy measurements to this level.

1.6.2.1 DART and NEPTUNE-Canada BPRs

Paroscientific BPRs form the backbone of the DART buoy system operated as part of the PTWCs (Pacific Tsunami Warning Centers) of NOAA. BPRs also form the backbone of the regional-scale (maximum 2700-m depth) tsunami and infrared gravity wave array of the Ocean Networks Canada (formerly NEPTUNE Canada) cabled observatory in the northeast Pacific. (www.neptunecanada.com).

DART II became operational in 2005 (Green, 2006). Although most DART buoys are located around the Pacific Rim, several buoys are now located in the Caribbean and off the east coast of the United State (Figure 1.32), with plans to expand to the Indian Ocean. Tsunami wave heights recorded by a bottom-anchored seafloor BPR (Figure 1.33) are sent in real time by an acoustic modem link to a nearby surface mooring, which then transmits the data via satellite to a shore site in the United States (Gonzalez et al., 1998). The BPR collects temperature and pressure at 15-s

FIGURE 1.32 DART buoy coverage for the world ocean in 2012 by contributing country. *(U.S. National Oceanic and Atmospheric Administration, NOAA.)*

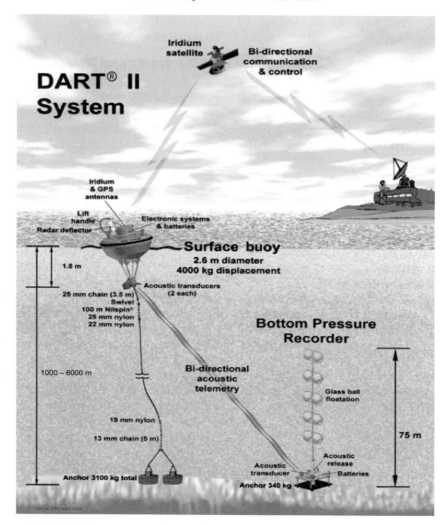

FIGURE 1.33   Cartoon showing the components of a typical DART II deployment system. *(NOAA.)*

intervals. The pressure data are corrected for temperature and the pressure converted to an equivalent sea surface height (height of the ocean surface above the seafloor) using a constant conversion of 670 mm/psia (pounds per square inch absolute). The system has two data reporting modes, standard and event. The system operates routinely in standard mode, in which four spot values of 15-s data are sent at 15-min intervals at scheduled transmission times. When the internal detection software identifies an event, the system begins event mode transmissions (Mofjeld, 2009). In the event mode, 15-s values are transmitted during the initial few minutes, followed by 1-min averages. Event mode messages also contain the time of the initial occurrence of the event. The system returns to standard transmission after 4 h of 1-min real-time transmissions if no further events are detected. Two-way communication via the Iridium commercial satellite communications system (Meinig et al., 2005) allows the Tsunami Warning Centers (TWCs) to set stations in event mode in anticipation of possible tsunamis or retrieve the high-resolution (15-s intervals) data in 1-h blocks for detailed analysis. The DART® buoys have two independent and redundant communications systems. The National Data Buoy Center (NDBC) distributes the data from both transmitters under separate transmitter identifiers. NDBC receives the data from the DART II systems, formats the data into bulletins grouped by ocean basin (see the NDBC - DART® GTS Bulletin Transmitter List, for a listing of the bulletin headers used for each transmitted identifier), and then delivers them to the National Weather Service Telecommunications Gateway (NWSTG) that then distributes the data in real time to the TWCs via NWS communications and nationally and internationally via the Global Telecommunications System. A comprehensive appraisal of DART is provided by Mungov et al. (2012).

The system maintains a measurement accuracy within ±1.0 cm of the observed tides deployed in similar depths off Hawaii within 100 nautical miles of an operational DART® station using a standard tide model for tidal adjustments between stations. Measurement resolution is 0.25 mm in water depths of 1000–6000 m. For 15-s data for 1 h duration, the time from a request for data to the receipt of data at the TWC must be less than 15 min. In event mode, the time from the start of the event to receipt of the data at the TWC's server must be less than 3 min.

The BPRs deployed within the NEPTUNE-Canada and VENUS cabled observatory networks of Ocean Networks Canada (ONC) are being used to study tsunamis, seafloor loading, tides, infragravity waves, seismic waves, and other ocean and earth phenomena off the west coast of Canada (Barnes et al., 2008). A low power high-precision signal period counter developed jointly by the Pacific Geoscience Centre and Bennest Enterprises (E. Davis, R. Meldrum,

J. Bennest, pers. com., Sidney, BC, 2013) is coupled to standard Paroscientific quartz pressure sensors to provide a resolution of better than 0.04 mm in 4000 m of water at a sampling rate of 1 Hz. The 2009-Samoan tsunami, the 2010-Chilean tsunami and the 2011 Tohoku tsunami originating off Japan were clearly recorded by the bottom-mounted NEPTUNE-Canada "tsunami array" of BPRs moored in depths of up 2700 m off the coast of Vancouver Island, by open-ocean DART stations located nearby, and by several tide gauges on the British Columbia—Washington coast (Thomson et al., 2009; Rabinovich et al., 2012; Thomson et al., 2013). These high quality, far-field tsunami records provided a comprehensive analysis of the events up to 11,000 km from the source area.

1.6.3 Satellite Altimetry

Conventional sea-level measurement systems are limited by the need for a fixed platform installation. As a result, they are only possible from coastal or island stations where they can be referenced to the land boundary. Unfortunately, there are large segments of the world's ocean without islands, so that the best hope for long-term global-scale sea-level measurements lies with satellite-borne radar altimetry. Early studies (Huang et al., 1978) using GEOS-3 altimeter data with its fairly low precision of 20–30 cm, demonstrated the value of such data for estimating the variability of the sea surface from repeated passes of the satellite radar. In this case, the difference between repeated collinear satellite passes eliminates the unknown contribution of the earth's geoid to the radar altimeter measurement. This same technique was employed by Cheney et al. (1983), using 1000 orbits of high quality SEASAT radar altimeter data.

With a known precision of 5–8 cm, the early radar data provided some of the first large-scale maps of mesoscale variability in the world's ocean. Satellite altimetry is now actively pursued by physical oceanographers interested in ocean circulation problems, including the formation and propagation of mesoscale eddies, and major current systems. The experience with GEOS-3 and SEASAT altimeter data demonstrated the great potential of these systems, which now provide sufficient accuracy to allow the specification of the mean ocean circulation related to the ocean surface topography. As with earlier applications, the primary concern is with the contribution of the earth's geoid to the satellite altimeter measurements. The geoid is known to have variations with space scales similar to the scales of sea-level fluctuations associated with the mean and mesoscale ocean circulation. In addition, satellite altimetry data must be corrected for: (1) variations in satellite orbit; (2) atmospheric effects, requiring knowledge of the intervening atmospheric temperature and water vapor profiles; and (3) sea state, which affects the shape of the reflected radar waveforms. Altimeters currently onboard TOPEX/Poseidon and newer series of satellites (combined with more exact processing methods) allow more precise determination of sea-level variations, to a level of about 2 cm, while assimilation of radar altimeter data into numerical ocean models has paved the way to improved mapping of mean sea level and of the geoid. Future altimeter missions, such as the Surface Water Ocean Topography Mission (SWOT), will also allow higher-resolution description of the ocean mesoscale variability.

Considerable headway has been made in the area of satellite altimetry due to the successful deployment of a number of spaceborne altimeters. The first to generate a lot of new data was the GEOSAT satellite first launched in 1985 by the U.S. Navy in an effort to more precisely map the influences of the geoid on missile tracks. After an 18-month "geodetic mission" the Navy was convinced by Dr. Jim Mitchell and others to put the satellite into an "exact repeat orbit" in November, 1986, using the same orbit as the previous SEASAT satellite (Tapley et al., 1982). The altimeter data from this orbit had already been

made public and thus the classified altimeter data from the geodetic mapping mission were already compromised for this orbit. By having the satellite operating in this orbit, scientists would be able to collect and analyze data on the ocean's height variability. Fortunately, the GEOSAT altimeter continued to function into 1989 providing almost three full years of repeated altimetry measurements. In addition, the navy has released the "crossover" data from the geodetic mission. In this mission, the track did not repeat but the crossovers between ascending and descending tracks provided valuable information on ocean height variability. Thus, it is possible to combine data from the earlier crossovers and repeat orbits from the "exact repeat mission" to form a nearly five-year time series of sea surface height variations. It should be stressed that without a detailed knowledge of the earth's geoid it is not possible to compute absolute currents and the main area of investigation provided by the GEOSAT data was in studying the ocean's height variability.

Considerable experience was gained in computing the various corrections that are needed to correct satellite altimeter data (Chelton, 1988). These include the ionospheric correction, the dry tropospheric correction, and the wet tropospheric correction. Added to these are the errors due to electromagnetic bias, antenna mispointing, antenna gain calibration, the inverse barometer effect, ocean/earth tides, and precise orbit determination. Since the GEOSAT satellite did not carry a radiometer to compute tropospheric water vapor, other operational satellite sensors were used to compute the atmospheric moisture to correct the altimeter pathlength (Emery et al., 1989a). Many experiments were conducted to better understand the electromagnetic bias correction (Born et al., 1982; Hayne and Hancock, 1982). Other corrections can be routinely computed from available sources, including the dry troposphere correction, which requires knowledge of the atmospheric pressure (Chelton, 1988).

GEOSAT data have been used to map both the large-scale and smaller scale regional circulations of the ocean. Miller and Cheney (1990) used GEOSAT data to monitor the meridional transport of warm surface water in the tropical Pacific during an El Niño event. Combining crossover and colinear data, the authors constructed a continuous time series of sea-level changes on a $2 \times 1°$ grid in the Pacific between $20°$ N and $20°$ S for the four-year period from 1985 to 1989. They concluded that the 1986–87 El Niño was a low-frequency modulation of the normal seasonal sea-level cycle and that a buildup of sea level in the western Pacific was not required as a precursor to an El Niño event. A similar analysis of colinear GEOSAT data for the tropical Atlantic (Arnault et al., 1990) showed good agreement between the satellite-sensed sea-level changes and those measured in situ using dynamic height methods. Using GEOSAT sea-level residuals computed

from a two-year mean, Vazquez et al. (1990) examined the behavior of the Gulf Stream downstream of Cape Hatteras. Comparisons with NOAA infrared satellite imagery show a fair agreement with the Gulf Stream path, but with some sea-level deviation maps not showing a clear location of the main stream. In the same geographic region Born et al. (1987) used a combination of GEOSAT altimetry and airborne XBT data, to map geoid profiles as the difference between the altimetric sea level and the baroclinic dynamic height. Many other oceanographers have used GEOSAT data to study a great variety of oceanographic circulation systems (e.g., Figure 1.34).

In December of 1992, the long awaited TOPEX/Poseidon (T/P) altimetric satellite was launched. Carrying two altimeters (one French and one U.S.) with a single antenna, TOPEX/Poseidon marked a significant step forward in altimetric remote sensing. The NASA altimeter

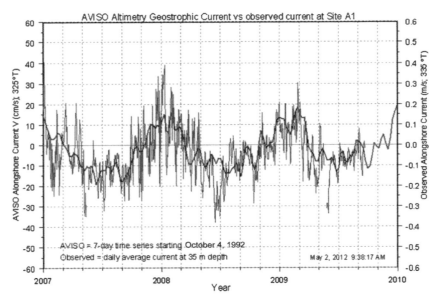

FIGURE 1.34 Comparison of the alongshore component of geostrophic current velocity derived from satellite altimetry (blue line) and the corresponding alongshore current observed by a current meter moored at 35-m depth (red line). The geostrophic currents are computed at 7-day time steps by AVISO; the current meter data are daily mean values for a nearby long-term mooring site located in 400 m of water on the continental slope off southwest Vancouver Island, British Columbia. The altimeter products were produced by Ssalto/Duacs and distributed by Aviso, with support from Cnes (http://www.aviso.oceanobs.com/duacs/).

is a dual-frequency altimeter, which is able to compensate for the influence of ionospheric changes. In early 2012 Jason-1 was placed in a non-repeat geodetic orbit which it completed and ceased operation in early 2013. The French altimeter was the first solid-state instrument to be deployed in space. In addition, there was a boresight microwave radiometer (TOPEX Microwave Radiometer) to provide real-time atmospheric water vapor measurements for the computation of wet troposphere corrections for the onboard altimeters. The resulting combination of data provided altimeter heights accurate to ±2.5 cm, providing important data for tidal analysis of global sea levels and other large-scale processes (e.g., Schrama and Ray, 1994; Ray 1998, Cherniawsky et al., 2001; Di Lorenzo et al., 2005; Foreman et al., 1998). A truly joint project, the satellite was built in the U.S. and launched by the French *Ariane* launch vehicle. In early 2002, the altimetric satellite Jason-1 (launched in December 2001) was placed into the TOPEX/Poseidon orbit and operated for six months just 90 sec behind T/P to intercalibrate the two altimeters. After this intercalibration period the older satellite shifted to an orbit midway between that of the Jason-1 orbital paths in order to improve the spatial resolution of altimeter coverage. Jason-1 had a repeat orbital period of 10 days and was capable of high-precision ocean altimetry measurements from the satellite to the ocean surface with accuracies of a couple of centimeters. A companion altimetric satellite OSTM/Jason-2—where OSTM stands for Ocean Surface Topography Mission—was launched in June 2008 and is now in a circular, non-sun-synchronous orbit at an inclination of 66° to earth's equator, which allows it to survey 95% of earth's ice-free ocean every 10 days. In 2009, Jason-1 was moved to the opposite side of earth from Jason-2, and flew over the same region of the ocean that Jason-2 flew over five days earlier. The ground tracks of Jason-1 fell midway between those of Jason-2, which are about 315 km apart at the equator. This combined mission provided twice the number of measurements of the ocean's surface for accurate observations of sea surface height variations that include changes in global sea level, the velocity of ocean currents, including those associated with mesoscale eddies, and changes in ocean heat storage (thermosteric effect). The combined mission helps pave the way for a future ocean altimeter mission that would collect much more detailed data with its single instrument than the two Jason satellites now do together. In early 2012, Jason-1 was placed in a non-repeat geodetic orbit which it completed and ceased operation in early 2013.

Altimetry data are available through the Web site AVISO (Archiving, Validation and Interpretation of Satellite Oceanographic data). Several data sets are available, including the mean sea surface height anomaly (msla) acquired from the Ssalto/Duacs project. AVISO presently distributes satellite altimetry data from TOPEX/Poseidon, Jason-1, ERS-1 and ERS-2, EnviSat, and DORIS precise orbit determination and positioning products. Altimetry is also provided by Cryosat-2 with an altimeter (Siral) working in an interferometric mode, with a high orbit inclination of 92° to satisfy scientific needs for observing the polar regions and ice sheets, and with an orbit non-sun-synchronous (commonly used for remote sensing satellites). The Chinese satellite HY-2A has a 14-day orbit. Figure 1.33 is an example of what can be achieved using data downloaded from AVISO. Here, we have compared the alongshore surface currents derived from altimetry with co-located alongshore currents observed at 35 m depth at long-term current meter mooring site A1 in 500 m of water off the west coast of Vancouver Island, British Columbia (cf., Thomson and Ware, 1996). The satellite node (at 48.5° N, 126.3° W) and the mooring location are within 7 km of one another. In general, the 10-day altimetry data reproduces the seasonal cycle of the alongshore currents over the continental slope but cannot reproduce the energetic higher frequency motions associated with local, as well as remote, wind forcing (cf. Hickey et al., 2003; Connolly et al., 2014).

1.6.4 Inverted Echo Sounder

As noted in Section 1.5, accurate depth measurements using acoustic sounders require corrections on the order of $\pm1\%$ for variations in sound speed introduced by changes in oceanic density. Rossby (1969) suggested that this effect could be used to advantage since it provided a way to measure variations in travel times of acoustic pulses sent from the sea floor due to changes in the depth of the thermocline. Moreover, the fact that travel times are integrated measurements means that they effectively filter out all but the fundamental mode of any vertical oscillations. This idea led to the development of the IES in which the round-trip travel time of regularly spaced 10 kHz acoustic pulses from the seafloor are now used to determine temporal variability in the integrated density structure of the ocean. The IES has been widely used in studies of the Gulf Stream, where its records are interpreted in terms of thermocline depth, heat content, and dynamic height (Rossby, 1969; Watts and Rossby, 1977). It has also been used in the equatorial Pacific and Atlantic although interpretation of the data is more uncertain because of a lack of repeated deep CTD casts to determine density variability (Chiswell et al., 1988).

Tidal period variability and large changes caused by El Niño-Southern Oscillation events are potentially serious problems in the interpretation of echo sounding data. In particular, the CTD data are needed to convert time series of acoustic travel time Δt between two depth levels (z_1, z_2) to a time series of dynamic height ΔD integrated over the pressure range p_1 to p_2 with an accuracy of $\pm0.01-0.04$ dynamic meters. The obvious similarity between these two parameters (Watts and Rossby, 1977) can be seen from the relations

$$\Delta t_{z1/z2} = 2 \int_{z_1}^{z_2} [1/c(S,T,p)]dz \tag{1.24}$$

and

$$\Delta D_{p2/p1} = \int_{p_1}^{p_2} [1/\rho(S,T,p) - 1/\rho(35,0,p)]dp$$

$$= 10^3 \int_{p_1}^{p_2} \delta dp$$

$$\tag{1.25}$$

where ρ is the water density, c is the speed of sound (which is dependent on the water density through specified salinity S, temperature T, and pressure p), and $z =$ depth (positive downward). Finally, δ, defined as

$$\delta = \alpha(S,T,p) - \alpha(35,0,p)$$
$$= 1/\rho(S,T,p) - 1/\rho(35,0,p) \tag{1.26}$$

is the specific volume anomaly. In these expressions, we use SI units with depth in meters, density in kilogram per meter cubed, pressure in decibars, and dynamic height in dynamic meters (1 dyn·m = 10 m^2/s^2).

Chiswell et al. (1988) compare time series of dynamic height from an IES with sea-level height ($z_1 = -\eta$) from a pressure sensor located 70 km away on Palmyra Island in the central equatorial Pacific. The spectra for the dynamic height variations determined from the IES closely resembled those from the pressure gauge. Significant coherence was found between the two signals at the 99.9% level of significance. Although, in principle, varying mixtures of vertical internal modes could produce a frequency dependence in the conversion of IES to dynamic height, the effect was not significant over the year-long data series. Wimbush et al. (1990) discussed moorings in 4325 m of water 72 km west of the subsurface pressure gauge in the Palmyra Lagoon (5°53' N, 162°05' W). The IES was set to 1/2-h sampling, each sample consisting of 20 pulses 10 s apart. Outliers were eliminated and the median value taken as representative of the acoustic travel time. According to Wimbush et al., a conventional IES without a pressure sensor adequately records synoptic-scale dynamic height oscillations with 20 to 100-day periods. Chiswell (1992) discussed 14-month records from five IESs deployed in February 1991 in a 50-km array in 4780 m of water near 23° N, 158° W north of Hawaii. The CTD and acoustic Doppler current profiler (ADCP) data collected during monthly surveys at the array site provided sufficient density data to calibrate the IES data in terms of dynamic height and geostrophic currents.

Wimbush et al. (1990) used the response method of Munk and Cartwright (1966) to determine the diurnal and semidiurnal tides, and filtered the data with a 40-h Gaussian low-pass filter to examine the residual (nontidal) motions. Chiswell has attempted to resolve the tidal motions through 36-h burst sampling of the density structure from three-hourly CTD profiles. IES deployments show that there is a linear relationship between dynamic height and travel time (Figure 1.35), with the calibration slope dependent on the particular T-S properties of the region. In this case, we can link variations in ΔD

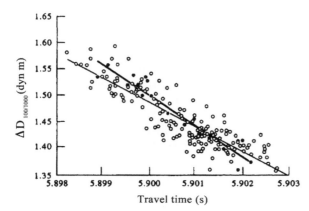

FIGURE 1.35 Dynamic height at 100 m relative to 1000 m ($\Delta D_{100/1000}$) from 186 shallow CTD casts plotted against corresponding travel time measured by IESs (open circles). Thin line is the least squares fit. Thick line and solid circles give $\Delta D_{100/1000}$ calculated from 17 deep casts plotted against the corresponding travel time from 4500 m to the surface, $T_{0/4500}$ ($r^2 = 0.93$ and slope $= -57.8$ dyn m/s). *(From Chiswell (1992).)*

(for depths shallower than the reference level p_{ref} used in the dynamic height calculation) to the acoustic travel time Δt_{ref}

$$\Delta D = m \Delta t_{ref} \qquad (1.27)$$

where total acoustic travel time to the bottom is

$$\Delta t = \Delta t_{ref} + \gamma H_2 \qquad (1.28)$$

in which $\gamma = 2/c_b$ and c_b is the average sound speed (assumed constant) between p_{ref} and the bottom. The depth H_2 is the depth range between the seafloor (pressure $= p_b$) and the reference pressure level, p_{ref}. Solving yields,

$$\Delta D = m \left[\Delta t - (\gamma / \rho'_b g') p_b \right] \qquad (1.29)$$

where gravity and bottom density are scaled as $g' = 0.1g$ and $\rho'_b = 10^{-3} \rho_b$, respectively. For oscillations in density having periods longer than about 20 days, the second term on the right-hand side of Eqn (1.29) may be dropped, whereby

$$\Delta D = m \Delta t \qquad (1.30)$$

Wimbush et al. (1990) find $m = -70$ dyn·m/s to convert Δt to ΔD, while Chiswell (1992) finds $m = -57.8$ dyn·m/s for Δt defined for $z = 0$–4500 m and ΔD at 100 m referenced to 1000 m. The high squared correlation coefficient, $r^2 = 0.93$, is based on 186 shallow (1000 m) and 17 deep (4500 m) CTD casts. The error in the slope using the deep casts is 4 dyn·m/s, with an RMS deviation of 0.017 dyn·m/s; for the shallow casts, the mean is 0.1 dyn·m/s with a deviation of 0.029 dyn·m/s. The travel times for the subtropical Pacific moorings of Chiswell correlate better with dynamic height measured below 100 m than with surface dynamic heights. This is because large variations in the temperature and salinity relation in the upper 100 m affect dynamic height more than they affect acoustic travel time (Chiswell et al., 1988). The tidal range of 0.08 dyn·m is relatively large compared with the seasonal range of 0.25 dyn·m and illustrates the need for detailed CTD sampling. Geostrophic currents have been derived from the array using the time series of dynamic height created from the multiple IES moorings. Aliasing of the records by high-frequency motions and a lack of CTD data to the depth of the IES remain problems for this method.

Today, researchers can purchase commercial IES combined with an optional Paroscientific pressure sensor (the instrument is designated a "PIES"). A combined IES, data logger, and acoustic release with both pressure and Aanderaa current velocity sensors is called a CPIES. PIES is a long-life sensor logging unit that transmits a wideband acoustic pulse to accurately measure the average sound velocity through a column of water from the seabed to the sea surface and back again. Units are commonly used to examine tides, currents, and long-term, in situ, changes in the thermal properties of the ocean along with coincident variations in barotropic pressure. The pressure sensor provides an accurate measurement of depth (distance to the surface). The sampling interval of PIES can be configured serially before deployment and also via its internal acoustic telemetry link. This telemetry link also allows recorded data to be transmitted to surface at data rates ranging from 100 to 6000 bits per second. PIES can be free-fall deployed. When the study is complete, its integrated acoustic release enables it to be commanded to disconnect from its tripod stand and return to the surface under its own buoyancy ready for collection by the surface vessel. Results from IESs can be found in Meinen and Watts (2000), Meinen et al. (2004), and Li et al. (2009). Numerous agencies (including NOAA/AMOL (Atlantic Oceanographic and Meteorological Laboratory)) have acquired these instruments from the University of Rhode Island (URI), the only known manufacturer of these systems.

1.6.5 Wave Height and Direction

Any discussion of sea-level would be incomplete without some mention of surface gravity wave measurement. Methods include: a capacitance staff which measures the change in capacitance of a conductor as the air–water interface moves up and down with passage of the waves; an upward-looking, high-frequency acoustic sounder or ADCP with a vertical-pointing transducer, which can be used to examine both the surface elevation and the associated orbital currents; a fixed graduated-staff attached to a drill platform, stuck in the sand or otherwise attached to the seafloor; satellite altimetry; a bottom-mounted pressure gauge with rapid sampling time; a shipborne Tucker wave-recorder system; and the waverider and directional waverider buoys. For brevity, we limit our presentation to the directional waverider since it represents reliable off-the-shelf technology. Commercial units include the Datawell Mark II directional waverider and the TRI-AXYS™ Directional Wave Buoy. Both companies offer a solar-powered unit.

The Datawell directional waverider is a spherical, 0.9-m diameter buoy for measuring wave height and wave direction. The buoy contains a heave-pitch-roll sensor, a three-axis fluxgate compass and two fixed "x" and "y" accelerometers. The directional (x, y, z) displacements in the buoy frame of reference are based on digital integration of the horizontal (x, y) and vertical (z) accelerations. Horizontal motions rather than wave slope are measured by this system. Vertical motions are measured by an accelerometer placed on a gravity-stabilized platform. The platform consists of a disk, which is suspended in a fluid within a plastic sphere placed at the bottom of the buoy. Accelerations are derived from the electrical coupling between a fixed coil on the sphere and a coil on the platform. A fluxgate compass is used to convert displacements from the buoy frame of reference to true earth coordinates. The operation of the solar-powered three accelerator TRIAXYS™ Directional Wave Buoy is similar to that of the Datawell buoy.

Displacement records are internally filtered at a high-frequency cutoff of 0.6 Hz. Onboard data reduction computes energy density, the prevailing wave direction, and the directional spread of the waves. Frequency resolution is around 0.01 Hz for waves in the range 0.033–1.0 Hz (periods of 30 to 1 s). Transmission of data is through the Argos satellite link, through a standard 27–40 MHz radio link, or via cellular to shore. The buoy will measure heave in the range ±20 m with 0.5% wave height resolution for wave periods of 1.6–30 s in the moored configuration. The direction range is 0–360° with a resolution of 1.5°. The manufacturers of the TRIAXYS buoy (AXYS Technologies, Inc.) claim better battery life and, because of solid-state accelerometers, a more linear response to heave as a function of frequency. Successful deployments in water depths of up to 300 m have been undertaken. The buoy is fitted with an Inmarsat D+ transceiver, which allows GPS data to be collected. Upon activation, the satellite system is also capable of secondary telemetry should the primary system be unavailable, as well as for a WatchMan500™ controller alarm should the buoy drift too far out of position. If the buoy

were to move beyond a defined radius, the end user will receive a series of alarm messages. The instrument collects, processes, and logs wave and SST data on the buoy, which is then transmitted via VHF, cellular, or satellite telemetry to a base station hosting the AXYS Data Management System. The TRIAXYS Waves plug-in data display allows for full presentation and archiving of data along with diagnostic utilities.

A crucial aspect of collecting reliable, long-term wave data is the mooring configuration. If not designed and moored correctly, there is little chance the mooring will survive the constant stresses of the wave motions. As illustrated in Figure 1.36, the recommended configuration consists of a single-point vertical mooring with two standard rubber shock cords and heavy bottom chain. This arrangement ensures sufficient symmetrical horizontal buoy response for small motions at low frequencies while the low stiffness of the rubber cords allows the waverider

FIGURE 1.36 Morning configuration for a Datawell Directional Waverider buoy on a shallow continental shelf. (*Modified after Datawell bv, 1992; Courtesy T.Fuhász and R. Kashino.*)

to follow waves up to 40 m high. Current velocities can be up to 2.5 m/s, depending on water depth (Datawell *bv*, 1992).

1.7 EULERIAN CURRENTS

The development of reliable, self-recording current meters is one of the major technological advances of modern oceanography. These sturdy, comparatively lightweight instruments are, in part, a by-product of the rapid improvement in electronic recording systems, which make it possible to record large volumes of digital data at high sampling rates (Baker, 1981). Although they can be used in either moored or tethered profiling modes, most current meters are used in time-series measurement of current speed and direction at fixed locations. (Such fixed-location measurements are called Eulerian measurements after the Swiss mathematician Leonhard Euler (1707–83) who first formulated the equations for fluid motion in a fixed frame of reference.) The development of reliable mooring technology and procedures also has played a major role in advancing the use of moored current meters and associated instrumentation. Acoustic release technology, which has proven so critical to oceanic research, will be discussed at the end of this section.

Most commercially available current meters have sufficient internal power and data storage to be moored for several months to several years. With the exception of current meters incorporated within real-time cabled observatory networks, which supply power through cables connected directly to a shore-based facility, the instrument's longevity in the ocean clearly depends on the selected sampling rate, the data storage capacity, the battery life, and the ambient water temperature. Greater power can be obtained from lithium batteries than from more conventional batteries but the user sometimes faces numerous transportation regulations and operational concerns with lithium batteries.

Operating time for all types of batteries decreases with water temperature. Despite their sophistication, most current meters are made to withstand a fair amount of abuse during deployment and recovery operations. Typical "off-the-shelf" current meters (and releases) can be deployed to depths of 1000–2000 m, and many manufacturers fabricate deep versions of their products with heavy-duty pressure cases and connectors for deployments to depths of up to 6000 m. Most modern current meters also allow for the addition of ancillary sensors for concurrent measurement of temperature, conductivity (salinity), water clarity (light attenuation and turbidity), pressure, and other scalars. Instrument failure is less likely than in the past but is certainly not uncommon. Leakage through connector ports, seals, and cables are major causes for failures. Damaged (or missing) O-rings, saltwater corrosion, or poorly tightened fasteners are also causes for instrument damage and data loss. Corrosion sometimes occurs as a result of electrical faults, the use of combined dissimilar metals, or crevice corrosion in stainless steel (as with some acoustic releases and instrument frames and hardware). Instruments are typically much more reliable than in the past, due, in part, to reduced size, and the use of integrated surface mounts, rather than discrete components, in the electronics. Battery life, rather than memory capacity, has become the limiting factor for instrument deployments.

Current meters differ in their type of speed and direction sensors, and in the way they internally process and record data. Although most oceanographers would prefer to work with the scalar components u, v of the horizontal current velocity vector, $\mathbf{u} = (u, v)$, current meters can directly measure only the speed ($|\mathbf{u}|$) and direction (θ) of the horizontal flow. (For now, we ignore the vertical velocity component, w.) It is because of this constraint that most current meter editing and analysis programs historically work with speed and direction. From a practical point of view, both the (u, v) and the $(|\mathbf{u}|, \theta)$

representations have their advantages, despite the difficulties with the discontinuity in direction at the ends of the interval of 0–360°.

Speed sensors can be of two types: *mechanical sensors*, which measure the current-induced spin of a rotor or paddle wheel; and *nonmechanical sensors*, which measure the current-induced change in: (1) a known electromagnetic field; (2) the difference in acoustic transmission times along a fixed acoustic path; or (3) in the case of many modern current meters, the Doppler frequency shift in gated acoustic pulses reflected off backscatters considered to be drifting passively in the water column. Despite these fundamental differences, all current meters have certain basic components that include speed sensors, a compass to determine orientation relative to the earth, built-in data processing algorithms, a digital storage device, and a source of power, such as a battery pack. Possible speed sensors include:

1. Propellers (with or without ducts).
2. Savonius rotors.
3. Acoustic detectors (sound propagation time or Doppler frequency shift).
4. Electromagnetic sensors (induced magnetic field).
5. Platinum resistors (flow-induced cooling).

Flow direction relative to the axes of the current meter is usually sensed using a separate vane or by configuring the speed sensors along two or three orthogonal axes. In all current meters, the absolute orientation of the instrument, relative to the earth's magnetic field, is determined by an internal compass. At polar latitudes where the horizontal component of the earth's magnetic field is weak, measurement of absolute current requires that the meter be positioned rigidly in a known orientation. Direction resolution depends on the type of compass used in the measurement; e.g., clamped potentiometer for the earlier Aanderaa Recording Current Meters (RCMs), optical disk for Marsh-McBirney electromagnetic current meters (ECMs), and flux

gate (Hall effect) compasses for the EG&G Vector Measuring Current Meter (VMCM), the InterOcean S4 current meter, and the Teledyne-RDI, Nortek, and SonTeck acoustic current meters (ACMs). For each deployment, compass direction must be corrected for the local deviation of the earth's magnetic field before the velocity data are converted to north–south and east–west components. The accuracy, precision, and reliability of a particular current meter are functions of the specific sensor configuration and the kind of processing applied to the data. Rather than comment on all the many possible variations, we will discuss a few of the more generic and successful configurations.

The problems and procedures associated with the use of these instruments, and the analysis of the resultant data, are sufficiently similar that the discussion should be instructive in the use of instruments not specifically mentioned. Current meter technology has advanced considerably over the past decade and several of the companies mentioned in earlier editions of this book are no longer manufacturing current meters for ocean science. Most manufacturers of ACMs and ECMs are still in operation; the high accuracy and durability provided by the single-point ECMs and ACMs, and profiling ADCPs built by these companies, have enabled them to dominate marine research and industry applications. However, much of the data in historical archives are from older type of current meters. For this reason, we have retained the sections on current meters published in previous editions of this book.

1.7.1 Early Current Meter Technology

One of the earliest forms of current measurement was the tilt of a weighted line lowered from a ship. The time it took an object dropped alongside a vessel to travel the length of the ship also provided a measure of the surface flow. (The term "knot" is from the use by Dutch sailors of a knotted line to measure the speed of their sailing vessel.) Although we like to think of the current meter as a recent innovation, the Ekman current meter was in use as early as the 1930s (Ekman, 1932). Although many different mechanical current meters were built in those days (see Sverdrup et al., 1942), few worked and most scientists went back to the Ekman meter. To measure the current, the instrument was lowered over the side of the ship to a specific depth, started by a messenger (a weight that is slid down the line), and then allowed several minutes before being stopped by a second messenger. The current speed for each time increment was determined by reading a dial that recorded the number of revolutions of an impeller turned by the current. A table was used to convert impeller revolutions to current speed. Current direction was determined from the distribution of copper balls that fell into a compass box below the meter at a fall rate that was a function of the rotation of the propellor. A profile from 10 to 100 m typically took about 30 min. Obvious problems with this instrument included low accuracy in speed and direction, limited endurance, and the need to work from a ship or other stationary platform. One of the first commercial current meters was the self-contained Geodyne 850 current meter built in the United States in the 1960s. The Geodyne was a large and bulky, vertically standing unit with a small direction vane and four-cup Savonius rotor. Burst sampling was permissible in the range of 60–660 s. The Nerpic CMDR current meter built in France in the 1960s was a torpedo-like device that oriented itself with the current flow and used an impellor-type rotor to measure current speed. In the original versions, data were recorded on punched paper tape. The Kaijo Denki current meter built in Japan in the 1970s was one of the earliest types of instruments that used differential acoustical travel time to measure flow velocity.

Although some oceanographers might disagree, the age of the modern current meter appears to have started with the Aanderaa RCM developed by Ivar Aanderaa in Norway in the early 1960s under sponsorship of the North

Atlantic Treaty Organization (NATO). The fact that many of these internally recording current meters still remain in operational condition attests to the instrument's durability. There was a time when many oceanographers considered the Aanderaa RCM4 (Figure 1.37) and its deep (>2000 m) counterpart, the RCM5, the workhorses of physical oceanography. It certainly was the most common and reliable current meter used to measure ocean currents. For this reason, there have been more studies, intercomparisons, and soul searching with this instrument than with any other type of meter.

1.7.2 Rotor-Type Current Meters

1.7.2.1 *The RCM Series of Current Meters*

The Geodyne and RCM4 current meters were the first current meters to use a Savonius rotor to measure current speed. This rotor consists of six axisymmetric, curved blades enclosed in a vertical housing, which is oriented normally to the direction of flow (Figure 1.37). Data collected by the RCM4 were recorded on a small 1/4-inch reel-to-reel magnetic tape. Today, data are mainly stored in solid state memory. Allowable sampling rate settings are 3.75×2^N min (e.g., 3.75, 7.5, 15, 30, 60 min) where N ($= 0$, 1, 2, ...) is an integer. Although shorter sampling periods are possible, they are not practical given the mechanical limitations of the rotor. Speed is obtained from the number of rotor revolutions for the entire sample interval while direction is the single direction recorded at the end of the sample period. Thus, speed is based on the average value for the recording interval while direction involves a single measurement. In the past, the number of revolutions per recorded data "count" was varied by changing the entire rotor counter module. More recent RCMs allow the investigator to set the number of revolutions per count (e.g., 2^M revolutions per count, where $M = 1$, 2, 3, ...) so that the speed range of the instrument can be adjusted for the flow conditions. For example, in the coastal tidal passes of British

(a)

FIGURE 1.37   (a) Aanderaa RCM4 current meter; (b) Exploded view of the encoder side of the Aanderaa RCM4 current meter. The reverse side contains a reel-to-reel 1/4" tape system for recording the data from the different channels. The recorder unit is attached to a directional vane. *(Courtesy G. Gabel.)*

Columbia and Alaska, the common upper range of 3 m/s for standard rotor settings was not always sufficient to measure peak tidal speeds; the peak speed of 7.5 m/s that occurs in Nakwakto Rapids at the entrance to the Seymour–Belize inlet system in coastal British Columbia is beyond the range of most modern current meters. The direction vane of the RCM4 is rigidly affixed to the pressure case containing the data logger. The unit is then inserted in the mooring line and the entire current meter allowed to orient in the direction of the current. Although the RCM does not average internally,

vector-averaged currents can be obtained through postprocessing of the data (Thomson et al., 1985). A meteorological package for surface applications also is available with the same data logging system (Pillsbury et al., 1974).

Part of the reason for the popularity of the RCM series of current meters has been their reliability, comparatively low cost, and relatively simple operation. Both calibration, and maintenance of the instruments, can be performed by individuals, with fairly limited electronics expertise. In more recent years, many of the other types of current meters, such as ECM and the

(b)

FIGURE 1.37 *(continued)*.

ACM, have advanced to the point that they require electronics expertise due to the advanced computer diagnostics available from the manufacturers of these instruments. Moreover, modern high maintenance instruments often need to be shipped back to the manufacturer for warranty-supported calibration. Another attractive feature of the RCM was (they are no longer supported by the manufacturer) the easy addition of sensors for measuring temperature, conductivity, and pressure (depth). The Aanderaa RCM7 introduced in the late 1980s could be purchased with standard temperature (-2 to $+35\,°C$), expanded temperature (i.e., over a narrower range such as $0-10\,°C$), conductivity (for salinity), and total pressure. The 0 to 5 V output from a Sea Tech transmissometer for measuring water clarity was readily incorporated in the instrument package. Thus, there was the potential to collect a wide range of parameters other than just currents alone.

The profiling Cyclesonde (van Leer et al., 1974; Baker, 1981) consists of a RCM4 current meter affixed to a buoyancy-driven platform, which makes repeated automatic round trips between the surface and some specified depth (<500 m) along a taut-wire mooring. The vertical cycling of the instrument is controlled by changing the density of the instrument package, by a few percent, using an inflatable bladder. Depending on the prescribed sampling interval and the duration of each round-trip (or depth of water sampled), the instrument can provide time series of currents, temperature, and salinity over periods of weeks to months at depths of every 10 m or so through the water column (Stacey et al., 1987; Webb and Pond, 1986).

Processing of RCM4 data includes four major steps: (1) tape transcription (quarter-inch tape to computer format); (2) calibration, or conversion to physical units; (3) error detection, spike removal, and interpolation; and (4) data analysis. The last point will be discussed in detail in later chapters of this book. The first three steps provide an example of the procedure required in producing useful data from moored current meters. The data in an RCM4 are recorded as 10-bit binary words (numbers from 0 to 1023) on 1/4-inch magnetic tape. For each cycle, six binary words are written on the tape. For the temperature channel, a near-linear calibration curve is applied to the measured value to convert it to temperature. The relationship between speed in physical units (e.g., cm/s) and rotor count is nearly linear so that speed also can be calculated from a linear calibration. Current speed for earlier versions was handled somewhat

differently since speeds had to be calculated as the difference between consecutive integers recorded on the appropriate data channel.

Tape translation is carried out by connecting a 1/4-inch tape recorder to a digital computer. With this setup, the digital data are transferred from the 1/4-inch tape to computer compatible format for further editing and analysis. To these raw character data is added "header" information such as the start and stop times of the particular mooring. As a check, one calculates the number of instrument cycles that should have occurred during the mooring period and this should equal the number of records in the raw data file. If this is not the case, then the data have timing errors, which must be corrected before processing can continue.

To convert the dated raw data to physical units (i.e., speed, direction) calibration constants are needed for the individual sensors. For most parameters the calibration values are found for each meter separately as quadratic fit to the calibration data. As has been mentioned above, this is not the case for the speed parameter for which a general curve can be used for all rotors, if currents are typically greater than 10 cm/s. Directions also are handled somewhat differently in that no formula is derived from the calibration data but rather a simple lookup table is developed for the calibration data from which compass readings can be converted directly into degrees from true or magnetic north.

1.7.2.2 The Vector Averaging Current Meter

As discussed by Baker (1981), one of the important data reduction techniques in oceanography was the introduction of the "burst sampling" scheme of Richardson et al. (1963) whereby short samples of densely packed data are interspersed with longer periods of no data. In continuous mode, the average current speed and instantaneous direction are recorded once per sampling interval. In burst mode, a rapid series of speed and direction measurements are averaged over a short segment of the sampling interval.

In vector-average mode, the instrument uses speed and direction to calculate the horizontal and vertical components of the absolute velocity during the burst. The instrument then separately averages each component internally to provide a single value of velocity vector for each burst. If enough is known about the spectrum of the flow variability, the burst samples can be used to adequately estimate the total energy in the various frequency bands. This procedure greatly reduces the amount of recording space needed to sample the currents. The vector-averaging current meter (VACM) introduced in the 1970s used both burst sampling and internal processing to compute the vector-average components of the current for each sampling period. Current speed was obtained using a Savonius rotor similar to that on the RCMs but direction was from a small vane that was free to rotate relative to the chassis of the current meter. Vectors were computed for every eight revolutions of the rotor and averaged over periods of from 4 to 15 min, depending on the selected sampling interval.

1.7.2.3 Problems with the Savonius Rotor

Because of its widespread use in the past, the Savonius rotor sensor needs to be covered in some detail. We begin by noting that a principal shortcoming of the RCM4/5 is its inability to record currents accurately in regions affected by surface wave motions. The problem with the Savonius rotor response is that it is omnidirectional and therefore responds excessively to oscillatory wave action. An intercomparison experiment using a mooring array shown schematically in Figure 1.38(a) demonstrated the differences between Savonius rotor measurements and those made with an ECM (Woodward et al., 1990). Even under moderate wave conditions, the near-surface moored RCM4 can have its speeds increased by a factor of two through wave pumping (Figure 1.38(b)). The effect of wave pumping on the Savonius rotor significantly increases the spectral energy at both low and high frequencies (Figure 1.38(c)). Hence,

the instrument is best suited to moorings supported with subsurface floats but is not suitable for mooring beneath surface buoys or in the upper ocean wave regime. Unlike the earlier Aanderaa current meters, VACMs provided accurate measurements when deployed in near-surface wave fields and from surface-following moorings (Halpern, 1978). In a comparison between Aanderaa and VACM measurements, Saunders (1976) concluded that "the Aanderaa instrument, excellent though it is on subsurface moorings, is not designed, nor should it be used, where wave frequency fluctuations are a significant fraction of the signal." In this, and a later paper, Saunders (1980) pointed out that the contamination of the Aanderaa measurements in near-surface applications is due also to a lag in the response of the direction vane to oscillatory flow.

In 1991, Aanderaa began manufacturing the vector-averaging RCM7 with a paddle-wheel rotor (Figure 1.37(a)) and internal solid state, E-prom modular memory. In earlier versions of the RCM7, the paddle-wheel rotor was partially shielded by a semicircle baffle, which was intended to reduce wave-induced "pumping". This was subsequently abandoned since the baffle was found to shed small-scale eddies, which interfere with the response of the paddle wheel in other operations. Field tests indicate that the vector-averaging RCM7 had only slightly better wave-region performance than the earlier RCMs (Figure 1.38(c)) and the overall improvements are marginal for most applications. During the selected recording interval, the number of rotor revolutions and compass direction are sampled 50 times per recording interval; e.g., every 12 s for a 10-min sampling interval. As with other vector-averaging current meters, the speed and direction are then resolved internally into east—west and north—south components and successive components are added and temporarily stored. When the selected recording interval has elapsed, the resulting average vector and its angle are calculated and

(a)

J-TEC anemometer

Surface buoy

Subsurface float

5 m   Marsh-McBirney

9 m   Marsh-McBirney

10 m   Aanderaa (Paddle)
        Aanderaa (Savonius)

Depth not to scale

Acoustic release

AML P.G.   Aanderaa P.G.
39 m   35 m   Anchor

FIGURE 1.38   (a) Mooring arrangement for comparison of current speed and direction from Aanderaa RCM4 (Savonius rotor) and RCM7 (paddle wheel) current meters and Marsh-McBirney (Electromagnetic) current meters moored at 10-m depth during September 1983 in an oceanic wave zone (Hecate Strait, British Columbia); (b) Winds were measured using a J-Tec vortex-shedding anemometer. In moderate wind-wave conditions, a surface or near-surface moored RCM4 with Savonius rotor can have its speeds increased by a factor of two through wave pumping. The paddle-wheel RCM7 behaves somewhat better; (c) Power spectra for current measurements in (a). *(Adapted from Woodward et al. (1990).)*

stored. A problem with the electronic memory is that data are lost if the instrument floods, as it often does when the instrument is hit by fishnet or tug boat lines. This was not the case for the 1/4-inch magnetic tapes used in the RCM4. Thomson (1977) reports finding a long lost RCM, that had lain on the bottom of Johnstone Strait, British Columbia for over three years. Although the metal components and circuit boards had turned to mush, the salt-encrusted tape contained a full record of error-free data.

Another problem common to all Savonius rotor current meters is that bearing friction results in fairly high threshold of the rotor and an improper response of the rotor to low current speeds. For the Aanderaa RCM4/5, this threshold level is about 2 cm/s and current measurements taken in quiescent portions of the

ocean will have many missing values where the currents were too slow to turn the rotors during the sampling interval. (A recent attempt to use mechanical RCM7s to measure dense, oxygen-rich water intruding over the shallow sill at the head of Effingham Inlet, a fjord with anoxic bottom water on the west coast of Vancouver Island, British Columbia failed to generate more than a few rotor turns over a period of many months, despite evidence of intrusive events in the coincident-moored CTD time series. In contrast, an acoustic Doppler current meter (ADCM) moored at the same location and depth the following year, yielded numerous weak, ≤ 1 cm/s, inflow events over the same months, consistent with CTD time series.) According to manufacturer specifications, the response is linear for current speeds between

FIGURE 1.38 *(continued)*.

2.5 and 250 cm/s so that once the rotor is turning it has acceptable response characteristics. In this range, accuracy is given as 1 cm/s, or 2% of the speed, whichever is greater. Accuracies for the other associated sensors are ±1% for pressure, ±0.3 °C for temperature, and ±0.05 psu for salinity. All of these accuracies are really "relative" values and regular calibration is required to insure reliable measurements. Such a calibration procedure is discussed in detail in Pillsbury et al. (1974) for RCM4s. We will only highlight some of the more important aspects of this

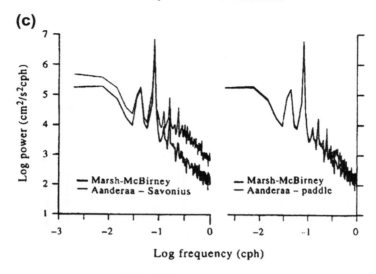

(c)

FIGURE 1.38 *(continued)*.

calibration in order to suggest problem areas where historical data from Aanderaa current meters may be subject to error.

As described by Pillsbury et al. (1974), calibration of the RCM4 compass is important because more compass failures occurred for a set of instruments than all other sensor failures combined. Careful calibration will reveal the several different kinds of compass failure. The compass calibration is performed for selected compass bearings by rotating the instrument through 360° on a pivoted stand. This operation is repeated 10 times. A reliable compass is one which repeats its calibration curve within 3°. From calibration work reported by Gould (1973), it is clear that there is a significant departure from linearity in most RCM4 compasses. The magnitude of the nonlinearity errors (approximately 1% of the scalar mean speed per degree of compass nonlinearity) means that many of the residual velocity values observed in the ocean could be introduced by a nonlinearity of 1° or 2° in the direction sensor. If such residual values are to be trusted, care must be taken to "calibrate out" instrument

nonlinearities in the data analysis procedure. Such precautions are particularly important if the current meter records are to be used to deduce shears from pairs of instruments or circulation patterns from horizontal current meter arrays.

Turning to the rotor, it was found that for speeds several centimeters per second above the threshold, the calibration of all rotors of a given type can be considered as equal. For calibration, this threshold was found to be roughly 10 cm/s, below which each rotor should be calibrated with its corresponding current meter. For mean speeds greater than 10 cm/s, a general calibration curve can be used for all instruments (Figure 1.39). This calibration curve is fitted by a line and used for all calibrations. Deviations from this line varied from 19% at 2 cm/s to less than 1% at 30 cm/s, with a mean value of 4%.

1.7.2.4 Vector Measuring Current Meter

To circumvent the nonlinear response problems of the RCM4, Weller and Davis (1980) developed the VMCM, which used two

FIGURE 1.39 A general calibration curve of current speed (cm/s) vs rotor counts in revolutions/second for mean speeds greater than 10 cm/s. This particular calibration curve has a linear relation for all calibrations.

orthogonal propeller current sensors with an accurate cosine response. This instrument produced negligible rectification and therefore accurately measured mean flow in the presence of unsteady oscillating flow. In laboratory tests, the VMCM performed well in the presence of combined mean plus oscillatory flow as compared with poorer performances by Savonius rotor/vane systems and by electromagnetic and acoustic sensors. The open fan-type rotors of the VMCM were highly susceptible to fouling by small filaments of weed and other debris.

1.7.3 Nonmechanical Current Meters

Nonmechanical current meters use electromagnetic or acoustic sensors to determine the current speed along two or more directional axes. Once the flow direction relative to the current meter is determined, absolute direction is found using a built-in magnetic compass. Single-point current meters measure the flow velocity in the immediate vicinity of the instrument, whereas profiling ACMs (addressed in the following section) provide flow measurements at specified distances from the transducer head.

1.7.3.1 ACM: *Differential Travel Time*

Differential travel time ACMs measure the difference in the time delay of short, high-frequency (megahertz) sound pulses transmitted between an acoustic source and receiver separated by a fixed distance, L. In all cases, the transducer and receiver are combined into one source-receiver unit. The greater the speed of the current component in the direction of sound propagation, the shorter the pulse travel time, and vice versa. For instance, suppose that the speed of sound in the absence of any current has a value c. The times for sound to travel simultaneously in opposite directions from two combined transducer–receiver pairs in the presence of an along-axis current of speed v is: $t_1 = L/(c + v)$ for transducer–receiver pair No. 1 and $t_2 = L/(c - v)$ for transducer–receiver pair No. 2. The velocity component along the transducer axis is therefore

$$v = L(t_2 - t_1)/(2t_1 t_2) \qquad (1.31)$$

A three-axis current meter determines the three-dimensional velocity by simultaneously measuring time differences along three orthogonal axes. This technology does not depend upon the presence of acoustic scatterers in the water for measuring currents and can, therefore, be relied upon to measure velocity in very clear water where scatterer-dependent Doppler currents meters may fail to give a continuous signal. The instruments also work close to the water surface and to the seafloor where ADCMs may give spurious acoustic reflections.

Examples of commercial ACMs include the three-axis MAVS-3 ACM available from NOBSKA (Woods Hole, USA) and a two-axis ACM available from FSI (Falmouth, USA). Earlier versions of ACMs are the SimTronix UCM 40 and the Neil Brown ACM current meters. Because of the rapid (≈ 1500 m/s) propagation of sound in water, these current meters are capable of high-frequency sampling and processing, with typical data rates of 25 Hz and higher. The instruments also can provide estimates of the sound velocity, c, along the two paths of length L between the sensors. More specifically, $c = 2L/t$, where $t = t_1 t_2 / (t_1 + t_2)$ is the effective time of propagation. Manufacturer specifications are as follows:

- *Speed accuracy:* ± 0.3 cm/s.
- *Speed resolution:* ± 0.03 cm/s (FSI claim a resolution better than 0.01 cm/s for their 2-D ACM).
- *Threshold speed:* not specified.
- *Speed range:* $0-2$ m/s
- *Compass direction:* accuracy $\pm 2°$; resolution $\pm 1°$.
- *Sampling rate:* infinity to 25 Hz.
- *Acoustic frequency:* 1.7 MHz.
- *Allowable tilt:* a true cosine tilt response up to $\pm 20°$.
- *Sound speed:* range of $1350-1600$ m/s and accuracy of ± 5 m/s.

NOBSKA was formed in 1997 to commercialize the MAVS Current Meter designed with federal funding by Woods Hole Oceanographic Institution to address the need for low flow, bottom boundary layer measurements. The product has evolved into a general purpose current meter that retains its ability to measure in low flow, clear water environments. MAVS-4 has been developed for the U.S. Ocean Observatories Initiative (OOI). Additional sensors include temperature, conductivity, and turbidity. Research results based on MAVS can be found in (Garcia-Berdeal et al., 2006; Johnston et al., 2006).

Because of their sophisticated technology, ACMs are often difficult to operate and maintain without dedicated technical support. For example, biofouling of the transducers can be a problem on any long-term mooring in the euphotic (near-surface light influenced) zone. The instruments also must undergo frequent recalibration due to problems with sensor misalignment and changes in the physical dimensions of the transducer—receiver pairs. As discussed by Weller and Davis (1980), this is a particular weakness of this type of ACM, which has proved difficult to calibrate due to drifts in the zero level and in the amplifier gain. In one comparison, they found that the background electrical noise of the ACM had the same level as the signal. As they point out, these problems are with the system electronics and have obviously proven to be solvable. Similar problems were encountered by Kuhn et al. (1980) in their intercomparison test using an early prototype model.

1.7.3.2 *Electromagnetic Current Meters*

ECMs such as the Marsh-McBirney 512 and the InterOcean S4 use the fact that an oceanic current behaves as a moving electrical conductor. As a result, when an ocean current flows through a magnetic field generated within the instrument, an electromotive force is induced, which is directly proportional to the speed of the ocean current and at right angles to both the magnetic field and the direction of the current (Faraday's law of electromagnetic induction). In general, the magnetic field may be that of the earth or the one generated by an electric current flowing through appropriately shaped coils (Figure 1.40). In 1832, Faraday tried to measure the flow of the Thames River using large electrodes positioned on either side but was unsuccessful because his galvanometers

FIGURE 1.40 Principle of the electromagnetic current meter. The instrument measures the electromotive force on an electric charge (the oceanic flow) moving through the magnetic field generated by the coil. This produces a voltage potential at right angles to both the magnetic induction field and the direction of flow.

were not sensitive enough to record a change in the electric current induced by the river in the presence of earth's magnetic field. Following the Second World War, the principle was used successfully to estimate the flow along the English Channel by measuring the potential difference between electrodes on either side using a telegraph cable for the distant electrode and the vertical component of the earth's magnetic field. In marine sciences, two-axis ECMs with an internal compass are used to determine the horizontal components of the flow velocity referenced to earth coordinates. The electrical voltage induced between the electrodes by the water motion through the magnetic field produced by the instrument gives the oceanic flow components relative to the instrument axes while the internal compass determines the orientation of the axes relative to the horizontal component of earth's magnetic field. ECMs such as the S4 measure the electrical potential generated across two pairs of exposed metal (titanium) electrodes located on opposite sides of the equatorial plane on the surface of a plastic sphere (Figure 1.41). The electrodes form orthogonal (x, y) axes that

detect changes in the induced electrical potential generated by the ocean current. The induced voltage potential (or EMF) $\mathbf{E}$ is found by Faraday's Law through the cross product.

$$\mathbf{E} = \int_0^\infty \mathbf{v} \times \mathbf{B} dL \qquad (1.32)$$

where $\mathbf{v}$, is the velocity of the flow past the electrodes, $\mathbf{B}$ is the strength of the applied magnetic field supplied by a battery-driven coil oriented along the vertical axis of the instrument, and L is the distance from the center of the coil. The magnetic field is directed vertically past the electrodes so that current flow parallel to the x-axis generates a voltage along the y-axis that is directly proportional to the strength of the water flow. The electric current induced by the voltage potential can be measured directly and converted to components of the flow velocity using laboratory calibration factors. Alternatively, a gain-controlled amplifier can be used to maintain a constant DC voltage at the logical output. The feedback current needed to maintain that

FIGURE 1.41 InterOcean S4 electromagnetic current meter, (a) View of the instrument showing electrodes; (b) Cut-away view of the electronics. The spherical hull has a diameter of 25 cm and the instrument weighs 1.5 kg in water. *(Courtesy, Mark Geneau, InterOcean.)*

electric current is directly proportional to the flow speed. As with the ACMs, InterOcean's specifications are characterized as follows (Marsh-McBirney no longer makes an oceanic ECM):

- *Speed accuracy:* ±2% of reading (with a minimum speed of 1 cm/s).
- *Speed resolution:* 0.3–3.5 mm/s, depending on range, for the S4A (2 Hz instrument) and 0.37–4.3 mm/s, depending on range, for the S4AH (5 Hz instrument).
- *Threshold speed:* 1–4 mm/s; equal to resolution and limited by noise.
- *Speed range:* standard is 0–3.5 m/s, but can be expanded to 0–7.5 m/s or reduced for higher resolution.
- *Compass direction:* accuracy of ±2° (within tilts of 5°) and resolution of 0.5°.
- *Allowable tilt:* cosine tilt response up to ±25°.

The standard S4A and deep S4AH ECMs have a fast response platinum temperature sensor, an inductive flow-through conductivity sensor, and high-resolution depth sensor. Additional add-ons include optical backscatter (suspended solids), transmissometer (turbidity meter), dissolved oxygen, and pH. Standard memory is 20 megabytes. The S4A also provides for two separate autonomous sampling cycles. This multi-tasking allows the instrument to collect data that normally requires several instruments running different cycles. Each "mode" allows for complete control of the sampling interval, averaging time (continuous or burst sampling), and parameters to record. All of the sensors available on the S4 can be programed to sample in either or both of the chosen sampling modes. Data can be averaged over regular intervals of a few seconds to tens of minutes, or set to burst sampling with a specified number of samples per burst at a given sampling interval. In addition, one can set the number of times velocity is sampled compared with conductivity and temperature. There is also an adaptive sampling option, which makes it possible to program the instrument to only save to memory those events exceeding a preset threshold. This feature in the S4 allows burst-mode recording of data over extended periods previously not possible due to memory limitations with normal burst-mode recording. This has particular application for current and wave data recording where the user is only interested in recording high level occurrence events. The surface of the S4 housing is grooved to maintain a turbulent boundary layer and prevent flow separation at higher speeds.

1.7.4 Acoustic Doppler Current Meters

ADCMs are nonmechanical current meters that measure flow speed and direction by transmitting high-frequency sound waves and then determining the Doppler frequency shift of the return signal scattered from assemblages of "drifters" in the water column. In a sense, the instrument "whistles" at a known frequency and listens for changes in the frequency of the echo. This technology makes it possible to profile the flow velocity as a function of distance from the instrument; profiling instruments, known as ADCPs or acoustic Doppler profilers (ADPs), are discussed later in this chapter. The Doppler acoustic technique relies on the fact that: (1) sound is reflected and/or scattered when it encounters marked changes in density; and (2) the frequency of the reflected sound is increased (decreased) in direct proportion to the rate at which the reflectors are approaching (or receding from) the instrument along the acoustic path. Principle (2) is responsible for the shift to lower frequency in the whistle from a passing train and is used by astronomers to measure the rate at which stars and galaxies are moving relative to the earth. The commonly observed "red shift" of starlight reveals that most distant galaxies are receding from the earth at an accelerating rate due to the expansion of the universe. Reflectors ensonified by ADCMs include "clouds" of planktonic organisms such as euphausiids, copepods and gellies, fish (with and without swim bladders), suspended particles, and discontinuities in water density. Buoyant wastewater plumes from coastal sewage outfalls and buoyant hydrothermal plumes rising from seafloor spreading regions are two common examples of density discontinuities that can be detected acoustically.

Unlike the single-point current meters discussed in the previous sections, which measure current time series at a fixed depth, profiling-type ADCMs (ADCPs) provide time series of the flow averaged over a suite of ensonified depth bins. ADCPs are like having a stack of current meters, albeit with the assumption that the flow is homogeneous over the individual volumes (cells) of water being ensonified. The transducers are tilted by 20–30° relative to the vertical axis of the instrument so that the volume of water being ensonified increases with distance from the transducer head. Commercial ADCMs—which includes both single-point and profiling instruments—are built by Aanderaa Instruments, Teledyne-RD Instruments, NortekUSA/Nortek International, and SonTek/YSI. We have worked with ACMs from all four manufacturers and found them all to be quality products with very similar flow accuracy and resolution. There are, however, differences in instrument reliability and customer service that the investigator should address before purchasing a particular instrument. Because it has been around longer, the Teledyne-RDI (formerly RDI) ADCP has been the focus of numerous comparisons and analyses (e.g., Pettigrew and Irish, 1983; Pettigrew et al., 1986; Flagg and Smith, 1989; Schott and Leaman, 1991). Teledyne-RDI makes a self-contained internally recording unit (Figure 1.42), a direct reading unit, and a vessel-mounted unit. The

FIGURE 1.42 A direct reading 150 kHz acoustic Doppler current meter with external Teledyne-RD-232 link manufactured by RD Instruments. Side view shows three of the four ceramic transducers. Each transducer is oriented at 30° to the axis of the instrument. The pressure case holds the system electronics and echo-sounder power boards.

instruments have generally been available at frequencies of 75, 150, 300, 600, and 1200 kHz; the more newly developed Broadband™ ADCP also includes a 2400 kHz unit. The choice of frequency is dependent on the particular application. Nortek also makes high quality ADCMs, specifically the 0.4–2.0 MHz Aquadopp for high-resolution single-point and medium range profiling measurements and the 190 and 470 kHz Continental for longer range profiler measurements. The lower frequency Continental has a range of 200–250 m compared to the range of around 500 m for the Teledyne-RDI 75 kHz Sentinal ADCP. Because ADCPs are well suited to oceanographic applications, we will consider this instrument in some detail.

The standard ADCM measures current by first estimating the relative frequency change, Δf, of backscattered echoes from a single transmit pulse (Gordon, 1996). The more recent Broadband Teledyne-RDI ADCP technology measures the current by determining the phase shifts ("time dilation") $\Delta \phi$ of backscattered echoes from a series of multiple transmitted pulses. The Aanderaa (Doppler current meter) DCM and Nortek Continental ADP profilers operate at higher frequencies than the Teledyne-RDI ADCPs, have fewer bins than the ADCP, and use three rather than four separate transducer beams to determine the 3-D flow. The use of four transducers provides redundancy in the flow calculations. Teledyne-RDI uses the redundancy to provide an error estimation for the calculated flow. Instantaneous flow estimates that have widely different values, can be ignored when calculating an ensemble average over some specified averaging time. A report on an intercomparison between a 614 kHz Broadband ADCP and two 607 kHz DCMs moored in 11.5 m of water in Øresund, Denmark has been prepared by the Danish Hydraulic Institute (Rørbaek, 1994).

Aside from some custom-built units, such as the 5-beam ADCP that has one beam pointing directly along the axis of the instrument (Gargett et al., 2004), the standard narrow-band ADCPs employ four separate transducers oriented in a Janus configuration with beams pointing at an angle of 30° to the plane of the transducers (Janus was the Roman god who looked both forward and backward at the same time). Broadband ADCP units, as well ADPs built by Nortek, employ three transducers at an angle of 20° to the plane of the transducers, enabling them to record values closer to "hard" reflectors such as the ocean surface or the seafloor. The 2-MHz Aanderaa RCM11 has two orthogonal transducers measuring the Doppler shift in the horizontal plane only, so that the vertical component of velocity is not measured. It is assumed that the small drifters reflecting the transmitted sound pulse are being carried passively by the current and their drift velocity has a near-uniform distribution over the volume of water being ensonified by the ADCP. For a narrowband ADCP with a transmit pulse having a fixed length of a few milliseconds, the Doppler frequency shift, Δf, of the backscattered signal is proportional to the component of relative velocity, $v \cos \theta$, along the axis of the acoustic beam between the backscatterers and the transducer head (Figure 1.43(a)).

For a given source frequency, f, and bin k (depth range $= D_k$) we find

$$v_k = \frac{1/2(\Delta f_k/f)c}{\cos(\theta_k)} \qquad (1.33)$$

where v_k is the relative current velocity for bin k at depth D_k, θ_k is the angle between the relative velocity vector and the line between the scatters and the ADCP beam, and c is the speed of sound at the transducer. The ADCP first determines the current velocity relative to the instrument by combining the observed values of frequency change along the axes of each of the acoustic beams (the instrument can only "see" along the *axis* of a given transducer, not across it; Figure 1.43(a)). Absolute velocity components in east−west and north−south coordinates, called "earth" coordinates, are then obtained using measurements from an internal magnetic compass.

The relative frequency shift, $\Delta f_k/f$ for bin D_k, is derived using the observed frequency of the returning echo (Figure 1.43(b)). To calculate the Doppler frequency shift caused by the moving scatterers, the ADCP first estimates the autocovariance function, $C(\tau)$, of the return echo using an internal hardware processing module. The slope of $C(\tau)$ as a function of time lag, τ, is then related to the frequency change due to the movement of the scatterer region during the time that

FIGURE 1.43 Principles of ADCP measurement. (a) Relative velocity, $v \cos \theta$, along the axis of the acoustic beam between the backscatterers and the transducer head; (b) auto-spectrum of returned acoustic signal showing the Doppler frequency shift for a given bin. *(RD Instruments (1989).)*

it was ensonified by the transmit pulse. Because of inherent noise in the instrumentation and the environment, as well as distortion of the backscattered signal due to differences in acoustic responses of the possible targets, the returned signal will have a finite spectral shape centered about the mean Doppler shifted frequency (Figure 1.43(b)). The spectral width (SW) of this signal has the form $SW = 500/D$, where D is the bin thickness in meters, and is a direct measure of the uncertainty of the velocity estimate

due to the finite pulse length, turbulence, and nonuniformity in scattering velocity. In the case of the standard Teledyne-RDI ADCP, depth cell lengths, D, can range from 1 to 32 m but are usually set at 4–8 m. Depth cell size for profiling ACMs ranges from 0.12 m for the higher frequencies to 32 m for the lower frequencies. Each acoustic beam of the Teledyne-RDI ADCP has a width of 2–4° (at the −3 dB or half-power point of the transducer beam pattern) so that the "footprint" over which the

TABLE 1.4　Teledyne-RD Instruments Acoustic Wavelengths (λ) and Depth Ranges (m) for Different Transducer Frequencies for the Low Power and High Power Settings

Freq. (kHz)	λ (mm)	Depth Range (m)		Standard Deviation (cm/s)			e_1 (dB/m)	e_2 (dB/m)
		Low	High	$N = 15$	$N = 30$	$N = 60$		
76.8	20	400	700	6.72	4.75	3.36	0.025	0.0221
153.6	10	240	400	3.36	2.38	1.68	0.039	0.0395
307.2	5	120	240	1.68	1.19	0.84	0.062	0.0726
614.4	2.5	60	60	0.84	0.59	0.42	0.139	0.1884
1228.8	1.25	25	25	0.42	0.30	0.21	0.440	0.6466

(low power is for self-contained units while high power is for either self-contained or externally powered units). Standard deviation for velocity of given frequency are for ensemble averages of N pings per ensemble, a depth cell size (bin length and length of transmit pulse) of 8 m and 30° beam angle orientation. For 20° angle multiply values by 1.5; for other depth cell sizes D (m) multiply values by 8/D. The values e_1 and e_2 are different published estimates of the absorption of sound at 4 °C, 35 psu and atmospheric pressure. At high frequencies, the range of transducers is limited by nonlinear dynamics (cavitation) and heat dissipation so that the ranges at high and low power output are the same. (From RD Instruments (1989).)

acoustic averaging is performed is fairly small. Nortek profilers have corresponding beam widths of 3.7° for 400-kHz system to 1.7° for the 2.0-MHz system. At a distance of 300 m, the footprint of an ADCP has a radius of 5—10 m. However, the horizontal separation between beams is roughly equal to the distance to the depth cell so that the assumption of horizontal uniformity of the current velocity is not always valid, especially for those cells farthest from the transducers.

Sidelobes of the transducer acoustic pattern can limit the reliability of the data. For the standard 4-beam 30 degree-angle ADCP, measurements taken over the last 15% [$\approx (1 - \cos 30)$] of the full-scale depth range are not valid if the ocean surface (or seafloor) are within the range of an upward (or downward) looking instrument. In general, the maximum range, R_{max}, of acceptable data for a vertically oriented ADCP within proximity to a "hard" reflecting surface such as the sea surface or sea floor is given by $R_{max} \approx H \cdot \cos \varphi$, where H is the distance from the ADCP to the reflecting surface and φ is the angle the transducers make with the instrument axis (for a 20-degree instrument, only 6% of the range [$\approx (1 - \cos 20)$] is lost near the sea surface or seafloor). For vessel-mounted systems

working in areas of rough or rapidly sloping bottom topography, a more practical estimate is $R_{max} \approx H(\cos \varphi - \alpha)$, where $\alpha \approx 0.05$ is a correction factor that accounts for differences in water depth during short (<10 min) ensemble averaging periods.

The higher the frequency, the shorter the distance an acoustic sounder can penetrate the water, but the greater the instruments ability to resolve velocity structure (Table 1.4). The 75- and 150-kHz Teledyne-RDI units are mainly used for surveys over depth ranges of 0—500 m while higher frequencies such as the 600 and 1200 kHz units are favored for examining flow velocity in shallow water of 25 to 50-m depth. Nortek profilers have an approximate range of 60—90 m for the low-frequency 400-kHz system to 4—10 m for the high-frequency 2.0-MHz system. As noted above, ADCPs employ four separate transducers each pointing at an angle of 20° or 30° to the plane of the transducers. Since only the current speeds along each of the beam axes can be estimated, trigonometric functions must be applied to the velocities to transform them into horizontal and vertical velocity components. The instrument provides one estimate of the horizontal velocity and two independent estimates of the vertical velocity. The ADCP senses the Doppler frequency shift in each 1-s acoustic "ping" by looking at the time-delayed gated signal returning from distinct "bins" (also depth cells or distance ranges) from the transducer along each of the four-beam axes. The resultant speed estimates are then converted within the instrument to common bin positions centered at $D_o + M \times D - D/2$ meters ($M = 1, 2, \ldots, 8$ to a maximum of 128 bins or cells) along the central axes normal to the plane of the transducers. Here, D is the cell width and D_o is a constant blanking length (see below). The three-transducer Nortek profilers also have a maximum of $M = 128$ cells. Since the different time delays t_k of each pulse correspond to different distances D_k from the transducers, the instrument provides estimates of the horizontal (u, v) and vertical (w) components of velocity averaged over adjoining depth ranges (or depth bins). As illustrated by Figure 1.44, the averaging consists of a linear weighting over twice the bin length, $D = z_{k+1} - z_k = c(t_{k+1} - t_k)$, where c is the sound speed. For the 4-m bin length selected in Figure 1.44, the triangular weighted average is over 8 m. The depth range of a particular bin covers the distance:

from : blank depth + (bin number)

$\times$ (bin length) − (bin length)/2

to : blank depth + (bin number) × (bin length)

+ (bin length)/2

A 4-m blanking is applied to the beginning of the beam to eliminate nonlinear effects near the transducer. The minimum length of the blank is frequency dependent but a larger value can be selected by the user. (For the Nortek Aquadopp profilers, the minimum blanking ranges from 1 m for the 0.4-MHz unit to 0.05 m for the 2.0-MHz unit.) For the particular setup shown in Figure 1.44, there are 15 1-s pings for each 20-s ensemble; bottom tracking is turned on every four pings. This option, together with machine processing "overhead" and time for transmission up the tow cable, uses up a segment of the total time available for each ensemble-averaging period.

The maximum range of the standard (single-transmit pulse) ADCP depends on the depth at which the strength of the return signal drops to the noise level. Depending on the rate of energy loss and heat dissipation, the instrument is generally capable of measuring current velocity to a range $R(m) = 250(300/f)$, where f is the frequency in kilohertz (kHz). The velocities (and backscatter intensity, which we discuss later in the section) from a series of pings are averaged to form an "ensemble" record. This saves on storage space in memory, reduces the amount of processing, and improves the error estimate for the velocity record. Each acoustic ping lasts about 1−10 ms and 10 or more separate pings, together with an equal number of compass readings, are typically used to calculate an ensemble-averaged velocity estimate for each recorded increment of time in the time series. The random error of the horizontal velocity (in meters per second) for each ensemble is given as

$$\sigma(m/s) = 1.6 \times 10^2 \Big/ \left(f D N^{1/2} \right)$$

where N is the number of individual 1-s pings per ensemble and D is the bin length in meters. For example, a 30-s ensemble-averaging period chosen during the instrument setup procedure, generally allows for about 20 pings plus 10 s of processing time. This overhead time is inherent to the system and must be taken into account when determining the error estimates. As indicated in Table 1.5, the standard deviation of the vertical and horizontal velocity estimates for this case is about 3 cm/s for $D = 8$ m and a 150-kHz transducer. The greater the number of pings used in a given ensemble, the greater the accuracy of the velocity estimate, with $\sigma \approx N_{-1/2}$. Tilt sensors are used to calculate changes in the orientation of the transducer axis and to ensure that data are binned into correct depth ranges. These sensors are limited to

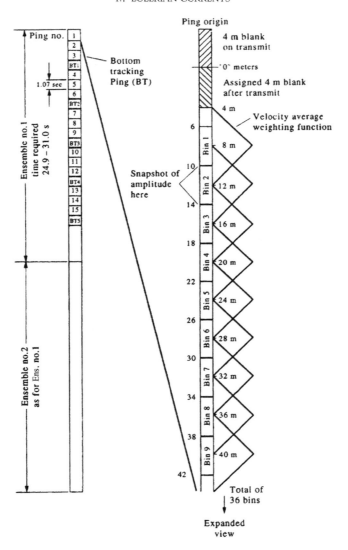

FIGURE 1.44 Allocation of depth bins and machine overhead for a narrow band (standard) 150 kHz ADCP having a bin length of 4 m, a blanking range of 4 m and a depth range of 36 bins. The instrument obtained 15 1-s pings for each 20-s ensemble and used the remaining time for internal processing and data transmission up an electrical cable. The information on the right is an expansion of the bin allocation for the first ping. A triangular weighting is used to determine the velocity for each bin. Similar results apply to the remaining pings for each of ensemble. A 4-m blanking is applied to the beginning of the beam to eliminate nonlinear effects the transducer. *(Courtesy, George Chase.)*

±20° for the ADCPs and ±30° for the Nortek ADPs so that for greater tilts, the velocity components cannot be determined accurately. Only three of the four beams of ADCPs are needed for each three-dimensional velocity calculation. The built-in redundancy provides for an "error velocity" estimate for each ensemble velocity, which involves subtracting the two independent

TABLE 1.5　Comparison of Hourly Time Series of Longshore Currents Over a 90-day Period from 308-kHz ADCP and Conventional Current Meters Off Northern California

Depth (m)	Moored Current Meter	Correlation Coefficient, r	Speed Difference (cm/s)	
			Mean	RMS
10	VACM	0.94	−3.7	8.1
20	VMCM	0.97	0.8	4.6
35	VMCM	0.98	0.2	2.7
55	VMCM	0.98	0.0	2.4
70	VMCM	0.98	0.3	2.2
90	VMCM	0.98	1.0	2.2
110	VMCM	0.98	0.5	1.9
120	VACM	0.97	−0.1	2.0

Results are found using the two-beam solution for the ADCP. VACM, Vector averaging current meter; VMCM, vector measuring current meter. (Adapted from Pettigrew and Irish (1986).)

estimates of the vertical velocity component for each ping. When the two vertical velocity estimates agree closely, the horizontal velocity components are most likely correct. In addition to the reliability check, the fourth beam serves as a backup should one of the transducers fail. Another measure provided by the ADCP is the "percent good" which is the percentage of pings that exceed the signal-to-noise threshold. Normally, the percent good rapidly falls below 50% at some depth and stays below that level. In practical terms, there usually is little difference in the data for assigned values of 25, 50, or 75 percent good.

Because Doppler current meters were originally designed for measuring currents from a moving platform, the ADCP records instrument heading, pitch, roll, and yaw. These data are then used to correct the measured velocities. In order to determine the true current velocity in "earth coordinates" from a moving vessel, the ADCP is capable of measuring the velocity of the instrument over the seafloor, providing that bottom is within range of the ADCP transducers and the bottom reflection exceeds the background noise level. A separate bin is used for this bottom tracking. The bottom tracking mode is usually turned on for a fraction of the total sampling time, uses a longer pulse length and provides a more accurate estimate of relative velocity than other bins. Modern shipboard GPS systems are accurate to better than ±10 m—the US provides GPS to the civilian community at performance levels specified in the GPS Standard Positioning Service Performance Standard of <7.8 m at the 95% confidence level—while accuracies of a few centimeters are available using GPS in combination with augmentation systems (it is the augmentation systems that improve the accuracy of military applications of GPS.) A ±10-m accuracy means that estimates of the ship speed taken at time increments of, say, 10–100 s can have errors as high as 10–100 cm/s, which are generally comparable to the kinds of current speeds we are trying to measure. Augmented Differential GPS (DGPS), which relies on error corrections transmitted from fixed land-based reference stations for which satellite positioning and timing errors have been calculated, is accurate to better than ±1 m (to ±0.1 m in some implementations). Shipboard systems working in this mode can be used to determine absolute currents to an accuracy of roughly ±1–10 cm/s by subtracting the accurately determined ship's velocity over the ground from relative currents measured by the ADCP (see note at the end of this section).

DGPS uses a network of fixed, ground-based reference stations to broadcast the difference between the positions indicated by the satellite systems and the known fixed positions. These stations broadcast the difference between the measured satellite pseudo-ranges and actual (internally computed) pseudo-ranges, and receiver stations may correct their pseudo-ranges by the same amount. The digital correction signal is typically broadcast locally over ground-based transmitters of shorter range. A similar system that transmits corrections from orbiting satellites instead of ground-based transmitters is called a Wide-Area DGPS (WADGPS) or Satellite-Based Augmentation System.

There are several factors that limit the accuracy of ADPs: (1) The accuracy of the frequency shift measurement used to obtain the relative velocity. This estimate is conducted by software within the instrument and strongly depends on the signal/noise ratio and the velocity distribution among the scatters; (2) the size of the footprint and the homogeneity of the flow field. For example, at a distance of 300 m from the transducers of the Teledyne-RDI Sentinal ADCP, the spatial separation between sampling volumes for opposite beams is 300 m so that they are seeing different parts of the water column, which may have different velocities; and (3) the actual passiveness of the drifters; i.e. how representative are they, in aggregate, of the in situ current? (Many species of zooplankton are active swimmers and vertical migrators.) In the shipboard system, the ADCP can track the bottom and obtain absolute velocity, provided the acoustic beam ranges to the bottom. Once out of range of the bottom, only the velocity relative to the ship or some level of no motion can be measured. As noted above, standard GPS positioning without the highly accurate (<±1 m)

differential mode cannot be used to obtain ship velocity since the accuracy of the standard mode (around 10 m) yields ship speed accuracies that are, at best, comparable to the absolute current speeds being measured. Erroneous velocity and backscatter data are commonly obtained from shipboard ADCP measurements due to vessel motions in moderate to heavy seas. In addition to exposure of the transducer head, the acoustic signal is strongly attenuated by air bubbles under the ship's hull or through the upper portion of the water column. Much better data are collected from a ship "running" with the seas than one lying in the trough or hove to in heavy seas. Our experience is that data collected in moderate to heavy seas are often unreliable and need to be carefully scrutinized. In deep water, zooplankton aggregations can lead to the formation of "false bottoms" in which the instrument mistakes the high reflectivity from the scattering layer as the seafloor.

The only way to improve velocity measurement accuracy with the standard single-pulse narrow-band ADCP is to lengthen the transmit pulse. A longer transmit pulse extends the length of the autocorrelation function and increases the number of lag values that can be used in the calculation of velocity. Since bin length is proportional to pulse length, this results in improved uncertainty in the velocity estimates. The trade-off is reduced depth resolution. By transmitting a series of short pulses, the newer Broadband™ ADCP circumvents these problems. Because of the multiple transmit pulses, the Broadband ADCP is capable of much better velocity resolution and higher vertical resolution. The time between pulses sets the correlation lags available for velocity computation while pulse length governs the size of the depth cells, as in the standard unit. Moreover, velocity is determined from differences in the arrival times of successive pulses. By increasing the effective bandwidth of the received signal by two orders of magnitude, the Broadband ADCP can reduce the variance of the velocity measurement by as

(a)

(b)

FIGURE 1.45 Alongshore currents measured by a Broadband RD Instruments Vessel-Mounted 300-kHz system along an eastward transect to the south of the Bass Point headland near Sydney, Australia (See Middleton et al., 1993). (a) Cross section of the flow for all depth bins, (y-axis) versus distance (x-axis) with red corresponding to northward flow and blue to southward flow; the black region denotes the seafloor; (b) Ensemble-average velocity at 5-m depth for the cross section in (a). Flow was northward in the wake of the headland and southward seaward of the point. The green vertical line to the right of the plot gives the flow speed scale in cm/s. The right panel gives information on the ship's course and speed from the bottom-track mode of the ADCP. *(Courtesy, Jason Middleton and Greg Nippard.)*

much as two orders of magnitude. Figure 1.45 is an example of currents (cross shore and alongshore) and acoustic backscatter obtained using a shipboard Broadband ADCP during a cross-shelf transect of the inner continental shelf east of Sydney, Australia. The new ADCP system offers "real-time" computer screen display for at-sea operations. The standard narrow-band ADCP uses a data acquisition system, that is no longer supported by Teledyne-RD Instruments, to display output of velocity components, beam-averaged backscatter intensity, percent good, and other ship related parameters such as heading and pitch and roll.

A further note on GPS measurements. There are currently a variety of chart datums that are used in the setup menu of a GPS and one must be sure to select that datum, which matches the chart being used for navigation. The general default datum is WGS-84 (World Geodetic Survey 1984) which applies to any region of the world. A commonly used datum in the eastern North Pacific and western North Atlantic is NAD-27 (North American Datum 1927) which was subsequently replaced by NAD-83 (North American Datum 1983). Other datums are WGS-72, Australian, Tokyo, European, and Alaska/Canada. *Selective Availability* is the name given by the United States Department of Defense for degradation of the GPS satellite constellation accuracy for civilian use. When disabled (as it was during the 1st Gulf War involving Iraq), GPS accuracy increases by about a factor of 10.

1.7.4.1 Deep-sea Observations Using an ADP in Low Scatter Environments

One of the prime concerns with ADCMs is the possible low signal-to-noise (S/N) level due to a lack of scatterers in the water column. A recent study by Thomson et al. (2012) provides some confidence that even a marginal S/N level can often yield high quality time-series data. In November 2005, a 2 MHz single-point Nortek Aquadopp ACM was deployed in an upward-looking configuration in 4386 m of water near the middle of the Middle America Trench off Costa Rica, roughly 0.72 km northwest of Ocean Drilling Program Borehole Site 1253. The ACM recorded three-dimensional current velocity, three-beam acoustic backscatter intensity, pressure, and temperature at an elevation of 21 m above bottom (mab) every 15 min based on a 1-Hz sampling rate and 2-min burst-averaging period. Currents were measured in earth coordinates based on an ensonified water volume within a radius of a few meters of the three acoustic transducers. The instrument was recovered by the DSV *Alvin* from the R.V. *Atlantis* in February 2009 and, with the exception of two missing data points, functioned flawlessly until its battery failed on 21 April 2007. According to the data, sensor resolutions were ±0.005 m/s for velocity, $\pm0.01\,°C$ for temperature, ±2 counts for backscatter, and ±0.013 m for depth. The depth

resolution was roughly 0.0003% of the full-scale pressure based on a background density of 1027.744 kg/m^3 from CTD data; the backscatter resolution was ~0.90 dB based on Nortek's specification of 0.40–0.47 dB/count (Lohrmann, 2001).

Linear interpolation was used to fill in the two missing 15-min values that occurred part way through the 523-day velocity time series. The current direction was then corrected for local magnetic declination and the current vectors rotated from a north and east reference frame to a principal component reference frame in which horizontal velocity components u, v are in the northeast (cross trench; 45° T) and northwest (along trench; 315° T) directions, respectively. As indicated in Figure 3 in the study, horizontal currents were strongest in the along-trench direction and reached 15-min average speeds of up to 0.3 m/s toward the northwest. Vertical velocities were also strong and commonly exceeded the instrument resolution of 0.005 m/s. The acoustic backscatter intensity had a range of 25 counts (~11 dB) superimposed on a background value of about 40 counts, with no suggestion of an acoustic noise threshold. Backscatter from the three beams was nearly identical.

1.7.4.2 Acoustic Backscatter

Although it was originally designed to measure currents, the ADCP has become a highly useful tool for investigating the distribution and abundance of zooplankton in the ocean. In particular, the intensity of backscattered sound waves for each depth bin—actually a "snapshot" of the intensity at a distance of two-thirds the way along the bin (Figure 1.44)—can be used to estimate the integrated mass of the backscatters over the "footprint" volume (width and thickness) of the original acoustic beams (Flagg and Smith, 1989). As with velocity, the instrument compensates for apparent changes in bin depth due to instrument tilt and roll. Calculation of the backscatter anomaly caused by plankton or other elements in the water column requires an understanding of the various factors causing dispersion and attenuation of the sound waves in water. Proper calibration of the acoustic signal as a function of acoustic range is essential for correct interpretation of the ADCP backscatter data. The measured backscatter intensity (also energy or amplitude squared) I_r is given by.

$$I_r/I_a = b \exp{(-2e_i z)}/z^2 + A_n \qquad (1.34)$$

where $b = I_0/I_a$ is the transducer gain, I_0 is the intensity of the ADCP transducer output, I_a is a reference intensity, e_i, is the absorption coefficient for water (cf. Table 1.4; $i = 1$, 2), $1/z^2$ is the effect of geometric beam spreading over the range z, and A_n is the relative noise level. The factor b arises because the ADCP does not record output intensity from the transducers, only relative intensity. The acoustic volume scattering strength, S_v, of the ADCP is then given by the logarithm of Eqn (1.34) as

$$S_v = 10 \log{(I_r/I_0)} - 10 \log{(b)} \qquad (1.35)$$

where the first term is the absolute acoustic scattering strength of the ADCP and the second term is an unknown additive constant. Since the later term is unknown, a relative measure of the volume scattering strength S_{vc} to some standard calibration region can be determined as $S_v' = S_v - S_{vc}$ (Thomson et al., 1991, 1992; Burd and Thomson, 1994, 2012). Thomson et al. (1992) use a vertically towed vehicle and are therefore able to calibrate their data relative to the near-uniform backscatter reference layer at intermediate depths (1000–1500 m) in the northeast Pacific. (The full sonar equation for the volume cross-scattering cross section, σ_b, from which we derive $S_v = \log{(\sigma_b)}$, can be found in Urick (1967)).

ADCMs do not measure directly the input or output of the acoustic backscatter intensity but rather the voltage from the so-called Automatic Gain Control (AGC), which is an internal adjustment, positive feedback circuit in the output device that attempts to keep the transducer output power constant. The average compensation voltage in the AGC is recorded and can be used to estimate the relative backscatter intensity. By incorporating a user exit program, the ensemble average AGC for each of the beams for each bin can also be recorded. As we will discuss later, this is proportional to the biomass (density × cross section) of the scatterers. The instrument also measures temperature—which it needs to calculate response correctly—and, in the case of ADCPs, the percent good, which is a measure of the number of reasonable pings per ensemble.

The speed of sound in water varies with temperature, salinity, and depth but is generally around 1500 m/s. Therefore, sound oscillations of 150 kHz (a common frequency used on shipboard systems and moored systems) have a wavelength of about 1 cm. Using the standard rule of thumb that the acoustic wave detects objects of about one-quarter wavelength, objects greater than 2.5 mm will reflect sound while objects less than this scatter the sound. The proportion of the sound beam transmitted, reflected, or scattered by the object is influenced by small contrasts in compressibility and density between the water and the features of the object. Organisms with a bony skeleton, scaly integument, and air bladder reflect/scatter more sound than an organism made up mostly of protoplasm such as salps and jellyfish (Flagg and Smith, 1989). Similarly, organisms that are aggregated into patches or layers return more scattered sound energy per unit volume (i.e., have a greater volume scattering strength) than uniform distributions of the same organisms.

A major problem with using the ADCP for plankton studies is common to all bioacoustical measurements; namely, determining the species composition and size distribution of the animals contributing to the acoustic backscatter. Invariably, in situ sampling using net tows is needed to calibrate the acoustic signal. If the ADCP is incorporated in the net system, the package has the advantage that the volume flow through each net can be determined accurately using the ADCP-measured velocity (Burd and Thomson, 1993). An attempt to calibrate the ADCP against net samples was conducted by Flagg and Smith (1989) who also pointed out problems with the response of the shipboard system to temperature fluctuations in the ADCP electronics. A more recent "calibration" (Burd and Thomson, 2012) compares volume scattering strength for a 153-kHz ADCP against 197 coincident, mixed-species zooplankton samples collected over six summers using a Tucker trawl net system towed to depths of up to 3000 m in the northeast Pacific. Results show that the acoustic backscatter data from the single-frequency ADCP mounted near the opening of the towed net system accounts for 84% of the variance in total net biomass, despite the extensive mix of faunal types, depth range, and broad spatial and temporal extent of the study.

1.7.5 Comparisons of Current Meters

As noted earlier, a major problem with the Savonius rotor is contamination of speed measurements by mooring motions (Gould and Sambuco, 1975). The contamination of the rotor speed is caused primarily by vertical motion or "rotor pumping" as the mooring moves up and down under wave action. In effect, the speed overestimates of the rotor result from its ability to accelerate about three times faster than it decelerates. Pettigrew et al. (1986) summarize studies on the ability of VMCMs and VACMs in laboratory tests to accurately measure horizontal flow in the presence of surface waves. For wave orbital velocities, W, of the same magnitude as the steady towing speed, U, of the current meter through the water (i.e. $W/U \approx 1$), the accuracy of the VACM depends on the ratio W/U. The percentage error increases as the ratio W/U increases and substantial overestimation of the true speed occurs for $W/U > 0.5$. The results for the VMCM differ significantly from those of the VACM. In particular, the VMCM underestimates the true velocity

by as much as 30% for $W/U \approx 1$, while for $W/U > 2$, speed errors do not appear to be strongly dependent on either W/U or on the relative orientation of the mean and wave current motions. For $W/U < 1/3$, the VMCM was within 2% of the actual speed. While vector averaging can reduce the effect of vertical motion on the recorded currents by smoothing out the short-term oscillatory flow, the basic sensor response is not well tuned to conditions in the wave zone or those for surface moorings. Intercomparisons of conventional current meters (Quadfasel and Schott, 1979; Halpern et al., 1981; Beardsley et al., 1981) have shown that VACM speeds are only slightly higher on surface moorings than on subsurface moorings and that contamination by mooring motion was only important for higher frequencies (>1 cph). At frequencies above 3–4 cph, ocean current spectra computed from VACM current meters did not flatten (i.e., not decrease with frequency) as much as spectra from other rotor equipped current meters. Near the surface this is due to horizontal motion of the mooring (Zenk et al., 1980), which is rectified by the Savonius rotor while at greater depths the surface float motion translates into vertical motion, which aliases the rotor speed due to rotor pumping. Further details can be found in Weller and Davis (1980), Mero et al. (1983), and Beardsely (1987).

Another problem with the Savonius rotor is that it does not have a cosine response to variations in the angle of attack of the flow due to interference of the support posts. In a study of rotor contamination, Pearson et al. (1981) conclude that Savonius rotor measurements, made from a mooring with a float 18 m below the sea surface, were not seriously contaminated by surface wave-induced mooring motion. In sharp contrast, Woodward et al. (1984) compared a standard Savonius rotor with a paddle-wheel rotor designed for wave-field applications, and an ECM. The electromagnetic speed sensors appeared to perform well in the near-surface wave field while the standard

Savonius rotor was severely contaminated by wave-induced currents (Figure 1.38).

Field comparisons (Halpern et al., 1981) demonstrated that above the thermocline (5 to 27-m depth) the VMCM, the VACM, and ACM all produced similar results for frequencies below 0.3 cph, regardless of mooring type. Above 4 cph, it was recommended that the VACM be used with a spar buoy surface float while both the VMCM and the ACM could be used with surface-following floats such as a donut buoy. In general, better quality measurements were made at depths from subsurface moorings than from surface moorings, indicating that even the VMCM data were contaminated somewhat by mooring motion.

The processing of current meter data is specific to the type of meter being used. It is interesting to read in current meter comparisons such as Beardsley et al. (1981) or Kuhn et al. (1980), the variety of processing procedures required to produce compatible data for the intercomparison of observations from different current meters. An important part of the data processing is the application of the instrument-specific calibration values to render measurements in terms of engineering units. In this regard, it is also important to have both a pre- and postexperiment calibration of the instrument to detect any serious changes in the equipment that might have occurred during the measurement period.

One of the earliest comparisons between a bottom-mounted ADCP and conventional mechanical current meters was conducted in 133 m of water near the shelf-break off northern California in 1982 (Pettigrew and Irish, 1983; Pettigrew et al., 1986). The 90-day time series of horizontal currents from a prototype upward-looking 308-kHz ADCP with 4-m bin length was compared with currents from a nearby (~300 m) string of VACMs and VMCMs. Despite the fact that only two of the beams could be used and the instrument had a 10-degree list, results show striking agreement between the

two sets of data (Table 1.5). Mean differences between corresponding acoustic and mechanical current meters were typically less than 0.5 cm/s while RMS differences were about 2 cm/s. Since acoustic currents were based on two beams tilted at 20° to the vertical, the relatively poor correlation at 10-m depth probably resulted from rotor pumping and overspeeding of the VACM rather than side-lobe contamination of the ADCP which would occur in the upper 6% of the depth range. Similar results were obtained by Schott (1986).

In 2002, the Institute of Ocean Sciences (Sidney, British Columbia) conducted a 4-month comparison of three, single-point ACMs (the Nortek Aquadopp, the Sontek Argonaut MD, and the Aanderaa RCM11) and one single-point ECM (the InterOcean S4) against Aanderaa RCM5 and RCM8 paddle-wheel current meters. Because they had been used in many studies prior to 2002, the half-shielded paddle-wheel RCMs provided the standard against which the other instruments were compared. Instruments pairs were roughly 2 m apart vertically and moored at roughly 2200-m depth at three separate locations within the axial valley of Endeavour Ridge in the northeast Pacific where currents are typically less than 10 cm/s. The results were never published and problems were experienced with all of the nonmechanical current meters. The principal finding was that the nonmechanical current meters were capable of measuring the very weak flows in the valley whereas rotor friction caused the RCM5/8s to erroneously record zero speeds for extended periods of time, for as much as a half semidiurnal tidal period. Subsequent experience with the nonmechanical instruments shows that quality, reliability, and durability have greatly improved since the time of this intercomparison. The results of this intercomparison helped in the design for the four long-term moorings presently installed at depths of around 2200 m within the axial valley of Endeavour Ridge as part of the Ocean Networks Canada cabled observatory. Each mooring consists of paired, real-time recording single-point Nortek Aquadopp 3000 2-MHz ADCMs and SBE 37 MicroCAT CTDs at 5, 50, 125, and 200 mab and an upward-looking Teledyne-RDI 75-kHz Long Ranger ADCP housed in a 45″ Flotation

Technologies syntactic foam float at 250 mab (Figure 1.46). Results show that the ADCP is providing current velocity and acoustic back-scatter intensity records for all 128 4-m bins over a depth range of 512 m above the top of

FIGURE 1.46 Schematic of one of four, bottom-anchored moorings located at around 2200-m depth within the axial valley of Endeavour Ridge, northeast Pacific. The mooring is connected by cable to the Ocean Networks Canada cabled observatory network, which supplies power and transmits 1-s data from the instruments to a Node, and hence by fiber optic cable to shore. The mooring release consists of a pull-pin that is operated by a remotely operated vehicle. Elevations are in meters above bottom (mab). *(Mooring design and image courtesy of David Spear, Tamás Juhász, and Lucius Perreault, Institute of Ocean Sciences.)*

the mooring, or ~1000 mab. There are also Flotation Technologies CF-12 12″ clamp-on football floats distributed along the length of the mooring to support the mooring and the cable when the system is on the surface during recovery.

1.7.6 Electromagnetic Methods

The dynamo interaction of moving, conducting seawater with the earth's stationary magnetic field induces electric currents in the ocean. These "motional" electric fields, whose existence in the ocean was first postulated by Faraday in 1832, produce a spatially smoothed measure of the water velocity at subinertial periods (periods longer than $1/f = 11.964$ h/sin(latitude)). For a given point on the seafloor, the electric fields are proportional to the vertically averaged, seawater-conductivity weighted water velocity averaged over a horizontal radius of a few water depths (Chave and Luther, 1990). Technologies that measure the horizontal electric field (HEF) yield direct observations of the barotropic transport in the overlying water column. Electric field measurements of transport are obtained from abandoned submarine communication cables or from self-contained bottom recorders. For a submarine cable, the motional HEF is integrated along the cable length.

According to theory (Sanford, 1971; Chave and Luther, 1990; Chave et al., 1992), the horizontal velocity vector field v^* is related to the HEF E_h by

$$E_h = F_z \mathbf{k} \times v^* \qquad (1.36a)$$

(sensor in a reference frame fixed to the seafloor) or

$$E_h = -F_z \mathbf{k} \times (V - v^*) \qquad (1.36b)$$

(sensor moving relative to seafloor), where F_z is the local vertical component of the geomagnetic field, $\mathbf{k}$ is a unit vector in the upward vertical direction, V is the vector sum of the horizontal velocities of the ocean relative to the earth and the sensor relative to the ocean, and

$$v^* = C \int_{-H}^{0} \sigma(z') v_h(z') dz' \Big/ \int_{-H}^{0} \sigma(z') dz' \qquad (1.37)$$

is the scaled (by the constant C) horizontal water velocity. The water velocity is averaged vertically over the water column of thickness H and weighted by the seawater conductivity, $\sigma(z)$. Equation (1.37) reduces to the scaled barotropic velocity, Cv when either the conductivity profile or the horizontal velocity is depth-independent. In the northern hemisphere, where F points into the earth, the north electric field is proportional to the west component of velocity while the east electric field is proportional to its north component. Neglecting the noise, we can solve Eqn (1.36a) to obtain

$$v^* = -\mathbf{k} \times E_h / F_z \qquad (1.38)$$

Since F is known to one part in 10^4 for the entire globe, measurement of E_h yields the horizontal flow field.

Measurement of the HEF is entirely passive, being based on naturally occurring fields, and hence has low power requirements and is nonintrusive. Motional electromagnetic may be used in a Eulerian configuration (bottom recorders or submarine cables) or a Lagrangian configuration (surface drifter, subsurface float, or towed fish). Equation (1.36b) shows that a relative velocity estimate is possible by measuring the HEF from a moving platform. On many instances, lack of a specific knowledge of v^* is not critical limitation since it is independent of depth by Eqn (1.37). The moving frame of reference Eqn (1.36b) is exploited by vertical profilers such as the electromagnetic velocity profiler and the expendable current profiler produced by Sippican. Horizontal profiles of the HEF can be obtained from a towed instrument and used with precise navigation to yield estimates of v^*

and the surface water velocity. The original form of such a towed instrument is the geomagnetic electrokinetograph (GEK) of von Arx (1950).

1.7.7 Other Methods of Current Measurement

There are numerous other ways to make Eulerian current measurements though not all have been successfully commercialized. For example, prior to the ADCP, scientists in Japan used towed electrodes at the ocean surface (the GEK) to routinely monitor the currents off the east coast of Japan. Goldstein et al. (1989) report on the use of SAR to measure surface currents from the phase-delay maps of aircraft-borne radar.

1.7.7.1 High Frequency Coastal Radar

High Frequency (HF) radar has become a wide spread technique for mapping surface currents along coastlines. This method uses short-baseline radars with frequencies ranging from 5 to 45 MHz to sense the backscatter from wind-generated capillary waves. The surface currents, which carry the capillary waves, give rise to a Doppler shift in the radar carrier frequency. This shift yields a radial velocity of the current along a radial line extending seaward from the radar shore station. For two radar stations installed along a shoreline, the radial velocities intersect, making it possible to compute a true surface current vector. The higher the frequency of the radar the higher the spatial resolution of the surface currents. This coverage is typically limited to the near shore region. Often these HF radars are operated at two different frequencies with that at 25 MHz giving a 2 km spatial resolution in a region extending about 50−60 km offshore. The 12 MHz band yields a 6 km spatial resolution over an area extending 150 km from shore.

One of the most widely used HF radar systems is CODAR, which is marketed under the name "SeaSonde." CODAR claims a surface current accuracy of ± 7 cm/s. These radars can also provide information on significant wave height, as well as wave direction and period for the radials emanating out from the shore stations to a range of about 3 km from the coast. Masson (1996) used a SeaSonde CODAR array to examine the effect of tidal currents on wind waves off the southern end of the Queen Charlotte Islands (now officially called Haida Gwaii) in northern British Columbia. There is an extensive array of CODAR stations along the west coast of the United States. Results from this system have been published by Kim et al. (2011). A comprehensive comparison between CODAR measured surface currents and surface current estimates from satellite altimetry has been carried out by Roesler et al. (2013).

1.7.7.2 Acoustic Correlation

Other recent techniques, such as the correlation sonar and acoustic "scintillation" flow measurements use pattern recognition and cross-correlation methods, respectively, to determine the current over a volume of ensonified water (Farmer et al., 1987; Lemon and Farmer, 1990). The acoustic scintillation method determines the flow in a turbulent medium by comparing the combined spatial and temporal variability of forward-scattered sound along two closely spaced parallel acoustic paths separated by a distance, Δx. Assuming that the turbulent field does not change significantly during the time it takes the fluid to travel between the two paths, the pattern of amplitude and phase fluctuations at the downstream receiver will, for some time lag, Δt, closely resemble that of the upstream receiver. Examination of the time delay in the peak of the covariance function for the two signals gives Δt, which then determines the mean velocity $v = \Delta x/\Delta t$ normal to the two acoustic paths. The technique has been used successfully to measure the horizontal flow in tidal channels and rivers, as well as the vertical velocity of a buoyant-plume rising from a deep-sea hydrothermal vent in the northeast Pacific (Lemon et al., 1996; Xu and Di Iorio, 2011; Di Iorio et al., 2012).

1.7.7.3 Displacement of Oceanic Features

Numerous papers have discussed the computation of surface currents from the displacements of patterns of SST in thermal AVHRR imagery. In the maximum cross-correlation (MCC) method, the cross correlation between successive satellite images is used to map the displacements due to the advection of the SST pattern (Emery et al., 1986). Wu (1991, 1993) has advanced a "relaxation labeling method" for computing sea surface velocity from sequential time-lapsed images. The method attempts to address two major deficiencies with the MCC method, namely: (1) the MCC approach is strictly statistical and does not exploit a priori knowledge of the physical problem; and (2) pattern deformation and rotation, as well as image noise, can introduce significant error into MCC vector estimates. The latter problem was addressed by Emery et al. (1992) who showed that rotation can be resolved using large search windows. (We note that the correlation method mentioned in the previous section is a form of feature displacement method. It is the "frozen" spatial structure of the dominant eddies rising within the turbulent plumes that enable the technology to determine the vertical velocity of buoyant plume.)

1.7.7.4 Other Doppler Current Measurements

Measurement of the vertical velocity, volume flux, and expansion rate of black smokers rising from hydrothermal vents can now be made using acoustic Doppler backscatter time series from bottom-mounted acoustic systems such as Cabled Observatory Vent Imaging Sonar (COVIS) (Jackson et al., 2003). In a recent study, Xu et al. (2013) connected the 400-kHz COVIS to the NEPTUNE-Canada (Ocean Networks Canada) Cabled Observatory to measure the flow velocity over a 10-m segment of a buoyant near-bottom plume at the Main Endeavour Field on the Juan de Fuca Ridge. Analysis is based on the covariance method of Jackson et al. (2003) in which the velocity component, v_r, in the direction of the acoustic line of sight is given by

$$v_r = \frac{c\Delta f}{2f} \tag{1.39a}$$

where

$$\Delta f = \frac{1}{2\pi\Delta t} angle \left[\sum_{n=1}^{N_p} \int_{t=0}^{T_w} E(t)E^*(t+\Delta t)dt \right] \tag{1.39b}$$

is the Doppler frequency shift from acoustic signals backscattered from particles and turbulence in the plume, c is the sound speed, and $f \sim 400$ kHz is the sonar frequency in the Doppler mode. The angle operator in Eqn (1.39b) calculates the phase angle in radians of a complex number. $E(t)$ is a demodulated complex signal corresponding to a given azimuthal beam and a given ping, whose amplitude and phase at the time are related to the amplitude of the acoustic backscatter and its phase shift relative to the transmitted pulses (* denotes the complex conjugate). The integral in Eqn (1.39b) estimates the autocorrelation function at the time lag, Δt. A rectangular window with length $T_w = 1$ ms is used to truncate the received signal. A summation over $N_p = 40$ pings at each elevation angle of the tiltable acoustic transducer reduces the uncertainty in the measurement caused by turbulence and background noise. The standard deviation, v_{std}, of v_r is calculated over the 40 pings and is used as a metric for uncertainty in the Doppler measurements. Over the roughly one-month proof-of-concept study, the method yielded temporal variations of the plume vertical volume flux of approximately $2-5$ m^3/s, centerline vertical velocities in the range $0.11-0.24$ m/s, and plume radius expansion rates of $0.082-0.21$ m per meter of vertical rise (Xu et al., 2013).

1.7.8 Mooring Logistics

In terms of accuracy and reliability, current meter data from surface and subsurface moorings cannot be divorced from the mooring itself. While many common mooring procedures are available, there is no single accepted technique nor is there agreement on the subsequent behavior of the mooring while in the water. Surface moorings with their flotation on the wavy surface of the ocean will behave differently than subsurface moorings over which the buoyancy is distributed vertically along the mooring line as in Figure 1.46. For the case of subsurface moorings, the addition of pressure sensors to most current meters gives confidence on the instrument depth (especially over steep or complicated bottom topography) and helps characterize mooring motion and determine its effect on the measured currents. Variations in the depth of the sensor can be calculated from the pressure fluctuations and used to estimate the depth and position of the moored instruments as a function of time. Also, models of mooring behavior have been developed which enable the user to predetermine line tensions and mooring motions based on the cross-sectional areas of the mooring components and estimates of the horizontal current profile. For example, the program SSMOOR distributed by Cable Dynamics and Mooring Systems in Woods Hole (Berteaux, 1990, 1991), uses a finite element technique to integrate the differential equilibrium equations for cables subjected to steady state currents. Factors taken into consideration include: the mooring wire (or rope) diameter, weight in water, and modulus of elasticity; and the shapes, cross sections, drag coefficients, weights, and centers of buoyancy of the recording instruments. Up to 10 current speeds can be specified for the current profile and as many as 20 instruments inserted in the anchoring line.

Mooring motions are largest when surface floats are used. For surface moorings in deep water, the length of the mooring line creates a relatively large "watch circle" that the surface float can occupy. This will add apparent horizontal motion to the attached current meters while, at depth, the surface wave and wind-driven fluctuations translate into mainly vertical oscillations of the mooring elements. Some intercomparison experiments have tried to use a variety of mooring types to test the effects of moorings alone. Zenk et al. (1980) compare VACM measurements from a taut-line surface mooring with a single-line spar buoy float and a more rigid two-line, H-shaped mooring. As expected, the H-shaped mooring was more stable and the other two exhibited much stronger oscillations. The current meters on the rigid H-mooring registered the greater current oscillations since the meters on the other, less restricted, moorings moved with the flow rather than measuring it.

In their current meter comparison, Halpern et al. (1981) discuss four different types of mooring buoyancy; three surface and one subsurface. The surface floats were: a toroid, a spar-buoy. and a torpedo-shaped float. They found that rotor pumping was much greater under the toroid than under the spar buoy and that the effect of rotor pumping on the resulting current spectra was significant at frequencies above 4 cph. While this was true for near-surface current meters, they also found that for deeper instruments the spar buoy float transmitted larger variations to the deeper meters making it a poor candidate for flotation in deep water current measurements. They found that both the VMCM and the ACM are less affected by the surface motions of a toroidal buoy. In a different comparison, Beardsley et al. (1981) tested an Aanderaa current meter suspended from a surface spar buoy, and found a significant reduction in the contamination of the measured signal by wave effects due to both currents and orbital motion with the spar buoy. Even with this flotation system, however, the Aanderaa current meter continued to register high current speeds compared with other sensors.

In an overall review of the recent history of current meter measurement, Boicourt (1982) makes the interesting observation that "results from current measurement studies are independent of the quality of the data". In making this claim, he remarks that often the required results are only qualitative, placing less rigorous demands on the accuracy of the measurements. He also points out that present knowledge of the high-frequency performance of most flow sensors is inadequate to allow definitive analysis of the current measuring system. In this regard, he states that ACMs and ECMs, with their fast velocity response sensors, hold great promise for overcoming the fundamental problems with mechanical current sensing systems, as observations have now shown. Finally, he calls for added research in defining the high-frequency behavior of common current meters.

Fieldwork by the Bedford Institute of Oceanography on Georges Bank in the western Atlantic has revealed another unwelcome problem with moored rotor-type current meters. Comparisons between currents measured by a subsurface array of Aanderaa current meters on the bank and a shipboard ADCP indicated that current speeds from the moored array were 20–30% lower than concurrent speeds from the profiler. To test the notion that the underspeeding was due to high-frequency mooring vibration caused by vortex shedding from the spherical floatation elements, an accelerometer was built into one of the subsurface moorings. Accelerations measured by this device confirmed that the current meters were being subjected to high-frequency side-to-side motions. Under certain flow conditions, the amplitudes of the horizontal excursions were as large as 0.5 m at periods of 3 s. Tests confirmed that the spherical buoyancy packages were the source of the motions. By enclosing the spherically shaped buoyancy elements in more streamlined torpedo-shaped packages, the mooring line

displacements were reduced to about 10% of what they were for the original configuration. Excellent agreement was found between the current meter and vessel-mounted ADCP current records.

In certain areas of the world (e.g., Georges Bank), the survivability of a mooring can have more to do with fishing activity than to environmental conditions. Also, in the early days of deep-sea moorings, the Scripps Institution of Oceanography lost equipment on surface moorings to theft and vandalism. Preventing mooring and data loss in such regions can be difficult and expensive. For fishery oceanography studies the dilemma is that, to be of use, the measurements must be obtained in areas where they are most vulnerable to fishnet fouling and fish-line entanglement. Damage to nets equates to lost fishing time and damaged or lost instrumentation. Aside from providing detailed information on the mooring locations in printed material handed out to commercial fishermen, or published in "Notices to Mariners", or provided on designated Web sites, fish processing companies and coast guard, the scientist may need to resort to closely spaced "guard buoys" in an attempt to keep fishermen and shipping traffic from subsurface moorings. Our experience is that a limited array of only three or so coast guard-approved buoys more than 0.5 km from the mooring is inadequate, and that certain operators will even use the buoys to guide their operations, thereby increasing the chance of damage.

1.7.9 Acoustic Releases

An acoustic release is a remotely controlled motorized linkage device that connects the expendable bottom anchor (often a set of used train wheels or specially designed concrete block) to the recoverable elements of a mooring (Figure 1.47). Modern acoustic releases are critical for free-fall deployment of moorings from ships and for reliable recovery of equipment on acoustic demand (Heinmiller, 1968). Operation of the

FIGURE 1.47   a). A deep-sea mooring using an IXBLUE SAS Oceano 2500S-Universal AR861 acoustic release (attached to a blue rope) being prepared for deployment off the deck of the Canadian Coast Guard Research Ship "John P. Tully". The yellow MF45 Series float manufactured by Flotation Technologies supports an upward-looking Teledyne-RDI LR75 KHz ADCP; a Seabird SBE 37 CTD has been attached to the release. Two train wheels form the anchor; (b). Details of an IXBLUE SAS Oceano 2500S-Universal AR861 acoustic release open on the bench. The hock release arm is on the left end and the acoustic transponder on the right end. The central part shows the electronic circuit boards and the battery packs. The stainless steel deep pressure case appears in the background. *(Photo courtesy of Tamás Juhász, Institute of Ocean Sciences, 2013).*

release requires a deck unit specially built for the particular type of release and a transducer for acoustic interrogation of the release. A ship's sounder can be used in place of the hand-held transducer provided it is of compatible frequency and has a wide beam. This is useful since it allows the technician to talk to the release from the ship's laboratory rather than by lowering a transducer over the side of the ship. However, for "acoustically noisy" ships, lowering a transducer over the side is often the only way to talk to the release. More advanced releases enable the user to measure the *slant range* from the ship to the release based on the two-way time delay. By taking into account the slant of the acoustic path, the user can determine the coordinates of the release. Long-life (three-year) acoustic "pingers" built into the releases also are used to locate the depth and position of moorings using triangulation procedures. This is particularly useful for those

moorings that fail to surface on command and must be dredged from the ship using a long line and grappling hook. In some acoustic releases, a rough estimate of the orientation of the release can be obtained remotely through changes in ping rate. For example, in the case of the InterOcean release, a doubled ping rate means that it is lying on its side rather being upright in the water column. Acoustic releases for most oceanic applications are manufactured by EdgeTech USA (Model 8242XS, CART), Inter-Ocean Systems, Inc. USA (1090E to 1500 m depth, 1090ED to 8000 m depth), Teledyne-Benthos USA (Model 866-A shallow and Model 865-A to 12,000 m depth), IXBlue SAS (France) Oceano Technologies (2500 Universal AR861 to 6000 m depth), and Ashtead Technology Offshore Division (Sonardyne Oceanographic Release Transponder, Type 7409). As a side note, the acoustic releases used successfully in a 3-year study of the deep currents in the Middle America Trench (Thomson et al., 2010), were 35 year-old Oceano USA RT181 releases that were still "chirping" in the laboratory many months after recovery.

Most modern releases use separate "load" and "release" codes so that the release can be remotely opened and closed. Some releases also provide a release code that signals the operator on the ship that the mooring has released and should be expected to surface in a time appropriate for the depth and net buoyancy of the mooring elements. Some releases also have a pinging mode, which allows the user to follow its rise using the ship's echo sounder (commonly 12 kHz). Releases typically can operate in situ for 2 years on alkaline batteries with another year in reserve. Using a lithium battery pack can extend their service life to 10 years. Thorough maintenance of the releases is an effective way to reduce mooring loss. As a result of improved technology and maintenance, the reliability of acoustic releases has improved to the point where they now function nearly 100% of the time.

As it is sometimes difficult to predict precisely where a mooring will surface, the time immediately after the release code has been sent can be quite tense. Spotting the mooring from the ship can be a real challenge, especially in rough weather. Attachment of a pressure-rated radio beacon and flashing light to the top float of the mooring can aide considerably with the recovery operation. If the mooring fails to surface after an appropriate time, a search can be initiated assuming that the mooring has surfaced and has not been spotted. Past experience has demonstrated the wisdom of having a dual release system with two acoustic releases side-by-side in a parallel harness often from different release manufacturers. A triangular bridle at the top connects the releases to a single point in the mooring line while a spreader bar connects the package to a single attachment point on the anchor chain. The extra cost can help avoid the need to dredge for the mooring if one of the releases should fail.

Dredging is a last (and often unrewarding) resort that can be extremely harmful to the mooring hardware, leading to severe damage to the current meters and other instruments on the mooring line. In addition, there is a correct dredging procedure and oceanographers new to the field should talk to more experienced colleagues for guidance. Safety issues include tangled mooring lines that need to be guided past the ship's propeller and carefully handled on deck. Tangled lines that are under load from the mooring components and/or still-attached anchor can be especially dangerous.

Moorings are sometimes inadvertently "hit" by fishing gear, towlines, surface vessels, or submarines. Improperly designed mooring components can also cause moorings to break apart. To facilitate the protection and recovery of fragmented moorings, technicians at the Institute of Ocean Sciences (IOS) in British Columbia have devised a protocol for mooring loss prevention, including: (1) the reduction in corrosion of mooring components by applying protective surface coatings, isolating dissimilar metals, using marine grade materials, and installing sacrificial anodes such as zinc blocks; (2) the use of distributed floatation such that any portion of the mooring remains positively buoyant should there be an unexpected break in the mooring line; (3) selecting mooring locations that minimize potential loss and aid recovery, while maintaining the integrity of the science program. In addition to the scientific program, consideration must be given to marine traffic, the nature of the seafloor, and proximity of submarine hazards such as transmission cables; (4) the use of simple acoustic beacons ("pingers") to mark the two ends of a mooring section. These can be recovered should the pinging mode of the acoustic release fail when the mooring arrives on the surface. In addition to providing recovery backup, the two pingers can be used to confirm that the mooring is still in position and intact prior to recovery. Pingers used at IOS transmit a 27-kHz frequency pulse, with 1 s-repetition rate, at 1/4 W output. They are mounted on an "in-line" strength member and have a conservative 2.5-year continuous operation life. They are simple, salt water activated, and have a limited range of roughly 0.5—1.0 nautical miles, depending on environmental conditions; and (5) equipping each mooring with a variety of locating beacons, for example VHF radio beacons, Argos Platform transmit terminal (PTT) satellite beacons, and XF xenon flashing lights. Iridium beacons serve the same purpose as Argos but with 2-way communication capability. Because of their expense and service fees, beacons are placed on the most valuable portion of the mooring only. As an example, IOS uses a Model 265 Beacon made locally by Oceanetic Measurement Ltd. The beacons are submersible to 1500-m depth, have either a pressure or salt-water switch, and a service life of about 1 month. The beacon signals when a mooring has risen to the surface and periodically updates a track of its latitude and longitude within 1-km lowest accuracy, to 300-m highest accuracy. The PTT is a "dumb" transmitter so that the number and quality of the position fixes are determined by the current satellite constellation. The VHF and XF units used by IOS are manufactured by Novatech, a branch of Metocean Ltd. In addition, Xeos Ltd. provides VHF, Argos, XF and Iridium beacons, with modern electronics and low power consumption.

1.8 LAGRANGIAN CURRENT MEASUREMENTS

A fundamental goal of physical oceanography is to provide a first-order description of the global ocean circulation and its temporal variability. The idea of following individual parcels of water (the Lagrangian perspective) is attractive since it permits investigation of a range of processes taking place within a tagged volume of water. Named after Joseph L. Lagrange (1736—1811), the French mathematician noted for his early work on fluid dynamics and tides, Lagrangian descriptions of flow can be used to investigate a broad suite of biophysical and geochemical processes ranging from the dispersion of substances discharged into the ocean from a point source to the productivity of a semienclosed marine ecosystem within a mesoscale eddy as it drifts across the ocean. Early Lagrangian measurements consisted of tracking some form of tracer such as a surface float or dye patch. While giving vivid displays of water motions over short periods of time, these techniques demanded considerable onsite effort on the part of the investigator. Initial technical advances were made more rapidly in the development of moored current meters, which yielded a strictly Eulerian picture of the current. However, improvements in tracking systems and buoy technology, since the 1970s, have made it possible to follow unattended surface and subsurface drifters for periods of many months to several years. Satellite-tracked surface buoys and acoustically tracked, neutrally buoyant SOFAR (SOund Fixing And Ranging or "Swallow") floats have been able to provide reliable, long-term, quasi-Lagrangian trajectories for many different parts of the world. (The trajectories are called quasi-Lagrangian since the drifters have a small "slip" of the order of 1—3 cm/s relative to the advective flow and because they do not move on true density surfaces. Surface drifters, for example, move on a two-dimensional plane rather than a three-dimensional density surface.) More recently, the more than 3000 vertically profiling floats deployed since 1999 as part of the international Argo program are providing oceanographers with an unprecedented volume of data on the temperature, salinity, and current velocity of the world ocean (http://www.argo.ucsd.edu). By December 2012, these drifters had yielded more than one million temperature and salinity profiles to depths of 2000 m.

Remotely tracked drifters provide a convenient and relatively inexpensive tool for investigating ocean variability without continued direct involvement by the investigator. In the case of the satellite-tracked buoys, the scientist can now dial-up the position of drifters or collect data from ancillary sensors on the buoys. The number of possible satellite positional fixes varies with latitude (Table 1.6). Time delays between the time that the data are collected by the spacecraft and the time they are available to the user is typically less than a few hours (Figure 1.48). This feature makes the drifters useful for tracking floating objects or oil spills. Oceanic platforms and satellite data transmission systems have become so reliable that both moored and drifting platforms are now used for the collection of a variety of oceanic and meteorological data, including SST, sea surface pressure, wind velocity, dissolved oxygen concentration, fluorescence, and mixed layer temperature. A new era of oceanographic data collection is in progress with less direct dependence on ships and more emphasis on data collection from autonomous platforms.

TABLE 1.6   The Mean Number of Satellite Passes per Day (24 h) for the Argos System for Different Numbers of Satellites in View

Transmitter Latitude (°North and South)	With Two Satellites	With Three Satellites	With Four Satellites
0	7	10	14
15	8	12	16
30	9	13	18
45	11	16	22
55	16	24	32
65	22	33	44
75	28	42	56
90	28	42	56

There are currently seven satellites that carry the Argos system. Because the number of passes scales linearly, the tabulated values can be extrapolated to give the mean number of passes for the seven-satellite system. (Table courtesy of Bill Woodward, President and CEO of CLS America, Inc. (2013).)

1.8.1 Drift Cards and Bottles

Until the advent of modern tracking techniques, estimates of Lagrangian currents were obtained by seeding the ocean surface with marked waterproof cards or sealed bottles and determining where these "drifters" came ashore. The card or bottle contained a note requesting that the finder notify the appropriate addressee of the time and location of recovery. To improve the chances of notification, a small token reward was usually offered (one Australian group gave out boomerangs). Although drift cards and bottles provide a relatively low-cost approach to Lagrangian measurements, they have major limitations. Because they float near the surface of the ocean, the movements of the cards and bottles are strongly affected by wind drag and wave-induced motions. In fact, much of what these type of drifters measure is wind and wave-induced drift ("windage") rather than underlying ocean currents (see section 1.8.1.1).

Moreover, even if the recovery rate was fairly high (1% is considered excellent for most drift card studies), the drifters provide, at best, an estimate of the lower bound of the mean current averaged over the time from deployment to recovery. Unless the card/bottle was recovered at sea, the scientist could never know if the drifter had recently washed ashore or had been lying on the beach for some time. In addition, the drifter provided no information on the current patterns between the deployment and recovery points.

1.8.1.1 Windage

Estimating windage is important when determining the drift of surface objects and has proven to be particularly important for predicting the trans-oceanic drift of the estimated 1.5 million tonnes of surface debris from the March 11, 2011-Tohoku-oki tsunami in the North Pacific Ocean (Cummins et al., 2012). The velocity, $\mathbf{U}$, of

**Argos mean data disposal time
May 2013**

FIGURE 1.48 Argos data disposal time (data throughput time) in minutes for May 2013. The time between the satellite observation and the time that the data are processed and available from Service Argos. The data timeliness is a function of several variables, including the number of satellites in the system and the number of antennae which are connected and receiving Argos data in real time. *(Courtesy of Bill Woodward, President and CEO of CLS America, Inc. (2013).)*

objects subject to windage can be separated into two components, $\mathbf{U} = \mathbf{U}_{water} + \mathbf{U}_{wind}$, where $\mathbf{U}_{water}$ is the horizontal velocity of the water at the ocean surface and $\mathbf{U}_{wind}$ is the velocity of the object relative to the water (windage). Windage results from direct forcing on an object by the wind blowing over the sea surface. Assuming a balance of wind force and ocean drag, the wind-driven component of velocity of an object can be estimated as

$$\mathbf{U}_{wind} = k\sqrt{\frac{A}{B}}\mathbf{U}_{10} \qquad (1.40)$$

where $\mathbf{U}_{10}$ is the wind velocity at a height of 10 m above the sea surface, A and B are, respectively, the surface area of the object above and below the water line, and $k \sim 0.025$ is a constant. According to this relation, an object with equal surface area above and below the water line $(A/B = 1)$ drifts relative to the water surface at 2.5% of the wind speed (as measured at the standard reference elevation of 10 m above the sea surface). This drift is in the direction of the wind, which may be different than the direction of the prevailing surface current. An object with high windage will be transported across the ocean more rapidly than an object with low windage.

1.8.2 Modern Drifters

Quasi-Lagrangian drifters can be separated into three basic types: (1) Surface drifters having a surface buoy that is tethered to a subsurface drogue at some specified depth (typically less than 300 m); (2) Subsurface, neutrally buoyant floats that are designed to remain on fixed subsurface density surfaces; and (3) Popup floats that cycle between subsurface density surfaces (where they remain for some fixed period of

FIGURE 1.49 (a) Examples of the basic drogue designs for satellite-tracked drifters: holey-sock drogue; parachute drogue; window-shade drogue; and Tristar drogue *(Niiler et al. (1987))*; (b) Schematic of the standard WOCE/TOGA holey-sock surface drifter showing pattern of holes in cloth panels *(From Sybrandy and Niiler (1990))*; (c) Cut-away view of foam-filled Plexiglass shell and PTT (satellite transmitter) used for the surface buoy. When complete, the surface float has an excess buoyancy greater than 7 kg. *(From Sybrandy and Niiler (1990).)*

FIGURE 1.49 *(continued)*.

time) and the ocean surface. Modern surface drifters have a radio frequency transmitter (called a PTT) for communication to a listening device while subsurface drifters may act either as a source or receiver of acoustic signals. Popup drifters use transmitters to report their locations to a passing satellite. Examples of the possible drogue configurations for modern satellite-tracked drifters are presented in Figure 1.49(a) along with the design for the standard holey-sock (World Ocean Circulation Experiment) WOCE/TOGA near-surface velocity drifter (Figure 1.49(b)). The purpose of the drogue is to reduce slippage between the drifter package and the water. The surface float contains the PTT, temperature, and atmospheric

(C)

FIGURE 1.49 (*continued*).

pressure sensors along with a drogue sensor and other electronics (Figure 1.49(c)); the purpose of the subsurface buoy is to reduce the "snap loading" on the drogue and cable by absorbing some of the shock from surface wave motion. In the case of the WOCE/TOGA holey-sock drifter, the ratio of the drogue cross-sectional area to the cross-sectional area of the other drifter components (such as the wire tether and subsurface float) is about 45:1, a relatively high drag-area ratio for typical drogues.

Trajectories from an early type of drifter deployed in the NORPAX experiment are found in McNally et al. (1983). Similar tracks from more modern drifters are presented in Figure 1.50. As examples, we have chosen trajectories from the North Atlantic near the Azores convergence zone (Figure 1.50(a)) and from the TOGA/ WOCE equatorial Pacific (Figure 1.50(b)). Note that, despite the extensive buoy coverage in these two cases, there are still regions unvisited by the drifters. The example shown in Figure 1.50(c) is a unique point-source deployment from the 106-mile site southeast of New York City. This site was the only ocean disposal site in the U.S.A. designated for dumping sewage sludge during the 1980s.

The essential technology for the above type of tracking was the development of a random access positioning system for polar-orbiting satellites that could simultaneously fix the positions of many platforms using the Doppler shift of the radio signals transmitted at regular intervals from the buoy. Early versions of the satellite tracking system were flown on the NIMBUS 6 (Kirwan et al., *1975*) and the French EOLE (Cresswell, 1976) satellites. Cresswell (1976) tested the accuracy of this system by examining the time series from a moored buoy and from an antenna mounted on top of a laboratory. For both sites, the uncertainties were less than 1.0 km. Similar RMS position fix errors were reported for the NIMBUS 6 systems by Kirwan et al. (1975) and by Richardson et al. (1981).

The early satellite-tracking systems have been replaced by the French Argos system (Collecte Localisation Satellite, CLS) carried onboard U.S. NOAA polar-orbiting weather satellites. As reported by Krauss and Käse (1984), this twin satellite system is capable of positional accuracies better than ±0.2 km. Location quality depends on a number of factors such as the quality of the ephemeris data (orbital parameters), the stability of the receiver oscillator and temperature control, the duration of the satellite pass, and the number of messages it receives from the drifter. Statistical information processed by Service Argos from thousands of fixed or slowly drifting platforms (Service ARGOS, 1992) indicates that quality locational fixes have 68% ($=\pm1\sigma$) accuracies of 150 m while standard fixes have 68% accuracies of ±350 m (here, σ is one standard deviation). As indicated by Table 1.6, the number of fixes per day is a function of latitude and the number of satellites available. Higher accuracy is possible when the platform is fixed over periods longer than 7 min during two successive satellite passes. The drifters themselves cost a few thousand dollars and are considered expendable. Typical tracking costs (for 2013) are of the order of $2500 per year for full

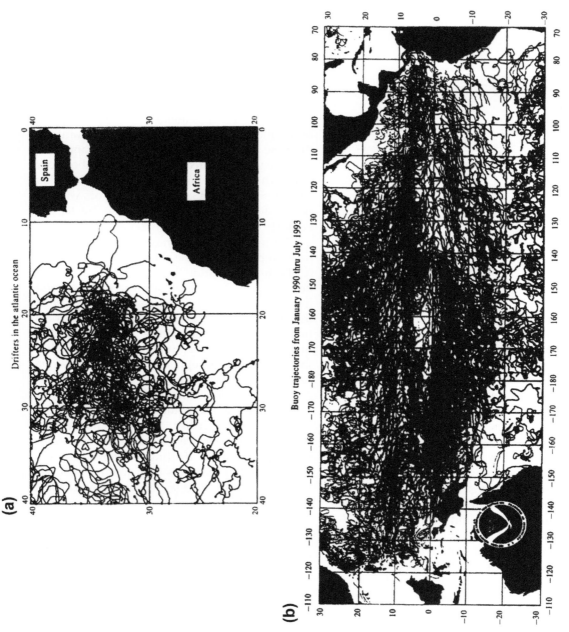

FIGURE 1.50 Trajectories from modern surface drifters with shallow (10–15 m) drogue depths. (a) Trajectories of 103 WOCE holey-sock drifters deployed near the Azores in the eastern North Atlantic from July 6, 1991 to October 25, 1993 (courtesy, Mayra Pazos, NOAA); (b) Trajectories of TOGA/WOCE holey-sock drifters for the Equatorial Pacific from January 1990 to July 1993 (courtesy, Mayra Pazos, NOAA); (c) Tracks of 66 holey-sock drifters centered at 10-m depth released from 106-mile site southeast of New York between October 1989 and June 1991. (Courtesy Paul Dragos, Battelle Ocean Sciences; Service Argos Newsletter 46, May 1993).

(c)

Trajectories of satellite-tracking drifting buoys
released at the 106-mile site October 1989 to June 1991

FIGURE 1.50 (*continued*).

tracking (no positional fixes omitted) and one-third of this for the one-third-duty cycle permitted by Service Argos (i.e., full-time tracking for 8 or 24 h followed by no tracking for 16 or 48 h, respectively; cf. Bograd et al., 1999a). Studies of surface and subsurface currents using Argos-tracked drifters can be found throughout the literature. Examples for the North Pacific are presented by Thomson et al. (1990; mesoscale eddies), Thomson et al. (1997; a basin scale North Pacific trajectory), Thomson et al. (1998; inertial and tidal currents), Bograd et al. (1999b; mean surface currents in the northeast Pacific), Rabinovich and Thomson (2001; diurnal shelf waves), and Rabinovich et al. (2002; anticylonic eddies).

1.8.3 Processing Satellite-tracked Drifter Data

Positional data obtained through satellite tracking need to be carefully examined for erroneous locations and the time of loss of drogue. In fact, one of the main problems with surface drifters, aside from the need for accurate positioning, is knowing if and when the drogue has fallen off. Strain sensors are often installed to sense drogue attachment, but they have proven unreliable. The tether linkage between the surface buoy and the drogue is the major engineering problem in designing robust and long-life drifters. Because of this problem, drifters often have a subsurface float to help absorb the snap loading on the drogue caused by surface waves and also ensure that the surface element is not constantly submerged in rough weather. An abrupt and sustained order-of-magnitude increase in the velocity variance derived from first differences of the edited positional data can be considered as evidence for drogue loss. The cubic spline routine in most software analysis packages works well for positional data provided the sampling interval is only a few hours (Bograd et al., 1999a). Although it is not recommended, the user can obtain the velocity components (u, v) directly from spline coefficients for the positional data.

1.8.4 Drifter Response

As with all Lagrangian tracers, it is difficult to know how accurately a drifter is coupled to the water and what effects external forces on the drifter's hull might have on its performance. In most applications, the coupling between the buoy and the water is greatly improved by the drogue. For shallow drifters with drogue depth centers less than 30 m, typical drogue-to-tether drag ratios are around 40:1. For deeper drogues (>100 m) the ratio decreases due to the added length of the tether. A smaller diameter wire can help offset the increased drag but at the expense of durability. There are as many different drogue designs as there are buoy hull shapes and it is difficult to get a consensus on the efficiencies of these drifter system elements. In a theoretical and experimental study, Kirwan et al. (1975) examined the effects of wind and currents on various hull and drogue types. They found that parachute drogues were more efficient than the common window-shade drogues, in strong contrast to the finding by Vachon (1973) that a bottom-ballasted window blind drogue was the most effective.

Subsequent studies by Dahlen and Chhabra (1983) have determined that a holey-sock drogue is more efficient than either the window shade or the parachute. This shape is easy to deploy and was selected for the standard drifter used in the WOCE program (Sybrandy and Niiler, 1990). Another innovative drogue called the Tristar developed by Niiler et al. (1987) uses a cross pattern of window shades with an additional horizontal plane (Figure 1.49(a)). The idea behind this drogue design is to reduce "sailing" of the drogue as is often occurs with a single window shade. Although easy to deploy (it goes into the water in a soluble box), this type of drifter is difficult to recover. For compatibility reasons, the standard drifter used in the WOCE Surface Velocity Program (WOCE-SVP) uses a holey-sock drogue.

In addition to the disagreements about which type of drogue is best, the field studies of Kirwan et al. (1978) reported that the wind drag correction formula, given by Kirwan et al. (1975), is much too large for periods of high wind. The subsequent conclusion was that drifter velocities uncorrected for wind drag are better indicators of the true prevailing surface currents than are those corrected for the influence of wind drag on the buoy hull. In this context, it should be recognized that Lagrangian drifting buoys respond to the integrated drag forces including the forces on the drogue and the direct forcing on the hull. The driving forces in the water column consist of a superposition of geostrophic currents plus wind- and tide-generated currents. To evaluate the role of wind forcing on drifter trajectories, McNally (1981) compared monthly mean drifter trajectories with the flow lines for mean monthly winds computed from Fleet Numerical Weather Central's (now the Fleet Numerical Meteorology and Oceanography Center, FNMOC) synoptic wind analysis. He found that the large-scale, coherent surface flow followed isobars of sea-level pressure and was $20-30°$ to the right of the surface wind in the North Pacific (Figure 1.51). Overall buoy speeds were 1.5% of the geostrophic wind speed during periods of strong atmospheric forcing (fall, winter, and spring). In the summer, mesoscale ocean circulation features, unrelated to the local wind, tended to determine the buoy trajectories.

McNally (1981) also compared trajectories among buoys with drogues at 30-m depth, buoys with drogues at 120 m, and buoys without drogues. He found that drifters drogued below 100-m depth behaved very differently from ones drogued at 30 m, but that those drogued at 30 m and those without drogues behaved similarly. Using the record from a drogue tension sensor (drogue on—off sensor), McNally found that it was not possible to detect from the trajectory alone when a buoy had lost its

FIGURE 1.51 Monthly average wind and buoy speeds over the ADS North Pacific region from June 1976 through July 1977. (a) Monthly average drifter speeds; (b) monthly average wind speeds; (c) monthly average difference angle between wind direction and drifter direction. Vertical bars denote ±1 standard deviation.

drogue. This result suggests a lack of vertical current shear in the upper 30 m where the flow apparently responds more directly to wind-driven currents than to baroclinic geostrophic flow. The result was supported by the poor correlation between mean seasonal dynamic height maps and the tracks of near-surface drifters reported by McNally et al. (1983). McNally (1981) also described an annual increase by a factor of 5 of the wind speed in the North Pacific while the drifter speeds increased by a factor of 3.5, somewhat surprising considering that the wind stress that drives the currents is proportional to the square of the wind speed. During this same time the mean seasonal dynamic height amplitude changed only slightly.

Emery et al. (1985) confirmed the lack of agreement between drifter tracks and the synoptic geostrophic current estimates, as well as the high correlation between drifter displacements and the geostrophic wind speed and direction (Figure 1.52). In a rather complex analysis of the wind-driven current derived from drifter trajectories, Kirwan et al. (1979) concluded that, while the drifter response is best described by a two-parameter linear system (consistent with the driving of the buoy by wave-driven Stokes drift), a combination of Ekman current plus Stokes drift also adequately described the resulting trajectories. Calculations by Emery et al. (1985), based on the nominal hull size, suggest that the Stokes drift component is relatively small and that the current in the surface Ekman layer is the primary driving mechanism for the mean drifter motion. That the angle this current makes to the wind is less than the 45° predicted by Ekman (1905), is expected since the real ocean conditions never seem to meet the conditions for Ekman's derivation. McNally (1981) found an average angle of 30° while Kirwan et al. (1979) reported an angle of 15°, both to the right of the wind for the northern hemisphere.

A search for the elusive Ekman spiral was conducted from November 20, 1991 to February 29, 1992 by Krauss (1993) using 10 satellite drifters drogued at five different levels within well-mixed homogeneous water of 80-m depth in the North Sea midway between England and Norway. The holey-sock drogues used in the study were 10-m long and centered at 5-m depth intervals from 7.5 to 27.5 m (Figure 1.53(a)). Results for the first four weeks of drift when the drifters were relatively close together revealed a clockwise turning and decay of the apparent wind stress with depth as required by Ekman-layer theory (Figure 1.53(b)). Here, the apparent wind stress is derived from the fluctuations in current velocity shear measured by the satellite-tracked drifters. Sea surface slopes needed to complete the calculations are from a numerical model. The observed amplitude decay of 0.90 and deflection of 10° near the surface are in close agreement with theory (apparent wind increases from 0° at the surface and is associated with an Ekman current that should be 45° to the right of the apparent wind). The angle increases to 41.6° in 25-m depth. The total current field is a superposition of barotropic currents due to sea-level variations and Ekman currents. The classical Ekman theory is unable to fully describe the observed deflection of the apparent wind (and Ekman current) to the right of the wind and its decay with depth. To be consistent with Ekman's theory, an eddy viscosity of 10^3 cgs units would be needed, which is well beyond the norm. However, as noted by Krauss, "… the deflections are a strong indication that some type of Ekman spiral dominates within the upper 30 m" In a related study, Lenn and Chereskin (2009), used repeated high-resolution profiles from a shipboard ADCP (see Section 1.7.4) to examine the Ekman spiral and transport in Drake Passage in the Southern Ocean. Based on data from 156 transects between South America and the Antarctic Peninsula from September 1999 to October 2006, the authors show that the mean Ekman spiral penetrates to roughly 100-m depth in the unstratified surface waters and is compressed vertically relative to theoretical predictions based on a constant eddy

FIGURE 1.52 (a) Comparison between monthly mean buoy and geostrophic wind directions; (b) Comparison between monthly mean buoy and geostrophic wind speeds. *(From Emery et al. (1985).)*

FIGURE 1.53 Test of Ekman's theory. The clockwise turning and decay of the apparent wind stress τ_D at depth D (m) relative to the observed surface wind stress. The apparent wind stress is derived from the current velocity shear dv_D/dz measured by satellite-tracked drifters (a) drogued at different depths during homogeneous winter conditions; (b) Histogram of the relative angle (in degrees) between the surface wind-stress vector and the calculated apparent wind-stress as a function of depth (surface wind minus apparent wind). Linear regression values (α,β) give apparent wind-stress as a function of surface wind stress. Offset results from the different time scales of the winds and the currents. Mean values given in upper right corner of figure. *(Courtesy W. Krauss (1994).)*

viscosity. The amplitude of the current was found to decay more rapidly than it rotated anticyclonically with depth. Fluctuations in the upper ocean stratification associated with diurnal buoyancy fluxes were thought to contribute to the Ekman spiral compression and the nonparallel shear—stress relation in the passage (see also Yoshikawa et al., 2007).

In an early study, McNally and White (1985) examined wind-driven flow in the upper 90 m using a set of buoys drogued at different depths. They found a sharp change in buoy behavior when the drogue entered the deepening surface mixed layer. This response was characterized by

a sudden increase in the amplitude of near-inertial motions with a downwind drifter velocity component three times that of the crosswind component. They also found that 80—90% of the observed crosswind component could be explained by an Ekman slab model. The large downwind response leads to surface currents, calculated from the buoy displacements, that are greater than 0° but less than 45° (about 30°) to the right of the wind. This behavior was true for all buoys with drogues above the upper mixed layer; once in the mixed layer all buoys behaved the same regardless of drogue depth.

In summary, the question of the relative coupling of drogued and undrogued drifting buoys to the water is still not completely resolved. Drifters measure currents, but which components of the flow dominate the buoy trajectories is still a topic of debate. Based on the recent literature, it appears that shallow drifters with drogue depths less than about 50 m are driven mainly by the wind-forced surface frictional Ekman layer, whereas deep drifters with drogue depths exceeding 100 m are more related to geostrophic currents. The likely percentage of contribution by these two current types depends on the type of drogue system that is used. A problem with trying to measure the deeper currents is that deeper drogue systems tend to fail sooner and it is difficult to access quantitatively the role of the drogue in the buoy trajectories. Drogue loss due to wave loading and mechanical decoupling of the surface buoy and the drogue is still the main technical problem to extending drifter life.

There are other problems with drifters worth noting. In addition to drogue loss and errors in positioning and data transmission, the transmitters submerge in heavy weather and loose contact with the passing satellite. Low drag ratios lead to poor flow response characteristics and, because of the time between Service Argos satellite passes, there is generally inadequate sampling of tidal and near-inertial motions (especially at low latitudes or for the one day on-two days off

(b)

FIGURE 1.53 (*continued*).

duty cycles) leading to aliasing errors. (Satellite positional sampling rate is not a problem for drifters tracked by GPS.) Drifters also have an uncanny tendency to go aground and to concentrate in areas of surface convergence.

1.8.5 The Global Drifter Program

The success of drifters as observational platforms led to the creation of the Global Drifter Program (GDP), a principal component of the Global Surface Drifting Buoy Array, which in the United States is part of NOAA's Global Ocean Observing System (GOOS). The objectives of the GDP are to:

1. Maintain a global 5 × 5° array of 1250 satellite-tracked surface drifting buoys to meet the need for an accurate and globally dense set of in situ observations of upper mixed layer currents, sea surface temperature,

salinity, atmospheric pressure, and wind velocity;

2. Provide a data processing system for scientific use of these data. At present, the data support short-term (seasonal to interannual) climate predictions as well as climate research and monitoring.

In the United States, the GDP is administered by NOAA's Atlantic Oceanographic and Meteorological Laboratory (AOML), which coordinates deployments, processes the data, archives the data, maintains META files describing each drifter deployed, develops and distributes data-based products and updates the GDP Web site (http://www.aoml.noaa.gov/phod/dac/gdp_objectives.php). AOML supervises the drifter industry, upgrades the technology, purchases drifters, and develops enhanced data sets. NOAA's GDP is part of an international program involving Argentina, Australia, Brazil, Canada, France, India, Italy, Korea, Mexico, New Zealand, South Africa, Spain, and the United Kingdom.

The drifting buoy configuration for the GDP is presented in Figure 1.54. Although there are a number of manufacturers of these drifters—so that individual buoys will look somewhat different—the basic configuration is the same as that in Figure 1.54. Deployment of the buoys (Figure 1.55) is easily carried out from volunteer observing ships or from the air. Because the GDP objective is to maintain a global array of 1250 buoys operating simultaneously, there is a need for repeated annual deployments to fill in gaps that arise from buoy failures or to cover areas from which buoys have been rapidly advected away.

As noted previously, the GDP Drifter Data Assembly Center has the goal of assembling and providing uniform quality control of SST and surface velocity measurements to help improve climate prediction models, which require accurate estimates of SST to initialize their ocean component. In addition to supporting numerical modeling, the drifters provide SST for the calibration and validation of satellite infrared SST which, in turn, allow for greater spatial coverage for climate prediction models. The status of the GDP drifter array as of May 2013 is shown in Figure 1.56. The red dots in this figure denote buoys that measure SST only (a total of 479), the blue dots denote those that also measure sea-level pressure (389), and the green dots buoys that also measure surface salinity. The majority of the buoys only measure SST, but a substantial number also measure atmospheric pressure. Both types of measurement are critical for weather and climate models. Salinity is a more recent requirement as oceanographers have only recently begun to measure salinity from space and need in situ verification on a global scale. A different view of the data distribution is given in Figure 1.57, which shows the growth of the GDP since 1988. Again, the data are binned according to the contributions of the buoys to SST, atmospheric pressure, wind, and salinity sensor measurements. The SST sensing buoys constitute a large majority of the total number of buoy days; atmospheric pressure sensing platforms began in 1994 and reached a maximum number in 2010. The addition of wind sensors began in 1997 and peaked in 2002. Salinity sensors were added in 2010 and their numbers have steadily increased through 2012.

An informative application of these drifter data is to map surface currents and the distribution of eddy kinetic energy (EKE) (Figure 1.58).

Figure 1.58 clearly highlights regions with strong zonal currents and strong poleward flowing western boundary currents such as the Kuroshio, Gulf Stream, and East Australian Current. Many of these same areas correspond to areas of high EKE with some additional areas of high EKE such as the Brazil—Malvinas Current confluence off the south-central east coast of South America. The EKE is substantially lower in the Antarctic Circumpolar Current where strong zonal currents do not exhibit the same

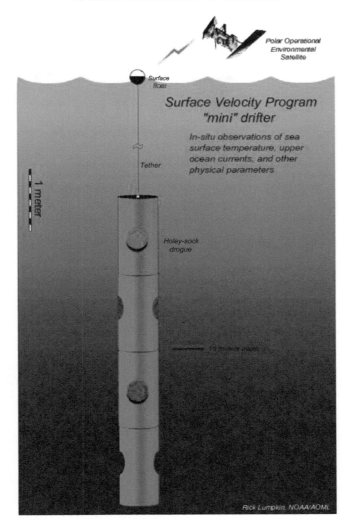

FIGURE 1.54 Drifting buoy configuration for the Global Drifter Program, GDP. *(Courtesy of Rick Lumpkin, NOAA/AOML.)*

high degree of mesoscale variability as other strong currents.

Because surface drifters closely track the two-dimensional flow of the ocean surface, they are ideal for studying the dispersion of surface particles, such as fish larvae, and buoyant pollutants, such as oil spills. Drifter observations of dispersion can also be used to quantify the effects of mesoscale variability on the mean transport

(Davis, 1991). Lumpkin et al. (2002) compared the Lagrangian and Eulerian length scales from surface drifters and altimetry, and found that they were proportionally related only in the most energetic part of the ocean (such as the Gulf Stream) where Lagrangian particles are advected around eddies at timescales much shorter than the Eulerian timescale. Here, "proportionally related" means that the two scales

VOS crew deploy next generation SVP drifter
Photo by: GDP

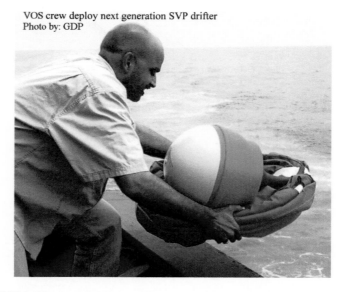

FIGURE 1.55   GDP drifter showing the surface float and drogue being deployed.

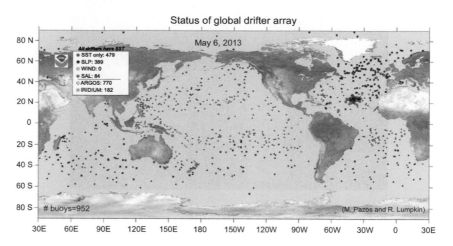

FIGURE 1.56   Status of the Global Drifter Program (GDP) drifter array as of May 6, 2013.

had different magnitudes but varied in a similar manner. Where the currents are weaker, the two scales were completely unrelated.

Bauer et al. (2002) separated mean and eddy drifter velocities in the tropical Pacific using optimized bicubic splines and found that eddy diffusivity in this region is strongly anisotropic: zonal diffusion is up to seven times greater than meridional diffusion due to the trapping of water parcels in coherent features such as equatorial and tropical instability waves. Two years later, Zurbas and Oh (2004)

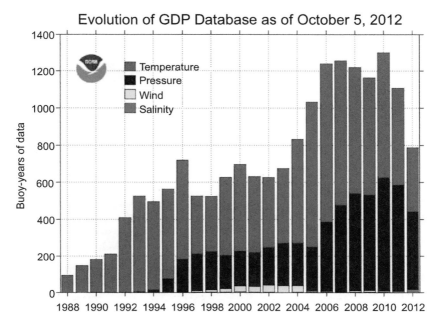

FIGURE 1.57   Growth of the Global Drifter Program drifter database in terms of sensor years of data for the period 1988 to 2012.

mapped the global surface diffusivity using a method designed to avoid the contamination of their eddy statistics by the time-varying mean flow shear. Their maps of apparent diffusivity (Figure 1.59) reveal enhanced values in zonal bands near 30° latitude in both the southern and northern hemispheres. Zurbas and Oh suggested that these features were due to meandering, eddy-rich eastward currents in the North Atlantic (Azores Current) and the North Pacific (North Subtropical Countercurrent) and to the westward drift of Agulhas retroflection eddies in the South Atlantic. In the South Pacific, they attributed high diffusivity values to the presence of a South Subtropical Countercurrent.

Some oceanic regions have been well sampled during parts of the year but poorly sampled at other times due to infrequent batch buoy deployments from research vessels or volunteer observing ships. In areas with strong seasonal fluctuations, such as the tropical Atlantic, this produces biased estimates of the mean flow when they are averaged in bins. Lumpkin (2003) addressed this problem by dividing the Lagrangian time series into spatial bins and, within each bin, decomposing the time series into a time-mean, annual and semiannual harmonics, and a residual component. This decomposition was performed using a Gauss-Markov method familiar in the oceanographic literature for its use in box inverse models (e.g., Ganachaud and Wunsch, 2000). This approach can resolve the amplitudes and phases of seasonal fluctuations throughout most of the tropical Atlantic with the present density of drifter observations (Lumpkin and Garraffo, 2005).

An important new application of drifter data is its synthesis with satellite altimetry. Altimetry is excellent at capturing the variations of sea-level height associated with geostrophic velocity anomalies but cannot yet map absolute sea level

FIGURE 1.58 Time averaged speed (top) and eddy kinetic energy (bottom) calculated from drifter observations. Values are shown at 1-degree spatial resolution. *(Courtesy of Rick Lumpkin, Director, GDP program, NOAA/AOML.)*

with sufficient accuracy to resolve time-mean currents. In addition, the sea level anomaly fails to account for the significant ageostrophic components of current variations, such as centrifugal effects that may account for differences in drifter-derived and altimetry-derived EKE on either side of the Gulf Steam front (Fratantoni, 2001). Moreover, gridded altimetric data tend to smooth the observations, which results in systematic underestimates of the EKE. Niiler et al. (2003) describe a method to synthesize Ekman-removed drifter speeds with gridded altimetric velocity anomalies in regions where they are significantly correlated. This method uses concurrent drifter and altimetry velocities to calibrate altimetry, making its amplitude consistent with the in situ drifter observations while using the time series from altimetry to correct for biased drifter sampling of mesoscale to interannual variations. Using this method applied to the global drifter data set, Niiler et al. (2004) produced a map of the absolute sea level for the period 1992–2002 (Figure 1.60).

1.8.6 Other Types of Surface Drifters

Before leaving this topic, it is appropriate to mention that while the satellite-tracked buoys are perhaps the most widely used type of surface follower for open waters, there are other buoy tracking methods being used in more confined coastal waters. A common method is to follow the surface buoy using ship's radar or radar from a nearby land-based station. More expensive buoys are instrumented with both radar reflectors and transponders to improve the tracking. The accuracies of such systems all depend on the ability of the radar to locate the

FIGURE 1.59   A map of the lateral diffusivity in the Pacific Ocean, combined with the mean current vectors in the surface layer, derived from drifters (*From Zurbas and Oh (2004).*)

FIGURE 1.60 Long-term absolute mean (1992–2002) sea level from satellite altimetry and drifter trajectories. Note: Elevations are in cm. *(From Niiler et al. (2004).)*

platform and also on the navigational accuracy of the ship. Fixes at several near-simultaneous locations are needed to triangulate the position of the drifter accurately. Data recording techniques vary from hand plotting on the radar screen to photographing the screen continuously for subsequent digital analysis. These techniques are manpower intensive when compared with the satellite data transmission, which provides direct digital data output.

In addition to radar, several other types of buoy tracking systems have been developed. Older drifters relied on radio wave navigation techniques such as LORAN or NAVSTAR (satellite navigation). For example, the subsurface drogued NAVocean and Candel Industries Sea Rover-3 Loran-C drifters had built-in Loran-C tracking systems that can both store and transmit the positional data to a nearby ship within a range of 25–50 km. Absolute positional accuracy in coastal regions was around 200 m but diminished offshore with decreased Loran-C accuracy. However, relative positional errors are considerably smaller. Based on time-delay transmission data from three regional Loran-C transmitters, Woodward and Crawford (1992) estimated relative position errors of a few tens of meters and drift speed uncertainty of ±2 cm/s for drifters deployed off the west coast of Canada. Once it is out of range of the ship, the Loran-C drifter could be lost unless it was also equipped with a satellite transmission system. Meteor-burst communication is a well-known technique that makes use of the high degree of ionization of the troposphere by the continuous meteor bombardment of the earth. A signal sent from a coastal master station skips from the ionosphere and is received and then retransmitted by the buoy up to several thousand kilometers from the source. Since the return signal is highly directional, it gives the distance and direction of the buoy from the master station. Buoys can also be positioned using VHF radio transmission via direction and range. The introduction of small, low-cost GPS receivers now makes it possible for buoy platforms to position themselves continuously to within several tens of meters. Provision for differential GPS (using a surveyed land-based shore station) has improved the accuracy to order of a few meters. Data are then relayed via satellite to provide a higher resolution buoy trajectory than is presently possible with Argos tracking buoys. Given the high positioning rate possible for GPS systems, it is the spatial accuracy of the fixes that limits the accuracy of the velocity measurements. Japanese scientists have claimed recently to have used a highly accurate mini-drifter GPS-Argos tracking system to study nearshore currents based on 10–20 min positional data (http://www.argos-system.org/web/en/55-news.php?item=511).

1.8.7 Subsurface Floats

New technological advances in subsurface, neutrally buoyant float design have improved interpretation and understanding of deep ocean circulation in the same way that surface drifters have improved research in the shallow ocean. In their earliest form (Swallow, 1955), subsurface quasi-Lagrangian drifters took advantage of the small absorption of low-frequency sound emitted in the sound channel (the sound velocity minimum layer) located at intermediate depths in the ocean. The sound-emitting drifters were tracked acoustically over a relatively short range from an attending ship. The development of the autonomous SOFAR float, which is tracked from listening stations moored in the sound channel (Rossby and Webb, 1970) has removed the burden of ship tracking and made the SOFAR float, a practical tool for the tracking of subsurface water movements. Although positional accuracies of SOFAR floats depend on both the tracking and float-transponder systems, the location accuracy of ±1 km given by Rossby and Webb (1970) is a representative value. In this case, neutrally buoyant SOFAR floats have a positional accuracy that is markedly lower than that of satellite-tracked drifting buoys. Using high-power 250-Hz sound sources, the early SOFAR floats are credited with the discovery of mesoscale variability in the ocean and for pioneering our understanding of Lagrangian eddy statistics (Freeland et al., 1975). The familiar "spaghetti-diagram" (Figure 1.61) is characteristic of the type of eddylike variability measured by SOFAR floats deployed in the upper ocean sound channel (Richardson, 1993).

SOFAR floats transmit low-frequency sound pulses, which are tracked from shore-listening positions or from specially moored "autonomous" listening stations. The need to generate low-frequency sound means that the floats are long (8 m) and heavy (430 kg), making them expensive to build and difficult to handle. Since greater expense is involved in sending sound signals than receiving them, a newer type of float

called the RAFOS (SOFAR spelled backwards) float has been developed in which the buoys listen for, rather than transmit, the sound pulses (Figure 1.62). In this configuration, the float acts as a drifting acoustic listening station that senses signals emanating from moored sound sources (Rossby et al., 1986). The positions of RAFOS floats in a particular area are then determined through triangulation from the known positions of the moored source stations. A typical moored sound source, which broadcasts for 80 s every two days at a frequency of 260 Hz, has a range of 2000 km and an average lifetime of three years (*WOCE Notes*, June 3, 1991)

Because RAFOS floats are much less expensive to construct than SOFAR floats and more difficult to locate (since they are not a sound source), RAFOS floats are considered expendable. The data processed and stored by each RAFOS

FIGURE 1.61 "Spaghetti-diagram" of all SOFAR float tracks from 1972 to 1989, excluding data from the POLYMODE Local Dynamics Experiment. Ticks on tracks denote daily fixes. Short gaps have been filled by linear interpretation. Plots are characteristic of the type of eddy-like variability measured by SOFAR floats deployed in the upper ocean sound channel. *(Courtesy, Phillip Richardson (1994).)*

FIGURE 1.62 Schematic of a RAFOS float. (*Courtesy, Thomas Rossby.*)

is primarily a satellite PTT and a ballast device that periodically comes to the surface and transmits its location data and "health" status (an update on its battery voltage and other parameters) to the Argos system (Davis et al., 1992). The only known points on the buoy trajectory are those obtained when the buoy is on the surface. As with the RAFOS buoys, the pop-up float sinks to its prescribed depth level after transmitting its data to a satellite system and continues its advection with the deep currents. The advantage of such a system is that it can be designed to survive for a considerable time using limited power consumption. Assuming that deep mean currents are relatively weak, the pop-up float is an effective tool for delineating the spatial pattern of the deep flow, which up to now has not been possible over large areas. The Autonomous Lagrangian Circulation Explorer (ALACE) described by Davis et al. (1992) and the Argo floats presently being maintained by the international oceanographic community (basically, modern versions of the earlier ALACE floats) drift at a preset depth (typically less than 2000 m) for a set period, for example 10 days, then rise to the surface for about a day to transmit their position to the satellite. The drifter then returns to a prescribed depth, which is maintained by pumping fluid to an external bladder that changes its volume and hence its buoyancy. Modern Argo floats (see section 1.8.7.1) are capable of making about 150 round-trips to depths of 2 km over a lifetime of about five years. Errors are introduced by surface currents when the device is on the surface. The floats also provide temperature and salinity profiles during ascent or descent. Additional sensors, such as dissolved oxygen sensors, have been incorporated into some of the floats.

As with the surface drifter data, the real problem in interpreting SOFAR float data is their fundamental "quasi-Lagrangian" nature (Riser, 1982). From a comparison of the theoretical displacements of true Lagrangian particles in simple periodic ocean current regimes with the

buoy as it drifts within the moored listening array must eventually be transmitted to shore via the Argos system or other satellite link. To do this, the RAFOS float must come to the surface periodically to transmit its trajectory information. After "uplinking" its data, the buoy again descends to its programed depth and continues to collect trajectory data. The cycle is repeated until the batteries run out.

The need for deep ocean drifters that are independent of acoustic tracking networks has led to the development of the "pop-up" float. The float displacements of real quasi-Lagrangian floats, Riser concludes that the planetary scale (Rossby wave) flows in his model contribute more significantly to the dispersion of 700-m depth SOFAR floats than do motions associated with near-inertial oscillations or internal waves of tidal period. Based on these model speculations, he suggests that while a quasi-Lagrangian drifter will not always behave as a Lagrangian particle it nevertheless will provide a representative trajectory for periods of weeks to months. For his Rossby wave plus internal wave model, Riser derived a correlation timescale of about 100 days. He also suggests that the residence of some floats in the small-scale (25 km) features, in which they were deployed, provides some justification for his conclusions.

For pop-up floats, problems in the interpretation of the positional data arise from: (1) interruptions in the deep trajectory every time the drifter surfaces; (2) uncertainty in the actual float position between satellite fixes; and (3) contamination of the deep velocity record by motions of the float on the surface or during ascent and descent. An essential requirement in the accurate determination of the subsurface drift is to find the exact latitude/longitude coordinates of the buoy when it first breaks the ocean surface and when first begins to re-sink. The ability to interpolate Argos fixes to these times is determined by the nature of the surface flow and the number satellite fixes. ALACE and Argo floats ascend more rapidly than they descend and spend little time at the surface. In a trial of an ALACE float to 1 km depth, the drifter spent 0.3 h in the upper 150 m, and 4 h between 150- and 950-m depth. Thus, according to Davis et al. (1992), most of the error was from vertical velocity shear at depths of 150 m and deeper, below the surface wind-driven layer. As an example, ALACE drifter 653 was launched at 2004 UTC on June 22, 1997 over the northern end of Loihi Seamount (18°56.229′ N, 155°15.347′ W). The float was equipped with temperature, conductivity, and pressure sensors, and ballasted to float at roughly 1350 dbar ($\approx$ 1335 m), coincident with the core of the hydrothermal plume emanating from the seamount. The float was programed to rise to the ocean surface every 15 days, spend roughly 24 h reporting its position and sensor data through Service Argos, and then submerge to its predetermined depth. Satellite positioning in the region had an accuracy of roughly 300 m at the 95% level of confidence, with generally seven to eight fixes per day at tropical latitudes. Location data provided averages of the mid-depth current at 15-day intervals and estimates of the surface currents at roughly 3-h intervals. Because the study was mainly interested in the response of the drifter trajectory to topographic features, no attempt was made to correct the observed drift for shear-induced velocity errors (cf. Thomson and Freeland, 1999 for further details).

1.8.7.1 Profiling Argo Floats

Argo floats are pop-up drifters with high-resolution temperature and conductivity (salinity) sensors that profile the upper 2000 m of the ice-free global ocean. In addition to temperature, conductivity, and pressure sensors, a fraction of the drifters are also equipped with dissolved oxygen sensors. As with other pop-up drifters, currents are determined from time-varying changes in positional data broadcast to satellites when the float is on the surface. This limits the temporal resolution of velocity time series to periods longer than at least twice the subsurface drift duration. The deployment of Argo drifters began in year 2000, supported by 31 different nations. By November 2007, Argo had achieved its initial goal of 3000 floats and by November 29, 2012, there were 3619 active floats in the world ocean gathering profiles at a rate of one about every 4 min—equal to 360 profiles a day or about 11,000 every month. By November 4, 2012, Argo had collected its one-millionth vertical profile. This is an impressive achievement considering that since the beginning of deep-sea oceanography in the late nineteenth century, ship-based observations have gathered just over half a million temperature and salinity profiles to a depth of 1 km and only 200,000 to 2 km (Gilbert, 2012). For Argo to be maintained at the 3000-float level, nations need to provide about 800 floats per year. Even with the 3000 floats, additional floats are needed because some areas of the ocean are over populated while others have gaps that need to be filled. Argo strives to maintain an average distance of 300 km between floats and works to provide a quality real-time data system that delivers 90% of profiles to users via two global data centers within 24 h. A delayed mode quality control system (DMQC) has been established and 60% of all eligible profiles have had DMQC applied. Float reliability has improved each year. Argo is now a major contributor to the World Climate Research Program's Climate Variability and Predictability Experiment (CLIVAR) project and to the Global Ocean Data Assimilation Experiment (GODAE). The Argo array is part of the Global Climate Observing System/Global Ocean Observing System (GCOS/GOOS). While continuing its core mission of monitoring ocean temperature and salinity, Argo is extending into ice-covered and shallower areas and partners are adding measurements of ocean biogeochemistry.

Argo floats are comprised of three subsystems: (1) Hydraulics, which control the buoyancy adjustment through an inflatable external bladder, allowing the float to surface and dive; (2) microprocessors which manage function control and scheduling; and (3) a data transmission system that controls communication with passing satellites. The floats have an approximate weight of 25 kg, a maximum operating depth of 2000 m, and crush depth of 2600 m. The three models presently in use are the PROVOR built by KANNAD (France) in close collaboration with IFREMER, the APEX float produced by Teledyne-Webb Research Corporation (USA) and the SOLO float designed and built by Scripps Institution of Oceanography (USA). Two float manufactures have new models; the

ARVOR is a new generation PROVOR float built by KANNAD while SOLO-II is a new generation SOLO built by MRV systems. The floats use temperature and salinity sensors manufactured by either SBE or by FSI. The temperature data are accurate to a few millidegrees over the float lifetime (see the Argo Web site for a discussion of salinity and oxygen data accuracy).

The autonomous battery-powered Argo floats drift at a specified "parking depth" pressure at which they are designed to be neutrally buoyant. At this pressure, the floats have a density equal to the ambient pressure and a compressibility that is less than that of seawater. The three types of floats operate in a similar fashion but differ slightly in their design characteristics. At 10-day intervals (a typical float duty cycle), the internal pumps direct fluid into an external bladder, causing the float to rise to the ocean surface over a period of about 6 h. A series of roughly 200 temperature-salinity-pressure measurements are obtained during the ascent and stored internally in the float. Polar-orbiting satellites determine the drifter position and retrieve the data from the float once it reaches the surface. The bladder then deflates and the float returns to its original density level. Floats are designed to make about 150 complete cycles. As illustrated by Figure 1.63, the present standard Argo mode consists of a "park and profile" cycle during which the float descends to a target depth of 1000 m to begin its drift. After 9.5 days, the float then descends to 2000 m from which it begins its temperature and salinity profile (Figure 1.64). Starting in 2010, 70% of the floats were profiling to depths greater than 1500 m. Another 20% profiled between depths of 1000 and 1500 m.

Data from most floats in Argo are received through Service Argos. In order to ensure error-free data reception and location in all weather conditions, the float must spend between 6 and 12 h at the ocean surface. Positions are accurate to ~±100 m depending on the number of satellites within range and the

FIGURE 1.63 Standard 10-day Argo cycling mode. The float first descends to a target depth of 1000 m to begin drifting. After 9.5 days, the float then descends to 2000 m from which it begins its temperature and salinity profile. After a 10-day submergence, the float surfaces to transmit data to a satellite. *(Courtesy, Denis Gilbert, Fisheries and Oceans Canada, The Maurice Lamontagne Institute (2012).)*

geometry of their distribution. Argo is also testing the use of GPS positioning and data communication through Iridium satellites. Iridium is an attractive option to Service Argos as it allows more detailed profiles to be transmitted within a shorter time at the surface. Iridium also allows for two-way communication. As of 2010, 250 floats had been deployed with Iridium antennas.

When on the surface, Argo floats are affected by wind drag, currents, and other factors. There also is potentially significant contamination by current shear during the slow ascent and descent phases of the float cycle. A study of velocity errors for the oceanic region southwest of Japan by Ichikawa et al. (www.jamstec.go.jp/J-ARGO/results/data_management/management/drifting_velocity/Drifting.pdf) has shown that surface drift and vertical shear can cause Argo drift velocities at the 1000-m parking depth to be overestimated by as much as 10−25%. However, the analysis of Ichikawa et al. invokes a model for the interaction between a profiling float and a model shear. The model shear used is representative of an area close to the Kuroshio, a Western Boundary Current region with high current shear. Much smaller shears of order 10% or less can be expected in less dynamic regions of the ocean.

1.8.8 Surface Displacements in Satellite Imagery

As noted briefly at the end of Section 1.7.7, well-navigated (geographically located) sequential satellite images can be used as "pseudo-drifters" to infer surface currents. The assumption is that the entire displacement of

FIGURE 1.64 The annual cycles in temperature (top) and salinity (below) in the Labrador Sea, North Atlantic Ocean. *(Courtesy, Igor Yashayaev, Fisheries and Oceans Canada, Bedford Institute of Oceanography (2012).)*

surface features seen in the imagery is caused by surface current advection. This displacement estimate method, (called the MCC) was applied successfully to sea ice displacements by Ninnis et al. (1986). Later, the same approach, was applied to infrared images of SST by Emery et al. (1986). The patterns and velocities of the SST-inferred currents were confirmed by the drifts of shallow (5-m drogue) drifters and by a CTD survey. Later studies (Tokamamkian et al., 1990; Kelly and Strub, 1992) have confirmed the utility of this method in tracking the surface displacements in different current regimes. When applied to the Gulf Stream (Emery et al, 1992), the MCC method reveals both the

prevailing flow and meanders. A numerical model of the Gulf Stream, used to evaluate the reliability of the MCC currents found that, for images more than 24-h apart, noise in this strong flow regime begins to severely distort the surface advection pattern.

The MCC method can also be applied to other surface features such as chlorophyll and sediment patterns mapped by ocean color sensors. In the future, it may be possible to combine ocean color tracking with infrared image tracking. Infrared features are influenced by heating and cooling, in addition to surface advection, while surface chlorophyll patterns respond to in situ biological activity. Since these two features should reflect the same advective patterns (assuming similar advective characteristics for temperature and color), the differences in calculated surface vectors should reflect differences in surface responses. Thus, by combining both color and SST it should be possible to produce a unique surface flow pattern that corrects for heating/cooling and primary biological production.

Another use of the MCC method is its application to SAR imagery. In this case, surface slicks result in the suppression of the SAR backscatter so that, provided that wind speeds are not too high to mask the oceanic features, the MCC method can be used to track the movements of the slick present in the surface layer (Qazi et al., 2013). Unlike infrared and ocean color images, which can be separated by as much as 24 h, the SAR images perform best in the MCC application when they are 30-min apart, as was the case with ENVISAT and ERS-2 satellites until mid-2009 (Qazi et al., 2013). The MCC-SAR surface currents are much more highly resolved than those computed from infrared or ocean color. In addition, the MCC-SAR currents can be computed much closer to the shoreline due to the higher spatial resolution of the SAR images.

1.8.9 Autonomous Underwater Vehicles

Autonomous underwater vehicles (AUVs) and Remotely Operated Vehicles (ROVs) provide sensor platforms for measuring oceanic water properties. ROVs such as the two-body 5000-m depth Remotely Operated Platform for Ocean Sciences (ROPOS, operated by The Canadian Scientific Submersible Facility) and the 6500-m depth Jason/Medea (designed by the Woods Hole Oceanographic Institution, Deep Submergence Laboratory) are highly maneuverable, unmanned tethered submersibles (decoupled from surface motion by intermediary controller package) that are "flown" from a ship to safely study and instrument many features of the world ocean including

hydrothermal venting systems at seafloor spreading regions of the deep ocean. They can be instrumented with CTDs, water sample carousels, high resolution still and video cameras, as well as robotic tools such as drills and mechanical manipulators. In contrast to the tethered ROVs, AUVs (including Gliders) are free of any surface vessel and therefore capable of conducting marine work during poor sea conditions at greater through-the-water speeds. The downside of all these devices is that they are expensive to purchase, highly expensive to operate, and labor intensive, requiring dedicated technical support, shiptime, and maintenance. Although many research groups have purchased AUVs and Gliders for oceanic surveys, most platforms are used by the military and by ocean industry. The Autosub6000 build by Underwater Systems Laboratory at the National Oceanography Centre (Southampton, United Kingdom) is a battery powered unit that supports magnetometer, turbidity, CTD, and electromagnetic EH sensors. It also has a 3-m altitude collision avoidance system and a 6000-m design limit. REMUS AUVs (REMUS 100, 600, and 6000 developed originally by the Woods Hole Oceanographic Institution and then transferred to HYDROID now a subsidiary of Kongsberg Marine) are also capable of extensive marine research.

Underwater gliders use small changes in buoyancy, similar to Argo drifters, in conjunction with wings, to convert vertical fall to forward motion, allowing the instrument to move horizontally at speeds of around 1 knot (0.5 m/s) with low power consumption. The concept for a glider with a buoyancy engine powered by a heat exchanger was introduced to the oceanographic community by Henry Stommel in a 1989 article in *Oceanography*. In the article, Stommel proposed the use of a glider, called *Slocum*, developed with research engineer Doug Webb. The glider name was for Joshua Slocum, who made the first solo circumnavigation of the globe by sailboat. Stommel and Webb proposed harnessing energy from the thermal gradient

FIGURE 1.65 A Slocum glider operated by Rutgers University, U.S.A.

between deep ocean water (2—4 °C) and surface water (near-atmospheric temperature) to achieve globe-circling range constrained only by battery power on board for communication, sensors, and navigational computers. By 2005, not only had a working thermal-powered glider (*Slocum Thermal*) been demonstrated by Webb Research, but they and other institutions had introduced battery-powered gliders (Figure 1.65) with impressive duration and efficiency, far exceeding that of traditional survey-class AUVs.

Although gliders have just sufficient speed to make headway against low-velocity flow regions of the ocean, the buoyancy-derived propulsion provides a marked increase in range and duration compared to vehicles operating with electrically driven propellers, thereby extending ocean surveys from hours to months, and over ranges of thousands of kilometers. The glide path consists of sawtooth-like up and down data profiles on temporal and spatial scales unavailable to other AUVs. A wide variety of glider designs are in use by the US Navy and ocean research organizations. Glider costs are typically about $100,000. On December 4, 2009, one of the gliders build by Webb Research became the first to complete a transatlantic journey, traveling from New Jersey on the east coast of the United States to the west coast of Spain in 221 days. The *Seaglider*

built by the University of Washington and the *Spray* built by the Scripps Institution of Oceanography have crossed the Gulf Stream from the mainland USA to Bermuda, and, together with the Webb *Slocum*, have conducted sustained, multivehicle collaborative monitoring of oceanographic variables in Monterey Bay off California. Wave gliders are different from the above gliders in that they use wave orbital motion to drive the platform. The blades that carry out this function are deployed about 6 m beneath the surface float and take advantage of wave action to pull the surface float forward (Figure 1.66). The combined float and driving system are shown together in Figure 1.67.

A rudder installed on the subsurface blades is used to steer the glider assembly while the surface float is equipped with solar panels to drive any sensors installed along with the GPS unit and Iridium communications system (Figure 1.68).

Wave gliders can carry a variety of sensors including wind speed and direction, air temperature, near-surface sea temperature, and temperature along the vertical tether, along with pressure. In past deployments, a wave glider was launched on the US west coast and traveled to Hawaii and then on to Australia and separately to Japan (http://liquidr.com/pacx/pacific-crossing.html).

FIGURE 1.66 Schematic of wave glider movement showing how the subsurface blades tip up and down to rectify the vertical wave motion and drive the surface float forward.

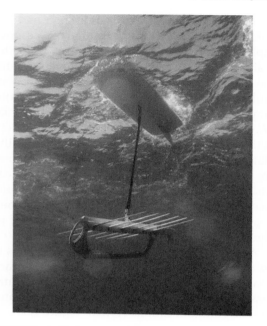

FIGURE 1.67 Underwater photograph showing wave glider subsurface motive blades along with the surface float. The rudder on the surface float is only for stability and does not direct the floats motion.

FIGURE 1.68 Photograph of wave glider float showing solar panels, GPS antenna, and Iridium antenna.

1.9 WIND

The surface wind stress is one of the primary mechanisms driving ocean variability over a broad range of frequencies. For example, turbulent mechanical wind mixing is a major factor in the deepening of the surface mixed layer while passing storm fronts are responsible for the generation of near-inertial currents, coastally trapped waves, the large-scale gyral circulation of the open ocean, and numerous other oceanic processes. Coastal upwelling and baroclinic undercurrents such as the California Undercurrent that form along eastern boundary regions of the world ocean respond to changes in the alongshore wind forcing (cf. Connolly et al., 2014). It is therefore not surprising to find a section on wind data in an oceanographic text. Moreover, we can state with some confidence that most of the scientific assessment of wind data over the ocean has been done by oceanographers searching for the best way to define the meteorological forcing field for oceanic processes. This is especially true of observationalists working on upper ocean dynamics and numerical modelers who require synoptic or climatological winds to drive their circulation models. It is not the intent of this book to discuss in detail the many types of available wind sensors and to evaluate their performance, as is done with the oceanographic sensors. Instead, we will briefly review the types of wind data available for ocean regions and make some general statements about the usefulness and reliability of these data.

Open-ocean wind data are of four basic types: (1) six-hourly geostrophic wind data computed from measured distributions of atmospheric sea surface pressure over the ocean; (2) directly measured wind data from ships and moored platforms (typically provided at hourly intervals); (3) 12-hourly wind speed and direction inferred from scatterometers flown on selected polar-orbiting satellites; and (4) six-hourly reanalysis wind velocity data derived from a blend of observations and numerical models. In addition to directly observed winds, several agencies use observations and atmospheric models to generate nowcasts, medium range (3–7 day) forecasts, and extended range (monthly to seasonal) forecasts of atmospheric pressure, wind velocity, and other parameters for various sectors of the world ocean. For example, up to one-week wind forecasts are provided for the northeast Pacific and west coast of North America by the Global Forecast System in the United States and the Global Environmental Multiscale Model of Environment Canada. Similar forecasts are generated by the European Centre for Medium-range Weather Forecasts (ECMWF) headquartered in the United Kingdom (http://www.ecmwf.int/), the Japan Meteorological Agency in Tokyo (http://www.jma.go.jp/en/week), the Bureau of Meteorology in Canberra (http://www.bom.gov.au/australia/), and the Indian Meteorological Department in Pune (http://www.imdpune.gov.in/).

Atmospheric pressure maps are prepared from combinations of data recorded by ships at sea, from moored or drifting platforms, and from ocean island stations. Analysis procedures have changed over the years with early efforts depending on the subjective hand contouring of the available data. More recently, there has been a shift to computer-generated "objective analysis" of the atmospheric pressure data. Since they are derived from synoptic weather networks, the pressure data are originally computed at six-hourly intervals (00, 06, 12, and 18 UTC). While some work has been done to correct barometer readings from ships to compensate for installation position relative to sea level, no systematic study has been undertaken to test or edit these data or analyses. However, in general, sea-level pressure patterns appear to be quite smooth, suggesting that the data are generally reliable. Objective analysis smooths the data and suppresses any noise that might be present.

It is not a simple process to conformally map a given atmospheric pressure distribution into a surface wind field. While the computation of the geostrophic wind velocity from the spatial gradients of atmospheric pressure is fairly straightforward, it is more difficult to extrapolate the geostrophic wind field through the sea-surface boundary layer. The primary problem is our imperfect knowledge of the oceanic boundary layer and the manner in which it transfers momentum from the wind to the ocean surface. While most scientists have agreed on the drag coefficient for low wind speeds (<5 m/s), there continues to be some disagreement on the appropriate coefficient for higher wind speeds and wave-current conditions (cf. Kara et al., 2007). Added to this is a lack of understanding of boundary layer dynamics and how planetary vorticity affects this layer. This leads to a lack of agreement on the backing effect and the resulting angle one needs to apply between the geostrophic wind vector and the surface wind vector. Thus, wind stress computations have required the a priori selection of the wind stress formulae for the transformation of geostrophic winds into surface wind stresses. The application of these stress calculations will therefore always depend on the selected wind stress relation and any derived oceanographic inferences are always subject to this limitation.

Anemometers installed on ships, moored buoys, or island stations provide another source of open-ocean wind data. The ship and buoy records are subject to problems arising from measuring the wind around structures and relative to a moving platform, which is itself being affected by the wind. These effects are difficult to estimate and even more difficult to detect once the data have been recorded or transmitted. Many of the earlier ship-wind data in climatological archives are based on wind estimates made by the ship's officers from their evaluation of the local sea state. (The Beaufort Scale was designed for the days of sailing vessels and uses the observed wave field to estimate the wind speed.) Analysis of the ship-reported winds from the Pacific (Wyrtki and Meyers, 1975a, b) has demonstrated that, with some editing and smoothing, these subjective data can yield useful estimates of the distribution of wind over the equatorial Pacific. Barnett (1983) has used objective analysis on these same data to produce an even more filtered set of wind observations for this region. Following a slightly different approach, Busalacchi and O'Brien (1981) reanalyzed the ship wind-data to fill in spatial gaps before applying the wind fields to oceanographic model studies.

There are now a large number of moored meteorological buoys in the world ocean that can be found at http://www.ndbc.noaa.gov/. The buoys typically have dual sensors (in case of instrument failures or damage) that provide hourly measurements of wind speed and direction, atmospheric pressure, air and water temperature, significant wave height and peak wave frequency. (Significant wave height is the average height of the highest 1/3 of all the measured waves over some specific period, typically 20 min) The data are noisy, have numerous spikes, and generally require considerable effort to "clean up". Meteorological buoys and their sensors are also subject to damage or loss during extreme wave conditions. Because such conditions generally occur in winter when it is difficult to service the platforms from ships, there are often gaps in the data series.

Included in other widely used sets of wind data are the synoptic wind fields produced by the Fleet Numerical Meteorological and Ocean Center (FNMOC)—formerly the Fleet Numerical Ocean Center (FNOC)—in Monterey, California. These analyses use not only ship, buoy, and island reports but also winds inferred from satellite-borne scatterometers and the tracking of clouds in sequences of visible and infrared satellite imagery. The latter technique uses the infrared image to estimate the temperature and, therefore, infer the elevation of the cloud mass being followed. By examining sequences of satellite images, specific cloud forms can be followed and the corresponding wind speed and direction computed for the altitude of the cloud temperature. As might be expected, this procedure is dependent not only on the accuracies of the satellite sensors but also on the interpretive skills of the operator. As a consequence, no real quantitative levels of accuracy can be attached to these data. Comparison between the FNMOC winds and coincident winds measured from an open-ocean buoy (Friehe and Pazan, 1978) showed excellent agreement in speed and direction over a period of 60 days. Although this single-point comparison is too limited to establish any uncertainty values for the FNMOC wind fields, the comparison provides some confirmation of the validity of techniques used to derive the FNMOC winds.

A wind product for the Pacific Ocean similar to the FNMOC winds is generated by the National Marine Fisheries Service (NMFS) in Monterey, California (Holl and Mendenhall, 1972; Bakun, 1973). In this product, the geostrophic "gradient" winds are first computed at a $3 \times 3°$ latitude–longitude grid spacing from spatial gradients in the six-hourly synoptic atmospheric pressure fields at the 500 or 800 mb surfaces. To obtain the surface-wind vectors in the frictional atmospheric boundary layer, the magnitudes of the calculated geostrophic wind vectors are reduced by a factor of 0.7 and the wind vectors rotated (backed) by 15°; here, "backed" refers to a counterclockwise motion in the northern hemisphere and a clockwise rotation in the southern hemisphere. (Some of the original work on this method can be traced to Fofonoff, 1960.)

Thomson (1983) compared winds computed by the NMFS with winds measured from moored buoys off the coast of British Columbia during the summers of 1979 and 1980 (Figure 1.69). In this comparison, it was concluded that winds computed from atmospheric pressure provided an accurate representation of the oceanic winds for timescales longer than several days but failed to accurately resolve short-term wind reversals associated with transient weather systems. Computed winds also tended to underestimate percentages of low and high wind speed. Similar results were

FIGURE 1.69 Comparison of observed and calculated oceanic winds for the period May to September 1980 on the west coast of Vancouver Island. Insert shows location of the moored buoys for 1979 and 1980 (triangles) and location (solid square) of grid point (49° N, 127° W) for the geostrophic winds. Observed winds are from anemometers on moored buoys; calculated winds are the six-hourly geostrophic winds provided by the National Marine Fisheries Service (NMFS) in Monterey, California. *(From Thomson (1983).)*

reported by Marsden (1987) for the northeast Pacific (including Ocean Weather Station P) and by Macklin et al. (1993) for the rugged coast of western Alaska. The poor correlation of observed and computed winds at short time-scales is thought to be due to the large (3 × 3°) spacing, the coarse 6-h sampling of the pressure field and the strong influence of orographic effects in mountainous coastal regimes. In Thomson's study, peak-computed winds were roughly 20° to the right of the observed peak inner-shelf winds, suggesting that the computed winds were representative of more offshore conditions or that the 15° correction for frictional

effects was too small. Spectra of observed winds were found to be dominated by motions at much larger wavelengths than were found in the computed values. The NMFS winds were found to contain a significant 24-h sea-breeze component in the inner shelf observed winds but not in the records farther offshore. Based on spectral comparisons it was concluded that the NMFS winds closely represented the actual winds for periods longer than two days (frequencies less than 0.02 cph) and only marginally matched actual winds for periods shorter than two days.

The Comprehensive Ocean-Atmosphere Data Set (COADS) and the international version

(ICOADS) are cooperative efforts between the National Center for Atmospheric Research (NCAR, USA) and the National Oceanic and Atmospheric Administration (NOAA, USA) to generate observed wind fields for the global ocean. COADS consists of global marine surface observations for the period 1784—1997 and monthly summary statistics of these observations for 1800—1997. The observations are primarily from ships (merchant, ocean research, fishing, and naval), moored platforms, and drifting buoys. Other sources of real-time and delayed mode data are included. Because coastal oceanographic studies rely on accurate nearshore wind data, Cherniawsky and Crawford (1996) compared monthly mean wind speeds and directions from buoys off the west coast of Canada to those from COADS for the period 1987—92. Differences between the 2 × 2° COADS and buoy winds were mainly due to inconsistencies in the ship recording methods. The effect of large ocean waves on buoy wind measurements was also a potential source of measurement error for winds greater than 7—10 m/s; above this range, buoy winds may be underestimated. Korăcin and Dorman (2001) used wind data generated by a 9-km grid, University of Washington Mesoscale Model 5 (MM5 regional model) to examine topographic influences on the summer marine boundary layer (MBL) along the California coast in June 1996. The modeled winds, which include orographic effects on the wind fields, and coastal buoy winds were deemed to be sufficiently similar for the authors to be confident in the MM5 results. The wind structure near coastal capes appears to be typically composed of an upstream convergence zone (compression bulge) and a downstream supercritical divergent flow (expansion fan) followed by a "deceleration zone". This flow structure undergoes marked diurnal variability, which affects the local wind divergence field and cloud formation. The authors further concluded that the overall MBL structure (including winds) is governed primarily by topography in the inner coastal zone lying within 100 km of shore. In a more recent study, Tinis et al. (2005) compared MM5 winds observed at coastal meteorological buoys from British Columbia to northern California in order to assess their suitability for use in regional biological ocean modeling (Pitcher et al., 2010). Two three-month study periods from 2003 were chosen: summer (July to September), which is most important for the growth of toxic algae off the Washington State coast, and fall (October to December) when downwelling favorable wind events force the onshore movement of potentially toxin-contaminated shelf water. Wind speeds determined by the MM5 model ranged from 81% to 101% of observed wind speeds. Mean winds were well modeled in summer but were, on average, 35° clockwise in the fall compared to buoy winds. Winds were strongest in the diurnal and 2—5 day bands in both seasons; spectral coherence between the model and observed winds in both of these frequency bands were highest (0.66—0.93) for the coast of Washington and northern Vancouver Island. In some nearshore regions, modeled winds were insufficiently accurate to represent the observed winds.

As the previous discussion indicates, a primary caution when using coastal wind data is that winds often need to be corrected for local orographic effects especially along mountainous coasts (Macklin et al., 1993). This is also true of winds generated by orographically sensitive regional numerical models such as the University of Washington MM5 winds. If the measured wind data are to be considered representative of the coastal ocean region, the wind sensor must be unobstructed along the direction of the wind. If not ideally situated, the measured wind data can still be used if the directional data are weighted to account for the bias due to local wind channeling by the topography. Marsden (1987) found good agreement between measured and calculated winds at the rugged but exposed anemometer site at Cape St. James on the central British Columbia coast, but relatively poor agreement for these winds at the protected coastal station at Tofino Airport 300 km to the south of the Cape.

As new in situ sensing methods evolve, emphasis is being placed on the ability to measure wind over the ocean. Sensors on polar-orbiting satellites are capable of systematically providing measurements over the entire globe. Moreover, sensors operating at microwave frequencies can make measurements of the ocean surface day and night and under nearly all weather conditions. Both active (radar) and passive (radiometer) microwave sensors have been shown capable of retrieving the ocean surface wind speed, with active microwave instruments being used to also retrieve the wind direction for winds stronger than 3 m/s. The U.S. Navy's WindSat mission, a space-based radiometer system, has also been shown capable of determining the wind direction using polarimetric and multilook observations for winds stronger than 7 m/s. The usefulness of microwave scatterometer data was clearly demonstrated during the SEASAT mission (Brown, 1983), which confirmed scatterometer accuracies of 1–2 m/s (speed) and 1–20° (direction). A study of SEASAT data (Thompson et al., 1983) has shown that radar backscatter from the ocean depends on surface wind stress for a wide range of transmitted wavelengths. These authors found that SEASAT SAR data, combined with simultaneous SEASAT scatterometer data, provided a good estimate of the coefficient of wind speed to wind stress. Hence, in the future, it may be possible to measure wind stress directly rather than infer it from wind or pressure measurements. In the past decade, a number of new systems have been deployed that are capable of measuring wind speed over the ocean. The GEOSAT altimeter discussed earlier is able to observe wind speed from the change in the shape of the altimeter waveform. While direction sensing is not possible, the altimeter is able to provide relatively accurate wind speeds

along the satellite subtracks every few kilometers (Witter and Chelton, 1991). This capability has been used by the U.S. Navy to routinely map the global wind field over the ocean. Comparisons of these winds with moored buoys and operational numerical model analyses have demonstrated the relative accuracy of these satellite winds. On the other hand, the presence of significant cloud liquid water presents major challenges for the passive polarimetric technique and thus limits its utility in supporting operational marine weather forecasting and warning. The development and refinement of instrumentation and algorithms for ocean surface wind retrieval is an ongoing process being conducted in both the active and passive remote sensing areas. Satellite-derived ocean surface wind products currently available include wind vector fields derived from QuikSCAT, ASCAT, OSCAT, WindSAT and ERS-2; wind speed fields are derived from Special Sensor Microwave Imager (SSM/I).

The QuikSCAT mission was launched in 1999 and ended on November 23, 2009. Surface wind fields were provided at 12.5 and 25-km resolution over that period. Ocean surface vector wind data from the European Advanced Scatterometer (ASCAT) system at 25 and 50-km resolution provides a partial replacement for QuikSCAT. The Oceansat-2 Scatterometer (OSCAT) now also provides surface ocean winds at 12.5 and 25-km resolution. ASCAT ocean winds are neutral stability winds referenced to 10-m elevation above the sea surface. These products are processed by NOAA/NESDIS utilizing measurements from ASCAT aboard the EUMETSAT METOP satellite. The current geophysical model function (GMF) being used is CMOD5.5, where the GMF relates the normalized radar cross section to the ocean surface wind speed and direction. OSCAT is a Ku-band conically scanning scatterometer system designed and built by the Indian Space Research Organization (ISRO)/Space Applications Center (SAC) and was launched aboard the Oceansat-2 satellite on September 23, 2009. The OSCAT ocean surface wind retrievals represent a 10-m neutral stability wind; OSCAT Level 1B and Level 2 products are provided to NOAA by ISRO on an orbit-by-orbit basis via EUMETSAT. The NOAA OSCAT wind retrievals are processed with the Scatterometer Wind Data Processor developed at the NESDIS/Center for Satellite Applications and Research and utilizing the OSCAT L1B data provided by ISRO.

The current geophysical model function being used is derived from NSCAT-2 and was provided by the Scatterometer Project at the NASA/Jet Propulsion Laboratory, where the GMF relates the normalized radar cross section to the ocean surface wind speed and direction. The wind vector retrievals flagged as potentially being contaminated by rain are colored in black. The current rain flag is underflagging for rain. For closer examination of the wind fields, the global image is further divided into $30 \times 20°$ bins between latitudes 80° N to 80° S and longitudes 180° W to 180° E, forming an HTML link map for the regions of interest. The global wind images display the available data from the previous 22 h up to the image creation time

(identical resolution and delay apply to the ASCAT wind data, the WindSat/Coriolis measurements, and the ERS-2 scatterometer wind data). The WindSAT wind retrievals are at a 10-m height assuming neutral stability and are derived from WindSat microwave brightness temperatures measurements. WindSat, the first spaceborne polarimetric microwave radiometer, was developed by the Naval Research Laboratory under sponsorship of the U.S. Navy and National Polar orbiting Operation Environmental Satellite System (NPOESS). For additional information on WindSat, visit the IPO WindSat Web site.

It is now possible to detect changes in the ambient acoustic noise level due to wind-driven surface effects. The exact mechanisms causing these acoustic noise variations are still being investigated but empirical data clearly suggest a linear relationship to wind-stress fluctuations.

The passive microwave sensor on the Defense Meteorological Satellite Program (DMSP) satellites, called the SSM/I, is able to sense wind speed but not direction (Figure 1.70; color plate). The SSM/I is a seven-channel four-frequency, linearly

Annual Mean 1992

SSMI Wind Speed Referenced to 10 m, m s⁻¹

FIGURE 1.70 Global annual mean of the SSMI (Special Sensor Microwave Imager) surface wind speed for 1991. *(Courtesy of David Halpern (From Halpern et al. (July 1993)). JPL Publication 93—10.)*

polarized microwave radar operating in a sun-synchronous orbit at an altitude of 860 km. Three of the four channels (19.3, 37.0, and 85.5 GHz) are dual-polarized while the 22.2 GHz channel is only vertically polarized, for a total of seven channels. The nearly 1400-km swath of the conically scanned SSM/I produces complete coverage between 87°36' S to 87°36 'N every three days per satellite (Halpern et al., 1993). There are usually at least two SSM/I operating. While the spatial resolution is poor due to the sensing capabilities at the microwave frequencies, algorithms have been developed that appear to produce reliable estimates of wind speed over the open ocean (Wentz et al., 1986; Gooberlet et al., 1990; Halpern et al., 1993). Wind speed accuracies are about ±2 m/s for the range of speeds between 3 and 25 m/s under rain-free conditions. Since the emissivity of land is very different from that of water, the SSM/I cannot be used to estimate wind speed within 100 km of land. Similarly, surface wind speed within 200 km of the ice edge cannot be computed from SSM/I data. However, wind speeds computed from the SSM/I compare reasonably well with open-ocean winds (Emery et al., 1994). Waliser and Gautier (1993) find that in the central and eastern equatorial Pacific, SSM/I wind-speed comparisons were well within the accuracies specified for the SSM/I. Biases (buoy-SSM/I) were generally less than 1 m/s and RMS differences were less than 2 m/s. However, in the western equatorial Pacific, biases were generally greater than 1−3 m/s and RMS differences closer to 2−3 m/s. According to Waliser and Gautier, "… there are still some difficulties to overcome in understanding the influences that local synoptic conditions (e.g., clouds/rainfall), and even background atmospheric and oceanic climatology effects, have on the retrieval of ocean-surface wind speeds from spaceborne sensors."

The most comprehensive spaceborne measurement of the wind field is made using a microwave scatterometer, which measures the radar scattering cross section of the sea surface at different incidence and azimuthal angles. The SEASAT scatterometer demonstrated the applicability of this instrument for the measurement of open-ocean wind speed and direction. Using a combination of fan-beam antennas, the scatterometer is able to compute both the wind speed and direction. As with many other satellite-borne systems, the scatterometer uses the Doppler shift of the received signal to compute the speed component while multiple fan-beam antennas (called sticks) are required to unambiguously resolve the wind direction. Since scattering cross section at radar frequencies is mostly related to the small wavelets that form when the wind acts on the sea surface, the scatterometer signal is actually related to the wind stress rather than to the wind speed. Unlike anemometers and other like instruments, no additional conversion from wind speed to wind stress is needed. The problem is that all historical calibration information is based on wind speed and direction, rather than wind stress. As a consequence, all present algorithms still convert the scatterometer measurements to wind speed and direction. Studies of SEASAT scatterometer data (Pierson, 1981; Guymer et al., 1981) have demonstrated the ability of the satellite scatterometer to reliably measure wind speed and direction relative to ship and buoy observations. Scatterometers flew on the European ERS-1 and ERS-2 satellites and on the NASA-Japanese ADEOS mission.

The ERS-2 scatterometer is used to generate ocean surface winds at 10-m height from satellite passes. The empirical model currently used by the ESA is referred to as CMOD4, which relates normalized radar cross section with wind speed and direction. Images may contain data up to 22 h previous to update time. SSM/I radiometer(s) data include: SSM/I Ocean Surface Wind Speeds, SSM/I Water Vapor, and SSM/I Rain Rate. The SSM/I ocean surface winds are for a height of 19.5 m as calculated from the ascending and descending satellite passes of SSM/I. SSM/I brightness temperatures are used by the U.S. Navy at the Fleet Numerical

Meteorology and Oceanography Center to calculate the wind data. Here, wind speed is calculated as a function of radiometric brightness temperature at specific frequencies and polarizations (Goodberlet et al. 1989; Andersen et al., 2006). All data for which TB37v-TB37hr < 50 or TB19h > 165 are rain flagged. The strictest rain flag is being used to yield wind speeds within 2 m/s.

Trenworth and Olson (1988) consider the surface wind field computed by the ECMWF to be the best winds for general operational global analyses. ECMWF forecast analyses of surface wind components at 10-m height are issued twice a day at 00 and 12 UTC. Numerical modelers examining large-scale circulation in the Pacific Ocean typically make use of the monthly mean and annual wind stress climatology provided by the Hellerman and Rosenstein (1983) wind fields. These data have problems near the equator where they tend to underestimate wind strength.

1.9.1 Reanalysis Meteorological Data

Many oceanographers now make use of so-called "reanalysis products" including the National Centers for Environmental Prediction-National Center for Atmospheric Research (NCEP-NCAR, USA) reanalysis and North American Regional Reanalysis (NARR) winds and other meteorological parameters, the European Reanalysis (ERA-interim) meteorological data sets, and the Japanese Reanalysis (JRA-25) meteorological data. NCEP-NCAR reanalysis are available through NOAA (http://www.cdc.noaa.gov/cdc/reanalysis/; Kistler et al., 2001; Kalnay et al., 1996; Reynolds et al., 2002). There are several versions of the reanalysis data series. Daily averages are found in the Reanalysis-1 data set, which contains time series of meteorological data (including surface fluxes) at 6-hr and $1.8 \times 2.0°$ resolutions. SST data are weekly. Gridded reanalysis fields are classified into three categories which

define the relative contributions from the observations and model used to derive the variable. Wind velocity and SST are classified as type-A variables since they rely most heavily on the data and are considered the most reliable. Type-B variables, such as surface air temperature and relative humidity, are influenced by both observation and model, and are less reliable. The least reliable type-C data, which includes surface heat fluxes, are derived solely from the model. The main data sets used in all mixed layer depth models (e.g. Turner and Kraus, 1967; Gaspar, 1988; Thomson and Fine, 2009) are the short and long-wave radiation fluxes, the latent heat flux, and the sensible heat flux. Bulk models also require time series of precipitation and wind speed.

Reanalysis data have been subjected to several quality reviews. For example the study by Kalnay et al. (1996) shows that monthly and annual mean heat fluxes from reanalysis agree favorably on a global scale with those derived from observational data sets; the study by Ladd and Bond (2002) indicates that the reanalysis shortwave radiation flux for the vicinity of Station "P" in the central northeast Pacific has a positive bias of roughly 20 W/m^2. Figure 1.71 provides a comparison between hourly winds observed at meteorological buoy C46206 moored on La Perouse Bank off the southwest coast of Vancouver Island and corresponding 3-hourly winds from a nearby NARR site for the period 29 April to 2 September 2012. As in Figure 1.69, the observed and reanalysis winds are in close agreement.

1.10 PRECIPITATION

Precipitation is one of the most difficult and challenging measurements to make over the ocean. Simple rain gauges installed on ships are invariably affected by the pitch and roll of the ship and by salt spray and wind flow over the ship's hull and superstructure. The short

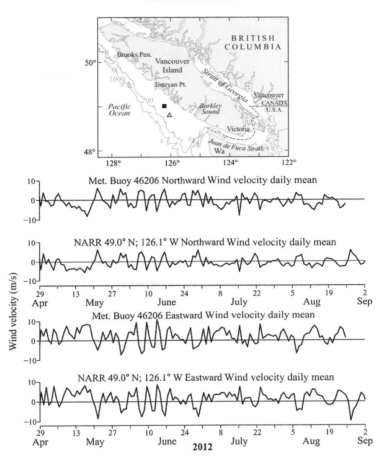

FIGURE 1.71 Comparison of observed and North American Regional Reanalysis (NARR) winds for the period April 29 to September 2, 2012 for the west coast of Vancouver Island, British Columbia. Wind vectors have been defined in terms of their northward (top two panels) and eastward components (bottom two panels). Insert shows location of the moored meteorological buoy C46206 (triangle) and location (solid square) of the grid point 49.0° N, 126.1° W for the reanalysis winds. Observed winds are from dual anemometers on the moored buoy; calculated winds are NCAR/NOAA. *(Figure courtesy, Roy Hourston, Fisheries and Oceans Canada (2013).)*

space and timescales of precipitation make it difficult to interpret point measurements. Rain gauges have two conflicting requirements that make use on shipboard difficult. First the gauge needs to be installed away from the ship influences, such as salt spray, which calls for positioning as high as possible on a mast. However, this conflicts directly with the second requirement, which calls for the regular maintenance of the gauge by ship's personnel. Few systematic studies have been made of precipitation measurements taken from ships, and little effort is made today to instrument ships to routinely observe rainfall over the ocean. A 25-year time series from Ocean Station P (Figure 1.72) in the northeast Pacific is one of a few in the open ocean (most others were taken at Ocean Weather Stations similar to Station P).

FIGURE 1.72 A 25-year time series (1956–81) of precipitation collected from Canadian Weather Ships at Ocean Station PAPA (50° N, 145° W) in the northeast Pacific. Solid line is from use of a Savitsky-Golay smoother (order = 13 months). *(Data courtesy, Sus Tabata.)*

Specialized rain gauges have been developed for use at sea, but they are easily damaged or stolen when mounted on buoys on the ocean surface.

One new technique is to infer rainfall from variations in the upper-ocean acoustic noise. While it may seem a bit confusing to interpret ocean upper-layer acoustic noise both in terms of rainfall and wind, the frequency signatures of the two noise-generating mechanisms are sufficiently different to be distinguishable. The unique characteristics of the sounds produced by different kinds of rainfall make it possible to use Acoustic Rain Gauges (ARGs) to identify and measure raindrop size, fall rate, and other properties of rainfall over the ocean. This area of the ocean surface contributing to the sound recorded by the ARG increases as the gauge is placed deeper in the ocean. Because rainfall can vary markedly over short distances, this spatially integrated rain measurement is considered preferable to a point measurement. Heavy rain can increase noise levels by up to 35 dB over frequencies ranging from roughly 1 kHz to greater than 50 kHz. Extreme rain events

produce loud signals that can reach as much as 50 dB above the background noise. Individual raindrops create underwater sound in two ways. The first sound is made by the impact of the raindrop hitting the ocean surface. Following the initial impact, sound can radiate from air bubbles trapped under water during the splash. These bubbles generally produce the louder sound. Raindrops of different sizes produce different sound intensity and frequency. Small raindrops (0.8–1.2 mm diameter) are surprisingly loud because they generate bubbles with every splash. Sound from these drops has frequencies between 13 and 25 kHz. Medium raindrops (1.2–2.0 mm) do not generate bubbles and are therefore remarkably quiet. Large (2.0–3.5 mm) and very large (>3.5 mm) raindrops trap larger bubbles, which can produce frequencies as low as 1 kHz.

Satellite techniques that relate the area of cold clouds to surface rainfall have had some success (Joyce and Arkin, 1997). However, infrared (IR) techniques underestimate warm rain, and lead to frequent false readings for certain anvil and thick cirrus clouds with cold IR brightness temperatures. A commonly used rainfall IR product is the Global Precipitation Index (GPI); an algorithm was by Joyce and Arkin (1997). The GPI overestimates rainfall from the large areas of cold clouds that form over the West Pacific Ocean (Liu et al., 2007) and underestimates certain other kinds of rainfall, but appears to do quite well in other selected regimes.

The 1987-launch of the SSM/I on one of the DMSP satellites provided a new opportunity to infer precipitation from microwave satellite measurements. While a precipitation algorithm was developed prior to the launch (Hollinger, 1989) later studies have improved upon this algorithm to formulate better retrievals of precipitation over both land and ocean. In the list of "environmental products" for the SSM/I the "precipitation over water" field shows a 25-km resolution, a range of 0–80 mm/h, an absolute accuracy of ±5 mm/h for quantization levels of 0, 5, 10,15, 20, and $\geq$25 mm/h. This algorithm utilized both the 85.5 and 37 GHz SSM/I channels, thus limiting the spatial resolution to the 25-km spot sizes of the 37 GHz channel.

A study by Spencer et al. (1989) employed only the two different polarizations of the 85.5-GHz channel, thus allowing the resolution to improve to the 12.5 km per spot size of this channel. This algorithm was compared with 15-min rain gauge data from a squall system in the southeast United States (Spencer et al., 1989). The 0.01″ (0.039 mm) rain gauge data were found to correlate well ($r^2 = 0.7$) with the SSM/I polarization corrected 85.5 GHz brightness temperatures. This correlation is surprisingly high considering the difference in the sampling characteristics of the SSM/I versus the rain gauge data. Portions of a rain system adjacent to the squall line were found to have little or no scattering signature in either the 85.5 or the 37 GHz SSM/I data due likely to the lack of an ice phase presence in the target area. This appears to be a limitation of the passive microwave methods to discern warm rain over land. Microwave sensors and precipitation radar have been used increasingly in recent years, with the potential to improve precipitation estimates from the surface and from space. Unlike IR, these techniques directly sense precipitation particles rather than cloud tops. However, significant difficulties remain. Rainfall retrieved from the precipitation radar suffers from uncertain attenuation correction, problems over complex terrain, and the limit of minimum detectable signal (Iguchi et al., 2000). Microwave retrievals over the ocean are thought to rival radar retrievals for accuracy, but retrievals over land are compromised because of variations of the surface emissivity (Spencer et al., 1989; Kummerow et al., 2001). The Tropical Rainfall Measuring Mission (TRMM) has led to a significant advancement in the quantification of moderate to intense rainfall. Despite this success, current rain measuring sensors lack sufficient

sensitivity and retrieval difficulties to detect and estimate light rainfall, especially over subtropical and high latitude oceans. Among various spaceborne sensors, CloudSat provides superior retrieval of light rainfall and drizzle. By complementing rain estimates from Cloud-Sat and precipitation radar aboard TRMM, it has been determined that the quasi-global (60° S to 60° N latitude) mean oceanic rain rate is about 3.05 mm/day, considerably larger than that obtained from any individual sensor product. Together with careful consideration of scaling issues, rainfall estimates from TRMM PR, TRMM TMI, AMSR-E, MHS, IR, and Cloud-Sat CPR sensors have been analyzed. Results show that the highest agreement among sensors in measuring the frequency and amount of rain occurs in the zone between 20° S and 20° N. However, toward higher latitudes and within the subtropical high-pressure regions, a majority of the sensors miss a significant fraction of the rainfall. This underestimation can exceed 50% of the total rain volume. The dual-frequency precipitation radar (DPR) with Ka/Ku-bands and a multichannel passive microwave radiometer on Global precipitation measurement "core" satellite extends the TRMM capability to measure light rain over 65° S to 65° N latitude, creating an unprecedented opportunity to improve global quantification and properties of precipitation.

1.11 CHEMICAL TRACERS

Oceanographers use a variety of chemical substances to track diffusive and advective processes in the ocean. These chemical tracers can be divided into two primary categories: *conservative* tracers such as salt and helium whose concentrations are affected only by mixing and diffusion processes in the marine environment; and *nonconservative* tracers such as dissolved oxygen, silicate, iron, and manganese whose concentrations are modified by chemical and biological processes, as well as by mixing and diffusion. The *conventional* tracers, temperature, salinity, dissolved oxygen and nutrients (nitrate, phosphate and silicate), have been used since the days of Wüst (1935) and Defant (1936) to study ocean circulation. More recently, *radioactive* tracers such as radiocarbon (^{14}C) and tritium (^3H) also are being used to study oceanic motions and water mass distribution. The observed concentrations of those substances, which enter from the atmosphere, must first be corrected for natural radioactive decay and estimates made of these substance's atmospheric distributions prior to their entering the ocean. If these radioactive materials decay to a stable daughter isotope, the ratio of the radioactive element to the stable product can be used to determine the time that the tracer was last exposed to the atmosphere. *Transient tracers*, which we will consider separately, are chemicals added to the ocean by anthropogenic sources in a short time span over a limited spatial region. Most transient tracers presently in use are radioactive. What is important to the physical oceanographer is that chemical substances that enter the ocean from the atmosphere or through the seafloor provide valuable information on a wide spectrum of oceanographic processes ranging from the ventilation of the bottom water masses, to the rate of isopycnal and diapycnal (cross-isopycnal) mixing and diffusion, to the downstream evolution of effluent plumes emanating from hydrothermal vent sites.

Until recently, many of these parameters required the collection and post-cast analysis of water bottle samples using some which are then subsampled and analyzed by various types of chemical procedures. There are excellent reference books presently available that describe in detail these methods and their associated problems (e.g., Grasshoff et al., 1983; Parsons et al., 1984). The book by Grasshof et al. (1983) also contains an excellent section on water samples and their application to chemical analyses. There are important concerns for the reliability of the chemical measurements regarding contamination of the sampling bottle or the subsampling procedure. Also, the volumes required for different chemical analyses vary greatly. A list of sample volumes for chemical observations as part of the WOCE can be found in Volume I of the WOCE Implementation Plan (WOCE, 1988). It is certain that the collection of these volumes will include both presently available "off-the-shelf samplers, sampling systems newly developed by private companies and sampling units designed and built by scientists." In any case, the precision and accuracy of these measurements depends, in part, on the sampling technique used.

Modern chemical "sniffers" (or chemical pumps) are being developed that allow for in situ analysis of samples (Lupton et al., 1993). The requirement for in situ chemical sampling of hydrothermal vents lead to the development of the submersible chemical analyzer (SCANNER) for analyses of Mn and Fe, the submersible system used to assess vented emissions (SUAVE) for Mn, Fe, Si, H_2S, and one of PO_4 or Cl, and the zero angle photon spectrometer (ZAPS) for detecting dissolved Mn to ambient seawater concentrations (≤ 1 nmol/1) (Lilley et al., 1995). The SCANNER and SUAVE systems comprise on-line colorimetric chemical detectors while ZAPS is a fiber-optic spectrometer, which combines solid-state chemistry with photomultiplier tube detection to make flow through in situ chemical measurements. Recent publications on the use of modern chemical sensors at hydrothermal vents can be found in the American Geophysical Union monograph "Mid-Ocean Ridges: Hydrothermal interactions between the lithosphere and oceans" (German et al., 2004) and the RIDGE 2000, Special Issue of *Oceanography* (Fornari et al., 2012).

For many chemical measurements, no single set of procedures applies so that groups, or individual scientists, must be responsible for their own data quality. It is impossible to evaluate after-the-fact the influences of sampling technique, sample history (storage, etc.) and analysis technique. It is therefore more difficult to attach levels of accuracy to these diverse methods. In this text, we will make some general comments regarding potential problems for each of the important parameters. For a more extensive discussion of chemical tracers, the reader is referred to Broecker and Peng (1982) and Charnock et al. (1988).

1.11.1 Conventional Tracers

1.11.1.1 Temperature and Salinity

If it were not for large-scale geographical differences in heat and buoyancy fluxes through the ocean surface from the overlying atmosphere, ocean temperatures and salinity would be nearly homogeneous, disrupted only by input from geothermal heating through the seafloor (Warren, 1970; Jenkins et al., 1978; Reid, 1982). In fact, below 1500-m depth the salinity range throughout the world ocean is only about 0.5 psu despite the regular deep-water formation at high latitudes (Warren, 1983). Temperature, salinity, and density distributions enable us to identify different water masses and track the movement of these water masses in the world oceans.

Atlases of temperature and salinity for the Atlantic Ocean, were produced by Wüst (1935) and Defant (1936) using data from the 1925–27 *Meteor Expedition*. These maps help define the depths of vertical mixing and upwelling in the upper ocean and reveal the extent of ventilation of deep and intermediate waters by sinking of cold, high salinity, high-density water from the Southern Ocean and the Labrador Sea.

Updated atlases for the Atlantic were presented in Fuglister (1960) and Worthington (1976). Similar maps for the Pacific Ocean were produced by Reid (1965) and Barkley (1968). Reid's atlas included distributions of dissolved oxygen and phosphate/phosphorous. An atlas of water properties for the North Pacific, was presented by Dodimead et al. (1963) and Favorite et al. (1976). Wyrtki (1971) provided conventional tracer data for the Indian Ocean obtained

from the International Indian Ocean Expedition. An atlas of the Bering Sea, is provided by Sayles et al. (1979). A summary of the global water mass distribution can be found in Emery and Meincke (1985). Surveys conducted during the World Ocean Circulation Experiment (1991–97) provide updated maps of conventional tracer distributions in the global ocean. As discussed in section 1.8.6.1, the international Argo program is presently yielding unprecedented volumes of high-resolution temperature and salinity data within the upper 2000 m of the World Ocean. Experimentalists and numerical modelers alike have enthusiastically endorsed these global datasets.

1.11.1.2 Dissolved Oxygen

Along with temperature and salinity, dissolved oxygen concentration is considered one of the primary scalar properties needed to characterize the physical attributes of marine and freshwater environments. Although it is not usually a conservative quantity, dissolved oxygen serves as a valuable tracer for mixing and ventilation throughout the water column and is a key index of water quality in regions of strong biological oxygen demand. This demand may arise from animal respiration, bacteria-driven decay, or nonorganic chemical processes (the discharge of pulp-mill effluent into the marine environment places a heavy burden on oxygen levels). Dissolved oxygen is widely used by physical oceanographers to delineate water-mass distributions, to estimate the timing and intensity of coastal upwelling processes and to establish the occurrence of deep-water renewal events in coastal fjords. In a study of the North Pacific, Reid and Mantyla (1978) found that dissolved oxygen gives the clearest signal of the subarctic cyclonic gyre in the deep ocean.

The apparent oxygen utilization (AOU) is the difference between the possible saturated oxygen content at a given pressure and temperature, and the actually observed oxygen content (Figure 1.73). Below the euphotic zone, this

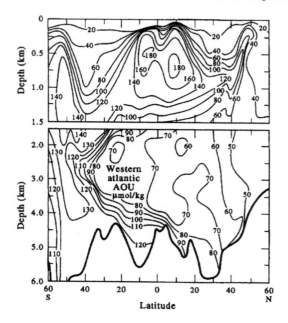

FIGURE 1.73 Vertical section of Apparent Oxygen Utilization (AOU) in mol/kg for the western basin of the Atlantic Ocean. (Figure 3.9 from GEOSECS program, Broecker and Peng, 1982.) The section is broken at 1500-m depth.

parameter provides an approximate measure of biological demand due to respiration and decay. It also is commonly used to trace water-mass movement and to determine the "age" (defined as the time away from exposure to the surface source) of oceanic water masses. Use of AOU suggests that the intermediate waters of the northeast Pacific have an age of several thousand years and are among the oldest (last to be ventilated) waters of the world ocean. Mantyla and Reid (1983) arrived at similar conclusions based on global distributions of potential temperature, salinity, oxygen and silicate. A more complete discussion of this parameter can found in Chapter 3 of Broecker and Peng (1982).

The "core-layer" method introduced by Wüst (1935) identified water masses, and their boundaries, on the basis of maxima or minima in temperature, salinity and dissolved oxygen content. In the ocean, dissolved oxygen levels are high near the surface where they contact the atmosphere but rapidly diminish to a minimum near 500−1000 m depth due to the decay of upper-ocean detritus. Oxygen values again increase with depth toward the bottom. Wyrtki (1962) discusses the relationship between the observed subsurface oxygen minimum in the North Pacific and the general circulation of the ocean, suggesting that it is to be a balance between upward advection, downward diffusion and in situ biological/chemical consumption. Miyake and Saruhashi (1967) argued that the effect of horizontal advection has a much greater effect on dissolved oxygen distributions than horizontal diffusion and biological consumption. In certain deep regions of the ocean, such as the Weddell Enderby Basin off Antarctica, the consumption of oxygen is below the detectable limit of the data so that oxygen may serve as a conservative chemical tracer (Edmond et al., 1979). Within coastal regions, narrow fjords often have one or more shallow cross-channel sills that greatly limit the exchange of offshore oceanic water with the bottom water in deep adjoining basins. The suboxic to anoxic conditions found in these inner basins of inlets are indicative of weak vertical mixing (stagnant bottom water) and infrequent dense water intrusions. In many of these deep anoxic basins, high levels of dissolved hydrogen sulfide (H_2S) can develop. Examples include Koljöfjord and Byfjord in western Sweden (Hansson et al., 2013) and Effingham Inlet on the west coast of Canada (Dallimore et al., 2005; Patterson et al., 2007, 2013).

When water bottle sampling was the only method for oceanographic profiling, the measurement of dissolved oxygen was only slightly more cumbersome and time consuming than the measurement of temperature and salinity. The advent of the modern CTD with its rapid temperature and conductivity responses left oxygen sampling behind. Thus, despite the importance of dissolved oxygen distributions to our understanding of chemical processes and biological consumption in the ocean, dissolved oxygen is far less widely observed than temperature or salinity. At present, there are three principal methods for measurement of dissolved oxygen: (1) water bottle sampling followed by chemical "pickling" and endpoint titration using the Winkler method (Strickland and Parsons, 1968; Hichman, 1978); (2) electronic sampling using a membrane covered polarographic "Clark" cell (Langdon, 1984); and (3) electronic sampling using luminescence-quenching technology (Thomson et al., 1988; Tengberg et al., 2006). The primary problems with standard water-bottle sampling of dissolved oxygen are the potential for sample contamination by the ambient air when the subsampling is carried out on deck, poor sampling procedure (such as inadequate rinsing of the sample bottles), and the oxidization effects caused by sunlight on the sample. Thus, laboratory procedures call for the immediate fixing of the solution after it is drawn from the water bottle by the addition of manganese chloride and alkaline iodide. During the pickling stage of the Winkler method, the dissolved oxygen in the sample oxidizes Mn(II) to Mn(III) in alkaline solution to form a precipitate MnO_2. This is followed by oxidation of added I^- by the Mn(III) in acidic solution. The resultant I_2 is titrated with thiosulfate solution using starch as an endpoint indicator. After the sample is chemically "fixed", the precipitate that forms can be allowed to settle for 10–20 min. At this stage, samples may be stored in a dark environment for up to 12 h before they need to be titrated. Parsons et al. (1984) give the precision of their recommended spectrophotometric method as $\pm 0.064/N$ (mg/1), where N is the number of replicate subsamples processed. The Winkler method is accurate to 1% provided the chemical analysis methods are rigorously applied. Another measure is the percentage saturation, which is the ratio of dissolved oxygen in the water to the amount of oxygen the water could hold at that temperature, salinity and pressure. Saturation curves closely follow those for dissolved oxygen.

In situ electronic dissolved oxygen sensors have been developed for use with profiling systems such as CTDs or as ancillary sensors attached to single-point current meters. All existing sensors use a version of the Clark cell, which operates on the basis of electro-reduction of molecular oxygen at a cathode, or are based on the ability of certain substances to quench fluorescence from a blue light source. The fluorescent substance is embedded in a gas permeable material that is exposed to the surrounding water. When used in a polarographic mode, the electric current supplied by the cathode in a Clark cell is proportional to the oxygen concentration in the surrounding fluid. To lessen the sensitivity of the device to turbulent fluctuations in the fluid, the electrode is covered with an electrolyte and membrane. Oxygen must diffuse down gradient through the membrane into the electrolyte before it can be reduced at the surface of the cathode. There are a number of drawbacks with the present systems. First of all, the diffusion of oxygen through the boundary layer near the surface of the probe is slow, limiting the response time of the cell to several minutes. Also, the electrochemical reaction within the cell consumes oxygen and stirring may be required to maintain the correct external oxygen concentration. Changes in the structure of the cell—due to alterations in the diffusion characteristics of the membrane as a result of temperature, mechanical stress and biofouling and to deterioration of the electrolyte and surfaces—require that the cell be recalibrated every several hours. The need for frequent re-calibration limits the use of the polarographic technique for profiling and mooring applications. Langdon (1984) uses a pulse technique to reduce the calibration drift. This improves long-term stability but time constants are still the order of minutes.

The Yellow Springs Instruments and Beckman polaragraphic dissolved oxygen sensors (Brown and Morrison, 1978) sense the oxygen content by the current in an electrode membrane combined with a thermistor for membrane temperature correction. The current through this membrane depends on the dissolved oxygen in the water and the temperature of the membrane. Samples of both membrane current and temperature are averaged every 1.024 s giving a resolution of 0.5 μA (microamps) with an accuracy of ±2 μA over a range of 0–25 μA. These in situ sensors have yet to be critically evaluated with reference to well-tested and approved methods. There are concerns with changes in the membrane over the period of operations and problems with calibration. Nevertheless, as measurement technology improves, an in situ oxygen sensor will be a high priority in that it saves considerable processing time and avoids errors possible with shipboard processing.

The Aanderaa Data Instruments (AADI) Oxygen Optode employs fluorescence quenching to obtain rapid and stable measurements of dissolved oxygen in a wide range of marine conditions (Tengberg et al., 2006). Optodes are now incorporated in profiling and moored CTD platforms and in single-point current meters. Although the use of fluorescence quenching for oxygen determination has been known since the 1930s (Kautsky, 1939) and widely used for in vivo measurement of the partial pressure of oxygen in blood (Peterson et al., 1984), the first application in oceanography was not reported until 1988 (Thomson et al., 1988). As with the modern Optode sensor, this prototype fluorescence-based dissolved oxygen sensor operated on the principle that the fluorescence intensity of an externally light-excited fluorophore is attenuated or "quenched" in direct relation to the concentration of dissolved oxygen in an ambient fluid (Figure 1.74(a)). Optimum results were obtained using a high-intensity blue-light source (wavelength of 450–500 nm) since

FIGURE 1.74 (a) Schematic of the first solid-state oceanic dissolved oxygen sensor. System uses blue light from (1) to excite a fluorophore in the sensor tip (9). The concentration of dissolved oxygen in the ambient fluid sensed by (6) is proportional to the degree of quenching of blue light fluoresced by the chemical-doped sensor; (b) Simultaneous profiles of oxygen in Saanich Inlet. YSI = YSI dissolved oxygen sensor. *(From Thomson et al. (1988).)*

this is the wavelength that most readily excites the known fluorophores. Results from a six-day time-series record of dissolved oxygen concentration from a moored instrument in Saanich Inlet in 1987 suggested that the technique can be used to build a rapid (<1 s) response profiling sensor (Figure 1.74(b)) with long-term stability and high (<0.1 ml/l) sensitivity. The fact that the oxygen spectra closely resembled the temperature spectra for the entire frequency band up to a period of 2 h suggested that the oxygen data were at least as stable as the thermistor on the Aanderaa RCM4 current meter that was used in the moored study. Since no blue-light source was available at the time, the prototype device relied on a high-power white-light source and a car battery to drive the system. Modern luminescence quenching sensors employ the needed blue-light source and a chemically-stable probe capable of withstanding the rigors of shipboard operations and high hydrostatic pressures. The lack of a commercial blue-light source with sufficient power (≈ 1 mW) to produce a strong fluorescence response is no longer an impediment to future improvements in this technology.

The AADI Optode for measuring dissolved oxygen became commercially available in 2002. According to the company's Web site, instruments have been used on autonomous Argo floats (e.g., Joos et al., 2003; Johnson et al., 2010) and gliders (Nicholoson et al., 2008), long-term monitoring in coastal environments with high bio-fouling (Martini et al., 2007), on coastal buoys (Jannasch et al., 2008), on Ferry box systems (Hydes et al., 2009), on profiling CTD instruments down to 6000 m, and in chemical sensor networks (Johnson et al., 2007). The manufacturer specifies an operating range of 0–500 µM (0–16 ml/l), a resolution of 1 µM, an accuracy of better than ± 8 µM (or 5% of the value, whichever is greater), and a 63% response time of 25 s (a response time of 8 s is also listed but it is not clear if this is the capability of the technology or the sensor).

1.11.1.3 Nutrients

Nutrients such as nitrate, nitrite, phosphate, and silicate are among the "old guard" of oceanic properties obtained on standard oceanographic cruises. One need only examine the early technical reports published by oceanographic institutions to appreciate the considerable effort that went into collection of these data on a routine basis. Oceanographers are again beginning to use these data on a routine basis to understand the distribution and evolution of water masses. However, there are a number of problems with nutrient collection that need to be heeded. To begin with, the data must be collected in duplicate (preferably triplicate) in small 10-mm vials and frozen immediately after the samples are drawn using a "quick freeze" device or alcohol bath. This is to prevent chemical and biological transformations of the sample while it is waiting to be processed. Careful rinsing of the nutrient vials is required as the samples are being drawn. Silicate must be collected using plastic rather than glass vials to prevent contamination by the glass silicate. Plastic caps must not be placed on too tightly and some space must be left in the vials for expansion of the fluid during freezing. Nutrient sample analysis is labor-intensive, time-consuming work. Although storage time can be extended to several weeks, we strongly recommend that nutrients be processed as soon as possible after collection, preferably on board the research ship using an autoanalyzer. With individual parameter techniques this is less likely to be possible than with more recent automated methods, which have been developed to handle most nutrients (Grasshof et al., 1983). These automated systems, which use colorimetric detection for the final measurement, need to be carefully standardized and maintained. Under these conditions, they are capable of providing high quality nutrient measurements on a rapid throughput basis.

Profiles of nutrients and dissolved oxygen for the North Pacific are presented in Figure 1.75. As first reported by Redfield (1958), the concentrations of nitrate, phosphate, and oxygen are closely linked except near source or sink regions of the water column. A weaker relationship exists between these variables and silicate. Nitrite

(a)

FIGURE 1.75 Plots of nitrate (N), phosphate (P) and dissolved oxygen (O) for the North Pacific. (a) Station 11 at 24°48.3′ N; 154°37.8′ W; (b) Station 50 at 38°30.3′ N; 152°00.3′ W; (c) Station 86 at 53°41.1′ N; 151°58.9′ W. (*Data from Martin et al. (1987).*)

only occurs in significant amounts near the sea surface where it is associated with phytoplankton activity in the photic zone and in the

detritus layer just below the seasonal depth of the mixed layer. Although the linear relationships between these parameters vary from region to region, the reason for the strong correlations is readily explained. Within the photic zone, phytoplankton fix nitrogen, carbon and other materials using sunlight as an energy source and chlorophyll as a catalyst. In regions of high phytoplankton activity such as midlatitudes in summer, the upper layers of the ocean are supersaturated in oxygen and depleted in nutrients. That is, there are sources and sinks for oxygen and nutrients. However, below the photic zone, bacterial decay and dissolution of detritus raining downward from the upper ocean leads to chemical transformations of oxidized products. This, in turn, leads to a reduction of oxygen compounds and corresponding one-to-one release of nitrate, phosphate and silicate. This linear relation would prevail throughout the ocean below the photic zone if were not for other sources and sinks for these chemicals. For example, we now know that silicate enters the ocean through resuspension of bottom sediments and from hydrothermal fluids vents from mid-ocean ridge systems (Talley and Joyce, 1992). Chemosynthetic production by bacteria in hydrothermal plumes is also a source/sink region as the analog to photosynthetic processes in the upper ocean.

It is generally thought that limitations in upper ocean nutrients, especially nitrates, combined with zooplankton predation (grazing) and turbulent mixing processes control primary (phytoplankton) productivity in the ocean. It is generally accepted that other nutrients such as the aeolian supply of iron compounds might ultimately control productivity in areas such as the equatorial and subarctic Pacific and the Southern Ocean where nitrate concentrations are high year-round but spring and fall blooms do not occur (Chisholm and Morel, 1991). These high nutrient, low chlorophyll (HNLP) regions have become the focus of increasing numbers of multi-disciplinary studies (Hamme et al., 2010; Parsons, 2012). From a climate change perspective, it is important to understand how to increase productivity in the upper ocean in order to increase sequestering of atmospheric CO_2 into the deep ocean, which holds about 50 times more CO_2 that, the atmosphere. This higher CO_2 concentration in the deep ocean is due to the "biological pump" whereby phytoplankton use photosynthesis to fix CO_2 at the ocean surface into organic carbon. When the plankton die, they sink into the deep ocean, where the organic carbon is consumed by other organisms and respired back to CO_2 to form dissolved carbonate.

In a classic paper, Redfield (1958) suggested that organisms both respond to and modify their external environments. His premise was that the nitrate of the ocean and the oxygen of the atmosphere are determined by the biochemical cycle and not conditions imposed on the organisms through factors beyond their control. Support for his thesis was derived from the fact that the well-defined nitrogen, phosphorous, carbon, and oxygen compositions of plankton in the upper ocean were almost identical to the concentrations of these elements regenerated from chemical processes in the 95% of the ocean that lies below the autotrophic zone. As pointed out by Redfield, the synthesis of organic material by phytoplankton leads to oceanic changes in concentration of phosphorous, nitrogen, and carbon in the ratio 1:16:106. During heterotrophic oxidation and remineralization of this biogenic material (i.e., decomposition of these organisms), the observed ratios are 1:15:105. Thus, for every phosphorous atom that is used by phytoplankton during photosynthesis in the euphotic zone, exactly 16 nitrogen atoms and 106 carbon atoms are used up. Alternatively, for every phosphorous atom that is liberated during decomposition in the deep ocean, exactly 15 nitrogen and 105 carbon atoms are liberated. The oxidation of these atoms during photosynthesis requires about 276 oxygen atoms while during decomposition 235 oxygen atoms are withdrawn from the water column for each atom of phosphorous that is added. If this process were simply one way, the primary nutrients would soon be completely depleted from the upper ocean. That is why life-supporting replenishment of depleted nutrients to the upper ocean through upwelling and vertical diffusion of deeper nutrient rich waters is such an important process to the planet. Brulandei et al. (1991) give a modern version of the Redfield ratios based on phytoplankton collected under bloom conditions as: C:N:P:Fe:Zn:Cu,Mn,Ni,Cd = 106:16:1:0.005:0.002:0.0004 (see also Martin and Knauer, 1973).

According to the above ratios, the formation of organic matter by phytoplankton in the surface autotrophic zone leads to the withdrawal of carbonate, nitrate, and phosphate from the water column. Oxygen is released as part of photosynthesis and the upper few meters of the ocean can be supersaturated in oxygen at highly productive times of the year. When the plants die and sink into the deeper ocean, decomposition by oxidation returns these compounds back to the seawater. Thus, increases in carbonate, nitrate, and phosphate concentrations below the euphotic zone are accompanied by a corresponding decrease in oxygen levels. This process leads to a rapid increase in nitrate and phosphate and a corresponding rapid decrease in oxygen within the upper kilometer or so of the ocean (Figure 1.75). Nitrate and phosphate reach subsurface maximums at intermediate depths and then begin to decrease slowly with depth to the seafloor. Oxygen, on the other hand, falls to a mid-depth minimum (the oxygen minimum layer) before starting to increase slowly with depth toward the seafloor. In the upper zone, the balance of chemicals is altered considerably by biological activity while near the coast the balance is altered by runoff, which provides a different ratio of nutrients. However, below the surface layer, the changes occur in the manner suggested by the Redfield ratios (Redfield et al., 1963). Note that the concentration of silicate is almost like that of the other nutrients, except that it doesn't reach a maximum at mid-depth and becomes more decoupled from the accompanying oxygen curve. This suggests a source function for silicate in the deep ocean. Indeed, there are two sources: resuspension and dissolution of siliceous material from rocks and other inorganic material on the seafloor and the injection of silicates into the ocean from hydrothermal venting along mid-ocean ridges and other magmatic source regions in the deep ocean.

The fact that carbon and oxygen concentrations greatly exceed the levels required by plankton while those of phosphorous and nitrogen were identical to those observed on average in the ocean (carbon is at least 10 times that needed for photosynthesis), prompted Redfield to suggest that phosphate and nitrate are limiting factors to oceanic primary productivity. It is thought that nitrate (NO_3) is the primary limiting factor although phosphorous limitation is still important in certain coastal areas. Airborne iron is also thought to be a limiting nutrient for primary productivity in the open ocean. Evidence for this is based on the year-round absence of phytoplankton blooms in the subarctic Pacific, equatorial Pacific and Southern Ocean despite the high near-surface concentrations of nitrate and phosphate. In these areas, autotrophic processes fail to exploit NO_3 and PO_4. The idea is that iron, or some other mineral, limits growth, which is not the case in areas served by aeolion transport from the land. Unfortunately, it not yet possible to sort out the effects of iron limitations from grazing by herbivorous zooplankton or from physical mixing in the surface layer which prevents stratification from confining the animals to a thin upper layer. The first open-ocean iron fertilization experiment (Ironex 1), conducted in October 1993 within the equatorial Pacific near the Galapagos Islands, showed that iron enrichment with $FeSO_4$ could dramatically increase surface productivity (Coale et al., 1998). Using sulfur hexafluoride to track the 64-km^2 iron-enriched area

containing 443 kg of iron sulfate solution, scientists found that the rate of growth and total mass of phytoplankton doubled over a period of three days. However, the iron soon precipitated out of solution as ultra-fine particles and sank, causing a sharp decrease in productivity levels. Since then, there have been numerous international iron fertilization experiments to examine carbon sequestering in the ocean. A recent experiment (LOHAFEX) was conducted in March 2009 in the southwestern Atlantic by Indian and German scientists onboard the research vessel *Polarstern*. Over a period of two and a half months, scientists fertilized a 300-km^2 cyclonic mesoscale eddy using six tonnes of dissolved iron, and followed the effects of the fertilization on the plankton continuously for 39 days. As in previous studies, the iron stimulated the growth of plankton, which doubled their biomass within the first two weeks by taking up carbon dioxide from the water. However, further plankton growth, and hence further drawdown of CO_2 in the upper ocean was prevented by increased grazing pressure from small crustacean zooplankton (copepods). Algal species, which regularly generate blooms in coastal regions including the Antarctic, were most heavily grazed (Wikipedia, 2013). At present, the question of iron enrichment and ocean productivity remains unresolved (Boyd et al., 2007).

1.11.1.4 Silicate

The oceanic distribution of many elements is determined by their involvement in the biochemical cycle. Nitrate and phosphate are associated with the labile tissue and protoplasm of surface plankton whereas silicate and alkalinity are linked to the refractory hard parts of the organisms such as the silica shells of diatoms. The term "silicate" applies to dissolved reactive silicate (monosilicic acid, $Si(OH)_4$; Iler, 1979) measured from water samples. Since most of the silicate undergoes dissolution in the water column rather than the seafloor (Edmond et al.,

1979), its distribution serves as tracer for water-mass mixing and advection. A north-south plot of silicate concentration in the Atlantic Ocean (from 80° S to 60° N), with the derived movement of principal water masses, is presented in Meckler et al. (2013). The advantage of silica over carbonates or other compounds is that siliceous sediments are found only in well-defined areas associated with surface upwelling and their distribution is not particularly dependent on depth. According to Edmond et al. (1979) the average flux of dissolved silica from the sediments to the deep ocean is about 3 $\mu mol/cm^2/year$, which is sufficient to make it a useful tracer of deep-sea flow. Large fluxes are observed in the Weddell-Enderby Basin off Antarctica and in the northern Indian Ocean. In the Meckler et al. (2013) study, a 550-thousand year opal sediment core formed by the siliceous diatom shells originating from the Northwest Africa upwelling system, is used to show degassing of CO_2 into the atmosphere off Antarctica at the time of major stadial event (relatively short cooling events, such as the Younger Dryas, that interrupt the warming leading out of a major glacial period).

In the extreme northeast Pacific (northeast of 45° N, 160° W), silicate concentration increases with depth to the bottom while in the equatorial Pacific no anomalies are observed despite the presence of opaline deposits. The increased silicate with depth in the northeast Pacific appears to be associated, in part, with westward advection of dissolved siliceous sediments deposited over the continental margin of the wind-induced upwelling domain that extends from British Columbia to Baja California along the west coast of North America. As noted by Edmond et al. (1979), the existence of silica sources at the seafloor makes it impossible to use global correlations with the extensive silicate distribution data to determine the distribution of other variables such as trace metals. Historical data, together with transect data collected along 47° N (Talley et al., 1988), further suggest that both the intermediate silica maximum in the

depth range 2000–2400 m and the near-bottom silica maximum in Cascadia Basin to the east of the Juan de Fuca Ridge (Figure 1.76) may be due, in part, to hydrothermal venting of high silicate waters (Talley and Joyce, 1992). The silica in the hydrothermal plumes emanating from the vents originates as silicates stripped from the crustal rocks by the high-temperature hydrothermal fluids. Other factors include vertical flux divergence of settling silicate particles, dissolution from opaline bottom sediments, and upslope injection from the bottom boundary layer.

Macdonald et al. (1986) point out that improper thawing of frozen silicate samples can result in a significant and variable negative bias in seawater determination of silicate. The problem arises from conversion of reactive silicate to a nonreactive, polymetric form in the frozen sample. This polymerization need not affect accuracy for frozen samples provided that sufficient thawing time is allowed for depolymerization to the reactive form. To control bias, the analyst must adjust the length of time between thawing and analysis. The appropriate "waiting time" varies according to the salinity of the sample, the silicate concentration and the length of time the sample was frozen. Waiting time increases with the time that the samples were frozen and with silicate/salinity ratio. For example, deep silicate samples collected from the northeast Pacific (salinity ≈ 35 psu; silicate ≈ 180 $\mu mol/l$) and stored for one to two months must be thawed for about 8 h before processing. This increases to 24 h for samples stored for more than five to six months. Macdonald et al. (1986) conclude that "If the objectives of sampling can accept a 5% negative bias and a slight loss of precision, then freezing is a simple method for storing a wide range of samples. However, samples should be analyzed as soon as practicable."

1.11.1.5 pH

The concern for increased acidification of the world ocean—arising from reduced

FIGURE 1.76   The meridional distribution of silica (micromoles per liter) in the North Pacific along approximately 152°W (Hawaiian region to Kodiak Island, Alaska). Mid-depth maximum values in excess of 180 μmol/l are emphasized. *(From Talley et al. (1991); Talley and Joyce (1992).)*

availability of carbonate ions (CO_3^{2-}) due to increased anthropogenic input of atmospheric CO_2 emissions (Feely et al., 2012)—has led to the need for accurate and reliable aquatic pH sensors. A primary concern is that changes in pH change the aragonite saturation state, Ω_{arag}, which is of importance for calcifying organisms. In thermodynamic equilibrium, $\Omega_{arag} > 1$ gives rise to the precipitation of aragonite while for $\Omega_{arag} < 1$, seawater is corrosive to calcium carbonate so that, in the absence of

biologically mediated protective mechanisms, dissolution will begin (Fabry et al., 2008; Espinosa, 2012). Marine organisms have been observed to reduce their calcification rates in undersaturated waters (Feely et al., 2004; Hoegh-Guldberg et al., 2007; Doney et al., 2009).

In aqueous solutions, pH is defined in terms of the proton (hydrogen ion, H^+) activity,

$$pH = -\log a_{H^+} \qquad (1.41a)$$

where the activity of a substance is proportional to the concentration

$$a_{H^+} = \gamma_{H^+}\left[H^+\right] \qquad (1.41b)$$

The activity coefficient, γ_{H^+}, varies with both total ionic strength (i.e., salinity) and the concentration of other ions in the fluid (Robie Macdonald, Institute of Ocean Sciences, Sidney, BC, pers. comm., 2012). This "activity" depends on is how many protons there are in the solution, including interactions with all the other ions in the solution. Activity is usually lower than concentration, but not always; a number of ionic interactions enhance activity, particularly for acid-base reactions. Both the old litmus paper test (colored acid-base indicator) and modern glass electrodes respond to "activity" (apparent concentration) rather than to absolute concentration. The historic gravimetric measurements generally reported concentrations, rather than activities. Most modern pH probes measure activity, but activity is defined operationally and is dependent on both the method used and the analyte solution; a glass electrode placed into a glacial freshwater pond measures something quite different from a spectrophotometric measurement, or even a measurement using that same electrode in the ocean (Robie Macdonald, pers. com., Fisheries and Oceans Canada, Institute of Ocean Sciences, Sidney, BC, 2012). There are presently two primary methods for measuring ocean pH: electrochemical and spectrophotometric (see *Guide to Best Practices for Ocean CO$_2$ Measurements* by Dickson et al., 2007, PICES Special Publication 3). Most in situ pH probes use electrochemical methods based on a glass combination electrode. In these probes (Figure 1.77), the liquid junction potential (essentially a "blank" signal resulting from the construction of the electrode and its interaction with the sample) is either minimized or very accurately defined, allowing for extremely precise pH measurements. However, when the electrode is placed in seawater, the high salinity changes the liquid junction potential, destabilizing the electrical signal and resulting in prohibitively long analysis times (in many

FIGURE 1.77 Schematic of a combination glass pH electrode. The liquid junction potential arises across the porous plug separating the exterior reference solution from the sample. *(David Reckhow, University of Massachusetts at Amherst, U.S.A.).*

situations, the electrode never attains a stable reading, although some threshold level of change is accepted as a stable measurement). In addition, glass electrodes respond not only to hydrogen ions but also to other positively charged ions such as sodium (Na^+), which occurs at extremely high concentrations in seawater, creating a substantial interference in the pH measurement. Most efforts to adapt electrochemical pH measurements for practical seawater applications have focused on ways to minimize the sodium interference and eliminate the liquid junction potential. While some new probes are better than the originals, their precision is still not adequate for many oceanographic applications.

The difficulties with electrochemical pH seawater measurements have led to the resurrection of older pH measurement methods based on dye molecules that change color in acids versus bases (the principle behind litmus paper). For example, *meta*-cresol purple, currently the most commonly used dye in seawater pH measurements, is yellow in its protonated (acidic) form and purple in its deprotonated (basic) form. To determine the pH of a solution, a small amount of the dye is added, and the absorbance measured at the maximum wavelengths (yellow and purple) of the two dye forms. The ratio of those absorbance values gives the precise pH of the solution. Because pK_a, the pH at which half the dye molecules are protonated and half are not, is about eight for *meta*-cresol purple (the same as for seawater), it is generally a good indicator for seawater analyses. However, other dyes are sometimes used for some special conditions, such as in estuaries or polar waters. While this method is very quick and easy to perform, its accuracy is dependent on the quality of the spectrophotometer used and the accuracy to which the pK_a of the dye has been determined for the sample type, taking into account the effects of salinity, temperature, and interfering ions.

Because pH is dependent on the analytical method used, careful and accurate calibration is critical. Standard buffer solutions of organic compounds of very specific and well-characterized pK_a are generally used as standards for pH analyses. The most commonly used buffers are those produced by the US National Bureau of Standards (NBS buffers) in purified water at pHs of 4, 7, and 10 (these are generally dyed pink, yellow, and blue, respectively). Because NBS standards are prepared in dilute solution, they are inappropriate for calibrating seawater pH analyses. For the spectrophotometric method, the dye stability constants that match the samples will not match the buffers, and in electrochemical analyses, using buffers that do not match the samples introduces an additional liquid junction potential and destabilizes the measurements. Although recipes for seawater standard buffers have been available for a couple decades, they have not been reproducible between laboratories. Truly standardized seawater buffers (produced by Andrew Dickson at Scripps Oceanographic Institution, who also produces standards for seawater alkalinity and total inorganic carbon) have only recently become available, and they have not yet been fully tested and implemented.

With respect to what buffers are used to calibrate the analysis, several different pH scales have been proposed for environmental analyses, based on how they've been calibrated. For seawater analyses, the appropriate scale to use depends on whether the dyes or electrodes are calibrated in NaCl or artificial seawater solutions, termed the "free" versus "total" hydrogen ion scales. In studies involving only high-salinity seawater samples, the "total" scale is preferred, but in studies of more varied environments such as estuaries or polar waters influenced by sea ice melt, the "free" scale can be preferable. Converting between the scales requires knowledge of or assumptions about the behavior of sulfate and fluoride ions in the samples.

Biological processes in closed systems can cause large changes in pH. Most of these issues are not important, if you are just trying to keep your swimming pool, aquarium, or possibly even fish farm in balance. However, when trying to look at interannual or climate change variations, including anthropogenic ocean acidification, extremely precise and accurate analyses are required in order to confidently identify the relatively small changes that might be occurring. For example, mixed layer pH is thought to be changing at about 0.002/year. The standard electrochemical method for seawater has an optimum precision of 0.004, while the spectrophotometric method is at about 0.001. In comparison, typical low price pH meters intended for use in low-salinity waters have a precision of about 0.01. As yet, successful use of a pH probe for long-term deployment on a mooring has not been reported. However, a number of manufacturers are now claiming to have pH sensors with sufficient stability for long-term deployment in seawater with precision adequate for climate studies. Their claims just have not yet been confirmed by independent deployments. In summary, it's safe to say that while we have made excellent progress on the precision of pH analyses in seawater, accuracy is still an unknown. However, even to get the necessary precision in seawater analyses requires specialized methods and equipment; the pool kit or the cheap hand-held pH probe are not going to give meaningful pH numbers in seawater.

1.11.2 Light Attenuation and Scattering

The light energy in a fluid is attenuated by the combined effects of absorption and scattering. In the ocean, absorption involves a conversion of light into other forms of energy such as heat; scattering involves the redirection of light by water molecules, dissolved solids and suspended material without the loss of total energy. Transmissometers are optical instruments that measure the clarity of water by measuring the fraction of light energy lost from a collimated light beam as it passes along a known pathlength (Figure 1.78). Attenuation results from the combined effects of absorption and shallow-angle Rayleigh (forward) scattering of the light beam by impurities and fine particles in the water. Water that is completely free of impurities is optically pure. Nephelometers (or turbidity meters) measure scattered light and respond primarily to the first-order effects of particle concentrations and size. Depending on manufacturer, commercially available nephelometers examine scattered light in the range from 90° to 165° to the axis of the light beam. Most instruments use IR light with a wavelength of 660 nm. Because light at this wavelength is rapidly absorbed in water (63% attenuation every 5 cm), there is little contamination of the source beam due to sunlight except within the top meter or so of the water column.

FIGURE 1.78 Exploded view of a Sea-Tech transmissometer. Red light of wavelength 660 nm passes from the light-emitting diode (LED) to the sensor over a fixed pathlength of 0.25 m.

The intensity $I(r)$ of a light beam of wavelength λ traveling a distance r through a fluid suspension attenuates as

$$I(r) = I_0 \exp(-cr) \qquad (1.42)$$

where I_0 is the initial intensity at $r = 0$ and $c = c(\lambda)$ is the rate of attenuation per unit distance. Attenuation of the light source occurs through removal or redirection of light beam energy by scattering and absorption. In the ocean, visible long-wave radiation (red) is absorbed more than visible short-wave radiation (blue and green) and what energy is left at long wavelengths undergoes less scattering than at short wavelengths. As a consequence, the ocean appears blue to blue-green when viewed from above. The exact color response depends on the scattering and absorption characteristics of the materials in the water including the dissolved versus the suspended phase—factors that are used to advantage in remote sensing techniques. For a fixed monochromatic light source, the clarity of the water, measured relative to distilled uncontaminated water, provides a quantitative estimate of the mass or volume concentration of suspended particles. Such material can originate from a variety of sources including terrigenous sediment carried into the coastal ocean by runoff, from current-induced resuspension of material in the benthic layer, or from detectable concentrations of plankton.

The "Secchi disk" is one of the simplest and earliest methods for measuring light attenuation in the upper layer of the ocean. A typical Secchi disk consists of a flat, 30-cm diameter white plate that is lowered on a marked line (suspended from the disk center) over the side of the ship. The depth at which the disk can no longer be seen from the ship is a measure of the amount of surface light that reaches a given depth and can be used to obtain a single integrated estimate of the extinction coefficient, $c(\lambda)$. The disk is still in use today. For example, Dodson (1990) used Secchi disk data from a series of lakes in Europe and the U.S.A. that suggest a direct relationship between the depth of day–night (diel) migration of zooplankton and the amount of light penetrating the epihelion. In this case, the zooplankton minimize mortality from visually feeding fish and maximize grazing rate. Despite its simplicity, there are a number of problems with this technique, notwithstanding the fact that it fails to give a measure of the water clarity as a function of depth and is limited to near-surface waters. In addition, the visibility of the disk will depend on the amount of light at the ocean surface (and type of light through cloud cover), on the roughness of the ocean surface, and the eyesight of the observer. Today, oceanographers rely on transmissometers and nephelometers to determine the clarity of the water as a function of depth.

A typical transmissometer consists of a constant intensity, single-frequency light source and receiving lens separated by a fixed pathlength, r_0. The attenuation coefficient in units of per meter is then found from the natural logarithm relation

$$c = -(1/r_0) \ln(I/I_0) \qquad (1.43)$$

in which r_0 is measured in meters, and I/I_0 is the ratio of the light intensity at the receiver versus that transmitted by the red (660 nm) LED. This choice of light wavelength is useful because it eliminates attenuation from dissolved organic substances consisting mainly of humic acids or "yellow matter" (also called "gelbstoff"; Jerlov, 1976). The Sea Tech transmissometer (Bartz et al., 1978) has an accuracy of $\pm 0.5\%$ and a small ($<1.03°$ or 0.018 radians) receiver acceptance angle that minimizes the complication of the collector receiving specious forward-scattering light. To obtain absolute values, the source and lens must be calibrated in distilled water and air since scatter can affect the results. As an example, a 0.25 m pathlength transmissometer which has a calibration value of $I_0 = 94.6\%$ in clean water and reading of 89.1% in the ocean corresponds to a light attenuation coefficient

$$c = -4 \ln(I/I_0) = -4 \ln(0.891/0.946)$$

$$= 0.240 \text{m}^{-1} \qquad (1.44)$$

Values of c in the ocean range from around 0.15/m for relatively clear offshore water for concentrations of particles as low as 100 μg/1 to around 21/m for turbid coastal water with particle concentrations of 140 mg/1 (Sea Tech user's manual). In studies of hydrothermal venting, measurement of water clarity is often one of the best methods to determine the location and intensity of the plume (Baker and Massoth, 1987; Thomson et al., 1992; Figure 1.79).

Problems with the transmissometer technique are: (1) drift in the intensity of the light source with time; (2) clouding of the lens by organic and inorganic material which affect the in situ calibration of the instrument; and (3) scattering, rather than absorption, of the light. If we ignore the influence of dissolved substances, the attenuation coefficient, c, depends on the concentration of the suspended material but also on the size, shape, and index of refraction of the material (Baker and Lavelle, 1984). Thus, a linear relationship between c and particle concentration C such that.

$$c = \alpha C + \alpha_o \qquad (1.45)$$

only occurs when the effects of size, shape and index of refraction are negligible or mutually compensating; here, α_o denotes the offset in c at zero particle concentration. After concentration, particle size is the next most important variable effecting clarity. Accurate estimates of concentration therefore require calibration in terms of the distribution of particle sizes and shapes in suspension as for example in Baker and Lavelle (1984). Laboratory results demonstrate that calibrations of beam transmissometer data in terms of particle mass or volume concentration are acutely sensitive to the size distribution of the particle population under study. There is also a trend of decreasing calibration slopes from environments where large particles are rare (deep ocean) to those where they are common (shallow estuaries and coastal waters). Theoretical attenuation curves agree more with observation when the natural particles are treated as disks rather than as spheres as in Mie scattering theory. The need for field calibration is stressed.

The results of Baker and Lavelle (1984) can be summarized as follows: (1) calibration of beam transmissometers is acutely sensitive to the size distribution of the particle population under study; (2) theoretical calculations based on Mie scattering theory and size distributions measured by a Coulter counter agree when attenuation for glass spheres is observed but underestimate the attenuation of natural particles when these particles are assigned an effective optical diameter equal to their equivalent spherical diameter deduced from particle volume measurements; (3) treating particles as disks expands their effective optical diameter and increases the theoretical attenuation slope close to the observed values; (4) there is a need to collect samples along with the transmissometer measurements, especially where the particle environment is nonhomogeneous.

Transmissometers are best used for measuring the optical clarity of relatively clear water whereas nephelometers are most suitable for measuring suspended particles in highly turbid waters. In murky waters, nephelometers have superior linearity over transmissometers while transmissometers are more sensitive at low concentrations. "Turbidity" or cloudiness of the water is a relative, not an absolute term. It is an apparent optical property depending on characteristics of the scattering particles, external lighting conditions and the instrument used. Turbidity is measured in nephelometer units referenced to a turbidity standard or in Formazin Turbidity Units (FTUs) derived from diluted concentrations of 4000-FTU formazin, a murky white suspension that can be purchased commercially. Since turbidity is a relative measure, manufacturers recommend that calibration involve the use of suspended matter from the waters to

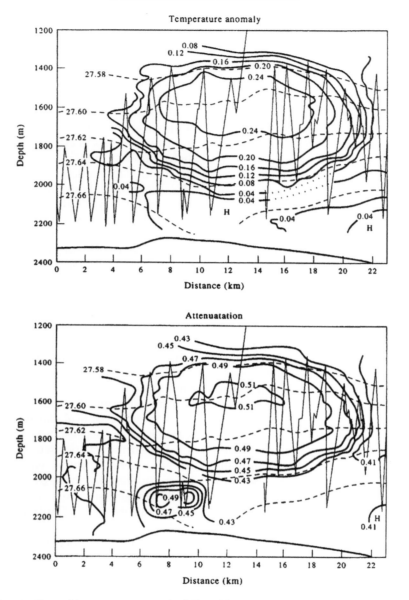

FIGURE 1.79   Cross sections of temperature anomaly (°C) and light attenuation coefficient (1/m) for the "megaplume" observed near the hydrothermal main site on the Cleft Segment of Juan de Fuca Ridge in the northeast Pacific in September 1986. Temperature anomaly gives temperature over the plume depth relative to the observed background temperature. Dotted line shows σ_θ surfaces and solid line the saw tooth track of the towed CTD path. *(From Baker et al. (1989).)*

be monitored. This is not an easy task if one is working in a deep or highly variable regime.

1.11.3 Oxygen Isotope: $\delta^{18}O$

The ratio of oxygen isotope 18 to oxygen isotope 16 in water is fractionated by differences in weight. The lighter element ^{16}O is more easily evaporated than ^{18}O and is therefore a measure of temperature; the higher the temperature the greater the $H_2^{18}O/H_2^{16}O$ ratio. In contrast to the variability in the surface ocean, average $H_2^{18}O/H_2^{16}O$ ratios for the deep ocean (>500-m depth) vary by less than 1%. This ratio (in percent) is expressed in conventional delta "δ" notation as

$$\delta^{18}O(\%) = \left(R_{std}/R_{sample} - 1 \right) \times 10^{10}$$

where $R = H_2^{18}O/H_2^{16}O$ is the ratio of the two main isotopes of oxygen and the subscript "std" refers to Standard Mean Ocean Water. The low variability in $\delta^{18}O$ values in waters in the deep sea has led to widespread use of oxygen isotopes as a paleothermometric indicator. These methods assume relatively little variation (about 1%) in the $\delta^{18}O$ values of deep ocean water over geological time. The $\delta^{18}O$ values of carbonate, silica, and phosphate precipitated by both living and fossil marine organisms, such as foramininferans, radiolarians, coccolithophorids, diatoms, and barnacles, have been used to estimate temperatures of the water in which the organism lived based on temperature-dependent equilibria between the oxygen in the water and the biomineralized phase of interest. The $\delta^{18}O$ values vary in space and time in different regions of the ocean. For example, shallow continental shelves are influenced by freshwater input, particularly at high latitudes. Thus, oxygen removed from seawater by organisms should reflect oceanic conditions at the time. Salinity and ^{18}O content are related in most ocean waters with similar processes influencing both in tandem.

According to Kipphut (1990), the $H_2^{18}O/H_2^{16}O$ ratio in seawater in the Gulf of Alaska shows only slight variation except near those coastal margins where there is significant input of freshwater from melting of large glaciers ($\delta^{18}O \approx -23\%$) and runoff from coastal precipitation ($\delta^{18}O \approx -10\%$). Precipitation is generally depleted in the heavier isotopes of oxygen because of isotopic fractionation processes, which occur during evaporation and condensation. Since the fractionation processes are temperature dependent, precipitation at higher latitudes and elevations shows progressively lower $H_2^{18}O/H_2^{16}O$ ratios. The ratio is a conservative property of water and when combined with salinity may be useful in determining distinct components of water masses. The isotope data south of Alaska suggest that the coastal waters in southwestern Alaska are derived from a combination of glacier melt and runoff from as far-east as south-central Alaska. If we add the freshwater added by runoff from the large rivers of northern British Columbia, the Alaska Coastal Current (Royer, 1981; Schumacher and Reed, 1986) is continuous feature flowing more than 1500 km from the southern Alaska Panhandle to Unimak Pass at the beginning of the Aleutian Island chain. Information regarding the important application of $\delta^{18}O$ observed in cores to the study of paleoclimate change are found in Chapter 6 of the 2007 report of the Intergovernmental Panel on Climate Change (Jansen et al., IPCC 2007).

1.11.4 Helium-3; Helium/Heat Ratio

Helium-3 (^3He) is an inert and stable isotope of helium whose residence time of about 4000 years in the ocean makes it a useful tracer for oceanic mixing times and deepsea circulation. There are two main sources in the ocean. In the upper mixed layer and thermocline, ^3He is produced by the β-decay of anthropogenic tritium; in the deep ocean, ^3He originates with mantle degassing of primordial helium from mid-ocean ridge hydrothermal vents. Anderson (1993) also argues that ^{3}He and neon from hotspot magmas and gases may reflect an extraterrestrial origin; specifically, subduction of ancient pelagic sediments rich in solar ^{3}He and neon originate with interplanetary dust particles now being recycled at oceanic hotspots. [For counterarguments see Hiyagon (1994) and Craig (1994)]. The distinct isotopic ratio of mantle helium (^{3}He/^{4}He $= 10^{-5}$) versus a ratio of 10^{-6} for atmospheric helium makes ^{3}He/^{4}He a useful tracer in the ocean. In a classic paper, Lupton and Craig (1981) showed that the ^{3}He/^{4}He ratio in the 2500-m deep core of the hydrothermal plume emanating from the East Pacific Rise at 15° S in the Pacific Ocean was 50 times higher than the ratio of atmospheric helium. The helium plume could be traced more than 2000 km westward from the venting region on the crest of the mid-ocean ridge (Figure 1.80). To quote the authors, "In magnitude, scale, and striking asymmetry, this plume is one of the most

remarkable features of the deep ocean, resembling a volcanic cloud injected into a steady east wind". Helium-3 is now used extensively as tracer for hydrothermal plumes in active spreading regions such as the Juan de Fuca Ridge in the northeast Pacific and the East Pacific Rise in the South Pacific.

Data collected during GEOSECS indicates that the deep Pacific is the oceanic region most enriched in ^{3}He with a mean ratio concentration δ^3He value of 17% compared with 10% in the Indian Ocean, 7% in the Southern Ocean and 2% in the Atlantic (Jamous et al., 1992). The core of the plume at the East Pacific Rise has a value of 50%. (Here, δ^3He(%) $= (R/R_a - 1) \times 100$, where $R = {}^3$He/^{2}He is the isotopic ratio of the sample and R_a is the atmospheric ratio.) The differences in concentration relate directly to the differences in hydrothermal input and inversely to the degree of deep-water ventilation. For example, there is a considerably greater hydrothermal activity in the Pacific than in the Atlantic while the Atlantic deep water is highly ventilated compared with the Pacific. Similarly, the values of δ^3He (≈ 28) at the bottom of the Black Sea reflect the presence of a strong source at the seafloor. In contrast, the strong correlation between dissolved oxygen concentration and ^{3}He in the Southern Ocean, (Figure 1.81) indicates that the distributions of these tracers in this region of the world ocean are mainly determined by ventilation processes.

Early vent-fluid samples taken from hydrothermal systems on the Galapagos Rift and at 21° N on the East Pacific Rise were found to have nearly equal ratios of ^{3}He to heat despite the considerable geographical separation of the sites and widely different fluid exit temperatures (≈ 20 and 350 °C, respectively). Here, "heat" is the excess amount of heat (in calories or joules) added to the ambient water by geothermal processes. By combining independent estimates of the mantle flux of ^{3}He within the ocean with the observed ratio ^{3}He/Heat $\approx 0.5 \times 10^{-12}$ cm^3 STP/cal, Jenkins et al. (1978) calculated a global

FIGURE 1.80   Cross section of $\delta(^3\text{He})$ over the East Pacific Rise at 15°S. The level of neutral plume buoyancy, as determined by the core depth of the ^{3}He plume, is about 400 m above the ridge crest. The ratio is defined as $\delta(^3\text{He}) = (R/R_{\text{ATM}} - 1) \times 100$ where $R = {}^3\text{He}/^4\text{He}$ and $R_{\text{ATM}} = 1.40 \times 10^{-6}$. (From Luplon and Craig (1981).)

FIGURE 1.81 Correlation between dissolved oxygen concentration O_2 and ^{3}He in the Southern Ocean indicates that the distributions of these tracers in the region of the world ocean are mainly determined by ventilation processes. Combination of GEOSECS and INDIGO-3 data. *(From Jamous et al. (1992).)*

signature of fluids vented into the oceans by mid-ocean ridge hydrothermal systems.

1.12 TRANSIENT CHEMICAL TRACERS

"Transient tracers" are anthropogenic compounds that are injected into the ocean over spatially limited regions within well-defined periods of time. The time "window" makes these compounds especially well-suited to studies of upper-ocean mixing and deep-sea ventilation. Transient tracers are commonly used to constrain solutions of global "box" models used to investigate climate-scale carbon dioxide fluxes within coupled atmosphere-ocean systems (Broecker and Peng, 1982; Sarmiento et al., 1988), and in generalized inverse models incorporating both data and ocean dynamics to determine oceanic flow structure (see Bennett, 1992). The timed release into the ocean may take place over a few hours, as in the case of rhodamine dye, or last longer than a century, as in the case of chlorofluorocarbons (CFCs). Injection of certain tracers, such as radiocarbon (^{14}C) greatly augments the natural distributions of these chemicals while for others, such as CFCs and tritium (^{3}H), the tracer is superimposed on an almost nonexistent background concentration. Because of the slow advection and mixing processes in the ocean, as well as the extensive research needed to measure the tracer distributions, most tracers are used in the study of seasonal to decadal scale oceanic variability. For all transient tracers, studies are limited by imperfect knowledge of the surface boundary conditions during water-mass formation. This is especially true of tracers entering from the atmosphere. Tritium and radiocarbon are radioactive isotopes whose observed concentrations must first be corrected for natural radioactive decay. Both tracers have widespread use in descriptive studies and large-scale numerical modeling of ventilation of the deep ocean and

oceanic hydrothermal heat flux of 4.9×10^{19} cal/year. An examination of the ^{3}He/Heat ratios in the 20-km wide megaplume observed in August 1986 on Juan de Fuca Ridge (Lupton et al., 1989) has shown that the ratios can vary by as much as an order of magnitude and that heat fluxes based on ^{3}He measurements must be taken with caution. Specifically, the ratio ^{3}He/Heat was found to vary with height within the megaplume formed during the hydrothermal event. The megaplume had lower helium values and five times the temperature anomaly as the near-bottom chronic venting regime. Since helium is extracted from the magma by the circulating fluids in the hydrothermal system, the relatively low ratios of ^{3}He/Heat in the megaplume presumably resulted from relatively high water-to-rock ratios and the youth of the hydrothermal fluid prior to its injection into the overlying ocean. Lupton et al. (1989) suggest that a value of $\approx 2 \times 10^{-12}$ cm^3 STP He/cal may be a reasonable estimate for the average ^{3}He/Heat the transformation of water masses over periods of decades. Our main purpose in this section is to provide a brief outline of the types of studies possible with transient tracers. Only results for the main tracers will be presented; secondary tracers such as krypton-85 and argon-39 are not discussed.

1.12.1 Tritium

During the late 1950s and early 1960s, large amounts of bomb-produced radiocarbon (^{14}C), strontium (^{90}Sr) and tritium (^{3}H) were released into the stratosphere during aboveground testing of thermonuclear weapons (Figure 1.82(a and b)). Of these, "bomb" tritium (the heaviest isotope of hydrogen) has an extensive database and is measurable to high precision and sensitivity. Tritium is incorporated directly in water molecules as HTO so that it is a true water-mass tracer. Most of the tritium was produced by tests conducted in the northern hemisphere and was eventually deposited onto the earth's surface north of 15 °N (Weiss and Roether, 1980; Broecker et al., 1986). Deposition into the oceans is through vapor diffusion and rainfall at a ratio of roughly 2:1 according to observational data. A study by Lipps and Hemler (1992) suggests that the ratio varies according to the type of rainfall. The large fronts across the Pacific and Atlantic oceans at subtropical latitudes impede lateral mixing and the southward transport of tritium. As a result, tritium with a half-life of 12.43 years serves as a useful tracer for water motions on timescales of decades. It is most useful when combined with measurements of its stable, inert daughter product ^{3}He. This combination helps determine the age of tritium entering the ocean and provides additional information on the distribution of tritium in the atmosphere before it entered the ocean (Jenkins, 1988). Most large-scale studies are based on the extensive tritium data collected in the North Pacific during the Geochemical Ocean Sections Study (GEOSECS: 1972–74) and Long Lines (1983–85). Roughly 0.3 l of seawater

FIGURE 1.82 Time series of bomb-produced elements released into the stratosphere during above-ground testing of thermonuclear weapons during the late 1950s and early 1960s: (a) radiocarbon (^{14}C) and (b) strontium (^{90}Sr) from measurements of atmospheric carbon dioxide and tritium (^{3}H) based on rain at Valencia Ireland. *(Adapted from Quay et al. (1983); Broecker and Peng (1982).)*

are required for the measurement of tritium by beta-decay counting.

Tritium in natural waters is expressed in "tritium units" (TU), which is the abundance ratio $^{3}H/^{1}H \times 10^{18}$. The ratio abundance corresponds to 7.09 disintegrations per min per kg of water. To remove the effect of normal radioactive decay from a data series, the tritium concentrations are corrected to a common reference of January 1, 1981. Thus, TU81N is the ratio of $^{3}H/^{1}H$ a sample would have as of 1981/01/01. The measurement error for "decay-corrected" data is 0.05TU or 3.5%, whichever is greater

(Van Scoy et al., 1991). Water having values less than 0.2TU81 are considered to reflect cosmogenic background levels or arise from dilution by mixing of bomb tritium. The fact that decay-corrected tritium is a conservative quantity that was added to a selected area of the world ocean in a relatively short period of time (Figure 1.83) makes it attractive as an oceanic tracer. Changes in the spatial distribution of tritium with time provide a measure of horizontal advection while depth penetrations on isopycnals that do not outcrop to the atmosphere are indicative of cross-isopycnal (diapycnal) mixing. Fine (1985) uses upper ocean tritium data from the GEO-SECS program to show that there is a net transport of $5 \times 10^6 \, \mathrm{m}^3/\mathrm{s}$ in the upper 300 m from the Pacific to the Indian Ocean through the Indonesian Archipelago. This contrasts with values of $1.7 \times 10^6 \, \mathrm{m}^3/\mathrm{s}$ obtained using hydrographic data (Wyrtki, 1961) and $5–14 \times 10^6 \, \mathrm{m}^3/\mathrm{s}$ from salt and mass balances (Godfrey and Golding, 1981; Piola and Gordon, 1984; Gordon, 1986).

Gargett et al. (1986) have examined the nine-year record of tritium from Ocean Station P (50° N, 145° W) in the northeast Pacific. Results suggest that the observed vertical distribution of tritium in this region is determined mainly through advection along isopycnals rather than by isopycnal or diapycnal diffusion in the density range of maximum vertical tritium gradient. Tritium data studied by Van Scoy et al. (1991) show evidence for wind-driven circulation to the depth of the dissolved oxygen minimum near 1000-m depth ($\sigma_t = 27.40$) in subpolar regions of the North Pacific. The authors conclude that, after two decades of mixing, advection along isopycnal surfaces appears to be the dominant process influencing the distribution of tritium in the North Pacific and that cross-isopycnal mixing in the subpolar region is important for ventilating the nonoutcropping isopycnals. According to Van Scoy et al. (1991), tritium has penetrated on average 100-m deeper into the ocean during the 10 years between the GEOSECS and Long Lines

FIGURE 1.83 Decay-corrected tritium (TU81) water column inventories over the world oceans based on results obtained as part of the GEOSECS program and NAGS expedition. (*Adapted from Broecker et al. (1986).*)

surveys. Depletions of tritium in the upper ocean are seen in the tropics and at high southern latitudes. Moreover, the above-background tritium levels observed on nonoutcropping isopycnals surfaces in the North Pacific indicate that

ventilation is still taking place despite the absence of deep convective mixing in this region. In the Atlantic Ocean, deep convection is the dominant mechanism for the invasion of surface waters into the deep ocean (Figure 1.84).

FIGURE 1.84　Cross section of tritium (TU81N) concentration in the western Atlantic Ocean (a) GEOSECS 1972; (b) TTO 1981. (From Östlund and Rooth (1990).) Results suggest that the observed vertical distribution of tritium in this region is determined mainly through advection along isopycals rather than by isopycnal or diapyncal diffusion in the density range of maximum vertical tritium gradient.

Tritium data are used to constrain circulation models for the world ocean. For example, tritium records combined with a three-box model of the Japan Sea—a comparatively isolated oceanic region with a mean depth of 1350 m—have yielded overturn times for the deep water of 100 years and overall residence times of 1000 years (Watanabe et al., 1991). Similar estimates for this region based on the same box-model constrained by ^{226}Ra and ^{14}C data yielded a turnover time of 300—500 years for deep water and 600—1300 years for the residence time (Harada and Tsunogai, 1986). Applications to larger oceanic basins are generally less successful. Memery and Wunsch (1990) found that the tritium data did not strongly constrain their circulation model for the North Atlantic and that large errors ($\approx 20\%$) in the input of tritium at the surface can be accommodated by relatively minor changes in the model circulation. According to Wunsch (1988), "Any uncertainty in the transient tracer boundary conditions and sparse interior ocean temporal coverage greatly weakens the ability of such tracers to constrain the ocean circulation". Although the authors still believe in the usefulness of tritium records, they suggest that chlorofluorocarbons will improve modeling capability since the atmospheric concentration of these compounds remains relatively high despite the 1988 Montreal Accord and are better known than for tritium.

Jenkins (1988) describes the use of the tritium-^{3}He age, which takes advantage of the radioactive clock of ^{3}He and the long timescale of tritium to measure the elapsed time since the Helium gas was in equilibrium with the atmosphere. Time scales for which this combined tracer is useful are 0.1—10 years.

1.12.2 Radiocarbon

Carbon-14 (^{14}C) dating requires prior knowledge of long-term variations in the ^{14}C/^{12}C ratio in the atmosphere. Because of the difficulties in separating radiocarbon produced from thermonuclear devices and cosmic rays, bomb-generated radiocarbon is a less useful tracer of upper ocean processes than is tritium. The problem of using radiocarbon data collected prior to 1958 together with tritium measurements to establish the prenuclear levels of radiocarbon is discussed by Broecker and Peng (1982). Once the prenuclear surface-water cosmic radiocarbon concentration is known for each locality, water column inventories for bomb-radiocarbon can be obtained from the depth profiles of ^{14}C/C, ^{3}H, and $\sum CO_2$ concentration obtained as part of the GEOSECS, NORPAX, and TTO programs (Broecker et al., 1985). Bomb-produced radiocarbon is delivered through a nearly irreversible process from the atmosphere to the ocean so that it is possible to estimate the amount of this isotope that has entered any given region of the ocean. As a result of this production, levels of $^{14}CO_2$ increased by about a factor of two in the northern hemisphere during the late 1950s and early 1960s. Measurement of radiocarbon by beta decay requires 200—250 l of seawater to give the desired accuracy of 3—4 ppt. Age resolution is 25—30 years for abyssal oceanic conditions for which the introduction of bomb-radiocarbon effects remain negligible. Radiocarbon has a half-life of 5680 years and decays at a fixed rate of 1% every 83 years. A rapid onboard technique for measuring radiocarbon using an accelerator mass spectrometer is described by Bard et al. (1988). This technique decreases the sample size by 2000 compared with that using the standard β-counting method.

By convention, radiocarbon assays are expressed as Δ^{14}C, which is the deviation in parts per 1000 (ppt) of the ^{14}C/^{12}C ratio from that of a hypothetical wood standard with δ^{13}C $= {}^{13}$C/^{12}C $= -25$ ppt and corrected from the actual δ^{13}C values (around 0 ppt for seawater to exactly -25 ppt to compare with the wood standard). The standard is a way to compare the observed ratio of carbon isotopes to the atmospheric value prior to the industrial revolution of about 1850. The quantity of ^{14}C in a sample of seawater is proportional to the actual uncorrected $^{14}C/^{12}C$ ratio $(1 + 0.001\delta^{14}C)$. More precisely

$$\Delta^{14}C = \delta^{14}C - 2(\delta^{13} + 25)(1 + \delta^{14}C/1000)$$

(1.46)

where

$$\delta^{14}C = 1000\left[(^{14}C/C)_{\text{sample}}\right.$$

$$\left. - (^{14}C/C)_{\text{standard}}/(^{14}C/C)_{\text{standard}}\right]$$

Pre-bomb $\Delta^{14}C$ values from corals collected in the early 1950s average around -50 (± 5) ppt (Druffel, 1989). Thus, any $\Delta^{14}C$ value above -50 ppt will indicate the presence of anthropogenic radiocarbon, mainly produced by the atmospheric nuclear testing in the early 1960s. The determination of inventories for bomb-produced radiocarbon in the ocean is much more complex than for bomb tritium. The reason is that the amount of natural tritium in the sea is negligible compared with the amount of bomb-produced tritium. In the case of radiocarbon, the delivery of isotopes to the ocean requires a better knowledge of wind speeds over the ocean and of the wind speed dependence of the CO_2 exchange rate.

The concentration of ^{14}C in the ocean is influenced by several processes. For example, bottom water formation in the Weddell Sea and the North Atlantic provides a direct input of surface water ^{14}C (Figure 1.85). Additional input of ^{14}C to the deep sea can occur by transport along isopycnals, by vertical mixing in the main oceanic thermocline, by lateral mixing of water masses and by upwelling in coastal and equatorial regions. Addition of CO_2 and ^{14}C comes from the dissolution of carbonate skeletons and the oxidation of organic materials from sinking particles. Stuiver et al. (1982) use radiocarbon data from GEOSECS to estimate abyssal (>1500 m) waters replacement times for the Pacific, Atlantic and Indian Oceans of 510, 275, and 250 years, respectively. The deep waters of the entire world ocean are replaced

on average every 500 years. Ostlund and Rooth (1990) found a relative decrease in the difference in $\Delta^{14}C$ between the surface and the northerly abyssal layers of the North Atlantic of 25–30%. If this were due to vertical diffusivity a high value of 10 cm^2/s would be required based on a scale depth of 1 km and 10 years between surveys. This is a factor of 10 too large so that high latitude injection processes must be responsible for the observed evolution below 1000-m depth. Measurements of $\Delta^{14}C$ from seawater and organisms from the Pacific coast of Baja California (Druffel and Williams, 1991) revealed the effects of coastal upwelling and bottom-feeding habits. Dilution of nearshore waters by upwelling accounts for reduced radioactive carbon levels observed near the coast while feeding on sediment-derived carbon explains the reduced levels of ^{14}C in sampled organisms relative to dissolved inorganic carbon in the water column. Broecker et al. (1991) have addressed the concerns about the accuracy of ventilation flux estimates for the deep Atlantic due temporal changes in the $^{14}C/C$ ratio for atmospheric CO_2. Despite the fact that $\Delta^{14}C$ values have declined from about 10 to 20 ppt over the past 300 years due to changes in the solar wind and the addition of $\Delta^{14}C$-free CO_2 to the atmosphere from fossil fuel burning, temporal effects have been considerably buffered in the ocean and errors in radiocarbon ages are too low by only 10–15%. The reason is that the northern and southern source waters for the Atlantic deep water have $\Delta^{14}C$ ratios, and hence relative time variability, considerably lower than the atmospheric ratio. For further details on the methodology, the reader is again referred to Chapter 6 of the 2007 IPCC report (Jansen et al., IPCC 2007).

1.12.3 Chlorofluorocarbons

Chlorofluorocarbons (CFCs) are a group of volatile anthropogenic compounds that until the 1988 Montreal Protocol found increasingly

FIGURE 1.85 Cross section of radiocarbon concentrations in the western Atlantic Ocean (a) GEOSECS 1972; (b) TTO 1981 (cf. Figure 1.84). Note that significant changes occur mainly in the deep waters north of 40° N. *(From Östlund and Rooth (1990).)*

widespread use in aerosol propellants, plastic foam blowing agents, refrigerants and solvents. Also known as chlorofluoromethanes (CFMs) and "Freons" (a Dupont trade name), most of these chemicals eventually find their way into the atmosphere where they play a primary role in the destruction of stratospheric ozone. The two primary compounds CFC-12 or F-12

(CF$_2$Cl$_2$) and CFC-11 or F-11 (CFCl$_3$) have respective lifetimes in the troposphere of 111 and 74 years. Although more than 90% of production and release of F-11 and F-12 takes place in the northern hemisphere, the meridional distributions of these compounds in the global troposphere are relatively uniform due to the high stability of the compounds and the rapid mixing that occurs in the lower atmosphere. The source function at the ocean surface differs by only about 7% from the northern hemisphere to the southern hemisphere (Bullister, 1989). During the period 1930–75 the ratio F-11/F-12 in the atmosphere and ocean surface increased with increasing uses of these chemicals (Figure 1.86). The regulation of CFC use in spray cans in the U.S.A. during the late 1970s decreased the rate of CFC-11 increase so that the ratio F-11/F-12 ratio in the atmosphere has remained nearly constant. As a consequence, measurements of the ratio provide information on when a particular water mass was last in contact with the atmosphere. In shelf waters the CFCs concentration is determined by rates of

mixed layer entrainment, gas exchange and mixing with source water. At a removal rate of about 1% per year from the atmosphere by stratospheric photolysis, CFCs will serve as ocean tracers well into the next century.

Since they are chemically inert in seawater, Chlorofluorocarbons are used to examine gas exchange between the atmosphere and ocean, ocean ventilation and mixing on decadal scales. The limit of detection of F-11 and F-12 in seawater volumes as small as 30 ml is better than 5×10^{-15} mol/kg seawater (Bullister and Weiss, 1988), or roughly three orders of magnitude higher than near-surface concentrations in the ocean. Modern techniques allow for processing of CFCs at sea with processing times of the order of hours. Gammon et al. (1982) examined the vertical distribution of CFCs at two offshore sites in the northeast Pacific. Using a one-dimensional vertical diffusion/advection model driven by an exponential surface source term, they obtained a characteristic depth penetration of 120–140 m. For the Gulf of Alaska station at 50° N, 140° W, vertical profiles of F-11 and F-12 gave consistent vertical diffusivities of order 1 cm^2/s and an upwelling velocity of 12–14 m/year. Woods (1985) used CFCs to estimate the transit time and mixing of Labrador seawater from its northern source region to the equator along the western Atlantic Ocean boundary. In a related study, Wallace and Lazier (1988) used CFCs and a simple convection model to examine recently renewed Labrador seawater formed by deep convection to depths greater than 1500 m following a severe winter in the North Atlantic. Their observed CFC levels of 60% saturation with respect to contemporary atmospheric concentrations suggest that deep convection took place too rapidly for air-sea gas exchange to bring CFC levels to equilibrium. Trumbore et al. (1991) have used CFCs collected in 1984 to examine recent deep-water ventilation and bottom water formation near the continental shelf in the Ross Sea in the Antarctic Ocean. Using CFC data combined with conventional

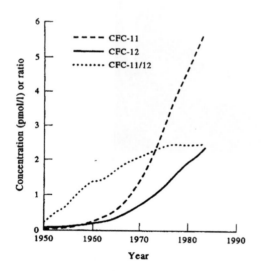

FIGURE 1.86 CFC-12 and CFC-11 concentrations in the upper ocean for T = −1 °C and S = 34.3 psu as function of time. *(From Trumbore et al. (1991).)*

(temperature, salinity, dissolved oxygen, and nutrient) tracer data in a time-dependent convection model they estimate shelf-water resident times of about three years for the Ross Sea. At the other end of the globe, Schlosser et al. (1991) use hydrographic and CFC data to suggest that formation of Greenland Sea Deep Water decreased in the 1980s. The dissolved F-12 concentration in Figure 1.87 illustrates several aspects of the circulation in the North Atlantic. In particular, we note the core of the Labrador Seawater mentioned earlier in this section, the presence of a lens of Mediterranean outflow water ("Meddy") at 22°N and the core of high CFC over the equator which is thought to be a longitudinal extension of flow from the western boundary near Brazil (Bullister, 1989).

1.12.4 Radon-222

Radon (^{222}Rn) is a chemically inert gas with a radioactive half-life of 3.825 days. It occurs naturally as a radionuclide of the ^{238}U series and is injected into the atmosphere by volcanic eruptions. The gas has proven particularly useful at timescales of a few days to weeks for examining the rate of gas transfer between the atmosphere and the ocean surface (Peng et al, 1979), in studies of water column mixing rates (Sarmiento et al., 1976), and for estimating the heat and chemical fluxes from hydro-thermal venting at mid-ocean ridges (Rosenberg et al., 1988; Kadko et al., 1990).

The new application of ^{222}Rn studies to hydrothermal venting regions has been especially successful (Rosenberg et al., 1988). In this case, it is assumed that there is a constant flux of radon into the effluent plume that typically rises several hundred meters above the venting region at depths of 2—3 km on the ridge axis. Typical venting regions have scales of 100 m and are spaced at several kilometers along the ridge axis. Waters exiting from black smokers can be up to 400 °C. At steady state, the amount of radon lost to radioactive decay at some point in the laterally spreading nonbuoyant plume is balanced by a supply of radon from the venting region. To obtain the total heat (or chemical species) issuing from the venting region, the observer first uses a submersible or towed sensor package to measure the ratio of radon to heat (or species) anomaly, ^{222}Rn/ΔT, in the plume near the vent orifice—before the radon in the plume has a chance to disperse or age. The observer then uses a towed sensor package to map the total inventory of radon in the spreading plume (Figure 1.88). Taking into account the effect of cold-water entrainment on the rising plume at Endeavor Ridge in the northeast Pacific (47°47′ N, 129°06′ W), Rosenberg et al. (1988) found an initial radon/ ΔT value of 0.03 dpm (disintegrations per minute—the standard unit of measurement for

FIGURE 1.87 Dissolved CFC-12 concentrations (10^{-12} mol/kg) along a North Atlantic section. *(From Bullister (1989).)*

FIGURE 1.88 Apparent age of the neutrally buoyant plume on the isopycnal surface $\sigma_\theta = 27.68$ (roughly 2100-m depth) at the Cleft Segment of Juan de Fuca Ridge in the northeast Pacific. Distribution based on Radon-222 data for September 1990. Depths in meters. Plume rises from the hydrothermal vent depth of 2280 m to approximately 2100 m. *(From Gendron et al. (1993).)*

radioactive materials) or 55 atoms per joule. They then used hydocast bottle data to estimate the standing crop of radon above 2100-m depth as ^{222}Rn(Total) $= 8 \times 10^{12}$ dpm. At steady-state, hydrothermal venting must be adding this much radon to the system so that the total heat emanating from the vents is.

$$^{222}\text{Rn (Total)}/(^{222}\text{Rn}/\Delta T)$$

$$= 3(\pm 2) \times 10^9 \text{ watts} \qquad (1.47)$$

which compares with estimates based on direct measurements of the total heat content anomaly of the plume in combination with local currents (Baker and Massoth, 1986, 1987; Baker et al., 1995). Gendron et al. (1993) have used ^{222}Rn to

examine time variability in hydrothermal venting on the Cleft Segment of Juan de Fuca Ridge and to estimate the age of the plume as a function of location relative to the known vent sites. They found that the hydrothermal flux decreased from 2.2 ± 0.3 GW in 1990 to 1.2 ± 0.2 GW in 1991 (1 GW $= 10^9$ W).

The estimates using radon-222 in the ocean are complicated by the fact that radon concentration is a function of both radioactive decay and dilution with ambient seawater. Similar estimates can be made using ^{3}He to heat ratios combined with the total inventory of ^{3}He in the ocean. The result (Jenkins, 1978) is a global hydrothermal heat flux of 4.9×10^{19} cal/year.

Baker and Lupton (1990) have used the ^{3}He/heat ratio as a possible indicator of magmatic/tectonic activity at ridge segments. The change from a ratio of 4.4×10^{-12} cm^3 STP/cal immediately following the megaplume eruption at Cleft segment to 1.3×10^{-12} cm^3 STP/cal two years later suggests that high ratios may be indicative of venting created or profoundly perturbed by a magmatic-tectonic event, while lower values may typify systems at equilibrium.

1.12.5 Sulfur Hexafluoride

In certain instances, there is a distinct advantage to a controlled and localized release of a chemical into the environment. Prefluorinated tracers such as sulfur hexachloride (SF$_6$) and perfluorodeclin (PFD) are among the new generation of deliberately released tracers used to measure mixing and diffusion rates in the ocean. These substances are particularly good at examining vertical mixing. Their appeal is that they are readily detectable conservative tracers that have no significant effect on the environment and no toxicity. A thorough description of the use of these tracers as well as rhodamine dyes can be found in Watson and Ledwell (1988). In the case of rhodamine dyes, the detection limit by fluorometers is set by the background fluorescence of natural substances in water, which is about one part in 10^{12} in the deep ocean, although Laane et al. (1984) and J.M. Suijlen (pers. comm., 2002) suggest that this limit can be increased to one part in 10^{16}. For SF$_6$ the background limit is set by dissolution from the atmosphere where the compound is present at one to two parts in 10^{12} by volume. Surface values in the ocean are roughly 5×10^{-17} and diminish to zero in deep water. The instrumental detection for SF$_6$ is limited to about 1/10 of the near-surface value (Watson and Liddicoat, 1985). PFD has no measurable background level in the ocean and is limited by instrumental detection to about one part in 10^{16}. For a release of 1 metric ton ($\equiv$ 1 tonne) at a given density level

in an experiment, these detection limits translate to maximum horizontal scales of 100 km for rhodamine dye, 1000 km for PFDs and basin scales for SF$_6$. Lifetimes for the tracers range from months to about a year. Despite their usefulness, the long-term prognosis for SF$_6$ and PFDs is limited as industrial injection of SF$_6$ into the atmosphere and medical use of PFDs will eventually increase background levels and take away from their ability to serve as tracers.

Rhodamine dye is used mainly in coastal studies and can be as affective as SF$_6$ as a conservative tracer for studying a broad range of mixing processes near the sea surface provided one uses two rhodamines with different photolytic decay rates (Suijlen and Buyse, 1994; Upstill-Goddard et al., 2001). SF$_6$ has been used successfully in WOCE. The North Atlantic Tracer Release Experiment (NATRE) was a large-scale WOCE-related study using SF$_6$ to examine the stirring and diapycnal mixing in the pycnocline of the North Atlantic. In May 1992, 139 kg of sulfur hexafluoride was released on the isopycnal surface 26.75 kg/m^3 (310 dbar) along with eight SOFAR floats and six pop-up drifters in the eastern subtropical Atlantic near 25.7° N, 28.3° W. To sample the tracer, investigators towed a vertical array of 20 integrating samplers at 0.5 m/s through the patch. A prototype 18-chamber sampler at the center of the array obtained a lateral resolution of about 360 m. The average profile increased from an RMS thickness of 6.8 m after 14 days to an RMS thickness of about 45 m by April 1993, yielding a diapycnal eddy diffusivity of 0.1–0.2 cm^2/s ($= 0.1–0.2 \times 10^{-4}$ m^2/s). To be successful, experiments like NATRE require the tracer to be injected on a constant density surface rather than a constant depth. Internal wave oscillations and other vertical motions would broaden the tracer concentration more than necessary if it were released at a constant depth. Care must be taken during injection to ensure the tracer's buoyancy is correct and that the turbulent wake of the injection apparatus is not excessive. A recent large-scale study of diapycnal mixing near 1500-m depth in the Antarctic Circumpolar Current using trifluoromethyl sulfur pentafluoride (CF_3SF_5) is presented by Ledwell et al. (2011). Released to the west of Drake Passage, the tracer indicates that the diapycnal diffusivity, averaged over 1 year and over tens of thousands of square kilometers, is $(1.3 \pm 0.2) \times 10^{-5} \, m^2/s$. The authors report that turbulent diapycnal mixing of this intensity is characteristic of the mid-latitude ocean interior, where the energy for mixing is thought to originate with internal wave breaking. Results support evidence that diapycnal mixing in the interior mid-depth ocean is weak and is likely too small to dictate the mid-depth meridional overturning circulation of the ocean.

1.12.6 Strontium-90

The distribution of bomb-produced ^{90}Sr in the ocean is quite similar to that of tritium. However as pointed out by Toggweiler and Trumbore (1985), ^{90}Sr has the virtue that the ratio $^{90}Sr/Ca$ incorporated into coral skeletons has the same value as this ratio in seawater. Corals average out seasonal variations in the ^{90}Sr content of seawater so that annual bands provide a time-averaged measure of the amount of strontium in the water. The results of Toggweiler and Trumbore (1985) suggest that waters move into the Indian Ocean via passages through the Indonesian Archipelago. In addition, the data suggest that there is a large-scale transport of water between the temperate and tropical North Pacific.

Data Processing and Presentation

2.1 INTRODUCTION

Most instruments neither measure oceanographic properties directly nor store the related engineering or geophysical parameters that the investigator eventually wants from the recorded data. For example, the temperature and conductivity data generated by Sea-Bird 911 conductivity-temperature-depth (CTD) instruments originate as measurements of the electrical resistivity of the sensors in response to their external environment. In turn, the sensors form the variable element in separate Wien bridge oscillator modules whose frequencies of oscillation change in response to the changes in electrical resistance. These changes are converted to an analog voltage which is then converted to a digital signal before being converted to temperature and conductivity using calibration against an accurately known standard in the laboratory. Pressure (depth) is determined from the electrical output from a mechanical strain gauge transducer with temperature compensation. In these sensors, voltages are obtained from pressure-induced changes in the dimension of a flexurally vibrating, load-sensitive resonator.

This progression from sensor electrical response, to the response of an oscillator circuit, to oceanic parameter is further complicated by the fact that all measurement systems alter their characteristics with time and therefore require repeated calibration to define the relationship between the measured and/or stored values and the geophysical quantities of interest. The usefulness of any observation depends strongly on the care with which the calibration and subsequent data processing are carried out. Data processing consists of using calibration information to convert instrument values to engineering units and then using specific formulas to produce the geophysical data. As an example, calibration coefficients are used to convert voltages collected in the different channels of a CTD to salinity, temperature, and depth. Salinity is a function of conductivity, temperature, and pressure, while depth is a function of pressure and temperature. These can then be used to derive such quantities as potential temperature (the pressure compression-corrected temperature) and steric height (the vertically integrated specific volume anomaly derived from the density structure).

Once the data are collected, further processing is required to check for errors and to remove erroneous values. In the case of temporal measurements, for example, a necessary first step is to check for timing errors. Such errors arise because of problems with the recorder's clock which cause changes in the sampling interval (Δt), or because digital samples are skipped

Data Analysis Methods in Physical Oceanography
http://dx.doi.org/10.1016/B978-0-12-387782-6.00002-8

during the recording stage. If N is the number of samples collected, then $(N - 1)\Delta t$ should equal the total length of the record, T. This points to the obvious need to keep accurate records of the exact start and end times of the data record and, where possible, inserting "time stamps" at regularly spaced intervals throughout the data series. When $T \neq (N - 1)\Delta t$, the investigator needs to conduct a search for possible missing records. Simultaneous, abrupt changes in recorded values on all channels often point to times of missing data. Gradual changes in the clock sampling rate (clock "speed") are more of a problem and one has often to assume some sort of linear change in Δt over the recording period. Abrupt shifts in clock speed are generally more easily determined. When either the start or the end time is in doubt, the investigator must rely on other techniques to determine the reliability of the sampling clock and sampling rate. For example, in the case of moored time series records obtained in regions with reasonable tidal motions, one can check that the amplitude ratios among the normally dominant K_1, O_1 (diurnal) and M_2, S_2 (semidiurnal) tidal constituents (Table 2.1) are consistent with previous observations. If they are not, there may be problems with the clock (or calibration of the signal amplitude). If the phases of the constituents are known from previous observations in the region, these can be compared with phases from the suspect instrument. For diurnal motions, each 1-h error in timing corresponds to a phase change of $15°$; for semidiurnal motions, the phase change is $30°$ per hour. Large discrepancies suggest timing problems with the data. In the case of the high data collection and transmission rates available for cabled ocean observatories, there is sometimes a tendency to insert too many time stamps

into the data stream. The finite time it takes to insert the time stamp can lead to disruptions in the data flow and a corresponding loss of cadence in the information flow.

Two types of errors must be considered in the editing stage: (1) large "accidental" errors or "spikes" that result from equipment failure, power surges, or other major data flow disruptions (including planktors such as salps and small jellyfish which partially impede the flow of seawater past or through the sensor) and (2) small random errors or "noise" that arise from changes in the sensor configuration, electrical and environmental noise, and unresolved environmental variability. The noise can be treated using statistical methods, while elimination of the larger errors generally requires the use of more subjective evaluation procedures. Data summary diagrams or distributions are useful in identifying the large errors as sharp deviations from the general population, while the treatment of the smaller random errors requires knowledge of the population density function for the data. It is often assumed that random errors are statistically independent and have a normal (Gaussian) probability distribution. A summary diagram can help the investigator evaluate editing programs that "automatically" remove data points whose magnitude exceed the record mean value by some integer multiple of the record standard deviation. For example, the editing procedure might be asked to eliminate data values for which $|x - X| > 3\sigma$, where X and σ are the mean and standard deviation of x, respectively. This is wrought with pitfalls, especially if one is dealing with highly variable or episodic systems. By not directly examining the data points in conjunction with adjacent values, the investigator can never be certain

TABLE 2.1   Frequencies (Cycles per Hour) for the Major Diurnal (O_1, K_1) and Semidiurnal (M_2, S_2) Tidal Constituents

Tidal constituent	O_1	K_1	M_2	S_2
Frequency (cph)	0.03873065	0.04178075	0.08051140	0.08333333

that he or she is not throwing away reliable values. For example, during the strong 1983—1984 El Niño, water temperatures at intermediate depths along Line P in the northeast Pacific exceeded the mean temperature by 10 standard deviations (10σ). Had there not been other evidence for basin-wide oceanic heating during this period, there would have been a tendency to dispense with these "abnormal" values.

In July 2009, technicians collecting high-resolution CTD data in the nearly landlocked Belize Inlet on the central coast of mainland British Columbia observed anomalous, small-scale (1—10 m) steplike structures in the temperature and salinity data from 70 to 210 m depth at several inner sites in the basin. The initial response on the deck of the ship was that these features were due to an instrument malfunction and that the CTD should be replaced with the backup system. However, on closer examination, the observed structures were recognized as thermohaline staircases, which turned out to be the first observation of double-diffusive features within a coastal fjord (Spear and Thomson, 2012). Salt-fingering staircases of ~10 m thickness were present between 70 and 140 m depth (the temperature and salinity minima were at 150 m depth), while diffusive convection staircases of ~1 m thickness were present between 160 and 210 m depth.

2.2 CALIBRATION

Before data records can be examined for errors and further reduced for analysis, they must first be converted to meaningful physical units. The integer format generally used in the past to save storage space and to conduct onboard instrument data processing is not amenable to simple visual examination. Binary and American Standard Code for Information Interchange (ASCII) formats are the two most common ways to store the raw data, with the storage space required for the more basic Binary format being about 20% of that for the integer values of ASCII format. Conversion of the raw data requires the appropriate calibration coefficients for each sensor. These constants relate recorded values to known values of the measurement parameter. The accuracy of the data then depends on the reliability of the calibration procedure as well as on the performance of the instrument and the number of individual measurements used to generate each output value. Very precise instruments with poor calibrations will produce incorrect, error-prone data. Common practice is to fit the set of calibration values by least squares quadratic expressions, yielding either functional (mathematical) or empirical relations between the recorded values and the appropriate physical values. This simplifies the postprocessing since the raw data can readily be passed through the calibration formula to yield observations in the correct units. We emphasize that the editing and calibration work should always be performed on *copies* of the original data; never work directly on the raw, unedited data.

In some cases the calibration data do not lend themselves to description in terms of polynomial expressions. An example is the direction channel in the older Aanderaa current meter data for which the calibration data consists of a table relating the recorded direction in raw 10-byte integer format (range, 0—1024) to the corresponding direction in degrees from the compass calibration (Pillsbury et al., 1974). Some thought should be given to producing calibration "functions" that best represent the calibration data. With the availability of modern computing facilities, it is no more burdensome to build the calibration into a table than it is to convert it to a mathematical expression. Most important, however, is the need to ensure that the calibration accurately represents the performance range and characteristics of the instrument. Unquestioned acceptance of the manufacturer's calibration values is not recommended for the processing of newly collected data. Instead, individual laboratory and/or field calibration may be needed for each instrument. In some cases this is not possible (for example, in the case of expendable bathythermograph (XBT) probes which come prepackaged and ready for deployment) and some overall average calibration relation must be developed for the measurement system regardless of individual sensors.

Some instruments are individually calibrated before and after each experiment to determine if changes in the sensor unit had occurred during its operation. In fact, most manufacturers of oceanic equipment ask that their equipment be sent back for regular scheduled calibration. Although these calibrations can, for the most part, be done by the purchaser themselves, the companies have both the equipment and the expertise to do the work efficiently, and to their specifications. For example, Sea-Bird Electronics will calibrate not only their sensors (temperature, conductivity, pressure, and oxygen) but also third-party add-ons such as transmissometers and fluorometers. These companies also have the ability to repair the instruments should the calibration reveal a problem with one of the sensors.

The conversion to geophysical units must take both pre- and postcalibrations into account. Often the pre- and postcalibration are averaged together or used to define a calibration trend line, which can then be used to transform the instrument engineering units to the appropriate geophysical units. Sometimes a postcalibration reveals a serious instrument malfunction and the data record must be examined to find the place where the failure occurred. Data after this point are eliminated (or modified to account for the instrumental problems) and the postcalibration information is not used in the conversion to geophysical values. Even if the instrument continues to function in a reasonable manner, the calibration history of the instrument is important to producing accurate geophysical measurements from the instrument.

Since each instrument may use a somewhat different procedure to encode and record data, it is not possible to discuss all the techniques employed. We therefore have outlined a general procedure only. Appendix A provides a list of the many physical units used today in physical oceanography. Although there have been many efforts to standardize these units, one must still be prepared to work with data in nonstandard units. This may be particularly true in the case of older historical data collected before the introduction of acceptable international units. These standard units also are included in Appendix A. To give an example of different units for the same quantity, pressure can be expressed as Pascals (Pa)—including hPa (hundreds of Pascal) or kPa (thousands of Pascal), bars (including mbars, where 1 mbar = 1 hPa), psi (pounds per square inch), mm of Mercury (mmHg, or torr), or meters of water (m H_2O). Oceanographers typically measure pressure as hPa (or mbar), while physicians taking a patient's systolic and diastolic blood pressure measure in mmHg.

2.3 INTERPOLATION

Data gaps or "holes" are a problem fundamental to many geophysical data records. This is particularly true of physical oceanographic data which are typically collected by instruments subjected to harsh environmental conditions. Gappy data are frequently the consequence of uneven or irregular sampling (in time and/or space), or they may result from the removal of erroneous values during editing, from sporadic recording system failures, or from scheduled or unscheduled system shutdowns. An example of gaps in spatial data is the one that occurs in infrared sea surface temperatures introduced by clouds in the infrared image. Infrequent data gaps, having limited duration relative to strongly energetic periods of interest, are generally of minor concern, unless one is interested in short-term episodic events rather than stationary periodic phenomena. Major difficulties arise if the length of the holes exceeds a significant fraction (1/3—1/2) of the signal of interest and the overall data loss rises beyond 20—30% (Sturges, 1983). Gaps have a greater effect on weak signals than on strong signals and the adverse effects of the gaps increases most rapidly for the smallest percentages of data lost. While some useful computational techniques have been developed for unevenly spaced data (Meisel, 1978, 1979) and there are even some advantages to having a range of Nyquist frequencies within a given data set (Press et al., 1992), most analysis methods require data values that are regularly spaced in time or space. As a consequence, it is generally necessary to use an interpolation procedure to create the required regular set of continuous data values as part of the data processing. Noteworthy exceptions are tidal analysis programs used to derive the tidal constituents of scalar and vector time series (Foreman, 1977, 1978; updated 2004; Pawlowicz et al., 2002). Because they are based on least squares analyses for sine and cosine Fourier components at specified frequencies, these programs have no difficulty working with gappy time series. The problem of interpolation and smoothing is discussed in more detail in Chapter 3.

2.4 DATA PRESENTATION

2.4.1 Introduction

The analysis of most oceanographic records necessitates some form of "first-look" visual display. Even the editing and processing of data typically requires a display stage, as, for example, in the exact determination of the start and end of a time series, or in the interactive removal and interpolation of data spikes and other erroneous values. A useful axiom is, "when in doubt, look at the data". In order to look at the data, we need specific display procedures. A single set of display procedures for all applications is not possible since different oceanographic data sets require different attributes. Often, the development of a new display method may be the substance of a particular research project. For instance, the advent of satellite oceanography has greatly increased the need for interactive graphics display and digital image analysis. The development of interactive edge and gradient detection software for determining the location of ocean features is still very much at the forefront of satellite-based Earth observation studies (Belkin and O'Reilly, 2009; Belkin et al., 2009; Williams et al., 2013). Our discussion begins with traditional types of data and analysis product presentations. These were developed as oceanographers sought ways to depict the ocean they were observing. The earliest shipboard measurements consisted of temperatures taken at the sea surface and soundings of the ocean bottom. These data were most appropriately plotted on maps to represent their geographical variability. The data were then contoured by hand to provide a smooth picture of the variable's distribution over the survey region. Examples of historical interest are the meridional sections of salinity from the eastern and western basins of the North Atlantic based on data collected during the German *Meteor* Expedition of 1925—1927 (Figure 2.1; Spiess, 1928). The water property maps from this expedition were among the first to indicate the north—south movements of water masses in the Atlantic basin.

As long as measurements were limited to the sea surface or sea floor, the question of the horizontal level for display was never raised. As oceanographic sampling became more sophisticated and the vertical profiling of water properties became possible, new data displays were required. Of immediate interest were simple vertical profiles of temperature and salinity, such as those shown in Figure 2.2. These property profiles, based on a limited number of sample bottles suspended from the hydrographic wire at

FIGURE 2.1   Latitudinal section of salinity in the western basin of the Atlantic Ocean. *(After Spiess (1928).)*

standard hydrographic depths, originally served to both depict the vertical stratification of the measured parameter and to detect any sampling bottles that had not functioned properly. The data points could then either be corrected or discarded from the data set. Leakage of the watertight seals, failure of the bottle to trip, and damage against the side of the ship are the major causes of sample loss. Leakage problems can be especially difficult to detect.

The data collected from a research vessel at a series of hydrographic stations may be represented as vertical section plots. Here, the discretely sampled data are entered into a two-dimensional (2D) vertical section at the sample depths and then contoured to produce the vertical structure along the section (Figure 2.3). Two things need to be considered in this presentation. First, the depth of the ocean, relative to the horizontal distances, is very small and vertical exaggeration is required to form readable sections. Second, the stratification can be separated roughly into two near-uniform layers with a strong density gradient layer (the pycnocline) sandwiched between. This two-layer system led early German oceanographers to introduce the terms "troposphere" and "stratosphere" (Wüst, 1935; Defant, 1936), which they described as the warm and cold water spheres of the ocean, respectively. Introduced by analogy to the atmospheric vertical structure, this nomenclature has

not been widely used in oceanography. The consequence of this natural vertical stratification, however, is that vertical sections are often best displayed in two parts, a shallow upper layer, with an expanded scale to show the considerable detail normally found in the upper ocean, and a deeper layer with a much more compressed vertical resolution due to the smaller structural variability.

Vertical profiling capability makes it possible to map quantities on different types of horizontal surfaces. Usually, specific depth levels are chosen to characterize spatial variability within certain layers. The near-vertical homogeneity of the deeper layers means that fewer surfaces need to be mapped to describe the lower part of the water column. Closer to the ocean surface, additional layers may be required to properly represent the strong horizontal gradients.

The realization by oceanographers of the importance of both along- and cross-isopycnal processes has led to the practice of displaying water properties on specific isopycnal surfaces. Since these surfaces do not usually coincide with constant depth levels, the depth of the isopycnal (equal density) surface also is sometimes plotted.

Isopycnal surfaces are chosen to characterize the upper and lower layers separately. Often, processes not obvious in a horizontal depth plot are clearly shown on selected isopycnal

FIGURE 2.2 Vertical profiles. (a) Temperature profiles for tropical (low) latitudes, midlatitudes, and polar (high) latitudes in the Pacific Ocean. (b) Salinity profiles for the Atlantic, Pacific, and tropical oceans for different latitudes. The dicothermal layer in (a) is formed from intense winter cooling followed by summer warming to shallower depths. Both salinity (solid line) and temperature (dashed line) are plotted for the tropics in (b). *(From Pickard and Emery (1992).)*

(sigma) surfaces. This practice is especially useful in tracking the lateral distribution of tracer properties such as the deep and intermediate depth silicate maxima in the North Pacific (Talley and Joyce, 1992) or the spreading of hydrothermal plumes that have risen to a density surface corresponding to their level of neutral buoyancy (Feely et al., 1994).

A plot relating one property to another is of considerable value in oceanography. Known as a "characteristic diagram" the most common is that relating temperature and salinity called the *TS* diagram. Originally defined with temperature and salinity values obtained from the same sample bottles, the *TS* relationship was used to detect incorrect bottle samples and to

FIGURE 2.3 Latitudinal cross-sections of (a) in situ temperature and (b) salinity for the Atlantic Ocean. Arrows denote direction of water mass movement based on the distribution of properties. Ant. Bott., Atlantic Bottom Water; Ant. Int., Antarctic Intermediate Water. *(From Pickard and Emery (1992).)*

define oceanic water masses. *TS* plots have been shown to provide consistent relationships over large horizontal areas (Helland-Hansen, 1918) and have recently been the focus of studies on the formation of water masses (McDougal, 1985a,b). Plots of potential temperature vs salinity (the $\theta-S$ relationship) or vs potential density (the $\theta-\sigma_\theta$ relationship) have been proved to be particularly useful in defining the maximum height of rise of hydrothermal plumes formed over venting sites along midocean ridges (Figure 2.4; Thomson et al., 1992).

Characteristic diagrams are not limited to *TS* plots. Various combinations of scalar quantities including temperature, salinity, dissolved oxygen, silicate, nitrate, phosphate, alkalinity,

and/or derived biogeochemical quantities have been used.

Except for some minor changes, plots of vertical profiles, vertical sections, horizontal maps, and time series, as discussed in the following sections, continue to serve as the primary display techniques for physical oceanographers. The development of electronic instruments, with their rapid sampling capabilities and the growing use of high-volume satellite data, may have changed how marine scientists display certain data but most of the basic display formats remain the same. Today, a computer is programmed to carry out both the required computations and to plot the results. Image formats, which are common with satellite data, require

FIGURE 2.4   Plot of mean potential temperature $(\bar{\theta})$ vs mean salinity $(\bar{S})$ for depths of 1500–2200 m at Endeavour Ridge in the northeast Pacific. The least squares linear fit covers the depth range 1500–1900 m, where $\bar{\theta} = -6.563 \cdot \bar{S} + 228.795$ °C. The abrupt change in the $\bar{\theta} - \bar{S}$ relationship at 1900 m depth marks the maximum height of rise of the hydrothermal plume. *(From Thomson et al. (1992).)*

further sophisticated interactive processing to produce images with accurate geographical correspondence. Despite this, the combination of vertical sections and horizontal maps continues to provide most investigators with the requisite geometrical display capability.

2.4.2 Vertical Profiles

Vertical profiles obtained from ships, buoys, aircraft, or other platforms provide a convenient way to display oceanic structure (Figure 2.2). One must be careful in selecting the appropriate scales for the vertical and the horizontal property axes. The vertical axis may change scale or vary

nonlinearly to account for the marked changes in the upper ocean compared with the relative homogeneity of the lower layers. The property axis needs to have a fine enough scale so as to define the small vertical gradients in the deeper layer without the upper layer going off-scale. When considering a variety of different vertical profiles together (Figures 2.5 and 2.6), a common property scale is an advantage, although consideration must be given to the strong dependence of vertical property profiles on latitude and season.

The development and regular use of continuous, high-resolution, electronic profiling systems now provides oceanic fine structure information previously not possible with standard

FIGURE 2.5 Time series of salinity profiles ("waterfall plot") taken in a highly stratified fjord. The effects of large internal waves can be seen around 0100 and 1300 on 12 November. *(From Farmer and Smith (1980).)*

hydrographic casts. Profiles from standard bottle casts required smooth interpolation between observed depths so that structures finer in scale than the smallest vertical sampling separation were missed. Vertical profiles from modern CTD systems are of such high resolution that they are generally either vertically averaged or sub-sampled to reduce the large volume of data to a manageable level for display. For example, with the rapid (≈ 10 Hz) sampling rates of modern CTD systems, parameters such as temperature and salinity, which are generated approximately every 0.01 m, are not presentable in a plot of reasonable size. For plotting purposes, the data are typically averaged to create files with sampling increments of 1 m or larger.

Studies of fine-scale (centimeter scale) variability require the display of full CTD resolution and will generally be limited to selected portions of the vertical profile. These portions are chosen to reflect the part of the water column that is of greatest concern for the study. Full-resolution CTD profiles reveal fine-scale structure in both T and S, and can be used to study mixing processes such as interleaving and double diffusion. Expressions of these processes are also apparent in full-resolution TS diagrams using CTD data. One must be careful, however, not to confuse instrument noise (e.g., those due to vibrations or "strumming" of the support cable caused by vortex shedding) with fine-scale oceanic structure. Processing should be used where possible to separate the instrument noise from the wave number, band-limited signal of mixing processes. Often, computer programs for processing CTD data contain a series of different display options that can be used to manipulate the stored high-resolution digital data. The abundance of raw CTD digital data, and the variety of in situ calibration procedures, make it difficult to interpret and analyze CTD records using a universal format. This is a fundamental problem in assembling a historical file of CTD observations. Hopefully, the statistics of CTD data that have been smoothed to a resolution comparable to that of

FIGURE 2.6 Time series of monthly mean profiles of upper ocean temperature at Ocean Weather Station "P", northeast Pacific (50° N, 145° W). Numbers denote the months of the year. *(From Pickard and Emery (1992).)*

traditional bottle casts are sufficiently homogeneous to be treated as updates to the hydrographic station data file. The increasingly wide use of combined CTD and rosette profiling systems has led to a dramatic decrease in the number of standard bottle casts. (A rosette system consists of a carrousel holding about 12 hydro bottles that can be "tripped" from the ship by sending an electric pulse down the conducting CTD support cable. The CTD is generally placed in the center at the lead end of the carrousel so that the sensors encounter water that is relatively undisturbed by the trailing rosette system.)

2.4.3 Vertical Sections

Vertical sections are a way to display vertically profiled data collected regionally along the track of a research vessel or taken from more extended crossings of an ocean basin (usually, meridionally or zonally). Marked vertical exaggeration is necessary to make oceanic structure visible in these sections. A basic assumption in any vertical section is that the structure being mapped has a persistence scale longer than the time required to collect the section data. Depending on the type of data collected at each station, the length of the section, and the speed of the vessel, shipboard collection times can

run from a few days to a few weeks. Thus, only phenomena with timescales longer than these periods are properly resolved by the vertical sections. Recognizing this fact leads to a trade-off between spatial resolution (between-station spacing) and the time to complete the section. Sampling time decreases as the number of profiles decreases and the samples taken approach a true *synoptic* representation (samples collected at the same time). Airborne surveys using expendable probes such as airborne XBTs from fixed-wing aircraft and helicopters yield much more synoptic information but are limited in the type of measurement that can be made and by the depth range of a given measurement. Although aircraft often have hourly charge-out rates that are similar to ships and generally are more cost-effective than ships on a per datum basis, operation of aircraft is usually the domain of the military or coast guard.

A major change in this regard is the advent of Unmanned Aerial Vehicles (UAVs) which has dramatically expanded in recent years. Rules controlling the application of such platforms are changing such that they can be considered as a useful platform for many oceanographic applications. These planes have the advantage that they can fly low and slow over the ocean for very long periods of time making them ideal for the collection of space–time maps of oceano-graphic surface properties. Most sensors deployed on these planes have focused on surface expressions of oceanographic phenomena looking at the optical and thermal infrared portions of the electromagnetic spectrum. Passive microwave sensors are now being designed for UAV applications. In addition, experimental programs are underway to develop packages that can be dropped from these aircraft to profile the upper portions of the ocean. For now, it is to be expected that most of the UAV sampling will be in mapping surface features of the ocean.

Fewer sample profiles means wider spacing between stations and reduced resolution of smaller, shorter term variability. There is a real danger of short timescale or space-scale variability aliasing quasisynoptic, low-resolution vertical sections. Thus, the data collection scheme must be designed to either resolve or eliminate (by filtering) scales of oceanic variability shorter than those being studied. With the ever-increasing interest in ocean climate, and at a time when the importance of mesoscale oceanic circulation features has been recognized, investigators should give serious consideration to their intended sampling program to optimize the future usefulness of the data collected.

Traditional bottle hydrographic casts were intended to resolve the slowly changing background patterns of the property distributions associated with the mean "steady-state" circulation. As a result, station spacings were usually too large to adequately resolve mesoscale features. In addition, bottle casts require long station times leading to relatively long total elapsed times for each section. The fact that these data have provided a meaningful picture of the ocean suggests that there is a strong component of the oceanic property distributions related to the steady-state circulation. For these reasons, vertical sections based on traditional bottle-cast station data provide useful definitions of the meridional and zonal distributions of individual water masses (Figure 2.1).

The importance of mesoscale oceanic variability has prompted many oceanographers to decrease their sample spacing. Electronic profiling systems, such as the CTD and CTD rosette, require less time per profile than standard bottle casts so that the total elapsed time per section has been reduced over the years despite the need for greater spatial resolution. Still, most oceanographic sections are far from being synoptic owing to the low speeds of ships and some consideration must be given to the definition of which time/space scales are actually being resolved by the measurements. For example, suppose we wish to survey a 1000-km oceanic section and collect a meager 20 salinity–temperature profiles to 2000 m depth along the way. At an average speed of 12 knots, steaming time alone will amount to about 2 days. Each bottle cast would take about 2 h and each CTD cast about 1 h. Our survey time would range from 3 to 4 days, which is just marginally synoptic by most oceanographers' standards. In more protected coastal waters, the advent of self-contained CTD systems has made it possible to use Hovercraft, large inflatables, or other small craft to move rapidly between stations. For example, scientists in the STRATOGEM (Strait of Georgia Ecosystem Modelling) project used hovercraft and large car ferries with as many as 14 daily scheduled sailings for their studies in the Strait of Georgia, British Columbia (Pawlowicz et al., 2007; Halverson and Pawlowicz, 2008; Riche, 2011).

Expendable profiling systems such as the XBT make it possible to reduce sampling time by allowing profile collection from a moving ship. Ships also can be fitted with an acoustic current profiling system, which allows for the measurement of ocean currents in the upper few hundred meters of the water column while the ship is underway. The depth of measurement is determined by frequency and is about 500 m for the commonly used Teledyne-RDI 150-kHz transducers. Most modern oceanographic vessels also have Shipboard ASCII Interrogation Loop (SAIL) systems for rapid (≈ 1 min) sampling of the near-surface temperature and salinity at the intake for the ship's engine cooling system. SAIL data are typically collected a few meters below the ship's waterline. Oceanographic sensor arrays towed in a saw-tooth pattern behind the ship provide another technique for detailed sampling of the water column. This method has wide application in studying near-surface fronts, turbulent microstructure, and hydrothermal venting. These technological improvements have lowered the sample time and increased the vertical resolution. Unfortunately, the instruments typically require considerable technical support and processing of the data can be highly labor intensive. Determining the locations of the data values also requires integration of the oceanic instrumentation with a reliable Global Positioning System.

As referred to earlier, it is common practice when plotting sections to divide the vertical axis into two parts, with the upper portion greatly expanded to display the larger changes of the upper layer. The smaller (closer spacing between lines) contour interval used in the upper part may be greater than that used for the weaker vertical gradients of the deeper layer. It is important, however, to maintain a constant contour interval within each layer to faithfully represent the gradients. In regions with particularly weak vertical gradients, additional contours may be added but a change in line weight, or type (dots, dashes, etc.), is customary to distinguish the added line from the other contours. All contours must be clearly labeled. Color is often very effective in distinguishing gradients represented by the contours. While it is common practice to use shades of red to indicate warm regions, and shades of blue for cold, there is no recommended color coding for properties such as salinity, dissolved oxygen, or nutrients. The color atlas of water properties for the Pacific Ocean published by Reid (1965) provides a useful color scheme.

In sections derived from bottle samples, individual data points are usually indicated by a dot or by the actual data value. In addition, the station number is indicated in the margin above or below the profile. Stations collected with CTDs usually have the station position indicated but no longer have dots or sample values for individual data points. Because of the high vertical resolution, only the contours are plotted.

The horizontal axis usually represents distance along the section and many sections have a small inset map showing the section location. Alternatively, the reader is referred to another map, which shows all section locations. Since many sections are taken along parallels of latitude or meridians of longitude, it is customary to include the appropriate latitude or longitude scale at the top or bottom of each section (Figure 2.3). Even when a section only approximates zonal or meridional lines, estimates of the latitude or longitude are frequently included in the x-axis label to help orient the reader. Station labels should also be added to the axis.

A unique problem encountered when plotting deep vertical sections of density is the need to have different pressure reference levels for the density determination to account for the dependence of seawater compressibility on temperature. Since water temperature generally decreases with pressure (greater depths), artificially low densities will be calculated at the greatest depths when using the surface pressure as a reference (Lynn and Reid, 1968; Reid and Lynn, 1971). When one wants to resolve the deep density structure, and at the same time display the upper layer, different reference levels are used for different depth intervals. As shown in Figure 2.7, the resulting section has discontinuities in the density contours as the reference level changes.

A final comment about vertical sections concerns the representation of bottom topography. The required vertical exaggeration makes it necessary to represent the bottom topography on an exaggerated scale. This often produces steep-looking islands and bottom relief. There is a temptation to ignore bottom structure, but as oceanographers become more aware of the importance of bottom topography in dictating

FIGURE 2.7   Cross-section of density (σ) (kg/m^3) across Drake Passage in 1976. *(From Nowlin et al. (1986).)*

certain aspects of the circulation, it is useful to include some representation of the bottom structure in the sections.

2.4.4 Horizontal Maps

In the introduction, we mentioned the early mapping of ocean surface properties and bottom depths. Following established traditions in map making, these early maps were as much works of art as they were representations of oceanographic information. The collection of hydrographic profiles later made it possible to depict property distributions at different levels of the water column (Figure 2.8). As with vertical sections, the question of sample time vs horizontal resolution needs to be addressed, especially where maps cover large portions of an ocean basin. Instead of the days to weeks needed to collect data along a single short section, it may take weeks, months, and even years to obtain the required data covering large geographical regions. Often, horizontal maps consist of a collection of sections designed to define either the zonal/meridional structure or cross-shore structure for near-coastal regions. In most cases, the data presented on a map are contoured with the assumption that the map corresponds to a

(a)

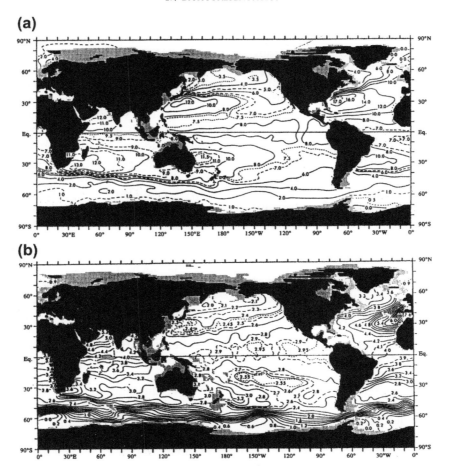

(b)

FIGURE 2.8 Horizontal maps of annual mean potential temperature in the world ocean at (a) 500 m; (b) 1000 m depth. *(From Levitus (1982).)*

stationary property distribution. For continental shelf regions, data used in a single map should cover a time period that is less than the approximately 10-day e-folding timescale of mesoscale eddies. In this context, the *"e-folding time"* is the time for the mesoscale currents to decay to $1/e^1 = 0.368$ of their peak values.

Much of what we know about the overall structure of the ocean, particularly the deep ocean, has been inferred from large-scale maps of water properties. A presentation developed by Wüst (1935) to better display the horizontal

variations of particular water masses is based on the *core layer* method. Using vertical property profiles, vertical sections, and characteristic (one property vs another property) diagrams, Wüst defined a core layer as a property extremum and then traced the distribution of properties along the surface defined by this extremum. Since each core layer is not strictly horizontal, it is first necessary to present a map showing the depth of the core layer in question. Properties such as temperature, salinity, oxygen, and nutrients also can be plotted along these layers in addition to the percentage of the appropriate water mass defined from the characteristic diagrams. A similar presentation is the plotting of properties on selected density surfaces. This practice originated with Montgomery (1938) who argued that advection and mixing would occur most easily along surfaces of constant entropy. Since these isentropic surfaces are difficult to determine, Montgomery suggested that surfaces of constant potential density would be close approximations in the lower layers and that sigma-*t* would be appropriate for the upper layers. Known as *isentropic analysis* because of its thermodynamic reasoning, this technique led to the practice of presenting horizontal maps on sigma-*t* or sigma-θ (potential density) surfaces. While it may be difficult to visualize the shape of the density surfaces, this type of format is often better at revealing property gradients. As with the core layer method, preparing maps on density surfaces includes the plotting of characteristic property diagrams to identify the best set of density surfaces. Inherent in this type of presentation is the assumption that diapycnal (cross-isopycnal) mixing does not occur. Sometimes steric surfaces or surfaces of thermosteric anomaly are chosen for plotting rather than density.

The definition and construction of contour lines on horizontal maps has evolved in recent years from a subjective hand-drawn procedure to a more objective procedure carried out by a computer. Hand analyses usually appear quite smooth but it is impossible to adequately define the smoothing process applied to the data since it varies with user experience and prejudice. Only if the same person contoured all the data, is it possible to compare map results directly. Differences produced by subjective contouring are less severe for many long-term and stationary processes, which are likely to be well represented regardless of subjective preference. Shorter term and smaller space-scale variations, however, will be treated differently by each analyst and it will be impossible to compare results. In this regard, we note that weather maps used

in 6-hourly weather forecasts are, in part, still drawn by hand since this allows for needed subjective decisions based on the accumulated experience of the meteorologist. Hand contouring by physical oceanographers connected the analyst much more closely to the data generating the contours. This was realistic when the volume of data was small and each point could be individually considered by the analyst. However, this is no longer possible with the large volumes generated by electronic sampling instruments. Objective analysis and other computer-based mapping procedures have been developed to carry out the horizontal mapping and contouring. Some of these methods are presented individually in later sections of this chapter. Since there is such a wide selection of mapping methods, it is not possible to discuss each individually. However, the reader is cautioned in applying any specific mapping routine to ensure that any implicit assumptions are satisfied by the data being mapped. The character of the result needs to be anticipated so that the consequences of the mapping procedure can be evaluated. For example, the mapping procedure called "objective analysis" or "optimum interpolation", is inherently a smoothing operation. As a consequence, the output gridded data may be smoothed over a horizontal length scale greater than the scale of interest in the study. One must decide how best to retain the variability of interest and still have a definable mapping procedure for irregularly spaced data.

2.4.5 Map Projections

One neglected aspect of mapping oceanographic variables is the selection of an appropriate map projection. A wide variety of projections has been used in the past. The nature of the analysis, its scale, and geographic region of interest dictate the type of map projection to use (Bowditch, 1977). Polar studies generally use a conic or other polar projection to avoid distortion of zonal variations near the poles. An example of a simple conic projection for the Northern Hemisphere is given in Figure 2.9. In this case, the cone is tangent at a single latitude (called a standard parallel), which can be selected by changing the angle of the cone (Figure 2.9(a)). The resulting latitude—longitude scales are different around each point and the projection is said to be nonconformal (Figure 2.9(b)). A conformal (= orthomorphic; conserves shape and angular relationships) conic projection is the Lambert conformal projection which cuts the Earth at two latitudes. In this projection, the spacing of latitude lines is altered so that the distortion is the same as along meridians. This is the most widely used conic projection for navigation since straight lines nearly correspond to great circle routes. A variation of this mapping is called the "modified Lambert conformal projection". This projection amounts to selecting the top standard parallel very near the pole, thus closing off the top of the map. Such a conic projection is conformal over most of its domain. Mention should also be made of the "polar stereographic projection" that is favored by meteorologists. Presumably, the advantages of this projection are its ability to cover an entire hemisphere, and its low distortion at temperate latitudes.

At mid and low latitudes, it is common to use some form of Mercator projection which accounts for the meridional change in the Earth radius by a change in the length of the zonal axis. Mercator maps are conformal in the sense that distortions in latitude and longitude are similar. The most common of these is the transverse Mercator or cylindrical projection (Figure 2.10). As the name implies it amounts to projecting the Earth's surface onto a cylinder which is tangent at the equator (equatorial cylindrical). This type of projection, by definition, cannot include the poles. A variant of this is called the oblique Mercator projection, corresponding to a cylinder which is tangent to the Earth along a line tilted with respect to the equator. Unlike the equatorial cylindrical this oblique projection can represent the poles (Figure 2.11(a)). This form of Mercator projection also has a conformal character, with equal distortions in lines of latitude and longitude (Figure 2.11(b)). The most familiar Mercator mapping is the universal transverse Mercator grid, which is a military grid using the equatorial cylindrical projection. Another popular midlatitude projection is the rectangular or equal-area projection, which is a cylindrical projection with uniform spacing between lines of latitude and lines of longitude. In applications where actual Earth distortion is not important, this type of equal-area projection is often used. Whereas Mercator projections are useful for plotting vectors, equal-area projections are useful for representing scalar properties. For studies of limited areas, special projections may be developed such as the azimuthal projection, which consists of a projection onto a flat plane tangent to the Earth at a single point. This is also called a gnomonic projection. Stereographic projects perform similar projections; however, where gnomonic projections use the center of the Earth as the origin, stereographic projections use a point on the surface of the Earth.

The effects of map projection on mapped oceanographic properties should always be considered. Often the distortion is unimportant since only the distribution relative to the provided geography (land boundaries) is important. In other cases, such as plots of Lagrangian trajectories, it is important to compare maps using the same projection from which it should be possible to roughly estimate velocities along the trajectories. Variations in map projections can also introduce unwanted variations in the displays of properties.

2.4.6 Characteristic or Property vs Property Diagrams

As noted in the introduction to this section, it is useful in many oceanographic applications to relate two simultaneously observed variables. Helland-Hansen (1918) first suggested the utility of plotting temperature (T) against salinity (S). He found that TS diagrams were similar over

(a)

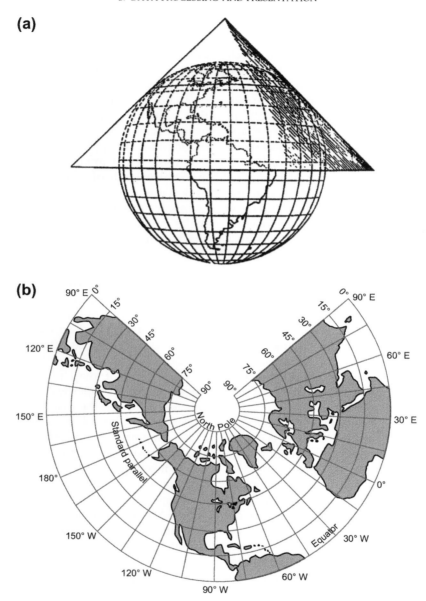

(b)

FIGURE 2.9 Example of a simple conic projection for the Northern Hemisphere. The single tangent cone in (a) is used to create the map in (b). *(From Bowditch (1977).)*

(a)

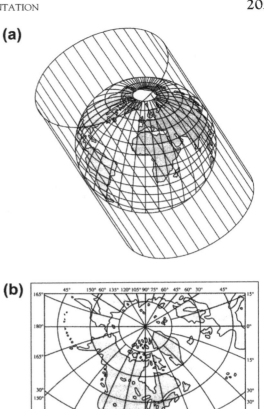

FIGURE 2.10 The transverse Mercator or cylindrical projection. *(From Bowditch (1977).)*

large areas of the ocean and remained constant in time at many locations. An early application of the *TS* diagram was the testing and editing of newly acquired hydrographic bottle data. When compared with existing *TS* curves for a particular region, *TS* curves from newly collected data quickly highlighted erroneous samples which could then be corrected or eliminated. Similar characteristic diagrams were developed for other ocean properties. Many of these, however, were not conservative and could not be expected to exhibit the constancy of the *TS* relationship (we will use *TS* as representative of all characteristic diagrams).

As originally conceived, characteristic diagrams such as the *TS* plots were straightforward

(b)

FIGURE 2.11 An oblique Mercator or oblique cylindrical projection that includes the poles. The cylinder in (a) is used to generate the transverse Mercator map of the western hemisphere in (b). *(From Bowditch (1977).)*

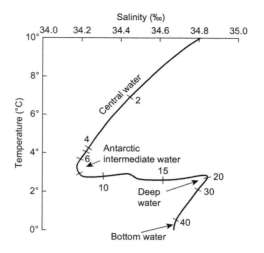

FIGURE 2.12 Temperature—salinity curve for the western basin of the South Atlantic at 41° S latitude. Depths are marked in hundreds of meters. *(Adapted from Tchernia (1980).)*

to construct. Pairs of property values from the same water bottle sample constituted a point on the characteristic plot. The connected points formed the *TS* curve for the station (Figure 2.12). Each *TS* curve represented an individual oceanographic station and similarities between stations were judged by comparing their *TS* curves. These traditional *TS* curves exhibit a unique relationship between *T*, *S*, and *Z* (the depth of the sample). What stays constant is the *TS* relationship, not its correspondence with Z. As internal waves, eddies, and other unresolved dynamical features move through a region, the depth of the density structure changes. In response, the paired *TS* value moves up and down along the *TS* curve, thus maintaining the water mass structure. This argument does not hold for near-surface layers where the water is being modified by wind mixing and heat and buoyancy fluxes with the atmosphere, or in frontal zones where the water mass is being modified by turbulent mixing and interleaving.

Temporal oceanic variability has important consequences for the calculation of mean *TS*

diagrams where *TS* pairs, from a number of different bottle or CTD casts, are averaged together to define the *TS* relationship for a given area or lapsed time interval. Perhaps the easiest way to present this information is in the form of a scatter plot (Figure 2.13) where the dots represent individual *TS* pairs. The mean *TS* relationship is formulated as the average of *S* over intervals of *T*. Depth values have been included in Figure 2.13 and represent a range of *Z* values spanning the many possible depths at which a single *TS* pair is observed. Thus, it is not possible to define a unique mean *T*, *S*, *Z* relationship for a collection of different profiles.

The traditional *TS* curve presented in Figure 2.13 is part of a family of curves relating measured variables such as temperature and salinity to density (sigma-*t*) or thermosteric anomaly ($\Delta_{S,T}$). The curvature of these lines is due to the nonlinear nature of the ocean's equation of state. In a traditional single-cast *TS* diagram, the stability of the water column, represented by the *TS* curve, can be easily evaluated. Unless one is in an unstable region, density should always increase with depth along the *TS* curve. Furthermore, the analysis of *TS* curves can shed important light on the advective and mixing processes generating these characteristic diagrams. We note that the thermosteric anomaly, $\Delta_{S,T}$, is used for *TS* curves rather than specific volume anomaly, $\delta_{S,T}$, since the pressure term that is included in $\delta_{S,T}$ has been found to be negligible for hydrostatic computation and can be approximated by $\Delta_{S,T}$, which lacks the pressure term.

The time variability of the *TS* relation is also a useful quantity. A simple extension of this characteristic diagram shown in Figure 2.14 reveals the monthly mean *TS* pairs for surface water samples over a year in the vicinity of the Great Barrier Reef. The dominant seasonal cycle of the physical system is clearly displayed with this format.

Another more widely used variation of the *TS* diagram is known as the volumetric *TS* curve.

FIGURE 2.13 Mean temperature–salinity curves for the North Pacific (10–20° N; 150–160° W). Also shown is the density anomaly $\Delta_{S,T}$. *(From Pickard and Emery (1992).)*

Introduced by Montgomery (1958), this diagram presents a volumetric census of the water mass with the corresponding TS properties. The analyst must decide the vertical and horizontal extent of a given water mass and assign to it certain TS properties. From this information, the volume of the water mass can be estimated and entered on the TS diagram (Figure 2.15). The border values correspond to sums across T and S values. Worthington (1981) used this procedure, and a three-dimensional plotting routine, to produce a volumetric TS diagram for the deep waters of the world ocean (Figure 2.16). The distinct peak in Figure 2.16 corresponds to a common deep water which fills most of the deeper parts of the Pacific. Sayles et al. (1979) used the method to produce a good descriptive analysis of Bering Sea water. This type of diagram has been made possible

with the development of computer graphics techniques, which greatly enhance our ability to display and visualize data.

In a highly site-specific application of TS curves, McDuff (1988) has examined the effects of different source salinities on the thermal anomalies produced by buoyant hydrothermal plumes rising from midocean ridges. In potential temperature–salinity (θ–S) space, the shapes of the θS curves strongly depend on the salinity of the source waters and lead to markedly different thermal anomalies as a function of height above the vent site.

2.4.7 Time Series Presentation

The graphical presentation of time series records is of particular importance in oceanography. Early requirements were generated by

FIGURE 2.14 Monthly mean temperature—salinity pairs for surface water samples over a year in the lagoon waters of the Great Barrier Reef. *(From Pickard and Emery (1992).)*

FIGURE 2.15 Volumetric temperature—salinity (T—S) curve in which the number of T—S pairs in each segment of the plot can be calculated. *(From Pickard and Emery (1992).)*

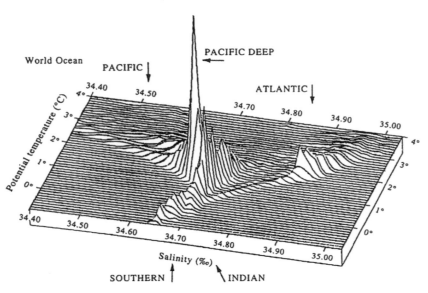

FIGURE 2.16 Three-dimensional volumetric *TS* diagram for the deep waters of the world ocean. The distinct peak corresponds to common deep water which fills most of the deeper parts of the Pacific. *(From Pickard and Emery (1992).)*

shore-based measurements of sea-level heights, sea surface temperature, and other relevant parameters. As ship traffic increased, the need for offshore beacons led to the establishment of light or pilot ships, which also served as platforms for offshore data collection. Some of the early studies, made by geographers in the emerging field of physical oceanography, were carried out from light ships. The time series of wind, waves, surface currents, and surface temperature collected from these vessels needed to be displayed as a function of time. Later, dedicated research vessels such as weather ships were used as "anchored" platforms to observe currents and water properties as time series. Today, many time series data are collected by moored instruments which record internally or send their data back to a shore station via satellites, radio telemetry, and electrical or fiber-optic bottom cables. The need for real-time data acquisition for operational oceanography and meteorology has created an increased interest in new methods of telemetering data. The development

of bottom-mounted acoustical modem systems, cable observatories such as the Monterey Accelerated Research Systems and Ocean Networks Canada, and satellite data collection systems such as Service Argos have opened new possibilities for the transmission of oceanographic data to shore stations and for the transmission of operational commands back to the offshore modules.

The simplest way to present time series information is to plot a scalar variable against time. The timescale depends on the data series to be plotted and may range in intervals from seconds to years. Scalar quantities can be displayed as time series plots on single or multiple vertical axes. Two-dimensional vector quantities (such as horizontal currents or surface winds) can be plotted as time series of speed (V) and direction (θ)—the two horizontal scalars actually measured by current meters and anemometers and more immediately available from the instrument—or as time series of two orthogonal components of velocity that need to be specified.

FIGURE 2.17   Time series of the low-pass filtered u (across-shelf, x) and v (along-shelf, y) components of velocity together with the simultaneously collected values of temperature (T) for the east coast of Australia immediately south of Sydney, August 31, 1983 to March 18, 1984. The axes for the stick vectors are rotated by $-26°$ from North so that "up" is in the alongshore direction. The current meter was at 137 m depth in a total water depth of 212 m. Time in Julian days as well as calendar days. *(Freeland et al. (1985).)*

Scalar time series of the u (x-direction) and v (y-direction) components of velocity are presented in Figure 2.17 along with simultaneously collected values of water temperature. Note that it is common practice in oceanography to rotate the u,v velocity components in order to align them with the dominant geographic or topographic orientation of the study region. This is especially true near coastal boundaries. The horizontal orthogonal axes can be in the cross-shore (x) and alongshore (y) directions or in the across-isobath (x) and along-isobath (y) directions. Over the continental shelf, the terms cross-shelf and along-shelf are used in place of cross-shore and alongshore. The vertical (z) component of current can also be plotted, although a separate vertical scale is generally needed because of the much weaker speeds compared to the horizontal speeds. We note that oceanographic convention has vectors of current (and wind) pointing in the direction the flow is *toward* whereas meteorological convention has wind vectors pointing in the direction the wind is *from*. To avoid

confusion, marine scientists are advised to use the oceanographic convention. Plots of the current velocity components derived from acoustic Doppler Current Profiler (ADCP) measurements are especially appealing since they show the flow as functions of depth and time. Figure 2.18 shows the daily mean alongshore component of the current at site A1 on the continental slope off southwest Vancouver Island, British Columbia (cf., Thomson and Ware, 2005). Current velocity was measured at 4-m bin intervals by an upward-looking 75-kHz ADCP moored in 500 m of water. Other scalar plots (in this case the alongshore component of wind stress from a nearby meteorological buoy, C46206) can be appended above or below the 2D current plot. The strong poleward currents in summer centered at 200 m depth represent the California Undercurrent (Thomson and Krassovski, 2010).

Another common approach is the use of the "stick plot" (Figure 2.19; also Figure 2.17) where each "stick" (i.e., vector) corresponds to a measured speed and direction at the specified

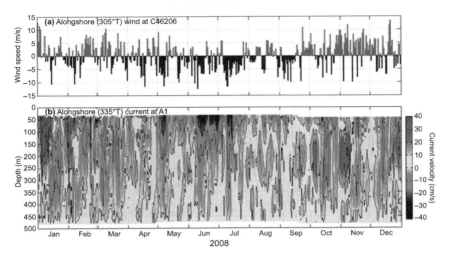

FIGURE 2.18 Time series (January 1–December 31, 2008) of the daily mean alongshore component of (a) wind stress at meteorological buoy C46206 and (b) current velocity at mooring site A1. The locations are within 20 km of one another off southwest Vancouver Island, British Columbia. The positive alongshore directions (305 and 335 °T, respectively) are given in "degrees True" compass bearing. Currents are measured every 4 m through the water column from an upward-looking 75-kHz ADCP moored in 500 m of water on the continental slope. Winds are measured about 20 km away on the continental slope. *(Data and analysis courtesy of Maxim Krassovski, Steve Mihály, Tamás Juhász and David Spear.)*

time. The length of the stick is scaled to the current speed. Direction may be relative to true north (pointed upward on the page) or the coordinate system may be rotated to align the axes with the dominant geographic or topographic boundaries (oceanographers in the Southern Hemisphere use true south instead of true north). The stick plot presentation is ideal for displaying directional variations of the measured currents. Rotational oscillations, due to the tides and inertial currents, are clearly represented. The once popular, but now less often used progressive vector diagram (PVD), can also be used to plot vector velocity time series (Figure 2.20). In this case, the time-integrated displacements along each of two orthogonal directions (x,y) are calculated from the corresponding velocity components (u,v) such that $(x,y) = (x_0,y_0) + \sum(u_i,v_i)\Delta t_i$ (where times Δt_i are for observations $i = 1, 2, ...$) to give "pseudo" downstream displacements of a parcel of water from its origin (x_0, y_0). The plot gives the vector sum of the individual current vectors plotting them head to tail for the period of interest. Distance in kilometers may be used in place of Earth coordinates, although there are distinct advantages to sticking with Mercator projections. Residual or long-term vector-mean currents are readily apparent in the PVD and rotational behavior also is well represented. The signature rotational motions of inertial and tidal currents can be easily distinguished in this type of diagram. The drawback is that the plots give the impression of a Lagrangian measurement with time (i.e., one that appears to be following the path of a fluid parcel) when in reality the data are from a single point (an Eulerian measurement) that cannot take into account the inevitable spatial inhomogeneities in the flow field; measurements at one location cannot inform us of the downstream trajectory of water parcels once they had crossed the recording location unless the flow is uniform in space.

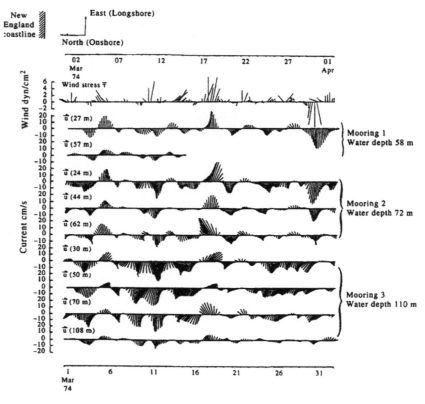

FIGURE 2.19 Vector (stick) plots of low-pass filtered wind stress and subtidal currents at different depths measured along the East Coast of the United States about 100 km west of Nantucket Shoals. East (up) is alongshore and north is cross-shore. Brackets give the current meter depth (m). *(Figure 7.11 from Beardsley and Boicourt (1981).)*

Yet another type of time series plot consists of a series of vertical profiles at the same locations as functions of time (Figure 2.21(a)). The vertical time series plot has a vertical axis much like a vertical section with time replacing the horizontal distance axis. Similarly, a time series of horizontal transects along a repeated survey line is like a horizontal map but with time replacing one of the spatial axes. Property values from different depth-time (z, t) or distance-time (x, t) pairs are then contoured to produce time series plots (Figure 2.21(b)) which look very similar to vertical sections and horizontal maps, respectively. This type of presentation is useful in depicting temporal signals that have a pronounced vertical structure such as seasonal heating and cooling. Other temporal changes due to vertical layering (e.g., from river a plume) are well represented by this type of plot.

2.4.8 Histograms

As oceanographic sampling matured, the concept of a stationary ocean has given way to the notion of a highly variable system requiring repeated sampling. Data display has graduated from a purely pictorial presentation to statistical representations. A plot format, related to

FIGURE 2.20 PVD constructed from the east–west and north–south components of velocity for currents measured every 10 min for a period of 50 days at a depth of 200 m in the Strait of Georgia, British Columbia. Plotted positions correspond to horizontal displacements of the water that would occur if the flow near the mooring location was the same as that at the derived location. *(From Tabata and Stickland (1972).)*

fundamental statistical concepts of sampling and probability, is the histogram or frequency-of-occurrence diagram. This diagram presents information on how often a certain value occurred in any set of sample values. As we discuss in the section on basic statistics, there is no set rule for the construction of histograms and the selection of a sample variable interval (called "bin size") is completely arbitrary. This choice of bin size will dictate the smoothness of the presentation but an appropriately wide enough interval must be used to generate statistically meaningful frequency-of-occurrence values.

2.4.9 New Directions in Graphical Presentation

Plotting oceanographic data has gone from a manpower-intensive process to one primarily carried out by computers. Computer graphics have provided oceanographers with a variety of new presentation formats. For example, all

the data display formats previously discussed can now be carried out by computer systems. Much of the investigator's time is spent ensuring that computer programs are developed, not only for the analysis of the data but also for the presentation of results. These steps are often combined, as in the case of objective mapping of irregularly spaced data. In this case, an objective interpolation scheme is used to map a horizontal flow or property field. Contouring of the output objective map is then done by the computer. Frequently, both the smoothing provided by objective analysis, and the computer contouring, can be performed by existing software routines. Sometimes problems with these programs arise, such as continuing to contour over land or the restriction to certain contour intervals. Both of these problems must be overcome in the computer routine or the data altered in some way to avoid the problems. For example, "masks" can be applied to avoid contouring over land surfaces.

FIGURE 2.21   Time series plots for: (a) Upper ocean temperature (°C); and (b) salinity (psu) and density (σ_t) at 10 m depth from repeated transects along Line P between Station P and the west coast of North America for the period January 1959 to December 1961. *(From Fofonoff and Tabata (1966).)*

2.4.9.1 *Three-Dimensional Displays*

In addition to computer mapping, the computer makes it possible to explore other presentations not possible in hand analyses. Three-dimensional plotting is one of the more obvious examples of improved data display possible with computers. For example, Figure 2.22 shows a three-dimensional plot of coastal bottom topography and a 2D projection (contour map) of the same field. One main advantage of the three-dimensional plot is the geometrical interpretation given to the plot. We can more clearly see both the sign and the relative magnitudes of the dominant features. A further benefit of this form of presentation is the ability to present views of the data display from different angles and perspectives. For example, the topography in Figure 2.22 can be rotated to emphasize the different canyons that cut across the continental slope. Any analysis, which outputs a variable as a function of two others can benefit from a three-dimensional display. A well-known oceanic example is the Garrett-Munk spectrum for internal wave variability in the ocean (Figure 2.23) in which spectral amplitude based on observational data is plotted as a function of vertical wave number (m) and wave frequency (ω). The diagram tells the observer what kind of spectral

FIGURE 2.22   Three-dimensional plot of water depth at 20-m contour interval off the southwest coast of Vancouver Island. The bottom plot is the 2D projection of the topography. (*Courtesy Gary Hamilton.*)

FIGURE 2.23 Garrett—Munk energy spectrum for oceanic internal waves based on different types of observations. Spectral amplitude (arbitrary units) is plotted against m (the vertical wave number in cycles per meter) and ω (the wave frequency in cycles per hour). Here, m^* is the wave number bandwidth, κ is the horizontal wave number, N is the buoyancy frequency, f is the Coriolis parameter, and $\gamma = (1 - f^2/\omega^2)^{1/2}$. MVC, moored vertical coherence and DLC, dropped lag coherence between vertically separated measurements. *(From Garvett and Munk (1979).)*

FIGURE 2.24 A Taylor diagram showing a statistical comparison of eight climate model estimates (models A through H) vs observations of the global pattern of mean annual precipitation. The dashed line labeled "observed" denotes the standard deviation of the observations. Green curves denote the RMS differences between the observations and the models. *(From Taylor (2005).)*

shape to expect from a specific type of profiling method.

2.4.9.2 Taylor Diagrams

Taylor (2001) introduced a method for graphically summarizing the statistical relationship between two related fields or sets of parameters. These "Taylor diagrams" are particularly useful for evaluating the relative skill of different numerical models against observations. In this case, the similarity between the modeled and observed fields is quantified in terms of their correlation coefficients (r), the degree of variation as measured by their standard deviations, and the centered root-mean-square (RMS) differences between the models and the observations. The mean values of the fields are removed before computing their higher order statistics so that the diagrams do not contain any bias

information. Only the centered pattern errors are applied.

An example of a Taylor diagram is presented in Figure 2.24. The purpose of the diagram is to show the relative skill with which eight climate models are capable of simulating the global pattern of annual mean precipitation (Taylor, 2005). The red dots with letters A—H show where the eight models are positioned within the 2D fields represented by three sets of curved lines for the three statistical parameters: (1) the circular lines radiating out from the origin (0, 0) and terminating on the x- and y-axes span the range of standard deviations (in millimeters/day) for the model simulations and for the data; the standard deviation for the single set of observations is shown by the dashed circular line (ending in "observed"); (2) the spokes emanating outward from origin span the full range (0—1) of possible correlation values that one might expect to find between the models and the observations; and (3) the last set of curved lines that span outward from where the dashed line of the observations intersects the

x-axis denotes the centered RMS difference between the simulated and observed values of the annual mean precipitation. The model which combines the highest correlation with the lowest RMS difference with the data (i.e., lies closest to the dashed line) would be considered as a possible "best performance" model. If the standard deviation of the model also resembles that of the observations, then the model is clearly performing very well.

In this example, Taylor (2005) considers model F, which has a pattern correlation of $r \sim 0.65$ and a centered RMS value of about 2.6 mm/day (based on the curved lines that radiate out from the dashed point on the x-axis), as a good candidate. The standard deviation of the simulated rainfall field is about 3.3 mm/day which is slightly greater than the standard deviation of 2.9 mm/day of the observations. Models A, B, and E lie closest to the dashed line (i.e., have standard deviations similar to the data) but only model A has a relatively low RMS error (roughly 2 mm/day) and only model A (and C) account for over 50% of the variance ($r > 0.71$) between the model and the data. Taylor (2001) notes that the diagram can be extended into a second quadrant to allow for negative correlation coefficients and that the statistics can be normalized by dividing both the RMS difference and standard deviations of the model results by the standard deviation of the observations.

The introduction of color into journal papers and online publications represents another important change in presentation method. As mentioned in the discussion of vertical sections, color shading has been used traditionally to better visually resolve horizontal and vertical gradients. Previously, most color presentations were restricted to atlas and report presentations and were not available in journal articles. New printing procedures have made color more affordable and much wider use is being made of color displays. One area of recent study where color display has played a major role is in the presentation of satellite and topographic images. Here, the use of false color enables the investigator to expand the dynamic range of the usual gray shades so that they are more easily recognizable by eye. False color is also used to enhance certain features such as sea surface temperature patterns and fronts inferred from infrared satellite images. The enhancements, and pseudocolor images, may be produced using a strictly defined function or may be developed in the interactive mode in which the analyst can produce a pleasing display. One important consideration in any manipulation of satellite images is to have each image registered to a ground map, which is generally called "image navigation" in oceanographic jargon. This navigation procedure (Emery et al., 1989b) can be carried out using satellite ephemeris data (orbital parameters) to correct for Earth curvature and rotation. Timing and spacecraft attitude errors often require the image to be "nudged" to fit the map projection exactly. An alternative method of image correction is to use a series of ground control points (GCPs) to navigate the image. GCPs are usually features such as bays or promontories that stand out in both the satellite image and the base map. In using GCP navigation a primary correction is made assuming a circular orbit and applying the mean satellite orbital parameters.

Access to digital image processing has greatly increased the investigator's capability to present and display data. Conventional data may be plotted in map form and overlain on a satellite image to show correspondence. This is possible since most image systems have one or more graphics overlay planes. Another form of presentation, partly motivated by satellite imagery, is the time sequence presentation of maps or images. Called "scene animation", this format produces a movie-style output which can be conveniently recorded on video. With widespread home use of video recorder systems, this form of data visualization is readily accessible to most people. A problem with this type of display is the present inability to publish videos

FIGURE 2.25 Perspective views of a mid-Holocene slide (Side A) and surrounding area in Douglas Channel, British Columbia using Fledermaus™ software. (a) Oblique view looking to the north showing several instability features adjacent to the slide and (b) oblique view looking to the east, showing several smaller slides at "n" at the top and back-slope of the main slide. The slide volume of 60 million m^3 was used by Thomson et al. (2012) to study landslide-generated tsunamis in the coastal region. The color code gives water depth in meters. *(From Conway et al. (2012).)*

or film loops. This greatly restricts the communication of results, which show the time evolution of a spatial field such as that shown by a series of geographically coincident satellite images. ADCP data from two or more nearby mooring sites also makes it possible to generate three-dimensional movies (using software like

QuickTime, VLC Media Player, or AVI player) of the currents, winds, and other properties simultaneously.

Digital image manipulation has also changed the way oceanographers approach data display. Using an interactive system the scientist-operator not only can change the brightness scale assignment (enhancement) but also can alter the orientation, the size (zoom in, zoom out), and the overall location of the output scene using a joystick, trackball, or mouse (digital tablet and cursor). With an interactive system, the three-dimensional display can be shifted and rotated to view all sides of the output. This allows the user to visualize areas hidden behind prominent features. Some of the most powerful applications have been developed by marine geologists and hydrographers whose display software and graphical information systems are used for navigation, marine resource mapping, and marine research (Figure 2.25).

As more oceanographers become involved with digital image processing and pseudocolor displays, there should be an increase in the variety of data and result presentations. These will not only add new information to each plot but also make the presentation of the information more interesting and "colorful". The old adage of a picture being worth a thousand words is often true in oceanography and the interests of the investigators are best served when their results can be displayed in some interesting graphical or image form.

Statistical Methods and Error Handling

3.1 INTRODUCTION

This chapter provides a review of some of the basic statistical concepts and terminology used in processing data. We need this information if we are to deal properly with the specific techniques used to edit and analyze oceanographic data. Our review is intended to establish a common level of understanding by the readers, not to provide a summary of all available procedures.

In the past, all collected data were processed and reduced by hand so that the individual scientist had an opportunity to become personally familiar with each data value. During this manual reduction of data, the investigator took into account important information regarding the particular instrument used and was able to determine which data were "better" in the sense that they had been collected and processed correctly. Within the limits of the observing systems, an accurate description of the data could be achieved without the application of statistical procedures. Individual intuition and familiarity with shipboard procedures took precedence in this type of data processing and analyses were made on comparatively few data. In such investigations, the question of statistical reliability was seldom raised and it was assumed that individual data points were a valid representation of the parameter being measured. The data values were considered to be "correct."

For the most part, the advent of the computer and electronic data collection methods has meant that a knowledge of statistical methods has become essential to any reliable interpretation of results. Circumstances still exist, however, for which physical oceanographers still assign considerable weight to the quality of individual measurements. This is certainly true of water sample data, such as dissolved oxygen, nutrients, and chemical tracers collected from bottle casts. In these cases, the established methods of data reduction, including familiarity with the data and knowledge of previous work in a particular region, still produce valuable descriptions of oceanic features and phenomena with a spatial resolution not possible with statistical techniques. However, for those more accustomed to having data collected and/or delivered on high density storage media, such as CD-ROM, optical disc, USB flash drive (also, thumb drive or key drive), or portable hard drive, statistical methods are essential to determining the value of the data and to decide how much of it can be considered useful for the intended analysis. This statistical approach arises from the fundamental complexity of the ocean, a multivariate system with many degrees of freedom in which nonlinear dynamics and sampling limitations make it difficult to separate scales of variability.

A fundamental problem with a statistical approach to data reduction is the fact that the

Data Analysis Methods in Physical Oceanography
http://dx.doi.org/10.1016/B978-0-12-387782-6.00003-X

ocean is not a stationary environment in which we can make repeated measurements. By "stationary" we mean a physical system whose statistical properties remain unchanged with time. In order to make sense of our observations, we are forced to make some rather strong assumptions about our data and the processes under investigation. Basic to these assumptions are the concept of randomness and the consequent laws of probability. Since each oceanographic measurement can be considered a superposition of the desired signal plus unwanted noise (due to measurement errors and unresolved geophysical variability), the assumption of random behavior often is applied to both the signal and the noise. We must consider not only the statistical character of the signal and noise contributions individually but also the fact that the signal and the noise can interact with each other. Only through the application of the concept of probability can we make the assumptions required to reduce this complex set of variables to a workable subset. Our brief summary of statistics will emphasize concepts pertinent to the analysis of random variables, such as probability density functions (PDFs) and statistical moments (mean, variance, etc.). A brief glossary of statistical terms can be found in Appendix B.

3.2 SAMPLE DISTRIBUTIONS

Fundamental to any form of data analysis is the realization that we are usually working with a limited set (or sample) of random events drawn from a much larger population. We use our sample to make estimates of the true statistical properties of the population. Historically, studies in physical oceanography were dependent on too few data points to allow for statistical inference and individual samples were considered representative of the true ocean. Often, an estimate of the population distribution is made from the sample set by using the relative frequency distribution, or histogram, of the measured data points. There is no fixed rule on how such a histogram is constructed in terms of ideal bin interval or number of bins. Generally, the more data there are, the greater the number of bins used in the histogram. Bins should be selected so that the majority of the measurements do not fall on the bin boundaries. Since the area of a histogram bin is proportional to the fraction of the total number of measurements in that interval, it represents the probability that an individual sample value will lie within that interval (Figure 3.1).

The most basic descriptive parameter for any set of measurements is the sample mean. The mean is generally taken over the duration of a time series (time average) or over an ensemble of measurements (ensemble mean) collected under similar conditions (Table 3.1). If the sample has N data values, $x_1, x_2, ..., x_N$, the sample mean is calculated as

$$\bar{x} = \frac{1}{N} \sum_{i=1}^{N} x_i \tag{3.1}$$

The sample mean is an unbiased estimate of the true population mean, μ. Here, an "unbiased" estimator is one for which the expected value, $E[x]$, of the estimator is equal to the parameter being estimated. In this case, $E[x] = \mu$ for which $\bar{x}$ is an unbiased estimator. The sample mean locates the center of mass of the data distribution such that

$$\sum_{i=1}^{N} (x_i - \bar{x}) = 0$$

that is, the sample mean splits the data so that there is an equal weighting of negative and positive values of the fluctuation, $x' = x_i - \bar{x}$, about the mean value, $\bar{x}$. The weighted sample mean is the general case of Eqn (3.1) and is defined as

$$\bar{x} = \frac{1}{N} \sum_{i=1}^{N} f_i x_i \tag{3.2}$$

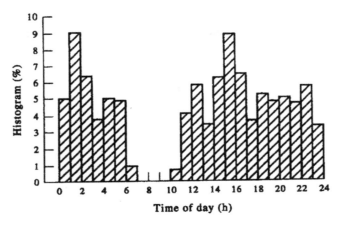

FIGURE 3.1 Histogram giving the percentage occurrences for the times of satellite position fixes during a 24-h day. Data are for satellite-tracked surface drifter #4851 deployed in the northeast Pacific Ocean from 10 December 1992 to 28 February 1993. During this 90-day period, the satellite receiver on the drifter was in the continuous receive mode.

where f_i/N is the relative frequency of occurrence of the ith value for the particular experiment or observational data set. In Eqn (3.1), $f_i = 1$ for all i.

The sample mean values give us the center of mass of a data distribution but not its width or how broadly the sample values are distributed. To determine how the data are spread about the mean, we need a measure of the sample variability or *deviation*, which is expressed in terms of the positive square root of the sample *variance*. For the data used in Eqn (3.1), the *sample variance* is the average of the square of the sample deviations from the sample mean, expressed as

$$s'^2 = \frac{1}{N} \sum_{i=1}^{N} (x_i - \bar{x})^2 \qquad (3.3)$$

The *sample standard deviation*, $s' = \sqrt{s'^2}$, the positive square root of Eqn (3.3), is a measure

of the typical difference of a data value from the mean value of all the data points. In general, these differ from the corresponding true *population variance*, σ^2, and the *population standard deviation*, σ. As defined by Eqn (3.3), the sample variance is a biased estimate of the true population variance. An unbiased estimator of the population variance, s, is obtained from

$$s^2 = \frac{1}{(N-1)} \sum_{i=1}^{N} (x_i - \bar{x})^2 \qquad (3.4a)$$

$$= \frac{1}{(N-1)} \left[\sum_{i=1}^{N} (x_i)^2 - \frac{1}{N} \left(\sum_{i=1}^{N} x_i \right)^2 \right] \qquad (3.4b)$$

where the denominator $N-1$ expresses the fact that we need at least two values to define a sample variance and standard deviation, s. The use of the estimators s vs s' is often a matter of debate among oceanographers, although it should be

TABLE 3.1 Statistical Values for the Data Set $x = \{x_i, i = 1,\dots, 9\} = \{-3, -1, 0, 2, 5, 7, 11, 12, 12\}$

Mean, $\bar{x}$	Biased Variance, s'^2	Unbiased Variance, s^2	Standard Deviation, s	Range	Median	Mode
5.00	30.22	34.00	5.83	15	5	12

noted that the difference between the two values decreases as the sample size increases. Only for relatively small samples $(N < 30)$ is the difference significant. Because s' has a smaller mean square (MS) error than s (suggesting a lower error than might be the case) and because s is an unbiased estimator when the population mean is known a priori, we recommend the use of Eqns (3.4a,b). However, a word of caution: if your hypothesis depends on the difference between s and s', then you have ventured onto shaky statistical ground supported by questionable data. We further note that the expanded relation Eqn (3.4b) is a more efficient computational formulation than Eqn (3.4a) in that it allows one to obtain s^2 from a single pass through the data. If the sample mean must be calculated first, two passes through the same data set are required rather than one, which is computationally less efficient when dealing with large data sets.

Other statistical values of importance are the range, mode, and median of a data distribution (Table 3.1). The *range* is the spread or absolute difference between the endpoint values of the data set while the *mode* is the value of the distribution that occurs most often. For example, the data sequence 2, 4, 4, 6, 4, 7 has a range of $|2 - 7| = 5$ and a mode of 4. The *median* is the middle value in a set of numbers arranged according to magnitude (the data sequence -1, 0, 2, 3, 5, 6, 7 has a median of 3). If there is an even number of data points, the median value is chosen midway between the two candidates for the central value. Two other measures, *skewness* (the third moment of the distribution and degree of asymmetry of the data about the mean) and *kurtosis* (a nondimensional number measuring the flatness or peakedness of a distribution) are less used in oceanography.

As we discuss more thoroughly later in this chapter, the shapes of many sample distributions can be approximated by a *normal* (also called a *bell* or *Gaussian)* distribution. A convenient aspect of a normal population distribution is

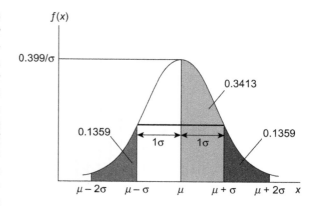

FIGURE 3.2 Normal distribution $f(x)$ for mean μ and standard deviation σ of the random variable X. Numerical values denote the relative areas for the different limit ranges. *(From Harriett and Murphy (1975).)*

that we can apply the following empirical "rule of thumb" to the data:

$\mu \pm \sigma$ spans approximately 68% of the measurements;

$\mu \pm 2\sigma$ spans approximately 95% of the measurements;

$\mu \pm 3\sigma$ spans most (99%) of the measurements.

The percentages are represented by the areas under the normal distribution curve spanned by each of the limits (Figure 3.2). We emphasize that the above limits apply only to normal distributions of random variables.

3.3 PROBABILITY

Most data collected by oceanographers are made up of samples taken from a larger unknown population. If we view these samples as random events of a statistical process, then we are faced with an element of uncertainty: "What are the chances that a certain event occurred or will occur based on our sample?" or "How likely is it that a given sample is truly representative of a certain population distribution?" (The last question might be asked of political pollsters who use small sample sizes to make sweeping statements about the opinions of the populace as a whole.) We need to find the best procedures for inferring the population distribution from the sample distribution and to have measures that specify the goodness of the inference. Probability theory provides the foundation for this type of analysis. In effect, it enables us to find a value between 0 and 1, which tells us just how likely is a particular event or sequence of events. A probability is a proportional measure of the occurrence of an event. If the event has a probability of zero, then it is impossible; if it has a probability of unity, then it is certain to occur. Probability theory as we know it today was initiated by Pascal and Fermat in the seventeenth century through their interest in games of chance. In the eighteenth century, Gauss and Laplace extended the theory to social sciences and actuarial mathematics. Well-known names like R.A. Fisher, J. Neyman, and E.S. Pearson are associated with the proliferation of statistical techniques developed in the twentieth century. The *frequentist* statistical techniques developed by these authors—which is the focus of this section—define probability in terms of the outcomes of a large number of near identical trials. The alternative *Baysian* statistical technique, named after Thomas Bayes, an eighteenth-century theologian and mathematician, views probability in terms of conditional probability in which the likelihood of a particular result or event depends on some prior probability, which is then updated when new information is obtained.

The *probability mass function*, $P(x)$, gives the relative frequency of occurrence of each possible value of a discrete random variable, X. Put another way, the function specifies the point probabilities $P(x_i) = P(X = x_i)$ and assumes nonzero values only at points $X = x_i$, $i = 1, 2, \ldots$. One of the most common examples of a probability mass function is the sum of the dots obtained from the roll of a pair of dice (Table 3.2). According to probability theory, the dice player is most

TABLE 3.2   The Discrete Probability Mass Function and Cumulative Probability Functions for the Sum of the Dots (Variable X) Obtained by Tossing a Pair of Dice

Sum of Dots (X)	Frequency of Occurrence	Relative Frequency	Probability Mass Function, $P(x)$	Cumulative Probability Function $F(x) = P(X \leq x)$
2	1	1/36	$P(x=2) = 1/36$	$F(2) = P(X \leq 2) = 1/36$
3	2	2/36	$P(x=3) = 2/36$	$F(3) = P(X \leq 3) = 3/36$
4	3	3/36	$P(x=4) = 3/36$	$F(4) = P(X \leq 4) = 6/36$
5	4	4/36	$P(x=5) = 4/36$	$F(5) = P(X \leq 5) = 10/36$
6	5	5/36	$P(x=6) = 5/36$	$F(6) = P(X \leq 6) = 15/36$
7	6	6/36	$P(x=7) = 6/36$	$F(7) = P(X \leq 7) = 21/36$
8	5	5/36	$P(x=8) = 5/36$	$F(8) = P(X \leq 8) = 26/36$
9	4	4/36	$P(x=9) = 4/36$	$F(9) = P(X \leq 9) = 30/36$
10	3	3/36	$P(x=10) = 3/36$	$F(10) = P(X \leq 10) = 33/36$
11	2	2/36	$P(x=11) = 2/36$	$F(11) = P(X \leq 11) = 35/36$
12	1	1/36	$P(x=12) = 1/36$	$F(12) = P(X \leq 12) = 1$
SUM	36	1.00		1.00

likely to roll a 7 (highest probability mass function) and least likely to roll a 2 or 12 (lowest probability mass function). The dice example reveals two of the fundamental properties of all discrete probability functions: (1) $0 \leq P(X = x)$; and (2) $\sum P(x) = 1$, where the summation is over all possible values of x. The counterpart to $P(x)$ for the case of a continuous random variable X is *the PDF*, $f(x)$, which we discuss, more in detail, later in this chapter. For the continuous case, the above fundamental properties become: (1) $0 \leq f(x)$; and (2) $\int f(x)dx = 1$ where the integration is over all x in the range $(-\infty, \infty)$.

To further illustrate the concept of probability, consider N independent trials, each of which has the same probability of "success" p and probability of "failure" $q = 1 - p$. The probability of success or failure is unity; $p + q = 1$. Such trials involve binomial distributions for which the outcomes can be only one of two events: for example, a tossed coin will produce a head or a tail; an XBT will work or it will not work. If X represents the number of successes that occur in the N trials, then X is said to be a discrete random variable having parameters (N, p). The term "Bernoulli trial" is sometimes used for X. The probability mass function that gives the relative frequency of occurrence of each value of the random variable X having parameters (N, p) is the binomial distribution.

$$p(x) = \binom{N}{x} p^x (1-p)^{N-x}, \quad x = 0, 1, ..., N$$

(3.5a)

where the expression

$$\binom{N}{x} = \binom{N}{N-x} = {}_NC_x \equiv N!/[(N-x)!x!]$$

(3.5b)

is the number of different *combinations* of groups of x objects that can be chosen from a total set of N objects without regard to order. The number of different combinations of x objects is always fewer than the number of *permutations*, ${}_NP_x$, of x objects $[{}_NP_x \equiv N!/(N-x)!]$. In the case of permutations, different ordering of the same objects

counts for a different permutation (i.e., *ab* is different than *ba*). As an example, the number of possible different batting orders (permutations) a coach can create among the first four hitters on a nine-person baseball team is $9!/(9-4)! = 9!/5! = 3024$. Each different ordering of the four players counts as a permutation. In contrast, the number of different groups of ball players a coach can put in the first four lead-off batting positions without regard to any particular batting order is $9!/[(9-4)!4!] = 9!/5!4! = 126$. The numbers

$$\binom{N}{x}$$

often are called *binomial coefficients* since they appear as coefficients in the expansion of the binomial expression $(a + b)^N$ given by the binomial theorem:

$$(a + b)^N = \sum_{k=0}^{N} \binom{N}{k} a^k b^{N-k}$$

(3.6)

The summed probability mass function

$$P(a \leq x \leq b) = \sum_{a}^{b} P(x)$$

for variable X over a specified range of values (a, b) can be demonstrated by a simple oceanographic example. Suppose there is a probability, $1 - p$, that a current meter will fail when moored in the ocean and that the failure is independent from current meter to current meter. Assume that a particular string of single-point meters will successfully measure the expected flow structure if at least 50% of the meters on the string remain operative. For example, a two-instrument string used to measure the barotropic flow will be successful if one current meter remains operative while a four-instrument string used to resolve the low mode, baroclinic flow will be successful if at least 2 instruments remain operative. We then ask: "For what values of p is a 4-instrument array preferable to a 2-instrument array?" Since each current meter is assumed to fail or function independently of the other meters, it follows that the number of functioning current meters is a binomial random variable.

The probability that a 4-instrument mooring is successful is then

$$P(2 \leq x \leq 4) = \sum_{k=2}^{4} \binom{4}{k} p^k (1-p)^{4-k}$$

$$= \binom{4}{2} p^2 (1-p)^2 + \binom{4}{3} p^3 (1-p)^1$$

$$+ \binom{4}{4} p^4 (1-p)^0$$

$$= 6p^2 (1-p)^2 + 4p^3 (1-p)^1 + p^4$$

Similarly, the probability that a 2-instrument array is successful is

$$P(1 \leq x \leq 2) = \sum_{k=1}^{2} \binom{2}{k} p^k (1-p)^{2-k}$$
$$= 2p(1-p) + p^2$$

From these two relations, we find that the 4-instrument string is more likely to succeed when

$$6p^2 (1-p)^2 + 4p^3 (1-p)^1 + p^4 \geq 2p(1-p) + p^2$$

or, after some factoring and simplification, when

$$(p-1)^2 + (3p-2) \geq 0$$

for which we find $3p - 2 \geq 0$, or $p \geq 2/3$. When compared to the 2-instrument array, the 4-instrument array is more likely to do its intended job when the probability, p, that the instrument works is $p \geq 2/3$. The 2-instrument array is more likely to succeed when $p \leq 2/3$.

In the previous example, specification of the probability p requires information on the rate of failure of the current meters. This, in turn, requires information on success rates from previous mooring programs that used these particular instruments. When examining these success rates, we would likely make the fundamental assumption, applicable to most of the data sets we collect, that each sample in our set of observations is an independent realization drawn from a random distribution. Individual events in this distribution cannot be predicted with certainty but their relative frequency of

occurrence, for a long series of repeated trials (samples), is often remarkably stable. We further remark that the binomial distribution discussed in the context of current meters is only one type of PDF. Other distribution functions will be discussed later in the chapter.

3.3.1 Cumulative Probability Functions

The probability mass function yields the probability of a specific event or probability of a range of events. From this function we can derive the *cumulative probability function, F(x)*—also called the cumulative distribution function, cumulative mass function, and probability distribution function—defined as that fraction of the total number of possible outcomes X (a random variable), which are less than a specific value x (a number). Thus, the distribution function is the probability that $X \leq x$, or

$$F(x) = P(X \leq x)$$

$$= \sum_{allX \leq x} P(x), \quad -\infty < x < \infty \tag{3.7a}$$
$$\text{(discrete random variable, } X)$$

$$= \int_{-\infty}^{x} f(x) dx \tag{3.7b}$$
$$= \text{(continuous random variable, } X)$$

The discrete cumulative distribution function for tossing a pair of fair dice (Table 3.2) is plotted in Figure 3.3. Since the probabilities P and f are limited to the range 0 and 1, we have $F(-\infty) = 0$ and $F(\infty) = 1$. In addition, the distribution function $F(x)$ is a nondecreasing function of x, such that $F(x_1) \leq F(x_2)$ for $x_1 < x_2$, where $F(x)$ is continuous from the right (Table 3.2).

It follows that, for the case of a continuous function, the derivative of the distribution function F with respect to the sample parameter, x

$$f(x) = \frac{dF(x)}{dx} \tag{3.8}$$

(a)   **(b)**

FIGURE 3.3   The discrete mass function $P(x)$ (a) and cumulative distribution function $F(x)$ (b) from tossing a pair of dice (see Table 3.2). *(From Harnett and Murphy (1975).)*

recovers the PDF, f. As noted earlier, the PDF has the property that its integral over all values is unity

$$\int_{-\infty}^{\infty} f(x)dx = F(\infty) - F(-\infty) = 1$$

In the limit $dx \to 0$ the fraction of outcomes for which x lies in the interval $x < x' < x + dx$ is equal to $f(x')dx$, the probability for this interval. The random variables being considered here are continuous so that the PDF can be defined by Eqn (3.8). Variables with distribution functions that contain discontinuities, such as the steps in Figure 3.3, are called discrete variables. A random variable is considered discrete if it assumes only a countable number of values. In most oceanographic sampling, measurements can take on an infinity of values along a given scale and the measurements are best considered as continuous random variables. The function $F(x)$ for a continuous random variable X is itself continuous and appears as a smooth curve. Similarly, the PDF for a continuous random variable X is continuous and can be used to evaluate the probability that X falls within some interval $[a, b]$ as

$$P(a \leq X \leq b) = \int_{a}^{b} f(x)dx \qquad (3.9)$$

3.4 MOMENTS AND EXPECTED VALUES

The discussion in the previous section allows us to determine the probability of a single event or experiment, or describe the probability of a set of outcomes for a specific random variable. However, our discussion is not concise enough to describe fully the probability distributions of our data sets. The situation is similar to Section 3.2 in which we started with a set of observed values. In addition to presenting the individual values, we seek properties of the data, such as the sample mean and variance to help us characterize the structure of our observations. In the case of probability distributions, we are not dealing with the *observed* mean and variance but with the *expected* mean and variance obtained from an infinite number of realizations of the random variable under consideration.

Before discussing some common PDFs, we need to review the computation of the parameters used to describe these functions. These parameters are, in general, called "moments" by analogy to mechanical systems where moments describe the distribution of forces relative to some reference point. The statistical concept of degrees of freedom is also inherited from the terminology of physical mechanical systems where the number of degrees of freedom specifies the motion possible within the physical constraints of the mechanical system and its distribution of forces. As noted earlier, the population mean, μ, and standard deviation, σ, define the first and second moments that describe the center and spread (distribution about the center) of the probability function. In general, these parameters do not uniquely define the PDF since many different PDFs can have the same mean and standard deviation. However, in the case of the Gaussian distribution, the PDF is completely described by μ and σ. In defining moments, we must be careful to distinguish between moments taken about the origin and moments taken about the mean (central moments).

When discussing moments it is useful to introduce the concept of expected value. This concept is analogous to the notion of weighted functions. For a discrete random variable, X, with a probability function $P(x)$ (the discrete analogue to the continuous PDF), the expected value of X is written as $E[X]$ and is equivalent to the arithmetic mean, μ, of the probability distribution. In particular, we can write the expected value for a discrete PDF as

$$E[X] = \sum_{i=1}^{N} x_i P(x_i) = \mu \qquad (3.10)$$

where μ is the population mean introduced in Section 3.2. The probability function $P(x)$ serves as a weighting function similar to the function f_i/N in Eqn (3.6). The difference is that f_i/N is the relative frequency for a single set of experimental samples whereas $P(x)$ is the expected relative frequency for an infinite number of samples from repeated trials of the experiment. The expected value, $E[X]$, for the sample which includes X, is the sample mean, $\bar{x}$. Similarly, the variance of the random variable X is the expected value of $(X - \mu)^2$, or

$$V[X] = E\left[(X - \mu)^2\right] = \sum_{i=1}^{N} (x_i - \mu)^2 P(x_i) = \sigma^2$$

$$(3.11)$$

In the case of a continuous random variable, X, with PDF $f(x)$, the expected value is

$$E[X] = \int_{-\infty}^{\infty} xf(x)dx \qquad (3.12)$$

while for any function $g(X)$ with a PDF $f(x)$, the expected value can be written as

$$E[g(X)] = \int_{-\infty}^{\infty} g(x)f(x)dx \text{ (continuous variable)}$$

$$(3.13a)$$

$$= \sum_{i=1}^{N} g(x_i)P(x_i) \text{ (discreate case)} \qquad (3.13b)$$

some useful properties of expected values for random variables are:

1. For c = constant; $\quad E[c] = c, \quad V[c] = 0$;
2. $E[cg(X)] = cE[g(X)], V[cg(X)] = c^2 V[g(X)]$;
3. $E[g_1(X) \pm g_2(X) \pm \ldots] = E[g_1(X)] \pm E[g_2(X)] \pm \ldots$;
4. $V[g(X)] = E[(g(X) - \mu)^2] = E[g(X)^2] - \mu^2$, (variance about the mean);
5. $E[g_1 g_2] = E[g_1]E[g_2]$;
6. $V[g_1 \pm g_2] = V[g_1] + V[g_2] \pm 2C[g_1, g_2]$.

Property (6) introduces the *covariance function* of two variables, C, defined as

$$C[g_1, g_2] = E[g_1 g_2] - E[g_1]E[g_2] \qquad (3.14)$$

where, using property (5), $C = 0$ when g_1 and g_2 are independent random variances. Using properties (1) to (3), we find that $E[Y]$ for the linear relation $Y = a + bX$ can be expanded to

$$E[Y] = E[a + bX] = a + bE[X]$$

while from (1) and (6) we find

$$V[Y] = V[a + bX] = b^2 V[X]$$

At this point, we remark that averages, expressed as expected values, $E[X]$, apply to ensemble averages of many (read, infinite) repeated samples. This means that each sample is considered to be drawn from an infinite ensemble of identical statistical processes varying under exactly the same conditions. In practice, we do not have a large number of repeated samples taken under identical conditions but rather time (or space) records having limited temporal (or spatial) extent. In using time or space averages as representative of ensemble averages, we are assuming that our records are *ergodic*. This implies that averages over an infinite ensemble can be replaced by an average over a single, infinitely long time series. An ergodic process is not to be confused with a stationary process for which the PDF of $X(t)$ is independent of time. In reality, time/space series can be considered stationary if major shifts in the statistical characteristics of the series occur over intervals that are long compared to the averaging interval so that the space/time records remain homogeneous (exhibit the same general behavior) throughout the selected averaging interval. A data record that is quiescent during the first half of the record and then exhibits large irregular oscillations during the second half of the record is not stationary.

3.4.1 Unbiased Estimators and Moments

As we stated earlier, $\bar{x}$ and s^2 defined by Eqns (3.2) and (3.4) are unbiased estimators of the true population mean, μ, and variance, σ^2. That is, the expected values of x and $(x - \bar{x})^2$ are equal to μ and σ^2, respectively. To illustrate the nature of the expected value, we will first prove that $E[\bar{x}] = \mu$. We write the expected value as the normalized sum of all $\bar{x}$ values

$$E[x] = E\left[\frac{1}{N}\sum_{i=1}^{N} x_i\right] = \frac{1}{N}\sum_{i=1}^{N} E[x_i]$$

$$= \frac{1}{N}\sum_{i=1}^{N}\mu = \mu$$

as required. Next, we demonstrate that $E[s^2] = \sigma^2$. We again use the appropriate definitions and write

$$E[s^2] = E\left[\frac{1}{N-1}\sum_{i=1}^{N}(x_i - \bar{x})^2\right]$$

$$= E\left[\frac{1}{N-1}\left\{\sum_{i=1}^{N}\left[(x_i - \mu)^2 - N(\bar{x} - \mu)^2\right]\right\}\right]$$

$$= \frac{1}{N-1}\left\{\sum_{i=1}^{N}E\left[(x_i - \mu)^2\right] - NE\left[(\bar{x} - \mu)^2\right]\right\}$$

$$= \frac{1}{N-1}\left\{\sum_{i=1}^{N}\sigma^2 - N\frac{\sigma^2}{N}\right\} = \frac{\sigma^2}{N-1}(N-1) = \sigma^2$$

where we have used the relations $x_i - \bar{x} = (x_i - \mu) - (\bar{x} - \mu)$, $E[(x_i - \mu)^2] = V[x_i] = \sigma^2$ (the variance of an individual trial) and $E[(x_i - \mu)^2] = V[x_i] = \sigma^2/N$ (the variance of the sample mean relative to the population mean). The last expression derives from the central limit theorem discussed in Section 3.6.

Returning to the discussion of statistical moments, we define the ith moment of the random variable X, taken about the origin, as

$$E\left[X^i\right] = \mu_i \tag{3.15}$$

Thus, the first moment about the origin ($i = 1$) is the population mean, $\mu = \mu_1$. Similarly, we can define the ith moment of X taken about the mean (called the ith central moment of X) as

$$E\left[(X - \mu)^i\right] = \mu_i \tag{3.16}$$

The population variance, σ^2, is the second ($i = 2$) central moment, μ_2.

3.5 COMMON PDFs

The purpose of this section is to provide examples of three common PDFs. The first is the uniform PDF given by

$$f(x) = \frac{1}{x_2 - x_1}, \quad x_1 \le x \le x_2$$
$$= 0, \quad \text{otherwise} \tag{3.17}$$

(a)

(b)

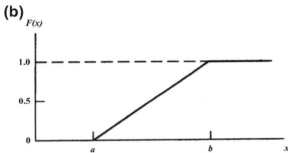

FIGURE 3.4 Uniform probability density distribution functions, (a) The probability density function, $f(x)$; and (b) the corresponding cumulative probability distribution function, $F(x)$. *(From Bendat and Piersol (1986).)*

(Figure 3.4) which is the intended PDF of random numbers generated by most computers and handheld calculators. The function is usually scaled between 0 and 1. The cumulative density function $F(x)$ given by Eqn (3.7b) has the form

$$F(x) = 0, \quad x < x_1$$

$$= \frac{x - x_1}{x_2 - x_1}, \quad x_1 \le x \le x_2$$

$$= 1, \quad x \ge x_2$$

while the mean and standard deviation of Eqn (3.17) are given by $\mu = (x_2 + x_1)/2$ and $\sigma = (x_2 - x_1)/2\sqrt{3}$.

Perhaps the most familiar and widely used PDF is the normal (or Gaussian) density function:

$$f(x) = \frac{e^{[-(x-\mu)^2/2\sigma^2]}}{\sigma\sqrt{2\pi}}, \quad \sigma > 0, \quad (3.18)$$

$$-\infty < \mu < \infty, \quad -\infty < x < \infty$$

where the parameter σ represents the standard deviation (or spread) of the random variable X about its mean value μ (Figure 3.2). For convenience, Eqn (3.18) is often written in shorthand notion as $N(\mu, \sigma^2)$. The height of the density function at $x = \mu$ is $0.399/\sigma$. The cumulative probability distribution of a normally distributed random variable, X, lying in the interval a to b is given by the integral (Eqn (3.9)).

$$P(a \le X \le b) = \int_a^b \frac{e^{[-(x-\mu)^2/2\sigma^2]}}{\sigma\sqrt{2\pi}} dx \quad (3.19)$$

which is the area under the normal curve between a and b. Since a closed form of this integral does not exist, it must be evaluated by approximate methods, often involving the use of tables of areas. We have included a table of curve areas in Appendix D (Table D.1). The normal distribution is symmetric with respect to μ so that areas need to be tabulated only on one side of the mean. For example, $P(\mu \le x \le \mu + 1\sigma) = 0.3413$ so by symmetry $P(\mu - 1\sigma \le x \le \mu + 1\sigma) = 2(0.3413) = 0.6826$. The latter is the value used in the rule of thumb estimates for the range of the standard deviation, σ. For the normal distribution, the tabulated values represent the area between the mean and a point z, where z is the distance from the mean measured in standard deviations. This leads to the familiar transform for a normal random variable X given by

$$Z = \frac{X - \mu}{\sigma} \quad (3.20)$$

called the standardized normal variable. The variable Z gives the distances of points measured from the mean of the normal random variable in terms of the standard deviation of the normal random variable, X (Figure 3.5). The standard normal variable Z is normally distributed with a mean of zero (0) and a standard deviation of unity (1). Thus, if X is described by the function $N(\mu, \sigma^2)$ then Z is described by the function $N(0, 1)$.

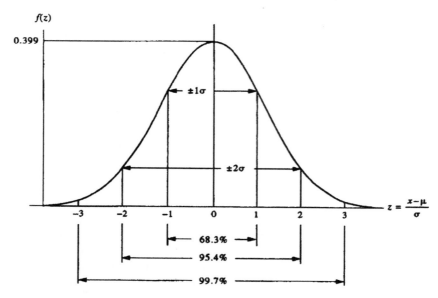

FIGURE 3.5   Distribution $f(z)$ for the standardized normal random variable, $Z = (X - \mu)/\sigma$ (cf. Figure 3.2). *(From Harriett and Murphy (1975).)*

Our third continuous PDF is the gamma density function, which applies to random variables, which are always nonnegative thus producing distributions that are skewed to the right. The gamma PDF is given by

$$f(x) = \frac{x^{\sigma-1}e^{-x/\beta}}{\beta^a \Gamma(\alpha)} \alpha, \quad \beta > 0; \quad 0 \leq x \leq \infty$$

$$= 0, \quad \text{elsewhere}$$

$$(3.21)$$

where σ and β are parameters of the distribution and $\Gamma(a)$ is the gamma function

$$\Gamma(a) = \int_0^\infty x^{\alpha-1}e^{-x}dx \qquad (3.22)$$

For any integer n,

$$\Gamma(n) = 1 \cdot 2 \cdot 3 \cdot \ldots \cdot (n-1) = (n-1)! \quad (3.23)$$

while for a continuous variable α

$$\Gamma(\alpha) = (\alpha-1)\Gamma(\alpha-1), \quad \text{for} \quad \alpha \geq 1 \quad (3.24)$$

where $\Gamma(1) = 1$. Plots of the gamma PDF for $\beta = 1$ and three values of the parameter α are presented in Figure 3.6. Since it is again impossible to define a closed form of the integral of the PDF in Eqn (3.21), tables are used to evaluate probabilities from the PDF. One particularly important gamma density function has a PDF with $\alpha = \nu/2$ and $\beta = 2$. This is the *chi-square random distribution* (written as χ_ν^2 and pronounced "ki square") with ν degrees of freedom (Appendix D, Table D.2). The chi-square distribution gets its name from the fact that it involves the square of normally distributed random variables, as we will explain shortly. Up to this point, we have dealt with a single random variable X and its standard normalized equivalent, $Z = (X - \mu)/\sigma$. We now wish to investigate the combined properties of more than one standardized independent normal variable. For example, we might want to investigate the distributions of temperature differences between reversing thermometers and a CTD (conductivity-temperature-depth) thermistor for

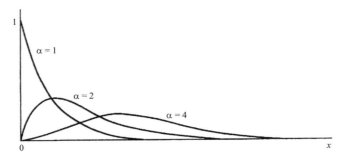

FIGURE 3.6 Plots of the gamma function for various values of the parameter α ($\beta = 1$).

a suite of CTD vs reversing thermometer inter-comparisons taken at the same location. Each cast is considered to produce a temperature difference distribution x_k with a mean μ_k and a variance σ_k^2. The set of standardized independent normal variables Z_k formed from the casts is assumed to yield ν independent standardized normal variables $Z_1, Z_2,..., Z_\nu$. The new random variable formed from the sum of the squares of the variables $Z_1, Z_2,..., Z_\nu$ is the chi-square variable χ_ν^2 where

$$\chi_\nu^2 = Z_1^2 + Z_2^2 + ... + Z_\nu^2 \qquad (3.25)$$

has ν degrees of freedom. For the case of our temperature comparison, this represents the square of the deviations for each cast about the mean. Properties of the distribution are

$$\text{Mean} = E[\chi_\nu^2] = \nu \qquad (3.26a)$$

$$\text{Variance} = E[(\chi_\nu^2 - \nu)^2] = 2\nu \qquad (3.26b)$$

We will make considerable use of the function χ_ν^2 in our discussion concerning confidence intervals for spectral estimates.

It bears repeating that PDFs are really just models for real populations whose distributions we do not know. In many applications, it is not important that our PDF be a precise description of the true population since we are mainly concerned with the statistics of the distributions as provided by the probability statements from the model. It is not, in general, a simple problem

to select the right PDF for a given data set. Two suggestions are worth mentioning: (1) use available theoretical considerations regarding the process that generated the data; and (2) use the data sample to compute a frequency histogram and select the PDF that best fits the histogram. Once the PDF is selected, it can be used to compute statistical estimates of the true population parameters.

We also keep in mind that our statistics are computed from, and thus are functions of, other random variables and are, therefore, themselves random variables. For example, consider sample variables $X_1, X_2,..., X_N$ from a normal population with mean μ and variance σ^2, then

$$\overline{X} = \frac{1}{N} \sum_{i=1}^{N} X_i \qquad (3.27)$$

is normally distributed with mean μ and variance σ^2/N. From this it follows that

$$Z = \frac{\overline{X} - \mu}{\sigma_x} = \frac{\overline{X} - \mu}{\sigma/\sqrt{N}} = \sqrt{N}\,\frac{\overline{X} - \mu}{\sigma} \qquad (3.28)$$

has a standard normal distribution $N(0, 1)$ with zero mean and variance of unity. Using the same sample, $X_1, X_2,..., X_N$, we find that

$$\frac{1}{\sigma^2} \sum_{i=1}^{N} (X_i - \overline{X})^2 = \frac{(N-1)s^2}{\sigma^2} = \chi_\nu^2 \qquad (3.29)$$

has a chi-square distribution (χ_ν^2) with $\nu = (N-1)$ degrees of freedom. (Only $N - 1$, not N, degrees of freedom are available since the estimator requires use of the mean which reduces the degrees of freedom by one.) Here, the sample standard deviation, s, is an unbiased estimate of σ. We also can use $(X - \overline{X})/(s/\sqrt{N})$ as an estimate of the standard normal statistic, $(X - \mu)/(\sigma/\sqrt{N})$. The continuous sample statistic $(X - \overline{X})/(s/\sqrt{N})$ has a PDF known as the *Student's t-distribution* (Appendix D, Table D.3) with $(N - 1)$ degrees of freedom. The name derives from an Irish statistician named W.S. Gossett who was one of the first to work on the statistic. Because his employer would not allow employees to publish their research, Gossett published his results under the name "Student" in 1908. Mathematically, the random variable t is defined as a standardized normal variable divided by the square root of an independently distributed chi-square variable divided by its degrees of freedom; viz $t = z/\sqrt{(\chi_\nu^2/\nu)}$, where z is the standard normal distribution. Thus, one can safely use the normal distribution for samples $\nu > 30$, but for smaller samples one must use the t-distribution. In other words, the normal distribution gives a good approximation to the t-distribution only for $\nu > 30$.

The above relations for statistics computed from a normal population are important for two reasons:

1. Often, the data or the measurement errors can be assumed to have population distributions with normal PDFs;
2. One is working with averages that themselves are normally distributed regardless of the PDF of the original data. This statement is a version of the well-known *central limit theorem*.

3.6 CENTRAL LIMIT THEOREM

Let $X_1, X_2, \ldots, X_i, \ldots$ be a sequence of independent random variables with $E[X_i] = \mu_i$ and $V[X_i] = \sigma_i^2$. Define a new random variable $X = X_1 + X_2 + \ldots + X_N$. Then, as N becomes large, the standard normalized variable

$$Z_N = \frac{\left(X - \sum_{i=1}^{N}\mu_i\right)}{\left(\sum_{i=1}^{N}\sigma_i^2\right)^{1/2}} \qquad (3.30)$$

takes on a normal distribution regardless of the distribution of the original population variable from which the sample was drawn. The fact that the X_i values may have any kind of distribution, and yet the sum X may be approximated by a normally distributed random variable, is the basic reason for the importance of the normal distribution in probability theory. For example, X might represent the summation of fresh water added to an estuary from a large number of rivers and streams, each with its own particular form of variability. In this case, the sum of the rivers and stream input would result in a normal distribution of the input of fresh water. Alternatively, the variable X, representing the success or failure of an AXBT launch, may be represented as the sum of the following independent binomially distributed random variables (a variable that can only take on one of two possible values).

$$\begin{aligned} X_i &= 1 \quad \text{if the } i\text{th cast is a success} \\ &= 0 \quad \text{if the } i\text{th cast is a failure} \end{aligned}$$

with $X = X_1 + X_2 + \ldots + X_N$. For this random variable, $E[X] = Np$ and $V[X] = Np(1 - p)$. For large N, it can be shown that the variable $(X - E[X])/\sqrt{V[X]}$ closely resembles the normal distribution, $N(0, 1)$.

A special form of the central limit theorem may be stated as: the distribution of mean values calculated from a suite of random samples X_i ($X_{i,1}, X_{i,2}, \ldots$) taken from a discrete or continuous population having the same mean μ and variance σ^2 approaches the normal distribution with mean μ and variance σ^2/N as N goes to infinity. Consequently, the distribution of the arithmetic mean

$$\overline{X} = \frac{1}{N}\sum_{i=1}^{N} X_i \qquad (3.31)$$

is asymptotically normal with mean μ and variance σ^2/N when N is large. Ideally, we would like $N \to \infty$ but, for practical purposes, $N \geq 30$ will generally ensure that the population of X is normally distributed. When N is small, the shape of the sample distribution will depend mainly on the shape of the parent population. However, as N becomes larger, the shape of the sampling distribution becomes increasingly more like that of a normal distribution no matter what the shape of the parent population (Figure 3.7). In many instances, the normality

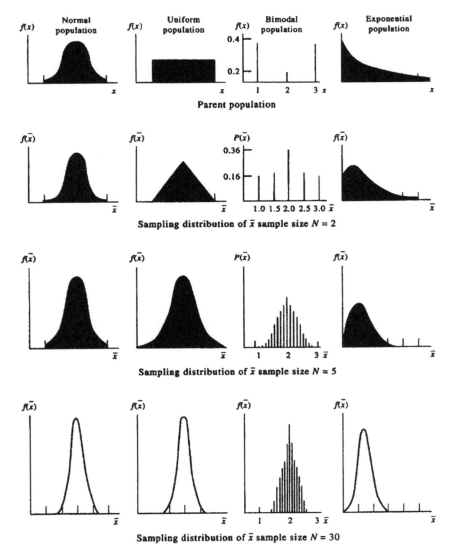

FIGURE 3.7 Sampling distribution of the mean $\bar{x}$ for different types of population distributions for increasing sample size, $N = 2$, 5, and 30. The shape of the sampling distribution becomes increasingly more like that of a normal distribution regardless of the shape of the parent population.

assumption for the sampling distribution for $\overline{X}$ is reasonably accurate for $N > 4$ and quite accurate for $N > 10$ (Bendat and Piersol, 1986).

The central limit theorem has important implications for we often deal with average values in time or space. For example, current meter systems average over some time interval, allowing us to invoke the central limit theorem and assume normal statistics for the resulting data values. Similarly, data from high-resolution CTD systems are generally vertically averaged (or averaged over some set of cycles in time), thus approaching a normal PDF for the data averages, via the central limit theorem. An added benefit of this theorem is that the variance of the averages is reduced by the factor N, the number of samples averaged. The theorem essentially states that individual terms in the sum contribute a negligible amount to the variation of the sum, and that it is not likely that any one value makes a large contribution to the sum. Errors of measurements certainly have this characteristic. The final error is the sum of many small contributions none of which contributes very much to the total error. Note that the sample standard error is an unbiased estimate (again in the sense that the expected value is equal to the population parameter being estimated) even though the component sample standard deviation is not.

To further illustrate the use of the central limit theorem, consider a set of independent measurements of a process whose probability distribution is unknown. Through previous experimentation, the distribution of this process was estimated to have a mean of 7 and a variance of 120. If $\overline{x}$ denotes the mean of the sample measurements, we want to find the number of measurements, N, required to a give a probability.

$$P(4 \leq \overline{x} \leq 10) = 0.866$$

where 4 and 10 are the chosen problem limits. Here, we use the central limit theorem to argue that, while we do not know the exact distribution of the specific variable, we do know that mean values are normally distributed. Using the standard variable, $z = (x - \mu)/(\sigma/\sqrt{N})$, substituting $\overline{x}$ for x, and using the fact that $\sigma = \sqrt{120} = 2\sqrt{30}$, we can then write our probability function as

$$P(4 \leq \overline{x} \leq 10)$$

$$= P\left[\frac{(4 - \mu)\sqrt{N}}{\sigma} < z < \frac{(10 - \mu)\sqrt{N}}{\sigma}\right]$$

$$= P\left[\frac{(4 - 7)\sqrt{N}}{2\sqrt{30}} < z < \frac{(10 - 7)\sqrt{N}}{2\sqrt{30}}\right]$$

$$= P\left[\frac{-3\sqrt{N}}{2\sqrt{30}} < z < \frac{3\sqrt{N}}{2\sqrt{30}}\right]$$

$$= 2P\left[z < \frac{3\sqrt{N}}{2\sqrt{30}}\right] - 1 = 0.866$$

from which we find

$$P\left[z < \frac{3\sqrt{N}}{2\sqrt{30}}\right] = 0.933$$

Assuming that we are dealing with a normal distribution, we can look up the value 0.933 in a table to find the value of the integrand to which this cumulative probability corresponds. In this case, $3/2\sqrt{N/30} = 1.5$, so that $N = 30$.

3.7 ESTIMATION

In most oceanographic applications, the population parameters are unknown and must be estimated from a sample. Faced with this estimation problem, the objective of statistical analysis is twofold: to present criteria that allow us to determine how well a given sample represents the population parameter; and to provide methods for estimating these parameters. An *estimator* is a random variable used to provide *estimates* of population parameters. "Good" estimators are those that satisfy a number of important criteria: (1) have average values that equal the parameter being estimated (*unbiasedness* property); (2) have relatively small variance (*efficiency* property); and (3) approach asymptotically the value of the population parameter as the sample size increases (*consistency* property). We have already introduced the concept of estimator bias in discussing variance and standard deviation. Formally, an estimate $\hat{\theta}$ of a parameter θ (here, the hat symbol (^) indicates an estimate), is an unbiased estimate provided that $E[\hat{\theta}] = \theta$; otherwise, it is a biased estimate with a bias $B = E[\hat{\theta}] - \theta$. An unbiased estimator is any estimate whose average value over all possible random samples is equal to the population parameter being estimated. An example of an unbiased estimator is the mean of the noise in an acoustic current meter record created by turbulent fluctuations in the velocity of sound speed in water; an example of a biased estimator is the linear slope of a sea-level record in the presence of a long-term trend (a slow change in average value). Other examples of unbiased estimators are $\bar{x}$ for $\hat{\theta}$, μ for $E[\hat{\theta}]$, and σ^2/N for $\sigma_{\hat{\theta}}^2$. The MS error of our estimate $\hat{\theta}$ is

$$E\left[\left(\hat{\theta} - \theta\right)^2\right] = V\left[\hat{\theta}\right] + B^2 \qquad (3.32)$$

The most efficient estimator (property 2) is the estimator with the smallest MS error. Since it is possible to obtain more than one unbiased estimator for the same target parameter, θ, we define the efficiency of an estimator as the ratio of the variances of the two estimators. For example, if we have two unbiased estimates $\hat{\theta}_1$ and $\hat{\theta}_2$, we can compute the relative efficiency of these two estimates as

$$\text{Efficiency} = V\left[\hat{\theta}_2\right]\Big/V\left[\hat{\theta}_1\right] \qquad (3.33)$$

where $V[\hat{\theta}_1]$ and $V[\hat{\theta}_2]$ are the variances of the estimators. A low value of the ratio would suggest that $V[\hat{\theta}_2]$ is more efficient while a high value would indicate that $V[\hat{\theta}_1]$ is more efficient. As an example, consider the efficiency of two familiar estimators of the mean of a normal distribution. In particular, let $\hat{\theta}_1$ be the median value and $\hat{\theta}_2$ be the sample mean. The variance of the sample median is $V[\hat{\theta}_1] = (1.2533^2\sigma^2/N)$ while the sample mean has a variance $V[\hat{\theta}_2] = \sigma^2/N$. Thus, the efficiency is

$$\begin{aligned} \text{Efficiency} &= V\left[\hat{\theta}_2\right]\Big/V\left[\hat{\theta}_1\right] \\ &= (\sigma^2/N)/(1.2533^2\sigma^2/N) \\ &= 0.6366 \end{aligned}$$

Therefore, the variability of the sample mean is 63% of the variability of the sample median, which indicates that the sample *mean* is a more efficient estimator than the sample *median.*

As a second example, consider the sample variances s'^2 and s^2 given by Eqns (3.3) and (3.4), respectively. The efficiency of these two sample variances is the ratio of s'^2 to s^2, namely,

$$\frac{1/N\sum_{i=1}^{N}(x_i - \bar{x})^2}{1/(N-1)\sum_{i=1}^{N}(x_i - \bar{x})^2} = \frac{N-1}{N} < 1$$

which indicates that s'^2 is a more efficient statistic than s^2.

We can view the difference $\hat{\theta} - \theta$ as the distance between the population "target" value and our estimate. Since this difference is also a random variable, we can ask probability-related questions, such as: "What is the probability

$$P\left(-b < \left(\hat{\theta} - \theta\right) < b\right)$$

for some range $(-b, b)$?" It is common practice to express b as some multiple of the sample standard deviation of σ_θ (e.g., $b = k\sigma_\theta$, $k > 1$). A widely used value is $k = 2$, corresponding to two standard deviations. Here, we can apply an important result known as *Tshebysheff's theorem,* which states that for any random variable Y, for $k > 0$:

$$P(|Y - \mu| < k\sigma) \geq 1 - \frac{1}{k^2} \qquad (3.34a)$$

or,

$$P(|Y - \mu| \geq k\sigma) \leq \frac{1}{k^2} \qquad (3.34b)$$

where $\mu = E[\widehat{Y}]$ and $\sigma^2 = V[\widehat{Y}]$. Applying this to the problem at hand, we find that for $k = 2$, $P(|\widehat{\theta} - \theta| < 2\sigma_\theta) \geq 1 - 1/(2)^2 = 0.75$. Therefore, most random variables occurring in nature can be found within two standard deviations ($\pm 2\sigma$) of the mean with a probability of 0.75. Note that the probability statement (Eqn (3.34a)) indicates that the probability is greater than or equal to the value of $1 - 1/k^2$ for any type of distribution. We can therefore, expect somewhat more than 75% of the values to lie with the range $(-2\sigma, 2\sigma)$. In fact, this is generally a conservative estimate. If we assume that oceanographic measurements are typically normally distributed, we find $P(|Y - \mu| < 2\sigma) = 0.95$, so that 95% of the observations lie within $\pm 2\sigma$. This is an important conclusion in terms of data editing methods that use criteria designed to select erroneous values from data samples based on probabilities.

3.8 CONFIDENCE INTERVALS

An important application of interval estimates for probability distribution functions is the formulation of *confidence intervals* for parameter estimates. These intervals define the degree of certainty that a given quantity θ will fall between specified lower and upper bounds θ_L, θ_U, respectively, of the parameter estimates. The confidence interval (θ_L, θ_U) associated with a particular confidence statement is usually written as

$$P(\theta_L < \theta < \theta_U) = 1 - \alpha, \quad 0 < \alpha < 1 \quad (3.35)$$

where α is called the *level of significance* (or confidence coefficient) for the confidence statement and $(1 - \alpha)100$ is the percent significance level for the variable θ. (The terms confidence coefficient, significance level, confidence level, and confidence are commonly used interchangeably). A typical value for α is 0.05, which means that 95% of the cumulative area under the probability curve (Eqn (3.35)) is contained

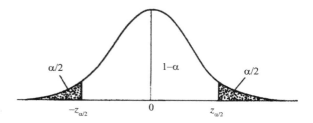

FIGURE 3.8 Location of the limits $(\theta_L, \theta_U) = (-Z_{\alpha/2}, + Z_{\alpha/2})$ for a normal probability distribution. For $\alpha = 0.05$, the cumulative area $1 - \alpha$ corresponds to the 95% interval for the distribution.

between the points θ_L and θ_U (Figure 3.8). For both symmetric and asymmetric probability distributions, each of the two points cuts off $\alpha/2$ of the total area under the distribution curve, leaving a total area under the curve of $1 - \alpha$; θ_L cuts off the left-hand part of the distribution tail and θ_U cuts off the right-hand part of the tail.

If θ_L and θ_U are derived from the true value of the variable θ (such as the population mean, μ), then the probability interval is fixed. However, where we are using estimates determined from a sample (for example, the mean $\overline{X}$) to determine the variable value, θ, the probability interval will vary from sample to sample because of changes in the sample mean and standard deviation. Thus, we must inquire about the probability that the true value of θ will fall within the intervals generated by each of the given sample estimates. The statement that $P(\theta_L < \theta < \theta_U)$ does not mean that the population variable θ has a probability of $P = 1 - \alpha$ of falling in the sample interval (θ_L, θ_U), in the sense that θ was behaving like a sample. The population variable is a fixed quantity. Once the interval is picked, the population variable θ is either in the interval or it is not (probability 1 or 0). For the sample data, the interval may shift depending on the mean and variance of the particular sample we select from the population. We should, therefore, interpret (Eqn (3.35)) to mean that there is a probability P that the specified random sample interval (θ_L, θ_U) contains the true population variable θ a total of $(1 - \alpha)$ 100% of the time. That is, $(1 - \alpha)$ is the fraction of the time that the true variable value θ is contained by the sample interval (θ_L, θ_U).

In general, we need a quantity, called a *pivotal quantity*, that is, a function of the sample estimator $\widehat{\theta}$ and the unknown variable θ, where θ is the only unknown. The pivotal quantity must have a PDF that does not depend on θ. For large samples $(N \geq 30)$ of unbiased point estimators, the standard normal distribution $Z = (\widehat{\theta} - \theta)/\sigma_\theta$ is a pivotal quantity. In fact, it is common to express the confidence interval in terms of Z. For example, consider the statistic $\widehat{\theta}$ with $E[\widehat{\theta}] = \theta$ and $V[\widehat{\theta}] = \sigma_\theta^2$; find the $100(1 - \alpha)\%$ confidence interval. To do this, we first define

$$P\left(-Z_{\alpha/2} < Z < Z_{\alpha/2} \right) = 1 - \alpha \qquad (3.36)$$

and then use the above relation $Z = (\widehat{\theta} - \theta)/\sigma_{\widehat{\theta}}$ to obtain

$$P\left(\widehat{\theta} - Z_{\alpha/2}\sigma_{\widehat{\theta}} < \theta < \widehat{\theta} + Z_{\alpha/2}\sigma_{\widehat{\theta}} \right) = 1 - \alpha \qquad (3.37)$$

This formula can be used for large samples to compute the confidence interval for θ once α is selected. Again, the significance level, $1 - \alpha$, refers to the probability that the population parameter θ will be bracketed by the given confidence interval. The meaning of these limits is shown graphically in Figure 3.8 for a normal population. We remark that if the population standard deviation σ is known it should be used in Eqn (3.37) so that $\sigma_{\widehat{\theta}} = \sigma$; if not, the sample standard deviation s can be used with little loss in accuracy if the sample size is sufficiently large (i.e., $N > 30$).

The three types of confidence intervals commonly used in oceanography are listed below. Specific usage depends on whether we are interested in the mean or the variance of the quantity being estimated.

3.8.1 Confidence Interval for μ (σ Known)

When the population standard deviation, σ, is known and the parent population is normally distributed (or $N > 30$), the $100(1 - \alpha)$ percent confidence interval for the population mean is given by the symmetrical distribution for the standardized normal distribution, z:

$$\bar{x} - z_{\alpha/2}\frac{\sigma}{\sqrt{N}} < \mu < \bar{x} + z_{\alpha/2}\frac{\sigma}{\sqrt{N}} \qquad (3.38)$$

As an example, we wish to find the 95% confidence interval ($\alpha = 0.05$) for μ, given the sample mean $\bar{x}$ and known normally distributed population variance, σ^2. Suppose that a thermistor installed at the entrance to the ship's engine cooling water intake samples every second for $N = 20$ s and yields a mean ensemble temperature $\bar{x} = 12.7\,°C$ for the particular burst. Further, suppose that the water is isothermal and that the only source of variability is instrument noise, which we know from previous calibration in the laboratory has a known noise level $\sigma = 0.5\,°C$. Since we want the 95% confidence interval, the appropriate values of z for the normal distribution are $z_{\alpha/2} = 1.96$ and $-z_{\alpha/2} = -1.96$ (see value for z for $\alpha/2 = 0.9750$ in Appendix D, Table D.1). Substituting these values into Eqn (3.38) along with $N = 20$, $\sigma = 0.5\,°C$, and $\bar{x} = 12.7\,°C$ we find

$$\left[12.7 - (1.96)0.5/\sqrt{20} \right]\,°C < \mu$$
$$< \left[12.7 + (1.96)0.5\sqrt{20} \right]\,°C$$

so that

$$12.48\,°C < \mu < 12.92\,°C$$

Based on our 20 data values, there is a 95% probability that the true mean temperature of the water will be bracketed by the interval (12.48 °C, 12.92 °C) derived from the random interval

$$\left(\bar{x} - z_{\alpha/2}\frac{\sigma}{\sqrt{N}}, \bar{x} + z_{\alpha/2}\frac{\sigma}{\sqrt{N}} \right)$$

3.8.2 Confidence Interval for μ (σ Unknown)

In most real circumstances, σ is not known and we must resort to the use of the sample standard deviation, s. Similarly, for small samples ($N < 30$), we cannot use the above technique but must introduce a formalism that works for any sample size and distribution, as long as the departures from normality are not excessive. Under these conditions, we resort to the variable $t = (\bar{x} - \mu)/(s/\sqrt{N})$, which has a student's t-distribution with $\nu = (N - 1)$ degrees of freedom. Derivation of the $100(1 - \alpha)\%$ confidence interval follows the same procedure used for the symmetrically distributed normal distribution, except that we must modify the limits. In this case

$$P\left[-t_{\alpha/2,\nu} < (\bar{x} - \mu)\bigg/\frac{s}{\sqrt{N}} < t_{\alpha/2,\nu}\right] = 1 - \alpha$$

$$(3.39)$$

This is easily arranged to give the $100(1 - \alpha)\%$ confidence interval for μ

$$\bar{x} - t_{\alpha/2,\nu}\frac{s}{\sqrt{N}} < \mu < \bar{x} + t_{\alpha/2,\nu}\frac{s}{\sqrt{N}} \qquad (3.40)$$

Note the similarity between Eqn (3.40) and the form Eqn (3.37) obtained for μ when σ is known. We return to our previous example of ship injection temperature and this time assume that $s = 0.5\,°C$ is a measured quantity obtained by subtracting the mean value $\bar{x} = 12.7\,°C$ from the series of 20 measurements. Turning to Appendix D (Table D.3) for the cumulative t-distribution, we look for values of $F(t)$ under the column for the 95% confidence interval ($\alpha = 0.05$) for which $F(t) = 1 - \alpha/2 = 0.975$. Using the fact that $\nu = (N - 1) = 19$, we find $t_{\alpha/2,\nu} = t_{0.025,19} = 2.093$. Substituting these values into Eqn (3.40) yields

$$\left[12.7 - 2.093\left(0.5/\sqrt{20}\right)\right]\,°C < \mu$$
$$< \left[12.7 + 2.093\left(0.5/\sqrt{20}\right)\right]\,°C$$
$$12.47\,°C < \mu < 12.93\,°C$$

There is a 95% chance that the interval ($12.47\,°C$, $12.93\,°C$) will bracket the true mean temperature. Because of the large sample size, this result is only slightly different than the result obtained for the normal distribution in the previous example when σ was known a priori.

3.8.3 Confidence Interval for σ^2

Under certain circumstances, we are more interested in the confidence interval for the signal variance than the signal mean. For example, to determine the reliability of a spectral peak in a spectral density distribution (or spectrum), we need to know the confidence intervals for the population variance, σ^2, based on our sample variance, s^2. To do this, we seek a new pivotal quantity. Recall from Eqn (3.29) that for N samples of a variable x_i from a normal population, the expression

$$\frac{1}{\sigma^2}\sum_{i=1}^{N}(x_i - \bar{x})^2 = \frac{(N-1)s^2}{\sigma^2} \qquad (3.41)$$

is a χ^2 variable with $(N - 1)$ degrees of freedom. Using this as a pivotal quantity, we can find the upper and lower bounds χ_U^2 and χ_L^2 for which

$$P\left[\chi_L^2 < \frac{N-1}{\sigma^2/s^2} < \chi_U^2\right] = 1 - \alpha \qquad (3.42)$$

or, upon rearranging terms,

$$P\left[\frac{(N-1)s^2}{\chi_U^2} < \sigma^2 < \frac{(N-1)s^2}{\chi_L^2}\right] = 1 - \alpha \quad (3.43)$$

Note that χ^2 is a skewed function (Figure 3.9), which means that the upper and lower bounds in Eqn (3.43) are asymmetric; the point $1 - \alpha/2$ rather than $-\alpha/2$ determines the point that cuts off $\alpha/2$ of the area at the lower end of the chi-square distribution.

From Eqn (3.43) we obtain the well-known $100(1 - \alpha)\%$ confidence interval for the variance σ^2 when sampled from a normal population

$$\frac{(N-1)s^2}{\chi_{\alpha/2,\nu}^2} < \sigma^2 < \frac{(N-1)s^2}{\chi_{1-\alpha/2,\nu}^2} \qquad (3.44)$$

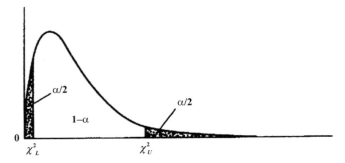

FIGURE 3.9 Location of the limits $(\theta_L, \theta_U) = (\chi_L^2, \chi_U^2)$ for a chi-square probability distribution. For $\alpha = 0.05$, the cumulative area $1 - \alpha$ corresponds to the 95% interval for the distribution. χ^2 is a skewed function so that the upper and lower bounds are asymmetric.

where the subscripts $\alpha/2$ and $1 - \alpha/2$ characterize the endpoint values of the confidence interval and $\nu = (N - 1)$ gives the degrees of freedom of the chi-square distribution. The smaller value of $\chi^2(\chi_L^2 = \chi_{1-\alpha/2,\nu}^2)$ appears in the denominator of the upper endpoint for σ^2 while the larger value of $\chi^2(\chi_U^2 = \chi_{\alpha/2,\nu}^2)$ appears in the denominator of the lower endpoint for σ^2. As an example, suppose that we have $\nu = 10$ in our spectral estimate of the eastward component of current velocity and that the background variance of our spectra near a distinct spectral peak is $s^2 = 10$ cm^2/s^2. What is the 95% confidence interval for the variance? How big would the peak have to be to stand out statistically from the background level? (Details on spectral estimation can be found in Chapter five.) In this case, $\alpha/2 = 0.025$ and $1 - \alpha/2 = 0.975$. Turning to the cumulative distribution $F(\chi^2)$ for $\nu = N - 1 = 9$ degrees of freedom in Appendix D (Table D.2), we find that $\chi_{0.025,9}^2 = 19.02$ for cumulative integral $F(\alpha/2 = 0.025)$ and that $\chi_{0.95,9}^2 = 2.70$ for cumulative integral $F(1 - \alpha/2 = 0.975)$. Thus, $P(2.70 < \chi_{\nu=9}^2 < 19.02) = 1 - \alpha = 0.95$. Substituting $N - 1 = 9$, $s^2 = 10$ cm^2/s^2, $\chi_{\alpha/2,9}^2 = 19.02$ for the value that cuts off $\alpha/2$ of the upper end area under the curve and $\chi_{1-\alpha/2,9}^2 = 2.70$ for the value that cuts off $1 - \alpha/2$ of the lower end area of the curve, Eqn (3.44) yields

$$9(10 \text{ cm}^2/\text{s}^2)/19.02 < \sigma^2 < 9(10 \text{ cm}^2/\text{s}^2)/2.70$$

$$4.7 \text{ cm}^2/\text{s}^2 < \sigma^2 < 33.3 \text{ cm}^2/\text{s}^2$$

Thus, the true background variance will lie between 4.7 and 33.3 cm^2/s^2. If a spectral peak has a greater range than these limits then it represents a statistically significant departure from background energy levels.

In most instances, spectral densities are presented in terms of the log of the spectral density function vs frequency or log-frequency (see Chapter 5). Dividing through by s^2 in Eqn (3.44) and taking the log, yields

$$\log (N - 1) - \log\left(\chi_{\alpha/2,\nu}^2\right) < \log\left(\sigma^2/s^2\right)$$
$$< \log (N - 1) - \log\left(\chi_{1-\alpha/2,\nu}^2\right)$$

or, upon subtracting $\log (N - 1)$ and rearranging the inequality

$$\log\left(\chi_{1-\alpha/2,\nu}^2\right) < \log (N - 1) - \log\left(\sigma^2/s^2\right)$$
$$< \log\left(\chi_{\alpha/2,\nu}^2\right)$$

The range R of the variance is then

$$R = \log\left(\chi_{\alpha/2,\nu}^2\right) - \log\left(\chi_{1-\alpha/2,\nu}^2\right) \qquad (3.45)$$

while the pivot point p_0 of the interval is

$$p_0 = \log (N - 1) - \log\left(\sigma^2/s^2\right) \qquad (3.46)$$

If we assume that the measured background value of s^2 is a good approximation to σ^2, so that $\sigma^2/s^2 = 1$, then $p_0 = \log(N - 1)$. The ranges between the maximum value and p_0, and the minimal value and p_0, are $\log(\chi^2_{\alpha/2,\nu}) - p_0$ and $\log(\chi^2_{1-\alpha/2,\nu}) - p_0$, respectively. Returning to our previous example for the 95% confidence interval, we find that

$$\log(2.70) < \log(9) < \log(19.02),$$
$$0.43 < 0.95 < 1.28$$

giving a range $R = 0.848$ with the pivot point at $p_0 = 0.95$.

3.8.4 Goodness-of-Fit Test

The goodness-of-fit statistical model is, as the model states, the best fit to data obtained as outcomes of an experiment. Measures of goodness-of-fit typically summarize the differences between observed values and the values expected from the model in question. When the set of outcomes for an experiment is limited to two outcomes (such as success or failure, on or off, and so on), the appropriate test statistic for the distribution is the binomial variable. However, when more than two outcomes are possible, the preferred statistic is the chi-square variable. In addition to providing confidence intervals for spectral estimates and other measurement parameters, the chi-square variable is used to test how closely the observed frequency distribution of a given parameter corresponds to the expected frequency distribution for the parameter. The expected frequencies represent the average number of values expected to fall in each frequency interval based on some theoretical probability distribution, such as a normal distribution. The observed frequency distribution represents a sample of values drawn from some probability distribution. What we want to know is whether the observed and expected frequencies are similar enough for us to conclude that they are drawn from the same probability

function (the "null hypothesis"). The test for this similarity using the chi-square variable is called a "goodness-of-fit" test.

Consider a sample of N observations from a random variable X with observed PDF $p_0(X)$. Let the N observations be grouped into K intervals (or categories) called *class intervals*, whose graphical distribution forms a frequency histogram (Bendat and Piersol, 1986). The actual number of observed values that fall within the ith class interval is denoted by f_i, and is called the *observed frequency* in the ith class. The number of observed values that we would expect to fall within the ith class interval if the observations really followed the theoretical probability distribution, $p(X)$, is denoted F_i, and is called the *expected frequency* in the ith class interval. The difference between the observed frequency and the expected frequency for each class interval is given by $f_i - F_i$. The total discrepancy for all class intervals between the expected and observed distributions is measured by the sample statistic

$$X^2 = \sum_{i=1}^{K} \frac{(f_i - F_i)^2}{F_i} \qquad (3.47)$$

where division by F_i transforms the sum of the squares into the chi-square-type variable, X^2.

The number of degrees of freedom, ν, for the variable X^2 is equal to K minus the number of different independent linear restrictions imposed on the observations. As discussed by Bendat and Piersol (1986), one degree of freedom is lost through the restriction that, if $K - 1$ class intervals are determined, the Kth class interval follows automatically. If the expected theoretical density function is normally distributed then the mean and variance must be computed to allow comparison of the observed and expected distributions. This results in the loss of two additional degrees of freedom. Consequently, if the chi-square goodness-of-fit test is used to test for normality of the data, the true number of degrees of freedom for X^2 is $\nu = K - 3$.

Equation (3.47) measures the goodness-of-fit between f_i and F_i as follows: when the fit is good (that is, f_i and F_i are generally close), then the numerator of Eqn (3.47) will be small and the hence the value of X^2 will be low. On the other hand, if f_i and F_i are not close, the numerator of Eqn (3.47) will be comparatively large and the value of X^2 will be large. Thus, the critical region for the test statistic X^2 will always be in the upper tail of the chi-square function (Figure 3.9) because we wish to reject the null hypothesis, whenever the difference between f_i and F_i, is large. More specifically, the region of acceptance of the null hypothesis (see Section 3.14) is

$$X^2 \leq \chi^2_{\alpha;\nu} \tag{3.48}$$

where the value of $\chi^2_{\alpha,\nu}$ is available from Appendix D (Table D.2). If X^2 is less than or equal to $\chi^2_{\alpha,\nu}$, the hypothesis that $p(X) = p_0(X)$ is accepted at the α level of significance (i.e., there is a $100\alpha\%$ chance that we are wrong in accepting the null hypothesis that our data are drawn from a normal distribution). However, if the sample value X^2 is greater than $\chi^2_{\alpha,\nu}$, the hypothesis is rejected at the level of significance. For example, suppose our analysis involves 15 class intervals and that the fit between the 15 estimates of f_i and F_i (where F_i is normally distributed) yields $X^2 = 23.1$. From tables for the cumulative chi-square distribution, $F(X) = p(X > \chi^2_{\alpha;\nu})$, we find that $p(X^2 > 21.03) = 0.05$ for $\nu = K - 3 = 12$ degrees of freedom. Thus, at the $\alpha = 0.05$ level of significance (95% certainty level), we cannot accept the null hypothesis that the observed values came from the same distribution as the expected values. In this case, our chances of being wrong are greater than $100\alpha\%$.

Chi-square tests for normality are typically performed using a constant interval width. Nonuniform distributions will yield different expected frequency distributions from one class interval to the next. Bendat and Piersol recommend a class interval width of $\Delta x \approx 0.4s$, where s is the standard deviation of the sample data. A further requirement is that the expected frequencies in all class intervals be sufficiently large that X^2 in Eqn (3.47) is an acceptable approximation to $\chi^2_{\alpha,\nu}$. A common recommendation is that $F_i > 3$ in all class intervals. When testing for normality, where the expected frequencies diminish on the tails of the distribution, this requirement is attained by letting the first and last intervals extend to $-\infty$ and to $+\infty$, respectively, so that $F_1, \ldots, F_K > 3$.

As an example of a goodness-of-fit test, we consider a sample of $N = 200$ surface gravity wave heights measured every 0.78 s by a Datawell waverider buoy deployed off the west coast of Canada during the winter of 1993–1994 (Table 3.3). The wave record spans a period of 2.59 min and corresponds to a time of extreme (5 m high) storm-generated waves. According to one school of thought (e.g., Phillips et al., 1993), extreme wave events in the ocean are part of a Gaussian process and the occurrence of maximum wave heights is related in a linear manner to the statistical distribution of the surrounding wave field. If this is true, then the heights of high-wave events relative to the background seas should follow a normal frequency distribution. To test this at the $\alpha = 0.05$ significance level, $K = 10$ class intervals for the observed wave heights were fitted to a Gaussian probability distribution. The steps in the goodness-of-fit test are as follows:

1. Specify the class interval width Δx and list the upper limit of the standardized values, $z_{\alpha/2}$, of the normal distribution that correspond to these values (as in Table 3.4; the lower bound for the first rank is $-\infty$). Commonly Δx is assumed to span 0.4 standard deviations, s, such that $\Delta x \approx 0.4s$; here we use $\Delta x \approx 0.5s$. For $\Delta x = 0.4s$, the values of $z_{\alpha/2}$, we want are $(\ldots, -2.4, -2.0, \ldots, 2.0, 2.4, \ldots)$ while for $\Delta x = 0.5s$, the values are $(\ldots, -2.5, -2.0, \ldots, 2.0, 2.5, \ldots)$.

2. Determine the finite upper and lower bounds for z_α from the requirement that $F_i > 3$. Since

TABLE 3.3   Wave Heights (mm) during a Period of Anomalously High Waves as Measured by a Datawell Waverider Buoy Deployed in 30 m Depth on the Inner continental shelf of Vancouver Island, British Columbia

4636	4840	4901	4950	4980	5014	5034	5060	5095	5130
4698	4842	4904	4954	4986	5014	5037	5066	5095	5135
4702	4848	4907	4955	4991	5015	5037	5066	5096	5135
4731	4854	4907	4956	4994	5017	5038	5069	5102	5145
4743	4856	4908	4956	4996	5020	5039	5069	5103	5155
4745	4867	4914	4956	4996	5020	5040	5071	5104	5157
4747	4867	4916	4959	4996	5021	5040	5072	5104	5164
<u>4749</u>	4870	4917	4960	<u>4997</u>	5023	5044	5073	5104	5165
4773	4870	4923	4961	4998	5024	5045	5074	5106	<u>5166</u>
4785	4874	4925	4963	5003	5025	5045	5074	5110	5171
4793	4876	4934	4964	5006	5025	5047	5074	<u>5111</u>	5175
4814	<u>4877</u>	4935	4964	5006	5025	5048	5078	5115	5176
4817	4883	4937	4966	5006	5025	5050	5079	5119	5177
4818	4885	4939	4966	5006	5028	5051	5080	5119	5181
4823	4886	<u>4940</u>	4970	5006	5029	5052	5081	5120	5196
<u>4824</u>	4892	4941	4971	5010	5029	<u>5053</u>	5086	5121	5198
4828	4896	4942	4972	5011	5029	5057	5089	5122	5201
4829	4897	4942	4974	5011	5030	5058	5091	5123	<u>5210</u>
4830	4898	4943	4977	5012	5031	5059	5093	5125	5252
4840	4899	4944	4979	5012	5032	5059	5094	5127	5299

The original $N = 200$ *data values have been rank ordered. The upper bounds of the K-class intervals have been underlined.(Courtesy, Diane Masson, Institute of Ocean Sciences, Sidney, B.C.).*

$F_i = NP_i$ (where $N = 200$ and P_i is the normal probability distribution for the ith interval), we require $P > 3/N = 0.015$. From tables of the standardized normal density function, we find that $P > 0.015$ implies a lower bound $z_{\alpha/2} = -2.0$, and an upper bound $z_{\alpha/2} = +2.0$. Note that for a larger sample, say $N = 2000$, we have $P > 3/2000 = 0.0015$ and the bounds become ± 2.8 for the interval $\Delta x = 0.4s$ and ± 2.5 for the interval $\Delta x = 0.5s$.

3. Calculate the expected upper limit, $x = sz_{\alpha/2} + \bar{x}$ (mm), for the class intervals and mark this limit on the data table (Table 3.3). For each upper bound, $z_{\alpha/2}$, in Table 3.4, find the corresponding probability density value. Note that these values apply to intervals so, for example, $P(-2.0 < x < -1.5) = 0.0668 - 0.0228 = 0.044$; $P(2.0 < x < \infty) = 0.0228$.

4. Using the value of P, find the expected frequency values $F_i = NP_i$. The observed frequency f_i is found from Table 3.3 by counting the actual number of wave heights lying between the marks made in step 3. Complete the table and calculate X^2. Compare to $\chi^2_{\alpha;\nu}$.

TABLE 3.4 Calculation Table for Goodness of Fit Test for the Data in Table 3.3

Class Interval	Upper Limit of Data Interval		P_i	$F_i = NP_i$	f_i	$F_i - f_i$	$\dfrac{(F_i - f_i)^2}{F_i}$
	z_α	$x = sz_\alpha + \bar{x}$					
1	−2.0	4767.4	0.0228	4.6	8	3.4	2.51
2	−1.5	4825.0	0.0440	8.8	8	0.8	0.07
3	−1.0	4882.5	0.0919	18.4	16	2.4	0.31
4	−0.5	4940.1	0.1498	30.0	23	7.0	1.63
5	0	4997.6	0.1915	38.3	33	5.3	0.73
6	0.5	5055.2	0.1915	38.3	48	9.7	2.46
7	1.0	5112.7	0.1498	30.0	35	5.0	0.83
8	1.5	5170.2	0.0919	18.4	18	0.4	0.01
9	2.0	5227.8	0.0440	8.8	9	0.2	0.00
10	∞	∞	0.0228	4.6	2	2.6	1.47
Totals			1.0000	200	200		10.02

The number of intervals has been determined using an interval width $\Delta x = 0.5s$ with z_α in units of 0.5 and requiring that $F_i > 3$. $N = 200$, $\bar{x}$ (mean) $= 4997.6$ mm, s (standard deviation) $= 115.1$ mm, and ν (degrees of freedom) $= k - 3 = 7$.

In the above example, $X^2 = 10.02$ and there are $\nu = 7$ degrees of freedom. From Appendix D (Table D.2), we find $P(X^2 > \chi^2_{\alpha,\nu}) = P(X^2 > \chi^2_{0.05,7}) = 14.07$. Thus, at the $\alpha = 0.05$ level of significance (95% significance level), we can accept the null hypothesis that the large wave heights measured by the waverider buoy had a Gaussian (normal) distribution in time and space.

3.9 SELECTING THE SAMPLE SIZE

It is not possible to determine the required sample size N for a given confidence interval until a measure of the data variability, the population standard deviation, σ, is known. This is because the variability of $\overline{X}$ depends on the variability of X. Since we do not usually know a priori the population standard deviation (the value for the true population), we use the best estimate available, the sample standard deviation, s. We also need to know the frequency content of the

data variable so that we can ensure that the N values we use in our calculations are statistically independent samples. As a simple example, consider a normally distributed, continuous random variable, Y, with the units of meters. We wish to find the average of the sample and want it to be accurate to within ± 5 m. Since we know that approximately 95% of the sample means of a normally distributed random variable will lie within $\pm 2\sigma_Y$ of the true mean, μ, we require that $2\sigma_Y = 5$ m. Using the central limit theorem for the mean, we can estimate σ_Y by

$$\widehat{\sigma}_Y = \frac{\sigma}{\sqrt{N}}$$

so that $2\sigma/\sqrt{N} = 2\sigma_Y = 5$ m, whereby $N = 4\sigma^2/25$ (assuming that the N observations are statistically independent). If σ is known, we can easily find N.

When we do not know σ, we are forced to use an estimate from an earlier sample within the range of measurements. If we know the sample range, we can apply the empirical rule for normal distributions that the range is approximately 4σ and take one-fourth the range as our estimate of σ. Suppose our range in the above example is 84 m. Then, $\sigma = 21$ m and

$$N = 4\sigma^2/25 = (4)(21 \text{ m})^2/(25 \text{ m}^2)$$
$$= 70.56 \approx 71$$

This means that, for a sample of $N = 71$ statistically independent values, we would be 95% sure (probability $= 0.95$) that our estimate of the mean value would lie within $\pm 2\sigma_Y = \pm 5$ m of the true mean.

One method for selecting the sample size for relatively large samples is based on Tshebysheff's theorem known as the "weak law of large numbers". Let $f(x)$ be a density function with mean μ and variance σ^2, and let $\overline{x}_N$ be the sample mean of a random sample of size N from $f(x)$. Let ε and δ be any two specified numbers satisfying $\varepsilon > 0$ and $0 < \delta < 1$. If N is any integer greater than $(\sigma^2/\varepsilon^2)\delta$ then

$$P[-\varepsilon < \overline{x}_N - \mu < \varepsilon] \geq 1 - \delta \qquad (3.49)$$

To show the validity of condition (Eqn (3.49)), we use Tshebysheff's inequality

$$P[g(x) \geq k] \geq \frac{E[g(x)]}{k} \qquad (3.50)$$

for every $k > 0$, random variable x, and nonnegative function $g(x)$. An equivalent formula is

$$P[g(x) < k] \geq 1 - \frac{E[g(x)]}{k} \qquad (3.51)$$

Let, $g(x) = [(\overline{x}_N - \mu) < \varepsilon]^2$ and $k = \varepsilon^2$ then

$$P[-\varepsilon < (\overline{x}_N - \mu) < \varepsilon]$$
$$= P[|\overline{x}_N - \mu| < \varepsilon]$$
$$= P\left[|\overline{x}_N - \mu|^2 < \varepsilon^2\right] \geq 1 - \frac{E\left[(\overline{x}_N - \mu)^2\right]}{\varepsilon^2}$$
$$= 1 - \frac{\sigma^2}{N\varepsilon^2} \geq 1 - \delta$$
$$\qquad (3.52)$$

For $\delta > \sigma^2/N\varepsilon^2$ or $N > \sigma^2/\delta\varepsilon^2$, the latter expression becomes

$$P[|\overline{x}_N - \mu| < \varepsilon] \geq 1 - \delta \qquad (3.53)$$

We illustrate the use of the above relations by considering a distribution with an unknown mean and variance $\sigma^2 = 1$. How large a sample must be taken in order that the probability will be at least 0.95 that the sample mean, $\overline{x}_N$, will lie within 0.5 of the true population mean? Given are: $\sigma^2 = 1$ and $\varepsilon = 0.5$. Rearranging the inequality (Eqn (3.53))

$$\delta \geq 1 - P[|\overline{x}_N - \mu| < 0.5] = 1 - 0.95 = 0.05$$

Substituting into the relation $N > (\sigma^2/\delta\varepsilon^2) = 1/(0.05)(0.5)^2$ shows that we require $N \geq 80$ independent samples.

3.10 CONFIDENCE INTERVALS FOR ALTIMETER-BIAS ESTIMATES

As an example of how to estimate confidence limits and sample size, consider an oceanographic altimetric satellite, where the altimeter is to be calibrated by repeated passes over a spot on the earth where surface-based measurements provide a precise, independent measure of the sea surface elevation. A typical reference site is an offshore oil platform having sea-level gauges and a location system, such as the multi-satellite global positioning system (GPS). For the TOPEX/POSEIDON satellite one reference site was an oil platform in the Santa Barbara channel off southern California (Christensen et al., 1994). Each pass over the reference site provides a measurement of the satellite altimeter bias, which is used to compute an average bias after repeated calibration observations. This bias is just the difference between the height measured by the altimeter and the height measured independently by the in situ measurements at the reference site. If we assume that our measurement errors are normally distributed with a mean of zero, then the uncertainty of the true mean bias,

$$\sigma_b = z s_b / \sqrt{N}$$

where z is the standard normal distribution, s_b is the standard deviation of the measurements, and N is the number of measurements (i.e., the number of calibration passes over the reference site).

Suppose we are required to know the true mean of the altimeter bias to within 2 cm, and that we estimate the uncertainty of the individual measurements to be 3 cm. We then ask: "What is the number of independent measurements required to give a bias of 2 cm at the 90%, 95%, and 99% confidence intervals?" Using the above formulation for the standard error, we find

$$N = \left(z_{\alpha/2} s_b / \sigma_b \right)^2 \qquad (3.54)$$

from which we can compute the required sample size. As before, the parameter α refers to the chosen significance level which, in the present case, correspond to $\alpha = 0.10$, 0.05, and 0.01. Now $\sigma_b = 2$ cm (required) and $s_b = 3$ cm (estimated), so that we can use the standard normal table for $z_{\alpha/2} = N(0, 1)$ in Appendix D (Table D.1) to obtain the values shown in Table 3.5. If we require the true mean to be 1.5 cm instead of 2.0 cm, the values in Table 3.5 become those in Table 3.6.

TABLE 3.5   Calculation of the Number of Satellite Altimeter Required to Attain a Given Level of Confidence in Elevation Using the Relation (3.54) for $\sigma_b = 2$ cm and $s_b = 3$ cm

Confidence Level (α)	Standard Normal Value (z_α)	Exact Number of Observations (N)	Actual Number of Observations
90%	1.645	6.089	7
95%	1.960	8.644	9
99%	2.576	14.931	15

TABLE 3.6   Calculation of the Number of Satellite Altimeter Observations Needed for a Given Level of Confidence in the Level Elevation Using the Eqn (3.54) for $\sigma_b = 1.5$ cm and $s_b = 3$ cm

Confidence Level (α)	Standard Normal Value (z_α)	Exact Number of Observations (N)	Actual Number of Observations
90%	1.645	10.82	11
95%	1.960	15.37	16
99%	2.576	26.54	27

Finally, suppose the satellite is in a 10-day repeat orbit so that we can only collect a reference measurement every 10 days at our ground site; we are given 240 days to collect reference observations. What confidence intervals can be achieved for both of the above cases if we assume that only 50% of the calibration measurements are successful and that the 10-day observations are statistically independent? Now, since we have only one calibration measurement every 10 days for 50% of 240 days we have

$$c = (0.5)(240 \text{ days}) (1 \text{ measurement}/10 \text{ days})$$

$$= 12 \text{ measurements}$$

Referring to the previous two tables, we see that for the first case (Table 3.5), where the mean bias was required to be 2.0 cm, we can achieve the 95% interval with 12 measurements; for the case where the mean bias is restricted to 1.5 cm (Table 3.6), only the 90% confidence interval is possible. Note that the values in the last column on the right are roundedup, integer versions of the calculated values of N.

3.11 ESTIMATION METHODS

Now that we have introduced methods to calculate confidence intervals for our estimates of μ and σ^2, we need procedures to estimate these quantities themselves. There are many different methods we could use but space does not allow us to discuss them all. We first introduce a very general technique, known as minimum variance unbiased estimation (MVUE), and then later discuss a popular method called the maximum likelihood method which leads to MVUE estimators.

Before introducing the MVUE procedure, we need to define two terms: *sufficiency* and *likelihood*. Let x_1, x_2,...,x_N be a random sample from a probability distribution with an unknown statistical parameter, θ (mean, variance, etc.). The statistic $U = g(x_1, x_2,..., x_N)$ is said to be sufficient for θ if the conditional distribution x_1, x_2,..., x_N, given U, does not depend on θ. In other words, once U is known, no other combination of x_1, x_2,..., x_N provides additional information about θ. This tells us how to check if our statistic is sufficient but does not tell how to compute the statistic.

To define likelihood, let $y_1, y_2,..., y_N$ be sample observations of random variables $Y_1, Y_2,..., Y_N$. For continuous variables, the likelihood $L(y_1, y_2,..., y_N)$ is the joint probability density $f(y_1, y_2,..., y_N)$ evaluated at the observations, y_i. Assuming that the Y_i are statistically independent

$$L\left(y_1, y_2, ..., y_N\right) = f\left(y_1, y_2, ..., y_N\right)$$
$$= f\left(y_1\right)f\left(y_2\right)...f\left(y_N\right) \quad (3.55)$$

where $f(y_i)$, $i = 1, 2,..., N$, is the PDF for the random variable Y_i.

As an oceanographic example, consider a record of daily average current velocities obtained using a single current meter moored for a period of one month ($N = 30$ days). Show that the monthly mean velocity, V, is a sufficient statistic for the population mean if the variance is known (in this case, estimated from the range of current values). Since the daily velocities are average values of shorter-term current velocity measurements (e.g., 30 min values), we can invoke the central limit theorem to conclude that the daily velocities are normally distributed. Hence, the PDF can be written as

$$f(V) = \frac{1}{\sigma(2\pi)^{1/2}} \exp\left[-\frac{1}{2\sigma^2}(V - \mu)^2\right]$$

We can write the likelihood L of our sample as

$$L = f(V_1, V_2, ..., V_{30})$$
$$= f(V_1)f(V_2)...f(V_{30})$$
$$= \frac{1}{\sigma(2\pi)^{1/2}} \exp\left[-\frac{1}{2\sigma^2}(V_1 - \mu)^2\right]$$
$$\times \frac{1}{\sigma(2\pi)^{1/2}} \exp\left[-\frac{1}{2\sigma^2}(V_2 - \mu)^2\right]...$$
$$\times \frac{1}{\sigma(2\pi)^{1/2}} \exp\left[-\frac{1}{2\sigma^2}(V_{30} - \mu)^2\right]$$
$$= \frac{1}{[\sigma(2\pi)]^{15}} \exp\left[-\frac{1}{2\sigma^2} \sum_{i=1}^{30} (V_i - \mu)^2\right]$$

Because σ is known from our range of current velocities, then L is a function of V and μ, only. Hence, V is a sufficient statistic for μ the population mean.

3.11.1 Minimum Variance Unbiased Estimation

For random variables $Y_1, Y_2,..., Y_N$, with PDF, $f(y)$, and unknown parameter θ, let one set of sample observations be $(x_1, x_2,..., x_N)$ and another be $(y_1, y_2,..., y_N)$. The ratio of the likelihoods of these two sets of observations can be written as

$$\frac{L(x_1, x_2, ..., x_N)}{L\left(y_1, y_2, ..., y_N\right)} \quad (3.56)$$

In general, this ratio will not be a function of θ if, and only if, there is a function $g(x_1, x_2, ..., x_n)$ such that $g(x_1, x_2, ..., x_n) = g(y_1, y_2, ..., y_N)$ for all choices of x and y. If such a function can be found, it is the minimum sufficient statistic for θ. Any unbiased estimator that is a function of a minimal sufficient statistic will be an MVUE; this means that it will possess the smallest possible variance among the unbiased estimators.

We illustrate what we mean with an example. Let $x_1, x_2, ..., x_N$ be a random sample from a normal population with the unknown parameters μ and σ^2. We want to find the MVUE of μ and σ^2. Writing the likelihood ratio we have

$$\frac{L(x_1, x_2, ..., x_N)}{L(y_1, y_2, ..., y_N)} = \frac{f(x_1, x_2, ..., x_N)}{f(y_1, y_2, ..., y_N)}$$

$$= \frac{\dfrac{1}{\sigma\sqrt{2\pi}}\exp\left[-\dfrac{1}{2\sigma^2}\displaystyle\sum_{i=1}^{N}(x_i - \mu)^2\right]}{\dfrac{1}{\sigma\sqrt{2\pi}}\exp\left[-\dfrac{1}{2\sigma^2}\displaystyle\sum_{i=1}^{N}(y_i - \mu)^2\right]}$$

$$= \exp\left\{-\frac{1}{2\sigma^2}\left[\sum_{i=1}^{N}(x_i - \mu)^2 - \sum_{i=1}^{N}(y_i - \mu)^2\right]\right\}$$

$$= \exp\left\{-\frac{1}{2\sigma^2}\left[\left(\sum_{i=1}^{N}x_i^2 - \sum_{i=1}^{N}y_i^2\right) -2\mu\left(\sum_{i=1}^{N}x_i - \sum_{i=1}^{N}y_i\right)\right]\right\}$$

For this ratio to be independent of μ, we must have

$$\sum_{i=1}^{N}x_i = \sum_{i=1}^{N}y_i \tag{3.57}$$

Similarly, for the ratio to be independent of σ^2, requires both Eqn (3.57) as well as

$$\sum_{i=1}^{N}x_i^2 = \sum_{i=1}^{N}y_i^2 \tag{3.58}$$

Thus, both $\sum x_i$ and $\sum x_j^2$ are minimum sufficient statistics for μ and σ^2. Since $\bar{x}$ is an unbiased estimate of μ

$$s^2 = \frac{1}{N-1}\sum_{i=1}^{N}(x_i - \bar{x})^2 \tag{3.59}$$

is an unbiased estimate of σ^2. Since both $\bar{x}$ and s^2 are functions of the minimal sufficient statistics

$$\sum_{i=1}^{N}x_i \quad \text{and} \quad \sum_{i=1}^{N}x_i^2$$

as expressed by Eqns (3.57) and (3.58), they also are MVUEs for μ and σ^2.

3.11.2 Maximum Likelihood

The procedure introduced earlier to compute the MVUE is complicated by the fact that one must find some function of the minimal sufficient statistic that gives the sought-after target parameter. Finding this function is generally a matter of trial and error. A more sophisticated procedure, the maximum likelihood method, often leads to the MVUE. Thus, an estimate that yields minimum variance equates to an estimate that has maximum likelihood of being correct.

The formal statement of this method is quite simple. Choose as estimates those parameter values that maximize the likelihood $L(y_1, y_2, ..., y_N)$. A simple example using discrete variables helps to illustrate the logic in the maximum likelihood method. This example is intended to give the reader an intuitive sense of the maximum likelihood method that can be more formerly defined and applied to continuous functions and PDFs. Assume we have a bag containing three marbles. The marbles can be black or white. We randomly sample two of the three and find that they are both black. What is the best estimate of the total number of black marbles in the bag? If there are actually two black and one white in the bag, the probability of sampling two black marbles is

$$\frac{\binom{2}{2}\binom{1}{0}}{\binom{3}{2}} = 1/3$$

where, as in Section 3.3, the binomial expression is

$$\binom{N}{r} = {_N}P_r/r! = N!/[r!(N-r)!] \qquad (3.60)$$

and ${_N}P_r$ is the number of permutations of N discrete variables sampled r at a time. In the above expression

$$\binom{2}{2}$$

indicates the first sample of two marbles, with both being black. The next term is the remaining unsampled marble (hence the 0 in the denominator) if it were white. Now if there are three black marbles in the bag the probability of sampling two blacks is

$$\frac{\binom{3}{2}\binom{0}{0}}{\binom{3}{2}} = 1$$

On this basis, it seems reasonable to choose three as the estimate of the number of black marbles in the bag in order to maximize the probability of the observed sample.

A more complex example can be used to illustrate the application of this method to our estimates of the mean, μ, and variance, σ^2, for a normal population. Again, let $y_1, y_2, \ldots, y_N$ be a random sample from a normal population with parameters μ and σ^2. We want to find the maximum likelihood estimators of μ and σ^2. To find the maximum likelihood, we need to write the joint PDF of the independent observations $y_1, y_2, \ldots, y_N$

$$L = f(y_1, y_2, \ldots, y_N)$$
$$= f(y_1)f(y_2)\ldots f(y_N)$$
$$= \frac{1}{\sigma(2\pi)^{1/2}}\exp\left[-\frac{1}{2\sigma^2}(y_1-\mu)^2\right]$$
$$\times \frac{1}{\sigma(2\pi)^{1/2}}\exp\left[-\frac{1}{2\sigma^2}(y_2-\mu)^2\right]\ldots$$
$$\times \frac{1}{\sigma(2\pi)^{1/2}}\exp\left[-\frac{1}{2\sigma^2}(y_N-\mu)^2\right]$$
$$= \frac{1}{\sigma^N(2\pi)^{N/2}}\exp\left[-\frac{1}{2\sigma^2}\sum_{i=1}^{N}(y_i-\mu)^2\right]$$
$$\tag{3.61}$$

We simplify this expression by taking the natural logarithm, $\ln(L)$, which we then differentiate to find the maximum. Specifically, we begin with

$$\ln(L) = -\frac{N}{2}\ln(\sigma^2) - \frac{N}{2}\ln(2\pi) - \sum_{i=1}^{N}\frac{(y_i-\mu)^2}{2\sigma^2}$$
$$\tag{3.62}$$

Taking derivatives of Eqn (3.62) with respect to μ and σ^2, we find

$$\frac{\partial[\ln(L)]}{\partial\mu} = \sum_{i=1}^{N}\frac{(y_i-\mu)}{\sigma^2} \qquad (3.63a)$$

$$\frac{\partial[\ln(L)]}{\partial\sigma^2} = -\frac{N}{2\sigma^2} + \sum_{i=1}^{N}\frac{(y_i-\mu)^2}{2\sigma^4} \qquad (3.63b)$$

Setting Eqns (3.63a,b) to zero and solving yields the required estimates of μ and σ^2. Beginning with Eqn (3.63a),

$$\sum_{i=1}^{N}\frac{(y_i-\mu)}{\sigma^2} = 0 \quad\text{or}\quad \mu = \frac{1}{N}\sum_{i=1}^{N}y_i = \bar{y} \quad (3.64)$$

Substituting $\bar{y}$ into Eqn (3.63b)

$$-N/\sigma^2 + \sum_{i=1}^{N}(y_i-\bar{y})^2/\sigma^4 = 0 \qquad (3.65a)$$

whereby the estimate for σ^2 becomes

$$\hat{\sigma}^2 = \sum_{i=1}^{N} (y_i - \bar{y})^2 / N = s'^2 \qquad (3.65b)$$

Thus, $\bar{y}$ and s'^2 are the maximum likelihood estimators of μ and σ^2. Although $\bar{y}$ is an unbiased estimate of μ, s'^2 is not unbiased for σ^2, as noted at the beginning of the chapter. However, s'^2 can easily be adjusted to the unbiased estimator

$$s^2 = \frac{1}{N-1} \sum_{i=1}^{N} (y_i - \bar{y})^2 \qquad (3.66)$$

Since the maximum likelihood method has widespread application, we present another simple example to illustrate its use. Let $y_1, y_2, ..., y_N$, be a random sample taken from a uniform distribution with $f(y_i) = 1/\theta = $ constant, $0 \le y_i \le \theta$, and $i = 1, 2, ..., N$. We want to find the maximum likelihood estimate of θ. Again, we write the likelihood, L, as the joint probability function

$$L = f(y_1, y_2, ..., y_N) = f(y_1)f(y_2)...f(y_N)$$
$$= (1/\theta)(1/\theta)...(1/\theta) = (1/\theta)^N$$
$$(3.67)$$

In this case, L is a monotonically decreasing function of θ and nowhere is $dL/d\theta = 0$. Instead, L increases monotonically as θ decreases and must be greater than or equal to the largest sample value, y_N. L is, therefore, not an unbiased estimate of θ. It can be adjusted to

$$\theta = \frac{(N+1)}{N} y_N \qquad (3.68)$$

which is unbiased. We note that if any statistic U can be shown to be a sufficient statistic for estimating θ, then the maximum likelihood estimator is always some function of U. If this maximum likelihood estimate can be found, and then adjusted to be unbiased, the result will generally be an MVUE.

To demonstrate the application of the maximum likelihood approach, assume that a random sample of size N is selected from the normal distribution (Eqn (3.18)) with μ and σ^2

as the mean and variance for each x (where we assume that the x_i values are independent). We ask: if $\bar{\theta} = (\theta_1, \theta_2) = (\mu, \sigma^2)$ is the parameter space for the PDF $f(x_1, x_2, ..., x_N)$, then what is the likelihood function? Also, find the maximum likelihood estimator $\hat{\theta}_1$ of θ_1, which maximizes the likelihood function and find the maximum likelihood estimator $\hat{\theta}_2$, which maximizes the likelihood function θ_2. We first write the PDF as

$$f(\bar{x}, \bar{\theta}) = \frac{1}{[\sigma^2(2\pi)]^{N/2}} \exp\left[\frac{1}{2\sigma^2} \sum_{i=1}^{N} (x_i - \mu)^2\right]$$

$$L(\bar{x}, \bar{\theta}) = \prod_{i=1}^{N} \left\{ \frac{1}{(2\pi\sigma^2)^{1/2}} \exp\left[\frac{-(x_i - \mu)^2}{\sigma^2}\right] \right\}$$

where L is the likelihood function written in terms of the product, Π, of the exponential. Taking the natural log of the above expression with respect to our estimated parameter, θ_1, and setting it equal to zero to find the maximum, we find

$$\ln(L) = -\frac{N}{2} \ln(\sigma^2) - \frac{N}{2} \ln(2\pi) - \sum_{i=1}^{N} \frac{(x_i - \mu)^2}{2\sigma^2}$$

where $\sigma > 0$ and $-\infty < \mu < \infty$. The derivative of this function with respect to θ_1 (which is μ) is

$$\frac{\partial L}{\partial \mu} = -\frac{1}{2\sigma^2} \sum_{i=1}^{N} (x_i - \mu)(-2)$$

$$= \frac{1}{\sigma^2} \sum_{i=1}^{N} (x_i - \mu) = 0$$

so that our estimate of μ is

$$\hat{\mu} = \bar{x} = \frac{1}{N} \sum_{i=1}^{N} x_i$$

Furthermore, the maximum likelihood estimator of θ_2 (which is σ^2) is given by

$$\frac{\partial L}{\partial \sigma^2} = -\frac{N}{2\sigma^2} - \frac{(-1)}{2\sigma^4} \sum_{i=1}^{N} (x_i - \mu)^2$$

$$= \frac{1}{2\sigma^2} \left[\frac{1}{\sigma^2} \sum_{i=1}^{N} (x_i - \mu)^2 - N \right] = 0$$

which yields the estimator

$$\hat{\sigma}^2 = \frac{1}{N} \sum_{i=1}^{N} (x_i - \hat{\mu})^2$$

For a normally distributed oceanographic data set, we can readily obtain maximum likelihood estimates of the mean and variance of the data. However, the real value of this technique is for variables that are not normally distributed. For example, if we examine spectral energy computed from current velocities, the spectral values have a chi-square distribution rather than a normal distribution. If we follow the maximum likelihood procedure, we find that the spectral values have a mean of ν, the number of degrees of freedom, and a variance of 2ν. These are the maximum likelihood estimators for the mean and variance. This example can be used as a pattern for applying the maximum likelihood method to a particular sample. In particular, we first determine the appropriate PDF for the sample values. We then find the joint likelihood function, take the natural logs, and then differentiate with respect to the parameter of interest. Setting this derivative equal to zero to find the maximum subsequently yields the value of the parameter being sought.

3.12 LINEAR ESTIMATION (REGRESSION)

Linear regression is one of a number of statistical procedures that fall under the general heading of linear estimation. Since linear regression is widely treated in the literature and is available in many software packages, our primary purpose here is to establish a common vocabulary for all readers. In our previous discussion and examples, we assumed that the random variables $Y_1, Y_2,..., Y_N$ were independent (in a probabilistic sense) and identically distributed, which implies that $E[Y_i] = \mu$ is a constant. Often this is not the case and the expected value $E[Y_i]$ of the variable is a function of some other parameter. We now consider the values y of a random variable, Y, called the dependent variable, whose values are a function of one or more *nonrandom* variables $x_1, x_2,..., x_N$, called independent variables (in a mathematical, rather than probabilistic sense).

If we model our random variable as

$$y = E[y] + \varepsilon = b_0 + b_1 x + \varepsilon \qquad (3.69)$$

we invariably find that the points y are scattered about the regression line $E[y] = b_0 + b_1 x$. The random variable ε in the right-hand term of Eqn (3.69) gives the departure from linearity and has a specific PDF with a mean value $\mu_\varepsilon = 0$. In other words, we can think of y as having a deterministic part, $E[y]$, and a random part, ε, that is randomly distributed about the regression line. By definition, simple linear regression is limited to finding the coefficients b_0 and b_1. If N independent variables $(x_1, x_2,..., x_N)$ are involved in the variability of each value y, we must deal with *multiple linear regression*. In this case, Eqn (3.69) becomes

$$y = b_0 + b_1 x_1 + b_2 x_2 + ... + b_N x_N + \varepsilon \qquad (3.70)$$

3.12.1 Method of Least Squares

One of the most powerful techniques for fitting a dependent model parameter y to independent (observed input) variable x_i ($i = 1, 2,..., N$) is the *method of least squares*. We apply the method in terms of linear estimation and will later readdress the topic in terms of more general statistical models. (Note: by "linear" we mean linear in the parameters $b_0, b_1,..., b_N$. Thus, $y = b_0 + b_1 x_i + \varepsilon$ is linear but $y = b_0 + \sin(b_1 x_1) + \varepsilon$ is not.) We begin with the simplest case, that of fitting a straight line to a set of points using the "best" coefficients b_0, b_1 (Figure 3.10). In a sense, the least squares procedure does what we do by eye—it minimizes the vertical deviations (residuals) of data points from the fitted line. The plots

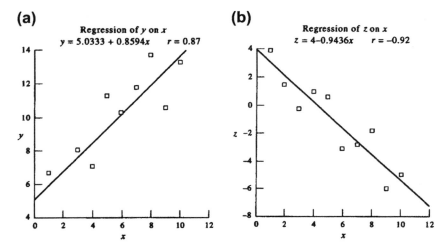

FIGURE 3.10 Straight-line (linear regression) fits to the sets of points in Table 3.7 using the "best" coefficients b_0, b_1. (a) Regression of y on x, for which $(b_0, b_1) = (5.0333, 0.8594)$; and (b) regression of z on x, for which $(b_0, b_1) = (4.0, -0.9436)$. r is the correlation coefficient.

in Figure 3.10 are for two separate data sets; $y = y(x)$ and $z = z(x)$. Let

$$y_i = \widehat{y}_i + \varepsilon_i \qquad (3.71)$$

where

$$\widehat{y}_i = b_0 + b_1 x_i \qquad (3.72)$$

is our estimator for the deterministic portion of the data and ε is the residual or error. To find the coefficients b_0, b_1 we need to minimize the sum of the squared errors (SSE) where SSE is the total variance that is not explained (accounted for) by our linear regression model given by Eqns (3.71) and (3.72)

$$SSE = \sum_{i=1}^{N} \varepsilon_i^2 = \sum_{i=1}^{N} (y_i - \widehat{y}_i)^2$$

$$= \sum_{i=1}^{N} \left[y_i - (b_0 + b_1 x_i) \right]^2 \qquad (3.73a)$$

$$= SST - SSR \qquad (3.73b)$$

in which

$$SST = \sum_{i-1}^{N} (y_i - \overline{y})^2 \quad \text{and}$$

$$SSR = \sum_{i-1}^{N} (\widehat{y}_i - \overline{y})^2 \qquad (3.73c)$$

Here, sum of squares total (SST) is the variance in the data, and sum of squares regression (SSR) is the amount of variance explained by our regression model. Minimization amounts to finding those coefficients that minimize the unexplained variance (SSE). Taking the partial derivatives of Eqn (3.73a) with respect to b_0 and b_1 and setting the resultant values equal to zero, the minimization conditions are

$$\frac{\partial SSE}{\partial b_0} = 0; \quad \frac{\partial SSE}{\partial b_1} = 0 \qquad (3.74)$$

Substituting Eqn (3.73a) into Eqn (3.74), we have for b_0

$$\frac{\partial SSE}{\partial b_0} = \frac{\partial}{\partial b_0}\left\{\sum_{i=1}^{N}[y_i - (b_0 + b_1 x_i)]^2\right\}$$

$$= -2\sum_{i=1}^{N}[y_i - (b_0 + b_1 x_i)]$$

$$= -2\left(\sum_{i=1}^{N}y_i - Nb_0 - b_1\sum_{i=1}^{N}x_i\right) = 0$$

$$(3.74a)$$

Now for b_1

$$\frac{\partial SSE}{\partial b_1} = \frac{\partial}{\partial b_1}\left\{\sum_{i=1}^{N}[y_i - (b_0 + b_1 x_i)]^2\right\}$$

$$= -2\sum_{i=1}^{N}(x_i)[y_i - (b_0 + b_1 x_i)]$$

$$= -2\left(\sum_{i=1}^{N}x_i y_i - b_0\sum_{i=1}^{N}x_i - b_1\sum_{i=1}^{N}x_i^2\right) = 0$$

$$(3.74b)$$

Once the mean values of y and x are calculated, these least squares equations can be solved simultaneously to find an estimate of the coefficient b_1 (the slope of the regression line); this is then used to obtain an estimate of the second coefficient, b_0 (the intercept of the regression line). In particular

$$\widehat{b_1} = \frac{\sum_{i=1}^{N}(x_i - \bar{x})(y_i - \bar{y})}{\sum_{i=1}^{N}(x_i - \bar{x})^2}$$

$$= \frac{\left[N\sum_{i=1}^{N}x_i y_i - \sum_{i=1}^{N}x_i\sum_{i=1}^{N}y_i\right]}{\left[N\sum_{i=1}^{N}x_i^2 - \left(\sum_{i=1}^{N}x_i\right)^2\right]} \quad (3.75a)$$

$$\widehat{b_0} = \bar{y} - \widehat{b_1}\bar{x} \quad (3.75b)$$

Several features of the regression values are worth noting. First, if we substitute the intercept

$\widehat{b_0} = \bar{y} - b_1\bar{x}$ into the line $\widehat{y} = b_0 + b_1 x$, we obtain

$$\widehat{y} = \bar{y} = b_1(x - \bar{x})$$

As a result, whenever $x = \bar{x}$, we have $\widehat{y} = \bar{y}$. This means: (1) that the regression line always passes through the point $(\bar{x}, \bar{y})$, the centroid of the distribution, and (2) that because the operation $\partial SSE/\partial b_0 = 0$ minimizes the error $\sum \varepsilon_i = 0$, the regression line not only goes through the point of averages $(\bar{x}, \bar{y})$ but it also splits the scatter of the observed points so that the positive residuals (where the regression line passes below the true point) always cancel exactly the negative residuals (where the line passed above the true point). The sample regression line is therefore an unbiased estimate of the population regression line. To summarize the linear regression procedure, we note that:

1. For each selected x (independent variable) there is a distribution of y (or z in the case of the examples shown in Figure 3.10) from which the sample (dependent variable) is drawn at random.
2. The population of y corresponding to a selected x has a mean μ that lies on the straight line $\mu = b_0 + b_1 x$, where b_0 and b_1 are regression parameters.
3. In each population, the standard deviation of y about its mean, $b_0 + b_1 x$, has the same value $(s_{xy} = s_\varepsilon, y = b_0 + b_1 x + \varepsilon)$. Note that ε is a random variable drawn from a normal population with $\mu = 0$ and $s = s_{xy}$.

Thus, y is the sum of a random part ε and a fixed part x; the fixed part determines the mean values of the y population samples, with one distribution of y for each x that we pick. The mean values of y lie on the straight line, $\mu = b_0 + b_1 x$, which is the population regression line. The regression parameter b_0 is the y mean for $x = 0$ and b_1 is the slope of the regression line. The random part, ε, is independent of x and y. To compute the regression parameters, we need values of $N, \bar{x}, \bar{y}, \sum x^2, \sum y^2$, and $\sum xy$. Earlier, we discussed the computational shortcuts to compute $\sum x^2$ and $\sum y^2$ without first computing the means of x and y. The same can be accomplished for xy using

$$\sum(x - \bar{x})(y - \bar{y}) = \sum xy - \sum x \sum y/N$$

As examples of linear regression, consider the data sets in Table 3.7 for dependent variables y_i and z_i which are both functions of the same independent variable x_i (for example, y_i, could be the eastward and z_i the northward component of velocity as functions of time x_i). We will compute the regression coefficients b_0, b_1 plus the sample variance s^2 and percent of explained variance (100 SSR/SST) for each data set.

To estimate the regression parameters, we must first compute the means of the three series

$$\bar{x} = 5.50; \quad \bar{y} = 9.78; \quad \bar{z} = -1.19$$

TABLE 3.7   Values for dependent variables y_i, z_i as function of x_i.

x_i	y_i	$\hat{y}_i$	z_i	$\hat{z}_i$
1.0	6.7	5.9	3.9	3.1
2.0	4.7	6.8	1.5	2.1
3.0	8.1	7.6	-0.2	1.2
4.0	7.1	8.5	1.0	0.2
5.0	11.3	9.4	0.6	-0.7
6.0	10.5	10.2	-3.1	-1.7
7.0	11.8	11.1	-2.8	-2.6
8.0	13.7	11.9	-1.8	-3.6
9.0	10.6	12.8	-6.0	-4.5
10.0	13.3	13.7	-5.0	-5.4

SST(y) = 80.64; SSR(y) = 61.11; SSE(y) = 19.53

SST(z) = 86.39; SSR(z) = 73.46; SSE(z) = 12.93

The estimated values $\hat{y}$ and $\hat{z}$ are derived from the linear regression analysis. Formulae at the bottom of the table are the sum of squares total (SST), sum of squares regression (SSR), and sum of the squared errors (SSE) to be derived in our regression analysis for N = 10

We then use the means to calculate the sums in Eqn (3.73a-c)

$$\sum_{i=1}^{10}(x_i - \bar{x})^2 = 82.50; \quad \sum_{i=1}^{10}(x_i - \bar{x})(y_i - \bar{y}) = 71.00;$$

$$\sum_{i=1}^{10}(x_i - \bar{x})(z_i - \bar{z}) = -77.85$$

$$\text{SST}(y) = \sum_{i=1}^{10}\left(y_i - \bar{y}\right)^2 = 80.64$$

$$\text{SST}(z) = \sum_{i=1}^{10}\left(z_i - \bar{z}\right)^2 = 86.36$$

For the regression of y on x ($\hat{y} = b_0 + b_1 x$) we find

$$b_0 = 5.05; \quad b_1 = 0.861; \quad s^2 = 2.44$$
$$100 \cdot \text{SSR}(y)/\text{SST}(y) = (100) \cdot 61.11/80.64$$
$$= 75.8\%$$

while for the regression of z on x ($\hat{z} = b_0 + b_1 x$), we have

$$b_0 = 4.00; \quad b_1 = -0.94; \quad s^2 = 1.62$$
$$100 \cdot \text{SSR}(z)/\text{SST}(z) = (100) \cdot 73.46/86.36$$
$$= 85.0\%$$

The ratio SSR/SST (variance explained/total variance) is a measure of the goodness of fit of the regression curves called the *coefficient of determination*, r^2 (*squared correlation coefficient*). If the regression line fits perfectly all the sample values, all residuals would be zero. In turn, SSE $= 0$ and SSR/SST $= r^2 = 1$. As the fit becomes increasingly less representative of the data points, r^2 decreases toward a possible minimum of zero.

3.12.2 Standard Error of the Estimate

The measure of the absolute magnitude of the goodness of fit of our estimate, $\hat{y}$, is the standard error of the estimate, s_ε ($=s_{yx}$), defined as

$$s_\varepsilon = [\text{SSE}/(N-2)]^{1/2}$$

$$= \left[\frac{1}{N-2} \sum_{i=1}^{N} (y_i - \widehat{y})^2 \right]^{1/2} \qquad (3.76)$$

The number of degrees of freedom, $N-2$, for s_ε is based on the fact that two parameters, b_0 and b_1 are needed for any linear regression estimate. If s_ε is from a normal distribution then approximately 68.3% of the observations will fall within $\pm 1 s_\varepsilon$ units of the regression line, 95.4% will fall within $\pm 2 s_\varepsilon$ units of the line and 99.7% will fall within $\pm 3 s_\varepsilon$ units of this line. For the examples of Table 3.7 (which includes estimates $\widehat{y}$ and $\widehat{z}$):

Variable y: $s_\varepsilon = [\text{SSE}(y)/(N-2)]^{1/2}$

$$= (19.53/8)^{1/2} = 1.56$$

Variable z: $s_\varepsilon = [\text{SSE}(z)/(N-2)]^{1/2}$

$$= (12.93/8)^{1/2} = 1.27$$

As a result, the $\pm 2 s_\varepsilon$ ranges are $\pm 2(1.56)$ and $\pm 2(1.27)$, respectively.

The standard error for the estimate of the y-intercept, $\widehat{b}_0$, of the regression line is given as

$$s_{b0} = s_\varepsilon \left(\frac{1}{N} + \frac{\bar{x}^2}{\sum_{i=1}^{N}(x_i - \bar{x})^2} \right)^{1/2} \qquad (3.77a)$$

and the standard error for the slope, $\widehat{b}_1$, of the regression line as

$$s_{b1} = \frac{s_\varepsilon}{\left[\sum_{i=1}^{N}(x_i - \bar{x})^2 \right]^{1/2}} \qquad (3.77b)$$

For small samples ($N < 30$), we can write the 90% confidence interval for the true value of the regression slope, b_1, as

$$\widehat{b}_1 - t_{0.05}s_{b1} \leq b_1 \leq \widehat{b}_1 + t_{0.05}s_{b1} \qquad (3.78)$$

where $t_{0.05}$ is the Student-t statistic for $\alpha/2 = 0.05$ and $N-2$ degrees of freedom. Turning to the regression line itself, it is useful to know how the confidence intervals for the regressional estimates of: (1) $\bar{y}$ (the mean of the dependent variable, y), and (2) y_i (a specific value of the dependent variable, y) vary for given values of the independent variable, x_i. Specifically, the confidence intervals for $\bar{y}$ and y_i are given by:

$$\widehat{\bar{y}} \pm t_{\alpha/2,N-2}s_\varepsilon \sqrt{\tilde{x}_i}; \quad \text{estimate of } \bar{y} \text{ given } \tilde{x}_i$$

$$\widehat{y}_i \pm t_{\alpha/2,N-2}s_\varepsilon \sqrt{1 + \tilde{x}_i}; \quad \text{estimate of } y_i \text{ given } \tilde{x}_i$$

$$(3.79a)$$

respectively, where

$$\tilde{x} = \frac{1}{N} + \frac{(x_i - \bar{x})^2}{\sum_{i=1}^{N}(x_i - \bar{x})^2} \qquad (3.79b)$$

Because of the dependence on x_i, these confidence limits would appear as hyperbolae in regression diagrams, such as Figure 3.10. The hyperbolae represent confidence belts for the different significance levels. Note the increasing uncertainty of making predictions for y for values of x far removed from the mean value, $\bar{x}$. Since the lines indicate that y must be within the confidence belt, higher significance levels have narrower belts. For all significance levels, estimates of $\bar{y}$ and y_i get worse as we move away from $\bar{x}$. Remember that these confidence belts are for the regression line itself and not for the individual points. Hence, in the case of the 95% confidence interval, if repeated samples of y_i are taken of the same size and the same fixed value of x, then 95% of the confidence intervals, constructed for the mean value of y and x, will contain the true value of the mean of y and x. If only one prediction is made for x, then the probability that the calculated interval will contain the true value is 95%.

3.12.3 Multivariate Regression

To extend the regression procedure to multivariate regression, it is best to formulate our linear estimation model in matrix terms. Suppose our model is of the form

$$Y = b_0 + b_l X_1 + b_2 X_2 + \ldots + b_k X_k + \varepsilon \quad (3.80)$$

and that we make N independent (probabilistic) observations $y_1, y_2, \ldots, y_N$ of Y. This means that we can write

$$y_i = b_0 + b_l x_{i1} + b_2 x_{i2} + \ldots + b_k x_{ik} + \varepsilon_i \quad (3.81)$$

where x_{ik} is the k th independent variable for the ith observation. Writing this in matrix form we have

$$Y = \begin{pmatrix} y_1 \\ y_2 \\ \ldots \\ \ldots \\ y_N \end{pmatrix}, \quad X = \begin{pmatrix} x_{10} & x_{11} & \cdots & x_{1k} \\ x_{20} & x_{21} & \cdots & x_{2k} \\ \ldots & \ldots & \ldots & \ldots \\ \ldots & \ldots & \ldots & \ldots \\ x_{N0} & x_{N1} & \cdots & x_{Nk} \end{pmatrix}$$

$$B = \begin{pmatrix} b_0 \\ b_1 \\ \ldots \\ \ldots \\ b_k \end{pmatrix}, \quad E = \begin{pmatrix} \varepsilon_1 \\ \varepsilon_2 \\ \ldots \\ \ldots \\ \varepsilon_N \end{pmatrix}$$

$$(3.82)$$

where the boldface letters indicate matrices. Using Eqn (3.82), we can represent the N equations relating y_i the independent variable x_{ij} as

$$Y = B \cdot X + E \quad (3.83)$$

If we restrict our analysis to the first two coefficients, Eqn (3.83) reduces to the simple straight-line fit model (Eqn (3.72)). In this case, the matrices for N observations become

$$Y = \begin{pmatrix} y_1 \\ y_2 \\ \ldots \\ \ldots \\ y_N \end{pmatrix}, \quad X = \begin{pmatrix} x_{10} & x_{11} & \cdots & x_{1N} \\ x_{20} & x_{21} & \cdots & x_{2N} \\ \ldots & \ldots & \ldots & \ldots \\ \ldots & \ldots & \ldots & \ldots \\ x_{N0} & x_{N1} & \cdots & x_{NN} \end{pmatrix}$$

$$B = \begin{pmatrix} b_0 \\ b_1 \end{pmatrix}, \quad E = \begin{pmatrix} \varepsilon_1 \\ \varepsilon_2 \\ \ldots \\ \ldots \\ \varepsilon_N \end{pmatrix}$$

$$(3.84)$$

Using these N observations in Eqn (3.83), the least squares equations are

$$N b_0 + b_1 \sum_{i=1}^{N} x_i = \sum_{i=1}^{N} y_i$$

$$b_0 \sum_{i=1}^{N} x_i + b_1 \sum_{i=1}^{N} x_i^2 = \sum_{i=1}^{N} x_i y_i \quad (3.85)$$

which we can solve for b_0 and b_1. We can generalize the procedure further by realizing that for $x_{i0} = 1$ in equation (3.81), we have

$$X' \cdot X = \begin{pmatrix} 1 & \cdots & \cdots & 1 \\ \cdots & \cdots & \cdots & \cdots \\ x_1 & \cdots & \cdots & x_N \end{pmatrix} \begin{pmatrix} 1 & x_i \\ \cdots & \cdots \\ 1 & x_N \end{pmatrix}$$

$$= \begin{pmatrix} N & \sum x_i \\ \cdots & \cdots \\ \sum x_i & \sum x_i^2 \end{pmatrix}$$

$$(3.86)$$

where X' is the transpose of the matrix X and, the sums are from 1 to N, and

$$X' \cdot Y = \begin{pmatrix} \sum_{i=1}^{N} y_i \\ \sum_{i=1}^{N} x_i y_i \end{pmatrix} \quad (3.87)$$

The least squares equations can then be expressed as

$$(X' \cdot X) \cdot B = X' \cdot Y \quad (3.88)$$

where

$$B = \begin{pmatrix} b_0 \\ b_1 \end{pmatrix} \quad (3.89)$$

Solving the above equations for B, we obtain

$$B = (X' \cdot X)^{-1} X' \cdot Y \quad (3.90)$$

3.12.4 A Computational Example of Matrix Regression

Since linear regression is widely used in oceanography, we will illustrate its use by a simple example. Suppose we want to fit a line to the data pairs consisting of the independent variable x_i and the dependent variable y_i given in Table 3.8. From these we find

$$\sum_{i=1}^{N} x_i = 0, \quad \sum_{i=1}^{N} y_i = 5, \quad \sum_{i=1}^{N} x_i y_i = 7,$$

$$\sum_{i=1}^{N} x_i^2 = 10$$

Substituting into Eqn (3.75a,b), we have

$$b_1 = \frac{\left[N \sum_{i=1}^{N} x_i y_i - \sum_{i=1}^{N} x_i \sum_{i=1}^{N} y_i \right]}{\left[N \sum_{i=1}^{N} x_i^2 - \left(\sum_{i=1}^{N} x_i \right)^2 \right]}$$

$$= \frac{[(5)(7) - (0)(5)]}{[(5)(10) - 10^2]} = 0.7$$

$$b_0 = \bar{y} - b_1 \bar{x} = 5/5 - (0.7)(0) = 1$$

This same problem can be put in matrix form (see the previous section)

$$\mathbf{Y} = \begin{pmatrix} 0 \\ 0 \\ 1 \\ 1 \\ 3 \end{pmatrix}, \quad \mathbf{X} = \begin{pmatrix} 1 & -2 \\ 1 & -1 \\ 1 & 0 \\ 1 & 1 \\ 1 & 2 \end{pmatrix}$$

$$\mathbf{X}' \cdot \mathbf{X} = \begin{pmatrix} 5 & 0 \\ 0 & 10 \end{pmatrix}, \quad \mathbf{X}' \cdot \mathbf{Y} = \begin{pmatrix} 5 \\ 7 \end{pmatrix},$$

$$(\mathbf{X}' \cdot \mathbf{X})^{-1} = \begin{pmatrix} 1/5 & 0 \\ 0 & 1/10 \end{pmatrix}$$

$$\mathbf{B} = (\mathbf{X}' \cdot \mathbf{X})^{-1} (\mathbf{X}' \cdot \mathbf{Y}) = \begin{pmatrix} 1/5 & 0 \\ 0 & 1/10 \end{pmatrix}$$

$$\times \begin{pmatrix} 5 \\ 7 \end{pmatrix} = \begin{pmatrix} 1 \\ 0.7 \end{pmatrix}$$

TABLE 3.8   Data Values Used in Least Squares Linear Fit of a Two-Coefficient Regression Model, $y_i = F(x_i)$

Data		Solution Values	
x_i	y_i	$(x_i)(y_i)$	x_i^2
−2	0	0	4
−1	0	0	1
0	1	0	0
1	1	1	1
2	3	6	4

so that by Eqn (3.89), $b_0 = 1$ and $b_1 = 0.7$.

An important property of the simple straight-line least-square estimators we have just derived is that b_0 and b_1 are unbiased estimates of their true parameter values. We have assumed that $E[\varepsilon] = 0$ and that $V[\varepsilon] = \sigma^2$; thus the error variance is independent of x and $V[Y] = V[\varepsilon] = \sigma^2$. Since σ^2 is usually unknown, we estimate it using the sample variance (Eqn (3.4)) given by

$$s^2 = \frac{1}{N-1} \sum_{i=1}^{N} (y_i - \bar{y})^2 \qquad (3.91)$$

However, if we use the output values, $\hat{y}_i$, from the least squares to estimate $\varepsilon_i(Y) = y_i - \hat{y}$, we must write Eqn (3.91) as

$$s^2 = \frac{1}{N-2} \sum_{i=1}^{N} (y_i - \hat{y}_i)^2 = \frac{1}{N-2} \text{SSE} \quad (3.92)$$

where SSE, given by Eqn (3.73a), represents the sum of the squares of the errors and the $N-2$ corresponds to the fact that two parameters, b_0 and b_1, are needed in the model. In matrix notation we can write the SSE as

$$\mathbf{SSE} = \mathbf{Y}' \cdot \mathbf{Y} - (\mathbf{B}' \cdot \mathbf{X}') \cdot \mathbf{Y} \qquad (3.93)$$

Using this with our previous numerical example, we write Eqn (3.93) as

$$(0 \quad 0 \quad 1 \quad 1 \quad 3) \begin{pmatrix} 0 \\ 0 \\ 1 \\ 1 \\ 3 \end{pmatrix} - (1 \quad 0.7) \begin{pmatrix} 1 & 1 & 1 & 1 & 1 \\ -2 & -1 & 0 & 1 & 2 \end{pmatrix} \begin{pmatrix} 0 \\ 0 \\ 1 \\ 1 \\ 3 \end{pmatrix}$$

$$= 11 - (1 \quad 0.7) \begin{pmatrix} 5 \\ 7 \end{pmatrix} = 11 - 9.9 = 1.1$$

Since $s^2 = \text{SSE}/(N-2)$, we have $s^2 = 1.1/(3) = 0.367$ as our estimator of σ^2

3.12.5 Polynomial Curve Fitting with Least Squares

The use of least-squares fitting is not limited to the straight-line regression model discussed thus far. In general, we can write our linear model as any polynomial of the form

$$Y = b_0 + b_1 x + b_2 x^2 + \ldots + b_N x^N + \varepsilon \quad (3.94)$$

The procedure is the same as with the straight line case except that now the **X** matrix has $N+1$ columns. Thus, the least-squares fit will have $N+1$ linear equations with $N+1$ unknowns, b_0, b_1, ..., b_N. These equations are called the *normal equations*.

3.12.6 Relationship between Least Squares and Maximum Likelihood

As discussed earlier, the maximum likelihood estimator is one that maximizes the likelihood of sampling a given parameter. In general, if we have a sample x_i from a population with the PDF $f(x_i, \theta)$, where θ is the parameter of interest, the maximum likelihood estimator $L(\theta)$ is the product of the individual independent probabilities.

$$L(\theta) = f(x_1, \theta) f(x_2, \theta) \ldots f(x_N, \theta) \quad (3.95)$$

If the errors all come from a normal distribution, this becomes from Eqn (3.61).

$$L(\theta) = \frac{\exp\left[-\sum_{i=1}^{N} (x_i - \theta)^2 \Big/ 2\sigma^2 \right]}{\sigma^N (2\pi)^{N/2}} \quad (3.96)$$

When this is maximized, it leads to the least-squares estimate

$$\widehat{\theta} = \frac{1}{N} \sum_{i=1}^{N} x_i = \overline{x}$$

In other words, the least-squares estimate of the mean of θ can be derived from a normal distribution using the maximum likelihood criterion. This value is found to be the average of the independent variable x.

3.13 RELATIONSHIP BETWEEN REGRESSION AND CORRELATION

The subject of correlation will be considered in more detail when we examine time series analysis methods. Our intension, here, is simply to introduce the concept in general statistical terms and relate it to the simple regression model just discussed. As with regression, correlation relates two variables but unlike regression it is measured without estimation of the population regression line.

The *correlation coefficient*, r, is a way of determining how well two (or more) variables covary in time or space. For two random variables x (x_1, x_2,..., x_n) and y (y_1, y_2,..., y_N) the correlation coefficient can be written as

$$r = \frac{1}{N-1} \sum_{i=1}^{N} \frac{(x_i - \overline{x})(y_i - \overline{y})}{s_x s_y} \quad (3.97a)$$

$$= C_{xy}/s_x s_y$$

where

$$C_{xy} = \frac{1}{N-1} \sum_{i=1}^{N} (x_i - \overline{x})(y_i - \overline{y}) \quad (3.97b)$$

is the *covariance* of x and y, and s_x and s_y are the standard deviations for the two data records as defined by Eqn (3.4). We note two important properties of r:

1. r is a dimensionless quantity since the units of the numerator and the denominator are the same;
2. the value of r lies between -1 and $+1$ since it is normalized by the product of the standard deviations of both variables.

For $r = \pm 1$, the data points (x_i, y_i) lie along a straight line and the samples are said to have a perfect correlation ($+$for "in-phase" fluctuations and minus ($-$) for $180°$ "out-of-phase" fluctuations). For $r \approx 0$, the points are scattered randomly on the graph and there is little or no relationship between the variables. The variables x_i, y_i in Eqn (3.97a,b) could be samples from two different, independent random variables or they could represent the independent (input) and dependent (output) variables of an estimation model. Alternatively, they could be samples from the same variable. Known as an *autocorrelation*, the latter is usually computed for increasing lag or shifts in the starting value for one of the time series. A lag of "m" means that the first m values of one of the series, say the x series, are removed prior to the calculation so that x_{m+1} becomes the new x_1, and so on.

Some authors prefer to use r^2 (the coefficient of determination discussed in Section 3.12.1 in the context of straight-line regression) rather than r (the correlation coefficient) since the squared value can be used to construct a significance level for r^2 in terms of a hypothesis test when the true correlation squared is zero. Writing

$$C_{xy}^2 / (s_x s_y)^2 = \text{SSR/SST} = r^2 \qquad (3.98)$$

we see that $r^2 =$ variance explained/total variance, as stated earlier. A value $r = 0.75$ means that a linear regression of y on x explains $r^2 = 56.25\%$ of the total sample variance. Our approach is to use r to get the sign of the correlation

and r^2 to estimate the joint variances. It is worth noting that a moderate value $r = 0.25$, for exmple, might seem significant (perhaps when comparing a current velocity record against a coincident wind stress record) until it is realized that it means that the variance in the wind stress only accounts for roughly $r^2 = 6.25\%$ of the variance in current velocity.

3.13.1 The Effects of Random Errors on Correlation

Before discussing the relationship between r and our simple regression model, it is important to realize that sampling errors in x_i and y_i can only cause r to decrease. This can be shown by writing our two variables as a combination of true values (α_i, β_i) and random errors (δ_i, ε_i). In particular

$$\begin{aligned} x_i &= \alpha_i + \delta_i \\ y_i &= \beta_i + \varepsilon_i \end{aligned} \qquad (3.99)$$

Using Eqns (3.97) and (3.99), we can write the correlation between x_i and y_i as

$$r_{xy} = \frac{s_\alpha s_\beta r_{\alpha\beta} + s_\beta s_\delta r_{\beta\delta} + s_\alpha s_\varepsilon r_{\alpha\varepsilon} + s_\delta s_\varepsilon r_{\delta\varepsilon}}{s_x s_y}$$

$$(3.100)$$

where for convenience we have dropped the index i. Since the random errors δ and ε are assumed to be independent of each other and of the variables α and β we know that

$$r_{\beta\delta} = r_{\alpha\varepsilon} = r_{\delta\varepsilon} = 0$$

so that Eqn (3.100) becomes

$$r_{xy} = \frac{s_\alpha s_\beta}{s_x s_y} r_{\alpha\beta} \qquad (3.101)$$

This result means that the ratio between the product of the true standard deviations (s_α, s_β) to the product of the measured variable (s_x, s_y) determines the magnitude of the computed correlation coefficient (r_{xy}) relative to the true value ($r_{\alpha\beta}$).

To determine Eqn (3.101), we expand the variances of x and y as

$$\left(s_x^2, s_y^2\right) = \frac{1}{N-1} \sum_{i=1}^{N} \left[(x_i - \bar{x})^2, (y_i - \bar{y})^2\right]$$

where, as usual, $\bar{x}, \bar{y}$ are the average values for samples x_i, y_i respectively. Expanding the numerator into its component terms through Eqn (3.99), and using the fact that the errors are independent of one another, and of x and y, yields

$$\sum_{i=1}^{N} (x_i - \bar{x})^2 = \sum_{i=1}^{N} \left[(\alpha_i - \bar{\alpha})^2 + \delta_i^2\right]$$
$$\sum_{i=1}^{N} (y_i - \bar{y})^2 = \sum_{i=1}^{N} \left[(\beta_i - \bar{\beta})^2 + \varepsilon_i^2\right]$$

Dividing through by $(N-1)$ and using the definitions for standard deviation, we find

$$s_x^2 = s_\alpha^2 + \frac{\sum_{i=1}^{N} \delta_i^2}{N-1}; \quad s_y^2 = s_\beta^2 + \frac{\sum_{i=1}^{N} \varepsilon_i^2}{N-1}$$
$$(3.102)$$

Since the second terms in each of the above expressions can never be negative $(N > 1)$, the observed variances s_x^2 and s_y^2 are always greater than the corresponding true variances. Applying this result to Eqn (3.101), we see that the calculated correlation, r_{xy}, derived from the observations is always smaller than the true correlation, $r_{\alpha\beta}$. Because of random errors, the correlation coefficient computed from the observations will be smaller than (or, at best, equal to) the true correlation coefficient.

3.13.2 The Maximum Likelihood Correlation Estimator

Returning to the relationship between correlation and regression, we note the maximum likelihood estimator of the correlation coefficient is, by Eqn (3.97a)

$$r = \left[\sum_{i=1}^{N} (x_i - \bar{x})(y_i - \bar{y})\right] \Bigg/ \left[\sum_{i=1}^{N} (x_i - \bar{x})^2 \sum_{i=1}^{N} (y_i - \bar{y})^2\right]^{1/2}$$
$$(3.103)$$

for a bivariate normal population (x_i, y_i). We can expand this using Eqn (3.97b) to derive

$$r = \frac{\left[N \sum_{i=1}^{N} x_i y_i - \sum_{i=1}^{N} x_i \sum_{i=1}^{N} y_i\right]}{\left\{\left[N \sum_{i=1}^{N} x_i^2 - \left(\sum_{i=1}^{N} x_i\right)^2\right]\left[N \sum_{i=1}^{N} y_i^2 - \left(\sum_{i=1}^{N} y_i\right)^2\right]\right\}^{1/2}}$$
$$(3.104)$$

Note that the numerator in Eqn (3.104) is similar to the numerator of the estimator for b_1 in Eqn (3.75a). For the case where the regression line passes through the origin in Eqn (3.75b), we have $b_1 = 0$ and our model is

$$\hat{y}_i = \hat{b}_1 x_i$$

and we can rewrite Eqn (3.75a) as

$$\hat{b}_1 = \frac{\sum_{i=1}^{N} (x_i y_i)}{\sum_{i=1}^{N} x_i^2}$$
$$(3.105)$$
$$= r s_y / s_x; \quad \text{or,} \quad r = \hat{b}_1 s_x / s_y$$

Thus, r can be computed from $\hat{b}_1$ and vice versa if the standard deviations of the sample variance x and y are known. Also, using the relationship between $\hat{b}_1$ and r we can write the variance of the parameter estimate in Eqn (3.105) as

$$s^2 = \frac{1}{N-2} \sum_{i=1}^{N} (y_i - \bar{y})^2 = \frac{1}{N-2} \text{SSE} \quad (3.106)$$

We can use this result to better understand the relationship between correlation and regression by writing the ratio of the regression variance in Eqn (3.106) to the sample variance for y alone; for large N, this becomes

$$\frac{s^2}{s_y^2} = \frac{(N-1)(1-r^2)}{N-2} \approx (1-r^2) \quad (3.107)$$

Thus, for N large, r^2 is that portion of the variance of y that can be attributed to its regression on x while $(1-r^2)$ is that portion of y's variance that is independent of x. Earlier it was noted that a computationally efficient way to calculate the variance was to use Eqn (3.4b) which required only a single pass through the data sample. A similar saving can be gained in computing the covariance by expanding the product

$$\sum_{i=1}^{N}(x_i - \bar{x})(y_i - \bar{y}) = \sum_{i=1}^{N}(x_i y_i)$$
$$- \frac{\left(\sum_{i=1}^{N} x_i\right)\left(\sum_{i=1}^{N} y_i\right)}{N}$$

$$(3.108)$$

3.13.3 Correlation and Regression: Cause and Effect

A point worth stressing is that a high correlation coefficient or a "good" fit of a regression curve $y = y(x)$ to a set of observations x, does not imply that x is "causing" y. Nor does it imply that x will provide a good predictor for y in the future. For example, the number of sockeye salmon returning to the Fraser River of British Columbia each fall from the North Pacific Ocean is often highly correlated with the mean fall sea surface temperature (SST) at Amphitrite Point on the southwest coast of Vancouver Island. No one believes that the fish are responding directly to the temperature at this point, but rather that temperature is a proxy variable for the real factor (or combination of factors) influencing the homeward migration of the fish. Of course, we are not saying that one should not draw inferences or conclusions from correlation or regression analysis (for example, SSTs often have large spatial and temporal correlation scales so that temperature may, indeed, be the main driving variable) but only that caution is advised when seeking cause-and-effect relationships between variables. We further remark that there is little point in drawing any type of line through the data unless the scatter about the line is appreciably less than the overall spread of the observations.

There is a tendency to fit trend lines to data with large variability and scatter even if a trend is not justified on statistical grounds. If $|r| < 0.5$, it hardly seems reasonable to fit a line for predictive purposes.

There is another important aspect of regression-correlation analysis that is worth stressing: although the value of the correlation coefficient or coefficient of determination does *not* depend on which variable (x or y) is designated as the independent variable and which is designated as the dependent variable, this distinction *is* very important when it comes to regression analysis. The regression coefficients a, b for the conditional distribution of y given x ($y = a_1 + b_1 x$) are different than those for the conditional distribution of x given y ($x = a_2 + b_2 y$). In general, $a_1 \neq -a_2/b_2$ and $b_1 \neq 1/b_2$ and so that the regression lines are different. In the first case, we are solving for the line shown in Figure 3.11(a), while in the second case we are solving for the line in Figure 3.11(b).

As an example, consider the broken lines in Figure 3.11(c) which show the two different linear regression lines for the regression of the observed cross-channel sea-level differences $y = \Delta\eta_c$, as measured by coastal tide gauges, and the calculated cross-channel sea-level difference $x = \Delta\eta_m$ obtained using concurrent current meter data from cross-channel moorings. The term $\Delta\eta_c$ is simply the difference in the mean sea level from one side of the 25-km-wide channel to the other, while $\Delta\eta_m$ is calculated from the current meter records assuming that the time-averaged along-channel flow is in geostrophic balance (Labrecque et al., 1994). The dotted line is the regression $\Delta\eta_c = a_1 + b_1\Delta\eta_m$ while the dashed line is the regression $\Delta\eta_m = a_2 + b_2\Delta\eta_c$ with $b_1 \neq b_2$. The correlation coefficient $r = 0.69$ is the same for the two regressions. The solid line in Figure 3.11(c) is the so-called *neutral* regression line for the two parameters (Garrett and Petrie, 1981) and might seem the line of choice since it

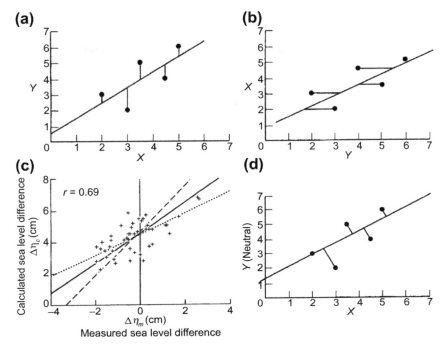

FIGURE 3.11 Straight-line regressions (a) y on x, and (b) x on y showing the "direction" along which the variance is minimized. (c) Scatter plot of $\Delta\eta_c$ vs $\Delta\eta_m$ for a cross-section of the 22-km-wide Juan de Fuca Strait separating Vancouver Island from Washington State. Plots give the regression of the observed cross-channel sea level differences $y = \Delta\eta_c$, as measured by coastal tide gauges, and the calculated cross-channel sea-level difference, $x = \Delta\eta_m$, obtained using concurrent current meter data from cross-channel moorings. The solid slopping line in plot (c) gives the bisector regression fit to the data (slope and 95% confidence level $= 0.96 \pm 0.37$); the dotted line (slope $= 0.66 \pm 0.14$) and the dashed line (slope $= 1.40 \pm 0.32$) are the standard slopes for $\Delta\eta_c$ vs $\Delta\eta_m$ and $\Delta\eta_m$ vs $\Delta\eta_c$, respectively. Here, $r = 0.69$ *(From Labrecque et al. (1994)).* (d) The "direction" along which the variance for the data points in (a) and (b) is minimized.

is not obvious which parameter should be the independent parameter and which should be the dependent parameter. Neutral regression is equivalent to minimizing the sum of the square distances from the regression line (Figure 3.11d).

In fisheries research, neutral regression is known as *geometric mean functional regression* (GMFR) and is commonly used to relate fish body proportions when there is no clear basis to select dependent and independent variables (Sprent and Dolby, 1980). For two variables with zero means, the slope estimator, b, is given by the square roots of the variance ratios.

$$b_{yx} = \text{sgn}\,(s_{xy}) \left[\frac{\sum_{i=1}^{N}(y_i - \bar{y})^2}{\sum_{i=1}^{N}(x_i - \bar{x})^2}\right]^{1/2};$$

regression $\hat{y}_i = \widehat{b}_{yx}x_i$

$$b_{xy} = \text{sgn}\,(s_{xy}) \left[\frac{\sum_{i=1}^{N}(x_i - \bar{x})^2}{\sum_{i=1}^{N}(y_i - \bar{y})^2}\right]^{1/2};$$

regression $\hat{x}_i = \widehat{b}_{xy}y_i$ (3.109)

where $\text{sgn}(s_{xy})$ is the sign of the covariance function $s_{xy} = \sum(x_i - \bar{x})(y_i - \bar{y})$ and $b_{yx} = 1/b_{xy}$ as required. Note that the slope b_{yx} lies midway between the slopes b_1 and b_2

$$b_1 = \frac{\sum_{i=1}^{N}(x_i - \bar{x})(y_i - \bar{y})}{\sum_{i=1}^{N}(x_i - \bar{x})^2};$$

regression line $\hat{y}_i = a_1 + \hat{b}_1 x_i$

$$b_2 = \frac{\sum_{i=1}^{N}(x_i - \bar{x})(y_i - \bar{y})}{\sum_{i=1}^{N}(y_i - \bar{y})^2};$$

regression line $\hat{x}_i = a_2 + \hat{b}_2 y_i$ (3.110)

given by Eqn (3.75a) for standard regression analyses (Figure 3.11(a)). The GMFR is then the geometric mean slope of the least-squares regression coefficient for the regression slope of y on x and the regression of x on y; $b_{yx} = [b_1/b_2]^{1/2}$. Since the slope from the GMFR is simply a ratio of variances, it is "transparent" to the determination of correlation coefficients or coefficients of determination. It is these correlations, not the slope of the line, that test the strength of the linear relationship between the two variables. Moreover, none of the standard linear regression models reduces to the GMFR slope estimate except under unlikely circumstances. According to Sprent and Dolby (1980), ad hoc use of the GMFR is not recommended when there are errors in both variables. The GMFR model, though appealing, rests on shaky statistical ground and its use remains controversial.

3.14 HYPOTHESIS TESTING

Statistical inference takes one of two forms. Either we make estimates of population variables, as we have done thus far, or we test hypotheses about the implications of these variables. Statistical inference in which we choose between two conflicting hypotheses about the value of a particular population variable is known as *hypothesis testing*.

Hypothesis testing follows scientific methodology from whose nomenclature the terms are borrowed. The investigator forms a "hypothesis,"

collects some sample data and uses a statistical construct to either reject or accept the original hypothesis. The basic elements of a statistical test are: (1) the *null hypothesis*, H_o (the hypothesis to be tested); (2) the alternate hypothesis, H_a; (3) the test statistic to be used; and (4) the region of rejection of the hypothesis. The active components of a statistical test are the test statistic and the associated rejection region, with the latter specifying the values of the test statistic for which the null hypothesis is rejected. We emphasize the point that "pure" hypothesis testing originated from early work in which the null hypothesis corresponded to an idea or theory about a population variable that the scientist hoped *would be rejected*. "Null" in this case means incorrect and invalid so that we could call it the "invalid hypothesis." In other words, the null hypothesis specified those values of the population variable, which it was thought did *not* represent the true value of the variable. This is a form of negative thinking and is the reason that many of us would rather think in terms of the *alternate hypothesis* in which we specify those values of the variable that we hope will hold true (the "valid" hypothesis). Regardless of which hypothesis is chosen, it is important to remember that the true population value under consideration must either lie in the test set covered by H_o or in the set covered by H_a. There are no other choices.

We restrict consideration of hypothesis testing to large samples ($N > 30$). In hypothesis testing, two types of errors are possible. In a type-1 error, the null hypothesis H_o is rejected when it is true. The probability of this type of error is denoted by α. Type-2 errors occur when H_o is accepted when it is false (H_a is true). The probability of type-2 errors is written as β. In Table 3.9, the probability $P(\text{accept } H_o | H_o \text{ is true}) = 1 - \alpha$ corresponds to the $100(1 - \alpha)\%$ confidence interval. Alternatively, the probability $P(\text{reject } H_o | H_o \text{ is false}) = 1 - \beta$ is the power of the statistical test since it indicates the ability of the test to determine when the null hypothesis is false and H_o should be rejected.

TABLE 3.9 The Four Possible Decision Outcomes in Hypothesis Testing and the Probability of Each Decision Outcome in a Test Hypothesis

		Possible Situation	
Action	Accept H$_o$	H$_o$ is true	H$_o$ is false
		Correct confidence level $1 - \alpha$	Incorrect decision; (Type-2 error); β
	Reject H$_o$	Incorrect decision (Type-1 error); α	Correct decision; power of the test $1 - \beta$
	Sum	1.00	1.00

For a parameter θ based on a random sample $x_1, \ldots, x_n$, we want to test various values of θ using the estimate $\hat{\theta}$ as a test statistic. This estimator is assumed to have an approximately normal sampling distribution. For a specified value of $\hat{\theta}(= \theta_0)$, we want to test the hypothesis, H$_o$, that $\hat{\theta}(= \theta_0)$ (written H$_o$: $\theta = \theta_0$) with the alternate hypothesis, H$_a$, that $\hat{\theta} > \theta_0$ (written H$_a$: $\theta > \theta_0$). An efficient test statistic for our assumed normal distribution is the standard normal Z defined as

$$Z = \frac{\left(\hat{\theta} - \theta\right)}{\hat{\sigma}_{\hat{\theta}}} \qquad (3.111)$$

where $\hat{\sigma}_{\hat{\theta}}$ is the standard deviation of the approximately normal sampling distribution of $\hat{\theta}$, which can then be computed from the sample. For this test statistic, the null hypothesis (H$_o$: $\theta = \theta_0$) is rejected for $Z > Z_\alpha$ where α is the probability of a type-1 error. Graphically, this rejection region is depicted as the shaded portion in Figure 3.12(a), which is called an "upper-tail" test. Similarly, a "lower-tail" test would have the shaded rejection region starting at $-Z_\alpha$ and corresponds to $Z < -Z_\alpha$ and $\theta < \theta_0$ (Figure 3.12b). A two-tailed test (Figure 3.12c) is one for which the null hypothesis rejection region is $|Z| > Z_{\alpha/2}$ and $\theta \neq \theta_0$. The decision of

which test alternative to use should be based on the form of the alternate hypothesis. If one is interested in parameter values greater than θ_0, an upper-tail test is used; for values less than θ_0, a lower-tail test is appropriate. If one is interested in any change from θ_0, it is best to use a two-tailed test. The following is an example for which a two-tailed test is appropriate.

Suppose that daily averaged currents for some mooring locations are available for the same month from two different years (e.g., January 2012 and January 2013). We wish to test the hypothesis that the monthly means of the alongshore component of the flow, V, for these two different years are the same. If the daily averages are computed from hourly observations, we invoke the central limit theorem and conclude that our sampling distributions are normally distributed. Taking each month as having 31 days, we satisfy the condition of a large sample ($N > 30$) and can use the procedure outlined above. Suppose we observe that for January 2012 the mean and standard deviation of the observed current is $V_{2012} = 23 \pm 3$ cm/s while for January 2013 we find a monthly mean speed $V_{2013} = 20 \pm 2$ cm/s (here, the standard deviations are obtained from the signal variances). We now wish to test the null hypothesis that the true (as opposed to our sampled) monthly mean current speeds were statistically the same for the two separate years. We use the two-tailed test to detect any deviations from equality. In this example, the *point estimator* used to detect any difference between the monthly mean records calculated from daily observed values is the sample mean difference, $\hat{\theta} - \theta_0 = V_{2012} - V_{2013}$. Our test statistic Eqn (3.111) is

$$Z = \frac{(V_{2012} - V_{2013})}{\left[s_{2012}^2/N_{2012} + s_{2013}^2/N_{2013}\right]^{1/2}}$$

which yields

$$Z = \frac{(23 - 20)}{[9/31 + 4/31]^{1/2}} = 17.06$$

FIGURE 3.12 Large-sample rejection regions (shaded areas) for the null hypothesis H_o: $\theta = \theta_0$, for the normally distributed function $f(\theta)$. (a) Upper-tail test for H_o: $\theta = \theta_0$, H_a: $\theta > \theta_0$; (b) lower-tail test with the rejection region for H_o: $\theta = \theta_0$, H_a: $\theta < \theta_0$; and two-tailed test for which the null hypothesis rejection region is $|Z| > Z_{n/2}$ and H_a: $\theta \neq \theta_0$.

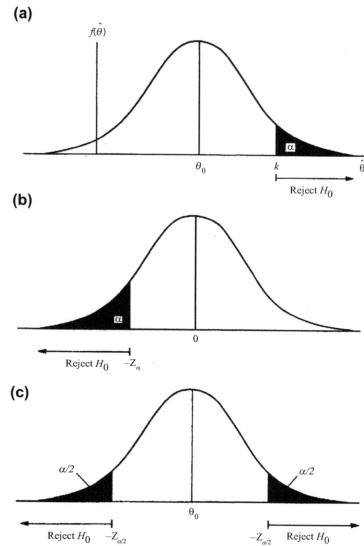

To determine if the above result falls in the rejection region, $Z > Z_\alpha$, we need to select the significance level α for type-1 errors. For the 95% significance level, $\alpha = 0.05$ and $\alpha/2 = 0.025$. From the standard normal table (Appendix D, Table D.1) $z_{0.025} = 1.96$. Our test value $Z = 17.06$ is greater than 1.96 so that it falls within the rejection region, and we must reject the hypothesis that the monthly mean current speeds are the same for both years. In most oceanographic applications hypothesis testing is limited to the null hypothesis and thus type-1 errors are most appropriate. We will not consider here the implementation of type-2 errors that lead to the acceptance of an alternate hypothesis as described in Table 3.9.

Turning again to satellite altimetry for an example, we note that the altimeter height bias, H_{bias}, discussed earlier in Section 3.10, is one of the error sources that contributes to the overall error "budget" of altimetric height measurements. Suppose that we wish to know if the *overall* height error H_T in the absence of the bias error, H_{bias}, is less than some specified amount, H_ε. We first set up the null hypothesis (H_0: $H_T - H_{bias} < H_\varepsilon$) that the overall height error in the absence of any bias is less than H_ε. At this point, we must also select a significance level for our test. A significance level of $1 - \alpha$ means that we do not want to make a mistake and reject the null hypothesis more than α $(100)\%$ of the time. We begin by defining our hypothesis limit, H_T, as

$$H_T = H_\varepsilon + \frac{Z_\alpha s_b}{\sqrt{N}} \qquad (3.112)$$

where the standard normal distribution Z_α for the bias error is given by Eqn (3.111) and s_b is the standard error (uncertainty) in our measurements. If the mean error of our measurements is greater than $H_T - H_{bias}$, then we reject H_0 and conclude that the height error in the absence of any bias error is greater than H_ε with a probability α of being wrong.

Suppose we set $H_\varepsilon = 13$ cm and consider $N = 9$ consecutive statistically independent satellite measurements in which each measurement is assumed to have an uncertainty of $s_b = 3$ cm. If the observed height error is 15 cm, do we accept or reject the null hypothesis for the probability level $\alpha = 0.10$? What about the cases for $\alpha = 0.05$ and $\alpha = 0.01$? Given our hypothesis limit $H_\varepsilon = 13$ cm and the fact that $N = 9$ and $s_b = 3$ cm, we can write Eqn (3.112) as $H_T = 13 + Z_\alpha$ cm. According to the results of Table 3.10, this means that we accept the null hypothesis that the overall error is less than 13 cm

TABLE 3.10   Test Results for the Null Hypothesis that the Overall Error H_T of Satellite Altimetry Data in the Absence of a Bias Error is Less than 13 cm (assuming normal error distribution)

Significance Level, a	Standard Normal Distribution, Z_α	Total Error Height, H_T	Decision
0.01	2.575	15.575 cm	Reject H_0
0.05	1.960	14.960 cm	Accept H_0
0.10	1.645	14.645 cm	Accept H_0

at the 5% and 10% probability levels but not at the 1% probability level (these are referred to as the 95%, 90%, and 99% significance levels, respectively).

3.14.1 Significance Levels and Confidence Intervals for Correlation

One useful application of null hypothesis testing is the development of significance levels for the correlation coefficient, r. If we take the null hypothesis as $r = r_0$, where r_0 is some estimate of the correlation coefficient, we can determine the rejection region in terms of r at a chosen significance level α for different degrees of freedom $(N - 2)$. A list of such values is given in Appendix E. In that table, the correlation coefficient r for the 95% and 99% significance levels (also called the 5% and 1% levels depending on whether or not one is judging a population parameter or testing a hypothesis) are presented as functions of the number of degrees of freedom.

For example, a sample of 20 pairs of (x, y) values with a correlation coefficient less than 0.444 and $N - 2 = 18$ degrees of freedom would not be significantly different from zero at the 95% confidence level. It is interesting to note that, because of the close relationship between r and the regression coefficient b_1 of these pairs of values, we could have developed the table for r values using a test of the null hypothesis for b_1.

The procedure for finding confidence intervals for the correlation coefficient r is to first transform it into the standard normal variable Z_r as

$$Z_r = \frac{1}{2}[\ln(1+r) - \ln(1-r)] \qquad (3.113)$$

which has the standard error

$$\sigma_z = \frac{1}{(N-3)^{1/2}} \qquad (3.114)$$

independent of the value of the correlation. The appropriate confidence interval is then

$$Z_r - Z_{\alpha/2}\sigma_z < Z < Z_r + Z_{\alpha/2}\sigma_z \qquad (3.115)$$

which can be transformed back into values of r using Eqn (3.113).

Before leaving the subject of correlations we want to stress that correlations are merely statistical constructs and, while we have some mathematical guidelines as to the statistical reliability of these values, we cannot replace common sense and physical insight with our statistical calculations. It is entirely possible that our statistics will deceive us if we do not apply them carefully. We again emphasize that a high correlation can reveal either a close relationship between two variables or their simultaneous dependence on a third variable. It is also possible that a high correlation may be due to complete coincidence and have no causal relationship behind it. The basic question that needs to be asked is "does it make sense?" A classic example (Snedecor and Cochran, 1967) is the high negative correlation (-0.98) between the annual birthrate in Great Britain and the annual production of pig iron in the United States for the years 1875–1920. This high correlation is statistically significant for the available $N - 2 = 43$ degrees of freedom, but the likelihood of a direct relationship between these two variables is very low.

3.14.2 Analysis of Variance and the F-Distribution

Most of the statistical tests we have presented to this point are designed to test for differences between two populations. In certain circumstances, we may wish to investigate the differences among three or more populations simultaneously rather than attempt the arduous task of examining all possible pairs. For example, we might want to compare the mean lifetimes of drifters sold by several different manufacturers to see if there is a difference in survivability for similar environmental conditions; or, we might want to look for significant differences among temperature or salinity data measured simultaneously during an intercomparison of several different commercially available CTDs. The *analysis of variance* (ANOVA) is a method for performing simultaneous tests on data sets drawn from different populations. In essence, ANOVA is a test between the amount of variation in the data that can be attributed to chance and that which can be attributed to specific causes and effects. If the amount of shared variability *between* samples is small relative to the shared variability *within* samples, then the null hypothesis H_o—that the variability occurred by chance—cannot be rejected. If, on the other hand, the ratio of these variations is sufficiently large, we can reject H_o. "Sufficiently large," in this case, is determined by the ratio of two continuous χ^2 probability distributions. This ratio is known as the *F-distribution*.

To examine this subject further, we need several definitions. Suppose we have samples from a total of J populations and that a given sample consists of N_j values. In ANOVA, the J samples are called J "treatments," a term that stems from early applications of the method to agricultural problems where soils were "treated" with different kinds of fertilizer and the statistical results compared. In the one-factor ANOVA model, the values y_{ij} for a particular treatment (input), x_j, differ from some common background value, μ, because of random effects; that is

$$y_{ij} = \mu + x_j + \varepsilon_{ij}; \quad \begin{array}{l} j = 1,2,...,J \\ i = 1,2,...,N_J \end{array} \qquad (3.116)$$

where the outcome y_{ij} is made up of a common (grand average) effect (μ), plus a treatment effect

(x_j), and a random effect, ε_{ij}. The grand mean, μ, and the treatment effects, x_j, are assumed to be constants while the errors, ε_{ij}, are independent, normally distributed, variables with zero mean and a common variance, σ^2, for all populations. The null hypothesis for this one-factor model is that the treatments have zero effect. That is, $H_0: x_j = 0$ ($j = 1, 2, ..., J$) or, equivalently, H_0: $\mu_1 = \mu_2 = ... = \mu_J$ (i.e., there is no difference between the populations aside from that due to random errors). The alternative hypothesis is that some of the treatments have a nonzero effect. Note that "treatment" can refer to any basic parameter we wish to compare such as buoy design, power supply, or CTD manufacturer. To test the null hypotheses, we consider samples of size N_j from each of the J populations. For each of these samples, we calculate the mean value $\overline{y}_j$ ($j = 1, 2, ..., J$). The grand mean for all the data is denoted as $\overline{y}$.

As an example, suppose we want to intercompare the temperature records from three types of CTDs placed in the same temperature bath under identical sampling conditions. Four countries take part in the intercomparison and each brings the same three types of CTD. The results of the test are reproduced in Table 3.11.

TABLE 3.11 Temperatures in °C Measured by Three Makes of CTD in the Same Calibration Tank

Measurement (i)	CTD Type 1 Sample $j = 1$	CTD Type 2 Sample $j = 2$	CTD Type 3 Sample $j = 3$
1	15.001	15.004	15.002
2	14.999	15.002	15.003
3	15.000	15.001	15.000
4	14.998	15.004	15.002
Mean $\overline{y}$ °C	$15.000 = \overline{y}_1$	$15.003 = \overline{y}_2$	$15.002 = \overline{y}_3$

Four instruments of each type are used in the test. The grand mean for the data from all three instruments is $\overline{y} = 15.002$ °C.

If H_0 is true, $\mu_1 = \mu_2 = \mu_3$ and the measured differences between $\overline{y}_1$, $\overline{y}_2$, and $\overline{y}_3$ in Table 3.11 can be attributed to random processes.

The treatment effects for the CTD example are given by

$$x_1 = \overline{y}_1 - \overline{y} = -0.002\,°C$$
$$x_2 = \overline{y}_2 - \overline{y} = +0.001\,°C$$
$$x_3 = \overline{y}_3 - \overline{y} = 0.0\,°C$$

where $\overline{y} = (\overline{y}_1 + \overline{y}_2 + \overline{y}_3)/3$. The ANOVA test involves determining whether the estimated values of x_j are large enough to convince us that H_0 is not true. Whenever H_0 is true, we can expect that the variability between the J means is the same as the variability within each sample (the only source of variability is the random effects, ε_{ij}). However, if the treatment effects are not all zero, then the variability between samples should be larger than the variability within the samples.

The variation within the J samples is found by first summing the squared deviations of y_{ij} about the mean value $\overline{y}_j$ for each sample, namely

$$\sum_{i=1}^{N_j} \left(y_{ij} - \overline{y}_j \right)^2; \quad j = 1, 2, 3$$

where N_j is the number of measurements in each sample. If we then sum this variation over all J samples, we obtain the *sum of squares within* (SSW).

Sum of squares within: SSW

$$= \sum_{j=1}^{J} \sum_{i=1}^{N_j} \left(y_{ij} - \overline{y}_j \right)^2 \tag{3.117}$$

Note that the sample lengths, N_j, need not be the same since the summation for each sample uses only the mean for that particular sample. Next, we will need the amount of variation between the samples (SSB). This is obtained by taking the squared deviation of the mean of the Jth sample, $\bar{y}_j$, and the grand mean, $\bar{y}$. This deviation must then be weighted by the number of observations in the Jth sample. The overall sum is given by

Sum of squares between: SSB

$$= \sum_{j=1}^{J} N_j \left(\bar{y}_j - \bar{y} \right)^2 \qquad (3.118)$$

To compare the variability within samples to the variability between samples, we need to divide each sum by its respective number of degrees of freedom, just as we did with other variance expressions, such as s^2. For SSB, the degrees of freedom (DOF) $= J - 1$ while for SSW

$$DOF = \left(\sum_{j=1}^{J} N_j \right) - J$$

The MS values are then:

Mean square between : $MSB = \dfrac{SSB}{J-1}$ (3.119a)

Mean square within : $MSW = \dfrac{SSW}{\left(\sum_{j=1}^{J} N_j \right) - J}$

$$(3.119b)$$

In the above example, $J - 1 = 2$ and $\sum_{j=1}^{J} N_j - J = 9$. The calculated values of mean square between (MSB) and mean square within (MSW) for our CTD example are given in Table 3.12. Specifically

$$SSW = \sum_{i=1}^{4} \left(y_{i1} - \bar{y}_1 \right)^2 + \sum_{i=1}^{4} \left(y_{i2} - \bar{y}_2 \right)^2$$

$$+ \sum_{i=1}^{4} \left(y_{i3} - \bar{y}_3 \right)^2$$

$$SSB = N_1 \left(\bar{y}_1 - \bar{y} \right)^2 + N_2 \left(\bar{y}_2 - \bar{y} \right)^2$$

$$+ N_3 \left(\bar{y}_3 - \bar{y} \right)^2 + N_4 \left(\bar{y}_4 - \bar{y} \right)^2$$

TABLE 3.12   Calculated Values of Sum of Squares and MS Values for the CTD Temperature Intercomparison

Type of Variation	Sum of Squares (°C²)	DOF	Mean Square (°C²)
Between samples (type of CTD)	20×10^{-6}	2	10×10^{-6}
Within samples (all CTDs)	18×10^{-6}	9	2×10^{-6}
Total	38×10^{-6}	11	(Ratio = 5.0)

$DOF = number\ of\ degrees\ of\ freedom$

where the total N_j (for $j = 1, \ldots, 4) = 12$. To determine if the ratio of MSB to MSW is large enough to reject the null hypothesis, we use the F-distribution for $J - 1$ and

$$\left(\sum_{j=1}^{J} N_j \right) - J$$

degrees of freedom.

Named after R. A. Fisher, who first studied it in 1924, the F-distribution is defined in terms of the ratio of two independent χ^2 variables divided by their respective degrees of freedom. If X_1 is a χ^2 variable with ν_1 degrees of freedom and X_2 is another χ^2 variable with ν_2 degrees of freedom, then the random variable

$$F(\nu_1, \nu_2) = \frac{X_1 / \nu_1}{X_2 / \nu_2} \qquad (3.120)$$

is a nonnegative chi-square variable with ν_1 degrees of freedom in the numerator and ν_2 degrees of freedom in the denominator. If $J = 2$, as in the CTD example above, the F-test is equivalent to a one-sided t-test. There is no upper limit to F, which like the χ^2-distribution is skewed to the right. Tables are used to list the critical values of $P(F > F_\alpha)$ for selected degrees of freedom ν_1 and ν_2 for the two most commonly used significance levels, $\alpha = 0.05$ and $\alpha = 0.01$. In ANOVA, the values of SSB and SSW follow χ^2-distributions. Therefore, if we let $X_1 = \text{SSB}$ and $X_2 = \text{SSW}$, then

$$F\left(J-1, \sum_{j=1}^{J} N_j - J\right) = \frac{[\text{SSB}/(J-1)]}{\text{SSW}/\left(\sum N_j - J\right)}$$

$$= \frac{\text{MSB}}{\text{MSW}}$$

(3.121)

When MSB is large relative to MSW, F will be large and we can justifiably reject the null hypothesis that the different CTDs (different treatment effects) measure the same temperature within the accuracy of the instruments. For our CTD intercomparison (Table 3.12), we have MSB/MSW = 5.0, $\nu_1 = 2$ and $\nu_2 = 9$. Using the values for the F-distribution for 2 and 9 degrees of freedom from Appendix D, Table D.4a, we find $F_\alpha (2, 9) = 4.26$ for $\alpha = 0.05$ (95% confidence level) and $F_\alpha (2, 9) = 8.02$ for $\alpha = 0.01$ (99% confidence level). Since, $F = 5.0$ in our example, we conclude that a difference exists among the different makes of CTD at the 95% confidence level, but not at the 99% confidence level.

3.15 EFFECTIVE DEGREES OF FREEDOM

To this point, we have assumed that we are dealing with random variables and each of the N values in a given sample are statistically independent. For example, in calculating the unbiased standard deviation for N data points, we assume there are $N - 1$ degrees of freedom. (We use $N - 1$ rather than N since we need a minimum of two values to calculate the standard deviation of a sample.) Similarly, in Sections 3.8 and 3.10, we specify confidence limits in terms of the number of samples rather than the "true" number of degrees of freedom of the sample. In reality, consecutive data values may not be independent. Contributions from low-frequency components and narrow band oscillations, such

as in inertial motions may lead to a high degree of correlation between values separated by large times or distances. The most common example of highly coherent narrow band signals are the tides and tidal currents, which possess a strong temporal and spatial coherence. If we want our statistics to have any real meaning, we are forced to find the *effective number of degrees of freedom* using information on the coherence and autocorrelation of our data.

The effects of coherent nonrandom processes on data series lead us into the question of data redundancy in multivariate linear regression. Our general model is

$$\widehat{y}(t_i) = \sum_{k=1}^{M} b_k x_k(t_i); \quad i = 1, ..., N \quad (3.122)$$

where the x_k represents M observed parameters or quantities at times t_i. The b_k are M linear-regression coefficients relating the independent variables $x_k(t_i)$ to the N model estimates, $\widehat{y}(t_i)$. Here, the x_k observations can be measurements of different physical quantities or of the same quantity measured at different times or locations.

The estimate $\widehat{y}$ differs from the true parameter by an error $\varepsilon_i = \widehat{y}(t_i) - y(t_i) = \widehat{y}_i - y_i$. Following our earlier discussion, we assume that this error is randomly distributed and is therefore uncorrelated with the input data $x_k(t_i)$. To find the best estimate, we apply the method of least squares to minimize the MS error.

$$\overline{\varepsilon^2} = \sum_{i=1}^{M} \sum_{j=1}^{M} b_i b_j \overline{x_i x_j} - 2 \sum_{j=1}^{M} b_j \overline{x_j y} + \overline{y^2} \quad (3.123)$$

In this case, the over bars represent *ensemble averages*. To assist us in our minimization, we invoke the Gauss–Markov theorem, which says that the estimator, given by Eqn (3.122), with the smallest MS error is that with coefficients.

$$b_k = \sum_{j=1}^{M} \left[\left\{ \overline{x_k x_j} \right\}^{-1} \overline{x_j y} \right] \quad (3.124)$$

where $\{\overline{x_k x_j}\}^{-1}$ is the k, j element of the inverse of the $M \times M$ cross-covariance matrix of the input variables (note: $\{\overline{x_k x_j}\}^{-1} \neq 1/\overline{x_k x_j}$). This MS product matrix is always positive definite unless one of the input variables x_k can be expressed as an exact combination of the other input values. The presence of random measurement errors in all input data make this "degeneracy" highly unlikely. It should be noted, however, that it is the partial correlation between inputs that increases the uncertainty in our estimator by lowering the degrees of freedom through a reduction in the independence of our input parameters. We can write the minimum least-square error $\overline{\varepsilon_o^2}$ as

$$\overline{\varepsilon_o^2} = \overline{y^2} - \sum_{i=1}^{M} \sum_{j=1}^{M} \overline{yx_j} \{\overline{x_k x_j}\}^{-1} \overline{yx_j} \qquad (3.125)$$

At this point, we introduce a measure of the reliability of our estimate called the *skill* (S) of the model. This skill is defined as the fraction of the true parameter variance explained by our linear statistical estimator; thus

$$S = \{\overline{y^2}\}^{-1} \sum_{i=1}^{M} \sum_{j=1}^{M} \left[\overline{yx_j} \{\overline{x_k x_j}\}^{-1} \overline{yx_j} \right] \qquad (3.126)$$

The skill value ranges from no skill ($S = 0$) to perfect skill ($S = 1$). We note that for the case ($M = 1$), S is the square of the correlation between x_1 and y.

The fundamental trade-off for any linear estimation model is that, while one wants to use as many independent input variables as possible to avoid interdependence among the estimates of the dependent variable, each new input contributes random measurement errors that degrade the overall estimate. As pointed out by Davis (1977) the best criterion for selecting the input data parameters is to use a priori theoretical considerations. If this is not possible, some effort should be made to select those inputs which contribute most to the estimation skill.

The conflicting requirements of limiting M (the observed parameters) and including all candidate input parameters is a dilemma. In considering this dilemma Chelton (1983) concludes that the only way to reduce the error limits on the estimated regression coefficients is to increase what are called the "effective degrees of freedom N^*." This can be done only by increasing the sample size of the input variable (i.e., using a longer time series) or by high-passing the data to eliminate contributions from unresolved, and generally coherent, low-frequency components. Since we are forced to deal with relatively short data records in which ensemble averages are replaced by sample averages over time or space, we need a procedure to evaluate N^*, the effective degrees of freedom.

In the case of real data, ensemble averages are generally replaced with sample averages over time or space so that the resultant values become estimates. Thus, the skill can be written as S given by Eqn (3.126). If we assume for a moment that the x_k input data are serially uncorrelated (i.e., we expand the data series into orthogonal functions), we can write the sample estimate of the skill as

$$\hat{S} = \sum_{i=1}^{M} \sum_{j=1}^{M} \frac{\overline{x_i x_j}^2}{\overline{x_j^2} \overline{y^2}} \qquad (3.127)$$

Following Davis (1978) we can expand this skill estimate into a true skill plus an artificial skill

$$\hat{S} = S + S_A \qquad (3.128)$$

The artificial skill, S_A, arises from errors in the estimates and can be calculated by evaluating the skill in Eqn (3.127) at a very long time (or space lag) where no real skill is expected. At this point, there is no true estimation skill and $\hat{S} = S_A$.

Davis (1976) derived an appropriate expression for the expected (mean) value of this artificial skill which relates it to the effective degrees of freedom N^*

$$\bar{S}_A = \sum_{k=1}^{M} (N_k^*)^{-1} \qquad (3.129)$$

where N_k^* is the effective degrees of freedom associated with the sample estimate of the covariance between the output y and input x_k of the model. Under the conditions that S (the true skill) is not large, that the record length N is long compared to the autocovariance scales of y and x, and that the N_k^* are the same for all N, we can write N^* as

$$N^* = \frac{N}{\left[\sum_{\tau=-\infty}^{\infty} C_{xx}(\tau)C_{yy}(\tau)\right] / \left[C_{xx}(0)C_{yy}(0)\right]}$$

$$= \frac{N}{\left[\sum_{\tau=-\infty}^{\infty} \rho_{xx}(\tau)\rho_{yy}(\tau)\right]}$$

(3.130a)

where $\rho_{\varsigma\varsigma}(\tau) = C_{\varsigma\varsigma}(\tau)/C_{\varsigma\varsigma}(0) = C_{\varsigma\varsigma}(\tau)/s_\varsigma^2$ is the normalized autocovariance function for any variable ζ (with variance s_ς^2), and

$$C_{\varsigma\varsigma}(\tau) = E[(\varsigma(t_i) - \bar{\varsigma})(\varsigma(\tau + t_i) - \bar{\varsigma})]$$
$$= \frac{1}{N'} \sum_{i=1}^{N'} [(\varsigma(t_i) - \bar{\varsigma})(\varsigma(\tau + t_i) - \bar{\varsigma})] \qquad (3.131)$$

where N' is the number of data values used in the summation for the particular lag, τ. A more complete form of this expression was given by Chelton (1983) as

$$N^* = \frac{N}{\left[\sum_{\tau=-\infty}^{\infty} (C_{xx}(\tau)C_{yy}(\tau) + C_{xy}(\tau)C_{yx}(\tau)) / C_{xx}(0)C_{yy}(0)\right]}$$

(3.132a)

$$= \frac{N}{\left[\sum_{\tau=-\infty}^{\infty} (\rho_{xx}(\tau)\rho_{yy}(\tau) + \rho_{xy}(\tau)\rho_{yx}(\tau))\right]}$$

(3.132b)

This expression now includes the cross-covariances between y and x (e.g., $C_{xy}(\tau)$ and $\rho_{xy}(\tau)$) and is not limited to cases where S is small.

In general, the true auto- and cross-covariances are not known and the computation of N^* requires the substitution of sample estimates over finite lags for the correlations in Eqn (3.132). The resulting effective degrees of freedom, N^*, can be used with standard tables to find the selected significance levels for S. In the ideal case, when all input variables are neither cross- nor serially correlated (and therefore independent) the effective number of degrees of freedom is N, the sample size. In general, however, input data series are serially correlated and $N^* \ll N$. The larger the time/space correlation scales in Eqn (3.132), the smaller the value of N^*. This means that it is the large scale, low-frequency components of the input data that lead to a decrease in the number of independent values in the data series.

The limitations of estimating regression characteristics for real data can be summarized as follows:

1. Accurate statistical results require the use of the effective number of degrees of freedom, N^*, with N^* generally much fewer than the total number of observations, N.
2. The accuracy of the estimated regression coefficients increases as N^* increases.
3. The accuracy of the regression coefficient decreases as the number of inputs N increases (measurement error is added).
4. The accuracy increases as the model skill increases and decreases as the input parameters become more correlated.

The above considerations emphasize the need for careful selection of the input data and the careful evaluation of the characteristics of these data. As pointed out by Davis (1977), a fundamental part of this selection process is the determination of the space and timescales to be studied. The methods used to extract this fundamental scale information from the input data can range from cross-spectral analysis (see Chapter 5) to a filtering of the data using preselected windows. Performing this filtering in the time domain rather than the frequency domain is often less complicated (see Chapter 6). The filtering process has the goal of eliminating scales that are not expected to contribute to the true correlation but which will add artificial correlation due to instrument and sampling errors.

Once the space and/or timescales are determined, selected or set by filtering, the next step is the selection of the input series to use in the estimate. At this stage, the dilemma arises between limiting the effects of errors and at the same time including as many as possible uncorrelated input variables to increase the degrees of freedom. Davis (1977) recommends using dynamical considerations to make this selection and shows how the data required for proper statistical estimation are generally those required to make the dynamical system well posed. However, he also mentions that, in general, the dynamics of most processes are not well enough understood and that specification data are not known with certainty. Nevertheless, some quantitative understanding of the physical system can serve as a useful guide to the selection of estimation data.

3.15.1 Simple Effective Degrees of Freedom

The calculation of the effective degrees of freedom given above is complex and, in many cases, will be too difficult to derive a quantitative solution. In these cases, a much more simple approach may be taken to estimate the effective degrees of freedom. If the correlation, r, between two variables is unity ($r = 1$) then the variables are perfectly correlated and there only 2 degrees of freedom, corresponding to the fact we are using two variables. If instead the two variables have a correlation of $r = 0$, then they are

completely uncorrelated and can be considered as independent variables. Under these conditions, there are as many degrees of freedom as there are data values, N. In general, r is between 0 and 1. Under these circumstances, one can use the correlation coefficient to estimate the effective number of degrees of freedom. In essence, the correlation coefficient represents a ratio of the independent to dependent values of the two variables. However, because totally dependent variables have a correlation of unity, corresponding to the minimum of degrees of freedom, it is clear that we cannot simply use the correlation coefficient to estimate the effective degrees of freedom, N^*. Instead, we need to subtract the correlation from unity (i.e., $1 - r$) and multiply this difference by the total number of data values in the series. The value $N^* = (1 - r)N$ is, then, a rough estimate of the effective degrees of freedom. While it is not as accurate as the complex procedure outlined above, this method gives a quick and useful estimate of the effective degrees of freedom. The topic of effective degrees of freedom is addressed again later in this chapter and in Chapter 5.

3.15.2 Trend Estimates and the Integral Timescale

Most oceanographic variability arises through a combination of random and nonrandom processes. The presence of tidal and low-frequency components means that data points in time or space series are not independent of one another. The data that we collect are not truly random samples drawn from random populations. There is invariably a nonzero correlation between values in the series that must be taken into account when we tally up the true number of independent samples or degrees of freedom we think we have in our system. This number is important when it comes to determining the confidence limits of linear regression slopes and parameter estimates. As an example, consider the confidence limits on the slope of the least squares linear regression $\hat{y} = b_0 + b_1 x$ (where, again, $\wedge$ denotes an estimator for the function y). From Eqn (3.40), the limits are

$$\pm \left(s_\varepsilon t_{\alpha/2,\nu} \right) \Big/ \left[(N-1)^{1/2} s_x \right] \qquad (3.133a)$$

Or, in terms of the estimator β_1 for b_1

$$b_1 - \frac{\left(s_\varepsilon t_{\alpha/2,\nu} \right)}{(N-1)^{1/2} s_x} < \beta_1 < b_1 + \frac{\left(s_\varepsilon t_{\alpha/2,\nu} \right)}{(N-1)^{1/2} s_x}$$

$$(3.133b)$$

where $\nu = N - 2$ is the number of degrees of freedom for the student's t-distribution at the $(1 - \alpha)$ 100% confidence level, and the standard error of the estimate, s_ε, is given by

$$s_\varepsilon = \left[\frac{1}{N-2} \sum_{i=1}^{N} (y_i - \hat{y})^2 \right]^{1/2}$$

$$= \left[\frac{1}{N-2} \mathrm{SSE} \right]^{1/2} \qquad (3.134)$$

The standard deviation for the x variable, s_x, is given by

$$s_x = \left[\frac{1}{N-1} \sum_{i=1}^{N} (x_i - \bar{x})^2 \right]^{1/2} \qquad (3.135a)$$

or,

$$(N-1)^{1/2} s_x = \left[\sum_{i=1}^{N} (x_i - \bar{x})^2 \right]^{1/2} \qquad (3.135b)$$

The question is: what do we use for the number of degrees of freedom if the N samples in our series are not statistically independent? The reason we ask this question is that the characteristic amplitudes of the fluctuations s_ε and s_x are calculated using all N values in our data series when we really should be using some sort of *effective* number of degrees of freedom N^* ($<N$), which takes into account the degree of correlation that exists between data points (see the previous section).

Suppose we decide to err on the conservative side by agreeing to work with that value of N^*, which makes the confidence limits $\pm (s_\varepsilon t_{\alpha/2,\nu}) / [(N^* - 1) s_x^2]^{1/2}$ as small as justifiably possible. This means that when we estimate the confidence limits for a regression slope for a given confidence coefficient, α, we know that we have probably been too cautious and that the confidence limits on the slope probably bracket those that we derive.

We begin by keeping s_ε as it is. If there are high frequency (possibly random) fluctuations superimposed on coherent low-frequency motions, retaining the high-frequency variability adds to the magnitude of s_ε. Had we low-pass filtered the data first and recomputed s_ε based on the true number of data points in our low-pass filtered record, we would expect s_ε to be somewhat smaller. By using s_ε as it is we are assuming that it is a fixed quantity no matter how we subsample or filter the data ($s_\varepsilon = $ constant). We do the same with s_x but now replace $N-1$ with $N^* - 1$, where $N^* < N$. This increases the magnitude of the confidence limits. All that remains is to assume that the number of degrees of freedom for the t-distribution are given by the effective number of degrees of freedom $\nu = N^* - 2$. This statistic has a larger value than for $\nu = N - 2$ so that, again, we are overestimating the magnitude of the confidence interval. This confidence interval is then given by

$$\pm \left(s_\varepsilon t_{\alpha/2,\nu} \right) \Big/ \left[(N^* - 1) s_x^2 \right]^{1/2} \qquad (3.136a)$$

that is,

$$b_1 - \frac{\left(s_\varepsilon t_{\alpha/2,\nu} \right)}{(N^* - 1)^{1/2} s_x} < \beta_1 < b_1 + \frac{\left(s_\varepsilon t_{\alpha/2,\nu} \right)}{(N^* - 1)^{1/2} s_x}$$

$$(3.136b)$$

with $\nu = N^* - 2$.

Our final task is to define the effective number of degrees of freedom, N^*, based on a knowledge of the autocovariance function $C(\tau)$ (Eqn (3.131)) as a function of lag τ. To do this, we must first find the integral timescale T for the data record

$$T = \frac{2}{C(0)} \sum_{k=0}^{m-1} \frac{\Delta\tau}{2} [C(\tau_k + \Delta\tau) + C(\tau_k)]$$

(discrete case)

(3.137a)

$$= \frac{2}{C(0)} \int_0^\infty C(\tau)d\tau \quad \text{(continuous case)} \quad (3.137b)$$

where m is the number of lag values incorporated in the summation, $\Delta\tau$ is the lag time step (which could be the sampling time step, Δt) and $\frac{1}{2}[C(\tau_k + \Delta\tau) + C(\tau_k)]$ is the mean value of C for the midpoint of the lag interval, $(\tau_\kappa, \tau_\kappa + \Delta\tau)$. The factor of 2 in Eqn (3.137a,b) accounts for the need to include negative values of τ in the summation and integral functions (the functions are symmetrical, so only the positive lags are considered). Once the integral timescale is known, the effective degrees of freedom are found by

$$N^* = \frac{N\Delta t}{T}$$

(3.138)

where Δt is the sampling increment and $N\Delta t$ is the total length (duration or distance) of the record. If, for example, $N = 120$, $\Delta t = 1$ h, and $T = 10$ h, then $N^* = 12$ ($\ll N$).

To find the autocovariance function, we let $\tau_\kappa = k\Delta\tau$ be the kth lag ($k = 0, 1, \ldots$), then

$$C(\tau_k) = \frac{1}{N - 1 - k} \sum_{i=1}^{N-k} (y_i - \bar{y})(y_{i+k} - \bar{y});$$

$$k = 0, \ldots, N_{\max}$$

(3.139a)

$$C(0) = \frac{1}{N - 1} \sum_{i=1}^{N} (y_i - \bar{y})^2 = s_y^2$$

(3.139b)

where $C(0)$ is the just the variance s_y^2 of the full data series. In both Eqns (3.139a) and (3.139b), the data start with the first value for $i = 1$; $N_{\max}$ is the maximum number of reasonable lag values (starting at zero lag and going to ($\ll N/2$) that can be calculated before the summation becomes erratic. In theory, we would like $C(\tau) \to 0$ as $\tau \to N$. In reality, however, the data series will contain low-frequency components, which will cause the autocovariance function to oscillate about zero or asymptote toward a nonzero value. It should also be obvious that the statistical significance of the summation becomes meaningless at large lag due to the fact that the statistic is based on fewer and fewer values as the lag becomes large. For example, at a lag $k = (N - 3)$ there are only four values that go into the summation and these are derived from neighboring points that are likely highly correlated. We can picture the integral timescale using Eqn (3.137b). Writing

$$T \cdot C(0) = \int_{-\infty}^{\infty} C(\tau)d\tau$$

FIGURE 3.13 Definition of the integral timescale, T. The area under the correlation curve $C(\tau)$ vs τ in (a) is equated to the rectangular region $T \cdot C(0)$ in (b). In practice, only a reasonable portion of the curve $C(\tau)$ is used to derive the area in (a).

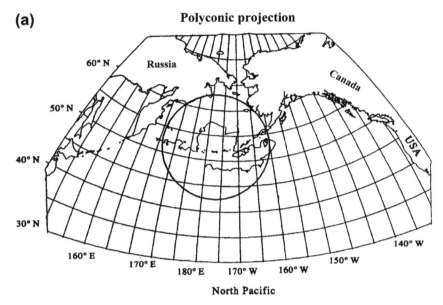

(a)   **Polyconic projection**

North Pacific

FIGURE 3.14 (a) Trajectory of a satellite-tracked drifter deployed to the south of the Aleutian Islands in the northeast Pacific and covering the period November 13, 1991 to July 30, 1993 based on a six-hourly sampling interval; (b, c) autocovariance functions and corresponding integral timescales for zonal (u) and meridional (v) velocities of the satellite-tracked drifter. *(Courtesy of Adrian Dolling.)*

we find that the area under the curve $C(\tau)$ for both positive and negative lag values, τ has been equated to the rectangular region $T \cdot C(0)$ (Figure 3.13). In essence, we take a reasonable portion of the curve $C(\tau)$, obtain its area and divide the integral (sum) by its value at zero lag, $C(0)$. An example of the autocovariance function and the integral timescale derived from it are shown in Figure 3.14 for satellite-tracked drifter data in the North Pacific.

3.16 EDITING AND DESPIKING TECHNIQUES: THE NATURE OF ERRORS

A major concern in processing oceanographic data is how to distinguish the true oceanic signal from measurement "errors" or other erroneous values. There are two very different types of measurement errors that can affect data. *Random errors*, usually equated with "noise," have random probability distributions and are generally small compared to the signal. Random errors are associated with inaccuracies in the measurement system or with real variability that is not resolved by the measurement system. The well-accepted statistical techniques for estimating the effects of such random errors are based largely on the statistics of a random population (see previous sections on statistics). Other errors that strongly influence data analysis are *accidental* errors. These errors are not representative of the true population and occur as a result of undetected instrument failures, misreading of scales, incorrect recording of data, and other human failings. In the following discussion, we will handle these two error types in reverse order since the large accidental errors must be removed first before techniques can be applied to treat the "statistical" (random) errors.

FIGURE 3.14   (*continued*).

One example of a large accidental error would be assigning an incorrect geographic location to an oceanographic measurement, which then transfers the observations to a region with which they have no direct relationship. Some of these errors, such as oceanographic stations on land, are easily detected, while others are less obvious. Another example of such errors would be biases in a group of measurements due to the application of incorrect sensor calibrations or undetected instrument malfunctions. An all too common error occurs during preparation of moored instruments when the operator inadvertently sets the start time using local time (or daylight saving time) while writing in the log that the time in UTC (Coordinated Universal Time). This typically happens when the person is rushed or is trying to work in rough seas on a rolling ship. Those working with the data will incorrectly interpret the time error as a phase shift.

A major goal of data processing is to remove or correct any errors in order to make the data set as self-consistent as possible. If we know the history of the data, meaning the details of its collection and reduction, we may be in a better position to understand the sources of these errors. If we have received the data from another source, or are looking at archived data, we may not have available the necessary details on the "pedigree" of the data and may have to come to some rather arbitrary decisions regarding its reliability. Considering the widespread use of computer-linked data banks, this is not a trivial problem. The question is how to ensure the necessary quality control yet ensure rapid dissemination and accessibility to data files.

3.16.1 Identifying and Removing Errors

There are two important axioms to follow when dealing with large erroneous values or "spikes":

1. To identify the large errors, it is necessary to examine all of the data in visual form and to get a "feel" for the data;

2. When large errors are encountered, it is usually best to eliminate them altogether rather than try to "correct" them and incorporate them back into the data set.

Of course, care must be taken not to reject important data points just because they do not fit either the previous data structure or our preconceived notion of the process. A good example is the determination of heat transport in the South Atlantic. Bennett (1976) suggested that the oceanic heat transport in this ocean is directed toward the equator, contrary to the widely accepted notion that oceanic heat transports are generally poleward. Stommel (personal communication) noted that, in his tabulation of property fluxes for the South Atlantic, Wüst (1957) conspicuously left out the flux of heat while treating other less easily computed transports such as those of nutrients and oxygen. Through an exchange of letters with a former student of Wüst's, Stommel learned that the heat content calculation indeed showed that heat is transported equatorward. Wüst considered this to be the wrong direction and the results were not published along with the other flux values. The point of this story is to illustrate the way in which our prejudice can lead us to reject significant results. In such cases, there is no hard rule as to how this decision is made and a great deal of subjectivity will always be inherent in this level of data interpretation. As for the heat flux in the Southern Ocean, present estimates show it is poleward but with a high degree of uncertainty. Moreover, mesoscale eddies contribute a significant fraction of the poleward flux (Volkov et al., 2010), an aspect of the circulation that could not be resolved by observations and numerical models.

The need to examine all the data to detect errors presents a difficult task because of the large numbers of values and the difficulty of looking at unprocessed data. In this case, it is more important to think of ways in which we can present the data so as to ask and answer the questions regarding consistency of the measurements. A compact overview of all the data is the best solution. This presentation may be as simple as a scatter diagram of the observations vs some independent variable, or a scatter diagram relating two concurrently measured parameters. While scatter diagrams cannot be used to resolve visually individual points, they do reveal groupings of points that relate to the physical processes expressed by the data. As an example, consider a temperature–salinity scatter diagram (Figure 3.15) computed using a large number of hydrographic data collected from bottle casts. Here, the groups of dots labeled "*a*," "*b*" refer to different water masses present in the 5° square

35–40 °N, 15–20 °W, where the data were collected. The data labeled "*c*" clearly represent a distinct water mass since the points lie along a line divergent from the rest of the scatter values. If we look at other similar *TS* scatter plots, we recognize that this line is consistent with the *TS* relationship from a corresponding square at this same longitude but south of the equator. Thus, it is likely that the latitude recorded was incorrect and that these data are simply misplaced. We correct this by eliminating the points "*c*" from our square. However, we cannot be sufficiently confident of our assumption to add the points to the other square even though the data coverage there is not very good.

FIGURE 3.15 TS relationship computed using a large number of hydrographic data collected from bottle casts. Groups labeled "a," "b" refer to different water masses present in the 5° square (35–40° N, 15–20° W) where the data were collected. The data labeled "c" clearly represent a distinct water mass since the points lie along a line divergent from the rest of the scatter values.

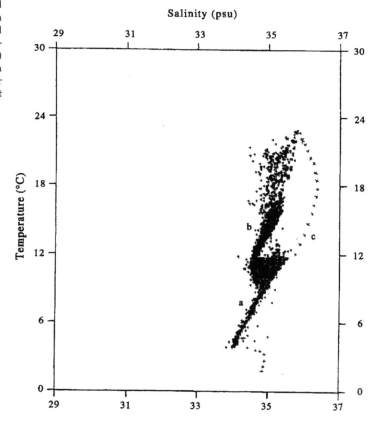

Often it is not possible to develop a simple summary presentation of all the data. In the case of current meter data, a time series presentation is the most appropriate way of looking at the data. As noted by Pillsbury et al. (1974), error detection using this technique is very time consuming. They note that this procedure can be used successfully for speed, pressure, salinity, and temperature but not for direction, which varies widely. This is due to the fact that direction is limited to the range 0–360° and shows no extreme values. Because of the wrap-around ("2π discontinuity") problem, in which $0° = 360°$ (or, alternatively, $-180° = +180°$), direction records tend to be very "spiky," especially in regions of strong tidal flow. A scatter diagram of speed vs direction can be used to detect systematic errors between the speed and

direction sensors and to pinpoint those times when the current speed is below the threshold recording level of the instrument. This would be displayed by the direction readings at speeds below threshold and would be easier to identify on the scatter plot than in the individual time series. The only way around the problem with the direction channel is to transform the recorded time series of speed and direction (U, θ) to orthogonal components of velocity (u, v). In particular, separate plots of the east–west (u) and the north–south (v) velocity components (or alongshore and cross-shore components for data collected near the coast) quickly reveal any erroneous values in the data (Figure 3.16).

Pillsbury et al. (1974) report that, for the now discontinued Aanderaa RCM4 and RCM5 current meters, there are several sources of large

FIGURE 3.16 A Plot of hourly data obtained from an Aanderra RCM4 current meter moored at 30 m-depth data in 250 m of water near the entrance to Juan de Fuca Strait (48°3.3′ N, 125°18.8′ W) during the period May 8–16, 1993. (a) Ambient pressure (instrument depth in meters); (b) East–west (u) component of velocity (m/s); (c) North–south (v) component of velocity (m/s); (d) velocity stick vector (m/s). Erroneous current velocity values ("spikes") stand out in the (u, v) records. Flow consisted of moderate tidal currents superimposed on a surface eatuarine overflow that weekend around May 13.

errors. We will discuss these as typical of the errors inherent in moored current meter data since data from many of these instruments remain in use. One source of error is the current meter's encoder, which might encounter a small electrical resistance. The probability of this occurring is considered small. A more likely error is due to nonuniformity in the 1/4-in magnetic tape which may have variations in the coating or carry residual magnetism. The tape transcriber is also a possible error source since it occasionally drops a bit. An error particular to the speed parameter where the speed is seen to abnormally increase, may be caused by nonuniformities in the speed potentiometer winding. A less frequent error type is that associated with clock and trigger malfunctions. Instances have been observed where a meter has cycled several times in rapid succession or conversely missed one or more cycles. These problems are addressed here under the section on timing errors. Direction errors are due to mechanical failures in the compass itself. In some cases the compass needle failed to contact the resistance ring around the compass while in others direction readings in one range all were recorded in a different range. Many of these compass problems were apparent in the raw data while others were only discovered later by looking at the direction histograms. Other problems with Aanderaa RCM4/5 current meters have been noted over the years. These can be minimized if the following protocol is observed (assuming that the instrument is operational and calibrated):

1. Use a new nonmagnetic battery and load test with a 100 Ω resistor to ensure that it meets the manufacturer's specification. Keep in mind that battery amp-hours decrease with increasing water temperature.

2. Do not overfill the supply spool with magnetic tape. Leave a 2 mm space so that the tape will not spill off the spool and jam the mechanical mechanism when the instrument is tilted or laid on its side.

3. Check the tape take-up spool clearance between pinch-rollers spring, circlip, and frame. Spin spool by hand. Check for space between the feed spool and pressure sensor (if installed). Wrap 20 turns of leader on the take-up spool and check the clutch tension.

4. Check that both spool nuts are in place and do not overtighten. Do not overtighten the nylon rotor pivot screw.

5. Ensure that no ferrous metal screws are used near the compass. Replace these with stainless steel or brass. Also, do not use a ferrous bar to balance the direction vane—it may be close enough to cause the compass to "stick" and ruin the directional data.

6. Inspect the O-ring for cuts or nicks and do not trap loose wiring under the ring seat when closing the case. Leakage of small amounts of water to the bottom of the instrument case can cause electrical malfunctions when the instrument tilts.

7. Do not jam a spinning rotor with tissue paper or other material prior to deployment. It is better to shield rotor from wind while on deck. Too often the instrument is recovered with the material still jammed in place.

8. It is essential to hand-record accurate times for the first and last data records. Make sure the time zone is recorded. Record the time the instrument enters the water on deployment and leaves the water on recovery. More problems can be linked to poor bookkeeping than any other cause.

9. Spin the rotor in multiples of 24 times (or some multiple of four) to ensure that sampling interval and rotor counter switch (if applicable) are correct.

We remark that items (6) and (8) apply to all moored instruments. Modern acoustic current meters have been known to have O-ring leaks and subsequent electrical failures, and the problem of recording time zone continues to be a problem despite the best efforts of technical protocols. Other problems include measurement errors due to low numbers of acoustic scatterers (poorly delineated Doppler frequency shift), gradual deterioration of acoustic transducers and their power output due to long-term effects of high pressures and damage from rough handling procedures on-deck, and insufficient battery power for the experiment duration and sampling rate.

A standard method for isolating large errors is to compute a histogram of the sample values. This amounts to completing step 1 in a goodness of fit calculation since a histogram is nothing more than a diagram showing the frequencies of occurrence of sample values. While this is a very straightforward procedure some thought must go into selecting the parameter intervals, or bins, over which the sample frequencies are calculated. If the bins are too large, the histogram will not resolve the character of the sample PDF and the effects of large error values will be suppressed by being grouped with more commonly occurring values. On the other hand, if the bins are too narrow, individual values take on more influence and the resulting distribution will not appear smooth. This makes it difficult to "see" the real shape of the distribution.

The use of a histogram in locating large errors is that it readily identifies the number of widely differing values that occur and shows whether these divergent values fit into the assumed PDF for the assumed variable. In other words, we can not only see how many values ("outliers") differ widely from the mean values, but also determine if the number of large values in the sample is consistent with the expected distribution of large values for the population. Thus, we have an added guideline for deciding whether the sample values should be retained or eliminated for subsequent analysis. Both PDFs and histograms use visual means of detecting large error values. It is possible to use more automated and objective techniques, such as eliminating all values that exceed a specified standard deviation (e.g., $\pm 3\sigma$). However, these approaches have the weakness

that they must first consider all data points, including the extreme values, as valid in order to determine decision levels for selecting or rejecting data. Here, we could use an iterative process in which the values outside the accepted range are omitted from each subsequent recalculation of the mean and standard deviation, until the remaining data have near constant statistics with each new iteration. Large errors, which are usually easy to spot using visual editing techniques, should be removed before proceeding to a more objective step involving the detection of less obvious random deviations. An objective technique for identifying outlier values is to compute a function, which selects extremes of the population, such as the first derivative of the measured variable with respect to an independent parameter. An example would be a time series of temperature measured from a line in a satellite image. After the extreme gradients are identified in the first derivative calculation, there is still the question of how widely the extremes should be allowed to differ from the rest of the population and whether a value should be considered as an error value or as simply as a maximum (or minimum) of the process being observed.

In making such a decision, it is necessary to have an estimate of the variability of the process. As discussed above, the dispersion (spread) of the population distribution is best represented by the variance or the standard deviation. If we are dealing with a normal population, we know that the standard deviation specifies the spread of the distribution and that 68% of the population values lie within $\mu \pm \sigma$ while 95% of these values are in the interval $\mu \pm 2\sigma$. Beyond $\mu \pm 3\sigma$ there is only 0.26% of the total frequency of occurrence, leaving 99.74% within this interval. Thus, it is again a matter of probabilities and significance level; and we must choose at what level we will reject deviations from the mean as errors. If we choose to discard all measurements beyond 2σ, we will have retained 95% of the sample population as our new sample population for which we will repeat out estimation of the statistics. This suggests that we will make our statistical estimate twice; first to decide what data to retain, and second to make statistical inferences about the behavior exhibited by the revised sample data. It is customary to use a much coarser subsampling interval, or to use broadly smoothed data, to compute the initial sample standard deviations for the purposes of editing the data. For our *TS* curve example (Figure 3.15), we might initially have used a computational interval of 1 or 2 °C to compute a standard deviation for the first-stage editing and then have used the newly defined sample population (original sample minus large deviations >2 °C) to recalculate the mean and standard deviation with a resolution of 0.1 °C, closer to the measurement accuracy for reversing thermometers. In statistical analysis we should not expect to exceed the inherent accuracy and resolution of our data. Modern computing facilities, and even pocket calculators, make it tempting to work with many decimal places despite the fact that higher place values are not at all representative of the ability of the instrument to make the oceanic measurement.

A form of two-step editing is used in the routine processing of CTD data, which is typically sampled at ≈25 samples per second per channel (≈25 Hz/channel). Since these instruments produce many more data than we are capable of examining, both smoothing and editing procedures are often built into the routine processing programs. The steps involved with processing calibrated CTD data should generally be as follows:

1. Write the data to a file for display on a computer screen using an interactive editing program written for the particular data set.
2. Examine all data for a given set of parameters by displaying the data simultaneously on a computer monitor; as a consistency check, it is important to know if large errors in one parameter, such as temperature, are

associated with some real feature in another parameter, such as salinity.
3. With the cursor, eliminate erroneous values collected near the ocean surface, where the probe rises in and out of the water with the roll of the ship.
4. Using the file in (3), calculate the pressure gradient vs depth for the data and eliminate those data values for which the depth is decreasing with time for adown-cast and increasing with time for an up-cast (wave action eliminator).
5. Using the file of (4), produce a hard-copy or computer screen plot of the entire profile plus an expanded version for the upper ocean (say 0–300 m depth).
6. On the copy, "flag" erroneous values and irregularities in all data channels.
7. Use the interactive screen display to eliminate "bad" data identified in (5). If gaps between data points are small, linearly interpolate between adjacent values.
8. Smooth the edited file by averaging values over a specified depth range. Typically,1-m averaged files are generated for profile data and 1-s averaged files for time series data.

Because of improved CTD technology in recent years, step (8) is often conducted first. This step is then eliminated or replaced with a larger averaging interval, such as 5 m.

Fofonoff et al. (1974) used a 1/2-s average (15 scans) to smooth the measured pressure series. From this smoothed set, a 1/10th decibar pressure series was generated. Even with the smoothing, the pressure was oversampled, with roughly two observations for each pressure interval. The goal of this computation was to produce a uniform pressure series that could be used to generate profiles of *T* and *S* with depth. Processing routines could be added that first sorted out spurious extreme *T* and *S* values, based on a running mean standard deviation, and which ensured that the pressure series was monotonically increasing. This would correct for small variations in the depth of the probe due to ship motion or strong current shear. Also, in making these editing decisions we should always keep in mind the instrument characteristics and not discard data well within the noise level of the measurement system.

When editing newly collected data, we should always consider what is already known from similar, or related measurements in order to detect obvious errors. A typical example is the use of *TS* curves to evaluate the performance of sample bottles in a hydro-cast. Since *TS* curves are known to remain relatively stationary for many areas, previously sampled *TS* curves for an area can be used to locate data points that may have been caused by the erroneous performance of a water sampler; these are generally due to inadvertent bottle "trips" in which the sampler likely closed before or after the desired depth was achieved. Prior *TS* curves also have served as a means of interpolating a particular hydrocast or perhaps providing salinities to match measured temperatures. This approach is limited, however, to those areas and those parameters for which a sufficient number of existing observations are available to define the mean state and variability. In many areas, and for many parameters, information is too limited for existing data to be of any real use in evaluating the quality of new measurements. As a matter of curiosity, it would be interesting to determine the numbers of deep hydrocast data that were unknowingly collected at hydrothermal venting sites and discarded because they were "erroneous." Anomalously high temperatures would be difficult to justify if one did not know about hydrothermal circulation and associated buoyant plumes. Similar comments apply to "anomalous" CTD profiles obtained within thermohaline staircases (double-diffusive features). As noted in Section 2.1, salt-fingering and diffusive convection generate small scale ($\sim$1−10 m) vertical structure that would appear to be highly erroneous if measured for the first time.

In contrast to large accidental errors, which lead to large offsets or systematic biases, random errors are generally small and normally distributed. These errors often are the result of inaccuracies in the instrumentation or data collection procedures and therefore represent the limit of our ability to measure the desired variable. Added to this is our inability to completely resolve the inherent variability in a particular parameter. This too may be a limitation of our instrument or of our sampling scheme. In either case, when we cannot directly measure a scale of oceanic variability that contributes to the alias of our measurement, the variability will form part of the uncertainty in the final calculated value.

The theory of random errors is well established (Scarborough, 1966). The fundamental approach is to treat the errors as random numbers with a normal PDF. Basic to this assumption is that positive and negative errors of the same size occur in about equal number and tend to cancel each other. This suggests that the appropriate way to treat data containing random errors is in terms of MS and root-mean-square (RMS) values. Another fundamental assumption is that the probability of an error occurring depends inversely on its magnitude; thus, small errors are more frequent than large ones. Following the first of these two assumptions, the PDF of the random errors might be written as

$$p(\varepsilon_x) = f(\varepsilon_x^2) \qquad (3.140)$$

where p is the PDF of the errors ε_x. The second characteristic requires that the probability decreases with increasing ε_x, so we can write for any real constant, k

$$p(\varepsilon_x) = C \exp(-k^2 \varepsilon_x^2) \qquad (3.141)$$

Using the fact that the integral under the curve of any PDF is unity, we solve for C and obtain

$$p(\varepsilon_x) = \frac{k}{\sqrt{\pi}} \exp(-k^2 \varepsilon_x^2) \qquad (3.142)$$

This expression is known as the probability equation or the error equation. A graph of the function gives the normal or Gaussian probability curve. The term k is a constant called the *index of precision* and sets both the amplitude and the width of the normal curve. As k increases, the normal curve becomes narrower and the errors get smaller, making the measurement more precise. (This description applies only for small random errors and not to systematic errors.)

3.16.2 Propagation of Error

Suppose we have a quantity, F, which is calculated from a combination of a number (N) of independently observed variables. For example, F might be oceanic heat transport computed from independent velocity and temperature profiles, x_i ($i = 1, 2$). We can estimate the combined random error of F as the SSE of the individual variables provided that the errors are independent of the variables and that they are all normally distributed. As a simple example, let F be a linear combination of our measurement variables, x_i

$$F = a_1 x_1 + a_2 x_2 + \ldots + a_N x_N \qquad (3.143)$$

where $a_1, \ldots, a_N$ are constants. The inverse of the squared error or *index of precision* (H) of F can be written

$$\frac{1}{H^2} = \frac{a_1^2}{h_1^2} + \frac{a_2^2}{h_2^2} + \ldots + \frac{a_N^2}{h_N^2} = \sum_{i=1}^{N} \frac{a_i^2}{h_i^2} \qquad (3.144)$$

where h_i is the error for the ith measurement, x_i.

A more generalized formula for error calculations for arbitrary F for which the contributing variables are uncorrelated is

$$\frac{1}{H^2} = \frac{(\partial F/\partial x_1)^2}{h_1^2} + \frac{(\partial F/\partial x_2)^2}{h_2^2} + \ldots + \frac{(\partial F/\partial x_N)^2}{h_N^2}$$

$$= \sum_{i=1}^{N} \frac{(\partial F/\partial x_i)^2}{h_i^2}$$

$$(3.145)$$

where partial derivatives $\partial F/\partial x_i$ are obtained from Taylor expansions of the function F in terms of the independent variables x_i. Specifically, $\partial F/\partial x_i = a_i$. To convert this expression to one in terms of relative errors, we use the fact that

$$\frac{1}{h^2} = \frac{r_e^2}{\rho_e^2} \qquad (3.146)$$

where r_e is the corresponding relative error and $\rho_e = 0.4769$ is a constant obtained from the error Eqn (3.142). Using this definition we can write our final error as

$$R_e = \left[(\partial F/\partial x_1)^2 r_1^2 + (\partial F/\partial x_2)^2 r_2^2 + \ldots \right.$$

$$\left. + (\partial F/\partial x_N)^2 r_N^2 \right]^{1/2} \qquad (3.147)$$

In this form, R_e is really only the RMS error that describes the equivalent combined error in the equation of interest. This Taylor expansion of the contributing error terms is known as the *propagation of errors formula*. It is limited to small errors and uncorrelated independent variables. Since these principles apply only to small random errors, it is necessary to use some data editing procedure to remove any large errors or biases in the measurements before using this formula. By using a MS formulation, we take advantage of the fact that small random errors can be expected to often cancel each other resulting in a far smaller MS error than would result if the measurement errors were simply added regardless of sign to yield a maximum "worst case error." The primary application of Eqn (3.147) is in determining the overall error in a quantity derived from a number of component variables all with measurement errors. This is a situation common to many oceanographic problems.

A more complicated propagation of error formula is needed if there is a nonzero correlation between the independent variables, x. In this case, we must also retain the covariance terms in any Taylor expansion of the small error terms. For example, the density ρ is a function of both temperature T and salinity S so that the errors (variances) in density σ_ρ^2 can be related to the measurement errors in temperature σ_T^2 and salinity σ_S^2 by

$$\sigma_\rho^2 = (\partial\rho/\partial T)^2 \sigma_T^2 + (\partial\rho/\partial S)^2 \sigma_S^2 \\ + 2[(\partial\rho/\partial T)\cdot(\partial\rho/\partial S)]C(T,S) \tag{3.148}$$

where $C(T, S)$ is the covariance between temperature and salinity fluctuations. Only when $C(T, S) = 0$ do we get the result (Eqn (3.147)). An example of a detailed error calculation for the measurement of flow through trawl nets towed at various angles through the water column is given in Burd and Thomson (1993).

3.16.3 Dealing with Numbers: the Statistics of Roundoff

Since we must represent all measurements in discrete digital form, we are forced to deal with the consequences of numerical roundoff, or truncation. The problem results from the limitations of digital computing machines. For example, the irrational fraction $1/3$ is represented in the computer as the decimal equivalent $0.3333...3$ with an obvious round-off effect. This may not seem to be a problem for most applications since most computers carry a minimum of eight decimal places at single precision. The large number of arithmetic operations carried out in a problem lasting for only a few seconds of computer processing time can, however, lead to large errors in due to roundoff and truncation errors. The case of greatest concern is when two nearly identical numbers are subtracted, requiring proper representation to the smallest possible digit. Such differences can easily occur unknowingly in a complicated computational problem. Rather than discuss procedures for estimating this roundoff error, we will discuss the nature of the problem and emphasize the need to avoid roundoff.

General floating-point values (decimal numbers) in a computer follow closely the so-called "scientific notation" and are represented as a mantissa (to the right of the decimal point) and an exponent (the associated power of 10). For example, in a three-digit system, the number 64.282 would be represented as 0.643×10^2 where the roundoff is accomplished by adding five in the thousands' decimal place and then truncating after the third digit. This process of rounding off results in a slight bias because it always rounds up when there is a 5 in the least significant digit. A way to overcome this bias is to use the last digit retained to determine whether to round up or down when the next digit is exactly 5. This rule, which leads to the least possible error, is to roundup if the next to the last digit retained is odd and to round down when it is even. This procedure can be summarized as follows. When rounding a number to k decimals:

1. if the $k + 1$ decimal is 0, 2, 4, 6, 8 then the k decimal is unchanged;
2. if the $k + 1$ decimal is 1, 3, 5, 7, 9 then the k decimal is increased by 1.

This system of rounding-off will result in errors that are generally less than 0.5×10^{-k} and maximum roundoff errors of 0.6×10^{-k}. In most applications, the effect of this roundoff bias is too small to justify the added numerical manipulation required to implement this even–odd roundoff scheme.

In computing systems, floating-point numbers are handled in a binary representation having 24 bits (word length is 32 bits but 8 bits are used for the exponent), which results in seven significant decimal digits. Called *single precision*, this level of accuracy is adequate for many computations. For those problems with repeated calculations, and the subsequent high probability of differencing two nearly identical numbers, a *double-precision* representation is used which has 56 binary bits leading to 16 significant decimal digits. Roundoff, in the case of double precision, results in very small biases, which can be completely ignored for most applications. Another approach to the problem of roundoff errors is to consider them to be random variables. In this way, statistical methods can be applied to better understand the effects of roundoff errors. Consider the roundoff of a single number x; for this number, all numbers occurring in the interval $x_0 - 1/2 < x < x_0 + 1/2$ (measured in units of the last digit) become that number. Thus, the roundoff has a uniform probability distribution in the last digit. We can write the corresponding PDF $f(x)$ for x as

$$f(x) = \begin{cases} 1 & \left(x_0 - \dfrac{1}{2}, \ x_0 + \dfrac{1}{2}\right) \\ 0 & \text{elsewhere} \end{cases} \qquad (3.149)$$

and note that

$$\int_{-\infty}^{\infty} f(x)\mathrm{d}x = 1 \qquad (3.150)$$

The most common measures of a PDF are its first two moments, the mean and the variance. The mean of $f(x)$ in Eqn (3.149) is x_0 and the variance is

$$V[f(x)] = \sigma^2 = \int_{x_0-1/2}^{x_0+1/2} [x - x_0]^2 f(x)\mathrm{d}x$$

$$= \int_{-1/2}^{+1/2} x'^2 \mathrm{d}x' = \frac{1}{12} \qquad (3.151)$$

Experimental tests have verified the uniform distribution of roundoff in computer systems. In fact, computers generate random numbers by using the overflow value of the mantissa.

We can represent roundoff as an additive random error (ε) superimposed on the true variable (x). In this case, we can write the computer representation of our variable (which we assume is free from measurement and sampling errors)

as $x + \varepsilon$. For a floating-point number system, it is better to use

$$x(1 + \varepsilon); |\varepsilon| < \frac{1}{2}(10^{-2}) \qquad (3.152)$$

for the variable with roundoff error ε. This formulation has the effect of focusing attention on the consequences of roundoff for every application in which it appears. For example, the product

$$x_1(1 + \varepsilon_1)x_2(1 + \varepsilon_2) = x_1 x_2(1 + \varepsilon_1 + \varepsilon_2 + \varepsilon_1\varepsilon_2) \qquad (3.153)$$

demonstrates how roundoff propagates during multiplication. Generally, the product $\varepsilon_1\varepsilon_2$ is sufficiently small to be ignored. However, in the above multiplication we must include the roundoff for this operation, whereby Eqn (3.153) becomes

$$x_1(1 + \varepsilon_1)x_2(1 + \varepsilon_2) = x_3(1 + \varepsilon_3)$$

$$= x_1 x_2(1 + \varepsilon_1 + \varepsilon_2 + \varepsilon) \quad |\varepsilon| < \frac{1}{2}(10^{-2});$$

$$\varepsilon_3 = \varepsilon_1 + \varepsilon_2 + \varepsilon \qquad (3.154)$$

Similar error propagation results are found for other arithmetical operations. We can extend this to a generalized product

$$y_1(1 + \varepsilon_1)y_2(1 + \varepsilon_2)...y_N(1 + \varepsilon_N) \qquad (3.155)$$

which becomes

$$y_1 y_2...y_N[1 + (\varepsilon_1 + \varepsilon_2 + ... + \varepsilon_N)] \qquad (3.156)$$

By the central limit theorem, the sum of N independent random numbers (the numbers (the roundoff errors)) approaches a normal distribution. The effect for the other operations is much the same; therefore, while individual roundoffs are from a uniform distribution, the result of many arithmetic roundoff operations tends toward a normal distribution. This also can be demonstrated experimentally.

As stated earlier, we will generally ignore roundoff as a source of error in the processing and analysis of oceanographic data. The above discussion has been presented here to make the reader aware of potential problems and provide some familiarity with the problems of using computing systems. In most data applications, the effects of roundoff error are small enough to be ignored. Only in the case of recursive calculations, where each computation depends on the previous one, do we anticipate large roundoff errors. This is usually a problem for numerical modelers who must deal with the repeated manipulation of computer-generated "data." In cases where roundoff errors are of some consequence, statistical methods can be used in which the errors can be treated as variables from a normal population.

3.16.4 Gauss—Markov Theorem

The term *Gauss—Markov process* is often used to model certain kinds of random variability in oceanography. To understand the assumptions behind this process, consider the standard linear regression model, $y = \alpha + \beta x + \varepsilon$, developed in the previous sections. As before, α, β are regression coefficients, x is a deterministic variable and ε a random variable. According to the Gauss—Markov theorem, the estimators α, β found from least squares analysis are the *best linear unbiased estimators* for the model for the following conditions on ε:

1. The random variable ε is independent of the independent variable, x;
2. ε has a mean of zero; that is $E[\varepsilon] = 0$;
3. Errors ε_j and ε_k associated with any two points in the population are independent of one another; the covariance between any two errors is zero; $C[\varepsilon_j, \varepsilon_k] = 0$, $j \neq k$;
4. ε has a finite variance $\sigma_\varepsilon^2 \neq 0$.

The estimators are *unbiased* since their expected value equals the population values (given 1 and 2) and they are *best* in that they are efficient (if 3 and 4 hold true), the variance of the least-squares estimators being smaller than any other linear unbiased estimator. A further assumption that is often made is that the errors, ε, are

normally distributed. In this case, the estimators of α, β, and μ using the least-squares requirements are identical to the estimators resulting from the use of maximum-likelihood estimation. This assumption, combined with the four previous assumptions, provide the rationale for the least-squares procedure.

3.17 INTERPOLATION: FILLING THE DATA GAPS

Most analysis procedures used in the physical sciences are designed for comparatively long and densely sampled series with equally spaced measurements in time or space. The wealth of information on time series analysis primarily applies to regularly spaced and abundant observations. There are two main reasons for this: (1) the mathematical necessity for long, equally spaced data for the derivation of statistically reliable estimates from modern analytical techniques; and (2) the fact that most modern measurement systems both collect and store data in digital format. Spectral estimates, for example, improve with increased duration of the data series in the sense that one is able to cover an increasing range of the dominant frequency constituents that make up the record. Digital sampling systems are considerably more economical than analogue recording systems in that they cut down on storage space, power consumption and postprocessing effort.

3.17.1 Equally and Unequally Spaced Data

Electronic systems now provide data at regularly spaced sampling increments. This includes data from autonomously recording instruments as well as data from instruments integrated into cabled observatory networks connected to a shore station, Any type of equipment failure generally leads to either data *gaps* or a premature termination of the record. The failure of electronic data logging systems—which in the case of cabled observatories also includes the hardware linking the instrument to the shore station—is but one source of gappy records in physical oceanography. Because of their very nature, shipborne measurements are a source of gappy records. Oceanographic research vessels are expensive platforms to operate and must be used in an optimal fashion. As a consequence, it is often impossible to collect observations in time or space of sufficient regularity and spacing to resolve the phenomenon of interest. Efforts are usually made to space measurements as evenly as possible but, for a variety of reasons, station spacings are often considerably greater than desired. Diminishing science budgets are having major impacts on maintaining historical sampling schedules for existing climate-based time series. Weather conditions, as well as ship and equipment problems, almost invariably lead to unwanted gaps in the data set. Sometimes equipment failures are not detected until the data are examined in the laboratory. In addition, editing out errors produces unwanted gaps in the data record.

The gap problem is even more severe when one is analyzing historical data or data collected from "platforms of opportunity." Historical data are a collection of many different sampling programs all of which had different goals and therefore very different sampling requirements. By its very nature, such collections of data will necessarily be irregularly spaced and variable in terms of accuracy and reliability. Further editing, dictated by the goals of the historical data analysis project, will add new gaps to the set of existing data series.

Monitoring stations, ships of opportunity, and satellite measurements frequently produce data series that are unevenly spaced. The geographic distribution of monitoring stations (e.g., Pacific island sea-level stations, Deepocean Assessment and Reporting of Tsunamis (DART) tsunami recording stations, and meteorological buoy stations) is far from uniform in terms of the spacing between stations. Thus, while the data series collected at each station, may themselves consist of evenly and densely spaced measurements in time, the space intervals between stations will be highly irregular. Open ocean buoys and current meter moorings also fit this classification of densely and evenly spaced temporal observations at widely and often irregularly spaced locations. Here again, any failure in the recording system, whether minor or catastrophic, will lead to gaps in the time series record. Often these gaps are quite large since unplanned recovery efforts are required to correct the problem. Such a correction effort assumes that the telemetering of data is available which, with the exception of data sent through satellites or through cabled observatory networks, is not widely available. Failures of onboard recording systems must wait until the scheduled servicing of the instrument which may then result in relatively large data gaps.

At the other end of the sampling spectrum, satellite observing systems provide dense and evenly spaced measurements that are often very irregular in time. A familiar source of temporal gaps in infrared image series is cloud cover. Both occasional and persistent cloud cover can interrupt a sequence of images collected to study changes of sea surface temperature. The effects of cloud cover apply also to satellite remote sensing in the optical bands. In addition to the cloud-cover problem, there are often problems with the onboard satellite sensing systems or associated with the ground receiving station that lead to gaps in time series of image data. Microwave sensing of the surface is not as sensitive to cloud attenuation but it is subject to sensor and ground-recording failure problems. Satellite radar data collected at the ocean surface are distorted by winds above a certain threshold speed. Off the Atlantic coast of North America, where there are strong oceanic features associated with the Gulf

Stream, the wind threshold speed is about 10 m/s whereas off the Pacific coast, where oceanic frontal structure is more subdued, the wind speed threshold is about 3 m/s (Williams et al., 2013).

Platforms of opportunity (usually merchant ships) produce uniquely irregular sets of measurements. Most merchant ships repeat the same course with minor adjustments for local weather conditions and season. A seasonal shift in course is generally seen at higher latitudes to take advantage of great circle routes during times of better weather. A return to lower latitudes is seen in winter data as the ships avoid problems with strong storms. Added to the seasonal track changes is the nature of the daily sampling procedure. Usually the ship takes measurements at some specified time interval which, due to variations in ship track, ship speed and weather conditions, may be at very different positions from sailing to sailing. Thus, the merchant ship data will be irregular in both space and time. Systems that operate continuously from ships of opportunity (e.g., injection SST) overcome this problem. These continuous measurements, however, are still subject to variations in ship track.

The net result of all these measurement problems is that oceanographers are often faced with short records of unequally spaced data. Even if the records are long they are often gappy in time or space. It is, therefore, necessary to interpolate these data to produce series of evenly spaced measurements. While some analysis procedures, such as least-squares harmonic analysis favored in tidal analysis (cf., Foreman et al.,1994 (updated 2009); Pawlowicz et al., 2002), apply directly to uneven or gappy data, it is more often the case that irregularly spaced data are interpolated to yield evenly spaced, regular data. These interpolated records can then be analyzed with familiar methods of time series analysis.

Interpolation also may be required with evenly spaced data if the subject dynamics apply to smaller space/timescales than are resolved by the measurements. Thus, the data points that are interpolated produce another set of regularly spaced points with a finer resolution. Many interpolation procedures have been developed that only apply to evenly spaced data.

3.17.2 Interpolation Methods

Interpolation techniques are needed for both irregularly spaced and evenly spaced data series. Before deciding which interpolation method is most effective, we need to consider the particular application. A series of appropriate questions regarding the selection of the best interpolation procedures are:

1. What samples (original data series, derivatives, etc.) should we use?
2. What class of interpolation function (linear, higher-order polynomial, cubic spline, etc.) best satisfies the dynamical restrictions of the analysis?
3. What mathematical criteria (exact data point matching, least-squares fit, continuity of slopes, etc.) do we use to derive the interpolated values?
4. Where do we apply these criteria?

Answers to these questions serve as guides to the selection of a unique interpolation procedure.

3.17.2.1 Linear Interpolation

The type of interpolation scheme to be employed depends on how many data points we want our interpolation curve (polynomial) to pass through (i.e., to "fit"). Increasing the number of points we want the curve to pass through, increases the order of the polynomial we need to do the fitting. The most straightforward and widely used interpolation procedure is that of *linear interpolation*. This consists of connecting a straight line between two data points and choosing interpolated values at the appropriate positions along that line. For a data series $y(x)$, this linear procedure can be written as

$$y(x) = y(a) + \frac{x-a}{b-a}[y(b) - y(a)]$$

$$= \frac{(b-x)y(a) + (x-a)y(b)}{b-a} \qquad (3.157)$$

where $x_{\text{start}} = a$ and $x_{\text{end}} = b$ are the times (positions) of the data collection at the start and end of the sampling increment being interpolated, and x represents the corresponding time (position) of the desired interpolated value within the interval $[a, b]$. This is the customary procedure for interpolating between values in most tables. The same formula can be applied to *extrapolation* (extending the data beyond the domain of the observations) where the point x lies beyond the interval $[a, b]$. Equation (3.157) is a special case of the Lagrange polynomial interpolation formula discussed in the next section.

3.17.2.2 *Polynomial Interpolation*

If we wish to interpolate between more than two points simultaneously, we need to use higher-order polynomials than the first-order polynomial (straight line) used in the previous section. For example, through three points we can find a unique polynomial of degree 2 (a quadratic); through four points, a unique polynomial of degree 3 (a cubic), and so on. The two methods described below are computationally robust in the sense that they yield reasonable results at most points. Polynomial interpolation techniques, such as Vandermonde's and Newton's methods are awkward to program and suffer from problems with roundoff error.

3.17.2.2.1 LAGRANGE'S METHOD

The Lagrange polynomial interpolation formula is a method for finding an interpolating polynomial $y(x)$ of degree N which passes through all of the available data points at the same time, (x_i, y_i); $i = 1, 2, ..., N+1$.

The general form for this polynomial, of which linear interpolation is a special case, is given as

$$y(x) = a_0 + a_1x + a_2x^2 + ... + a_Nx^N = \sum_{k=0}^{N} a_k x^k$$

$$= \sum_{i=1}^{N+1} \left[y_i \left(\prod_{\substack{k=1 \\ k \neq i}}^{N+1} \frac{x - x_k}{x_i - x_k} \right) \right]$$

$$(3.158)$$

where Π is the product function. Note that in the product function, the ith term—corresponding to the particular data point, x_i, in the denominator—is not included when calculating the product for the term involving x_i. Even though k ranges from 1 to $N+1$, Π uses only N terms and the final polynomial is of order N, as required.

The goal of the Lagrange interpolation method is to find an Nth degree polynomial which is constrained to pass through the original $N+1$ data points and which yields a "reasonable" interpolated value for any position x located anywhere between the original data points. To see that the polynomial passes through the original data points, note that the ith product function, Π_i, defined for the data point x_i in the denominator is constructed in such a way that $\Pi_i(x_j; x_i) = \delta_{ij}$ whenever $x = x_j$ is one of the data values (δ_{ij} is the Kronecker delta function). This means that $\Pi_i(x_j; x_i) = 0$ for all x_j except for the specific value $x = x_i$ found in the original data series that matches the term in the denominator. In the latter case, $\Pi_i(x_j; x_i) = 1$ and $y_i\Pi_i(x_j; x_i) = y_j$.

The general polynomial we seek is constructed as a sum of the product functions in Eqn (3.158) which can be expanded to give

$$y(x) = \sum_{i=1}^{N+1} y_i[Q_i(x)/Q_i(x_i)] \qquad (3.159)$$

in which

$$Q_i(x) = (x - x_1)(x - x_2)...(x - x_{i-1})(x - x_{i+1})$$
$$...(x - x_{N+1})$$

$$(3.160)$$

is the product of all the factors except the ith one. For any x, Eqn (3.159) can be expanded to give the interpolating polynomial

$$y(x) = y_1\frac{(x - x_2)(x - x_3)...(x - x_{N+1})}{(x_1 - x_2)(x_1 - x_3)...(x_1 - x_{N+1})}$$
$$+ y_2\frac{(x - x_1)(x - x_3)...(x - x_{N+1})}{(x_2 - x_1)(x_2 - x_3)...(x_2 - x_{N+1})} + ...$$
$$+ y_N\frac{(x - x_1)(x - x_2)...(x - x_N)}{(x_{N+1} - x_1)(x_{N+1} - x_3)...(x_{N+1} - x_N)}$$

$$(3.161)$$

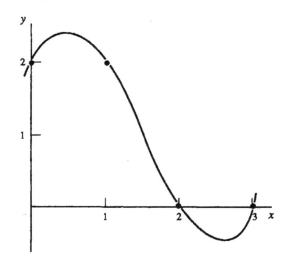

FIGURE 3.17 Use of Lagrange's method to fit a third-order (cubic) polynomial through the data points (x_i, y_i) given by (0, 2), (1, 2), (2, 0), and (3, 0).

Note that, for the original data points, $x = x_i$, the polynomial yields the correct output value $y(x_i) = y_i$, as required.

In the Lagrange interpolation method, the calculation is based on all the known data values. If the user wants to add new data to the series, the whole calculation must be repeated from the start. Although the above formula can be applied directly, programing improvements exist that should be taken into account (Press et al., 1992). Use of Neville's algorithm for constructing the interpolating polynomial is more efficient and allows for an estimate of the errors resulting from the curve fit.

As an example of this interpolation method, consider four points (x_i, y_i), $i = 1,..., 4$ given as (0, 2), (1, 2), (2, 0) and (3, 0) through which we wish to fit a (cubic) polynomial. Substituting these values into Eqn (3.161), we obtain

$$y(x) = 2\frac{(x - 1)(x - 2)(x - 3)}{(0 - 1)(0 - 2)(0 - 3)}$$
$$+ 2\frac{(x - 0)(x - 2)(x - 3)}{(1 - 0)(1 - 2)(1 - 3)} + 0 + 0$$
$$= \frac{2}{3}x^3 - 3x^2 + \frac{7}{3}x + 2$$

The resulting third-order curve is plotted in Figure 3.17.

3.17.2.3 Spline Interpolation

In recent years, the method that has received the widest general acceptance is the spline interpolation method. Splines, unlike other polynomial interpolations, such as the Lagrange polynomial interpolation formula, apply to a series of segments of the data record rather than the entire data series. This leads to the obvious question to ask in selecting the proper interpolation procedure: Do we want a single, high-order polynomial for the interpolation over the entire domain, or would it be better to use a sequence of lower-order polynomials for short segments and sum them over the domain of interest? This integration is inherently a smoothing operation but one must be careful of discontinuities, or sharp corners, where the segments join together. Spline functions are designed to overcome such discontinuities, at least for the lower-order derivatives. It is because discontinuities are allowed in higher-order derivatives that splines are so effective locally. Constraints placed on the interpolated series in one region have only very small effects on regions far removed. As a result, splines are more effective at fitting nonanalytic distributions characteristic of real data. The term "spline" derives from the flexible drafting tool used by naval architects to draw piecewise continuous curves.

Splines have other favorable properties such as good convergence, highly accurate derivative approximation, and good stability in the presence of roundoff errors. Splines represent a middle ground between a purely analytical description and numerical finite difference methods, which break the domain into the smallest possible intervals. The piecewise approximation philosophy represented by splines has given rise to finite element numerical methods.

With spline interpolation, we approximate the interpolation function $y(x)$ over the interval $[a, b]$ by dividing the interval into subregions with the requirement that there be continuity of the function at the joints. We can define a spline function, $y(x)$, of degree N with values at the joints.

$$a = u_0 \leq u_1 \leq u_2 \ldots \leq u_N = b \qquad (3.162)$$

and having the properties:

1. In each interval $u_{i-1} \leq x \leq u_i$ $(i = 1, \ldots, m)$, the function $y(x)$ is a polynomial of degree not greater than N;
2. At each interior joint, $y(x)$ and its first $N - 1$ derivatives are continuous.

The spline function in widest use is the cubic spline $(N = 3)$. To give the reader familiarity with the spline interpolation technique, we will develop the cubic spline equations and work through a simple example. Consider a data series with elements (x_i, y_i), $i = 1, \ldots, N$. Since we are working with a cubic spline interpolation, the first two derivatives $y'(x)$ and $y''(x)$ of the interpolation function, $y(x)$, can be defined for each of the points x_i while the third derivatives $y'''(x)$ will be a constant for all x. Here, the prime symbol denotes differentiation with respect to

the independent variable x. We write the spline function in the form

$$y(x) = f_i(x); \quad x_i \leq x \leq x_{i+1}, \ i = 1, \ldots, N - 1 \qquad (3.163)$$

and specify the following conditions at the junctions of the segments:

1. continuity of the spline function:

$$f_i(x_i) = y(x_i) = y_i, \quad i = 1, 2, \ldots, N - 1;$$
$$f_{i-1}(x_i) = y(x_i) = y_i, \quad i = 2, 3, \ldots, N; \qquad (3.164a)$$

2. continuity of the slope:

$$f'_{i-1}(x_i) = f'_i(x_i), \quad i = 1, 2, \ldots, N - 1 \qquad (3.164b)$$

3. continuity of second derivative:

$$f''_{i-1}(x_i) = f''_i(x_i), \quad i = 1, 2, \ldots, N - 1 \qquad (3.164c)$$

Since $y'''(x) = $ constant, $y''(x)$ *must be linear, so that*

$$f''_i(x_i) = y''_i \frac{(x_{i+1} - x)}{(x_{i+1} - x_i)}$$
$$= y''_{i+1} \frac{(x - x_i)}{(x_{i+1} - x_i)} \qquad (3.165)$$

Integrating twice and selecting integration constants to satisfy the conditions, Eqn (3.164a,b) on $f_i(x_i)$ and $f_{i-1}(x_i)$ gives

$$f_i(x_i) = y_i \frac{(x_{i+1} - x)}{(x_{i+1} - x_i)} + y_{i+1} \frac{(x - x_i)}{(x_{i+1} - x_i)}$$
$$- \frac{(x_{i+1} - x_i)^2}{6} y''_i \left\{ \frac{(x_{i+1} - x)}{(x_{i+1} - x_i)} - \left[\frac{(x_{i+1} - x)}{(x_{i+1} - x_i)} \right]^3 \right\}$$
$$- \frac{(x_{i+1} - x_i)^2}{6} y''_{i+1} \left\{ \frac{(x - x_i)}{(x_{i+1} - x_i)} - \left[\frac{(x - x_i)}{(x_{i+1} - x_i)} \right]^3 \right\} \qquad (3.166)$$

which uniquely satisfies the continuity condition for the second derivative but not, in general, for the first derivative (slope). To ensure continuity of the slope at the seams, we expand Eqn (3.165) by differentiation to get

$$f'_i(x_i) = \frac{(y_{i+1} - y_i)}{(x_{i+1} - x_i)} - \frac{(x_{i+1} - x_i)}{6}(2y''_i + y''_{i+1})$$

(3.167a)

$$f'_{i-1}(x_i) = \frac{(y_i - y_{i-1})}{(x_{i+1} - x_i)} - \frac{(x_{i+1} - x_i)}{6}(y''_{i-1} + 2y''_i)$$

(3.167b)

We then set Eqns (3.167a) and (3.167b) to be equal in order to satisfy slope continuity Eqn (3.164b), whereby

$$(x_i - x_{i-1})y''_{i-1} + 2[(x_{i+1} - x_{i-1})]y''_i + (x_{i+1} - x_i)y''_{i+1}$$

$$= 6\frac{(y_{i+1} - y_i)}{x_{i+1} - x_i} - \frac{(y_i - y_{i-1})}{x_i - x_{i-1}}, \quad i = 2, ..., N-1$$

(3.168)

which must be satisfied at $N - 2$ points by the N unknown quantities, y''_i We require two more conditions on the y''_i that we get by specifying conditions at the end points x_1 and x_N of the data sequence. After specifying these end values, we have $N - 2$ unknowns, which we find by solving the $N - 2$ equations. There are two main ways of specifying the end points: (1) we set one or both of the second derivatives, y''_i and y''_N at the end points to be zero (this is termed the *natural cubic spline)* so that the interpolating function has zero curvature at one or both boundaries; or (2) we set either y''_i and y''_N to values derived from Eqn (3.167) in order that the first derivatives of the interpolating function, y'_i, take on specified values at one or both of the termination boundaries.

As a general example, we consider the spline solution for six evenly spaced points with the data interval $h = x_{i+1} - x_i$ and function d_i defined in terms of y_i, as

$$d_i = \frac{(y_{i+1} - 2y_i + y_{i-1})}{2h^2}$$

(3.169)

We can write the Eqn (3.166) for these six equally spaced points in matrix form as

$$\begin{pmatrix} 4 & 1 & 1 & 0 \\ 1 & 4 & 1 & 0 \\ 0 & 1 & 4 & 1 \\ 0 & 0 & 1 & 4 \end{pmatrix} \begin{pmatrix} y''_2 \\ y''_3 \\ y''_4 \\ y''_5 \end{pmatrix} = \begin{pmatrix} 12d_2 - y''_1/h \\ 12d_3 \\ 12d_4 \\ 12d_5 - y''_6/h \end{pmatrix}$$

(3.170)

If we want to specify y'_i rather than y''_i, we need an equation relating both. If the end conditions are not known, the simplest choice is $y''_i = 0$ (the *natural spline* noted above). Another, and smoother choice (in the sense of less inflection or curvature at the interpolated point) is $y''_1 = 0.05 y''_2$ Although spline interpolation is a global, rather than a local, curve (altering a y''_i or an end condition affects the overall spline), the dominant diagonal terms in Eqn (3.170) cause the effects to rapidly decrease as the distance from the altered point increases.

We should point out the method of splines offers no advantage over polynomial interpolation when applied to either the approximation of well-behaved mathematical functions or to curve fitting when the experimental data are dense. "Dense" means that the number of data points in a subregion is more than an order of magnitude larger than the number of inflection points in the fitted curve and that there are no abrupt changes in the second derivative. The advantage of splines is their inherent smoothness when dealing with sparse data.

As a numerical example of spline fitting, we consider the six-point fitting of the points represented in Eqn (3.170) for the 11 data points in Table 3.13. Using a general polynomial fit yields the curve in Figure 3.18. Here, all but the last of the first six points lie on a near straight line. Due to this single point, the polynomial curve oscillates with an amplitude that does not decrease. In contrast, the spline amplitude (Figure 3.19) for the same 11 values reduces each cycle by a factor of 3.

TABLE 3.13　Data Pairs (x_i, y_i) Used for Interpolation Schemes in Figures 3.18 and 3.19

i	x_i	y_i
1	0	16
2	14	19
3	27	36
4	33	48
5	41	53
6	48	90
7	62	119
8	74	120
9	89	96
10	99	71
11	114	36

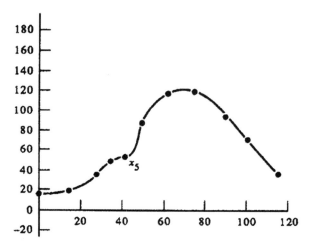

FIGURE 3.19　Cubic spline fit to the data values in Table 3.13. Amplitude of each cycle is reduced by a factor of three compared to Figure 3.18.

FIGURE 3.18　A general six-point polynomial fit to the data values in Table 3.13. Due to a single point, the polynomial curve oscillated with amplitude that does not decrease with x.

Often the first or second derivatives of the interpolated function are important. In Figure 3.18, we see that fitting a polynomial to sparse data can result in large, unrealistic changes in the second derivatives. The spline fit to the same points (Figure 3.19) using the end-point conditions $y_1'' = y_N'' = 0$ demonstrates the smoothness of the spline interpolation. In essence, the spline method sacrifices higher-order continuity to achieve second derivative smoothness.

Spline interpolation is generally accomplished by computer routines that operate on the data set in question. Computer routines solve for the spline functions by solving the equation

$$\sum_{i=1}^{N} \left[(g(x_i) - y_i)/\delta y_i \right]^2 = S \qquad (3.171)$$

where $g(x_i)$ is composed of cubic parabolas.

$$g(x) = a_i + b_i(x - x_i) + c_i(x - x_i)^2 + d_i(x - x_i)^3 \qquad (3.172)$$

for the interval $x_i \le x \le x_{i+1}$. The terms δy_i are positive numbers that control the amount of smoothing at each point; the larger is δy_i, the more closely the spline fits at each data point. A good choice for δy_i is the standard deviation of the data values.

The S term also controls smoothing, resulting in more smoothing when S increases. As S gets smaller, smoothing decreases and the splines fit the data points more closely.

When $S = 0$ the data points are fitted exactly by the interpolating spline functions. A recommended value of S is $N/2$, where N is the number of data points. An even smoother interpolation can be achieved using splines under tension. Tension is introduced into the spline procedure to eliminate extraneous inflection points. An iterative procedure is usually used to select the best level for the tension parameter.

3.17.3 Interpolating Gappy Records: Practical Examples

Gaps or "holes" occur frequently in geophysical data series. Gaps in a stationary time series are, of course, analogous to gaps in a homogeneous spatial distribution. Small gaps are of little concern and linear interpolation is recommended for filling the gaps. If the gaps are large (of the size of the integral time or space scale), it is generally better to work with the existing short data segments than to "make up" data by pushing interpolation schemes beyond their accepted limitations. For the gray area between these two extremes, one wants to know how large the data loss can be and still permit reasonable use of standard interpolation techniques and processing methods. The problem of gappy data in oceanography was addressed by Thompson (1971) who suggested that a random sampling of data points might be an optimally efficient approach. Further insight into the problem of missing data can found in Davis and Regier (1977) and Bretherton and McWilliams (1980). In this section, we present two examples of how to deal with gappy data. One is a straightforward analysis by Sturges (1983), who used monthly tide gauge data to investigate what

happens to spectral estimates when one punches holes in the data set. The other is a practical guide to the interpolation of satellite-tracked Lagrangian drifter data with its inherently irregular time steps.

3.17.3.1 Interpolating Gappy Records for Time Series Analysis

Sturges (1983) used a Monte Carlo technique to poke holes at random in a known time series of monthly mean sea level. The original record had a "red" spectrum, which fell off as f^{-3} at high frequencies (f) and contained a single major spectral peak at a period of 12 months. A total of 120 months of data were used in the analysis. The idea was to reconstruct the gappy series using a cubic spline interpolation method and see how closely the spectrum from the interpolated time series resembled that of the original time series. Data loss was limited to less than 30% of the record length and, for any individual experiment, the holes were all the same length. However, different hole lengths were used in successive runs. The only stipulation was that the length of the data segment before the next gap be at least as long as the gap itself. The program was not allowed to eliminate the first and last data points.

Cross-spectra were computed between the original time series and the interpolated gappy series. For a specified hole size, holes were generated randomly in the data series, the cross-spectra computed and the entire process repeated 1000 times. The magnitudes of the resulting cross-spectra provided estimates of how much power was lost or gained during the interpolation while the corresponding phases was interpreted as the error introduced by the interpolation process (Figure 3.20). Several important conclusions arise from Sturges' analysis:

1. Gaps have a more adverse effect on weak spectral components (spectral peaks) than on strong ones embedded in the same background spectrum;

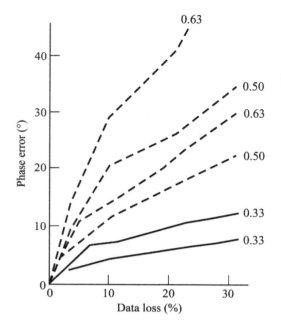

FIGURE 3.20 Absolute phase error in degrees (°) expressed as a function of the data lost (%) between the original sea level time series and the same series but with randomly generated holes that have been filled in using a cubic spline fit. Each line shows the ratio Δ/T, where Δ is the length of the random gap and T is the spectral period of interest; for example, the value 0.5 means that the holes were four units (months) long and that the period was 8 units long. Results are shown for the 90% and 99% confidence limits (lower and upper curves for each ratio value, respectively). *(From Sturges, 1983.)*

2. The phase can be estimated to roughly 10° uncertainty at the 99% confidence level for data losses of up to 30% for a strong spectral signal; the requirement is that the gaps are kept to about 1/3 (0.33) of the period of the signal being examined. If the gaps increase to 1/2 (0.5) of the period, maintenance of a 10° phase uncertainty at the 99% level requires that the data loss is less than 5% (in Figure 3.20, the *x*-axis indicates the percentage of data removed then filled with a cubic spline algorithm when examining the phase errors along the *y*-axis);

3. Although correlation functions can be computed for gappy data, it is much more difficult to compute the cross-correlation function for these data.

According to Sturges' analysis, the adverse effects of gaps depends on the length of gaps relative to the length of data set and on the magnitudes of the dominant spectral components in the signal.

3.17.3.2 Interpolating Satellite-Tracked Positional Data

The analysis of positional (latitude, longitude) time series collected through the Service Argos satellite-tracking system illustrates some of the problems that may arise with standard interpolation procedures. Because the times that polar-orbiting satellites pass over an oceanic region change through the day and because drifters move relative to the orbits of the satellite, the times between satellite fixes are irregular. At midlatitudes, times between locational fixes can range from less than an hour to as long as 10 h. Typical average times between fixes are around 2–3 h (Thomson et al., 1997). The challenge is to generate regularly spaced time series of latitude (*x*) and longitude (*y*) from which one can derive regularly spaced time series of drifter zonal velocity ($u = \Delta x/\Delta t$) and meridional velocity ($v = \Delta y/\Delta t$). This challenge is especially problematic where a "duty cycle" has been programed into the drifter transmitter to reduce the number (and cost) of transmissions to the passing NOAA satellites. A commonly used duty cycle, consisting of one day of continuous transmission followed by two days of no transmission, results in large data gaps that can make calculation of mean currents difficult in regions having strong currents in the inertial and tidal frequency bands. The duty cycle of 8 h continuous transmission followed by 16 h of silence is superior for midlatitude regions with strong inertial or tidal frequency variability.

Because of strong inertial motions in the upper layer of the open ocean and strong tidal motions over continental margins, sampling intervals of 3–4 h, or less, are preferable. A typical time step of 6 h used in many analyses of satellite-tracked drifters is inadequate to resolve inertial motions except in regions equatorward of 30° latitude where the inertial period $T = 1/f_{inertial}$ exceeds 24 h (at 50° latitude, $T \approx 16.5$ h; see *Coriolis frequency*). To generate time series at a reasonably short time step, say 3 h, we need to interpolate between irregularly spaced data points. To do this, we use a cubic spline interpolation for each of the positional records. After the correct start and end times for the oceanic portion of the record have been determined, the first step in the process is to remove any erroneous points from the "raw" data by calculating speeds over adjacent time steps, t_i; e.g., $u_i = (x_{i+1} - x_i)/(t_{i+1} - t_i)$. One then omits any unrealistic velocity values that exceed some threshold value (say 5 m/s). This "edited" record needs to undergo further editing by averaging successive data positions for which the time step Δt is less than an hour. The reason for this is quite simple: Because positional accuracies Δx and Δy are about 350 m roughly 63%

of the time, velocity errors are roughly $\Delta x / \Delta t > 0.1$ m/s when $\Delta t < 1$ h. Such error values are comparable to mean ocean currents and need to be eliminated from the records. Drifters located using GPS transmitters have much smaller positional errors and, therefore, better velocity resolution. The time series also need to be examined for drogue-on, drogue-off. If a reliable strain sensor is built into the drogue system, it can be used to determine if and when the drogue fell off. Otherwise, one needs to calculate the speed-squared from the raw data and look for sudden major "jumps" in speed that signal loss of the drogue (Figure 3.21). We recommend this approach for all modern-day drifters since strain gauge sensors appear to be unreliable. At the time this edition of the book was being written, drogue loss and not battery or transmitter failure, was the primary cause of drifter "failure" in the open ocean. Argo drifters are not drogued, but drift at depth where the vertical current shear is generally weak, thereby reducing slippage and associated velocity error.

Provided there are more than about six accurate satellite fixes per day, the edited positional records can be interpolated to regularly spaced 3-h time series using a cubic spline interpolation

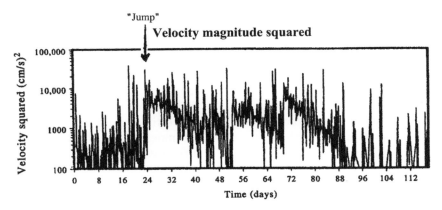

FIGURE 3.21 Sudden "jumps" in the speed-squared values from edited satellite-tracked drifter velocity data collected in the north Pacific near 50° N between 133 and 142° W longitude during the period September 4 to December 30, 1990. the "jump" indicates rapid acceleration of the drifter following probable loss of its drogue.

algorithm. In general, the spline curve will be well behaved and the fit will resemble the kind of curve one would draw through the data by eye. Inertial and tidal loops in the trajectory will be fairly well resolved. Spurious results will occur where data gaps are too large to properly condition the spline interpolation algorithm. Assuming that the spline interpolation of positions looks reasonable, the next step is to calculate the velocity components from the rate of change of position. It is tempting to equate the coefficient for the linear term in the cubic spline interpolation to the "instantaneous" velocity at any location along the drifter trajectory. That would be a mistake. Although trajectories can look quite smooth, curvatures can be large and resulting velocities unrealistic. In fact, use of the spline coefficients to calculate instantaneous velocity components leads to an increase in the kinetic energy of the motions. The reader can verify this by artificially generating a continuous time series of position consisting of a linear trend and time varying inertial motions. The artificial position record is then decimated to 3-hourly values and a cubic spline interpolation scheme applied. Using instantaneous velocity values at the 3-hourly time steps derived from the interpolation, one finds that the kinetic energy in most frequency bands is increased relative to the original record. The recommended procedure is to calculate the two horizontal velocity components (u, v) from the central differences between three consecutive values of the 3-hourly positional data. From first differences, the velocity components at each point "i" are then: $u_i = (x_{i+1} - x_i)/(t_{i+1} - t_i)$ and $v_i = (y_{i+1} - y_i)/(t_{i+1} - t_i)$ for simple two-point differences or for the recommended centered values, $u_i = (x_{i+1} - x_{i-1})/(t_{i+1} - t_{i-1})$ and $v_i = (y_{i+1} - y_{i-1})/(t_{i+1} - t_{i-1})$. In summary, for those oceanic regions subject to pronounced inertial and tidal frequency motions, we have recommended the use of cubic spline interpolation to generate 2–4-hourly time series for position but simple linear interpolation of positional data to generate the corresponding time series

for velocity. The interpolation requires more than six to eight satellite fixes through the day to be successful.

Trajectories with data gaps that are long relative to the local inertial period require special consideration. For gaps associated with a transmitter duty cycle of 8 h "on" followed by 16 h "off," we can obtain accurate daily mean positional values by least-squares fitting a time-varying continuous function to successive segments of the irregular data and then averaging the resulting function over successive 24-h periods. This filtering processes is as follows (see Bograd et al., 1999):

1. Use least squares to fit a specified function, $\xi(t)$, to several (N) successive 8-h days of zonal (or meridional) trajectory data. The general model has the form $\xi(t) = a + bt + ct^2 + dt^3 + a_1 \sin(2\pi ft + \phi_1) + a_2 \sin(2\pi f_2 t + \phi_2)$ where a, b, c, d, a_1, ϕ_1, a_2 and ϕ_2 are the unknown coefficients, f is the local Coriolis frequency and f_2 the semidiurnal frequency (~ 0.081 cph). The phases ϕ_1, ϕ_2 for the two frequencies will vary from segment to segment. We suggest that four to five days ($N = 4$ or 5) of data be used for each segment fit. Shorter segments will have too few data for an accurate least-squares fit; longer segments will result in too much smoothing of the intermittent inertial and tidal motions;

2. Repeat the least-squares operation for each segment of length N days, shifting forward in time by one day after each set of coefficients is determined. This yields one estimate for the first day $\xi_1 = \xi (t = t_1)$, two estimates for the second day, ξ_2, three estimates for the third day and four estimates for all other days until near the end of the record when the number of estimates again falls to unity for the last record, Average all the values in each daily segment for each of the multiple curves $\xi_i(t)$ ($i = 1, ...,$ up to N) to get the average daily latitude $\xi_x(t)$ and longitude $\xi_y(t)$;

3. The pairs of coefficients a_1, ϕ_2 and a_2, ϕ_2 can be used to give rough reconstructions of the inertial and semidiurnal tidal motions, respectively. However, expect the phases to fluctuate considerably from segment to segment due to natural variability in the phases of the motions and from contamination by adjacent frequency bands.

For the duty cycle consisting of one day "on" followed by two days "off," the model is less useful (except at equatorial latitudes) and requires a much longer data segment (say 12 days instead of four) for each least-squares analysis.

3.17.3.3 *Interpolation Records from Nearby Stations*

Provided that the spatial scales of the processes being examined are large compared to the separation between sampling sites, short gaps in the time series at one location can be filled using an identical type of time series from a nearby location. For example, missing hourly tide heights at one coastal tide gauge station can be filled using hourly tide heights from an adjacent station further along the coast. To do this, we first use coincident data segments to determine the relative amplitudes and phases of the time series at the two locations.

A simple cross-correlation analysis can be used to determine the peak time lag between the series while the relative amplitudes can be obtained from the ratio of the standard deviations of the two series. Gaps in one time series (series 1) are then filled by applying the appropriate time lag and amplitude factor to the uninterrupted data series (series 2). Because tide gauges are generally in relatively protected embayments, each of which has its own particular frequency response characteristics, it is not possible to apply the tidal constituents for one site to an adjacent site even if the tidal constituents in the offshore region are nearly identical. A more sophisticated approach would be to first obtain the complex transfer function $H_{12}(\omega) = |H_{12}(\omega)|\exp[i\phi_{12}(\omega)]$ as a function of frequency ω for the two coincident time series. The missing time series values at site 1 could then be filled using the amplitudes $|H_{12}(\omega)|$ and phase differences $\phi_{12}(\omega)$ of the transfer function applied to the uninterrupted data series.

3.18 COVARIANCE AND THE COVARIANCE MATRIX

Covariance, like variance, is a measure of variability. For two variables, the covariance is a measure of the joint variation about a common mean. When extended to a multivariate population, the relevant statistic is the covariance matrix. As we shall see, it is equivalent to what will be introduced later as the "mean product matrix." The covariance and covariance matrix are the fundamental concepts behind the spatial analysis techniques discussed in the next chapter.

3.18.1 Covariance and Structure Functions

The covariance $C(Y_1, Y_2)$, also written as $\text{cov}[Y_1, Y_2]$, between variables Y_1, Y_2 is

$$C(Y_1, Y_2) = E[(Y_1 - \mu_1)(Y_2 - \mu_2)] \quad (3.173)$$

where $\mu_1 = E[Y_1]$ and $\mu_2 = E[Y_2]$. A positive covariance indicates that Y_2 and Y_1 increase and decrease together while a negative covariance has Y_2 decreasing as Y_1 increases, and vice versa. We can expand Eqn (3.173) into a more convenient computational form

$$C(Y_1, Y_2) = E[Y_1 Y_2] - E[Y_1]E[Y_2] \quad (3.174)$$

Note, that if Y_1, Y_2 are independent random variables, then $C[Y_1, Y_2] = 0$.

For a two-dimensional isotropic velocity field, $u_i(y)$, the covariance tensor $C(r)$, also called the *structure function* from earlier studies of turbulence, takes the form

$$C_{ij}(r) = \left\langle u_i(\mathbf{y})u_j(\mathbf{y}+\mathbf{r}) \right\rangle$$
$$= \sigma^2 \frac{[f(r) - g(r)]r_i r_j}{r^2 + g(r)\delta_{ij}} \qquad (3.175)$$

where $\langle \cdot \rangle$ denotes an ensemble average, $r \equiv |\mathbf{r}|$, $\mathbf{y} = (y_1, y_2)$ is the position vector, $f(r)$ and $g(r)$ are, respectively, the one-dimensional longitudinal and transverse correlation functions, and $\sigma^2 = \langle u_i(\mathbf{y})^2 \rangle$. The longitudinal and transverse correlation functions are

$$f(r) = \langle u_L(\mathbf{y})u_L(\mathbf{y}+\mathbf{r}) \rangle \qquad (3.176a)$$

$$g(r) = \langle u_P(\mathbf{y})u_P(\mathbf{y}+\mathbf{r}) \rangle \qquad (3.176b)$$

where $u_L(\mathbf{y})$ and $u_p(\mathbf{y})$ are the velocity fluctuations parallel and perpendicular to $\mathbf{r} = (r_1, r_2)$. The velocity fluctuations are normalized so that the correlations equal unity at $r = 0$. If the two-dimensional flow field is horizontally nondivergent, homogenous and isotropic, then $C_{ij}(r) = 0$ and

$$g(r) = \frac{\mathrm{d}}{\mathrm{d}r}[rf(r)] \qquad (3.177)$$

Freeland et al. (1975) have used Eqn (3.177) to test for two-dimensional, nondivergent, homogenous, and isotropic low-frequency velocity structure in SOFAR (SOunding Fixing and Ranging) float data collected in the North Atlantic. Stacey et al. (1988) used this relation to test for similar flow structure in the Strait of Georgia, British Columbia. Although close to the error limits in certain cases, the observed structure is generally consistent with horizontal, nondivergent, homogeneous, and isotropic flow (Figure 3.22). The dotted lines in Figure 3.22 are the analytical functions

$$f(r) = (1 + br)e^{-br} \qquad (3.178a)$$

$$g(r) = (1 + br - b^2 r^2)e^{-br} \qquad (3.178b)$$

3.18.2 A Computational Example

If Y_1, Y_2 have a joint PDF

$$f(y_1, y_2) = \begin{cases} 2y_1, & 0 \le y_1 \le 1; \ 0 \le y_2 \le 1 \\ 0, & \text{elsewhere} \end{cases}$$
$$(3.179)$$

what is the covariance of Y_1, Y_2? We first write the expected value of Y_1, Y_2 as

$$E[Y_1 Y_2] = \int_0^1 \int_0^1 y_1 y_2 f(y_1, y_2)\, dy_1 dy_2$$

$$= \int_0^1 \int_0^1 y_1 y_2 (2y_1)\, dy_1 dy_2$$

$$= \int_0^1 \frac{1}{3} y_2 \left(2y_1^3\right)\Big|_0^1 dy_2 = \int_0^1 \frac{2}{3} y_2 dy_2 = \frac{2}{3} \frac{y_2^2}{2}\Big|_0^1$$

$$= \frac{1}{3}$$

Recall that, for discrete variables

$$E[g(Y_1, ..., Y_k)] = \sum_{y_k} \cdots \sum_{y_1} g(y_1, ..., y_k) P(y_1, ..., y_k)$$

or for continuous variables

$$E[g(Y_1, ..., Y_k)] = \int_{y_k} \cdots \int_{y_1} g(y_1, ..., y_k) f(y_1, ..., y_k)$$
$$dy_1 ... dy_k$$

For this example, we find $E[Y_1 Y_2] = 1/3$. Now

$$E[Y_1] = \int_0^1 \int_0^1 y_1 (2y_1)\, dy_1 dy_2 = \int_0^1 \frac{2}{3} y_1^3\Big|_0^1 dy_2$$

$$= \frac{2}{3} y_2\Big|_0^1 = \frac{2}{3}$$

and $E[Y_2] = 1/2$, so that $\mathrm{cov}[Y_1 Y_2] = E[Y_1 Y_2] - \mu_1 \mu_2 = 1/3 - (2/3)(1/2) = 0$. Therefore, Y_1 and Y_2 are independent. Of course, we could have anticipated this result since $f(y_1, y_2)$ in Eqn (3.179) is independent of y_2.

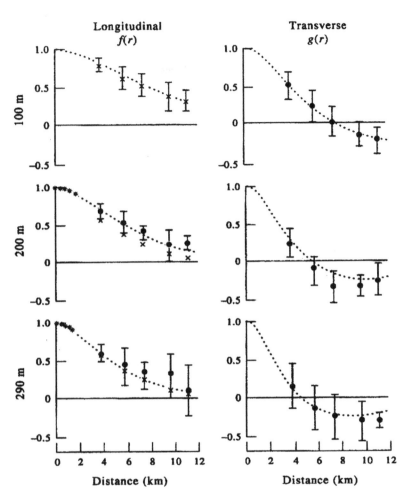

FIGURE 3.22 Longitudinal and transverse correlations at 100, 200, and 280/290 m depths. The dots are measured average values and error bars are the standard deviations. The mean and trend were removed from each time series before calculation of the correlations. The crosses are predicted values of $f(r)$ calculated using Eqn (3.177) by drawing straight line segments between the average values of $g(r)$ and integrating under the curve. (From Stacey et al. (1988.))

3.18.3 Multivariate Distributions

In the case of multivariate distributions, the covariance becomes the *covariance matrix*. If we have n measurements (samples) of N variables (Y), we can describe this as N random variables having a joint N-dimensional PDF.

$$f_{1,2,...,N}(Y_1, Y_2, ..., Y_N) \qquad (3.180)$$

If the random variables, Y, are mutually independent, the joint PDF can be factored in the usual way as

$$f_{1,2,...,N}(Y_1, Y_2, ..., Y_N) = f_1(Y_1)f_2(Y_2)...f_N(Y_N) \qquad (3.181)$$

An important multivariate PDF is the multivariate normal PDF.

$$f_Y(Y) = \frac{1}{(2\pi)^{N/2}|\mathbf{W}|^{1/2}}$$
$$\times \exp\left[-\frac{1}{2}(Y - \mu)^T\mathbf{W}^{-1}(Y - \mu) \right]$$

where $Y^T = (Y_1, Y_2,..., Y_N)$, $\mu^T = (\mu_1, \mu_2,..., \mu_N)$, are the row vectors and $\mathbf{W}^{-1}$ is the inverse of the covariance matrix $\mathbf{W}$

$$\begin{pmatrix} \sigma_1^2 & \sigma_1\sigma_2\rho_{12} & \sigma_1\sigma_3\rho_{13} & ... & \sigma_1\sigma_N\rho_{1N} \\ \sigma_2\sigma_1\rho_{12} & \sigma_2^2 & \sigma_2\sigma_3\rho_{23} & ... & \sigma_2\sigma_N\rho_{2N} \\ ... & ... & ... & ... & ... \\ \sigma_N\sigma_1\rho_{N1} & \sigma_N\sigma_2\rho_{N2} & ... & ... & \sigma_N^2 \end{pmatrix}$$

(3.182)

where $\rho_{ij} = C_{ij}/\sigma_i\sigma_j$ is the correlation function; we can also write this as

$$\mathbf{W} = \begin{pmatrix} V[Y_1] & C[Y_1Y_2] & C[Y_1Y_3] & ... & C[Y_1Y_N] \\ C[Y_2Y_1] & V[Y_2] & C[Y_2Y_3] & ... & C[Y_2Y_N] \\ ... & ... & ... & ... & ... \\ C[Y_NY_1] & C[Y_NY_2] & C[Y_NY_3] & ... & V[Y_N] \end{pmatrix}$$

(3.183)

Note that the covariance $C[Y_iY_j] = C[Y_jY_i]$ and therefore $\mathbf{W}$ is symmetric ($\mathbf{W} = \mathbf{W}^T$). In addition, $\mathbf{W}$ is positive semidefinite; that is, $|\mathbf{W}|$ and all its principal minors are nonnegative. Another way to show this is

$$V[\lambda^T Y] = E\left[\lambda^T (Y - \mu)(Y - \mu)^T\lambda\right] = \lambda^T |\mathbf{W}|$$

(3.184)

which will always be nonnegative for any λ.

3.19 THE BOOTSTRAP AND JACKKNIFE METHODS

Many data series in the natural sciences are nonreproducible and the researcher is left with only one set of observations with which to work. With only one realization of a series, it is impossible to compare it with a related series to determine if they are drawn from the same, or from different, populations. There are numerous oceanographic examples, including tsunami oscillations recorded by a coastal tide gauge, a single seasonal cycle of monthly mean currents at a mooring location, and a trend in long-term temperature data from a climate monitoring station. Marine biologists face similar limitations when analyzing groups of animal species caught in nets or bottom grab samples. The problem is that empirical observations are prone to error and any interpretation of an event must be devised based on statistical measures of the probability of the event. A fundamental measure for testing the validity of any property of a data set is its variance. Parametric statistical models have been developed that help the investigator decide the degree of faith to be placed in a given statistic. However, data and model are often nonlinear so that it is not usually possible to find an analytical expression for model variance in terms of the data variance.

The *parametric* statistical methods presented in the previous sections were institutionalized long before the time of modern digital computers when use of analytical expressions greatly simplified the laborious hand calculation of statistical properties. During the past few decades, *nonparametric* statistical methods have been developed to take advantage of the increasing computational efficiencies of computers. An advantage of the new methods is that they permit investigations of the statistical properties of a sample, which do not conform to a specific analytical model. Equally importantly, they can be applied to small data sets while still providing a reliable estimation of confidence limits on the statistic of interest. "Bootstrapping" and "jackknifing" are two of the more commonly used methods that could not be used effectively until the invention of the digital computer. Both are resampling techniques in which artificial data sets are generated by selection of points from an original set of data. Specifically, we start with a single realization of an "experiment" and from that one set of experimental data we create a multitude of new artificial realizations of the experiment without having to repeat the observations. These realizations are then used to estimate the reliability of the particular statistic of interest, with the underlying assumption that the sample data are representative of the entire population.

In the bootstrapping method, random samples selected during the resampling process are replaced before each new sample is created. As a consequence, any data value has the possibility of being drawn many times. The name bootstrap arises from the expression "to lift oneself up by one's bootstraps." In jackknifing, artificial data sets are created by selectively and systematically removing samples from the original data set. The statistics of interest are recalculated for each resulting truncated data set and the variability among the artificial samples used to describe the variability of these statistics. "Cross-validation" is an older technique. The idea is to split the data into two parts and set one part aside. Curves are fitted to the first part and then tested against values in the second part. Cross-validation consists of determining how well the fitted curves predict the values in the portion of data set aside. The data can be randomly split in many ways and many times in order to obtain the needed statistical reliability. For additional information on this technique, the reader is referred to Efron and Gong (1983).

3.19.1 Bootstrap Method

Introduced by Efron in 1977 (Diaconis and Efron, 1983), bootstrapping provides freedom from two limiting factors that have constrained statistical theory since its beginning: (1) the assumption of normal (Gaussian) data distributions; and (2) the focus on statistical measures whose theoretical properties can be analyzed mathematically. As with other nonparametric methods, bootstrapping is insensitive to assumptions made with respect to the statistical properties of the data and does not need an analytical expression for the connection between model and data statistical properties. Resampling techniques are based on the idea that we can repeat a particular experiment by constructing multiple data sets from the one

measured data set. Application of the resampling procedure must be modeled on a testable hypothesis so that the resulting probability can be used to accept or reject the null hypothesis. The methods can be applied just as well to any statistic, simple or complicated. A *bootstrap sample* is a "copy" of the original data that may contain a certain value (datum, x_n) more than once, once, or not at all (i.e., the number of occurrences of x_n lies between 0 and N, where N is the number of independent data points). Introductions into the bootstrapping procedure can be found in Efron and Gong (1983), Diaconis and Efron (1983), and Tichelaar and Ruff (1989). Nemec and Brinkhurst (1988) apply the method to testing the statistical significance of biological species cluster analysis for which there are duplicate or triplicate samples for each location. Connolly et al. (2009) use bootstrapping methodology to test three different models for species abundance distributions of Indo-Pacific corals and reef fishes.

Suppose that we have N values of a scalar or vector variable, x_n ($1 \leq n \leq N$), whose statistical properties we wish to investigate in relation to another variable. This could be a univariate variable, such as sea-level height $x_n = \eta(t_n)$ at a single location over a period of N time steps, t_n, or the structure of the first mode empirical orthogonal function $\phi_1(x_n)$ as a function of location, x_n. Alternatively, we could be dealing with a bivariate variable (x_{1n}, x_{2n}) such as water temperature vs dissolved oxygen content from a series of vertical profiles. Results apply to any other set of measurements whose statistics we wish to determine. We may want to compare means and standard deviations (variances) of different records to see if they are significantly different. Alternatively, we might want to place confidence limits on the slope of a line derived using a standard least-squares fit to our bivariate data (x_{1n}, x_{2n}), or, determine how much confidence we can have in the coefficients we obtained from the least-squares fit of an annual cycle to a single set of 12 monthly mean current records from a mooring location. Note that if there is a high degree of correlation among the N data values, the N are not statistically independent samples and we are faced with the usual problem of dealing with an effective number of degrees of freedom N^* for the data set.

The procedure is to equate each of our N independent data points with a number produced by a random number generator. We can do this by assigning each of the data values to separate uniform-width bins lying along the line $(-1, +1)$, or $(0, 1)$, depending on the random number generator being used. For N values, there will be N uniform-width bins on the line and each bin will be equated with one of the N data values (Figure 3.23). The bin width is $2/N$ if the line -1 to $+1$ is used. A random number generator such as a Monte Carlo scheme is used to randomly select sequences of N bins corresponding to the multiple bootstrap samples. Suppose that the random number generator picks a number, r, from the range $-1 \leq r \leq 1$. If this number falls into the range of bin k, corresponding to the range $[2(k-1)/N] - 1 \leq r_k \leq (2k/N) - 1$, for $k = 1,\ldots, N$, then the data value x_k assigned to bin k is taken to be one of the samples we need to make up our bootstrap data set. In Figure 3.23, there are 10 data values and 10 corresponding random number segments of length 0.2, with datum value x_1 assigned to the range -1.0 to -0.8, x_2 assigned to -0.8 to -0.6, and so on. Since bootstrapping works with replacement, it is quite possible that the random number generator will come up with the same bin several times, or not at all. The first N data values from our resampling constitute the first bootstrap sample. The process is then repeated again and again until hundreds or thousands of bootstrap samples have been generated. Diaconis and Efron (1983) discuss making a US billion bootstrap samples. They also take another approach. Instead of generating one bootstrap sample at a time by equating bins along the real line $(-1, 1)$ with N samples, they generate all the needed multiple copies of all the N data values (say one million copies of each of the original data values or data points) and place them all in a rotating "lotto" bin. They then reach in and pull out all the requisite number of N-value bootstrap samples from the shuffled points, being careful to throw each data point back into the bin before selecting the next value. This requires some sort of label for each value in the bin based on a random selection process that can identify a data point that has been selected.

Although bootstrapping has yet to find widespread application in the marine sciences, there are several noteworthy examples in the literature. Enfield and Cid (1990) examined the stationarity of different groupings of El Niño recurrence rates based on the chronology of Quinn et al. (1987). For example, group 1 consisted of all strong (S) and very strong (VS) events for the period 1525–1983, while groups 4 and 5 consisted of S/VS events for times of high and low solar activity for this period. Groups 6–10 contained different samples of intensities for the modern period of 1803–1987. Maximum likelihood

FIGURE 3.23 The assignment (binning) of observed data values x_n ($n = 1,\ldots, 10$) to 10 range values of the random number, r_k ($k = 1,\ldots, 10$). For each bootstrap sample of 10 values, 10 random numbers are selected and located according bin range. The datum values x_n assigned to each range are then used to form the bootstrap sample.

estimation was used to fit a two-parameter Weibull distribution $f(t)$ to each sample group,

$$f(t) = \left(\beta t^{\beta-1} / \tau^{\beta}\right) \exp\left[-(t/\tau)^{\beta}\right] \quad (3.185)$$

where β and τ are, respectively, the shape (peakedness) and timescale (RMS return interval) parameters, and t is the random variable for the return interval. For each group, only a single distribution could be fitted. To derive estimates of the mean and standard deviations of the parameters for each group, 500 bootstrap samples were generated and the Weibull parameters obtained for each sample. As indicated by Figure 3.24, this number of samples provides good convergence to the mean value for the Weibull distribution fit for each group. The distribution of El Niño return events for bootstrap samples for all intensities for the "early modern" period 1803–1891 is shown in Figure 3.25 along with its corresponding Weibull distribution. Enfield and Cid (1990) use the resampling analysis to show that, for the groups associated with times of low solar activity and those associated with times of high solar activity, there is comparatively little overlap between the bootstrap-derived frequency histograms and mean return timescales, τ (years) (Figure 3.26). These results suggest that there is a statistical

difference in the return times for the two groups and that return times are nonstationary.

Much of the present evidence for global warming is based on Northern Hemispheric annual surface air temperature records over the past 100 years (Jones et al., 1986; Hansen and Lebedeff, 1987; Gruza et al., 1988; Mann et al., 1998; IPCC, 2007). Interest in the reliability of the means and trends of these records (labeled H, J, and G) prompted Eisner and Tsonis (1991) to examine differences in means and trends of pairs of these records for the three global mean temperature curves. The data sets have been constructed using different averaging methods and different observational data bases. Data set H contains only observations from land stations whereas data set J uses both land-and ship-based observations. Averages for set H are derived using equal area boxes over the globe whereas data set G is constructed by visual inspection of anomalies from sea-level temperature analyses. The usual assumption is that these time series are representative of the same population, a result that appears to be supported by the statistically significant correlation $r > 0.79$ among the different curves. As pointed out by Eisner and Tsonis, however, the presence of trends in these data means that the linear cross-correlation coefficient may not be a reliable measure of the

Estimation of the population mean distribution parameters (Bootstrap method: resampling with replacement)

$\tau = 9.56$ years
$\sigma_{\tau} = \pm 0.95$ years

Mean (years)

Number of bootstrap samples

FIGURE 3.24 Estimation of the population mean distribution parameters (mean return time in years) using the bootstrap method for El Niño events taking place during times of low solar activity for the period 1525–1983. τ is the return time and σ_{τ} its standard deviation. *(From Enfield and Cid (1990)).*

FIGURE 3.25 Histogram of El Niño return times for all events between 1803 and 1987 (group #7) derived using the bootstrapping resampling technique. The solid curve is the Weibull distribution fitted to the histogram. The modal and mean return intervals (3.3 and 3.8 years, respectively) are the derived from the MLE-estimated population parameters. *(From Enfield and Cid (1990).)*

FIGURE 3.26 Histograms and fitted Weibull distributions obtained using the bootstrapping method. Plots show the occurrences of El Niño events for the times of low and high solar flare activity for (a) Strong and very strong El Niño events, only; (b) All El Niño events.

covariability of the records. Two questions can be addressed using the bootstrapping method: (1) are the three versions of the temperature records significantly different that we can say they are not drawn from the same population? (The null hypothesis is false.); and (2) are the trends in the three records sufficiently alike that they are measuring a true rise in global temperature?

Because of the strong linear correlation in the records, the authors work with difference records. A difference record is constructed by subtracting the annual (mean removed) departure record of one data set from the annual departure record of another data set. Although not zero, the cross-correlation for the difference records is considerably less than those for the original departure records, showing that differencing is a form of high-pass filtering that effectively reduces biasing from the trends. The average difference for all 97 years of data used in the analyses (the difference record H-J relative to the years 1951–1980) is −0.05 °C, indicating that the hemispheric temperatures of Jones et al. (1986) are slightly warmer than those of Hansen and Lebedeff (1987). Similar results were obtained for H-G and J-G. To see if these differences are statistically significant, 10,000 bootstrap samples of the difference records were generated. The results (Figure 3.27(a)) suggest that all three hemispheric temperature records exhibit significantly different nonzero means. The overlap in the distributions is quite minimal. The same process was then used to examine the trends in the difference records. For the H-J record, the trend is +0.15°C/century so that the trend of Hansen and Lebedeff is greater than that of Jones *et al.* As indicated in Figure 13.27(b), the long-term trends were distinct. On the basis of these results, the authors were forced to conclude that at least two of the data sets do not represent the true population (i.e., the truth). More generally, the results bring into question the confidence one can have that the long-term temperature trends obtained

from these particular data are representative of trends over hemispheric or global scales.

Biological oceanographers often have difficult sampling problems that can be addressed by bootstrap methods. For example, the biologist may want to use cluster analyses of animal abundance for different locations to see if species distributions differ statistically from one sampling location (or time) to the next. Cluster analyses of ecological data use dendrograms—linkage rules which group samples according to the relative similarity of total species composition—to determine if the organisms in one group of samples have been drawn from the same or different statistical assemblages of those of another group of samples. Provided there are, at least, replicates for most samples, bootstrapping can be used to derive tests for statistical significance of similarity linkages in cluster analyses (Burd and Thomson, 1994). In a more recent study, Connolly et al. (2009) derive and use several alternative bootstrap analyses to test the ability of three different species abundance models to characterize ecological assemblages of Indo-Pacific corals and reef fishes. For further information on this aspect of bootstrapping, the reader is referred to Nemec and Brinkhurst (1988). Finally, in this section, we note that it is possible to vary the bootstrap size by selecting samples smaller than N, the original size of the data set, to compare various estimator distributions obtained from different sample sizes. This allows one to observe the effects of varying sample size on sample estimator distributions and statistical power.

3.19.2 Jackknife Method

Several other methods are similar in concept to bootstrapping but differ significantly in detail. The idea, in each case, is to generate artificial data sets and assess the variability of a statistic from its variability over all the sets of artificial data. The methods differ in the way they generate the artificial data. Jackknifing

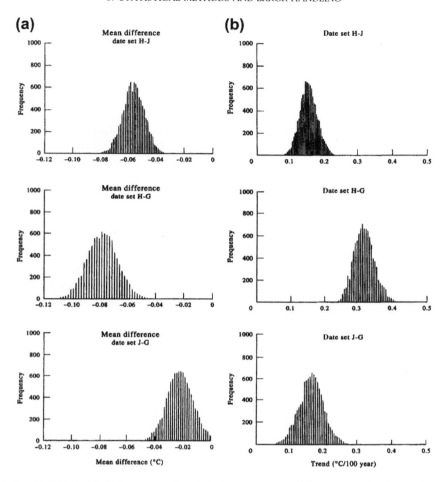

FIGURE 3.27 Bootstrap-generated histograms of global air temperature difference records obtained by subtracting the temperature records of Jones et al. (1986) (J), Hansen and Lebedeff (1987) (H), and Gruza et al. (1988) (G). (a) Frequency distributions of the mean differences plotted for 10^4 bootstrap samples. The x-axis (ordinate) gives the number of times the bootstrap mean fell into a given interval. All three distributions are located to the left of zero mean difference. (b) Same as (a), but for slope (trend) of the temperature difference curves. All three distributions are separated from zero indicating significant differences between long-term surface temperature trends given by each of the three data sets. *(From Elsner and Tsonis (1991).)*

differs from bootstrapping in that data points are not replaced prior to each resampling. This technique was first proposed by Maurice Quenouille in 1949 and developed by John Tukey in the 1950s. The name "jackknife" was used by Tukey to suggest an all-purpose statistical tool.

A jackknife resample is obtained by deleting a fixed number of data points (j) from the original set of N data points. For each resample, a different group of j values is removed so that each resample consists of a distinct collection of data values. In the "delete-j" jackknife sample, there will be $k = N - j$ samples in each new

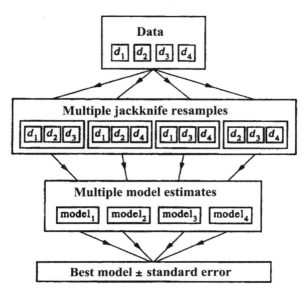

FIGURE 3.28 Schematic representation of the jackknife. The original data vector has four components (samples), labeled d_1 to d_4. The data are resampled by deleting a fixed number of components (here, one) from the original data to form multiple jackknife resamples (in case, four). Each resample defines a model estimate. The multiple model estimates are then combined to a best model and its standard deviation. *(From Tichelaar and Ruff (1989).)*

truncated data set. The total number of new artificial data that can be generated is

$$\binom{N}{j}$$

which the reader will recognize as $_NP_j = N!/(N-j)!$, the number of permutations of N objects taken j at a time. Consider the simple delete-1 jackknife. In this case, there are $N-1$ samples per artificial data set and a total of $_NP_j = N$ new data sets that can be created by systematically removing one value at a time. As illustrated by Figure 3.28, an original data set of four data values will yield a total of four distinct delete-1 jackknife samples, each of size three (3), which can then be used to examine various statistics of the original data set. The sample average of the data derived by deleting the ith datum, denoted by the subscript (i), is

$$\bar{x}_{(i)} = \frac{N\bar{x} - x_i}{N-1} = \frac{1}{N-1} \sum_{j \neq 1}^{N} x_j \qquad (3.186)$$

where

$$\bar{x} = \frac{1}{N} \sum_{i=1}^{N} x_i$$

is the mean found using all the original data. The average of the N jackknife averages, $\bar{x}_{(j)}$, is

$$\bar{x}^* = \frac{1}{N} \sum_{i=1}^{N} \bar{x}_{(i)} = \bar{x} \qquad (3.187)$$

The last result, namely that the mean of all the jackknife samples is identical to the mean of the original data set, is easily obtained using Eqn (3.186). The estimator for the standard deviation, σ_j, of the delete-1 jackknife is

$$\sigma_j = \sum_{i=1}^{N} \left[\left(\bar{x}_{(i)} - \bar{x}^* \right)^2 \right]^{1/2} \qquad (3.188a)$$

$$= \frac{1}{N-1} \sum_{i=1}^{N} \left[(x_i - \bar{x})^2 \right]^{1/2} \qquad (3.188b)$$

where Eqn (3.188b) is the usual expression for the standard deviation of N data values. Our expression differs slightly from that of Efron and Gong (1983), who use a denominator of $1/[(N-1)N]$ instead of $1/(N-1)^2$ in their definition of variance. The advantage of Eqn (3.188a) is that it can be generalized for finding the standard deviation of any estimator θ that can be derived for the original data. In particular, if θ is a scalar, we simply replace $x_{(i)}$ with $\theta_{(i)}$ and $x*$ with $\theta*$ where $\theta_{(i)}$ is an estimator for θ obtained for the data set with the ith value removed. Although the jackknife requires fewer calculations than the bootstrap, it is less flexible and at times less dependable (Efron and Gong, 1983). In general, there are N jackknife samples for the delete-1 jackknife as compared with

$$2N-1 P_N = \binom{2N-1}{N}$$

bootstrap points.

Our example of jackknifing is from Tichelaar and Ruff (1989), who generated $N = 20$ unequally spaced data values y_i that follow the relation $y_i = c x_i + \varepsilon_i$ ($c = 1.5$, exactly), where ε_i is a noise component drawn from a "white" spectral distribution with a normalized standard deviation of 1.5 and mean of zero. The least squares estimator for the standard deviation of the slope is

$$\widehat{\sigma} = \sum_{i=1}^{N}\left[(y_i - \widehat{c}x_i)^2\right] \Bigg/ \left[(N-1)\sum_{i=1}^{N} x_i^2\right]$$

(3.189)

where $\widehat{c} = \sum y_i x_i / \sum x_i^2$. Two jackknife estimators were used: (1) the delete-1 jackknife, for which the artificial sample sizes are $N-1 = 19$; and (2) the delete-half ($N/2$) jackknife for which

normalized frequency of occurrence distribution, f, for the delete-1 jackknife which yields $\widehat{c} = 1.518 \pm 0.0136$; (c) As in (b) but for the delete-half jackknife for which $\widehat{c} = 1.517 \pm 0.0141$; (d) The corresponding bootstrapping estimate, for which $\widehat{c} = 1.517 \pm 0.0132$. Note the subtle difference in distribution between (b) and (c). The dashed line is analytical distribution of $\widehat{c}$ *(From Tichelaar and Ruff (1989).)*

FIGURE 3.29 Use of the bootstrapping technique to estimate the reliability of a linear regression line. (a) A least-squares fit through the noisy data, for which the estimated slope $\widehat{c} = 1.518 \pm 0.0138$ (± 1 standard error); (b) The

the sample sizes are $N - N/2 = 10$. In both cases, the jackknife resamples had equal weighting in the analysis. For the delete-half jackknife, a Monte Carlo determination of 100 subsamples was used since the total samples $_{20}P_{10} = 20!/10!$ is very large. The results are presented in Figure 3.29. The last panel gives the corresponding result for the bootstrap estimate of the slope using 100 bootstrap samples. Results showed that the bootstrap standard error of the slope was slightly lower than those for both jackknifing estimates.

4

The Spatial Analyses of Data Fields

A fundamental problem in oceanography is how best to represent spatially distributed data (or statistical products computed from these data) in such a way that dynamical processes or their effects can best be visualized. As in most aspects of observational analysis, there has been a dramatic change in the approach to this problem due to the increased abundance of digital data and the ability to process them. Prior to the use of digital computers, data displays were constructed by hand and "contouring" was an acquired skill of the descriptive analyst. Hand contouring is still practiced by a few scientists today although, more likely, the data points being contoured are averaged values produced by a computer. In other applications, the computer not only performs the averaging but also uses objective statistical techniques to produce both the gridded values and the associated contours. This can lead to some strange results, as computers will contour values even when it is not justified by the data values or distribution.

The purpose of this section is to review data techniques and procedures designed to reduce spatially distributed data to a level that can be visualized easily by the analyst. We will discuss methods that address both spatial fields and time series of spatial fields since these are the primary modes of data distribution encountered by

the oceanographer. Our focus is on the more widely used techniques, which we present in a practical fashion, stressing the application of the method for interpretive applications.

4.1 TRADITIONAL BLOCK AND BULK AVERAGING

A common older method for deriving a gridded set of data is simply to average the available data over an arbitrarily selected rectangular grid. This averaging grid can lie along any chosen surface but is most often constructed in the horizontal or vertical plane. Because the grid is often chosen for convenience, without any consideration to the sampling coverage, it can lead to an unequal distribution of samples per grid "box." For example, because distance in longitude varies as the cosine of the latitude, the practice of gridding data by 5° or 10° squares in latitude and longitude may lead to increasingly greater spatial areas to be covered at low latitudes. Although this can be overcome somewhat by converting to distances using the central latitude of the box (Poulain and Niiler, 1989), it is easy to see that inhomogeneity in the sampling coverage can quickly nullify any of the useful assumptions made earlier about the Gaussian

313

nature of sample populations or, at least, about the set of means computed from these samples. This is less of a problem with satellite-tracked drifter data since satellite ground tracks converge with increasing latitude, allowing the data density in boxes of fixed longitude length to remain nearly constant.

With markedly different data coverage between sample regions, we cannot always compare the values computed in these squares with the same level of statistical confidence. At best, one must be careful to consider properly the amount of data being included in such averages and be able to evaluate possible effects of the variable data coverage on the mapped results. Each value should be associated with a sample size indicating how many data points, N, went into the computed mean. This will not dictate the spatial or temporal distributions of the sample data field but will, at least, provide a sample size parameter, which can be used to evaluate the mean and standard deviation at each point. While the standard deviation of each grid sample is composed of both spatial and temporal fluctuations (within the time period of the grid sample), it does give an estimate of the inherent variability associated with the computed mean value.

Despite problems with nonuniform data coverage, it has proven worthwhile to produce maps or cross sections with simple grid-averaging methods since they frequently represent the best spatial resolution possible with the existing data coverage. The approach is certainly simple and straightforward. Besides, the data coverage often does not justify more complex and computer-intensive data reduction techniques. Specialized block-averaging techniques have been designed to improve the resolution of the corresponding data by taking into account the nature of the overall observed global variability and by trying to maximize the coverage appropriately. For example, averaging areas in offshore regions are frequently selected which have narrow

meridional extent and wide zonal extent, taking advantage of the generally stronger meridional gradients observed in the ocean. Thus, an averaging area covering 2° latitude by 10° longitude may be used to better resolve the meridional gradients, which dominate the open ocean (Wyrtki and Meyers, 1975). This same idea may be adapted to more limited regions, such as continental margins, if the general oceanographic conditions are known. If so, the data can be averaged accordingly, providing improved resolution perpendicular to strong frontal features. For example, the averaging areas in the Benguela Current System off southwest Africa or the California Current System off the west coast of North America would be narrower in offshore directions at right angles to the coast than in the alongshore direction. A further extension of this type of grid selection would be to base the entire averaging area selection on the data coverage. This is difficult to formalize objectively since it requires the subjective selection of the averaging scheme by an individual. However, it is possible in this way to improve resolution without a substantial increase in sampling (Emery, 1983).

All of these bulk- or block-averaging techniques make the assumption that the data being considered in each grid box are statistically homogeneous and isotropic over the region of study. Under these assumptions, area sample size can be based strictly on the amount of data coverage (number of data values) rather than having to know details about processes represented by the data. Statistical homogeneity does not require that all the data were collected by the same instrument having the same sampling characteristics. Thus, grid-square averaging can include data from many different instruments, which generally have the same error limits.

One must be careful when averaging different kinds of measurements, even if they are of the same parameter. It is very tempting, for example, to average mechanical bathythermograph (MBT) temperatures with newer expendable bathythermograph (XBT) temperatures to produce temperature maps at specific depths. Before doing so, it is worth remembering that XBT data are likely to be accurate to 0.1 °C, as reported earlier (at least in temperature), while MBT data are decidedly less accurate and less reliable. Another marked difference between the two instruments is their relative vertical coverage. While most MBTs stopped at 250 m depth, XBTs are good to 500–1800 m, depending on probe type. Thus, temperature profiles from MBTs can be expected to be different from those collected with XBTs. Any mix of the two will necessarily degrade the average to the quality of the MBT data and bias averages to shallow (<300 m) depths. In some applications, the level of degraded accuracy will be more than adequate and it is only necessary to state clearly and be aware of the intended application when mixing the data from these instruments. Also, one can expect distinct discontinuities as the data make the transition from a mix of measurements at shallower levels to strictly XBT data at greater depths. This same argument holds when mixing XBT temperature profiles with much more accurate but less plentiful Conductivity-Temperature-Depth (CTD) profiles.

Other important practical concerns in forming block averages have to do with the usual uneven geographic location of oceanographic measurements. Consider the global distribution of all autumn oceanographic research measurements up to 1970 of the most common oceanographic observation, temperature profiles (Figure 4.1). It is surprising how frequently these observations lie along meridians of latitude or parallels of longitude. This makes it difficult to assign the data to any particular 5° or 10° square when the border of the square coincides with integer values of latitude or longitude. When the latter occurs, the investigator must decide to which square the borders will be assigned and be consistent in carrying this definition through the calculation of the mean values.

As illustrated by Figure 4.1, data coverage can be highly nonuniform. In this example, some areas were frequently sampled while others were seldom (or never) occupied. There was, and still is, a distinct concentration of measurements off the west coast of the U.S. near the location of Scripps Institution of Oceanography and there is an abundance of observations around Japan reflecting that country's interest in oceanographic research. Such nonuniformity in data coverage is a primary factor in considering the representativeness of simple block averages. It certainly brings into question the assumptions of homogeneity (spatially uniform sampling distribution) and isotropy (uniform sampling regardless of direction) since the sample distribution varies greatly with location and may often have a preferred orientation. The situation becomes even more severe when one examines the quality of the data in the individual casts represented by the dots in Figure 4.1. In order to establish a truly consistent data set in terms of the quality of the observations (i.e., the depth of the cast, the number of samples, the availability of oxygen and nutrients, and so on), it is generally necessary to reject many of the available hydrographic casts. The situation is changing with the advent of major oceanographic initiatives such as Argo, which has a dedicated objective to "… provide a quantitative description of the changing state of the upper ocean and the patterns of ocean climate variability from months to decades, including heat and freshwater storage and transport." (http://www.argo.ucsd.edu/About_Argo.html).

The question of data coverage depends on the kind of scientific questions the data set is being asked to address. For problems not requiring high-quality hydrographic or CTD profile stations, a greater number of observations are available, while for more restrictive studies requiring a higher accuracy, far fewer casts would match the qualifications. This is also true for other types of historical data but is less true of newly collected data. However, even now, one must

FIGURE 4.1   The global distribution of all temperature profiles collected during oceanographic surveys in the fall up to 1970. Sample is most dense along major shipping routes.

ensure that all observations have a similar level of accuracy and reliability. Variations in equipment performance, such as sensor response or failure, must be compensated for in order to keep the observations consistent. Also, changes in instrument calibration need to be taken into account over the duration of a sampling program. For example, transmissometer lenses frequently become matted with a biotic film that reduces the amount of light passing between the source and receiver lenses. A nonlinear, time-dependent calibration is needed to correct this effect.

Despite the potential problems with the block-averaging approach to data presentation, much information can be provided by careful consideration of the data rather than the use of more objective statistical methods to judge data quality. The shift to statistical methods represents a transition from the traditional oceanographic efforts of the early part of the twentieth century when considerable importance was given to every measurement value. In those days, individual scientists were personally responsible for the collection, processing, and quality of their data. Then, it was a simple task to differentiate between "correct" and "incorrect" samples without having to resort to statistical methods to indicate how well the environment had been observed. In addition, earlier investigations were primarily concerned with defining the mean state of the ocean. Temporal variability was sometimes estimated but was otherwise ignored in order to emphasize the mean spatial field. With today's large volumes of data, it is no longer possible to "hand check" each data value. A good example is provided by satellite-sensed information, which generally consists of large groupings of data that are usually treated as individual data values.

In anticipation of our discussion of filtering in Chapter 5, we point out that block averaging corresponds to the application of a box-car-shaped filter to the data series. This type of filter has several negative characteristics such as a slow filter roll-off and large side lobes, which can distort the information in the original data series.

4.2 OBJECTIVE ANALYSIS

In a general sense, *objective analysis* is an estimation procedure, which can be specified mathematically. The form of objective analysis most widely used in physical oceanography is that of least squares optimal interpolation, more appropriately referred to as *Gauss—Markov smoothing*, which is essentially an application of the linear estimation (smoothing) techniques discussed in Chapter 3. Since it is generally used to map spatially nonuniform data to a regularly spaced set of gridded values, Gauss—Markov smoothing might best be called "Gauss—Markov mapping." The basis for the technique is the Gauss—Markov theorem, which was first introduced by Gandin (1965) to provide a systematic procedure for the production of gridded maps of meteorological parameters. If the covariance function used in the Gauss—Markov mapping is the covariance of the data field (as opposed to a more ad hoc covariance function based on historical information, as is often the case), then Gauss—Markov smoothing is optimal in the sense that it minimizes the mean square error of the objective estimates. A similar technique, called "Kriging" after a South African engineer Daniel G. Krige, was developed in mining engineering (see Section 4.3). Oceanographic applications of objective analysis are provided by Bretherton et al. (1976), Freeland and Gould (1976), Bretherton and McWilliams (1980), Hiller and Käse (1983), Bennett (1992), and others.

The two fundamental assumptions in optimal interpolation are that the statistics of the subject data field are stationary (unchanging over the sample period of each map) and homogeneous (the same characteristics over the entire data field). A further assumption often made to simplify the analysis is that the statistics of the second moment, or covariance function, are isotropic (the same structure in all directions). Bretherton et al. (1976) point out that if these statistical characteristics are known, or can be estimated for some existing data field (such as a climatology based on historical data), they can be used to design optimum measurement arrays to sample the field. Since the optimal estimator is linear and consists of a weighted sum of all the observations within a specified range of each grid point, the objective mapping procedure produces a smoothed version of the original data field that will tend to underestimate the true field. In other words, if an observation point happens to coincide with an optimally interpolated grid point, the observed value and interpolated value will probably not be equal due to the presence of noise in the data. The degree of smoothing is determined by the characteristics of the signal and error covariance functions used in the mapping and increases with increasing spatial scales for a specified covariance function.

The optimal gridding process is complicated by several factors, including the fact that the ability of the interpolation procedure to generate representative maps decreases as the number of measurements being used goes down. In physical oceanography, observations are often very sparse. As indicated in Figure 4.1, oceanographic observations are generally not evenly spaced and are frequently few in number. This inhomogeneity is characteristic of many regions of the ocean and drives the need for objective techniques for mapping the variables of interest.

The general problem is to compute an estimate $\widehat{D}(\mathbf{x}, t)$ of the scalar variable $D(\mathbf{x}, t)$ at a position $\mathbf{x} = (x, y)$ from irregularly spaced and inexact observations $d(\mathbf{x}_n, t)$ at a limited number of data positions $\mathbf{x}_n$ ($n = 1, 2, ..., N$). These observations are "inexact" in that they are subject to various types of errors, such as instrumental noise and environmental variability, and to "errors" associated with the least-squares fit analysis used in the optimum interpolation procedure. Implementation of the

procedure requires *a priori* knowledge of the variable's covariance function, $C(\mathbf{r})$, and uncorrelated error variance, ε, where $\mathbf{r}$ is the spatial separation between positions. For isotropic processes, $C(\mathbf{r}) \rightarrow C(r)$, where $r = |\mathbf{r}|$. Although specification of the covariance matrix should be founded on the observed structure of oceanic variables, selection of the mathematical form of the covariance matrix is hardly an "objective" process even with reliable data (cf. Denman and Freeland, 1985). In addition to the assumptions of stationarity, homogeneity, and isotropy, an important constraint on the chosen covariance matrix is that it must be positive definite (no negative eigenvalues). Bretherton et al. (1976) report that objective estimates computed from nonpositive-definite matrices are not optimal and the mapping results are poor. In fact, nonpositive-definite covariance functions can yield objective estimates with negative expected square error. One way to ensure that the covariance matrix is positive definite is to fit a function, which results in a positive definite covariance matrix to the sample covariance matrix calculated from the data (Hiller and Käse, 1983). This results in a continuous mathematical expression to be used in the data weighting procedure. In attempting to specify a covariance function for data collected in continental shelf waters, Denman and Freeland (1985) further required that $\partial^2 C / \partial x^2$ and $\partial^2 C / \partial y^2$ be continuous at $r = 0$ (to ensure a continuously differentiable process) and that the variance spectrum, $S(k)$, derived from the transform of $C(\mathbf{r})$ be integrable and nonnegative for all wavenumbers, $\mathbf{k}$ (to ensure a realizable stochastic random process).

Calculation of the covariance matrix requires that the mean and "trend" be removed from the data (the trend is not necessarily linear). In three-dimensional space, this amounts to the removal of a planar or curvilinear surface. For example, the mean density structure in an upwelling domain is a curved surface, which is shallow over the outer shelf and deepens seaward. Calculation of the density covariance matrix for such a region first involves removal of the curved mean density surface (Denman and Freeland, 1985). Failure to remove the mean and trend would not alter the fact that our estimates are optimal but it would redistribute variability from unresolved larger scales throughout the wavenumber space occupied by the data. We would then map features that have been influenced by the trend and mean.

As discussed later in the section on time series, there are many ways to estimate the trend. If ample good-quality historical data exist, the trend can be estimated from these data and then subtracted from the data being investigated. If historical data are not available, or the historical coverage is inadequate, then the trend must be computed from the sample data set itself. Numerous methods exist for calculating the trend and all require some type of functional fit to the existing data using a least-squares method. These functions can range from straight lines to complex higher-order polynomials and associated nonlinear functions. We note that, although many candidate oceanographic data fields do not satisfy the conditions of stationarity, homogeneity, and isotropy, their anomaly fields do. In the case of anomaly fields, the trend and mean have already been removed. Gandin (1965) reports that it may be possible to estimate the covariance matrix from existing historical data. This is more often the case in meteorology than in oceanography. In most oceanographic applications, the analyst must estimate the covariance matrix from the data set being studied. Regardless of the approach used, the practice of dealing with anomalies rather than with the observations themselves can make the optimum interpolation procedure considerably more accurate.

In the following, we present a brief outline of objective mapping procedures. The interested reader is referred to Gandin (1965) and Bretherton et al. (1976) for further details. As noted

previously, we consider the problem of constructing a gridded map of the scalar variable, $D(\mathbf{x}, t)$, from an irregularly spaced set of scalar measurements, $d(\mathbf{x}, t)$, at positions $\mathbf{x}$ and times t. The notation $\mathbf{x}$ refers to a suite of measurement sites, x_n ($n = 1, 2, \ldots$), each with distinct (x, y) coordinates. We use the term "variable" to mean directly measured oceanic variables as well as calculated variables such as the density or streamfunction derived from the observations. Thus, the data, $d(\mathbf{x}, t)$, may consist of measurements of the particular variable we are trying to map or they may consist of some other variables that are related to D in a linear way. The former case gives

$$d(\mathbf{x}, t) = D(\mathbf{x}, t) + \varepsilon(\mathbf{x}) \qquad (4.1)$$

where the ε are zero-mean measurement errors, which are not, correlated with the gridded variable D or its anomaly. In the latter case

$$d(\mathbf{x}, t) = F[D(\mathbf{x}, t)] + \varepsilon(\mathbf{x}) \qquad (4.2)$$

in which F is a linear function that acts on the function D in a linear fashion to give a scalar (Bennett, 1992). For example, if $D(\mathbf{x}, t) = \Psi(\mathbf{x}, t)$ is the streamfunction, then the data could be current meter measurements of the zonal velocity field, $u(\mathbf{x}, t) = F[\Psi(\mathbf{x}, t)]$, where

$$d(\mathbf{x}, t) = u(\mathbf{x}, t) + \varepsilon(\mathbf{x}) = -\frac{\partial \Psi(\mathbf{x})}{\partial y} + \varepsilon(\mathbf{x}) \quad (4.3)$$

and $\partial \Psi / \partial y$ is the gradient of the streamfunction in the meridional direction.

To generalize the objective mapping problem, we assume that mean values have *not* been removed from the original data prior to the analysis. If we consider the objective mapping for a single "snap shot" in time (thereby dropping the time index, t), we can write linear estimates $\widehat{D}(\mathbf{x})$ of $D(\mathbf{x})$ as the summation over a weighted set of the measurements d_i ($i = 1, \ldots, N$)

$$\widehat{D}(\mathbf{x}) = \overline{D}(\mathbf{x}) + \sum_{i=1}^{N} b_i(d_i - \overline{d}) \qquad (4.4)$$

where the overbar denotes an expected value (mean), $d_i = d(\mathbf{x}) = d(x_i)$, and $1 \leq i \leq N$ is shorthand notation for the data values, and the $b_i = b(\mathbf{x}) = b(x_i)$ are, as yet unspecified, weighting coefficients at the data points x_i. The selection of the N data values is made by restricting these values to some finite area about the grid point. Often called "Cressman Weights," the b_i depends only on the distance between the grid point and the location of the observation and not on the observation itself. The search radius selected defines the length scale over which an observation is used in the interpolation to a specified grid point. This search radius is chosen by the user and can be made to vary in space depending on the data coverage and the inherent horizontal physical scales. The estimates of the parameters b_i in Eqn (4.4) are found in the usual way by minimizing the mean square variance of the error $e(\mathbf{x})^2$ between the measured variable, D, and the linear estimate, $\widehat{D}$, at the data location. In particular,

$$\overline{e(\mathbf{x})^2} = \overline{\left[D(\mathbf{x}) - \widehat{D}(\mathbf{x})\right]^2} \qquad (4.5)$$

which on substitution of Eqn (4.4) yields

$$\overline{e(\mathbf{x})^2} = \overline{[D(\mathbf{x}) - \overline{D}(\mathbf{x})]^2}$$
$$+ \sum_{i=1}^{N}\sum_{j=1}^{N} b_i b_j \overline{(d_i - \overline{d})(d_j - \overline{d})} \qquad (4.6)$$
$$- 2\sum_{i=1}^{N} b_i \overline{(d_i - \overline{d})(D - \overline{D})}$$

Note, that if the mean has been removed, we can set $\overline{D}(\mathbf{x}) = \overline{d}(\mathbf{x}) = 0$ in Eqn (4.6). The mean square difference in Eqn (4.6) is minimized when

$$b_i = \sum_{j=1}^{N}\left\{ \left[\overline{(d_i - \overline{d})(d_j - \overline{d})}\right]^{-1} \overline{(d_j - \overline{d})(D - \overline{D})} \right\} \qquad (4.7)$$

To calculate the weighting coefficients in Eqn (4.7), and therefore the grid-value estimates in Eqn (4.4), we need to compute the covariance

matrix by averaging over all possible pairs of data taken at points x_i, x_j; the covariance matrix is

$$\overline{(d_i - \overline{d})(d_j - \overline{d})} = \overline{(d(x_i) - \overline{d})(d(x_j) - \overline{d})} \quad (4.8)$$

We do the same for the interpolated value

$$\overline{(d_j - \overline{d})(D - \overline{D})} = \overline{(d(x_j) - \overline{d})(d(x_k) - \overline{D})}$$
$$(4.9)$$

where x_k is the location vector for the grid point estimate $\widehat{D}(x_k)$. This is a key step in optimum interpolation whereby the computation of the covariance function at the grid value $\widehat{D}(x_k)$ depends only on the distances between the measurement locations and the positions of the grid values. We should also note that since the covariance is for the anomaly or difference from the mean it could also be considered as the "error covariance."

In general, we need a series of measurements at each location so that we can obtain statistically reliable expected values for the elements of the covariance matrices in Eqns (4.8) and (4.9). The expected values in the above relations could be computed as ensemble averages over spatially distributed sets of measurements. Typically, however, we have only one set of measurements for the specified locations x_i, x_j. As a consequence, we need to assume that, for the region of study, the data statistics are homogeneous, stationary, and isotropic. If these conditions are met, the covariance matrix for the data distribution (for example, sea surface temperature (SST)) depends only on the distance r between data values, where $r = |x_j - x_i|$. Thus, we have elements i, j, of the covariance matrix given by

$$\overline{(d_i - \overline{d})(d_j - \overline{d})} = C(|x_j - x_i|) + \overline{\varepsilon^2}$$
$$(4.10)$$
$$\overline{(d_j - \overline{d})(D - \overline{D})} = C(|x_j - x_k|) + \overline{\varepsilon^2}$$

where $C(|\mathbf{r}|) = \overline{d(\mathbf{x})d(\mathbf{x} + \mathbf{r})}$ is the data covariance matrix as a function of the separation distance and the mean square error $\varepsilon(\mathbf{x})^2$ implies that this estimate is not exact and there is some error in the estimation of the correlation function from the data. This is referred to as the "mean square mapping error" of the interpolation. We note that this is not the same error in Eqn (4.6) that we minimize to solve for the weights in Eqn (4.7). The matrix can now be calculated by forming pairs of observed data values separated into bins according to the distance between sample sites, x_i. These are then averaged over the number of pairs that have the same separation distance to yield the product matrix

$$\overline{(d_i - \overline{d})(d_j - \overline{d})}$$

This computation requires us to define some "bin interval" for the separation distances so that we can group the product values together. To ensure that the resulting covariance matrix meets the condition of being positive definite, a smooth function satisfying this requirement can be fitted to the computed raw covariance function. This fitted covariance function is used for

$$\overline{(d_j - \overline{d})(D - \overline{D})}$$

and to calculate

$$\overline{\left[(d_i - \overline{d})(d_j - \overline{d})\right]}^{-1}$$

The weights b_i are then computed from Eqn (4.7). It is a simple process to then compute the optimal grid value estimates from Eqn (4.4). Note that, for the case where the data can provide no help in the estimate of D (that is, for $\varepsilon(\mathbf{x}) \to \infty$), then $b_i = 0$ and the only reasonable estimate is $\widehat{D}(\mathbf{x}) = \overline{D}$, the mean value. Similarly, if the data are error free (such that $\varepsilon(\mathbf{x}) \to 0$), then $\widehat{D}(x_i) = D(x_i)$ for all x_i ($i = 1, \ldots, N$). In other words, the estimated value and the measured data are identical at the measurement sites (within the limits of the inherent noise in the data values) and the estimator simply interpolates between the observations.

As with other interpolation methods, optimum interpolation is subject to sampling inhomogeneities. Consequently, for regions with very few observations, optimum interpolation can return a discontinuous field. The method also assumes, sometimes incorrectly, that all measurement values have the same error variance since the weighting calculation is based on distance only.

The critical step in the objective mapping procedure is the computation of the covariance matrix. We have described a straightforward procedure to estimate the covariance matrix from the sample data. As with the estimate of the mean or overall trend, it is often possible to use an existing set of historical data to compute the covariance matrix. This is frequently the case in meteorological applications where long series of historical data are available. In oceanography, however, the covariance matrix typically must be computed from the sample data. Where historical data are available, it is important to recognize that using these data to estimate the covariance matrix for use with more recently collected data is tantamount to assuming that the statistics have remained stationary since the time that the historical data were collected.

Bretherton et al. (1976) suggest that objective analysis can be used to compute the covariance matrix. In this case, they start with an assumed covariance function, $\widehat{F}$ which is then compared with a covariance function computed from data with a fixed distance x_o. The difference between the model $\widehat{F}$ and the real F computed from the data is minimized by repeated iteration.

To this point, we have presented objective analysis as it applies to scalar fields. We can also apply optimal (Gauss–Markov) interpolation to vector fields. One approach is to examine each scalar velocity component separately so that for n velocity vectors we have $2n$ velocity components

$$d_r = u_1(\mathbf{x}_r); \quad d_{r+n} = u_2(\mathbf{x}_r) \tag{4.11}$$

where u_1 and u_2 are the x, y velocity components at x_r. If the velocity field is nondivergent, we can introduce a scalar streamfunction $\Psi(\mathbf{x})$ such that

$$u_1 = -\frac{\partial \Psi}{\partial y}; \quad u_2 = \frac{\partial \Psi}{\partial x} \qquad (4.12)$$

and apply scalar methods to Ψ.

Once the optimal interpolation has been executed, there is a need to return to Eqn (4.6) to compute the actual error associated with each optimal interpolation. To this end, we note that we now have the interpolated data from Eqn (4.4). Thus, we can use $\hat{D}$ computed from Eqn (4.4) as the value for D in Eqn (4.6). The product in the last term of Eqn (4.6) is computed from the covariance in Eqn (4.9). In this way, it is possible to compute the error associated with each optimally interpolated value. Frequently, this error field is plotted for a specific threshold level, typically 50% of the interpolated values in the mapped field (see following examples). It is important to retain this error estimate

as part of the optimal interpolation since it enables the investigator to assess the statistical significance of individual gridded values.

4.2.1 Objective Mapping: Examples

An example of objective mapping applied to a single oceanographic survey is provided by the results of Hiller and Käse (1983). The data are from a CTD survey grid occupied in the North Atlantic about midway between the Azores and the Canary Islands (Figure 4.2). At each CTD station, the geopotential anomaly at 25 db (dbar) relative to the anomaly at 1500 db (written 25/1500 db) was calculated and selected as the variable to be mapped. The two-dimensional correlation function for these data is shown in three-dimensional perspective in Figure 4.3(a). A series of different correlation functions were examined and an isotropic, Gaussian function that was positive definite,

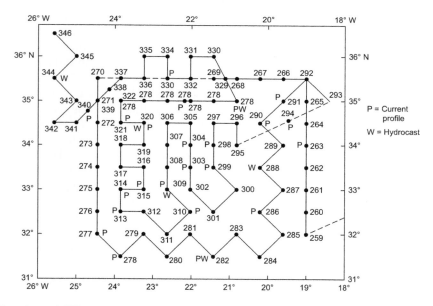

FIGURE 4.2   Locations of CTD stations taken in the North Atlantic between the Azores and the Canary Islands in spring 1982 (experiment POSEIDON 86, Hiller and Käse, 1983). Also shown are locations of current profile (P) and hydrocast (W) stations.

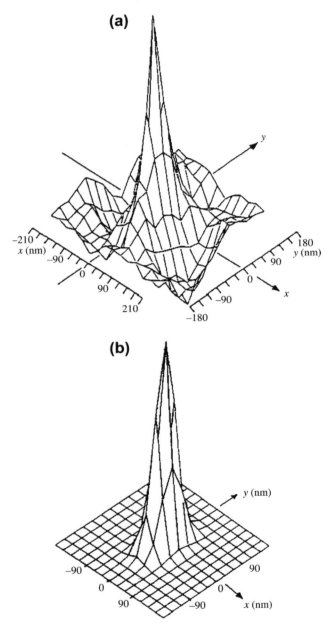

FIGURE 4.3 The two-dimensional correlation function $C(\mathbf{r})$ for the geopotential anomaly field at a pressure of 25 dbar referenced to 1500 dbar (25/1500 dbar) for the data collected at stations shown in Figure 4.2 (1 dbar $= 1 \, \text{m}^2/\text{s}^2$). Here, $\mathbf{r} = (x, y)$, where x, y are the eastward and northward coordinates, respectively. Distances are in nautical miles. (a) The "raw" values of $C(\mathbf{r})$ based on the observations; (b) A model of the correlation function fitted to (a). *From Hiller and Käse (1983).*

was selected as the best fit (Figure 4.3(b)). Using this covariance function, the authors obtained the objectively mapped 25/1500 db geopotential anomaly shown in Figure 4.4(a). Removal of a linear trend gives the objective map shown in Figure 4.4(b) and the associated root mean square (RMS) error field shown in Figure 4.4(c). Only near the outside boundaries of the data domain does the RMS error increase to around 50% of the geopotential anomaly field (Figure 4.4(b)).

As an example of objective mapping applied to a vector field, Hiller and Käse (1983) examined a limited number of satellite-tracked drifter trajectories that coincided with the CTD survey in space and time. Velocity vectors based on daily averages of low-passed finite difference velocities are shown in Figure 4.5(a). Rather than compute a covariance function for this relatively small sample, the covariance function from the analysis of the 25/1500 db geopotential anomaly was used. Also, an assumed error level, $\overline{\epsilon^2}$, was used rather than a computed estimate from the small sample. With the isotropic correlation scale estimated to be 75 km, the objective mapping produces the vector field in Figure 4.5(b). The stippled area in this figure corresponds to the region where the error variance exceeds 50% of the total variance.

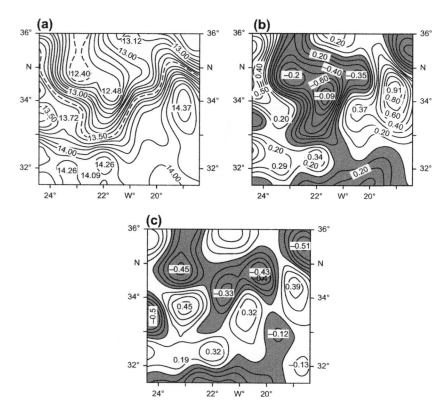

FIGURE 4.4 Objective analysis of the geopotential anomaly field 25/1500 dbar (m^2/s^2) using the correlation function in Eqn (4.2b). (a) The approximate center of the frontal band in this region of the ocean is marked by the 13.5 dbar pressure isoline; (b) Same as (a) but after subtraction of the linear spatial trend; (c) Objectives analysis of the residual mesoscale perturbation field 25/1500 dbar after removal of the composite mean field. *(After Hiller and Käse (1983).)*

FIGURE 4.5 Analysis of the velocity field for the current profile collected in the grid in Figure 4.2. (a) The input velocity field; (b) Objective analysis of the input velocity field with correlation scale $\lambda = 200$ km assumed noise variance of 30% of the total variance of the field. This approach treats mesoscale variability on scales less than 200 km as noise, which is smoothed out. In the shaded area, the error variance exceeds 50% of the total variance.

Due to the paucity of data, the area of statistically significant vector mapping is quite limited. Nevertheless, the resulting vectors are consistent with the geopotential height map in Figure 4.4(a).

Another example is provided by McWilliams (1976) who used dynamic height relative to 1500-m depth and deep float velocities at 1500 m to estimate the streamfunction field. The isotropic covariance function for the random fluctuations in streamfunction $\Psi' = \Psi - \overline{\Psi}$ at 1500-m depth is

$$C(r) = \overline{\Psi'(\mathbf{x}, z, t)\Psi'(\mathbf{x} + \mathbf{r}, z, t)}$$

$$= \overline{\Psi'^2}(1 - \varepsilon^2)(1 - \gamma^2 r^2)\exp\left(-\frac{1}{2}\delta^2 r^2\right)$$

$$(4.13)$$

where $\mathbf{r}$ is a horizontal separation vector, $r = |\mathbf{r}|$, ε is an estimate of relative measurement noise ($0 \leq \varepsilon \leq 1$), and γ^{-1}, δ^{-1} are decorrelation length scales found by fitting Eqn (4.13) to prior data. Denman and Freeland (1985) discuss the merits of five different covariance functions fitted to geopotential height data collected over a period of three years off the west coast of Vancouver Island. As discussed in Chapter 1, the widely used global SST data generated by National Oceanic and Atmospheric Administration (NOAA) (Reynolds and Smith, 1994) are based on optimal interpolation on a 1×1 degree grid. The weekly and monthly analyses use 7 days of in situ (ship and buoy) and satellite SST records. Error statistics show that the SST RMS data errors from ships are almost twice as large as the data errors from buoys or satellites, and that the average e-folding spatial scales for the error are 850 km in the zonal direction and 615 km in the meridional direction. The analysis also includes a preliminary step that uses Poisson's equation to correct any satellite biases relative to the in situ data. Reynolds and Smith (1994) demonstrate the importance of this correction using data following the 1991 eruptions of Mt Pinatubo. For other examples, the reader is referred to Bennett (1992).

As a final point, we remark that the requirement of isotropy is easily relaxed by using direction-dependent covariance matrices, $C(r_1, r_2)$, whose spatial structure depends on two orthogonal spatial coordinates, r_1 and r_2 (with $r_2 \geq r_1$). For example, the map of light attenuation coefficient at 20-m depth obtained from transmissometer profiles off the west coast of Vancouver Island (Figure 4.6) uses an exponentially decaying, elliptical shaped covariance matrix

$$C(r_1, r_2) = \exp\left[-a\Delta x^2 - b\Delta y^2 - c\Delta x\Delta y\right]$$

$$(4.14a)$$

where

$$a = \frac{1}{2}\left\{[\cos(\pi\phi/180)/r_1]^2 + [\sin(\pi\phi/180)/r_2]^2\right\}$$

$$b = \frac{1}{2}\left\{[\sin(\pi\phi/180)/r_1]^2 + [\cos(\pi\phi/180)/r_2]^2\right\}$$

$$c = \cos(\pi\phi/180)\sin(\pi\phi/180)[r_2^2 - r_1^2]/(r_1 r_2)^2$$

$$(4.14b)$$

Here, Δx and Δy are, respectively, the eastward and northward distances from the grid point to the data point, and ϕ is the orientation angle (in degrees) of the coastline measured counterclockwise from north. In this case, it is assumed that the alongshore correlation scale, r_2, is twice the across-shore correlation scale, r_1. The idea here is that, like water-depth changes, alongshore variations in coastal water properties such as temperature, salinity, geopotential height, and log-transformed phytoplankton chlorophyll-a pigment concentration occur over longer length scales than across-shore variations.

In their introduction to optimal interpolation, Barth et al. (2008; http://modb.oce.ulg.ac.be/wiki) construct an artificial data field based on a spatially inhomogeneous sampling density (Figure 4.7). The white area in this figure corresponds to a barrier in the original data field

FIGURE 4.6 Objective analysis map of light attenuation coefficient (per meter) at 20-m depth on the west coast of Vancouver Island obtained from transmissometer profiles. The covariance function $C(r_1, r_2)$ given by the ellipse is assumed to decay exponentially with distance with the longshore correlation scale $r_2 = 50$ km and cross-shore correlation scale $r_1 = 25$ km. Here, (r_1, r_2) is written as (R_1, R_2) in the figure.

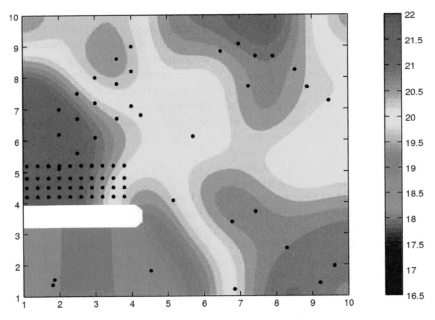

FIGURE 4.7 Optimal interpolation (colored region) of a data field and the sampling points (dots) of the original data field. The white region denotes a physical barrier. *From Barth et al. (2008).*

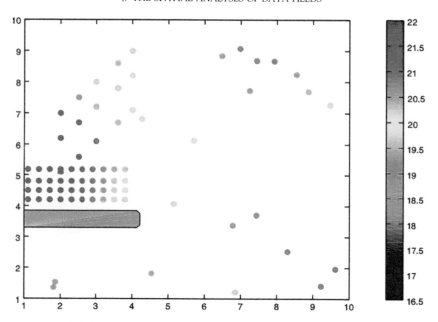

FIGURE 4.8 Actual values of the original field at the data locations shown in Figure 4.7. *From Barth et al. (2008).*

(the colored region). Figure 4.8 presents the actual data values from the original data field for the data locations specified in Figure 4.7.

A comparison between the interpolated data field (colored map) in Figure 4.7 and the original data field in Figure 4.8, which was used to generate the interpolated map, clearly reveals that much of the structure in the original field is not captured by the gridded sample data. It is also clear that this data sample has some very different spatial characteristics in that there is a regular grid just above the barrier with very dense sampling while much of the remaining data field is relatively sparsely sampled, with rather large regions not sampled at all. Moreover, the real data values would have errors associated with them, including instrumental errors, which may be random or have a bias. Representative errors arise when the observations do not precisely coincide with what we are attempting to determine. For example, Barth et al. (2008) remark that the investigator might

desire a monthly average whereas the measurements being used are instantaneous or averaged over a short period of time. In addition, not all measurements are likely to have been collected at the same time. Added to these possible sources of error are a list of other errors, including human error, transmission or recording errors, and instrumental malfunctions.

4.3 KRIGING

Kriging is another optimal interpolation technique and originated with the Master's thesis of Daniel G. Krige who pioneered the distance-weighted average gold grades at the Witwatersrand reef complex mine in South Africa. The method was developed to improve mapping the topography of a geographic region that had unevenly spaced data points. The goal was to estimate topographic values for grid locations for which there were no data values.

The theory was further developed by the French mathematician Georges Matheron. As with objective analysis discussed previously, Kriging belongs to the family of linear least-squares estimation algorithms. Again, the goal is to estimate the values of an unknown real-valued function from observations at known, but irregularly, spaced locations. There are several types of Kriging, including ordinary Kriging and detrended Kriging (Beers and Kleijnen, 2002). In the latter, the Kriging algorithm uses linear regressional analysis to "detrend" the data.

In general, spatial interpolation methods such as linear, spline, inverse distance, and triangular interpolation estimate values at given locations as the weighted sum of data values that surround a specific location. Almost all of these methods assign decreasing weights to the data with increasing separation distance from the interpolation point. Similar to objective analysis, Kriging assigns weights according to a moderately data-dependent weighting function rather than an arbitrary function. Kriging is, however, still an interpolation method. As such, it is subject to many of the basic limitations of other interpolation methods, including the propensity to underestimate the highs and overestimate the lows compared to the actual property distribution. In addition, interpolation methods are highly data-driven, whereby a dense distribution of observations will yield a better representation of reality than a sparse data distribution. Any interpolation method will perform well in data clusters but perform poorly in the gaps between these clusters.

There are some advantages to Kriging compared to other methods, such as the fact that it provides an estimation error (Kriging variance) for the estimate of a given variable. The availability of this estimation error can be used in the assimilation of interpolated fields into numerical simulation models. It also helps to compensate for the effects of data clustering by assigning lower weights for clustered values than for isolated data points.

4.3.1 Mathematical Formulation

All Kriging estimators are variants of the basic linear regression estimator $Z^*(u)$ defined as (Bohling, 2005)

$$Z^*(u)_i - m(u) = \sum_{\alpha=1}^{n(u)} \lambda_\alpha [Z(u_\alpha) - m(u_\alpha)] \quad (4.15)$$

where the u are location vectors corresponding to an estimation point (u) and one of the neighboring data points, u_α, indexed by α. The value n(u) is the number of data points in a neighborhood of u that are used for the estimation of $Z^*(u)$, while m(u) and $m(u_\alpha)$ are the expected values of Z(u) and $Z(u_\alpha)$, respectively. The values $\lambda_\alpha(u)$ are the Kriging weights assigned to data values $Z(u_\alpha)$ used to estimate Z(u).

We treat Z(u) as a random field with a trend component, m(u), and a residual component, $R(u) = Z(u) - m(u)$. The Kriging method estimates the residual at u as a weighted sum of residuals at surrounding data points. Kriging weights are derived from the semivariogram, which is another form of the covariance function. The use of distinct functions is one of the fundamental differences between the Kriging and optimum interpolation methodology.

The semivariogram and covariance functions quantify the correlation structure of a spatial field and provide measures of how well variables that are spaced closer together correlate compared to variables that are spaced further apart. Both measure the degree of statistical correlation as a function of separation distance, s. The semivariogram is defined as

$$\gamma(s_i, s_j) = \frac{1}{2} \text{var}[Z(s_i) - Z(s_j)] \quad (4.16)$$

where var is the variance of Z. If two locations, s_i and s_j, are close in terms of the distance measure $d(s_i, s_j)$, they are expected to have similar values so that the difference $Z(s_i) - Z(s_j)$ should be small. As locations s_i and s_j become farther apart, they become less similar and the difference $Z(s_i) - Z(s_j)$ is expected to increase

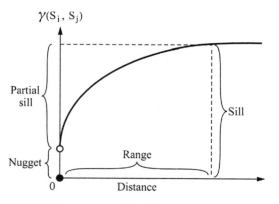

$\gamma(s_i, s_j)$

FIGURE 4.9 A typical semivariogram showing an increase in the semivariogram with separation distance.

(Figure 4.9). Semivariogram functions can also be thought of as dissimilarity functions. As illustrated in Figure 4.9, the height that the semivariogram reaches when it levels out is called the "sill" while the height discontinuity (and y-offset) at zero distance is called the "nugget effect." The nugget effect can itself be divided into a measurement error and a micro-scale variation, either of which can be zero. The distance at which the semivariogram reaches its sill height is called the "range."

There is a direct relationship between the semivariogram and the covariance function, C, whereby

$$\gamma(s_i, s_j) = \text{sill} - C(s_i, s_j) \qquad (4.17)$$

This equivalence makes it possible to use either the semivariogram or the covariance function to perform the interpolation. One important criterion is that the Kriging interpolated values have nonnegative Kriging standard errors, which means that only some functions can be used as the semivariogram or covariance functions. In those cases where the data-driven semivariogram has negative values, a nonnegative function is fit to the semivariogram. This is similar to the step in objective analysis where a nonnegative function is fit to the covariance function.

The goal of Kriging is to determine the weights λ_α that minimize the variance of the Kriging estimator

$$\sigma_E^2(u) = \text{var}[Z^*(u) - Z(u)] \qquad (4.18)$$

under the unbiased constraint that the expected value E

$$E[Z^*(u) - Z(u)] = 0 \qquad (4.19)$$

The random field $Z(u)$ is decomposed into a residual, R, and trend, m, component such that $Z(u) = R(u) + m(u)$, where the residual component is treated as a random field with a stationary mean of zero (0) and a stationary and isotropic covariance function, which is a function of the separation distance, h, but not of the position, u. Thus, we can write

$$E[R(u)] = 0 \qquad (4.20a)$$

$$\text{Cov}[R(u), R(u + h)] = E[R(u), R(u + h)]$$
$$= C_R(h)$$
$$(4.20b)$$

where $C_R(h)$ is the residual covariance function that is generally derived from the input semivariogram model, $C_R(h) = C_R(0) - \lambda(h) = \text{sill} - \lambda(h)$. Thus, the semivariogram entered into a Kriging program should represent the residual component of the variable.

There are three different types of Kriging: simple, ordinary, and Kriging with a trend. These all differ in their treatments of the trend component, m(u). For simple Kriging, we assume that the trend component is a known constant such that, m(u) = m, whereby

$$Z_{SK}^*(u) = m + \sum_{\alpha=1}^{n(u)} \lambda_\alpha^{SK}(u)[Z(u_\alpha) - m] \qquad (4.21)$$

This estimate is automatically unbiased since $E[Z(u_\alpha) - m] = 0$, so that $E[Z_{SK}^*(u)] = m = E[Z(u)]$.

The estimation error $Z_{SK}^*(u) - Z(u)$ is a linear combination of a random variable representing residuals at the data points, u_α, and the estimation point, u:

$$Z_{SK}^*(u) - Z(u) = [Z_{SK}^*(u) - m] - [Z(u) - m]$$

$$= \sum_{\alpha=1}^{n(u)} \lambda_\alpha^{SK}(u)R(u_\alpha) - R(u) = R_{SK}^*(u) - R(u)$$

$$(4.22)$$

Using rules for the variance of a linear combination of random variables, the error variance is then given by

$$\sigma_E^2(u) = \text{var}[R_{SK}^*(u)] + \text{var}[R_{SK}(u)]$$

$$-2\text{Cov}[R_{SK}^*(u), R_{SK}(u)]$$

$$= \sum_{\alpha=1}^{n(u)} \sum_{\beta=1}^{n(u)} \left[\lambda_\alpha^{SK}(u)\lambda_\beta^{SK}(u)C_R(u_\alpha - u_\beta) \right]$$

$$+ C_R(0) - 2 \sum_{\alpha=1}^{n(u)} \lambda_\alpha^{SK}(u)C_R(u_\alpha - u)$$

$$(4.23)$$

To solve for the Kriging weights, which is clearly a principal step in the entire process, we minimize the error variance in Eqn (4.23) by taking the derivative of this equation with respect to the Kriging weights and then setting each expression to zero. This leads to the following system of equations:

$$\sum_{\beta=1}^{n(u)} \left[\lambda_\beta^{SK}(u)C_R(u_\alpha - u_\beta) \right] = C_R(u_\alpha - u)$$

$$\alpha = 1, \ldots, n(u) \qquad (4.24)$$

Because of the constant mean, the covariance function for $Z(u)$ is the same as that for the residual component $C(h) = C_R(h)$, so that we can write the simple Kriging systems of equations in terms of $C(h)$:

$$\sum_{\beta=1}^{n(u)} \left[\lambda_\beta^{SK}(u)C(u_\alpha - u_\beta) \right] = C(u_\alpha - u)$$

$$\alpha = 1, \ldots, n(u) \qquad (4.25)$$

or

$$K\lambda_{SK}(u) = k \qquad (4.26)$$

where K is the covariance matrix between the measured data points with elements $K_{i,j} = C(u_i - u_j)$, and k is the vector of covariances between the data points and the estimation point whose elements are given by $k_i = C(u_i - u)$. Here, it is assumed that the covariance vector is simply a function of the distance between the data point and the estimation point. The vector λ_{SK} represents the weights for simple Kriging for the surrounding data points.

Assuming a positive definite covariance matrix, we can solve for the Kriging weights as:

$$\lambda_{SK}(u) = K^{-1}k \qquad (4.27)$$

With these weights, we can compute both the Kriging estimate and the Kriging variance from Eqn (4.23), which now reduces to

$$\sigma_{SK}^2(u) = C(0) - \lambda_{SK}^T(u)$$

$$= C(0) - \sum_{\alpha=1}^{n(u)} \lambda_\alpha^{SK}(u)C(u_\alpha - u) \qquad (4.28)$$

The result is a set of weights for estimating the value of the variable at the location of interest based on the measured values from a set of neighboring data points. The weighting for each data point generally decreases with increasing separation distance between a specific point and the interpolation grid location.

Bohling (2005) applies simple Kriging to a porosity field (Figure 4.10) where the x and y axes correspond to the eastward and northward directions (eastings and northings), respectively, and are measured in meters. From this field, Bohling derives a spherical semivariogram (Figure 4.11), which has a zero nugget (i.e., it intersects the origin) and a sill of 0.78 with a range of 4141 m. Since the

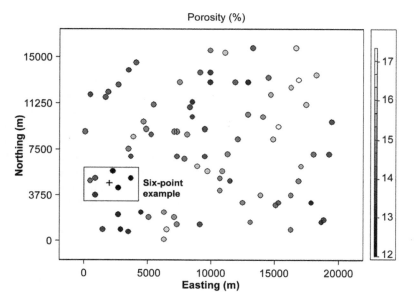

FIGURE 4.10 Point values of porosity (in %) mapped in eastward and northward (easting and northing) geographical coordinates. The plus sign (+) denotes an interpolation grid-point to be determined using the surrounding six observed values in the box. The gray-scale porosity code (%) is shown on the right. *(After Bohling (2005).)*

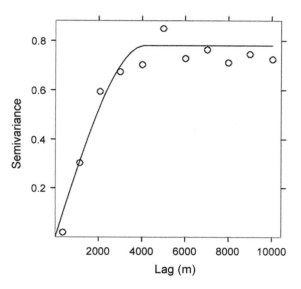

FIGURE 4.11 Semivariogram for the porosity field in Figure 4.10. *(After Bohling (2005).)*

study uses a spherical semivariogram, the covariance function is given by

$$C(h) = C(0) - \gamma(h)$$
$$= 0.78 \cdot \left[1 - 1.5(h/4141) + 0.5(h/4141)^3\right]$$
(4.29)

for separation distance h up to 4141 m, and then zero beyond that range. For the small sample of six points marked in Figure 4.10, the matrix of distances between pairs of data points is given in Table 4.1.

From this set of data, the covariance matrix is

$$\mathbf{K} = \begin{bmatrix} 0.78 & 0.28 & 0.06 & 0.17 & 0.40 & 0.43 \\ 0.28 & 0.78 & 0.43 & 0.39 & 0.27 & 0.20 \\ 0.06 & 0.43 & 0.78 & 0.37 & 0.11 & 0.06 \\ 0.17 & 0.39 & 0.37 & 0.78 & 0.37 & 0.27 \\ 0.40 & 0.27 & 0.11 & 0.37 & 0.78 & 0.65 \\ 0.43 & 0.20 & 0.06 & 0.27 & 0.65 & 0.78 \end{bmatrix}$$

TABLE 4.1 Matrix of Separation Distances (in Meters) between Pairs of Points for the Distribution of Six Points Presented in Figure 4.10

	Point 1	Point 2	Point 3	Point 4	Point 5	Point 6
Point 1	0	1897	3130	2441	1400	1265
Point 2	1897	0	1281	1456	1970	2280
Point 3	3130	1281	0	1525	2800	3206
Point 4	2441	1456	1523	0	1523	1970
Point 5	1400	1970	2800	1523	0	447
Point 6	1265	2280	3206	1970	447	0

where the matrix elements have been rounded off to two decimal places. Note the high correlation between point 5 and 6, which are separated by only 447 m. The corresponding vector of Kriging weights is

$$\begin{bmatrix} \lambda_1 \\ \lambda_2 \\ \lambda_3 \\ \lambda_4 \\ \lambda_5 \\ \lambda_6 \end{bmatrix} = \mathbf{K}^{-1}\mathbf{k} = \begin{bmatrix} 0.1475 \\ 0.4564 \\ -0.0205 \\ 0.2709 \\ 0.2534 \\ -0.0266 \end{bmatrix}$$

We note that data point 6 is assigned a very small weight relative to data point 1 even though they are both about the same distance from the interpolation point (+) and have about the same variance. This is because data point 6 is "shielded" by nearby data point 5 with which it is very strongly correlated and which already has a very strong influence on the estimation point. The covariances and, hence the Kriging weights, are determined entirely by the data configuration and the covariance model, and not by the actual data values themselves.

Using this procedure, and adding back in a mean field to the interpolated porosity values, yields the estimated porosity distribution (in %) using simple Kriging (Figure 4.12). The standard deviation of the interpolated field is presented in Figure 4.13, which clearly lines up along bands determined by the data distribution, as one might expect from this type of interpolation method.

The preceding results illustrate several important characteristics of the Kriging method (Bohling, 2005). Specifically: (1) kriged surfaces, as with most interpolated surfaces, are smooth,

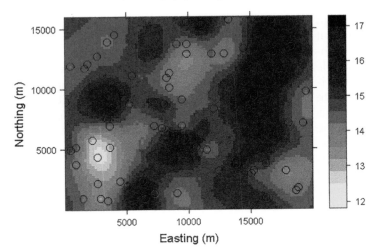

FIGURE 4.12 The result of applying simple Kriging to the porosity field in Figure 4.10. Porosity values (%) are denoted by the color bar on the right.

FIGURE 4.13 Standard deviation of the porosity field (in %; see bar on the right) interpolated using simple Kriging.

Simple kriging standard deviation

and often much smoother than the actual surface that would be generated if there were more data available; (2) "bull's-eyes," which are caused by local extremes in the data values, are inevitable; and (3) the error map for the interpolated surfaces (as illustrated by the standard deviation in Figure 4.13) is driven primarily by the data distribution rather than by the actual data values at the different locations.

In the case of "ordinary Kriging," the mean data field is no longer assumed to be constant (spatially uniform) over the entire domain but instead is assumed constant in the local neighborhood of each estimation point. In simple Kriging, we equate $C_R(h)$ with $C(h)$, which is the covariance matrix of the data itself due to the presence of a constant mean. While this equality does not hold for ordinary Kriging, the practice is to assume it is true on the assumption that the semivariogram, from which $C(h)$ is derived, effectively filters out the influence of large-scale trends in the mean. Both ordinary Kriging and simple Kriging are interpolation methods that follow naturally from a semivariogram analysis and both procedures tend to filter out trends in the mean.

When applied to the porosity data in Figure 4.10, ordinary Kriging yields an interpolation field which resembles the simple Kriging field presented in Figure 4.12. The resultant fields are similar and differ only in highly detailed features. The standard deviation fields agree closely, again primarily because they are a function of the data locations more than they are a function of the actual data values. Kriging with a trend (also known as "universal Kriging") resembles ordinary Kriging except that, instead of fitting a local mean in the neighborhood of the estimation point, a linear or higher-order trend is fitted to the (x, y) coordinates of the data points. Including this form of model to the Kriging system, is akin to an extension to a local mean for ordinary Kriging. In fact, ordinary Kriging can be thought of as universal Kriging with a zeroth-order trend model.

Other specialized Kriging methods, such as "indicator Kriging," uses indicator functions rather than a data field for the interpolation process. "Co-Kriging" is a form of multivariate Kriging that uses the covariance between two different variables. "Disjunctive Kriging" is a nonlinear generalization of Kriging and "lognormal Kriging" interpolates positive data, which has been transformed through the use of logarithms. Mathematically, Kriging is closely related to regressional analysis but involves the interpolation of a single variable while regression is the relationship between multiple data sets. It can also be shown that Kriging results in the best linear unbiased estimate. It is linear in the sense that all estimated values are weighted linear averages of the data values, unbiased because the mean error is zero, and best because it is designed to minimize the variance of the estimation errors.

Kriging that uses a semivariogram with a pronounced nugget will create discontinuities, with the interpolated surface oscillating up or down as it tries to "grab" any data point that happens to coincide with a grid node (estimation point). In such cases, it is possible to use "factorial Kriging," where the nugget is filtered out. Another option is to fit a semivariogram that does not have a nugget.

The method for selecting the appropriate neighboring data points can have at least as much influence on the estimates as the interpolation algorithm itself. The simplest approach is the nearest neighbor criterion; more complex methods may involve quadrant and octant searches, which look for data points within a certain distance in each quadrant or octant surrounding the estimation point.

4.4 EMPIRICAL ORTHOGONAL FUNCTIONS

The previous section dealt with the optimal smoothing of irregularly spaced data onto a gridded map. In other studies of oceanic variability, we may be presented with a large data set from a grid of time-series stations, which we wish to compress into a smaller number of independent pieces of information. For example, in studies of climate change, it is necessary to deal with time series of spatial maps, such as surface temperature. A useful obvious choice would involve a linear combination of orthogonal spatial "predictors," or modes, whose net response as a function of time would account for the combined variance in all of the observations. The signals we wish to examine may all consist of the same variable, such as temperature, or they may be a mixture of variables such as temperature and wind velocity or current and sea level. The data may be in the form of concurrent time-series records from a grid (regular or irregular) of stations $x_i(t)$, $y_i(t)$ on a horizontal plane or time-series records at a selection of depths on an $x_i(t)$, $z_i(t)$ cross section. Examples of time series from cross-sectional data include those from a string of current meters on a single mooring or from moorings of upward-looking bottom-mounted Acoustic Doppler Current Profilers (ACDPs) strung across-channel.

A useful technique for compressing the variability in this type of time-series data is *principal component analysis* (PCA). In oceanography, the method is commonly known as *empirical orthogonal function* (EOF) analysis. The EOF procedure is one of a larger class of inverse techniques and is equivalent to a data reduction method widely used in the social sciences known as *factor analysis (FA)*. The first reference we could find to the application of EOF analysis to geophysical fluid dynamics is a report by Edward Lorenz (1956) in which he develops the technique for statistical weather prediction and coins the term "EOF."

As discussed by Preisendorfer (1988), one of the essential aspects of the PCA method was developed by an Italian geometer, Beltrami, in 1873. He formulated a modern form of the resolution of a general square matrix into its singular value decomposition (SVD), which stands at the core of PCA. This same discovery was made independently by the French algebraist, Jordan, in 1874. PCA appears to have made its first appearance in the United States as an exercise in abstract algebra when Sylvester (1889) considered the problem of the reduction of a square matrix into its SVD. A decade later, Pearson (1901) recast linear regressional analysis into a new form to avoid the common asymmetrical relationship between "dependent" and "independent" variables. In his paper, Pearson introduced a clear geometric visualization of PCA in Euclidean space. The first application of PCA to meteorology appears to have been made at the Massachusetts Institute of Technology (MIT) by G.P. Wadsworth and his colleagues in 1948. The goal of their study was to develop a short-term prediction method for sea level atmospheric pressure over the northern hemisphere. In test calculations over the North Atlantic, Wadsworth was faced, in 1944, with the daunting task of hand-calculating the 91 eigenvalues of a 91×91 matrix. Confronted with this unmanageable numerical task, Wadsworth dropped the PCA approach and went on to use theoretical orthogonal functions (Tshebyschev polynomials) to complete the project. It is interesting to note that, about the same time, a completely independent use of PCA in meteorology was being carried out by Fukuoka (1951).

When the Whirlwind general-purpose computer became available at MIT in the 1950s, E.N. Lorenz, starting with the work of Wadsworth and colleagues, undertook prediction studies of the 500-mb height anomaly for January (1947–1952) for a grid of 64 points covering the mainland United States, Southern Canada, and portions of the surrounding oceans. Lorenz (1956) is now a classic in the field of statistical-dynamical approaches to weather prediction. The Statistical Forecasting Project at MIT under Lorenz's direction produced some outstanding early applications of PCA to short-range forecasting. Applications of PCA to oceanographic data sets began to appear about a decade after Lorenz's work. Trenberth (1975) related southern hemisphere atmospheric oscillations to SST observations. PCA studies based on SSTs in the Pacific by Barnett and Davis also appeared in the 1970s along with similar work by Weare et al. (1976). An interesting idea involving the use of extended Emprical Orthogonal Functions (EEOFs) for moving pattern detection in tropical Pacific Ocean temperatures is explored in Weare and Nasstrom (1982).

The advantage of EOF analysis is that it provides a compact description of the spatial and temporal variability of data series in terms of orthogonal functions, or statistical "modes." Usually, most of the variance of a spatially distributed series is in the first few orthogonal functions whose patterns may then be linked to possible dynamical mechanisms. It should be emphasized that no direct physical or mathematical relationship necessarily exists between the statistical EOFs and any related dynamical modes. Dynamical modes conform to physical constraints through the governing equations and associated boundary conditions (LeBlond and Mysak, 1979); EOFs are simply a method for partitioning the variance of a spatially distributed group of concurrent time series. They are called "empirical" to reflect the fact that they are defined by the covariance structure of the specific data set being analyzed (as shown below).

In oceanography and meteorology, EOF analysis has found wide application in both the time and frequency domains. Conventional EOF analysis can be used to detect standing oscillations only. To study propagating wave phenomena, we need to use lagged covariance matrix (Weare and Nasstrom, 1982), or complex PCA in the frequency domain (Wallace and Dickinson, 1972; Horel, 1984). Our discussion, in this section, will focus on space/time domain applications. Readers seeking more detailed descriptions of both the procedural aspects and their applications are referred to Lorenz (1956), Davis (1976), and Preisendorfer (1988).

The best analogy to describe the advantages of EOF analysis is the classical vibrating drum problem. Using mathematical concepts presented in most undergraduate texts, we know that we can describe the eigenmodes of drumhead oscillations through a series of two-dimensional orthogonal patterns. These modes are defined by the eigenvectors and eigenfunctions of the drumhead. Generally, the lowest modes have the largest spatial scales and represent the most dominant (most prevalent) modes of variability. Typically, the drumhead has, as its largest mode, an oscillation in which the whole drumhead moves up and down, with the greatest amplitude in the center and zero motion at the rim where the drum is clamped. The next highest mode has the drumhead separated in the center with one side 180° out of phase with the other side (one side is up when the other is down). Higher modes have more complex patterns with additional maxima and minima. Now, suppose we had no mathematical theory, and were required to describe the drumhead oscillations in terms of a set of observations; we would look for the kinds of eigenvalues in our data that we obtain from our mathematical analysis. Instead of the analytical or dynamical solutions that can be derived for the drum, we wish to examine "empirical" solutions based strictly on a measured data set. Since we are ignorant of the actual dynamical analysis, we call the resulting modes of oscillation EOFs.

EOFs can be used in both the time and frequency domains. For now, we will restrict ourselves to the spatial domain application and consider a series of N maps at times $t = t_i$ $(1 \leq i, N)$, each map consisting of scalar variables, $\psi_m(t)$, collected at M locations, $\mathbf{x}_m (1 \leq m \leq M)$. One could think of N weather maps available every 6 h over a total period of $6N$ h, with each map showing the sea surface pressure $\psi_m(t) = P_m(t)$ $(1 \leq m \leq M)$ recorded at M weather buoys located at mooring sites $\mathbf{x}_m = (x_m, y_m)$. The subscript m refers to the spatial grid locations in each map. Alternatively, the N maps might consist of pressure data $P(t)$ from $M - K$ weather buoys and velocity component records $u(t)$, $v(t)$ from $K/2$ current meter sites. Or, again the time series could be from $M/2$ current meters on a moored string. Any combination of scalars is permitted (this is a statistical analysis not a dynamical analysis). The goal of this procedure is to write the data series, $\psi_m(t)$, at any given location, $\mathbf{x}_m$, as the sum of M orthogonal spatial functions $\phi_i(\mathbf{x}_m) = \phi_{im}$ such that

$$\psi(\mathbf{x}_m, t) = \psi_m(t) = \sum_{i=1}^{M} [a_i(t)\phi_{im}] \qquad (4.30)$$

where $a_i(t)$ is the amplitude of the ith orthogonal mode at time $t = t_n$ $(1 \leq n \leq N)$. Simply put, Eqn (4.30) states that the time variation of the dependent scalar variable, $\psi(\mathbf{x}_m, t)$, at each location, $\mathbf{x}_m$, results from the linear combination of M spatial functions, ϕ_i, whose amplitudes are weighted by M time-dependent coefficients, $a_i(t)$ $(1 \leq I \leq M)$. The weights, $a_i(t)$, tell us how the spatial modes, ϕ_{im}, vary with time. There are as many (M) basis functions as there are stations for which we have data. Put another way, we need as many modes as we have time-series stations so that we can account for the combined variance in the original time series at each time, t. If we want, we can also formulate the problem as M temporal functions whose amplitudes are weighted by M spatially variable coefficients. Whether we partition the data as spatial or temporal orthogonal functions, the results should be identical.

Since we want the spatial functions, $\phi_i(\mathbf{x}_m)$, to be orthogonal, so that they form a set of basis functions, we require that

$$\sum_{m=1}^{M} [\phi_{im}\phi_{jm}] = \delta_{ij} \text{ (orthogonality condition)}$$

$$(4.31)$$

where the summation is over all observation locations and δ_{ij} is the Kronecker delta

$$\delta_{ij} = \begin{cases} 1, & j = i \\ 0, & j \neq i \end{cases} \qquad (4.32)$$

It is worth remarking that two functions are said to be orthogonal when the sum (or integral) of their product over a certain defined space (or time) is zero. Orthogonality in Eqn (4.31) does not mean $\phi_{im}\phi_{jm} = 0$ for each m. For example, in the case of continuous sines and cosines, $\int \sin\theta \cos\theta \, d\theta = 0$ when the integral is over a complete phase cycle, $0 \leq \theta \leq 2\pi$. By itself, the product $\sin\theta \cdot \cos\theta = 0$ only if the sine or cosine term happens to be zero.

There is a multitude of basis functions, ϕ_i, that can satisfy Eqns (4.30) and (4.31). Familiar examples are sine, cosine, and Bessel functions. The EOFs are determined uniquely among the many possible choices by the constraint that the time amplitudes $a_i(t)$ are uncorrelated over the sample data. This requirement means that the time-averaged covariance of the amplitudes satisfies

$$\overline{a_i(t)a_j(t)} = \lambda_t \delta_{ij} (\text{uncorrelated time variability})$$

(4.33)

in which the overbar denotes the time-averaged value and

$$\lambda_i = \overline{a_i(t)^2} = \frac{1}{N}\sum_{n=1}^{N}\left[a_i(t_n)^2\right] \qquad (4.34)$$

is the variance in each orthogonal mode. If we then form the covariance matrix $\psi_m(t)\psi_k(t)$ for the known data and use Eqn (4.33), we find

$$\overline{\psi_m(t)\psi_k(t)} = \overline{\sum_{i=1}^{M}\sum_{j=1}^{M}\left[a_i(t)a_j(t)\phi_{im}\phi_{jk}\right]}$$

$$= \sum_{i=1}^{M}[\lambda_i \phi_{im}\phi_{ik}]$$

(4.35)

Multiplying both sides of Eqn (4.35) by ϕ_{ik}, summing over all k and using the orthogonality condition Eqn (4.31), yields

$$\sum_{k=1}^{M}\overline{\psi_m(t)\psi_k(t)}\phi_{ik}$$

$$= \lambda_t \phi_{im} \ (i\text{th mode at the }m\text{th location;}$$
$$m = 1, \ldots, M)$$

(4.36)

Equation (4.36) is the canonical form for the *eigenvalue problem*. Here, the EOFs, ϕ_{im}, are the *i*th *eigenvectors* at locations x_m, and the mean-square time amplitudes

$$\lambda_i = \overline{a_i(t)^2}$$

are the corresponding *eigenvalues* of the mean product, $\mathbf{R}$, which has elements

$$R_{mk} = \overline{\psi_m(t)\psi_k(t)}$$

This is equal to the covariance matrix, $\mathbf{C}$, if the mean values of the time series $\psi_m(t)$ have been removed at each site, x_m. The total of M EOFs corresponding to the M eigenvalues of Eqn (4.36) forms a complete basis set of linearly independent (orthogonal) functions such that the EOFs are uncorrelated modes of variability. Assuming that the record means $\psi_m(t)$ have been removed from each of the M time series, Eqn (4.36) can be written more concisely in matrix notation as

$$\mathbf{C}\phi - \lambda\mathbf{I}\phi = 0 \qquad (4.37)$$

where the covariance matrix, $\mathbf{C}$, consists of M data series of length N with elements

$$C_{mk} = \overline{\psi_m(t)\psi_k(t)}$$

$\mathbf{I}$ is the unity matrix, and ϕ are the EOFs. Expanding Eqn (4.37) yields the eigenvalue problem

$$\begin{pmatrix} \overline{\psi_1(t)\psi_1(t)} & \overline{\psi_1(t)\psi_2(t)} & \cdots & \overline{\psi_1(t)\psi_M(t)} \\ \overline{\psi_2(t)\psi_1(t)} & \overline{\psi_2(t)\psi_2(t)} & \cdots & \overline{\psi_2(t)\psi_M(t)} \\ \cdots & \cdots & \cdots & \cdots \\ \overline{\psi_M(t)\psi_1(t)} & \overline{\psi_M(t)\psi_2(t)} & \cdots & \overline{\psi_M(t)\psi_M(t)} \end{pmatrix}$$

$$\times \begin{bmatrix} \phi_1 \\ \phi_2 \\ \cdots \\ \phi_M \end{bmatrix} = \begin{bmatrix} \lambda\,0\ldots0 \\ 0\,\lambda\ldots0 \\ \cdots \\ 0\ldots\lambda \end{bmatrix}\begin{bmatrix} \phi_1 \\ \phi_2 \\ \cdots \\ \phi_M \end{bmatrix} \qquad (4.38a)$$

corresponding to the series of linear system of equations

$$\left[\overline{\psi_1(t)\psi_1(t)} - \lambda\right]\phi_1 + \overline{\psi_1(t)\psi_2(t)}\phi_2 + \ldots + \overline{\psi_1(t)\psi_M(t)}\phi_M = 0$$

$$\overline{\psi_2(t)\psi_1(t)}\phi_1 + \left[\overline{\psi_2(t)\psi_2(t)} - \lambda\right]\phi_2 + \ldots + \overline{\psi_2(t)\psi_M(t)}\phi_M = 0$$

$$\cdots$$

$$\overline{\psi_M(t)\psi_1(t)}\phi_1 + \overline{\psi_M(t)\psi_2(t)}\phi_2 + \ldots + \left[\overline{\psi_M(t)\psi_M(t)} - \lambda\right]\phi_M = 0$$

(4.38b)

The eigenvalue problem involves diagonalization of a matrix, which in turn amounts to finding an axis orientation in M space for which there are no off-diagonal terms in the matrix. When this occurs, the different modes of the system are orthogonal. Since each $\mathbf{C}$ is a real symmetric matrix, the eigenvalues λ_i are real. Similarly, the eigenvectors (EOFs) of a real symmetric matrix are real. Because $\overline{C(x_m, x_k)}$ is positive, the real eigenvalues are all positive.

If Eqn (4.37) is to have a nontrivial solution, the determinant of the coefficients must vanish; that is

$$det \begin{vmatrix} C_{11} - \lambda & C_{12} & \ldots & C_{1M} \\ C_{21} & C_{22} - \lambda & \ldots & \ldots \\ \ldots & \ldots & \ldots & \ldots \\ C_{M1} & \ldots & \ldots & C_{MM} - \lambda \end{vmatrix} = 0$$

(4.39)

which yields an Mth order polynomial, $\lambda^M + \alpha\lambda^{M-1} + \ldots$, whose M eigenvalues satisfy

$$\lambda_1 > \lambda_2 > \ldots > \lambda_M \qquad (4.40)$$

Thus, the "energy" (more specifically, the variance) associated with each statistical mode is ordered according to its corresponding eigenvector. The first mode contains the highest percentage of the total variance, λ_1; of the remaining variance, the greatest percentage is in the second mode, λ_2, and so on. If we add up the total variance in all the time series, we obtain

$$\sum_{m=1}^{M} \left\{ \frac{1}{N} \sum_{n=1}^{N} [\psi_m(t_n)]^2 \right\} = \sum_{j=1}^{M} \lambda_j$$

Sum of variances in the data
= sum of variance in the eigenvalues   (4.41)

The total variance in the M time series equals the total variance contained in the M statistical modes. The final piece of the puzzle is to derive the time-dependent *amplitudes* of the ith statistical mode

$$a_i(t) = \sum_{m=1}^{M} \psi_m(t)\phi_{im} \qquad (4.42)$$

Equation (4.36) provides a computational procedure for finding the EOFs. By computing

the mean product matrix, $\overline{\psi_m(t)\psi_k(t)}(m, k = 1, \ldots, M)$ or "scatter matrix" $\mathbf{S}$ in the terminology of Preisendorfer (1988), the eigenvalues and eigenvectors can be determined using standard computer algorithms. From these, we obtain the variance associated with each mode, λ_j, and its time-dependent variability, $a_i(t)$.

As outlined by Davis (1976), two advantages of a statistical EOF description of the data are: (1) the EOFs provide the most efficient method of compressing the data; and (2) the EOFs may be regarded as uncorrelated (i.e., orthogonal) modes of variability of the data field. The EOFs are the most efficient data representation in the sense that, for a fixed number of functions (trigonometric or other), no other approximate expansion of the data field in terms of $K < M$ functions

$$\widehat{\psi}_m(t) = \sum_{m=1}^{K} a_i(t)\widehat{\phi}_{im} \qquad (4.43)$$

can produce a lower total mean-square error

$$\sum_{m=1}^{K} \left[\overline{\psi_m(t) - \widehat{\psi}_m(t)} \right]^2 \qquad (4.44)$$

than would be obtained when the $\widehat{\phi}_i$ are the EOFs. A proof of this is given in Davis (1976). Also, as discussed later in this section, we could just as easily have written our data $\psi(\mathbf{x}_m, t)$ as a combination of orthogonal temporal modes, $\phi_i(t)$, whose amplitudes vary spatially as $a_i(\mathbf{x}_m)$. Since this is a statistical technique, it does not matter whether we use time or space to form the basis functions. However, it might be easier to think in terms of spatial orthogonal modes that oscillate with time rather than temporal orthogonal modes that oscillate in space.

As noted above, EOFs are ordered by decreasing eigenvalue so that, among the EOFs, the first mode, having the largest eigenvalue, typically accounts for a considerable fraction of the variance of the data. Thus, with the inherent efficiency of this statistical description, only a few empirical modes generally are needed to describe the fundamental variability in a very large data set. Often it may prove useful to employ the EOFs as a filter to eliminate unwanted scales of variability. A limited number of the first few EOFs (those with the largest eigenvalues) can be used to reconstruct the data field, thereby eliminating those scales of variability not coherent over the data grid and therefore less energetic in their contribution to the data variance. An EOF analysis can then be made of the filtered data set to provide a new apportionment of the variance for those scales associated with most of the variability in the original data set. In this application, EOF analysis is much like standard Fourier analysis used to filter out scales of unwanted variability. In fact, for homogeneous time series sampled at evenly spaced increments, it can be shown that the EOFs are Fourier trigonometric functions.

The computation of the eigenfunctions, $a_i(t)$, in Eqn (4.42) requires the data values, $\psi_m(t)$, for all of the time series. Often these time series contain gaps, which make it impossible to compute $a_i(t)$ at those times for which the data are missing. One solution to this problem is to fill the gaps in the original data records using one of the procedures discussed in the previous chapter on interpolation. While this will provide an interpolation consistent with the covariance of the subject data set, these optimally estimated values of $\psi_m(t)$ often result in large expected errors if the gaps are large or the scales of coherent variability are small.

An alternative method, suggested by Davis (1976), that can lead to a smaller expected error is to estimate the EOF amplitude at time, t, directly from the existing values of $\psi_m(t)$, thus eliminating the need for the interpolation of the original data. Conditions for this procedure are that the available number of sample data pairs is reasonably large (gaps do not dominate) and that the data time series are stationary. Under these conditions, the mean product matrix $\overline{\psi_m(t)\psi_k(t)}(m, k = 1, ..., M)$ will be approximately the same as it would have been for a data set without gaps. For times

when none of the $\psi_m(t)$ values are missing, the coefficients $a_i(t)$ can be computed from Eqn (4.42). For times t when data values are missing, $a_i(t)$ can be estimated from the available values of $\psi_m(t)$

$$\widehat{a}_i(t) = b_i(t) \sum_{j=1}^{M'} \psi_j(t)\phi_{ij} \qquad (4.45)$$

where the summation over j includes only the available data points, $M' \leq M$. From Eqns (4.37), (4.43) and (4.45), the expected square error of this estimate is

$$\overline{[a_i(t) - \widehat{a}_i(t)]^2} = b_i^2(t) \sum_{j=1}^{M'} \left(\lambda_j \gamma_{ji}^2\right)$$
$$+ \lambda_i[1 + b_i(t)(\gamma_{ii} - 1)]^2 \quad (4.46)$$

where

$$\gamma_{ji} = \sum_k \phi_j(k)\phi_i(k) \qquad (4.47)$$

and the summation over k applies only to those variables with missing data. Taking the derivative of the right-hand side of Eqn (4.46) with respect to b_i, we find that the expected square error is minimized when

$$b_i(t) = (1 - \gamma_{ii})\lambda_i \left/ \left[(1 - \gamma_{ii})^2\lambda_i + \sum_j \lambda_j \gamma_{ji}^2\right]\right.$$
$$(4.48)$$

Applications of this procedure (Davis, 1976, 1978; Chelton and Davis, 1982, Chelton et al., 1982) have shown that the expected errors are surprisingly small even when the number of missing data is relatively large. This is because the dominant EOFs in geophysical systems generally exhibit large spatial scales of variability, leading to a high coherence between grid values. As a consequence, contributions to the spatial pattern from the most dominant EOFs at any particular time, t, can be reliably estimated using a relatively small number of sample grid points.

4.4.1 Principal Axes of a Single Vector Time Series (Scatter Plot)

A common technique for improving the EOF analysis for a set of vector time series is to first rotate each data series along its own customized principal axes. In this new coordinate system, most of the variance is associated with a major axis and the remaining variance with a minor axis. The technique also provides a useful application of PCA. The problem consists of finding the principal axes of variance along which the variance in the observed velocity fluctuations $\mathbf{u}'(t) = (u_1'(t), u_2'(t))$ is maximized for a given location; here u_1' and u_2' are the respective east–west and north–south components of the wind or current velocity obtained by removing the respective means $\overline{u}_1$ and $\overline{u}_2$ from each record; i.e., $u_1' = u_1 - \overline{u}_1$, $u_2' = u_2 - \overline{u}_2$. The amount of data "scatter" is a maximum along the major axis and a minimum along the minor axis (Figure 4.14). We also note that principal axes are defined in such a way that the velocity components along the principal axes are uncorrelated.

The eigenvalue problem Eqn (4.37) for a two-dimensional scatter plot has the form

$$\begin{vmatrix} C_{11} & C_{21} \\ C_{12} & C_{22} \end{vmatrix} \begin{vmatrix} \phi_1 \\ \phi_2 \end{vmatrix} = \begin{vmatrix} \lambda & 0 \\ 0 & \lambda \end{vmatrix} \begin{vmatrix} \phi_1 \\ \phi_2 \end{vmatrix} \qquad (4.49)$$

where the C_{ij}, are components of the covariance matrix, $\mathbf{C}$, derived from the current velocity data, and (ϕ_1, ϕ_2) are the eigenvectors associated with the two possible values of the eigenvalues, λ. To find the principal axes for the scatter plot of u_2' vs. u_1', we set the

FIGURE 4.14 The principal component axes for daily averaged velocity components u, v measured by a current meter moored at 175 m depth on the west coast of Canada. Here, the north–south component of velocity, $v(t)$, is plotted as a scatter diagram against the east–west component of current velocity, $u(t)$. Data cover the period October 21, 1992–May 25, 1993. The major axis along 340 °T can be used to define the longshore direction, $\mathbf{v}'$.

determinant of the covariance matrix Eqn (4.49) to zero

$$det|C - \lambda I| = det\begin{vmatrix} C_{11} - \lambda & C_{12} \\ C_{21} & C_{22} - \lambda \end{vmatrix}$$

$$= det\begin{vmatrix} \overline{u_i'^2} - \lambda & \overline{u_1'u_2'} \\ \overline{u_2'u_1'} & \overline{u_2'^2} - \lambda \end{vmatrix} = 0$$ (4.50a)

where (for $i = 1, 2$) the elements of the determinant are given by

$$C_{ii} = \overline{u_i'^2} = \frac{1}{N} \sum_{n=1}^{N} [u_i'(t_n)]^2$$ (4.50b)

$$C_{ij} = \overline{u_i'u_j'} = \frac{1}{N} \sum_{n=1}^{N} [u_i'u_j'(t_n)]$$ (4.50c)

Solution of Eqn (4.50) yields the quadratic equation

$$\lambda^2 - \left[\overline{u_1'^2} + \overline{u_2'^2}\right]\lambda + \overline{u_1'^2}\,\overline{u_2'^2} - \left(\overline{u_1'u_2'}\right)^2 = 0$$ (4.51)

whose two roots $\lambda_1 > \lambda_2$ are the eigenvalues, corresponding to the variances of the velocity fluctuations along the major and minor principal axes. The orientations of the two axes differ by 90° and the principal angles θ_p (those along which the sum of the squares of the normal distances to the data points u_1', u_2' are extrema) are found from the transcendental relation

$$\tan(2\theta_p) = \frac{2\overline{u_1'u_2'}}{\overline{u_1'^2} - \overline{u_2'^2}}$$ (4.52a)

or

$$\theta_p = \frac{1}{2}\tan^{-1}\left[\frac{2\overline{u_1'u_2'}}{\overline{u_1'^2} - \overline{u_2'^2}}\right]$$ (4.52b)

where the principal angle is defined for the range $-\pi/2 \leq \theta_P \leq \pi/2$ (Freeland et al., 1975; Kundu and Allen, 1976; Preisendorfer, 1988). As usual, the multiple $n\pi/2$ ambiguities in the angle that

one obtains from the arctangent ($\tan^{-1}$) function must be addressed by considering the quadrants of the numerator and denominator in Eqn (4.52a,b). Preisendorfer (1988; Figure 2.3) outlines the nine different possible cases. Proof of Eqn (4.52) is given in Section 4.4.5.

The principal variances (λ_1, λ_2) of the data set are found from the determinant relations Eqns (4.50a) and (4.51) as

$$\left.\begin{matrix} \lambda_1 \\ \lambda_2 \end{matrix}\right\} = \frac{1}{2}\left\{\left(\overline{u_1'^2} + \overline{u_2'^2}\right)\right.$$
$$\left. \pm \left[\left(\overline{u_1'^2} - \overline{u_2'^2}\right)^2 + 4\left(\overline{u_1'u_2'}\right)^2\right]^{1/2}\right\}$$
   (4.53)

in which the $+$ sign is used for λ_1 and the $-$ sign for λ_2. In the case of current velocity records, λ_1 gives the variance of the flow along the major axis and λ_2 the variance along the minor axis. The slope, $s_1 = \phi_2/\phi_1$, of the eigenvector associated with the variance λ_1 (i.e., the slope of the eigenvector in the north–south east–west Cartesian coordinate system) is found from the matrix relation

$$\begin{vmatrix} \overline{u_1'^2} - \lambda & \overline{u_1'u_2'} \\ \overline{u_2'u_1'} & \overline{u_2'^2} - \lambda \end{vmatrix}\begin{vmatrix} \phi_1 \\ \phi_2 \end{vmatrix} = 0$$ (4.54a)

Solving Eqn (4.54a) for $\lambda = \lambda_1$, gives

$$\left(\overline{u_1'^2} - \lambda_1\right)\phi_1 + \left(\overline{u_1'u_2'}\right)\phi_2 = 0$$
$$\left(\overline{u_2'u_1'}\right)\phi_1 + \left(\overline{u_2'^2} - \lambda_1\right)\phi_2 = 0$$
   (4.54b)

so that

$$s_1 = \left|\lambda_1 - \overline{u_1'^2}\right|\bigg/\overline{u_1'u_2'}$$ (4.54c)

with a similar expression for the slope, s_2, associated with the variance $\lambda = \lambda_2$; the products of the slopes $s_1 \cdot s_2 = -1$. If $\lambda_1 \gg \lambda_2$, then $\lambda_1 \approx \overline{u_1'^2} + \overline{u_2'^2}$, and $s_1 \approx \overline{u_1'^2}/\overline{u_1'u_2'}$. The usefulness of PCA is that it can be used to find the main orientation of fluid flow at any current meter or anemometer site, or within a "box" containing velocity variances derived from Lagrangian drifter trajectories (Figure 4.15). Since the mean and low frequency currents in relatively shallow waters are generally "steered" parallel to the coastline or local bottom contours, the major principal axis is often used to define the "alongshore" direction while the minor axis defines the "cross-shore" direction of the flow. In the case of prevailing coastal winds, the major axis usually parallels the mean orientation of the coastline or coastal mountain range that steers the surface winds. Although defining the cross-shore direction is vital to the estimation of cross-shore fluxes, reliable estimates are often difficult to obtain. This is especially true in regions where the alongshore

component of flow is strong and highly variable. In such cases, small "errors" in the specified orientation of the axes can lead to marked relative changes in the cross-shore flux estimates.

4.4.2 EOF Computation Using the Scatter Matrix Method

There are two primary methods for computing the EOFs for a grid of time series of observations. These are: (1) The scatter matrix method, which uses a "brute force" computational technique to obtain a symmetric covariance matrix **C**, which is then decomposed into eigenvalues and eigenvectors using standard computer algorithms (Preisendorfer, 1988); and (2) the computationally efficient SVD method, which derives all the components

FIGURE 4.15 Principal axes of current velocity variance (kinetic energy) obtained from surface satellite tracked drifter measurements off the coast of southern California during 1985−86. For this analysis, data have been binned into 200×200 km^2 boxes solid border denotes the region for which there were more than 50 drifter days and more than two different drifter tracks. *From Poulain and Niiler (1989).*

of the EOF analysis (eigenvectors, eigenvalues, *and* time-varying amplitudes) without computation of the covariance matrix (Kelly, 1988). The EOFs determined by the two methods are identical. The differences are mainly the greater degree of sophistication, computational speed, and computational stability of the SVD approach.

Details of the covariance matrix approach can be found in Preisendorfer (1988). This recipe, which is only one of several possible procedures that can be applied, involves the preparation of the data and the solution of Eqn (4.37) as follows:

1. Ensure that the start and end times for all M time series of length N are identical. Typically, $N > M$.
2. Remove the record mean and linear trend from each time-series record $\psi_m(t)$, $1 \leq m \leq M$ such that the fluctuations of $\psi_m(t)$ are given by $\psi'_m(t) = \psi_m(t) - [\overline{\psi_m(t)} + b_m(t - \bar{t})]$ where b_m is the slope of the least-squares regression line for each location. Other types of trend can also be removed.
3. Normalize each demeaned, detrended time series by dividing each data series by its standard deviation $s = \left[1/(N-1) \sum (\psi_{m'})^2 \right]^{1/2}$ where the summation is over all time, t $(t_n: 1 \leq n \leq N)$. This ensures that the variance from no one station dominates the analysis (all stations get an equal chance to contribute). The M normalized time-series fluctuations, ψ'_m, are the data series that we use for the EOF analysis. The total variance for each of the M eigenvalues $= 1$; thus, the total variance for all modes, $\sum \lambda_i = M$.
4. Rotate any vector time series to its principal axes. Although this operation is not imperative, it helps maximize the signal-to-noise ratio for the preferred direction. For future reference, keep track of the means, trends, and standard deviations derived from the M time series records.

5. Construct the $M \times N$ data matrix, $\mathbf{D}$, using the M rows (locations $\mathbf{x}_m$) and N columns (times t_n) of the normalized data series

$$\text{Time} \rightarrow$$
$$\mathbf{D} = \left. \begin{pmatrix} \psi'_1(t_1) & \psi'_1(t_2) & \cdots & \psi'_1(t_N) \\ \psi'_2(t_1) & \psi'_2(t_2) & \cdots & \psi'_2(t_N) \\ \cdots & \cdots & \cdots & \cdots \\ \psi'_M(t_1) & \psi'_M(t_2) & \cdots & \psi'_M(t_N) \end{pmatrix} \right\} \text{Location} \downarrow$$

$$(4.55)$$

and from this derive the symmetric covariance matrix, $\mathbf{C}$, by multiplying $\mathbf{D}$ by its transpose $\mathbf{D}^T$

$$\mathbf{C} = \frac{1}{N-1} \mathbf{D}\mathbf{D}^T \qquad (4.56)$$

where $\mathbf{S} = (N-1)\,\mathbf{C}$ is the scatter matrix defined by Preisendorfer (1988), and

$$\mathbf{C} = \begin{pmatrix} C_{11} & C_{12} & \cdots & C_{1M} \\ C_{21} & C_{22} & \cdots & C_{2M} \\ \cdots & \cdots & \cdots & \cdots \\ C_{M1} & \cdots & \cdots & C_{MM} \end{pmatrix} \qquad (4.57)$$

The elements of the real symmetric matrix $\mathbf{C}$ are

$$C_{ij} = C_{ji} = \frac{1}{N-1} \sum_{n=1}^{N} \left[\psi'_i(t_n)\psi'_j(t_n) \right] \qquad (4.58)$$

The eigenvalue problem then becomes

$$\mathbf{C}\boldsymbol{\phi} = \lambda\boldsymbol{\phi} \qquad (4.59)$$

where scalar values λ are the eigenvalues and $\boldsymbol{\phi}$ the eigenvectors.

At this point, we remark that we have formulated the EOF decomposition in terms of an $M \times M$ "spatial" covariance matrix whose time-averaged elements are given by the product $(N-1)^{-1}\mathbf{D}\mathbf{D}^T$ Eqn (4.56). We could just as easily have formed an $N \times N$ "temporal" covariance matrix whose spatially averaged elements are given by the product $(M-1)^{-1}\mathbf{D}^T\mathbf{D}$. The mean values we remove in preparing the two data sets are slightly different since preparation of $\mathbf{D}$ involves time averages while preparation of $\mathbf{D}^T$ involves spatial averages. However, in principle, the two problems are identical, and the percentage of the total time-series variance in each mode depends on whether one computes the spatial EOFs or temporal EOFs. As we further point out in the following section, another difference between the two problems is how the singular values are grouped and which is identified with the spatial function and which with the temporal function (Kelly, 1988). The designation of one set of orthogonal vectors as EOFs and the other as amplitudes is quite arbitrary.

Once the matrix $\mathbf{C}$ has been calculated from the data, the problem can be solved using "canned" programs from one of the standard statistical or mathematical computer libraries for the eigenvalues and eigenvectors of a real symmetric matrix. In deriving the values listed in Tables 4.2–4.7, we have used the double-precision program DEVLSF of the International Math and Science Library (IMSL). The program outputs the eigenvalues λ in increasing order. The time varying amplitudes $a_i(t) = \sum_{m=1}^{M}\psi_m(t)\phi_{im}$ (see Eqn (4.42)) follow from the relationship Eqn (4.66). The principal axes are derived in Table 4.4 but not used to rotate the

TABLE 4.2 Data Matrix $\mathbf{D}^T$ Components of Velocity (cm/s) at Three Different Sites at 1700-m Depth in the Northeast Pacific

Time (Days)	Site 1 (u_1)	Site 1 (v_1)	Site 2 (u_2)	Site 2 (v_2)	Site 3 (u_3)	Site 3 (v_3)
1	−0.3	0.0	0.4	−0.4	−0.8	−1.4
2	−0.1	0.3	0.4	−0.3	−1.1	0.0
3	−0.1	−0.4	0.0	−0.5	0.0	−2.5
4	0.2	0.6	0.0	−0.6	−0.7	0.4
5	0.3	−0.1	−0.6	−0.3	0.0	−0.3
6	0.5	0.0	0.9	−0.6	0.6	0.3
7	0.2	0.2	−0.1	−0.7	1.2	−2.8
8	−0.5	−0.9	0.0	−0.6	0.0	−1.8

Records Start September 29, 1985 and are Located Near 48° N, 129° W. For Each of the Three Stations We List the East–West (u) and North–South Component (v). The Means and Trends Have Not Yet Been Removed.

TABLE 4.3 Means, Standard Deviations, and Linear Trends for Each of the Time Series Components for Each of the Three Current Meter Sites Listed in Table 4.2

Component	Mean (cm/s)	Standard Deviation (cm/s) Raw Data	Standard Deviation (cm/s) Trend Removed	Trend (cm/s/day)
u_1 (east-west)	0.025	0.333	0.328	0.024
v_1 (north-south)	−0.037	0.457	0.418	−0.075
u_2 (east-west)	0.125	0.443	0.433	−0.038
v_2 (north-south)	−0.500	0.151	0.114	−0.040
u_3 (east-west)	−0.100	0.762	0.503	0.233
v_3 (north-south)	−1.012	1.278	1.250	−0.108

Means have been removed from the time series prior to the calculation of the standard deviations. The standard deviations have been calculated in two ways: with no trend removal (raw data) and with the linear trend removed.

TABLE 4.4 Principal Axes for the Current Velocity at Each Site in Table 4.2

Station ID	Angle θ (°)	Major axis (cm/s)	Minor axis (cm/s)
Site 1	59.9	0.226	0.054
Site 2	−3.3	0.172	0.020
Site 3	−69.7	1.574	0.362

The angle θ is measured counterclockwise from east. Axes lengths are in cm/s. Values have been derived using (4.52) and (4.53) without removal of the trends. Results can also be derived using eigenvalue analysis but will differ slightly from those in the table since EOF calculations generally involve removal of a linear trend. It is preferable to use the original data series when determining the principal axes.

coordinate system as we had recommended in Step four above. Moreover, we used the statistics derived for the Raw Data in Table 4.3 to obtain the results in Table 4.4, but used the Trend Removed data in Table 4.3 to calculate the results in Tables 4.5 to 4.7. Thus, the principal components (Table 4.4) are based on the non-detrended data since we argue that this is most representative of the actual flow orientation. To obtain λ in decreasing order of importance, we have had to invert the eigenvalue output.

TABLE 4.5   Eigenvalues and Percentage of Variance in Each Statistical Mode Derived from the Data in Table 4.2

Eigenvalue No.	Eigenvalue	Percentage
1	2.2218	37.0
2	1.7495	29.2
3	1.1787	19.6
4	0.6953	11.6
5	0.1498	2.5
6	0.0048	0.1
Total	6.0000	100.0

For each eigenvector or mode, the program normalizes all values to the maximum value for that mode. The amplitude of the maximum value is unity ($+=1$). Since there are M eigenvalues, the data normalization process gives a total EOF variance of $M(\sum \lambda_i = M)$. The canned programs also allow for calculation of a "performance index" (PI), which measures the error of the eigenvalue problem Eqn (4.59) relative to the various components of the problem and the machine precision. The performance of the eigenvalue routine is considered "excellent" if $PI < 1$, "good" if $1 \leq PI \leq 100$, and "poor" if $PI > 100$. As a final analysis, we can conduct an *orthogonality check* on the EOFs by using the relation Eqn (4.31). Here we look for significant

TABLE 4.6   Eigenvectors (EOFs) ϕ_1 for the Data Matrix in Table 4.2

Station ID	Mode 1	Mode 2	Mode 3	Mode 4	Mode 5	Mode 6
Site 1 u_1	1.000	−0.032	−0.430	0.479	−0.599	−0.969
Site 1 v_1	0.958	−0.078	−0.162	−0.966	1.000	0.085
Site 2 u_2	0.405	0.230	1.000	0.910	0.517	−0.295
Site 2 v_2	−0.329	−0.898	−0.525	1.000	0.784	−0.111
Site 3 u_3	0.349	1.000	−0.474	0.812	0.124	0.907
Site 3 v_3	0.654	−0.964	0.263	0.190	−0.539	1.000

Modes are Normalized to the Maximum Value for Each Mode.

TABLE 4.7   Time Series of the Amplitudes, $a_t(t)_1$ for Each of the Statistical Modes

Time	Mode 1	Mode 2	Mode 3	Mode 4	Mode 5	Mode 6
Day 1	0.798	−0.773	0.488	0.089	0.091	0.124
Day 2	−0.076	1.258	0.402	0.126	0.595	−0.089
Day 3	1.153	−1.582	−0.458	0.275	−0.492	−0.094
Day 4	−1.531	0.759	0.363	−1.585	−0.382	0.000
Day 5	0.097	1.647	−2.099	0.509	−0.128	0.039
Day 6	−2.169	−0.142	1.084	1.296	−0.171	0.008
Day 7	−0.721	−1.921	−0.866	−0.534	0.503	0.004
Day 8	2.450	0.754	1.085	−0.176	−0.017	0.008

departures from zero in the products of different modes; if any of the products

$$\sum_{m=1}^{M} \left[\phi_{im} \phi_{jm} \right]$$

are significantly different from zero for $i \neq j$, then the EOFs are not orthogonal and there are errors in the computation. A computational example is given in Section 4.4.4.

4.4.3 EOF Computation Using Singular Value Decomposition

The above method of computing EOFs requires use of covariance matrix, **C**. This becomes computationally impractical for large, regularly spaced data fields such as a sequence of infrared satellite images (Kelly, 1988). In this case, for a data matrix **D** over N time periods (N satellite images, for example), the covariance or mean product matrix is given by Eqn (4.56)

$$\mathbf{C} = \frac{1}{N-1} \mathbf{D}\mathbf{D}^T \qquad (4.60)$$

where $\mathbf{D}^T$ is the transpose of the data matrix **D**. If we assume that all of the spatial data fields (i.e., satellite images) are independent samples, then the mean product matrix is the covariance matrix and the EOFs are again found by solving the eigenvalue problem

$$C\phi = \phi\Lambda \qquad (4.61)$$

where ϕ is the square matrix whose columns are eigenvectors and Λ is the diagonal matrix of eigenvalues. For satellite images, there may be $M = 5000$ spatial points sampled $N = 50$ times, making the covariance matrix a 5000×50 matrix. Solving the eigenvalue problem for ϕ would take $\max\{O(M^3), O(MN^2)\}$ operations. As pointed out by Kelly (1988), the operation count for the SVD method is $O(MN^2)$, which represents a considerable savings in computations over the traditional EOF approach if M is large. This is primarily true for those cases where M, the number of locations in the spatial data matrix, D, are far greater than the number of temporal samples (i.e., images).

There are two computational reasons for using the SVD method instead of the covariance matrix approach (Kelly, 1988): (1) The SVD formulation provides a one-step method for computing the various components of the eigenvalue problem; and (2) it is not necessary to compute or store a covariance matrix or other intermediate quantities. This greatly simplifies the computational requirements and provides for the use of canned analysis programs for the EOFs. Our analysis is based on the double-precision program DLSVRR in the IMSL. The SVD method is based on the concept in linear algebra (Press et al., 1992) that any $M \times N$ matrix, D, whose number of rows M is greater than or equal to its number of columns, N, can be written as the product of three matrices: an $M \times N$ column-orthogonal matrix, U, an $N \times N$ diagonal matrix, S, with positive or zero elements, and the transpose (V^T) of an $N \times N$ orthogonal matrix, V. In matrix notation, the SVD becomes:

$$D = U \begin{pmatrix} s_1 & & & \\ & s_2 & & \\ & & \cdots & \\ & & & s_N \end{pmatrix} V^T \qquad (4.62)$$

For oceanographic applications, the data matrix, D, consists of M rows (spatial points)

and N columns (temporal samples). The scalars $s_1 \geq s_2 \geq \ldots \geq s_N \geq 0$ of the matrix S, called the *singular values* of D, appear in descending order of magnitude in the first N positions of the matrix. The columns of the matrix V are called the left singular vectors of D and the columns of the matrix U are called the right singular vectors of D. The matrix S has a diagonal upper $N \times N$ part, S', and a lower part of all zeros in the case when $M > N$. We can express these aspects of D in matrix notation by rewriting Eqn (4.62) in the form

$$D = [U|0]\begin{vmatrix} S' \\ \hline 0 \end{vmatrix} V^T \qquad (4.63)$$

where $[U|0]$ denotes a left singular matrix and S' denotes the nonzero part of S, which has zeros in the lower part of the matrix (Kelly, 1988).

The matrix U is orthogonal, and the matrix V has only N significant columns, which are mutually orthogonal such that,

$$\begin{aligned} V^T V &= I \\ U^T U &= I \end{aligned} \qquad (4.64)$$

Returning to Eqn (4.62), we can compute the eigenvectors, eigenvalues, and eigenfunctions of the PCA in one single step. To do this, we prepare the data as before following steps one to five in Section 4.4.2. We then use commercially available programs such as the double-precision program DLSVRR in the IMSL to solve Eqn (4.63). The elements of matrix U are the eigenvectors while those of matrix S are related to the eigenvalues $s_1 \geq s_2 \geq \ldots \geq s_N \geq 0$. To obtain the time-dependent amplitudes (eigenfunctions), we require a matrix A such that

$$D = UA^T \qquad (4.65)$$

which, by comparison with Eqn (4.62), requires $A^T = SV^T$ whereby

$$A = \left(A^T\right)^T = VS^T \qquad (4.66)$$

Hence, the amplitudes are simply the eigenvectors of the transposed problem multiplied by the transpose of the singular values, $\mathbf{S}$. Solutions of Eqn (4.62) are identical (within round-off errors) to those obtained using the covariance matrix of the data, $\mathbf{C}$. We again remark that the only difference between the matrices $\mathbf{U}$ and $\mathbf{V}$ is how the singular values are grouped and which is identified with the spatial function and which with the temporal function. The designation of $\mathbf{U}$ as EOFs and $\mathbf{V}$ as amplitudes is quite arbitrary.

The decomposition of the data matrix $\mathbf{D}$ through SVD is possible since we can write it as a linear combination of functions $F_i(x)$, $i = 1, \ldots, M$ so that

$$\mathbf{D} = \mathbf{F}\boldsymbol{\alpha} \tag{4.67a}$$

or

$$\begin{pmatrix} D(x_1, t_j) \\ D(x_2, t_j) \\ \ldots \\ \ldots \\ D(x_N, t_j) \end{pmatrix} = \begin{pmatrix} F_1(x_1) \ldots F_N(x_1) \\ F_1(x_2) \ldots F_N(x_2) \\ \ldots \\ \ldots \\ F_1(x_N) \ldots F_N(x_N) \end{pmatrix} \begin{pmatrix} \alpha_1(t_j) \\ \alpha_2(t_j) \\ \ldots \\ \ldots \\ \alpha_N(t_j) \end{pmatrix}$$

$$\tag{4.67b}$$

where the α_i are functions of time only. The functions F are chosen to satisfy the orthogonality relationship

$$\mathbf{F}\mathbf{F}^T = \mathbf{I} \tag{4.68}$$

so that the data matrix $\mathbf{D}$ is divided into orthogonal modes

$$\mathbf{D}\mathbf{D}^T = \mathbf{F}\mathbf{a}\mathbf{a}^T\mathbf{F}^T = \mathbf{F}\mathbf{L}\mathbf{F}^T \tag{4.69}$$

where $\mathbf{L} = \mathbf{a}\mathbf{a}^T$ is a diagonal matrix. The separation of the modes arises from the diagonality of the $\mathbf{L}$ matrix, which occurs because $\mathbf{D}\mathbf{D}^T$ is a real and symmetric matrix and $\mathbf{F}$ a unitary matrix. To reduce sampling noise in the data matrix $\mathbf{D}$, one would like to describe it with fewer than M functions. If $\mathbf{D}$ is approximated by $\tilde{\mathbf{D}}$, which uses only K functions ($K < M$), then the K

functions which best describe the $\mathbf{D}$ matrix, in the sense that

$$(\tilde{\mathbf{D}}\text{-}\mathbf{D})^T(\tilde{\mathbf{D}}\text{-}\mathbf{D})$$

is a minimum, are the EOFs, which correspond to the largest valued elements of the traditional EOFs, found earlier.

4.4.4 An Example: Deep Currents Near a Mid-Ocean Ridge

As an example of the different concepts presented in this section, we again consider the eight days of daily averaged currents ($N = 8$) at three deep current meter sites in the northeast Pacific near the Juan de Fuca Ridge (Table 4.2). Since each site has two components of velocity, $M = 6$, the data all start on the same day and have the same number of records. Following the five steps outlined in Section 4.4.2, we first removed the average value from each time series, but did not remove the trend. We then calculated the standard deviation for each time series and used this to normalize the time series so that each normalized series has a variance of unity. For convenience, we write the transpose of the data matrix, $\mathbf{D}^T$, where columns are the pairs of components of velocity (u, v) and rows are the time in days.

Time-series plots of the first three eigenmodes are presented in Figure 4.16. The PI for the scatter matrix method was 0.026, which suggests that the matrix inversion in the eigenvalue solutions was well defined. A check on the orthogonality of the eigenvectors suggests that the SVD gave vectors, which were slightly more orthogonal than the scatter matrix approach. For each combination (i, j) of the orthogonality condition Eqn (4.31), the products $\Sigma_{i,j}[\phi_{im}, \phi_{jm}]$ were typically of order 10^{-7} for the SVD method and 10^{-6} for the scatter matrix method. Similar results apply to the orthogonality of the eigenmodes given by Eqn (4.33).

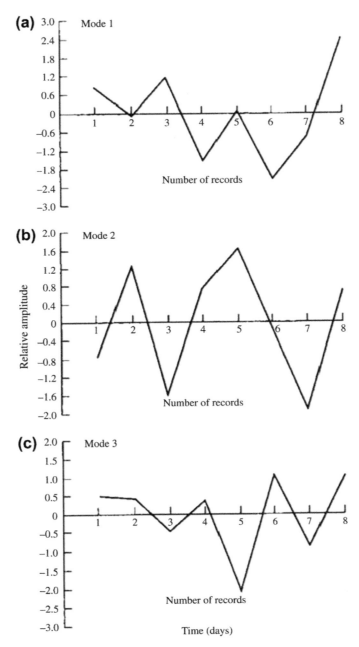

FIGURE 4.16 Eight-day time series for the first three EOFs for current meter data collected simultaneously at three sites at 1700-m depth in the northeast Pacific in the vicinity of Juan de Fuca Ridge, 1985. Modes 1, 2, and 3 presented in (a), (b) and (c), respectively, account for 37.0, 29.2, and 19.6% of the variance, respectively.

Before closing this section, we remark that we also could have performed the above analysis using complex EOFs of the form

$$\psi_m(t) = u_m(t) + iv_m(t)$$

in which case $M = 3$. This formulation not only allows the EOF vectors to change amplitude with time, as in our previous decomposition using $2M$ real EOFs, but also to rotate in time.

4.4.5 Interpretation and Examples of EOFs

In interpreting the meaning of EOFs, it is worth keeping in mind that, while EOFs offer the most efficient statistical compression of the data field, empirical modes do not necessarily correspond to true dynamical modes or modes of physical behavior. Often, a single physical process may be spread over more than one EOF. In other cases, more than one physical process may be contributing to the variance contained in a single EOF. The statistical construct derived from this procedure must be considered in light of accepted physical mechanisms rather than as physical modes themselves. It often is likely that the strong variability associated with the dominant modes is attributable to several identifiable physical mechanisms. Another possible clue to the physical mechanisms associated with the EOF patterns can be found in the time-series coefficients $a_i(t)$. The temporal variability of certain processes might resemble the time series of the EOF coefficients, which would then suggest a causal relationship not readily apparent in the spatial structure of the EOF.

One way to interpret EOFs is to imagine that we have displayed the data as a scatter diagram in an effort to discover if there is any inherent correlation among the values. For example, consider two parameters such as SST and sea-level pressure (SLP) measured at a number of points over the North Pacific.

This is the problem studied by Davis (1976) where he analyzed sets of monthly SST and SLP over a period of 30 years for a grid in the North Pacific. If we plot $x = $ SST against $y = $ SLP in a scatter diagram, any correlation between the two would appear as an elliptical cluster of points. A more common example is that of Figure 4.14 where we plotted the north–south (y) component of daily mean current, v, against the corresponding east–west (x) component, u, for a continental shelf region. Here, the mean flow tends to parallel the coastline, so that the scatter plot again has an elliptical distribution. To take this distribution into account, we redefine our coordinate system by rotating $x(u)$ and $y(v)$ through the counterclockwise angle θ to the principal axes representation x', y' (u', v') discussed in Section 4.4.2. These transformations are given by

$$
\begin{aligned}
x' &= x \cos\theta + y \sin\theta \\
y' &= -x \sin\theta + y \cos\theta
\end{aligned}
\tag{4.70}
$$

where (u', v') are found by simply replacing x with u and y with v in the above equations. Note that, in the case of currents, θ is measured in the counterclockwise direction from east in this coordinate system (which confuses things somewhat since east is 90° in terms of true compass bearing; north is 0° $\equiv$ 360° T). What we have done in this rotation is to formulate a new set of axes that explains most of the variance, subject to the assumption that the variance does not change with time. Since the axes are orthogonal, the total variance will not change with rotation. Let $V = \overline{x'^2} = N^{-1}\sum x'^2$ be the particular variance we want to maximize (as usual, the summation is over all N values of the time series). Note that we have focused on x' whereas the total variance is actually determined by r^2, where r is the distance of each point from the origin. However, we can expand $r^2 = x^2 + y^2$ and associate the variance with a given coordinate. In other words, if we maximize the variance associated with x', we will minimize the variance associated with y'. Using our summation convention, we can write

$$V = \overline{x'^2} = \overline{x^2}\cos^2\theta + 2\overline{xy}\sin\theta\cos\theta + \overline{y^2}\sin^2\theta \tag{4.71}$$

and

$$\frac{\partial V}{\partial\theta} = 2\left(\overline{y^2} - \overline{x^2}\right)\sin\theta\cos\theta + 2\overline{xy}\cos 2\theta \tag{4.72}$$

We maximize Eqn (4.72) by setting $\partial V/\partial\theta = 0$, giving Eqn (4.52a), which we previously quoted without proof

$$\tan(2\theta_p) = \frac{2\overline{xy}}{\overline{x^2} - \overline{y^2}} \tag{4.73}$$

From Eqn (4.73), we see that if

$$\overline{xy} \ll \max\left(\overline{x^2}, \overline{y^2}\right)$$

then $\tan(2\theta_p) \to 0$ and $\theta_P = 0$, or $\pm 90°$, and we are left with the original axes. If $\overline{x^2} = \overline{y^2}$ and $\overline{xy} \neq 0$, then $\tan(2\theta_p) \to \pm\infty$ and the new axes are rotated $\pm 45°$ from the original axes.

We now find the expression for V. Since $\sec^2(2\theta) = 1 + \tan^2(2\theta)$

$$\cos 2\theta = \left(\overline{x^2} - \overline{y^2}\right) / \pm D$$
$$\sin 2\theta = \left[1 - \cos^2(2\theta)\right]^{1/2} = 2\overline{xy}/\pm D \tag{4.74}$$

where

$$D = \left[\left(\overline{x^2} - \overline{y^2}\right)^2 + 4\overline{xy}^2\right]^{1/2} \tag{4.75}$$

Then, using the identities

$$\cos^2\theta = \frac{1}{2}(1 + \cos 2\theta),$$
$$\sin^2\theta = \frac{1}{2}(1 - \cos 2\theta) \tag{4.76}$$

we can write the variance as

$$V = \overline{x^2}\frac{(1 + \cos 2\theta_p)}{2} + \overline{y^2}\frac{(1 - \cos 2\theta_p)}{2} + \overline{xy}\sin 2\theta_p$$
$$= \frac{1}{2}\left\{\left(\overline{x^2} + \overline{y^2}\right) \pm \left[\left(\overline{x^2} - \overline{y^2}\right)^2 + 4\overline{xy}^2\right]^{1/2}\right\} \tag{4.77}$$

The two roots of this equation correspond to a maximum and a minimum of V. For a new axis for which $\overline{x'^2}$ is a maximum, we will find $\overline{y'^2}$ a minimum. This follows automatically from the fact that the total variance is conserved. However, we can confirm this mathematically by computing $\partial^2 V/\partial\theta^2 = 0$. From Eqn (4.72), and using Eqns (4.74) and (4.75), we find maximum (minimum) values from

$$\partial^2 V\Big/\partial\theta_p^2 = 2\left(\overline{y^2} - \overline{x^2}\right)\cos 2\theta_p - 4\overline{xy}\sin 2\theta_p$$
$$= -2\left[\left(\overline{x^2} - \overline{y^2}\right)^2 + 4\overline{xy}^2\right]\Big/\pm D$$
$$= \pm 2D = 0 \tag{4.78}$$

The positive sign in Eqn (4.78) corresponds to a maximum (since Eqn (4.77) is negative); the negative sign corresponds to a minimum. Solving Eqn (4.78) using Eqn (4.75), yields the relationship between the variances in the x and y variables. It so happens that the variance solutions given by Eqn (4.78) are also the eigenvalues of the covariance matrix. Thus, we can return to our previous methods where we used the covariance matrix to compute the EOFs.

A published example of EOF analysis is presented by Davis (1976), who examined monthly maps of SST and SLP for the years 1947–74. The SLP data were originally obtained from the Long-Range Prediction Group of the U.S. National Meteorological Center (NMC) as one-month averages on a 5°-diamond-shaped grid (i.e., 20° N–140° W, 20° N–150° W, ...,

25° N—145° W, 25° N—155° W, etc.). The data were transferred to a regular 5°-square grid using linear interpolation from the four nearest diamond grid points to fill in the square grid. The SST data were obtained from the U.S. National Marine Fisheries Service in the form of monthly averages over 2° squares. Because this grid spacing is not a submultiple of 5°, and because sometimes data were missing, the following data analysis scheme was employed. The 2° data were subjectively analyzed to produce maps contoured with a 1 °F contour interval. During this stage, missing values were filled in where feasible. The corrected values were then linearly interpolated onto a 1° grid and 25 values were averaged to formulate area averages on the chosen 5° grid coincident with the SLP data. The ship data originated as ship injection temperatures and are subject to all of the problems discussed earlier in the section on SST.

Before carrying out the EOF analysis, the SST and SLP data sets were further averaged onto a grid with a 5°-latitude spacing and a 10°-longitude spacing (Figure 4.17). In those cases where some SST values were missing, the available observations were used to compute the grid average. Even then there were some 5° × 10° regions with missing data in the SST fields. Both fields were then converted to anomalies using the mean of the 28-year data set as the reference field. Thus, each of the individual monthly maps was transformed into an anomaly map, corresponding to the deviation of local values from the long-term mean.

The standard deviations of both the SLP and SST anomaly fields are shown in Figure 4.18. It is interesting to note some of the basic differences between the variability of these two fields. The SLP field has its primary variability in the central northern part of the field just off the tip of the Aleutian Islands. Here, the Aleutian Low dominates the pressure field in winter and becomes the source of the main variability in the SLP data. In contrast, the SST field has near-uniform variance levels except in the Kuroshio Extension region off of northeast Japan where a maximum associated with advection from the Kuroshio is clearly evident.

To compute the EOFs from the anomaly fields, Davis (1976) used the covariance (scatter) matrix method presented in Section 4.4.2. The fraction of total variance accounted for by the EOFs for both the SST and SLP data is presented in Figure 4.19 as a function of the number of EOFs. The steep slope of the SLP curve means that fewer SLP EOFs are needed to express the variance. The slope of the SST EOF curve is consistently below that for the SLP EOF series. As a consequence, Davis presented only the first six SLP EOFs (labeled P_1-P_6 in Figure 4.20) but felt it necessary to present the first eight SST EOFs (labeled T_1-T_8 in Figure 4.21). The SLP EOFs exhibited fairly simple, large-scale patterns with P_1 having the same basic shape as the SLP

FIGURE 4.17 The grid of sea surface temperature (SST) and sea-level pressure (SLP). The 10° longitude by 5° latitude SLP averages are centered at grid intersections and SST averages are centered at crosses. *From Davis (1976).*

FIGURE 4.18 Standard deviation of: (a) Sea level pressure anomaly (mb); and (b) Sea surface temperature anomaly (°C) for the North Pacific. The anomalies are departures from monthly normal values. Variances are averaged over all months of the 28-year record (1947–1974). *From Davis (1976).*

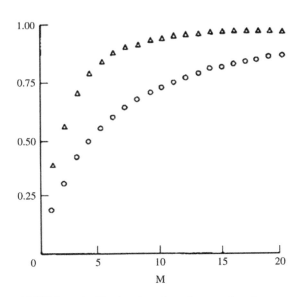

FIGURE 4.19 The fraction of total sea surface temperature (circles, o) and sea-level pressure (triangles, Δ) anomaly variance accounted for by the first M empirical orthogonal functions. *From Davis (1976).*

standard deviation (Figure 4.18). The structural sequence for the first three SLP EOFs was: For P_1, a single maximum; for P_2, two meridionally separated maxima; and for P_3, two zonally separated maxima. Higher modes appear to be combinations of these first three with an increasing number of smaller maxima.

The SST maps obtained by Davis were considerably more complicated than the SLP maps, with large-scale patterns dominating only the first three modes of the temperature field. As with the SLP modes, the sequence seems to be from a central maximum (T_1), to meridionally separated maxima (T_2), and then to zonally separated maxima (T_3). The higher-order EOFs have a number of smaller maxima with no simple structures. The overall scales are much shorter than those for the SLP EOFs. This turns out to be true for the time scales of the EOFs, with the SLP time scales being much shorter than those computed for the SST EOFs.

FIGURE 4.20 The six principal empirical orthogonal functions P_1-P_6 describing the sea level pressure anomalies. Function numbers are written to the left of each panel. *From Davis (1976).*

FIGURE 4.21 The eight principal empirical orthogonal functions T_1-T_8 describing the sea level pressure anomalies. Function numbers are written to the left of each panel. *From Davis (1976).*

The goal of the EOF analysis by Davis (1976) was to determine if there is some direct statistical connection between the SLP and SST anomaly fields. By using the EOF procedure he was able to present the primary modes of variability for both fields in the most compact form possible. This is the real advantage of the EOF procedure. In terms of the two anomaly fields, Davis found that there were connections between the variables. First, he found that SST anomalies could be predicted from earlier SST anomaly fields. This is a consequence of the persistence of individual SST patterns as well as the fact that some patterns appear to evolve from earlier patterns through advective processes. Davis also concluded that it was possible to specify the SLP anomaly on the basis of the coincident SST anomaly field. Finally, it was not possible to statistically predict the SST field from the simultaneous SLP field. These conclusions, would have been difficult to arrive at without using the EOF procedure, are consistent with the much greater heat capacity and persistence ("memory") of SST anomalies compared to SLP anomalies.

4.4.6 Variations on Conventional EOF Analysis

Conventional principal component (EOF) analysis is limited by a number of factors, including the dependence of the solution on the domain of analysis, the requirement for orthogonal spatial modes, and the lumping together of variability over all frequency bands. In addition, the method can detect standing waves but not progressive waves. Over the years, several authors have developed what might be called "variations" on the standard EOF theme. For the most part, the methods differ in the types of variances they insert into the algorithms used to determine the EOFs (principal components). Given that EOF analysis is a strictly statistical method, it is irrelevant how the variance is derived, provided that the type of variance

used in the analysis is the same for all spatial locations. All that required is that the matrix $\mathbf{D}$, derived from statistical averages (such as the covariance, correlation, and cross-covariance functions) of the gridded time series is a Hermitian matrix.

Departure from standard EOF analysis can have numerous forms. For example, one may choose to work in the frequency domain instead of the time domain by using spectral analysis to calculate the spectral "energy" density for specific frequency bands. In this case, the matrix $\mathbf{D}$ is complex, consisting of the cross spectra between the gridded time series over a specific frequency band. The spectral densities represent the data variances, which are used to determine the EOFs. Thus, the method is equally at home with variances obtained in the time or frequency domains. Regardless of variance-type, principal component methods are simply techniques for compressing the variability of the data set into the fewest possible number of modes.

Returning to the time domain, suppose that we are examining the statistical structure of alongshore wind and current fluctuations over the continental shelf and that we have reason to believe that current response to wind forcing is delayed by one or more time steps in the combined data series. A delay of half a pendulum day (≈ 12 h at mid-latitudes) is not unreasonable. From a causal point of view, the best way to examine the EOF modes for the combined wind and current data is to first create new time series in which the wind records are lagged (shifted forward in time) relative to the current records. Suppose we want a delay of one time step, then, alongshore wind velocity values $V_k(t_j)$ at site k at times t_j ($j = 2, 3, \ldots$) get replaced with the earlier records at times t_{j-1}. That is, $V_k(t_j) \rightarrow V_k(t_{j-1}) = V_k^*(t_j)$, while the current velocity record remains unchanged, $v_k(t_j) = v_k^*(t_j)$. In this case, the asterisk (*) denotes the new time series. Optimal empirical modes are those for which the wind and current records are properly "tuned" with the correct time lags. For large spatial regions with variable wind response times, this can get a little tricky so caution is advised.

4.5 EXTENDED EMPIRICAL ORTHOGONAL FUNCTIONS

EEOFs are an extension of the traditional spatial EOFs and are formulated to deal not only with spatial but also with temporal correlations in the space-time data sets. This departure from conventional EOF analysis was presented by Kundu and Allen (1976) who combined the zonal (u) and meridional (v) time series of currents into complex time series $U = u + iv$, where each scalar series is defined for times t_j and locations x_k. The method was applied to current data collected during the Coastal Upwelling Experiment (CUE-II) off the Oregon coast in the summer of 1973. The complex covariance matrix obtained from these time series were then decomposed into complex eigenvectors by solving a standard complex eigenvalue problem. Unlike the scalar approach to the problem, this complex EOF technique can be used to describe rotary current variability within selected frequency bands.

A further variation on conventional EOF analysis, which is related to complex EOF analysis, was provided by Denbo and Allen (1984). Using a technique we describe in Chapter 5, the current fluctuations in each of the time series (u, v) records collected during CUE-II were decomposed into clockwise (S^+) and counterclockwise (S^-) rotary spectra. The spectra (corresponding to the variance per unit frequency range) for the dominant spectral components, which is typically S^- in the ocean, were then decomposed into EOFs by solving the standard complex eigenvalue problem. Known as *rotary empirical orthogonal function analysis*, the method is best suited to flows with strong rotary signals such as continental shelf waves and near-inertial motions, but is not well suited to highly rectilinear flows such as those in tidal channels for which

S^+ and S^- are of comparable amplitude (see Hsieh, 1986; Denbo and Allen, 1986).

The first use of *complex EOFs* in the frequency domain was described by Wallace and Dickinson (1972) and subsequently used by Wallace (1972) to study long-wave propagation in the tropical atmosphere. Early oceanographic applications are provided by Hogg (1977) for long waves trapped along a continental rise and by Wang and Mooers (1977) for long, coastal-trapped waves (CTWs) along a continental margin. In this approach, complex eigenvectors are computed from the cross-spectral matrices for specified frequency bands. This is the most general technique for studying propagating wave phenomena. As noted by Horel (1984), however, EOF analysis in the frequency domain can be cumbersome if applied to time series in which the power of a principal component is spread over a wide range of frequencies as a result of nonstationarity in the data. Horel presents a version of complex EOF analysis in the time domain in which complex time series of a scalar variable are formed from the original time series and their Hilbert transforms. The complex eigenvectors are then determined from the cross-correlation or cross-covariance matrices derived from the complex time series. The Hilbert transform $u_m^H(t)$ of the original time series $u_m(t)$ represents a filtering operation in which the amplitude of each spectral component remains unchanged but the phase of each component is shifted by $\pi/2$. Because of this 90° shift in phase, the Hilbert transform is also known as the quadrature function. Expanding the scalar time series

$$u_m(t) = \sum_{\omega} [a_m(\omega) \cos(\omega t) + b_m(\omega) \sin(\omega t)]$$

(4.79)

as a Fourier series over all frequencies, ω, the Hilbert transform $u_m^H(t)$ is

$$u_m^H(t) = \sum_{\omega} [b_m(\omega) \cos(\omega t) - a_m(\omega) \sin(\omega t)]$$

(4.80)

In practice, the Hilbert transform can be derived directly from the coefficients of the Fourier transform of $u_m(t)$, although with the usual problems caused by aliasing and truncations effects. The complex covariance matrix $r_{mk} = \overline{U_m(t)U_k(t)^*}$ obtained for the series $U_m(t) = u_m(t) + iv_m(t)$ and its complex conjugate, $U_k(t)^*$, are shown to be useful for identifying traveling and standing wave modes; here, (u, v) are the zonal and meridional components of velocity. In the extreme case where the data set is dominated by a single frequency, the frequency domain EOF technique and complex time domain EOF technique are identical. According to Merrifield and Guza (1990), the Hilbert transform complex EOF only makes sense if the frequency distribution in the original time series $(u_m(t), v_m(t))$ is narrow band.

EEOFs were used in meteorological studies by Weare and Nasstrom (1982), and later applied to find propagating features in the upper atmosphere (Kimoto et al., 1991; Plaut and Vautard, 1994). In EEOF analysis, the state vector at time t, used in the traditional EOFs is "extended" to include temporal variations as (Hannachi, 2004)

$$\mathbf{X}_t = (X_{t,1}, \dots, X_{t+M-1,1}; X_{t,2}, \dots, X_{t+M-2,2}; \dots;$$

$$X_{t,p}, \dots, X_{t-1,p})$$

(4.81)

where $t = 1, \dots, n - M + 1$. The parameter M is known as the window-length or delay parameter, and p refers to the number of eigenvalues or spatial EOFs in the data set. The data matrix becomes

$$\mathbf{X} = \begin{pmatrix} \mathbf{x}_1 \\ \mathbf{x}_2 \\ \dots \\ \mathbf{x}_{n-M+1} \end{pmatrix}$$

(4.82)

From Eqn (4.80) we can see that time is now included with the spatial dimension. We can now write the data matrix as

$$\mathbf{X} = \begin{pmatrix} \mathbf{x}_1^1 & \mathbf{x}_1^2 & \mathbf{x}_1^p \\ \mathbf{x}_2^1 & . & . \\ . & . & . \\ \mathbf{x}_{n-M+1}^1 & \mathbf{x}_{n-M+1}^2 & \mathbf{x}_{n-M+1}^p \end{pmatrix}$$

(4.83)

which is similar to the data matrix in Eqn (4.38a) except that now the elements of the data matrix are vectors rather than scalars. This new data matrix is of the order $(n - M + 1)pM$. The covariance matrix of Eqn (4.83) is

$$\mathbf{C} = \frac{1}{n - M + 1} \mathbf{X}^T \mathbf{X}$$

$$= \begin{pmatrix} C_{11} & C_{12} & \cdots & C_{1M} \\ C_{21} & C_{22} & \cdots & \cdots \\ \cdots & \cdots & \cdots & \cdots \\ C_{M1} & \cdots & \cdots & C_{MM} \end{pmatrix}$$

(4.84)

where each C_{ij} ($1 \le i, j \le M$) is a lagged covariance matrix between gridpoint i and gridpoint j given by

$$C_{ij} = \frac{1}{n - M + 1} \sum_{k=1}^{n-M+1} \mathbf{x}_k^{i\,T} \mathbf{x}_k^j$$

(4.85)

For large values of n (as compared to the window length M) the covariance matrix is approximately a diagonal-constant matrix where each descending diagonal from left to right is constant. This is generally the case when we deal with daily observations or even monthly averages derived from more frequently sampled data.

EEOFs are the EOFs for extended versions of the data matrix given by Eqn (4.83), corresponding to the eigenvectors of the covariance matrix given by Eqn (4.84). The EEOFs can be computed directly by computing the eigenvalues/eigenvectors of Eqn (4.83) using the singular-value-decomposition method. With this formulation we can write

$$\mathbf{X} = \mathbf{VAU}^T$$

(4.86)

where $\mathbf{U} = (u_{ij}) = (\mathbf{u}_1, \mathbf{u}_2, ..., \mathbf{u}_d)$ represents the matrix of the d EEOFs, or the left singular vectors, of $\mathbf{X}$, and $d = Mp$ is the number of new variables represented by the number of columns in the data matrix. The diagonal matrix $\mathbf{A}$ contains the singular values $a_1, ..., a_d$ of $\mathbf{X}$ and $\mathbf{V} = (\mathbf{v}_1, \mathbf{v}_2, ..., \mathbf{v}_d)$ is the matrix of the right singular vectors, or the extended principal components.

In summary, conventional EOF analysis in the time domain works best when the variance is dominated by standing waves and spread over a wide range of frequencies and wavenumbers. Frequency domain EOF analysis should be used when the dominant variability within the data set is concentrated into narrow frequency bands. Rotary spectral EOF analysis is best used for data sets in which the variance is in narrow frequency bands and dominated by either the clockwise or counterclockwise rotating component of velocity. Complex time domain PCA allows for the detection of propagating wave features (if the process has a narrow frequency band) and the identification of these

motions in terms of their spatial and temporal behavior. However, regardless of which method is applied, the best test of a method's validity is whether the results make sense physically and whether the variability is readily visible in the raw time series.

4.5.1 Applications of EEOFs

In a study of tropical disturbances using data from the Tropical Ocean Global Atmosphere—Coupled Ocean-Atmosphere Response Experiment (TOGA COARE), Fraedrich et al. (1997) used windowed, vertical time delay EEOFs to examine connections between different modes of vertical and temporal variability in the tropical atmosphere. At the same time, the authors wanted to determine the dominant modes of variability for the wind and diabatic heating in the equatorial Western Pacific during TOGA-COARE. Focus was on observations from the TOGA-COARE Intensive Observing Period (IOP), which consisted of a set of rawinsonde soundings at seven

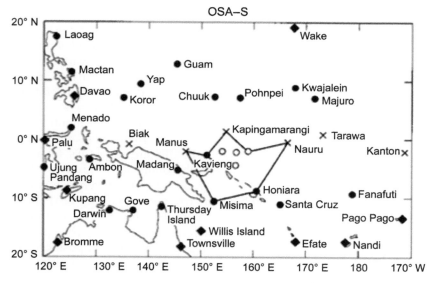

FIGURE 4.22 The OSA-S rawinsonde array (Solid line) of TOGA COARE stations over which the heating, drying, and wind time series are calculated. *From Frank et al. (1996).*

stations in the equatorial Pacific (Figure 4.22). The data were extracted from the Australian Bureau of Meteorology Tropical Analysis and Prediction System. Additional ship and island rawinsondes were collected during the IOP and were added to the data set. Some of the data were acquired from the National Center for Atmospheric Research (NCAR) along with a large number of surface reports, mostly from the National Climate Center of the Bureau of Meteorology.

All soundings were quality controlled to remove highly suspicious values. In order to remove the effects of surface winds measured at island stations, surface wind speeds were estimated as 72% of the wind speeds at the 850-hPa pressure level and all data averaged over 24 h to produce daily mean fields from the six hourly sonde launches. Only data from the 11 standard levels (surface, 1000, 850, 700, 500, 400, 300, 250, 200, 150, and 100 hPa) were used. Missing daily values (roughly 5% of the data set) were filled using a horizontal least squares fit (Frank, 1979). This procedure produced a complete data set at all stations at the 11 standard levels for all 120 days of the IOP.

The four variables used in this study were daily averages of zonal and meridional winds, along with heat and moisture fluxes in and out of the atmosphere. Since the heat and moisture budgets were based on daily averages over a fairly large area, they are representative of the net effects of many of the clouds and mesoscale connective systems that were incorporated into the averaged calculations. EEOF analysis was applied to the data, which was able to delineate the temporal evolution of the spatial patterns in these data sets. Consider a system described by a vector of K components ordered according to height and evolving with time. The vector series is weighted using a sliding window of length W. This generates a new vector series whose components constitute a height—time section, KXW, which evolves in time and represents the states of the system in KXW-dimensional phase space. Fraedrich et al. (1997) note that the following

should be considered when applying this method:

1. Using a sliding window results in some smoothing. The degree of smoothing depends on the window length, whereby a longer window provides a smoother reconstruction. An optimal choice of the window length (W) is the half-period of the longest signal. This resolves the signal with sufficient detail while providing sufficient data smoothing to increase the signal-to-noise ratio.
2. Wavelike oscillations are represented by a pair of EEOFs with similar eigenvalues, with the associated EEOF patterns shifted by quarter of a wavelength. In this way, EEOFs are able to properly represent propagating waves, whereas standard EOFs can only represent standing waves.
3. The advantage of EEOFs is their ability to: (1) represent dominant internally coherent patterns in both space and time; and (2) to distinguish oscillatory components with a variety of frequencies. These oscillatory components are characterized by pairs of EEOFs in quadrature and which are associated with eigenvalues of similar magnitude.

The first two EEOFs of the covariance matrix of the (u, v) height—time series are presented in the top frame of Figure 4.23. The two EEOFs explain 48.3% of the variance in the wind (Table 4.8). Here, we note that, for each mode, the variations in u are much greater than those in v.

In Figure 4.23, the wind EEOFs are contoured and the corresponding u,v vectors are plotted over these contour fields. EEOF1 shows a vertical wavenumber-one structure with the zonal wind varying most near the tropopause and near 800 hPa. EEOF2 is similar to EEOF1 but its phase is in quadrature with EEOF1. This suggests that this pair of EEOFs represent a single mode of variation. As indicated in Figure 4.23, this variation is approximately twice the

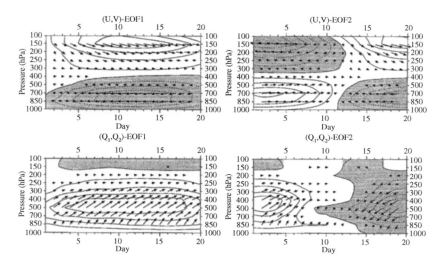

FIGURE 4.23 The first EOF pairs of the (u, v) and (Q_1, Q_2) height–time series with a 20-day window. Top panels: The vectors represent wind direction and strength. The contours are u-component (shaded negative); units are in meters per second. Bottom panels: The vectors represent (Q_1, Q_2) components with positive Q_1 upward and positive Q_2 to the right. The contours are Q_1 component (shaded negative); the units are in degrees per day.

window length, or about 40 days, and likely dictated by the 40–60 day Madden–Julian oscillation known to dominate these time-space series.

The first two modes of the heat-flux (Q_1, Q_2) height–time series are shown in the lower panel of Figure 4.23 and their contributions to the variance given in Table 4.1. As this table shows, the first two EEOFs explain 41% of the variance in

TABLE 4.8   Percent Variance (Left) and the Running Total (Right) for each of the two pairs of columns Contributed by the First Six EOFs in the Height–Time Delay Analysis of the Total (u, v) and (Q_1, Q_2) Variances

EOF	(u, v)	Wind	(Q_1, Q_2)	Heating
1	27.3	27.3	28.4	28.4
2	21.0	48.3	12.2	40.6
3	7.7	56.0	7.9	48.5
4	7.1	63.1	7.6	56.1
5	6.8	69.9	4.7	60.8
6	3.4	73.3	3.2	64.0

the time-space series of heat flux. As with the wind EEOFs, the two heat-flux modes are similar to each other and shifted in phase by roughly 90°. Again, the time scale is approximately twice the window size, or about 40 days.

Variations in Q_1 and, to a lesser extent, in Q_2 are dominated by variability in the vertical velocity field, the vertical structure of these quantities corresponds to an internal wavenumber-one structure in divergence. A similar pattern persists for the high-order EEOFs (not shown) suggesting that there is no significant difference in vertical heating as a function of wavenumber.

One unique aspect of the EEOF analysis is the ability to view the phase relationship between the EEOF modes. As noted earlier EEOF1 and EEOF2 appear to be similar but phase shifted. This can be seen more clearly in Figure 4.24 where EEOF1 is plotted against EEOF2. In the top-left panel, it is clear that EEOF structure is similar and the phase shift results in an almost circular rotation between the paired EEOF modes for the wind components. Comparison with the lower left panel reveals that this relationship is relatively independent of window size as the only change here is an increase in window size from 20 to 30 days. It is also interesting that the heating EEOFs are similar for the two window sizes and while they are noisier than for the wind they also exhibit this rotational behavior consistent with a shift in phase between the two EEOF modes. The period of the oscillations is the time required for a complete revolution in these diagrams. For the wind (Figure 4.24(a)) the vertical axis is crossed at 40 and 80 days, while the horizontal axis is crossed

at days 49 and 88 days resulting in an overall period of 39—40 days. Similarly the heating diagrams yield a period of about 41 days.

In this study the EEOF analysis has revealed a dominant pattern of space-time variation in both wind and heating—drying with a period of approximately 40 days. This result is very consistent with the temporal period discussed for TOGA COARE by Gutzler et al. (1994) and McBride et al. (1995). All of these authors associate this variation with the Madden—Julian oscillation. The structure of this oscillation as seen in EEOF1 and EEOF2 is dominated by zonal

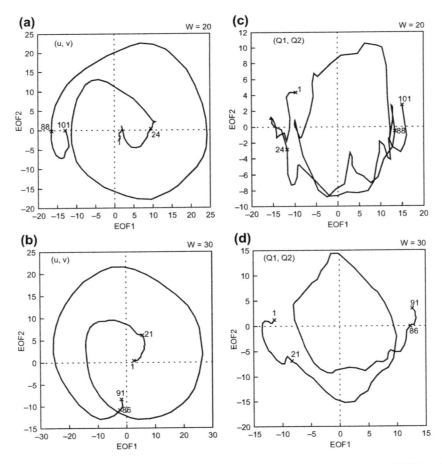

FIGURE 4.24 Dial of the principal components of the first height—time delay eigenvector pair (EOF 1 and EOF 2) for the two window lengths ($W = 20$, 30 days): (a), (b) the u, v winds and (c), (d) the (Q_1, Q_2) heating. Particular day numbers indicated are discussed in the text.

wind variations with a vertical structure of the first internal mode. This zonal wind pattern is consistent with an equatorially trapped Kelvin wave but the node is not totally a Kelvin wave.

In their analysis of Pacific SSTs anomalies, Weare and Nasstrom (1982) present the first and/or most important EEOF at three different times (Figure 4.25). The features are consistent with the standard EOF fundamental modes, which have been associated with El Niño warming of the eastern equatorial Pacific (Weare, 1981). The features in Figure 4.25 show remarkable persistence over time, which also agrees

with previous studies of equatorial Pacific behavior. Along with this strong persistence, Figure 4.25 suggests a shift of the maximum variability along the equator westward from the South American coast during the 6-month period examined in this study, which also agrees with previous studies of areal-averaged data in this region (Weare, 1982).

The second most important EEOF of Pacific SST anomalies (Figure 4.26) shows greater differences in the six-month sequence than are apparent in the first mode in Figure 4.25. As time processes, there is an extension westward

FIGURE 4.25 Most important "extended" empirical orthogonal function of monthly departures of tropical Pacific Ocean surface temperature for times t, $t+3$, and $t+6$ months reading from top to bottom.

FIGURE 4.26 Second most important function as in Figure 4.25.

along the equator of the positive values near Peru at a speed of about 0.25 m/s. This speed is in good agreement with estimates of the ocean current speeds in this region (Wyrtki, 1977). During this same period, a large negative feature appears to recede northward at a somewhat slower speed. Figure 4.26 suggests that SST anomalies associated with the Peru-South Equatorial Current system are often opposite in sign to changes in the California—North Equatorial Current system. Since both of these systems advect relatively cold water, this 180° (out-of-phase) relationship suggests that the southern currents weaken while the northern currents strengthen, and vice versa.

4.6 CYCLOSTATIONARY EOFs

In EOF analysis, a set of orthogonal eigenfunctions is found from a spatial covariance function. Data are decomposed into the sum of a set of individual modes composed of a single spatial pattern and a corresponding amplitude time series, referred to as the "loading vectors" (LVs) and "principal component time series" (PCTS), respectively. These empirically derived basis functions can provide useful insight into the physical processes behind the data and serve as a useful tool for developing statistical methods. The underlying assumption in EOF analysis is that the data being analyzed are stationarity so that the covariance function of the data does not depend on time. By definition, the spatial patterns represented by the EOF LVs are time independent (stationary) so that only the amplitudes of these stationary patterns vary in time, as described by the PCTS. However, geophysical variables, including climate-related variables, are rarely stationary even after the removal of cyclic components like the annual and semiannual cycles. Subsequently, physical inferences based on EOFs of climate signals can be misleading and potentially erroneous. The spatial patterns

of many phenomena in geophysics and climate science show the presence of seemingly random fluctuations in addition to a deterministic component such as the annual cycle. Such signals change in time with well-defined periods (deterministic components) in addition to fluctuating at longer timescales, and are thus best described by time-dependent covariance functions. These signals are said to be periodically correlated or cyclostationary. Because physical systems are generally not stationary but evolve and change over time, a suitable representation of this time-dependent response is important for the extraction of physically meaningful modes and their space-time evolutions from the data.

The decomposition in terms of a set of basis functions is often useful in understanding the complicated response of a physical system. When decomposed into simpler, basic patterns, insight can be gained into the nature of variability of a given system. While theoretical basis functions have been studied extensively, exact theoretical basis functions are very difficult to find and in general, computational basis functions are sought instead. Perhaps the simplest and most common computational basis functions are EOFs. Consider a simple system defined by:

$$T(x,t) = B(x,t)S(t) \qquad (4.87)$$

where $B(x, t)$ is a deterministic physical process that is modulated by a stochastic time series process, $S(t)$. It follows that the mean and the space-time covariance function are given by:

$$\mu(x,t) = \langle T(x,t)\rangle = B(x,t)\langle S(t)\rangle = B(x,t)\mu_s \qquad (4.88)$$

$$C(x,t;x',t') = \langle T(x,t)T(x',t')\rangle$$
$$= B(x,t)B(x',t')R_s(\tau) \qquad (4.89)$$

where, μ_s and R_s are the mean and autocovariance function of the stochastic component, $S(t)$, respectively. Thus, the first two moment statistics are time dependent in the presence of a time-dependent physical process $B(x, t)$. This time dependence of the statistics is due, in theory, to the physical component $B(x, t)$ and not to the stochastic component $S(t)$, which is assumed to be stationary over the time scales of interest.

There are many observational examples that suggest geophysical processes and the corresponding statistics are time dependent. In EOF analysis, the response characteristics of a physical process are assumed to be stationary and therefore not dependent on time. In light of the evidence from geophysical observations, the assumption of stationarity restricts the investigator's ability to interpret certain physical signals. The question arises, then, how can time-dependent characteristics be properly accounted for when attempting to compute basis functions that are assumed to be representative of the variability of a physical signal? If the covariance function is time dependent, computational eigenfunctions can be given by the solution of the Karhunen–Loeve equation (Loeve, 1978):

$$\int_D \int_T C(x, t; x', t') B_n(x', t') dt' dx' = \lambda_n B_n(x, t)$$

$$(4.90)$$

where D and T are space and time domains, respectively. Unfortunately, the solution to Eqn (4.90) is computationally intensive and not practical to derive. To address this problem, a simplification can be introduced known as the assumption of cyclostationarity. Equation (4.87) can be rewritten under the assumption that the response characteristics of the physical process are periodic in time. That is:

$$B(x, t) = B(x, t + d) \qquad (4.91)$$

where d is the "nested" periodicity of the process. The periodicity assumption is valid in many cases since many observed physical processes oscillate

with a well-defined period. The two moment statistics, can then be shown to be periodic:

$$\mu(x, t) = \langle T(x, t) \rangle = \langle T(x, t + d) \rangle = \mu(x, t + d)$$

$$(4.92)$$

$$C(x, t; x', t') = \langle T(x, t + d) T(x', t' + d) \rangle$$
$$= C(x, t + d; x', t' + d) \qquad (4.93)$$

Derivations of the above moment statistics requires that the stochastic component, $S(t)$, in Eqn (4.87) be stationary. A physical process that satisfies Eqns (4.92) and (4.93) is said to be "cyclostationary." The stationary case, for which $d = 1 \cdot \Delta t$ shows that stationarity is a special case within the cyclostationary framework. With the assumption of cyclostationarity, finding eigenfunctions as solutions to Eqn (4.89) becomes computationally tractable. The resulting eigenfunctions are also periodic in time with the same period as the corresponding statistics, leading to the definition of cyclostationary empirical orthogonal functions (CSEOFs) (Kim et al., 1996; Kim and North, 1997). In CSEOF analysis, space-time data are written as:

$$T(x, t) = \sum_n B_n(x, t) P_n(t) \qquad (4.94)$$

$$B_n(x, t) = B_n(x, t + d) \qquad (4.95)$$

where $B(x, t)$ are CSEOF loading vectors (LVs) and $P(t)$ are corresponding principal component time series (PCTS). Each eigenfunction represents not just one spatial pattern but also multiple spatial patterns, which repeat themselves in time. For example, when using monthly data and a nested period (d) of one year, the resulting LVs will be composed of 12 separate spatial patterns, one for each month of the year. In contrast to EOF analysis, the temporal variation of the data in CSEOF analysis has two distinct components: the time-dependent physical process, $B(x, t)$, and the stochastic undulation of the physical processes, $P(t)$. For example, when considering the annual seasonal signal, $B(x, t)$

represents the spatial pattern varying over the course of the year, while $P(t)$ represents the interannual variability in the amplitude of the seasonal signal.

While the assumption of periodic statistics is reasonable for many geophysical variables, it can be difficult to prove this periodicity and subsequently choose the nested period, d, for the CSEOF decomposition. The nested period must be determined based on an *a priori* physical understanding of the process being investigated. In many cases, there exists an obvious choice for the nested period. For instance, if one were studying the annual seasonal cycle, the nested period would obviously be one year. Sometimes, however, the period of the physical process of interest is not as obvious. For example, the El Niño-Southern Oscillation (ENSO) signal in the Equatorial Pacific does not have a well-defined period, but instead, has cyclicity of somewhere in the approximate range of two and five years. There is also the problem of selecting the nested period if one is studying several different geophysical signals, with a range of periods. In general, the nested period should be selected as the least common multiple of the periods of signals of interest. For instance, if there is a data set in which semiannual, annual, and biennial periodic signals are all present, the nested period should be set at two years. The semiannual cycle LVs would simply repeat four times, while the annual cycle LVs would repeat twice.

The concept of CSEOF analysis was first developed as a way to describe climatic time series with well-defined periods but unpredictable amplitude fluctuations. One of the first published applications involved performing a CSEOF decomposition of the globally averaged surface air temperature field (Kim et al., 1996). More recently, CSEOFs have been used to extract the annual cycle from the tropical Pacific SST field (Kim and Chung, 2001). This analysis was able to accurately explain the detailed structure and temporal modulation of the annual cycle. Similar work has been completed

on the satellite altimetry sea level record, demonstrating the ability to use CSEOFs to extract not only the modulated annual cycle but also the ENSO signal that is present in sea level data (Hamlington et al., 2011).

In the study by Hamlington et al. (2011), the CSEOF technique was applied to the quarter-degree resolution AVISO multiple altimeter-gridded global data set composed of sea level measurements spanning 1993–2008. Using CSEOF analysis it was possible to extract the time-variant (modulated) annual cycle in the sea level data. In order to distinguish the variability associated with the annual cycle, a nested period of 12 months was used in the CSEOF analysis. The annual cycle is described by the first CSEOF mode as it is the dominant signal in the sea level data. The top panels in Figure 4.27 show the time-dependent LVs associated with the annual cycle, while the bottom panel shows the PCTS of the CSEOF mode. The PCTS does not exhibit the annual cycle period of 12 months, which is instead described by the LVs that are required to have a period of one year. The PCTS describes the longer timescale fluctuations of the annual cycle. The fluctuations of the annual PCTS reflect variations of the strength of the annual cycle about some mean amplitude. The amplitude of the annual cycle varies within 20% of the mean, with values that are less than the mean, indicating weaker-than-normal annual cycles and values greater than the mean, indicating stronger-than-normal annual cycles.

The PCTS shows interannual variability that can be related to ENSO. The El Niño phase of ENSO tends to weaken the annual cycle and a weak negative correlation is observed between the PCTS and an ENSO index, such as the multivariate ENSO index (MEI) (Wolter and Timlin, 1998). The MEI can be understood as a weighted average of the main ENSO features contained in six different variables and serves as a tool for monitoring ENSO events. In addition to its relationship with ENSO, the CSEOF mode also

FIGURE 4.27   CSEOF mode 1 representing the modulated annual cycle (MAC) from the AVISO satellite altimetry data set. (a) The panel subplots show the monthly time-dependent CSEOF LVs (color images), the PCTS, and (b) the reconstructed mode's contribution to global mean sea level (GMSL).

contains oscillations with the period of six months, which are likely due to the semiannual cycle present in sea-level time series. By combining the LVs and the PCTS and then averaging, a global mean sea level (GMSL) time series associated with the annual cycle can be formed (Figure 4.27(b)). The 12-month periodicity is clear from this figure, as well as the time-varying amplitude of the GMSL annual cycle that is produced by the CSEOF analysis.

In addition to the ability to extract a modulated annual cycle, another advantage of the CSEOF technique comes from the potential to extract physically interpretable modes with less, albeit still significant, variability. The second CSEOF mode is shown in Figure 4.28. The

top panels show the temporally varying LVs while the bottom panel shows the associated PCTS. After plotting the MEI with the global mode associated with the second CSEOF mode as seen in Figure 4.28, the significance of this mode becomes clear. With a correlation of 0.80 between the MEI and global mean time series of the CSEOF mode and by looking at the spatial patterns of the LVs, it is clear that the second CSEOF represents the ENSO variability in the data set. The ability to extract a mode directly related to the ENSO signal represents one of the strengths of the CSEOF method.

In summary, the usual assumption of stationarity when applying EOF analysis is often

FIGURE 4.28 CSEOF mode 2 captures the ENSO signal in the AVISO satellite altimetry data. (a) The panel subplots show the monthly time-dependent CSEOF LVs (color images), the PCTS; and (b) and the reconstructed mode's contribution to global mean sea level (GMSL). *(from Hamlington et al., 2011).*

not justifiable. Cyclostationarity may be a better assumption for a wide range of geophysical processes, including climate signals such as those discussed in the above example using sea level data. CSEOF analysis finds computational modes of a cyclostationary process. By accounting for the time-dependent response of geophysical signals, clearer and more interpretable information can be extracted regarding the underlying physical processes of a data set. As with any other analysis technique, CSEOF is based on underlying assumptions, which lead to limitations on the capabilities of the analysis. However, when used in the right context, the CSEOF technique can be a significant improvement over commonly used techniques founded in the assumption of stationarity.

4.7 FACTOR ANALYSIS

As discussed in Preisendorfer (1988), Factor Analysis (FA) can be considered as the generalization of Principal Component Analysis (PCA) and linear regression analysis. For any given $n \times p$ data set $\mathbf{Z}$, we can perform either a PCA or an FA. The PCA is the simpler of the two analyses, while FA is the conceptually more complex of the two analysis methods. FA is a form of a linear statistical model that can hypothesize about phenomena underlying the data set **Z**, while PCA make no hypotheses about the linearity of the underlying phenomena, nor is it basically statistical in character. In spite of these fundamental differences between PCA and FA, their algebraic forms look alike. They are, however, methodologies for reaching quite different goals. If the analyst is interested in isolating the sources of the data's variability, he or she will use PCA. If instead, the investigator wishes to study the sources of data covariability, he or she will use FA.

FA itself is a statistical method used to describe variability among observed correlated variables in terms of a potentially lower number of unobserved variables, called *factors*. FA searches for potential joint variations. The observed variables are modeled as linear combinations of the potential factors and errors. The information gained about the interdependencies between observed variables can be used to reduce the set of variables of a data set. FA originated in psychometrics and is used largely in the behavioral and social sciences, marketing, product management, operations research, and other applied sciences that deal with large quantities of data.

4.8 NORMAL MODE ANALYSIS

In the previous sections, we were concerned with the partition of data variance into an ordered set of spatial and temporal statistical modes and maps. The eigenvalue problem associated with EOF modes was solved with little consideration given to the underlying physics of the oceanic system. In contrast, normal mode decomposition takes into account the physics and associated boundary conditions of the fluid motion. A common approach is to separate the vertical and horizontal components of the motion and to isolate the forced component of the response from the freely propagating response. As illustrations of these techniques, we consider two basic types of normal mode, eigenvalue problem:

1. The calculation of vertical normal modes (eigenfunctions), $\psi_k(z)$, for a stratified, hydrostatic fluid with specified top and bottom boundary conditions; and
2. the derivation of the cross-shore orthogonal modes (eigenfunctions), $\phi_k(x, z)$, for CTWs over a variable depth, stratified ocean with or without a coastal boundary.

The first problem can be solved without including the earth's rotation, f, while the second problem requires specification of f. Both eigenvalue problems yield solutions only for certain eigenvalues, λ_k, of the parameter, λ.

4.8.1 Vertical Normal Modes

A common oceanographic problem is to find the amplitudes (a_k) and phases (θ_k) of a set of K orthogonal basis functions, or modes, by fitting them to a profile of M ($>K$) observed values of amplitude and phase. For instance, one might have observations from $M = 5$ depths and want to find the modal parameters (a_k, θ_k) for the first three theoretical modes, $k = 1, 2, 3$, derived from an analysis of the equations of motion. Once the set of theoretical modes are derived, they can be fitted using a least-squares regression technique to observations of the along-channel (or cross-channel) current amplitude and phase. This yields the required estimates, (a_k, θ_k), for $k = 1, 2, 3$.

To obtain the vertical normal modes for a nonrotating fluid ($f = 0$), we assume that the pressure, p, density, ρ, and horizontal and vertical components of velocity (u, v) and w, respectively, can be separated into vertical and horizontal variables. This separation of variables has the form

$$[u(\mathbf{x}, t), v(\mathbf{x}, t), p(\mathbf{x}, t)/\rho_0] = \sum_{k=0}^{\infty} p_k(x, y, t)\psi_k(z)$$

(4.96a)

$$w = \sum_{k=0}^{\infty}\left[w_k \int_{-H}^{z} \psi_k(z)dz \right]$$

(4.96b)

$$\rho = \sum_{k=0}^{\infty} \rho_k \frac{d\psi_k(z)}{dz}$$

(4.96c)

where $k = 0, 1, 2, \ldots$ is the vertical mode number and the variables without subscripts are functions of $(\mathbf{x}, t) = (x, y, t)$. Substituting these expressions into the usual equations of motion (see LeBlond and Mysak, 1979; Kundu, 1990), we obtain the *Sturm–Liouville equation*

$$\frac{d}{dz}\left(\frac{1}{N^2}\frac{d\psi_k}{dz} \right) + \frac{1}{c_k^2}\psi_k = 0 \qquad (4.97)$$

where $N(z) = [-(g/\rho)d\rho/dz]^{1/2}$ is the Brunt–Väisälä frequency, c_k^2 is the separation constant and $1/c_k^2$ the eigenvalues, λ_k.

In the case of a rotating fluid (i.e., $f \neq 0$), we assume $N(z)$ is uniform with depth and replace simply N^2/c_k^2 in Eqn (4.97) as follows:

$$N^2/C_k^2 \to (N^2 - \omega^2)/gh_k, \quad k = 1, 2, \ldots \quad (4.98a)$$

where h_k is an "equivalent depth," ω is the wave frequency

$$gh_k = (\omega^2 - f^2)/(l^2 + q^2) = c_k^2 - f^2/l^2 \quad (4.98b)$$

and (l, q) are the wavenumbers in the horizontal (x, y) directions. Wavelike solutions are possible provided that $f^2 < \omega^2 < N^2$. For a rectangular channel of width L, the cross-channel wavenumber $q \to q_m = m\pi/L$ and solutions must be considered for both k, $m = 1, 2, \ldots$ (Thomson and Huggett, 1980). For both the rotating and nonrotating case, solutions to the eigenvalue problem Eqn (4.97) are subject to specified boundary conditions at the seafloor ($z = -H$) and the upper free surface ($z = 0$) of the fluid. These end-point boundary conditions are:

$$\frac{d\psi_k}{dz} = 0 \text{ (i.e., } w = 0) \text{ at } z = -H \quad (4.99a)$$

$$\frac{d\psi_k}{dz} + \frac{N^2}{g}\psi_k = 0 \left(\text{i.e., } \frac{\partial p}{\partial t} = \rho g w \right) \text{ at } z = 0$$

(4.99b)

Modal analysis of the type described by Eqns (4.97)–(4.99) is valid only for an inviscid hydrostatic fluid in which oscillations occur at frequencies much lower than the local buoyancy frequency, N, and for which the vertical length scale is much smaller than the horizontal length scale. In addition, the ocean must be of uniform depth and have no mean current shear. (For sloping bottoms, the horizontal cross-slope velocity component, u, is linked to the vertical boundary, w, through the bottom boundary condition $u = -w\, dH/dx$ and separation of variables is not possible.) The method can be applied to an ocean with zero rotation or with rotation that changes linearly with latitude, y. Solutions to Eqn (4.97) are obtained for specified values of $N(z)$ subject to the surface and bottom boundary conditions. Although the individual orthogonal modes propagate horizontally, the sum of a group of modes can propagate vertically if some of the modes are out of phase.

Analytical solutions: Simple analytical solutions to the Sturm–Liouville equation are obtained with and without rotation when $N = $ constant (density gradient uniform with depth). Assuming the rigid lid condition (i.e., no surface gravity waves so that $w = 0$ at $z = 0$), the vertical shapes of the orthogonal eigenfunctions $\psi_k(z)$ in Eqn (4.97) are given by

$$\psi_k(z) = \cos(k\pi z/H), \quad k = 0, 1, 2, \ldots \quad (4.100)$$

where $k = 0$ is the depth-independent barotropic mode, and $k = 1, 2, \ldots$ are the depth-dependent baroclinic modes. The kth mode has k zero crossings over the depth range $-H \leq z \leq 0$ and satisfies the boundary conditions $w = 0$ (cf. Eqn (4.99a)). Phase speeds (eigenvalues) of the modes are given by

$$c_o = (gH)^{1/2}, \quad k = 0 \text{ (barotropic mode)}$$
$$(4.101a)$$

$$c_k = NH/k\pi, \quad k = 1, 2, \ldots \text{ (barotropic mode)}$$
$$(4.101b)$$

In general, $N(z)$ is nonuniform with depth and, for a given k, the solutions will have the form

$$c_k = (gh_k)^{1/2} \qquad (4.102)$$

where the "equivalent depth" h_k is used in analogy with H in Eqn (4.101a). For an ocean of depth $H \approx 2500$ m and buoyancy frequency $N \approx 2 \times 10^{-3}/\text{s}$, the eigenvalue for the first baroclinic mode has a phase speed $c_1 \approx 1.6$ m/s and the equivalent depth $h_k = c_1^2/g \approx 0.26$ m. For a 400-m deep tidal channel, we find $N \approx 5 \times 10^{-3}$ m/s, $c_1 \approx 0.8$ m/s, and $h_k \approx 0.06$ m.

General solutions: To solve the general eigenvalue problem Eqns (4.97)–(4.99) for variable buoyancy frequency, $N(z)$, we resort to numerical integration techniques for ordinary differential equations with two-point boundary conditions. That is, given the start and end values of the function $\psi_k(z)$, and variable coefficient $N(z)$ we seek values at all points within the domain $(-H \leq z \leq 0)$. Fortunately, there exist numerous packaged programs for finding the eigenvectors and eigenvalues of the Sturm–Liouville equation for specified boundary conditions. For example, the NAG routine D02KEF (Nag Library Routines, 1986) and MatLab find the eigenvalues and eigenfunctions (and their derivatives) of a regular singular second-order Sturm–Liouville system of the form

$$\frac{d}{dz}\left[F(z)\frac{d\psi_k}{dz}\right] + G(z; \lambda)\psi_k = 0 \qquad (4.103)$$

together with boundary conditions

$$z_{a2}\psi_k(z_a) = z_{a1}F(z_a)d\psi_k(z_a)/dz \qquad (4.104a)$$

$$z_{b2}\psi_k(z_b) = z_{b1}F(z_b)d\psi_k(z_b)/dz \qquad (4.104b)$$

for real-valued functional coefficients F and G on a finite or infinite range, $z_a < z < z_b$. Provision is made for discontinuities in F and G and their derivatives. The following conditions hold on the function coefficients:

1. The function $F(z)$, which equals $1/N^2(z)$ in the case of Eqn (4.97), must be nonzero and of one sign throughout the closed interval $z_a < z < z_b$. This is certainly true in a stable oceanic environment where $N^2 > 0$; for $N^2 < 0$, the fluid is gravitationally unstable and vertical modes are not possible;

2. $\partial G/\partial \lambda$ must be of constant sign and nonzero throughout the interval $z_a < z < z_b$ and for all relevant values λ, and must not be identically zero as z varies for any relevant value of λ.

Numerical solutions to the Sturm–Liouville equation are obtained through a Pruefer transformation of the differential equations and a shooting method. (The shooting method and relaxation methods for the solution of two-point boundary value problems are described in *Numerical Methods* (Press et al., 1992)). The computed eigenvalues are correct to a certain error tolerance specified by the user. Eigenfunctions $\psi_k(z)$ for the problem have increasing numbers of inflection points and zero crossings within the domain $z_a < z < z_b$ as the eigenvalue increases. When the final estimate of λ_k is found by the shooting method, the routine D02KEF integrates the differential equation once more using that value of λ_k and with initial conditions chosen such that the integral

$$I_k = \int_{z_a}^{z_b} [\psi_k(z)]^2 \partial G/\partial \lambda(z; \lambda)dz \qquad (4.105)$$

is roughly unity. When $G(z, \lambda)$ is of the form $\lambda w(z) + \psi(z)$, which is the most common case, I_k represents the square of the norm of ψ_k induced by the inner product

$$\overline{\psi_k(z)\psi_m(z)} = \int_{z_a}^{z_b} \psi_k(z)\psi_m(z)w(z)dz \quad (4.106)$$

with respect to which the eigenfunctions are mutually orthogonal if $k \neq m$. This normalization of ψ for $k = m$ is only approximate but typically differs from unity by only a few percent.

If one is working with observed density (σ_t) profiles for the region of interest, a useful approach is to solve the Sturm–Liouville equation using an analytical expression for $N(z)$ by fitting a curve of the type $\sigma_t(z) = [\rho(z) - 1]10^3 = \sigma_o \exp[a/(z + b)]$ or other exponential form, to the data. The eigen (modal) analysis is fairly insensitive to small changes in density so that, even though changes in $N(z)$ are large in the upper oceanic layer, it is usually possible to get by with a simple analytical curve fit. Alternatively, we can specify the actual density on a numerical grid for which modes are to be calculated. Once $N(z)$ is available, we can use numerical methods to solve Eqn (4.97) subject to the boundary conditions Eqn (4.99), allowing for specified error

bounds or degree of convergence on the final boundary estimate. Based on the analytical solutions Eqn (4.100), the investigator can expect solutions ψ_k to resemble cosine functions whose vertical structure has been distorted by the nonuniform distribution of density along the vertical profile. There is a direct analogy here with the modes of oscillation of a taut string clamped at either end and having a nonuniform mass distribution along its length.

The normal modes are normalized relative to their maximum value and then fitted to the data in a least-squares sense (Table 4.9). If there are M current meters on a mooring string, the maximum possible number of normal baroclinic modes is $M - 1$. By comparing the normal modes with the data, we can derive the absolute values of the barotropic mode and a maximum of $M - 1$ baroclinic modes. Solutions to the least-squares fitting are described in (Press et al., 1992).

4.8.2 An Example: Normal Modes of Semidiurnal Frequency

Suppose that the along-axis semidiurnal currents, v, in a tidal channel have the form $v_m = a_m \cos(\omega t + \theta_m)$, where t is the time, and

TABLE 4.9 Model Amplitudes (cm/s) and Phases (Degrees Relative to 120° W Longitude) for Johnstone Strait M_2 Tidal Currents Computed from Nine-day Current Meter Records

Site	M	K_v (cm²/s)	Before (a_0, θ_0)	After (a_0, θ_0)	Before (a_1, θ_1)	After (a_1, θ_1)	Before (a_2, θ_2)	After (a_2, θ_2)
CM13	3	15	42, 55°	42, 55°	12, 172°	25, 171°	—	19, −10°
CM14	3	8	35, 51°	35, 51°	11, 169°	15, 171°	—	8, −4°
CM15	3	13	32, 35°	32, 36°	18, 175°	12, 166°	—	7, −31°
CM02	5	0	36, 42°	NC	21, 220°	NC	9, 13°	NC
CM04	4	7	29, 45°	50, 24°	13, 215°	79, 174°	2, −34°	70, 0°

Column two gives the number of current meters (CM) on the string. The first column for the barotropic mode (a_0, θ_0) and each of the two baroclinic modes (a_k, θ_k), $k = 1, 2$, gives the amplitude and phase (a, θ) before and after the bottom current meters is included in the analysis. The bottom current is included after its amplitude and phase are corrected for bottom boundary layer friction. The vertical eddy viscosity K_v is that value which gives the minimum ratio between the first and second baroclinic modes when the frictionally corrected bottom current meter is included. NC means "no change", implying perfect modal fit for all depths with and without bottom current meter record. At CM04, no near-surface current meter was deployed and the records were only five days long and therefore suspect.

a_m, θ_m ($m = 1, ..., M$) are the observed current amplitude and phase, respectively. In terms of tidal current ellipses, we can think of v as the major axis of the current ellipse for each current meter on the mooring line. The oscillations have frequency $\omega = \omega_{M_2}$ corresponding to M_2 semidiurnal tidal currents and the phase, θ, is referenced to some specific time zone or meridian of longitude so that we can intercompare values for different current meters and for the astronomical surface tide. The values a_m, θ_m for the different current meter records can be determined using harmonic analysis techniques (Foreman, 1976; Powlowicz et al., 2002) provided the measured data are at hourly (or other equally spaced) intervals over a period of seven days or longer so that the M_2 and K_1 constituents are separable. We next rewrite the above expression for v in the usual way as $v_m = A_m\cos(\omega t) + B_m\sin(\omega t)$, where $\tan\theta_m = A_m/B_m$ and $a_m^2 = (A_m^2 + B_m^2)$. This allows us to examine the sine and cosine components separately. The observed magnitudes A_m and B_m at each current meter depth z_m, $m = 1, ..., M$ are then used to compute the amplitudes and phases of the basis functions $\psi_k(z_m)$, for a maximum of K different modes ($K < M$). At best, we can obtain the amplitudes and phases of the barotropic mode ($k = 0$) and up to $M - 1$ baroclinic modes.

Details of the modal analysis at semidiurnal frequency using current meter data from a tidal channel are presented by Thomson and Huggett (1980). The first step is to obtain an exponential functional fit (Figure 4.29(a)) to the observed mean density structure, $N(z)$. This structure is then used with the local water depth H (assuming a flat bottom), the Coriolis parameter, f, and the wave frequency, ω, to calculate the theoretical dynamic modes (Figure 4.29(b)). A finite sum of these theoretical modes, $\Sigma\psi_k(z)$, is then least-squares fitted to the observed cosine component $A_m(z)$ to obtain estimates of the contributions A_k from each mode, k. This operation is repeated for the sine component B_k. (Recall that the maximum total of barotropic plus

baroclinic modes allowed in the summation is fewer than the number of current meter records per mooring string and that the vertical structure of each mode is found through the products (A_k, B_k)$\psi_k(z)$ where the coefficients are constant.) Using the relationships $\tan\theta_k = A_k/B_k$ and $a_k^2 = (A_k^2 + B_k^2)$, we obtain the amplitudes and phases of the various modes. In their analysis, Thomson and Huggett (1980) typically had only three reliable current meter records per mooring string. Normally, this would be enough to obtain the first two baroclinic modes. However, the bottom current meter in most instances was within a few meters of the bottom and therefore strongly affected by benthic boundary layer effects. To include a mode-2 solution in the estimates, the observed phase and amplitude of the bottom current meter record had to be adjusted for frictional effects via the added term $\exp(-z')$ $\cos(\omega t + \theta - z')$, where $z' = (z + H)/\delta$, and $\delta \approx (2K_v/\omega)^{1/2}$ is the boundary layer thickness for eddy viscosity K_v.

Since K_v is not known *a priori*, the final solution required finding that value of K_v which minimized the ratio formed by the first mode calculated with and without the bottom current meter included in the analysis (Table 4.9). In the case where five current meters were available, Thomson and Huggett found that there was no difference in the value of the second mode estimate with and without inclusion of the bottom current meter record in the analysis, suggesting that the three-mode decomposition was representative of the actual current variability with depth.

4.8.3 Coastal-Trapped Waves

Stratified or nonstratified oceanic regions characterized by abrupt bottom topography adjacent to deeper regions of uniform depth support the propagation of trapped ocean waves with frequencies, ω, which must be lower than the local inertial frequency, f. Trapped sub-inertial motions ($\omega < f$) typically are

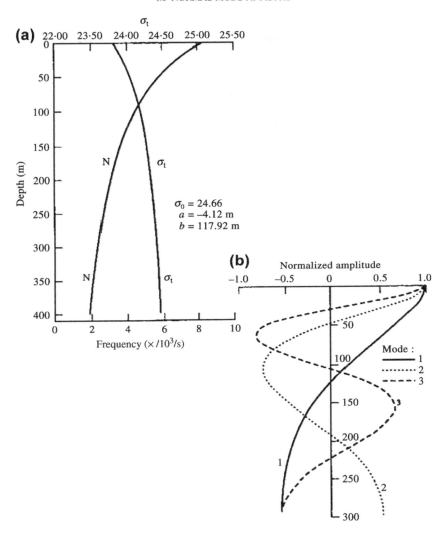

FIGURE 4.29 Baroclinic modes for semidiurnal frequency (ω_{M_2}) in a uniformly rotating, uniform depth channel (a) The mean density structure (σ_t) and corresponding buoyancy frequency $N(z)$ used to calculate the eigenvalues; (b) Eigenvectors for the first three baroclinic modes. The barotropic mode (not plotted) has a magnitude of unity at all depths. Phase speeds for the modes fitted to the current meter data are $c_1 \approx 34$ cm/s; $c_2 \approx 20$ cm/s. *From Thomson and Huggett (1980).*

found along continental margins where the coastal boundary is bordered by a marked change in water depth consisting of a shallow (<200 m) continental shelf, a steep continental slope, and a deep (>2000 m) more weakly sloping continental rise. The alongshore wavelengths vary from tens to thousands of kilometers while the cross-shore trapping scale is determined by the density structure and length scale for the cross-shore topography. For baroclinic waves, the *internal deformation radius*, $r = NH/f$, provides an estimate of the cross-shelf trapping scale while the *stratification parameter*, $S = (N^2_{max}H^2_{max})/f^2L^2$, characterizes the importance of stratification for a shelf-slope region of width L. For a mid-latitude ocean of depth $H \approx 2500$ m and buoyancy frequency $N \approx 2 \times 10^{-3}/s$, we find $r \approx 50$ km. For wide shelves ($L > 100$ km), the motions are confined mainly to the continental slope, while for narrower shelf regions, the motions extend to the coast where they "lean" up against the coastal boundary. For $S \gg 1$ the CTWs are strongly baroclinic, while for $S \ll 1$, they are mainly barotropic (Chapman, 1983; Connolly et al., 2014). The case $S \approx 1$ corresponds to barotropic shelf waves modified by stratification.

In addition to continental shelf regions, CTWs can occur along mid-ocean ridges and in oceanic trenches (where they are known as *trench waves*; cf., Mysak et al., 1979), as well as around isolated seamounts and islands (Pizarro and Shaffer, 1998). Phase propagation, in all cases, is with the coastal boundary to the right of the direction of propagation in the Northern Hemisphere and to the left of the direction of propagation in the Southern Hemisphere. For strongly baroclinic waves, energy propagation is always in the direction of phase propagation; for barotropic motions, short waves can propagate energy in the opposite direction to phase propagation.

The general CTW solutions consists of a Kelvin wave mode ($k = 0$), for which the cross-shore velocity component is identically zero at the coast ($U \equiv 0$ at $x = 0$), together with a hierarchy of higher mode shelf waves ($k = 1, 2, \ldots$) whose cross-shore velocity structures have increasing numbers of zero crossings (sign changes) normal to coast. The first shelf wave mode will have one zero crossing in sea surface elevation ζ over the continental margin; the second mode will have two crossings, and so on. For the current component, U, the first mode shelf wave will have no zero crossing in the cross-shelf direction, the second mode will have one crossing, and so on. The condition of no normal flow through the coastal boundary requires $U = 0$ at $x = 0$.

Computer programs that calculate the frequencies and cross-shore modal structure of CTWs of specified wavelength are available in reports written by Brink and Chapman (1987) and Wilkin (1987). We confine ourselves to a general outline of the programs for the interested reader. Practical difficulties with the numerical solutions to the equations are provided in these comprehensive reports. The programs of Brink and Chapman use linear wave dynamics in which the water depth, $h(x)$, is assumed to be a function of the cross-shore coordinate, x, alone. Similarly, the buoyancy frequency, $N(z)$, is a function of depth alone. The one profile that can be used in the analysis is best obtained by least-squares fitting a function (such as a polynomial or exponential) to a series of observed profiles. The wave parameters such as velocity, pressure, and density are assumed to be sinusoidal in time (t) and alongshore direction (y) such that for any particular wave parameter, ξ, we have

$$\xi(x, y, t) = \xi_0(x)\exp[i(\omega t + ly)] \qquad (4.107)$$

where ω is the wave frequency and l is the alongshore wavenumber. This gives rise to a two-dimensional eigenvalue problem in (ω, l) of the form

$$L[\xi_0(x; \omega, l)] = 0 \qquad (4.108)$$

where L is a linear operator. The problem is solved for arbitrary forcing and a fixed l. In particular, for a given wavenumber, k, the frequency ω is varied until the algorithm finds the free-wave mode resonance. Resonance is defined as the frequency at which the square of the spatially integrated wave variable

$$I_o = \int_0^\infty \xi_0^2 dx, \quad \text{or } I_p = \int_0^\infty \int_{-h}^0 (p^2 dz) dx \quad (4.109)$$

is at a maximum. The suite of programs tackle the following problems for which the user provides the bottom profile, $h(x)$, a mean flow profile (if needed) and a selection of boundary conditions:

1. The program BTCSW yields the dispersion curves $\omega = \omega(l)$ (the frequency as a function of wavenumber), the cross-shore modal structure for velocity component $U(x)$ and/or sea surface elevation $\zeta(x)$, and wind coupling coefficients for barotropic CTWs—including continental shelf waves and trench waves—for arbitrary topography and mean alongshore current. Options for the long-wave and rigid-lid approximations are included in the program. The user can specify one of two geometries corresponding to topography with and without a coastal boundary. The outer boundary at $x = x_{max}$ is set as $-2L$, where L is the width of the typographically varying domain in the cross-shore direction. Thus, about half the domain has a flat bottom. The outer boundary condition is specified as $\partial U / \partial x = 0$. To obtain solutions for both ζ and U, the depth at the coast should be given a nonzero value $h(0) \geq 1$ m.

2. For wave frequencies $\omega \leq 0.9f$, the program BIGLOAD2 yields dispersion curves $\omega = \omega(l)$, the horizontal modal structure, and wind-coupling coefficients for an ocean with continuous, horizontally uniform stratification, and arbitrary topography. Density in the model has the form $\rho^*(x, y, z, t) = \rho_0(z) + \rho(x, y, z, t)$, where ρ_0 is background density and ρ is the density perturbation. Since $\rho \ll \rho_0$, the Boussinesq approximation is assumed throughout (i.e., the small density perturbations are ignored in momentum terms involving the fluid inertia and Coriolis acceleration but are retained in vertical buoyancy terms where they multiply, g, the acceleration due to gravity; cf. LeBlond and Mysak, 1978). The program allows for the component of the β-effect normal to the coast and for both the free surface and rigid lid boundary conditions at the ocean surface. Solutions

are obtained using the coordinate transformation $\theta = z/h(x)$ and assuming a linear bottom friction drag. A total of 17 vertical and 25 horizontal grids (rectangles) are generated so that the vertical resolution is much better near shore than in deep water. Problems with singularities are avoided by setting $h(x) \geq 1$ m at the coast, $x = 0$. The program does not work well when the shelf-slope width (or width of a trench at the base of the shelf) is small relative to the internal deformation radius for the first mode in the deep ocean. Spurious features appear in unexpected places and force the user to increase the density of horizontal grids over regions of rapidly varying topography. In addition, a spurious mode occurs in the pressure equation for $\beta = 0$ at the local inertial frequency, $\omega = f$, making the overall solution suspect. As noted by the authors, the user will have difficulty finding the barotropic Kelvin wave parameters.

3. The program CROSS is used to find baroclinic coastal-trapped modes for $\omega \leq f$ for arbitrary stratification and uniform depth.

4. The program BIGDRV2 is used to obtain the velocity, pressure, and density fluctuations over a continental shelf-slope region of arbitrary depth, stratification, and bottom friction and is driven by an alongshore wind stress of the form $\tau_x(t, y) = \tau_0 \exp[i(\omega t + ly)]$. Specification of a linear friction coefficient of zero ($r = 0$) results in a divide-by-zero error. As a result, inviscid solutions should not be attempted. As with (2), solutions are obtained on a 25×17 stretched grid. In practice, it is generally best to start a study of coastally trapped waves using BTCSW since it gives first-order insight into the type of modal structure one can expect. However, if the barotropic dispersion curves do not fit the data (e.g., observations reveal strong diurnal-period shelf waves but the first-mode dispersion curves consistently remain below

the diurnal frequency band for realistic topography), then density and mean currents should be introduced using BIGLOAD2 and CROSS.

The Brink and Chapman programs have been used by Crawford and Thomson (1984) to examine free wave propagation along the west coast of Canada and by Church et al. (1986) and Freeland et al. (1986) to examine wind-forced CTWs along the southeast coast of Australia (Figure 4.30). In all cases, model results are compared with alongshore sea-level records and current meter observations from cross-shore mooring lines. The cross-shore depth profiles, $h(x)$, and associated

buoyancy frequencies, $N^2(z)$, used in the Australian model are presented in Figures 4.31(a, b). From these input parameters, the program was used to generate eigenvalues and eigenfunctions for the first three CTW wave modes (Figure 4.32) and the theoretical dispersion curves (Figure 4.33) relating wave frequency, ω, to alongshore wavenumber, l. The slopes of the (ω, l) curves give the phase speeds c_k for the given modes ($k = 1, 2, 3$) listed on the figure.

Wilkin (1987) presents a series of FORTRAN programs for computing the frequencies and cross-shore modal structure of free CTWs in a stratified, rotating channel with arbitrary bottom topography. The programs solve the linearized,

FIGURE 4.30 Southwest coast of Australia showing the locations of the tide gauge stations (■) and current meter lines (0, 1, 2, 3) occupied during the Australian Coastal Experiment (ACE). *From Freeland et al. (1986).*

FIGURE 4.31 Parameters used in determining the coastal-trapped wave eigenfunctions at Cape Howe Stanwell Park and Newcastle; (a) The cross-shore depth profiles $h(x)$; (b) The $N(z)^2$ profiles. Below 600 dbar ($\approx$ 590 m) all curves are similar so that only is drawn. *From Church et al. (1986).*

inviscid, hydrostatic equations of motion using the Boussinesq approximation. The Brunt–Väisälä frequency $N(z)$ is a function of the vertical coordinate only. As with Brink and Chapman (1987), the eigenvalue problem is solved using resonance iteration and finite difference equations. The cross-shore perturbation fields returned by the model include velocity, pressure, and density.

The difference with Wilkin's model is that it uses a staggered horizontal (Arakawa "C") grid for which the usual horizontal Cartesian coordinates (x, y) have been mapped to orthogonal curvilinear coordinates (ξ, η). Instead of using finite differencing, the vertical structures of the modes are determined through modified sigma coordinates with expansion of the field variables in terms of

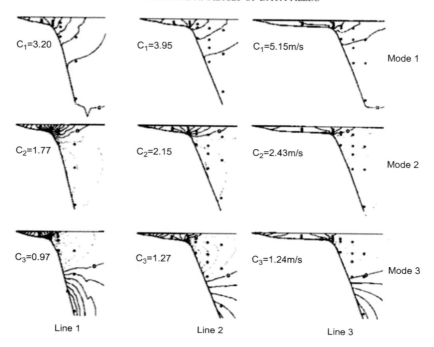

FIGURE 4.32 The eigenfunctions $U(x; z)$ for the first three baroclinic longshore current modes for the three lines in Figures 4.15 and parameters in Figure 4.16. The contouring is in arbitrary units. Phase speeds c_k (eignvalues) of each mode for each of the three lines also are shown. *From Church et al. (1986).*

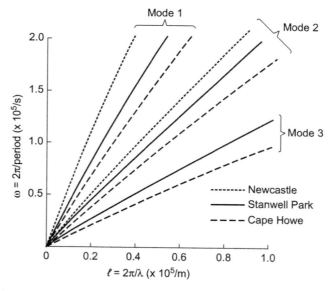

FIGURE 4.33 The theoretical dispersion curves $\omega = \omega(l)$ relating the longshore wavenumber, l, to the wave frequency, ω (here λ is the wavelength). Curves correspond to the first three baroclinic modes for each mooring location. For mode 3, the dispersion curve at Stanwell Park and Newcastle are almost identical. The slopes of the lines are the theoretical phase speeds, c_k. *From Church et al. (1986).*

Chebyshev polynomials of the first kind. The program has the option of specifying wavenumber, l, and searching for the corresponding free wave frequency, $\omega(l)$, as in Brink and Chapman, or specifying ω and searching for l. For reasons explained by Wilkin, the model is designed to be compatible with the primitive equation ocean circulation model developed by Haidvogel et al. (1988).

In the curvilinear coordinate system, a line element of length ds in the Wilkin model satisfies

$$ds^2 = dx^2 + dy^2 = d\xi^2/dm^2 + d\eta^2/dn^2$$

$$(4.110)$$

and the metric coefficients m, n are defined by

$$m = \left[(\partial x/\partial \xi)^2 + (\partial y/\partial \xi)^2\right]^{-1/2} \quad (4.111a)$$

$$n = \left[(\partial x/\partial \eta)^2 + (\partial y/\partial \eta)^2\right]^{-1/2} \quad (4.111b)$$

The velocity perturbations for time-dependent solutions of the form $\exp(-i\omega t)$ are then

$$U = \frac{1}{f^2 - \omega^2}\left(i\omega m \frac{\partial \phi}{\partial \xi} - fn \frac{\partial \phi}{\partial \eta}\right) \quad (4.112a)$$

$$V = \frac{1}{f^2 - \omega^2}\left(i\omega n \frac{\partial \phi}{\partial \eta} - fm \frac{\partial \phi}{\partial \xi}\right) \quad (4.112b)$$

$$w = \frac{i\omega}{N^2}\frac{\partial \phi}{\partial z} \quad (4.112c)$$

where (U, V, w) are the velocity components and $\phi = p/\rho_0$ is the perturbation pressure. Solutions are then sought for the resulting pressure equation

$$mn\frac{\partial}{\partial \eta}\left(\frac{n}{m}\frac{\partial \phi}{\partial \eta}\right) + (f^2 - \omega^2)\frac{\partial}{\partial z}\left(\frac{1}{N^2}\frac{\partial \phi}{\partial z}\right)$$
$$+ mn\frac{\partial}{\partial \xi}\left(\frac{m}{n}\frac{\partial \phi}{\partial \xi}\right) = 0$$

$$(4.113)$$

For a straight coastline, $m\partial/\partial \xi = \partial/\partial x$ and we arrive at the usual solutions for alongshore (x-direction) propagation of progressive waves of the form $F(y) \exp[i(lx - \omega t)]$.

The Wilkin model is less general than the Brink and Chapman model in that application of the rigid-lid approximation does not allow for the barotropic (long wave) Kelvin wave solution and a "slippery" solid wall is placed at the offshore boundary. The new vertical coordinate variable, σ, is defined by

$$\sigma = 1 + 2z/h(\eta) \quad (4.114)$$

so that the ocean surface is located at $\sigma = 1$ and the (now flattened) seafloor at $\sigma = -1$. Application of this model to the west coast of New Zealand (South Island) is presented by Cahill et al. (1991). Modes 1 and 2 of the alongshore current for the northern portion of this region based on Wilkin's program CTWEIG are reproduced in Figure 4.34. Similar results for the southern region are presented in Figure 4.35. Notice that the CTWs are nearly barotropic over the shallow shelf immediately seaward of the coast in both sections but are more baroclinic in the offshore region off the southwest coast.

4.9 SELF ORGANIZING MAPS

The Self Organizing Map (SOM) provides a method for extracting spatial patterns from high dimensional data sets by clustering the data into much lower dimensional arrays of orderly and smoothly connected mapping units. Arrays are commonly one or two-dimensional. Examples of high dimensional data sets include daily time series of SST measured at grid locations within a coastal upwelling region and daily time series of current velocity measured at mooring sites strung across the continental shelf. In the first case, a simple two-map system might have one map showing near-uniform SST values over the entire domain (such as might occur during calm conditions in the middle of winter) and a second map showing cold water near the coast and warmer water offshore (indicative of strong wind-driven upwelling in summer). In the second example, one map unit might consist of a near-spatially uniform southeastward flow characteristic of upwelling favorable wind conditions while the

FIGURE 4.34 The alongshore velocity structure of coastal-trapped waves for the northwestern shelf-slope region of South Island, New Zealand, (a) Mode 1; (b) Mode 2. Contour lines when multiplied by 10^{-7} correspond to the alongshore velocities in m/s for unit energy flux in watts. Negative values are dashed. Current meter locations are given by the dots. Here, c is the phase speed of the mode. *From Cahill et al. (1991).*

FIGURE 4.35 As for Figure 4.34 but for the shelf-slope region off the southwestern tip of South Island. Note the change in depth and offshore distance scale in the two figures. This line is roughly 500 km to the south of the line in Figure 4.34.

second map would consist of near unidirectional northwestward flow typical of downwelling favorable wind conditions. The incorporation of intervening map units within the array make it possible to examine the transitions between the two "end member" states. SOM of this kind were used successfully to study SST patterns (Lui et al., 2006) and current velocity structure (Liu and Weisburg, 2005) on the West Florida Shelf. Figure 4.36 shows the 12 map units

FIGURE 4.36 A 3 × 4 Self Organizing Map of 2-day low-pass filtered velocity data at three depth levels (see legend) from October 1998 through September 2001 on the West Florida Shelf. Adjoining map units show similar features, but similarity diminishes with separation between maps. The relative frequency of occurrence of each pattern (map unit) is shown in the right corner of each map. *Adapted from Liu and Weisberg (2005).*

obtained by Liu and Weisburg for the velocity structure on the shelf; Figure 4.37 gives the percentage of time that a particular map unit was "selected" as best matching the input data.

The SOM is an ordered, nonlinear, artificial neural network (ANN) technique with unsupervised learning (no instructor or teacher required) used for mapping high-dimensional spatially and temporally varying input data into a much smaller number of map units (also called, elements, nodes, archetypes, or neurons) of a regular low-dimensional array (Kohonen, 1982, 2001). For oceanographic applications, SOMs provide a pattern recognition and classification tool that

clusters input data into an array of gradually changing maps, typically two-dimensional, that can reveal the most commonly occurring structural features embedded in the large scale data sets. As illustrated by Figure 4.36, adjoining units in the array share similar structures and features but this similarity diminishes with increased separation between the map units.

As with other AANs, the SOM consists of numerical algorithms that simulate the processing capability of the brain, in which a network of interconnected units (cells, neurons) processes information or input data in parallel rather than sequentially. During unsupervised training, the

FIGURE 4.37 Temporal changes of the Best Matching Unit (BMU) for the 3 × 4 Self Organizing Map in Figure 4.36. The tick marks (x) on the vertical axis range from 1 to 12 and correspond to the pattern numbers in the SOM. The light gray lines connect adjoining tick marks to form a time series. *From Liu and Weisberg (2005).*

networks learn to generate their own classifications of the training (input) data without external help. This assumes that class membership is broadly defined by the input patterns sharing common features, and that the network will be able to identify those features across the range of input patterns (Bullinaria, 2004). SOMs can deal with noisy and gappy data and require no prior knowledge or requirements about the data, such as the need for normal distribution or equality of parameter variances. The learning is unsupervised in that the training of the network is entirely data-driven and no prior instructions (other than the dimensionality of the output maps) are given to the algorithms on how they are to reach their mapping goal. In supervised learning, target results for the input data vectors are provided to "train" the algorithm (e.g., Bayesian spam email filtering, Support Vector Machines).

SOM algorithms were developed in the early 1980s with first applications in image analysis and biological cluster analysis (Kohonen, 1982, 2001). The technique has now been used over a wide range of disciplines and was first applied to climate variability by Hewitson and Crane (1994). More recent climate applications can be found in Hsu et al. (2002) and Reusch et al. (2007). Oceanic examples of SOMs include applications to large-scale surface fields such as SLP (Hewitson and Crane, 2002), SST (Lui and Weisberg, 2006), QuikSCAT winds (Risien et al., 2004), and NCEP/NCAR reanalyses data (Tennant, 2004). SOMs have also been applied to analyses of coastal ocean current patterns derived from moored ADCPs (Lui and Weisberg, 2005) and estuarine circulation as defined by a single ADCP (Cheng and Wilson, 2006). *Richardson* et al. (2002) used the SOM to identify characteristic chlorophyll-a profiles in vertical measurements obtained in the Benguela upwelling system. The first public-domain general-purpose SOM software package SOM_PAK was released in 1990 by the Laboratory of Computer and Information Science of the Helsinki University of Technology. The package was then implemented by the same group in MatLab as a Toolbox (Vesanto et al., 2000). Software for SOM analysis is currently available as part of the Neural Network Toolbox in MatLab and as "kohonen" in the R programming language.

4.9.1 Basic Formulation

The SOM consists of J map units or elements that are typically arranged in a two-dimensional grid with a specified weight vector, $\mathbf{w}_j$, assigned to each map unit (some authors use $\mathbf{m}_j$ for the weights). SOM requires the analyst to specify the number of units in the array and to provide an initial guess for the weight vectors to be used at the first step of the clustering. In their study of currents on the West Florida Shelf, Liu and Weisberg (2005) specified a 3×4 $(J = 12)$ SOM consisting of 3 units down by 4 units across (Figure 4.36). The weights can be initialized randomly (as was done for the Florida Shelf study) or the analyst can use prior knowledge, such as output from PCA or the record mean values for each site, to specify the initial weights. Initializing using random weights means that the algorithm will take a bit longer to learn how to cluster the data. After their initial specification, the weights change following a learning rule (presented below) that specifies how to calculate new weights at each step in the clustering process. At a given time step, the jth weight vector for the jth map unit will have the form

$$\left(w_{j,1}, w_{j,2}, ..., w_{j,n} \right) \qquad (4.115)$$

where n is also the number of elements in the input vector created from the data. The input vector is also referred to as the "training vector" from ANN terminology.

Following specification of the SOM size (but see note * at the end of this paragraph) and initial weighting values, the incremental self-organizing (sequential training) algorithms are ready for the input of the p data vectors, $\mathbf{x}$, constructed from the high-dimensional data set. These input vectors are of length n and have the form

$$\left.\begin{matrix} \left(x_{1,1}, x_{1,2}, ..., x_{1,n} \right) \\ \left(x_{2,1}, x_{2,2}, ..., x_{2,n} \right) \\ \cdots\cdots\cdots\cdots\cdots \\ \left(x_{i,1}, x_{i,2}, ..., x_{i,n} \right) \\ \cdots\cdots\cdots\cdots\cdots \\ \left(x_{p,1}, x_{p,2}, ..., x_{p,n} \right) \end{matrix}\right\} p \text{ distinct input (training) vectors}$$

$$(4.116)$$

where vector elements are real numbers. The data (training) vectors are input into the SOM algorithm and the *activation* of each unit for the specific input vector is calculated. The activation function is typically a preselected function of the

Euclidian distance, $D^2(t)$, between the input vector and the weight vector for that particular unit (the squared differences between the vectors on a component by component basis). For each map unit, j, the SOM algorithm calculates $D^2(t)$ as

$$D_j^2(t) = \sum_{k=1}^{n} \left(x_{i,k} - w_{j,k}(t) \right)^2 \quad j = 1, ..., m;$$

$$i = 1, ..., p \qquad (4.117)$$

where the weights and the elements, $x_{i,k}$, of the input vectors have the forms Eqns (4.115) and (4.116), respectively. The unit whose weight vector shows the highest activation (i.e., minimum Euclidean distance, D^2, also written as *argmin* $(\|\mathbf{x}_k - \mathbf{w}_i\|)$ for the particular input vector is selected as the "winner" or "best matching unit" (BMU) for the particular SOM map unit. In effect, the winner (BMU), determined from the minimum of Eqn (4.117), is a measure of how closely a given input data vector, $\mathbf{x}_k$, matches the weight vector $\mathbf{w}_i$ for each of the map units in the array. From a neurological point of view, this class of unsupervised system is a type of competitive learning, whereby the neurons compete amongst themselves to be activated, with the result that only one neuron is activated at any one time. This activated neuron is called a "winner-takes-all neuron" or simply the "winning neuron". Such competition can be induced/implemented by having lateral inhibition connections (negative feedback paths) between the neurons. The result is that the neurons are forced to organize themselves. (* The subjective aspect of selecting the number of map units has led to the formulation of Growing Hierarchical Self Organizing Maps, GHSOM, whereby the optimal number of map units is determined through a more objective SOM procedure; see Section 4.9.5 below.)

The next step after specifying the initial weights and undertaking the first sequential mapping step, is to modify the "winner" weight, $\mathbf{w}_j(t)$, so that it more closely resembles the input data that was presented to it during the previous time step. Specifically, the weight vector of the winner $w_j(t + \Delta t)$ at $t = t + 1 \cdot \Delta t$ is moved toward the presented vector by a fraction of the Euclidian distance as determined by the time-diminishing learning rate, $\alpha(t)$, and the neighborhood function, $\varepsilon_{qi}(t)$. That is,

$$\mathbf{w}_j(t + \Delta t) = \mathbf{w}_j(t) + \alpha(t) \cdot \varepsilon(t) \left[\mathbf{x}_i(t) - \mathbf{w}_j(t) \right],$$

$$i = 1, \ldots, p$$

$$(4.118)$$

where the learning rate has one of the following forms (Liu and Weisberg, 2005):

$$\alpha(t) = \begin{cases} \alpha_o(1 - t/T), & \text{linear} \\ \alpha_o(0.05/\alpha)^{t/T}, & \text{power} \\ \alpha_o/(1 + 100t/T), & \text{inverse} \end{cases} \quad (4.119)$$

Here, α_o is the initial learning rate and T is the training duration. In the SOM MatLab Toolbox, the defaults are a linear function and $\alpha_o = 0.5$ for an initial training session and 0.05 for further fine tuning (Liu and Weisberg, 2005). Users can also specify the initial and final values of α or specify other time decreasing functions. The winner's activation will be even higher the next time the same input vector is presented to the algorithm. In addition to the learning rate, the weight vectors of units in the neighborhood of the winner are modified according to the decreasing spatial-temporal neighborhood function, $\varepsilon_{qi}(t)$, where

$$\varepsilon_{qi}(t) = \begin{cases} H(r_i - D_{qi}) & \text{bubble} \\ \exp(-D_{qi}^2/2r_i^2) & \text{Gaussian} \\ \exp(-D_{qi}^2/2r_i^2)\delta(r_i - D_{qi}) & \text{cut-Gaussian} \\ \max(0, \ 1 - (r_i - D_{qi})^2) & \text{Epanechikov} \end{cases}$$

$$(4.120)$$

Here, $H(a)$ is the Heaviside step function defined as: $H(a) = 0$ if $a < 0$, and $= 1$ if $a \geq 0$. As an example, the initial weight vector, $w_j(t_{start})$, at the start of a three unit SOM array might look something like

$$\left. \begin{array}{l} (w_{1,1}, w_{1,2,}, \ldots, w_{1,n}) \\ (w_{2,1}, w_{2,2,}, \ldots, w_{2,n}) \\ (w_{3,1}, w_{3,2,}, \ldots, w_{3,n}) \end{array} \right\}$$

$$= \begin{cases} (0.496, 0.877, \ldots, 0.317) \\ (0.169, 0.714, \ldots, 0.843) \\ (0.522, 0.069, \ldots, 0.589) \end{cases} \quad (4.121)$$

where the elements $w_{i,k}$ are random numbers between 0 and 1 that we selected using the RANDM function on a Hewlett Packard 30S hand calculator. Note that our choice of values could easily represent randomly selected current speeds in meters per second. During the analysis, each of the p vectors in the input data will fall into one of the m clusters or map units corresponding to the output vector, $\mathbf{y} = (y_1, y_2, \ldots, y_m)$ of length m (Figure 4.38). Here, the number of best matching units, m, can be smaller than, greater than, or equal to, the length, n, of the input vector for each time step, p. There is one weight vector of length n associated with each output unit, $\mathbf{y}$.

Two measures commonly used to assess the quality of the SOM analysis, are the *quantization error* (QE) and the *topological error* (TE) (Vesanto et al., 2000). The quantization error is the Euclidean distance between the Best Matching units (winners) and the corresponding input data. The average quantization error, along with

Network architecture

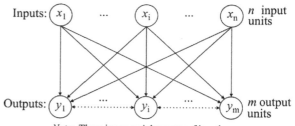

Inputs: x_1 $\cdots$ x_i $\cdots$ x_n n input units

Outputs: y_1 $\cdots$ y_i $\cdots$ y_m m output units

Note: There is one weight vector of length n associated with each output unit

FIGURE 4.38 Network architecture showing input x (of length n) and outputs y (of lengths m). Arrows denote the interconnectivity between input and output. *Adapted from http://genome.tugraz.at/MedicalInformatics2/SOM.pdf.*

its standard deviation, can be used to indicate how well the map units fit the data. The topological error is the ratio of occurrences of map units after all input data have been applied, and where the second-best matching unit is not a direct topological neighbor of the best matching unit. Thus, it gives an indication of how well the map units are topologically ordered.

4.9.2 SOM vs PCA

SOM analysis is often compared to, or used in conjunction with, PCA (see Section 4.4). Both methods effectively reduce the dimensionality of the original data and both search for spatial patterns that aid in the interpretation of the structure and dynamical features embedded in the data. In the analysis of a two-dimensional scalar data field, such as SST or along-channel velocity field from moored current meters, both methods can be used to isolate a set of representative spatial fields.

The EOFs describe a set of modes of variability in the data, which may or may not be related to the actual physical modes of variability. Often, only the first few modes and their respective Principal Component Time Series (PCTS) are sufficient to give a reasonable reconstruction of the original detrended data. The EOFs are found by maximizing the variability explained by each EOF while enforcing the integrated orthogonality condition among the EOFs. This results in "orthogonal ordering", whereby the dominant component of the flow, such as the estuarine circulation, determines the first mode. All subsequent modes are orthogonal to this first mode, which can be problematic in that there may be no particular reason why the physical processes should adhere to orthonormal conditions. Thus, the physical dynamics of the system, other than perhaps the one described by the dominant first mode, can be lost in combinations of the higher modes. Techniques have been developed to circumvent some of these shortcomings (see Sections 4.4–4.6), such as rotating the eigenvectors

which make up the EOFs, performing EOF analysis after removing certain modes, or using a non-orthogonal (oblique) basis functions (Richman, 1986; von Storch and Zwiers, 1999). However, all methods introduce a degree of subjectivity and can lead to non-zero correlation among the PCTS.

Lui et al. (2006) compare EOF analysis to SOM analysis using an artificial time series made up of a progressive sinusoidal wave plus noise. The EOF analysis is able to pick out the sinusoidal signal. However, for a more complex signal made up of an admixture of sine, step, sawtooth, and cosine waves, the leading mode was not representative of any of the signals but of a complex hybrid signal. Furthermore, even though the sine and cosine signals are orthogonal functions, they did not represent any of the higher modes in the time series.

4.9.3 The Self Organizing Map (SOM)

As noted previously, the SOM is a nonlinear, feed-forward neural network with the ability to objectively downscale high dimensional input data into a set of neurons (map units) ordered in a user-selected output space. The user-selected output space in which the map units (also cells, nodes and archtypes) are organized can be rectangular, cylindrical, toroidal or any other two (or higher) dimensional structure. The purpose of the output space is to provide a graphical view of the connections among the map units. Although SOMs have been used as a cluster analyzing tool, the map units obtained using the SOM approach are topologically ordered, unlike in traditional clustering. Thus, at each stage of the processing, each piece of incoming information is kept in its proper context or neighborhood, and map units (neurons) dealing with closely related pieces of information are kept close together so that they can interact via short synaptic connections. Thus, similar map units are assembled together within a region of output space whereas dissimilar map units are forced apart in output space. Because of the connections among map units and the topological ordering, SOM yields a continuum of states representing the original data. In SOM analysis, the map units and their time series are built using the concept of the Best Matching Unit (BMU). The BMU is determined by comparing the map units with the original input data, measuring the similarity (by Euclidean distance) and constructing a time series of the ranking of the map units with respect to their similarity. In this way, SOM and PCA can result in similar analysis products.

Topological ordering in SOM occurs as a result of a neighborhood function. However, in order to effect topological ordering, each neighbor of the winning map unit is also nudged closer to the input data by some smaller degree determined by a neighborhood function, $\varepsilon_{qi}(t)$, given by Eqn (4.120). With only two map units, which we consider as part of the discussion in the section that follows, the complexities in determining neighbors, such as the shape of the map unit array and the lattice of connections between the neurons within that array, do not have to be considered. Furthermore, since the topological ordering has no meaning in the two-map case, the effect of the neighborhood function is to "smooth" or minimize the separation polarization between the two map units. Because the smoothing tends to move the map units closer together, it also increases the quantization error of the mapping, or how well each map unit fits the data it is representing. In order to achieve the best representation of the input data by two map units, it is best to set the neighborhood function to zero. This allows the SOM software to produce results much like a cluster analysis and is better defined as unsupervised vector quantization.

4.9.4 Application to Estuarine Circulation in Juan de Fuca Strait

Juan de Fuca Strait (Figure 4.39) is a partially mixed tidal channel connecting the freshwater catchment basins of the Strait of Georgia and

Puget Sound to the Pacific coast of British Columbia (Canada) and Washington State (USA). The channel is 160 km long, 25–40 km wide, and has a maximum depth of 200 m. Estuarine circulation in the strait prevails 90% of the time in summer and 55% of the time in winter while "transient" wind-forced regimes occur roughly 10% of the time in summer and 45% of the time in winter (Thomson et al., 2007). If we consider the estuarine and transient flow regimes as the end states of a bimodal system, we can use the unsupervised learning capacity of SOM to delineate the "archetypical" structure of the two states and to determine those times in a current meter data set when one or the other of the two states dominates the circulation in the channel. The simple case of two neurons or map units is degenerate with respect topological ordering in SOM analysis. Whether the initial state is positioned to the left or right, upper or lower panel has no physical meaning and is solely determined by the initialization and the order that input data are presented to the map units. After examining the two-map system, we can progress to a system of multiple map units that characterize the transition between the two basic states.

4.9.4.1 Observations

Following Mihály and Thomson (unpublished report), we use a subset of the current vectors, $\mathbf{x}(t)$, obtained from arrays of ADCPs and single-point current meters moored across central Juan de Fuca Strait from 1998 to 2005 (Figure 4.39). The two criteria used in the selection of the subsets were: (1) sufficient coverage to resolve along-channel currents at the scale of the internal deformation, $r_D \sim 10$ km, in the cross-channel direction; and (2) the availability of continuous high quality data throughout the selected analysis period. High resolution cross-channel coverage is essential for delineating the variability of the highly horizontally and vertically sheared flows in the strait. Similarly, pattern-recognition algorithms require high

FIGURE 4.39 Map of the Juan de Fuca Strait and adjoining regions. The square in the upper panel marks the position of a meteorological buoy (C46206) and the open triangle marks the National Centers for Environmental Prediction/National Center for Atmospheric Research (NCEP/NCAR) site "126W49NG". Solid circles in the bottom panel show locations of single-point current meter (CM) moorings; circles with triangles denote ADCP or ADCP + CM moorings. *(Modified after Thomson et al. (2007).)*

quality, long duration records to identify archetypical flow structures. Because the SOM analysis is sensitive to changes in cross-channel coverage and data quality, close attention was paid to variability in data reliability and spatial coverage when interpreting results in terms of dynamical processes.

We selected observations from five ADCP deployments periods covering the period May 2002–May 2005 (Figure 4.40; Table 4.10). The data span two winter periods, two summer periods, and a separate continuous period encompassing both winter and summer months. All cross-channel configurations typically include

upward-looking 150 or 300 kHz ADCPs at stations 2B, 2C, and 2E augmented by a single-point current meter mooring at station 2A with instruments at nominal depths of 25 and 75 m. An additional current meter was sometimes deployed at a nominal depth of 145 m below the ADCP at station 2E on the US side of the strait. The ADCPs were able to achieve maximum vertical ranges in summer but not in winter when there are fewer suitably sized zooplankton scatterers in the strait. This seasonal reduction in scatterers, combined with normal diel migration, resulted in a lowering of backscatter energy from the surface bins during daytime and the

FIGURE 4.40 Yearly time lines for moorings in Figure 4.39 for years 2002–2005. ADCP refers to moorings having an upward-looking (bottom- or mid-depth mounted) acoustic doppler current profiler (ADCP); CM refers to moorings having either Aanderaa RCM4 or InterOcean S4 single-point current meters only. The mid-depth ADCPs often had single-point CMs moored below the ADCP. Gaps denote periods of mooring recovery and servicing, times of instrument damage or data loss, or no moored instrumentation. *(Modified after Thomson et al. (2007).)*

subsequent loss of reliable velocity estimates within the upper two to three ADCP bins. Because of instrument malfunction and other availability factors, mooring configurations changed slightly from deployment to deployment. For example, for the first series used in this study (summer of 2002), the 150 kHz ADCP at the central site 2C was replaced with three single-point current meters at depths of 28, 78 and 153 m to delineate flow in the three distinct depth ranges normally covered by the ADCP. During the following deployment (winter of 2002/03), the single-point current meters were replaced with an upward-looking 75 kHz ADCP. In the winter deployment of 2004–2005, the 75 kHz ADCP at the central mooring was replaced with a 300 kHz ADCP so that its configuration matched that at near-shore stations 2E and 2B.

The ADCPs typically sampled every 15 min, whereas the single-point current meters had sample intervals of 30 or 60 min. All time series were converted to hourly samples following low-pass filtering and resampling at hourly intervals. Because one of the goals of the SOM analysis was to understand the wind-forced circulation, a low-pass 30-h Kaiser-Bessel filter was applied to the hourly records to remove

TABLE 4.10 Times of the Five ADCP-single Point Current meter Deployments Periods Used in the Development of Self Organizing Maps for Alongshore Currents in Central Juan de Fuca Strait (Figure 4.39)

Summer 1	Winter 1	Summer/Winter	Summer 2	Winter 2
May–Sep. 2002	Nov. 2002–May 2003	July 2003–Jan. 2004	May–Sep. 2004	Sep. 2004–April 2005

the tides (see Chapter six regarding filters). The filtered data were then resampled at 12:00 h Coordinated Universal Time (UTC) to provide values for the daily mean circulation. The along-channel direction was chosen to be along the first principal component direction as determined for each current vector time series (see Section 4.4.1). The along-channel data from each mooring of the array was then interpolated vertically using a one-dimensional Hermite cubic spline. In order to represent a partial slip bottom boundary condition, the interpolating spline was forced to be zero at a depth that was arbitrarily chosen as 10% greater than the bottom depth. Near the surface, the velocity at the top-most bin or top-most current meter was extrapolated to the surface. These velocity estimates were then interpolated in two dimensions onto a 10 m vertical by 100 m horizontal cross-channel grid. The grid domain was made approximately 10% wider than the channel width and deeper than the chart depth in order to allow for slippage at the boundary resulting (we assume) in a more realistic bottom boundary layer.

4.9.4.2 Archetypical Flows for All Data

We begin the analysis by deriving a two-element (two map unit) SOM representing the archetypical flow conditions for the along-channel current velocity in central Juan de Fuca Strait for all five deployment periods listed in Table 4.10. The left panel in Figure 4.41 is representative of the cross-channel structure during a fully established estuarine circulation regime in summer; the right panel is representative of the flow structure during a major wind-forced transient event in winter, when strong southerly winds drive an intense (~ 1 m/s) eastward flow called the "Olympic Peninsula Countercurrent" on the US side of the strait (Thomson et al., 2007). Analyses conducted using a neighborhood weight function for the two-map array indicates a modest improvement of the average quantization error of about 10% compared to

when no neighborhood function is used. Quantization errors for the individual deployment periods presented in Table 4.10, as well as the Euclidean distance between the map units, shows that the map units for the summer periods are closely spaced (i.e., they share many topological similarities), whereas those for winter deployments—including the "all data" vectors for the period in column five of Table 4.10—the map units are spaced much farther apart, indicating greater topological differences. For the linear learning rate $\alpha(t) = \alpha_0(1 - t/T)$ from Eqn (4.119), the Epanechikov neighborhood function Eqn (4.120) based on the default initial radii also results in the same map units, indicating that use of the Epanechikov function for small map unit arrays is equivalent to using no neighborhood function. Applying a decreasing Gaussian neighborhood weight function Eqn (4.118), beginning at one and proceeding to 0 with sequentially changing weights, also results in the same set of map units when there are sufficient iterations. This implies that, with judicial selection of the neighborhood function parameters in the weight vectors, it is possible to balance the benefits of topological ordering against minimization of the quantization error.

According to the analysis, days for which the estuarine flow regime was the Best Matching Unit occurred 79% of the time; days when transient flow regimes were dominant accounted for the remaining 21% of the time. This objective finding is very close to a subjective analysis of the same period reported in Thomson et al. (2007) who found, by visually inspecting each daily mean flow period, that 78% were representative of an estuarine flow regime and 22% were representative of a transient flow regime. The time line for the frequency of occurrence of a specific Best Matching Unit (Figure 4.42) shows a clear demarcation between winter and summer. The SOM analysis based on seasons indicates that estuarine (transient) regime prevails 93% (7%)—97% (3%) of the time in the summer and 56% (44%)—74% (26%) of the time in the winter.

FIGURE 4.41 A two-map unit SOM representing the archetypical flow conditions for the along-channel current velocity, *u*, in central Juan de Fuca Strait for all five deployment periods listed in Table 4.10. Outflow toward the ocean is colored blue and inflow from the ocean is colored red. The light white line denotes $u = 0$. The left panel is representative of the cross-channel structure during a fully established estuarine circulation regime in summer (peak surface outflow is 14.8 cm/s, peak near bottom inflow is 7.8 cm/s); the right panel is representative of the flow structure during a major wind-forced transient event in winter, when there is a moderately intense eastward flowing Olympic Peninsula Countercurrent on the US side of the channel (peak near-bottom outflow is 8.3 cm/s, peak surface inflow is 29.8 cm/s). *Courtesy, Steve Mihály, Ocean Networks Canada.*

Corresponding values obtained from a subjective inspection of the daily observations over roughly the same periods (Thomson et al., 2007) are 92% (8%)–93% (7%) over the three summers and 59% (41%)–70% (30%) for the three winters. Both methods indicate that the winter of 2004–2005 had a greater portion of time with estuarine flow conditions, with the SOM and subjective analyses yielding 74% and 70% estuarine conditions, respectively.

The bimodal SOM map structure can be used to estimate the duration of the transient flow events. Over the entire three year data set, there were 60 occasions of contiguous transient flows

with a maximum duration of 9 days. If we assume it takes one day to return to estuarine conditions, there were two transition events with 14 day duration and one each of 10 and 11 day durations. The longer durations tended occur earlier in the winter season (October through December), but otherwise the occurrences of transient events did not reveal any other identifiable pattern during the winter season.

4.9.4.3 One Dimensional Linear SOM Analysis

In order to capture the transition between the two fundamental archetypical flows (from

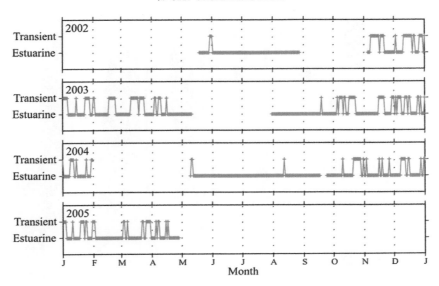

FIGURE 4.42 Frequency of occurrence of Best Matching Units for the two-map unit SOM in Figure 4.41 for observation years 2002–2005. *Courtesy, Steve Mihály, Ocean Networks Canada.*

estuarine to transient and back to estuarine), the bimodal SOM outlined above was expanded along a single dimension to form a linear array. To achieve this, we first examined the mean quantization error and the polar (Euclidian) range and distance for arrays of two to ten linear map units for the overall data and each of the five time periods presented in Table 4.10. The parameters used in the SOM analysis to obtain the error metrics are the same as those used for the initial two map unit array described above. The errors fall into two groupings: the smaller grouping of mean quantization error $(MQE = \langle QE \rangle)$ and Euclidean distances represent periods which only span the summer when there is very little transient flow activity and a smaller range of expected flow structures; the larger grouping of errors coincides with times when there are sufficient numbers of transient events to require many more map units to delineate the continuum of features in the data. As expected, the MQE diminishes with additional map units. From three map units and upward, the polar distance (the sum of the inter-map unit distances) has a linear increasing trend, whereas the polar range (the distance between the poles), typically begins to level off after five or six map units. This deficit between the polar range and polar distance is a measure of how well the linear array is representing the features in the data. The final metric, the topological error (TE), shows that, in general, there is no error until about six map units, and the behavior of the topological error is similar to that of the polar deficit in that it increases with increasing numbers of map units. Both metrics help define the ability of a linear array to represent the input data and, therefore, are an indication of the linearity of the input data in multidimensional space.

The six unit linear array (Figure 4.43) gives a reasonable representation of the daily mean along-channel flow in Juan de Fuca Strait and how the flow transitions between the two archetypical flow regimes consisting of the "pure" estuarine mode on the extreme left and the "pure" transient mode on the extreme right. Here, outflow to the ocean is colored blue and

FIGURE 4.43 A one-dimensional (linear), six unit SOM representing the transition in the daily along-channel flow, u, in Juan de Fuca Strait from runoff-driven estuarine circulation (1) to wind-driven transient circulation (6). Outflow toward the ocean is colored blue and inflow away from the ocean is colored red. The light white line denotes $u = 0$; u is positive for inflow. Numbers in each panel denote the number of times a particular map is the BMU; the fraction of the input data that is clustered into a particular map; and the peak values of the outflow (cm/s)/inflow (cm/s). *Courtesy, Steve Mihály, Ocean Networks Canada.*

inflow from the ocean is colored red. Contiguous pairs of map units have varying degrees of similarity. The values in the individual map units indicate: (1) the number of times that a particular map unit is the best matching unit (BMU); (2) the percentage of input data falling into the map unit; and (3) the ratio of maximum inflow (eastward current, positive) to the maximum outflow (westward current, negative). The strong inflow along the US side of the channel in the right-hand panels indicates the presence of the surface intensified "Olympic Peninsula Countercurrent" that forms during major southerly wind events (Thomson et al., 2007). Note that, while we attribute the transitional maps to a shift in the fundamental flow regime, some of the difference may actually be due to changes in instrumentation and data quality during the different deployment periods.

As in the case for Figure 4.43, the judicial selection of the parameters of the neighborhood function can result in near perfect topological ordering and a minimization of the quantization error (error in Euclidean distance). Large radii, r_i in Eqn (4.120) result in stiffer arrays, which then result in much larger quantization errors. Conversely, minimum values of the quantization error (D_j) can be reached by allowing the final radius to go to zero or by using no ordering at all. Allowing the radii to diminish to zero, or using no neighborhood function, results in a minimum achievable average D_j of around 289 cm/s when applied to the complete data set of along-channel flow in Juan de Fuca. However, topological errors remain unacceptably high in both cases, and the map units do not appear to span a reasonable continuum of data patterns. It is found that using a decreasing radius beginning at two and ending at 0.9 within a Gaussian shaped neighborhood function results in topological errors of less than 1% and an average quantization error of 310 cm/s.

4.9.4.4 Comments on Computations

SOM analysis generally begins with specification of the number of map units as well as the space in which they reside and their interconnection. In the oceanographic and meteorological references we have cited, space has been exclusively limited to a one or two dimensional rectangular grid, with the map dimensions chosen subjectively. As in spatial EOF modes, the individual map units have the same dimensions as the input data. After the array of map units have been initialized to some value, each input data field is compared to the array of map units. As in EOF analysis, both the input data and the map units are converted to vector form and the Euclidean distance measured between the input data and each map unit. The closest map unit is identified as the winner and is moved closer to the input data vector by a factor defined by a learning rate, which is specified by the user. At this point, the steps resemble traditional cluster algorithms and are a form of unsupervised vector quantization. In SOM analysis not only is the winning map unit vector moved closer to the presented input vector but the map units which surround the winning map unit are also incrementally adjusted toward the input vector in inverse proportion to their distances from the winning map unit by use of a neighborhood function. The result is that the SOM array becomes topologically ordered, whereby map units that resemble one another are moved closer together and map units that differ from one another are pushed further apart. This relationship between the winning map units and its neighbors also forces the SOM to produce more map units in regions where there is high input data density as well as forcing a continuum of map units over the entire data set.

To begin a SOM analysis, the map units are first initialized to a set of values. For the SOM Toolbox for MatLab 5 (Vesanto et al., 2000) used in the Juan de Fuca Strait study, there are two options, a random initialization based on the span of the input data and initialization with the leading EOFs. For the EOF initialization, in the case of the simplest one-dimensional SOM, the map units are initialized as the leading EOF. If the map unit array had been two-dimensional, the first two modes would have been used. Extensive testing of these two initializations (S. Mihláy, person. Com., 2008) indicated that both initializations resulted in the same set of map units; the randomly initialized map units took substantively longer to converge. Tests on larger (10 × 10) two-dimensional arrays with a small neighborhood function took as many as 9600 iterations for map units initialized randomly to reach a stable solution. In contrast, for an EOF initialized set of map units with the same neighborhood function, 480 iterations were more than sufficient. However, it should be noted that the number of iterations are strongly affected by the parameters of the neighborhood function; when the neighborhood function is strong, many fewer iterations are needed and the smoothing brings the SOM to stability much faster. Furthermore, if random initialization is used without a neighborhood function for larger arrays, depending on the algorithm, a few map units will remain random and not converge. In this case, unsupervised vector quantization does not necessarily converge to a stable solution.

Within the MatLab software packages, two algorithms are available to perform the analysis. Sequential incremental SOM (the traditional method), and batch SOM. In the sequential algorithm, each input vector is presented to the map units and, using a learning rate, the map units are moved incrementally toward the input data. In the batch method, Voronoi tessellation is used to move the map units closer to the data, and no explicit learning rate needs to be specified. Testing the two algorithms using the alongshore flow data indicated that the batch SOM was faster by an order of magnitude. Furthermore, the batch algorithm with a specified set of parameters always resulted in the same solution, whereas the sequential incremental SOM algorithm did tend toward the batch algorithm solution upon increased iterations, but repeated sequential algorithms did not always result in the same solution. The main benefit of the sequential algorithm is that it is amenable to a limited set of mathematical analyses, and hence has been studied to examine the behavior of the SOM as a neural network (Kohonen, 2001). Because of its speed and repeatability, Mihály and Thomson (unpublished) chose to use the batch algorithm for all the SOM analyses.

As with the bimodal analysis, Mihály and Thomson used the faster batch algorithm to derive the six map unit presented in Figure 4.43. This approach provides sufficient iterations so that the map units reach an absorbing state after which there is no change with each iteration. For the global topology, "sheet" is initially chosen (see Vesanto et al., 2000). This places the map units in a two dimensional array where the interconnections between the map units can be made either with a hexagonal or rectangular lattice. In the limiting case of a linear array, the choice of lattice does not change the interconnections between the map units. Using the sheet topology, either map unit at the ends of the one-dimensional array are only connected in varying degrees to map units toward the center of the array. This has the tendency to enhance the polarization between the two end map units. Circular topologies can be chosen so that the two outer end map units of a sheet array are connected to each other (cylindrical) or both the sides and the top and bottom can be connected (toroidal). In addition to these parameters, the number of map units and the nature of interconnection between the map units through a neighborhood function need to be chosen. These provide a strong subjective input into the analysis that cannot be avoided.

The six map units in Figure 4.43 describe a continuum of along-channel flow states in the daily current velocity time series. In order to quantify the degree of similarity between map unit neighbors, we can use the Euclidean distance metric. To gauge the global similarity of the map units in the array along the linear continuum, some interpretation is necessary. Since each map unit can be represented as a point in multi-dimensional space (defined by a position vector equal in length to the number of grid points of the map unit), the distance between any two map units is easily determined. However, since the magnitude of vector addition is only equal to scalar addition if the vectors have the same "direction" in multidimensional space, it is very unlikely that the distances between map units will sum to the distance between the first and last map units. The vector defining the distance between the "polar" (the two outside) map units is a straight line, but the positions of the map units between the two pole map units will describe a curve in multi-dimensional space. The difference between these two distances is a measure of the linearity of the trajectory of map units transitioning from one polar mode to the other from a particular SOM analysis. Therefore, in addition to quantization and topological error metrics, we can define two other measures specific to a one dimensional map array to assess SOM analysis. We define the polar range as the Euclidean distance between the first and last map unit, and we define the polar distance as the sum of the Euclidean distances between contiguous map units from the first map unit to the last map unit in the array. With these four metrics—mean quantization error, topological error, polar range, and distance—we can add some insight and objectivity in selecting the number of map units in a linear array as well as assessing the linear arrays adequacy in describing the evolution of the underlying data.

4.9.4.5 Summary

The stages of the SOM algorithm can be summarized as follows:

1. **Initialization**—Choose random values for the initial weight vectors, $\mathbf{w}_j$;
2. **Sampling**—Draw a sample training input vector, $\mathbf{x}$, from the input space;
3. **Matching**—Find the winning neuron, $J(\mathbf{x})$, with weight vector closest to input vector;
4. **Updating**—Apply the weight update equation $\mathbf{w}_j(t + \Delta t) = \mathbf{w}_j(t) + \alpha(t) \cdot \boldsymbol{\varepsilon}(t)[\mathbf{x}_i(t) - \mathbf{w}_j(t)]$;
5. **Continuation**—keep returning to step two until the feature maps stop changing.

4.9.5 Growing Hierarchical Self Organizing Maps

Growing hierarchical Self Organizing Maps (GHSOM) were designed to remove some of the subjectivity of choosing the SOM topology. Rules are made so that an initial SOM can be grown by row and column, as well as hierarchically, such that a single map unit can spawn new layers of SOM. According to Liu et al. (2006), the GHSOM improves the basic SOM by (1) providing an incrementally growing version of the SOM, which eliminates the need for the user to directly specify the initial size of the map beforehand; and (2) by enabling the SOM to adapt to hierarchical structures in the input data. Briefly, the steps are:

Step 1

Prior to the training process, a single map (therefore, by definition, not a Self Organizing Map) is created. The weight vector for this single map is the mean of all input vectors (e.g., the mean flow or the mean SST). This one-unit map is deemed "layer 0", and has a mean quantization error, $\langle QE \rangle_0$, given by

$$\langle QE \rangle_0 = \sum \|\mathbf{x}_k - \mathbf{w}_i\| \qquad (4.122)$$

where the double vertical lines denote the Euclidean distance. The only reason for creating this map is to obtain the above quantization error, which is then used in the next step.

Step 2

Below layer 0, a new 2×2 SOM, "layer 1" (Figure 4.44), is created. The mean QE of these four maps, $\langle QE \rangle_1$, is compared with $\langle QE \rangle_0$ in Eqn (4.122) with the requirement that

$$\langle QE \rangle_1 > \tau_1 \langle QE \rangle_0 \qquad (4.123)$$

where τ_1 is a number between 0 and 1. If τ_1 is set to 1, the above inequality will never be fulfilled and the "growing" will stop at this point; i.e., the result will be layer 1, the 2×2 map unit. If the inequality is fulfilled, the map unit with the greatest mean quantization error $\langle QE \rangle$ is found and defined as the error unit. Next, the most dissimilar adjacent neighbor of the error unit is identified and a new row or column of map units is inserted between the error unit and the most

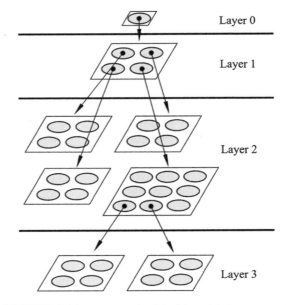

FIGURE 4.44 An example of the hierarchical structure of the Growing Hierarchical Self Organizing Maps (GHSOM). The single map from "layer 0" is expanded in the 2×2 map units. All of the four units in the first-layer SOM are expanded in the second layer. Only two units in one of the second layer SOMs are further expanded in the third layer.

dissimilar neighbor. In this way, the layer begins to grow in breadth; τ_1 is therefore called the "breadth controlling parameter". If τ_1 is set to zero, the map would grow to infinity, so the idea is to choose a reasonable value for τ_1. The default value in the MatLab toolbox is 0.3, which indicates that the mean or cumulative QE must be 70% greater than that for the preceding QE for the map to stop growing. Mihály and Thomson (unpublished) found that a higher value (0.6) improved the analysis.

Step 3

If the inequality Eqn (4.123) is still not satisfied, the strategy is to pick map units, which might be expanded further into a new hierarchical layer. A second inequality is used:

$$QE_i > \tau_2 \langle QE \rangle_0 \qquad (4.124)$$

where τ_2 is smaller than τ_1 ($0 < \tau_2 \leq \tau_1 < 1$); here, τ_2 is called the "depth controlling parameter" and has a MatLab default value 0.03 (Mihály and Thomson used a lower value of 0.005). In this process, the quantization error of each individual map is compared to the quantization error of the "layer 0" map unit. As a consequence, each single map unit that satisfies the above inequality is expanded in a subsequent layer. In general, decreasing the parameter τ_1, grows SOM arrays with greater breadth, while decreasing the parameter τ_2, grows SOMs with more layers in the GHSOM hierarchy. Choosing the appropriate parameters re-introduces a strong degree of subjectivity back into the analysis. Using GHSOM for a single layer illustrates the hierarchical aspect of the process while preserving the objectiveness of the approach.

4.10 KALMAN FILTERS

In 1960, Rudolf E. Kalman published his now famous paper describing a recursive solution to the discrete-data linear filtering problem of trying to estimate the state y, which is governed by the linear stochastic difference equation,

$$y_k = Ay_{k-1} + Bu_{k-1} + \text{random (white) noise} \qquad (4.125)$$

The matrix A relates the state at the present step, k, with that at the previous step, $k-1$, and the matrix B relates the optional control input (u, such as the Navier–Stokes equations of motion) to the state y. Kalman's original interest was in determining the orbits of planets from limited Earth observations but the filter soon found extensive use in autonomous and assisted navigation, with both civilian and military applications. The Kalman Filter is an efficient optimal estimator (a set of mathematical equations) that provides a recursive computational methodology for estimating the state of a discrete-data controlled process from measurements that are typically noisy, while providing an estimate of the uncertainty of the estimates. Because the filter is recursive, new measurements can be processed as they are received from external sensors. Unlike recursive filters, the method doesn't need to store all previous measurements nor reprocess all data at each time step. If the noise associated with all processes and data contributing to the state have Gaussian (normal) distributions, then the Kalman filter obtains solutions by minimizing the mean square error of the estimated parameters. If the Probability Density Functions of the variables are not known, and only the mean and standard deviation of the noise are available, the Kalman filter is still the best linear estimator; some non-linear estimators may be better than the linear estimators.

Kalman filters have a number of advantages. Specifically, the filter provides reliable practical results due to its optimality, it is designed for real time processing, and equations involving the measurements do not need to be inverted as part of the solution. Moreover, the method is easy to formulate and implement, given a basic understanding of the underlying mathematics. As with other filters, the purpose of the Kalman filter is to determine the "best" estimate of state parameters, Y, from noisy input data. The Kalman filter not only determines values of the parameters within the constraints of the measurement uncertainty but it also takes into account the error of the measurements relative to the error of the predictions. In essence, Kalman filters fuse prediction and measurement based on a weighted difference between an actual measurement and a measurement prediction. The method proceeds from an *a priori* estimate, $\widehat{y}_k^-$, of the state y_k at step k that is based on information of the process prior to step k (hence the minus sign superscript), to an *a posteriori* estimate of the state, $\widehat{y}_k$, at step k that incorporates newly acquired measurements, z_k (Welch and Bishop, 2006) Specifically,

$$\widehat{y}_k = \widehat{y}_k^- + K\left(z_k - H\widehat{y}_k^-\right) \qquad (4.126)$$

where the new measurement, z_k, is given in terms of the actual state of the process, y_k,

$$z_k = Hy_k + v_k \qquad (4.127)$$

plus some uncertainty, characterized by a random (white noise) variable, v_k, having a Gaussian probability distribution. Here, H is a matrix linking the state to the measurement, matrix K is *gain* or *blending factor*, and the difference $z_k - H\widehat{y}_k^-$ is the measurement *residual* or *innovation*. The errors of the *a priori* and *a posteriori* estimates are then, respectively,

$$e_k^- = y_k - \widehat{y}_k^- \qquad (4.128a)$$
$$e_k = y_k - \widehat{y}_k \qquad (4.128b)$$

with corresponding error covariance functions (traditionally written as P rather than Cov as we did earlier)

$$P_k^- = E\left[e_k^- e_k^{-^T}\right] \qquad (4.129a)$$
$$P_k = E\left[e_k e_k^T\right] \qquad (4.129b)$$

The key to solving Eqn (4.126) is to find that matrix K that minimizes, in the usual least squares sense, the a posteriori error covariance in Eqn (4.129b). This is accomplished by substituting Eqn (4.126) into Eqn (4.128b), which is then substituted into Eqn (4.129b), deriving the indicated expected values, taking the derivative of the trace of the result with respect to K, setting the result to zero (corresponding to minimization), and then solving for K. One expression for K that minimizes Eqn (4.129b) is (Welch and Bishop, 2006)

$$K_k = \frac{P_k^- H^T}{(HP_k^- H^T + R)} \qquad (4.130)$$

where R is the measurement error covariance. Note that as R approaches zero, $K_k \to 1/H$, so the gain K weights the residual between the observation and the initial estimate, $K(z_k - H\widehat{y}_k^-) \to (y_k - \widehat{y}_k^-)$, more favorably. In contrast, as the *a priori* estimate error covariance P_k^- approaches zero, $K_k \to 0$ so that the gain K weights the new estimate more favorably than the data.

As an example, consider the problem of determining the precise location of a sailboat participating in a trans-oceanic race. We assume that the boat is equipped with a Global Positioning System (GPS) unit that provides an estimate of the vessel's position within a few meters but that the calculated positions are noisy, with values that "dance" around while remaining within a few meters of actual position. (Early in our careers, we depended on Loran-C navigation for positioning in offshore regions, for which fixed positions could jump around by as much as several "cables" over periods of several minutes; a cable is 1/10 of a nautical mile.) Because the heading and speed of the boat are known, the boat's location can be computed from the GPS information provided the influence of the surface current and wind on the boat are known or can be estimated.

This "dead-reckoning" method—which, in the not-so-distant past, was one of the only ways sailing ships could determine their position at sea—yields smooth estimates of the vessel's position but which drift over time. As in other applications, the Kalman filter can be thought of as operating in two distinct phases of dead-reckoning: *predict* and *update*. During the prediction phase, the boat's position can be modified by the external forces acting on the boat (the dynamic or "state transition" model). In addition to calculating a new estimate of the vessel's position, a new covariance function can also calculated. Next, in the update phase, a measurement of the boat's position is obtained from the GPS, a measurement that, once again, has a degree of uncertainty. The covariance between the newly predicted position and that of the prediction from the previous phase determines how much the new measurement will affect the updated prediction. Ideally, if the dead-reckoning estimates tend to drift away from the real position, the GPS measurement will help "nudge" the position estimate back toward the real position, while not perturbing the system to the point that the predictions become rapidly changing and noisy. We have summarized the above procedure in Figure 4.45 where the first prediction phase is based on previously available data and precedes the collection of new data in the measurement phase. The variances of the two steps are related.

In Figure 4.45, the conditions for the state and its variance at the previous data step are $\widehat{y}_{k-1}$ and σ_{k-1}, respectively; the prediction for the state and the variance at the next step are $\widehat{y}_k^-$ and σ_k^-.

We use the dynamical model (i.e., physical system), together with the initial conditions, to make this prediction. We then take the measurement z_k and compute the corrected state, $\widehat{y}_k$ (with variance σ_k), by blending the prediction and the residual, which is always a case of merging two Gaussian variables. The result is an optimal estimate with a smaller variance.

The blending factor, K, is controlled by knowledge gained from the measurements or from the prediction. As noted earlier, if the prediction is more certain, then the prediction error covariance P_k decreases to zero, K also approaches zero and the filter weights the prediction more heavily than the residual, $z_k - H\widehat{y}_k^-$. In contrast, if the measurements are more certain that the prediction, then the measurement covariance (R) approaches zero as K approaches H^{-1} and the filter weights the residual more heavily than the prediction. The computational basis for the Kalman filter equations are outlined in Figure 4.46, which shows both the prediction and measurement phases of the Kalman filter.

In Figure 4.46, the recursive nature of the filter equations is indicated by the arrows at the top and bottom of the boxes. The boxes represent the prediction and measurement steps of the Kalman filter. The basic assumptions behind the filter are:

1. The model used to predict the "state" needs to be a linear function of the measurements;
2. The model error and the measurement error (measurement noise) must both be Gaussian with zero means.

FIGURE 4.45 The sequence of the Kalman equations for estimation of the variable y.

The Kalman equations

Make prediction based on previous data: $\widehat{y}^-; \sigma^-$

↓

Take measurement: z_k, σ_z

↓

Optimal estimate $(\widehat{y})$ = Prediction + (Kalman gain) * (Measurement − Prediction)

Variance of estimate = Variance of prediction * (1 − Kalman gain)

The set of Kalman filtering equations in detail

FIGURE 4.46 The Kalman filter equations (see text for details and functions).

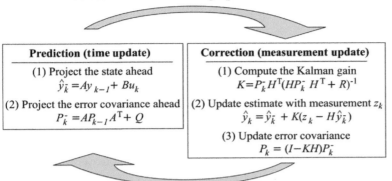

Prediction (time update)

(1) Project the state ahead
$$\hat{y}_{\bar{k}} = Ay_{k-1} + Bu_k$$

(2) Project the error covariance ahead
$$P_{\bar{k}} = AP_{k-1}A^\mathrm{T} + Q$$

Correction (measurement update)

(1) Compute the Kalman gain
$$K = P_{\bar{k}}H^\mathrm{T}(HP_{\bar{k}}H^\mathrm{T} + R)^{-1}$$

(2) Update estimate with measurement z_k
$$\hat{y}_k = \hat{y}_{\bar{k}} + K(z_k - H\hat{y}_{\bar{k}})$$

(3) Update error covariance
$$P_k = (I - KH)P_{\bar{k}}$$

Even if the noise is not Gaussian, and we know only the mean and standard deviation of the noise and not its probability distribution, the Kalman filter is still the best linear estimator. For highly non-Gaussian distributions, non-linear estimators may be preferable.

It worth emphasizing that the Kalman filter combines a system's dynamics (i.e., the physical laws of motion) known to control the inputs to that system with multiple sequential measurements to form an estimate of the systems varying quantities (its state) that is better than the estimate obtained by using any one measurement alone. Noisy sensor data, approximations to the physical equations that describe how a system changes with time, and unaccounted for external factors, introduce uncertainty in the system's state. The Kalman filter averages a prediction of the system's state with a newly acquired measurement whose contribution is determined by a series of weights. These weights, which also take into account the "trustworthiness" of the data relative to the prediction, are calculated from the covariance function, a measure of the estimated uncertainty of the prediction of the system's state. As a result, the new state estimate lies between the predicted and the measured state, and has a smaller estimated uncertainty than either of the two alone. This process is repeated for every time step, with the new estimate and its covariance forming the prediction used in the subsequent iteration. Hence, the recursive character of the Kalman filter, which requires only the last "best guess" rather, then the entire history of a system's state to calculate the new state.

Since the certainty of the measurements is often difficult to specify precisely, it is common to discuss the filter's behavior in terms of its gain. The Kalman gain depends on the relative uncertainty of the measurements and the current state estimate, and it can be adjusted to achieve a particular desired performance of the filter. For a high gain, the filter places more weight on the measurements and follows them more closely; for a low gain, the filter conforms more to the model predictions smoothing out noise and decreasing the responsiveness of the system. At the extremes, a gain of zero causes the measurements to be ignored completely while a gain of unity causes the estimate of the state to be ignored entirely.

Implementation of the Kalman filter is often difficult to achieve in practice due to the requirement for a good estimate of the noise covariance matrices. A study by Furrer and Bengtsson (2007) used Monte Carlo methods to estimate the Kalman filter variants. Another promising approach is the Autocovariance Least Squares (ALS) technique that uses the autocovariance of the data to estimate the noise covariance. A study by Odelson et al. (2005) demonstrates that the noise covariances estimated in this way are unbiased and converge to the true values with increasing sample size. They also add positive semi definiteness constraints to these covariances. Abdel-Hafez (2008) uses the same technique to estimate the Global Positioning System (GPS) measurement noise-covariance matrix.

The Kalman filter is known to be optimal since: (a) the model perfectly matches the real system; (b) the measurement noise is white; and (c) the covariances of the noise are exactly known. We have already discussed methods to estimate the error covariance's. Once these are known, it is useful to estimate the performance of the Kalman filter itself; i.e., to determine whether it is possible to further improve the state estimation quality. We also know that if the Kalman filter works optimally, the output prediction error is also white noise. This white noise character properly reflects the state estimation quality. To evaluate the performance of the Kalman filter, it is necessary to inspect the "whiteness" of the predictions.

An interesting example of Kalman filter application is given by www.cs.cornell.edu/Courses/cs4758/2012sp/.../MI63slides.pdf. This example concerns tracking the fluid level in a tank being filled. For this problem, we can write the analysis sequence as:

Predict:

$$\widehat{y}_{t|t-1} = A_t\widehat{y}_{t-1|t-1} + B_tu_t \qquad (4.131a)$$

$$P_{t|t-1} = A_tP_{t-1|t-1}A_t^T + Q_t \qquad (4.131b)$$

where: $\widehat{y}$ is the estimated state; A is the state transition matrix (i.e., the matrix describing the transition between states); u represents the optional control variables on the state; B is the control matrix (i.e., the matrix mapping the control variables to the state variables); P is the state error covariance matrix (i.e., error of the estimation);

and Q is the process covariance error matrix (i.e., the error due to the process).

Subscripts are: $t|t$ = the current time, $t-1|t-1$ = the previous time, and $t|t-1$ is time at intermediate steps. The Kalman filter removes noise from the system by assuming a pre-defined model of the system. This model should be defined by the following:

1. Understand the situation: Examine the problem and break it down in to the mathematical basics.
2. Model the state process: Start with a basic model, which may not be perfect at first, but can be refined later.
3. Model the measurement process: Analyze how to measure the process. The measurement space may not be in the same space as that of the state (e.g., using an electrical diode to measure weight, an electrical reading does not directly translate to weight).
4. Model the noise: This needs to be done both for the state and the measurements processes. The basic Kalman filter assumes Gaussian white noise so that the variance and the covariance (error) functions are meaningful (i.e., make sure that the error you model is suitable for the situation).
5. Test the filter: This step is often overlooked. Use synthetic data if needed. See if the filter is behaving as it should.
6. Refine filter: Try to change the noise parameters (filter), as these are the easiest to change. If necessary go back further and rethink the situation.

As an example, consider the water in a tank (Figure 4.47) and assume there is a measurement parameter that gives the water level of the tank as it fills or empties. The goal is to estimate the unknown level of water in this tank. In this example, the water level measurements are provided by a float in the tank. The tank could be: (a) Filling, emptying or static (level is increasing, decreasing or constant); or (b) sloshing around or

FIGURE 4.47 Water level in a tank. Measurements are provided by the float.

static (level changes from side to side or is flat and constant). We first consider the most basic model in which the fluid in the tank is level (horizontal) and the surface is constant ($L = c$). Using Eqn (4.10.7), we can write the state variable as a scalar so that $\hat{y} = y$, where y is now the estimate of L. Since we are assuming a constant model, there is no time variation and $y_{t+1} = y_t$, so that $A = 1$ for any $t \geq 0$. Both B and $u = 0$.

We now need to model the measurement process, z, which for simplicity we assume is precisely the water level in the tank. Finally, we need to model the noise. Again, for simplicity, we assume that there is noise in the measurement, and that $R = r$ in the expression Eqn (4.130) for the Kalman gain. The process is a scalar so we can assume that $P = p$; since the process is not well defined, we adjust the noise so that $Q = q$. To test the filter, we apply Eqn (4.131) whereby

$$y_{t|t-1} = y_{t-1|t-1} \qquad (4.132a)$$

$$p_{t|t-1} = p_{t-1|t-1} + q_t \qquad (4.132b)$$

The update is

$$y_{t|t} = y_{t|t-1} + K_t(z_t - y_{t|t-1}) \qquad (4.133a)$$

$$K_t = p_{t|t-1}(p_{t|t-1} + r)^{-1} \qquad (4.133b)$$

$$p_{t|t} = (1 - K_t)p_{t|t-1} \qquad (4.133c)$$

where, the new data value is z_t. The filter is now completely defined. To put some numbers into this model, we first assume that the true level is $L = 1$. We initialize the state with an arbitrary number, with an extremely high variance, as it is completely unknown. Specifically, $y_0 = 0$ and $p_0 = 1000$. If one initializes with a more meaningful value, the filter solution will convergence faster. We choose the system noise as $q = 0.0001$ since we believe that we have a realistic and accurate model. Thus,

Predict 1:

$$y_{1|0} = 0$$

$$p_{1|0} = 1000 + 0.0001$$

The hypothetical measurement we obtain from the float is $z_t = z_1 = 0.9$, which differs from the true water level value of $L = 1.0$ because of noise. Assuming a measurement noise of $r = 0.1$, we can write:

Update 1:

$$K_1 = (1000 \times 0.0001)(1000 \times 0.0001)^{-1}$$

$$= 0.9999$$

$$y_{1|1} = 0 + 0.9999(0.9 - 0) = 0.8999$$

$$p_{1|1} = (1 - 0.9999)(1000 \times 0.0001) = 0.1$$

As indicted by this first update step, the initialization value $y_{1|0} = 0$ has been brought close to the true value, $L = 1$, of the system. In addition, the error variance, $p_{1|1}$, has diminished to a more reasonable value.

If we do another step, we have:

Predict 2:

$$y_{2|1} = 0.8999$$

$$p_{2|1} = 0.100 + 0.001 = 0.1001$$

And, assuming a second measurement $z_2 = 0.8$ (again differing from unity because of noise), we find:

Update 2:

$$K_2 = 0.1001(0.1001 + 0.1)^{-1} = 0.5002$$

$$y_{2|2} = 0.8999 + 0.5002(0.8 - 0.8999) = 0.8499$$

$$p_{2|2} = (1 - 0.5002)0.1001 = 0.0500$$

If we continue this process, we eventually obtain the results as shown in Table 4.11.

From this table, we can see that the model works successfully. After stabilization at about time step 4, the estimated state is within 0.05 of the true value, while the specified measurements are between 0.8 and 1.2 (only within 0.2 of the true value). The results are plotted in Figure 4.48 where the purple line represents the values estimated by the Kalman filter; the yellow line is the true value (a constant) and the dark blue line are the input measurements.

As shown, the Kalman filter provides an optimal estimate of the true values even when the measurements are very noisy (a 20% errors in measurements resulted in only a 5% inaccuracy in the Kalman filter estimate). Hence, the Kalman filter has achieved its purpose.

We now examine a more realistic case in which the tank fills at a constant rate, f, such that the water level $L_t = L_{t-1} + f$. We assume that $f = 0.1$ per unit time and start with $L_0 = 0$.

We will also assume that the measurement and the process noise are constant with time (i.e., $q_t = 0.001$ and $r_t = 0.1$). Table 4.11 now takes the form shown in Table 4.12.

We see that, over time, the estimated state stabilizes (i.e., the variance becomes very small). While the estimate reduces the noise, it dramatically underestimates the true value, L, which is much closer to the measured values, z (Figure 4.49).

It is clear from Figure 4.49 that the estimates systematically underestimate the true and even the measured values. This is not a very attractive result even though the noise level has been greatly reduced. There are two possible causes for this problem: (a) the model we have chosen; and/or (b) the reliability of our process model (our chosen q value). The easiest corrective approach is to change the q value. A valid question is "Why did we chose $q = 0.0001$ to begin with?". The answer is that we thought our model was a good estimation of the true process and that the error level would be quite small. Apparently, our model was not as good as we had anticipated so we need to relax this requirement. Specifically, we now assume there

TABLE 4.11   Kalman Filter Applied to the Water Tank Problem (Figure 4.47)

Time (t)	Predict		Update			
	$y_{t|t-1}$	$p_{t|t-1}$	z_t	K_t	$y_{t|t}$	$p_{t|t}$
3	0.8499	0.0501	1.1	0.3339	0.9334	0.0334
4	0.9334	0.0335	1	0.2509	0.9501	0.0251
5	0.9501	0.0252	0.95	0.2012	0.9501	0.0201
6	0.9501	0.0202	1.05	0.1682	0.9669	0.0168
7	0.9669	0.0169	1.2	0.1447	1.0006	0.0145
8	1.0006	0.0146	0.9	0.1272	0.9878	0.0127
9	0.9878	0.0128	0.85	0.1136	0.9722	0.0114
10	0.9722	0.0115	1.15	0.1028	0.9905	0.0103

The Predict and Update Parameters are Explained in the Text.

FIGURE 4.48 Kalman filter estimates (purple line) of a constant water level (yellow line) in a tank. The measured values (z) input to the filter are shown in dark blue.

TABLE 4.12 Kalman Filter Applied to the Water Tank problem (Figure 4.47) for The Specific Values Presented in the Text (Specifically, $f = 0.1$ per Unit Time and Initial Water Level $L_0 = 0$)

	Predict		**Measurement and Update**				**Actual**
Time (t)	$y_{t\|t-1}$	$p_{t\|t-1}$	z_t	K_t	$y_{t\|t}$	$p_{t\|t}$	L
0	—	—	—	—	0	1000	0
1	0.000	1000×0.0001	0.11	0.9999	0.1175	0.100	0.1
2	0.1175	0.1001	0.29	0.5002	0.2048	0.0500	0.2
3	0.2048	0.0501	0.32	0.3339	0.2452	0.0334	0.3
4	0.2452	0.0335	0.50	0.2509	0.3096	0.0251	0.4
5	0.3096	0.0252	0.58	0.2012	0.3642	0.0201	0.5
6	0.3642	0.0202	0.54	0.1682	0.3945	0.0168	0.6

The Last Column Gives the Actual Water Level.

is a greater error with our process model and set $q = 0.01$. Application of the new q-value to the previous model results in the values plotted in Figure 4.50.

As indicated by Figure 4.50, changing q greatly improved the accuracy of the estimate, although the estimates are still significantly below the true and measured values. If we

FIGURE 4.49 Kalman filter estimates (purple line) of a water tank filling at a constant rate $f = 0.1$ per unit time with constant error $q = 0.0001$ per unit time. The yellow line shows the true water level, L, and the dark blue line the water level (z_t) measured by the float.

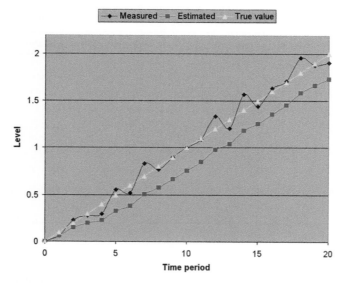

FIGURE 4.50 As with Figure 4.49, but with a much smaller constant error $q = 0.01$ for the process model.

FIGURE 4.51 As with Figure 4.49, but with an even much smaller constant error $q = 0.1$ for the process model.

increase q once again by setting $q = 0.1$, we obtain the results plotted in Figure 4.51, which shows an even greater improvement. The Kalman filter estimate begins to track the noisy measurements, an intentional outcome of increasing q, but still has a bit less noise than the measured points. The filter estimates now approach the true value more effectively than the estimates using smaller noise values.

It is clear that further increases in the process noise level will cause the estimated values to increasingly match the measured values, giving no benefit to using the Kalman filter. The lesson here is that a poorly defined model will not provide a good filter estimate. However, increasing the estimated error will allow the Kalman filter estimate to rely more on the measurement values, while still allowing some noise removal.

The primary use of Kalman filtering in physical oceanography is for the assimilation of oceanographic data into numerical models (Pham et al., 1998). This application uses the Extended Kalman Filter (EKF), which applies

to non-linear state systems or measurement equations. Here, a Kalman filter is a linearized version of these equations, which continues to yield an optimal solution. The key to this form of data assimilation is to approximate the error covariance of the data to be assimilated into the numerical model. This procedure amounts to making no correction in those directions for which the error is the most attenuated by the system. This has the added benefit of improving the filter stability. These "directions of correction" evolve with time according to the model evolution, which is a primary feature of this filter that distinguishes it from other sequential assimilation methods. Pham et al. (1998) suggest a method for initializing the filter based on EOFs discussed earlier in this chapter. They examine assimilation of wind stress forcing into a simple quasi-geostrophic (QG) model for a square ocean domain. Although this is an unrealistic test case, the result of this assimilation method are very encouraging.

In a study of coastal ocean problems, Chen et al. (2009) compared the reduced rank Kalman filter (RRKF) with the ensemble Kalman filter (EnKF) and the ensemble square-root Kalman filter (ENSKF) in three idealized regimes: (a) A flat bottom circular shelf driven by tidal forcing at the open boundary; (b) a linear slope continental shelf with river discharge; and (c) a rectangular estuary with tidal flushing intertidal zones and freshwater discharge. They used the unstructured grid Finite-Volume Coastal Ocean Model (FVCOM). Model run comparisons showed that the success of the data assimilation method depends on sampling location, assimilation methods (univariate or multivariate covariance approaches), and the nature of the dynamical system. In general, for these applications, the EnKF and EnSKF work better than RRKF, particularly for time dependent cases with large perturbations. In EnKF and ENSKF, multivariate covariance methods should be used to avoid the appearance of unrealistic numerical oscillations. Since the coastal ocean features multiscale dynamics in both time and space, an individual case-by-case approach should be used to determine the most efficient and reliable data assimilation approach for different dynamical systems.

4.11 MIXED LAYER DEPTH ESTIMATION

Much of this chapter has focused on methods for constructing smoothed oceanic fields from large-scale spatial and temporal data. Some methods take into account the dynamics of the physical system, while others take a strictly statistical approach. Little has been said about methods designed to delineate physical "breaks" in oceanic distributions. Here, we gave a brief overview of techniques used to define the depth of the surface mixed layer—the top layer of the ocean characterized by uniform to nearly uniform vertical water property structure—which

is of interest to a broad variety of oceanic studies including upper ocean productivity, air-sea exchange processes, and climate variability (cf. Curry and Roy, 1989; Robinson et al., 1993; Wijesekera and Gregg, 1996; and Kara et al., 2000a,b). Methods for determining temporal "regime shifts" in the ocean are presented in Chapter 5.

Surface windstress, convective cooling, breaking waves, current shear, and other turbulent processes in the upper ocean generate a surface layer characterized by uniform to near-uniform density, active vertical mixing, and high turbulent dissipation. The depth of this "mixing layer" is ultimately determined by a balance between the destabilizing effects of mechanical mixing and the stabilizing effects of surface buoyancy flux. As a result of temporal variations in these opposing affects, the mixing layer may be imbedded in a deeper "mixed layer" of almost identical density to the mixing layer and representing the time-integrated response to previous mixing events. Mixed layer depth (MLD) can vary by tens of meters over a diurnal cycle and by over 100 m over an annual cycle (cf. Large et al., 1994).

In the absence of direct turbulent dissipation measurements, mixed layer depth is derived from oceanic profile data using a variety of proxy variables. Methods for estimating the MLD from CTDs and other profiling instrumentation data fall into four broad categories: (1) Threshold methods, which find a predefined step in the surface profile (Price et al. 1986; Lukas and Lindstrom 1991; Peters et al. 1988) or find a critical gradient for the upper layer (Lukas and Lindstrom, 1991; Holte and Talley, 2009); (2) least-squares regression methods, which fit two or more line segments to near-surface profiles (Papadakis, 1981, 1985); (3) integral methods, which calculate a depth-scale for the upper layer based on integral properties of the water column such as conservation of mass (Ladd and Stabeno, 2012; Freeland et al., 1997, 2013); and (4) a split-and-merge algorithm that fits straight line segments based on a specified error minimization (Thomson and Fine, 2003, 2009). The methods have fundamentally different approaches. Methods one and four consider the mixed layer to be a physically distinct entity whose depth can be determined from the observed density structure independently of any integral conservation constraints. In contrast, method three views the mixed layer as the upper component of a two layer approximation to a continuous density profile who derivation must satisfy certain conservation requirements such as conservation of total mass.

4.11.1 Threshold Methods

In air-sea interaction studies (e.g., Wijesekera and Gregg, 1996; Smyth et al., 1996a,b), the depth of the surface mixed layer, D, is defined as that depth, z, at which the potential density difference $\Delta\sigma_\theta(z) = \sigma_\theta(z) - \sigma_\theta(z_0)$ in the upper ocean exceeds a specified threshold value, typically 0.01 kg/m^3 (Figure 4.52(a)); here, z_o is a reference depth (generally in the range $z_o = 0$, the ocean surface, to 10 m depth) and $\sigma_\theta(z) = \rho_\theta(z) - 1000$ kg/m^3 is the density anomaly for measured potential density, ρ_θ. Because the threshold method is comparatively simple—depth estimates can be made by hand without analytical computations—and because the threshold difference of 0.01 kg/m^3 generally yields diurnal depth estimates similar to those from turbulent dissipation measurements, the threshold difference method with the threshold 0.01 kg/m^3 has become the *de facto* standard for many mixed layer depth studies. In ocean-climate studies, where focus is on the variability in MLD averaged over periods of months and longer, threshold values in excess of 0.125 kg/m^3 are more commonly used for monthly mean data (cf. Table 1 in Kara et al., 2000b). A drawback with the 0.01 kg/m^3 threshold method is that it often neglects the underlying water of near-identical density encompassing high chlorophyll, nutrient, and particle concentrations

(e.g., Robinson et al., 1993; Washburn et al., 1998) and ignores the fact that "… the retreat of turbulent mixing to shallower depths proceeds faster than the erosion of the stratification at the base…" (Kara et al., 2000b).

Because oceanographers have access to more temperature profile data than salinity (and hence density) profile data, and because salinity measurements tend to be noisier than temperature measurements, the mixed layer depth is commonly linked to a step-like change in water temperature, with specified steps in the range 0.01−0.5 °C (e.g., Levitus, 1982; Weller and Plueddemann, 1996; Kara et al., 2000). However, where possible, it is better to use sigma theta (σ_θ) as an estimator for MLD, primarily because it is the density structure, which directly affects the stability and degree of turbulent mixing in the water column. *In situ* density (sigma-t, σ_t) is less reliable because over-turning turbulence can result in adiabatic changes of 0.04 kg/m^3 over depths of 10 m (Schneider and Müller, 1990).

The less frequently used threshold gradient method defines mixed layer depth as the depth at which the density gradient, $\partial\sigma_\theta/\partial z$, first exceeds 0.01 kg/m^4 (Figure 4.52(b)) or other specified level. The density gradient method is considered a less consistent estimator of MLD than the density difference approach (Schneider and Müller, 1990).

4.11.2 Step-Function Least Squares Regression Method

Problems with profile approximations have led to the formulation of customized curve-fitting algorithms for oceanic profiles, including the use of least-squares linear approximations (Papadakis, 1981; Freeland et al., 1997) and "form oscillators" (Papadakis, 1985). Papadakis (1981) uses a three-segment linear fit and the Newtonian approximation method to find a minimum variance solution to the general mixed layer depth problem. Freeland et al. (1997) uses a

FIGURE 4.52 Estimation of mixed layer depth (D) for different methods for the density profile obtained for a CTD station LH07 off the west coast of Vancouver Island, July 19, 1997 (see Figure 4.54 for station location). (a) Threshold difference method with standard density step $\Delta\sigma_\theta(z) = \sigma_\theta(z) - \sigma_\theta(0) = 0.01$ kg/m^3; (b) threshold gradient method for $\partial\sigma_\theta/\partial z = 0.01$ kg/m^4; (c) two-segment least-squares method for $z_a = 10$ m, $z_b = 200$ m, constant $\sigma_{\theta1}$ and $\sigma_{\theta2}$, and $\sigma(D) = (\sigma_{\theta1} + \sigma_{\theta2})/2$; and (d) integral depth-scale, D, with $z_a = 10$ m and $z_b = 200$. In (c), each of the paired shaded regions have equal positive (+) and negative (−) areas. In (d), the darkly shaded rectangle has sides of length D and $L_\sigma = \sigma_\theta(z_b) - \sigma_\theta(z_a)$, with the area $D \times L$ of the rectangle equal to the vertical integral $I = \int_{z_b}^{z} [\sigma_\theta(z) - \sigma_{\theta b}]dz$ denoted by the shaded region to the right of the density profile.

From Thomson and Fine (2003).

two-segment least-squares approach to obtain a time series of winter mixed layer depth at Ocean Station "P" (50° N, 145° W) in the northeast Pacific. The two-segment case can be solved analytically and is useful for pre-CTD data. The three-segment case requires special techniques (e.g., Papadakis, 1985) and solutions can be unstable.

The two-segment approach (Figure 4.52(c)) seeks a step-like least-squares approximation to a continuous water density profile, $\sigma_\theta(z)$, with z positive downward from the surface such that

$$\sigma_\theta(z) = \begin{cases} \sigma_{\theta_1} & 0 \le z_a < z < D \\ \sigma_{\theta_2} & D < z < z_b \end{cases} \quad (4.134)$$

where z_a is a near-surface depth, D is the estimated mixed layer depth (more realistically, the pycnocline depth rather than the base of the uniformly mixed surface layer), $z_b = 200 - 500$ m is an arbitrary depth below the depth of seasonal mixing, and $\sigma_{\theta_1}, \sigma_{\theta_2}$ are constant potential densities for the mixed layer and intermediate layer, respectively. Minimizing the integral

$$\Phi = \int_{z_a}^{D} [\sigma_\theta(z) - \sigma_{\theta 1}]^2 \, dz + \int_{D}^{z_b} [\sigma_\theta(z) - \sigma_{\theta 2}]^2 dz \quad (4.135)$$

with respect to $\sigma_{\theta_1}, \sigma_{\theta_2}$ and D leads to the solution

$$F(D) = \sigma_\theta(D) - \frac{1}{2}\left[\frac{\int_{z_a}^{D} \sigma_\theta(z)dz}{D - z_a} + \frac{\int_{D}^{z_b} \sigma_\theta(z)dz}{z_b - D} \right]$$

$$= 0, \quad (4.136)$$

which can be solved numerically. Two-segment approximations are computationally stable. Increasing the number of segments or "steps"

can improve the approximation to the profile data but typically leads to greater complexity. The quality of the least-squares fit varies from profile to profile depending on the structure of the underlying layering.

4.11.3 Integral Depth-Scale Method

A simple estimate of the mixed layer depth, D, is the integral depth-scale (Figure 4.52(d)), also called the "trapping depth" (Price et al., 1986), where

$$D = \frac{\int_0^{z_b} z N_b^2(z)dz}{\int_0^{z_b} N_b^2(z)dz} = \frac{\int_{z_a}^{z_b} (\sigma_{\theta b} - \sigma_\theta)dz}{\sigma_{\theta b} - \sigma_{\theta a}} \quad (4.137)$$

where z_a and z_b are a near-surface depth and an arbitrary reference depth (e.g., $z_a = 0$ and $z_b \sim 250$ m) and $\sigma_{\theta b} = \sigma_\theta(z_b)$, $\sigma_{\theta a} = \sigma_\theta(z_a)$ (cf. Freeland et al., 1997). Here,

$$N(z) = \left(-\frac{g}{\rho_0} \frac{d\rho_\theta}{dz} \right)^{1/2} \quad (4.138)$$

is the buoyancy (Brunt-Väisälä) frequency, g is the acceleration of gravity, and ρ_0 is a reference density. Unlike the threshold methods, which often require only the upper portion of a density profile, the step-function and integral-depth approaches require specification of a deep reference density that is much deeper than the mixed layer depth.

A second integral approach to fitting a two-layer function to a continuous profile finds the mixed layer depth (in effect, the pycnocline depth) by requiring that a parameter, ψ_{fit}, proportional to the potential energy of the two-layer system, be equal to corresponding value, ψ_o, for the original continuous profile of the water column. Once again, we let $\sigma_{\theta_1}, \sigma_{\theta_2}$ be the constant densities of the two layers (mixed and reference layers, respectively) and consider estimates made relative to a deep reference level, $z_b = H = 250$ db (~ 250 m). Following Ladd and Stabeno (2012), we can write

$$\Phi_o = \int_0^H \left[\sigma_\theta(z) - \overline{\sigma_\theta} \right] z \, dz \qquad (4.139a)$$

where

$$\overline{\sigma_\theta} = \frac{1}{H} \int_0^H \sigma_\theta(z) dz \qquad (4.139b)$$

is the depth-averaged density. Setting the integrals in Eqn (4.139a,b) equal to the corresponding values obtained for a two-layer representation, specifically,

$$\Phi_{fit} = \frac{1}{2} \left[D^2 \left(\sigma_{\theta_1} - \overline{\sigma_\theta} \right) + \left(H^2 - D^2 \right) \left(\sigma_{\theta_2} - \overline{\sigma_\theta} \right) \right]$$
$$(4.140a)$$

and

$$\overline{\sigma_\theta} = \frac{\left[D \sigma_{\theta_1} + (H - D) \sigma_{\theta_2} \right]}{H} \qquad (4.140b)$$

yields an estimate for the mixed layer depth,

$$D = \frac{2\Phi_o}{H \left(\overline{\sigma_\theta} - \sigma_{\theta_1} \right)} \qquad (4.141)$$

Solution to Eqn (4.141) requires that we specify a value for the upper layer density, σ_{θ_1}. In his study of winter mixed layer depth in the northeast Pacific, Freeland (2013) arbitrarily sets σ_{θ_1} equal to the mean density of the well-mixed upper layer averaged over the top 60 m. This is considered to be the maximum depth that can be chosen without approaching the shallowest mixed layer depth (~ 90 m) ever observed in winter in this region. The method does not work for summer months when solar heating and reduced wind mixing can cause the mixed layer depth in the northeast Pacific to shoal to 20 m or less. The method was used to examine temporal variability in the winter mixed layer depth and pseudo potential energy of the water column, Ψ, at Ocean Station P in the northeast Pacific.

4.11.4 The Split-and-Merge Algorithm

The threshold difference method often yields instances where the estimated mixed layer depth differs markedly from the visually estimated mixed layer depth (i.e., the depth to the top of the first well-defined pycnocline estimated from density plots). This, and the fact that depth estimates from the step-function and integral-scale methods are more representative of the main pycnocline depth than the surface mixed layer depth, led to formulation of the split-and-merge algorithm for estimating mixed layer depth (Thomson and Fine, 2003, 2009). First developed by Pavlidis and Horowitz (1974) to estimate the optimal decomposition of plane curves and waveforms, the method can also be used to approximate other structural features in the water column.

The split-and-merge algorithm approximates a specified curve using piecewise polynomial functions in which the breakpoints in the fitted curve (locations of subset boundaries and changes in slope) are adjusted to fit the available data (Figure 4.53). The algorithm provides profile decomposition by defining the locations of the breakpoints separating the different segments, the piecewise approximation parameters, and the error of the approximation. In general, the fitted segments are disjointed but can be made to satisfy continuity requirement by a retrospective local adjustment of the approximating curves. The method addresses the problem of fitting a series of segments to a profile, $\phi(z)$, where z is depth and ϕ represents temperature, salinity, density, fluorescence, or other profile variable. Given a set of points $S = \{z_i, \phi_i\}$, $i = 1, 2, ..., N$, we seek the minimum number, n, such that S is divided into n subsets $S_1, S_2, ..., S_n$ in which the data points for each subset are approximated by a polynomial of order at most $m - 1$ with an error norm less than some specified quantity, ε. The algorithm merges adjacent fitted segments with similar approximating coefficients and splits those

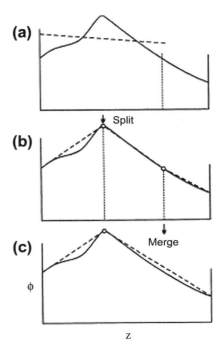

FIGURE 4.53 Illustration of the split-and-merge algorithm for which an optimum segmentation can be found in one iteration. (a) The initial fit (dashed line) to the left-hand segment of the curve $\phi(z)$ is split at the break-point to (b) form three separate segments. (c) The two right-hand segments are then merged to form one segment. *From Thomson and Fine (2003).*

segments with unacceptable error norms. A least-squares method, or similar approach, is used to fit the curve to the specified data set. Because fitting higher order polynomials has its own set of problems, the method is generally restricted to piecewise linear approximations, for which $m = 2$. The uppermost segment (the mixed layer) is assumed to have a near-uniform vertical distribution so that $m = 1$ for this segment.

The split-and-merge method removes the need for careful *a priori* choice of the number of segments, n, as well as the need to specify the initial segmentation. In this way, it is straightforward to obtain segmentation where the error

norm on each segment (or over all segments) does not exceed a specified bound. Computational experience indicates that the local minima found by the split-and-merge algorithm are close to the global minima for computational time of order N. For a specified error norm, the mixed layer depth, D, is equated with the uppermost segment of the piecewise-linear fit. To avoid scaling problems for profile variables, including the need for a different error norm for each variable, the variables z and $\phi(z)$ are normalized such that

$$z^* = \frac{z - z_{min}}{z_{max} - z_{min}}; \phi^*(z) = \frac{\phi(z) - \phi_{min}}{\phi_{max} - \phi_{min}} \quad (4.142)$$

where subscripts denote the maximum and minimum values of the given variable over the depth range of interest and $0 \leq (z^*, \phi^*(z)) \leq 1$. Following other methods (see Table 4.13), the split-and-merge algorithm uses a non-zero starting depth (e.g., $z_{min} = 2.5$ m), to avoid problems associated with prop-wash or turbulent flow past the hull of the ship during station keeping, and $z_{max} \geq 150$ m to ensure capture of the mixed layer depth regardless of season. As with the threshold method, the split-and-merge algorithm requires a predefined error norm. To determine the sensitivity of MLD estimates to the specified error norm, Thomson and Fine (2003) examined MLD estimates over a wide range of error values, from 0.001 to 0.03, and found that results for the norm $\varepsilon = 0.01$ typically gave values that were closest to "visual" MLD estimates for CTD profiles collected in July 1997 along ship survey lines off the west coast of Vancouver Island, Canada (Figure 4.54). For the density profile data examined, this error norm yielded the same mean mixed layer depth, $\overline{D}$, as the threshold method for $\Delta\sigma_\theta = 0.03$ kg/m^3; the error norm $\varepsilon = 0.003$ yielded the same $\overline{D}$ as the "standard" threshold method for which $\Delta\sigma_\theta = 0.01$ kg/m^3. Because the mean MLD estimates determined by the threshold and split-and-merge methods are nearly identical, a second-order statistic (the standard error) was

TABLE 4.13   Selected Definitions for Mixed Layer Depth and Other Upper Ocean Features

Source	Name	Definition
Price et al. (1986)	Trapping depth, D_T	$D_T = \Delta T^{-1} \int\limits_{z_i}^{z} T dz$
Peters et al. (1989)	Mixed layer depth, MLD	$\Delta\sigma_\theta = 0.01$ kg/m^3 (relative to $z = 0$ m)
Schneider and Müller (1990)	Mixed layer depth, MLD	$\Delta\sigma_\theta = 0.01$ kg/m^3 (relative to $z = 2.5$ m) $\Delta\sigma_\theta = 0.03$ kg/m^3 (relative to $z = 2.5$ m)
Wijffels et al. (1994)	Mixed layer depth, MLD top of thermocline, TTC	$\Delta\sigma_\theta = 0.01$ kg/m^3 (relative to $z = 2.5$ m) $\partial\sigma_\theta/\partial z = 0.01$ kg/m^4
Brainerd and Gregg (1995)	Mixed layer depth, MLD	$\Delta\sigma_\theta = 0.005 - 0.5$ kg/m^3 $\partial\sigma_\theta/\partial z = 0.0005 - 0.05$ kg/m^4
Smyth et al. (1996a,b)	Diurnal mixed layer, DML Upper ocean layer, UOL	$\Delta\sigma_\theta = 0.01$ kg/m^3 $\sigma_\theta < 22$ kg/m^3 (top of pycnocline)
Weller and Plueddemann (1996)	Mixed layer depth, MLD Isopycnal layer depth, ILD Seasonal thermocline depth, STD	$\Delta T = 0.01$ °C (relative to $z = 2.25$ m) $\Delta\sigma_\theta = 0.03$ kg/m^3 (relative to $z = 10$ m) $\Delta\sigma_\theta = 0.15$ kg/m^3 (relative to $z = 10$ m)
Wijesekera and Gregg (1996)	MLD MLD$_1$ MLD$_2$ MLD$_3$	$\Delta\sigma_\theta = 0.01$ kg/m^3 (relative to $z = 0$ m) $\partial\sigma_\theta/\partial z = 0.01$ kg/m^4 $\partial\sigma_\theta/\partial z = 0.025$ kg/m^4 $\partial\sigma_\theta/\partial z = 0.01$ psu/m
Skyllingstad et al. (1999)	MLD	$\Delta\sigma_\theta = 0.01$ kg/m^3 (relative to $z = 0$ m)
Thomson and Fine (2003, 2009)	Split-and-merge	Error norm $\varepsilon = 0.01$

Here, T is Temperature, S is Salinity, σ_θ is Potential Density. The "Trapping Depth", D_T (Price et al., 1986) is Analogous to the Integral Depth Scale. Modified After Thomson and Fine (2003).

used as a measure of the relative "performance" or predictive skill of a given method.

4.11.5 Comparison of Methods

A comparison among the various mixed layer depth estimators (the split-and-merge method, two threshold-type methods, the two- and three-step regression methods, and the integral-scale method) by Thomson and Fine (2003) demonstrates that, in the absence of turbulent dissipation measurements, only the threshold and split-and-merge methods provide accurate estimates of mixed layer depth. Estimates from the regression and integral methods are more representative of the permanent pycnocline depth than the surface mixed layer depth. The threshold and split-and-merge methods give nearly identical MLD results over a wide range of threshold and error norm values. For the threshold method, MLD estimates increase with increased threshold value, while for the split-and-merge method, the MLD estimate closely captures the "visual" mixed layer depth and remains unchanged over wide range of norm values. Where the upper layer has a weak density gradient, the ability of the threshold method to estimate the MLD is strongly dependent on the threshold value. In contrast, the split-and-merge algorithm readily finds an MLD that corresponds closely to the visual

FIGURE 4.54 Top portion of 1-m average density (σ_θ) profiles derived from CTD casts collected in July 1997 off the west coast of Vancouver Island (see map). Horizontal bars denote mixed layer depth derived using the split-and-merge algorithm. Profiles of σ_θ typically range from 22 to 25 kg/m^3. A scale of $3\sigma_\theta$ units is shown in the Line B profile plots. *From Thomson and Fine (2003).*

estimate. When the upper layer has a "significant" density gradient, both methods have difficulty estimating a mixed layer depth; there is no well-defined mixed layer and depth estimates invariably depend on the specified error values. In such cases, integral methods give a better approximation to a "two-layer" vertical structure.

Table 4.14 presents a statistical summary of the mean depths and standard deviations for

TABLE 4.14   The Mean Mixed Layer Depth (MLD_{sub}) and Standard Error (Standard Deviation/$\sqrt{N}$, $N =$ Number of Samples) for Different Estimation Methods (Depths in Meters)

	$MLD_{S\&M}$	MLD_{TH}	MLD_{GR}	D_σ	D_2	D_3
Mean (m)	(12.51 ± 0.32)	(12.43 ± 0.33)	(12.57 ± 0.38)	(60.4 ± 0.6)	(60.9 ± 0.7)	(35.60 ± 0.46)

Subscripts "S&M", "TH", and "GR" denote the split-and-merge, threshold difference, and threshold gradient methods, respectively. D_σ is the integral depth scale, D_2 is the MLD from the two-step function, and D_3 for the three-step-function approach.

five methods for all density profiles collected over two decades off the west coast of Vancouver Island. Results for the threshold difference method ("*TH*") are based on the standard threshold $\Delta\sigma_\theta = 0.01$ kg/m^3 and those for the threshold gradient method ("*GR*") on a gradient $\Delta\sigma_\theta/\Delta z = 0.006$ kg/m^4. The specified error norm $\varepsilon = 0.003$ for the split-and-merge method coincides with the threshold difference method value of $\Delta\sigma_\theta = 0.01$ kg/m^3. As the tabulated results indicate, the mean mixed layer depths obtained using the threshold and split-and-merge methods are markedly different from those based on the regression and integral methods. Mean depths from the threshold and split-and-merge methods (mean MLD ≈ 12 m) are a factor of five smaller than those for the regression and integral methods (mean MLD ≈ 60 m), and most accurately match "by-eye" estimates of the uniform-density surface mixed layer. For a more recent study of mixed layer depth estimation methods, the reader is referred to Esfahani (2014).

4.12 INVERSE METHODS

4.12.1 General Inverse Theory

General inverse methods have become a sophisticated analysis tool in the earth sciences. For example, in the field of geophysics, a goal of this technique is to infer the internal structure of the earth from the measurement of seismic waves. The essence of the geophysical *inverse problem* is to find an earth structure

model, which could have generated the observed acoustic travel-time data. This is in contrast to the *forward problem* which uses a known input and an understood physical system to predict the output. In the inverse problem, the input and output are known and the result is the *model* required to translate one set of data into the other.

In oceanography, inverse methods are used for a variety of applications, including the inference of absolute ocean currents using known tracer distributions and geostrophic flow dynamics (Wunsch, 1978, 1988). Other applications include the use of underwater acoustic travel times to determine the average temperature of the global ocean for long-term climate studies (Worchester et al., 1988) or the use of tsunami observations from bottom pressure recorders such as DART, cabled observatories, or coastal tide gauges, to help define the seismic source regions for major *trans*-oceanic tsunamis (Satake, 1993; Fine et al., 2005; Hayashi et al., 2012). A study by Mackas et al. (1987) and Masson (2006) used inverse techniques to determine the origins and mixing of water masses for the coast of British Columbia. In these oceanographic applications, the "solutions" are what we previously called the "models" in the geophysical problem. The kernel functions are formulated from the physics of the problem in question and the result is found by matching the "solution" to the input data. A cursory look at the problem is provided in this section. The interested reader is referred to Bennett (1992) for detailed insight into the theory and application of inverse methods in oceanography.

In general, the inverse problem takes the form

$$e(t) = \int_a^b C(t, \xi) m(\xi) d\xi \qquad (4.143)$$

where $e(t)$ are the input data, $m(\xi)$ is the model and $C(t, \xi)$ is the kernel function for the variable ξ. The kernel functions are determined from the relevant physical equations for the problem and are assumed to be known (Oldenburg, 1984). It is the judicious selection of these kernel functions that makes the inverse problem a complex exercise requiring physical insight from the oceanographer. In order to extract information about the model, $m(\xi)$, we will restrict our consideration of inverse theory to linear inverse methods applied to a set of observations. This is referred to as "finite dimensional inverse theory" by Bennett (1992). In his discussion of this form of inverse theory, Bennett suggests that it applies to:

1. An incomplete ocean model, based on physical laws but possessing multiple solutions.
2. Measurements of quantities not included in the original model but related to the model by additional physical laws.
3. Inequality constraints on the model fields or the data.
4. Prior estimates of errors in the physical laws and the data.
5. Analysis of the level of information in the system of physical laws, measurements, and inequalities.

Equations (4.143) is a *Fredholm equation* of the first kind. Inverse theory is centered on solving this equation in such a way as to extract information about the model, $m(\xi)$, when information is available for the data, $e(t)$. It is important to realize that the inverse problem cannot be solved unless the physics and the geometry of the problem are known (i.e., Eqn (4.143) has been set up). It is, therefore, impossible to consider a solution

to the inverse problem unless the forward problem can be solved. The physics of the forward problem may be ill-posed, in which case not all of the solutions will match or, if they do, it is a coincidence and not a solution to Eqn (4.143). Thus, the basic questions to ask regarding a solution of the inverse problem are: (1) Does a solution exist? In other words, is there an $m(\xi)$ which produces $e(t)$?; (2) How does one construct a solution?; (3) Is the solution unique?; and (4) How is the nonuniqueness appraised?

The answers to the above questions will depend on the data, $e(t)$. In theory, there exist three types of data:

1. An infinite amount of accurate data;
2. a finite amount of accurate data;
3. a finite amount of inaccurate data.

In reality, only option (3) occurs as we are forced to work with observations, which contain a variety of measurement and sampling errors. While perfect data are limited to the realm of the mathematical, it is often instructive to consider analytic "inverses". For example, the analytical inverse to

$$x(f) = \int_{-\infty}^{\infty} x(t) e^{-i2\pi ft} dt \qquad (4.144)$$

is

$$x(t) = \frac{1}{2\pi} \int_{-\infty}^{\infty} x(f) e^{i2\pi ft} dt \qquad (4.145)$$

Similarly, the inverse of

$$\phi(x) = 2/\lambda \int_x^a \left[r\varepsilon(r) \Big/ (r^2 - x^2)^{1/2} \right] dr \qquad (4.146)$$

is

$$\varepsilon(r) = -\lambda/\pi \int_r^a \left[(d\phi/dx) \Big/ (x^2 - r^2)^{1/2} \right] dx$$

$$(4.147)$$

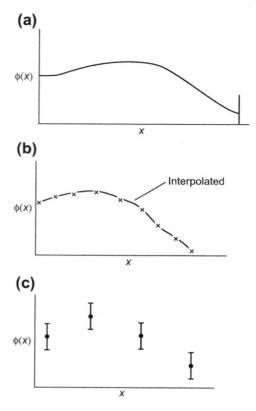

(a)

$\phi(x)$

x

(b)

Interpolated

$\phi(x)$

x

(c)

$\phi(x)$

x

FIGURE 4.55 Three examples of the function $\phi(x)$ required for the inverse solution, $\varepsilon(r)$, of Eqn (4.147). Analytical (a) and digital (b) versions of $\phi(x)$ for which inversion is readily possible, (c) A typical "observed" version of $\phi(x)$, consisting of four mean values (plus standard deviations) for which inversion is considerably less accurate.

In the second case, we require knowledge of $d\phi/dx$ to find $\varepsilon(r)$, which is easy to do for ideal continuous data (Figure 4.55(a)), or even for a finite sample of accurate data (Figure 4.55(b)). If, however, we have a finite sample of inaccurate data (Figure 4.55(c)), we have difficulty estimating $d\phi/dx$.

The problem of dealing with a limited sample of inaccurate measurements is the most common obstacle to the application of inverse methods. Usually, these inaccuracies can be treated as additive noise superimposed on the true data and,

therefore, can be handled with statistical techniques. These additive errors have the effect of "blurring" or distorting our picture of the solution (model). Unfortunately, one cannot conclude that if the error noise is small that the model distortions also will be small. The reason for this is that most geophysical kernel functions act to smooth the model, thus changing the length scale of the response for both the forward and inverse problems. In other words, the solution obtained with inaccurate data using the inverse procedure may be very different from the model, which actually generated the data. In addition, particular solutions to the model are not unique and a wide variety of solutions is equally possible.

In most oceanographic applications of inverse methods, we are primarily interested in finding a model, which reproduces the observations. Here, the fundamental problem is the nonuniqueness of any inverse solution, which is one of infinitely many functions that can reproduce a finite number of observations. This nonuniqueness becomes more severe when the data are inaccurate, as they must be in any practical oceanographic application. The key to the application of inverse methods in oceanography is to select the "correct" (by which we mean the most probable or the most reasonable) inverse model-solution.

Inverse construction in oceanography may take the form of parametric modeling. In this case, we write our model as $m = f(a_1, a_2, \ldots, a_N)$ and a numerical scheme is sought to find appropriate values of the parameters, a_i ($i = 1, \ldots, N$). Parameterization is justified when the physical system actually has this form and depends on a number of input parameters. The model is solved by collecting more than N data points and finding the parameters through a least-squares minimization of

$$\phi = \sum_{i=1}^{N} (e_i - e'_i)^2 \qquad (4.148)$$

where

$$e_i' = f(a_1, a_2, \ldots, a_N; \varepsilon_i) \qquad (4.149)$$

In Eqn (4.149) ε_i is the ith kernel function.

4.12.2 Inverse Theory and Absolute Currents

As reviewed by Bennett (1992), an important application of inverse theory to ocean processes was the computation of absolute currents for large-scale ocean circulation. In the 1970s, two different approaches to this problem were proposed. The first by Stommel and Schott (1977) was called the "beta spiral" technique, which demonstrated that the vertical structure of large-scale, open-ocean velocity fields could be explained using simple equations expressing geostrophy and continuity (conservation of mass). The second method, introduced by Wunsch (1977), showed that reference velocities could be estimated simultaneously around a closed path in the ocean. The resultant absolute velocities were consistent with geostrophy and the conservation of heat and salt at various levels. As a guide to oceanographic applications of inverse techniques, we provide succinct reviews of both applications.

4.12.2.1 *The Beta Spiral Method*

Insightful reviews of the Stommel and Schott (1977) beta spiral method are provided by Olbers et al. (1985) and Bennett (1992). The basic equations for this application are the usual linearized beta (β)-plane equations for horizontal geostrophic flow (u, v) in a Boussinesq fluid

$$-\rho_0 f v = -\partial p / \partial x \qquad (4.150a)$$

$$\rho_0 f u = -\partial p / \partial y \qquad (4.150b)$$

the hydrostatic equation

$$0 = -\partial p / \partial z - \rho g \qquad (4.151)$$

which relate pressure perturbations, $p(\mathbf{x}, t)$, to density fluctuations, $\rho(z, t)$, and the conservation of mass (or continuity) relation

$$\nabla \cdot \mathbf{u} + \partial w / \partial z = 0 \qquad (4.152)$$

In these equations, f is the Coriolis parameter, u, v, and w are, respectively, the eastward (x), northward (y) and upward (z) components of current velocity, and $\rho = \rho(x, y, z)$ is the density perturbation about the mean density $\rho_0 = \rho_0(z)$. Following Bennett (1992), we will reserve vector notation for horizontal fields and operators ($\mathbf{x} = (x, y)$, $\mathbf{u} = (u, v)$, etc.).

Using the above equations, we can derive the well-known "thermal wind" relations, whose vertically integrated velocity components are

$$u(\mathbf{x}, z) = u_0(\mathbf{x}) + (g/f\rho_0) \int_{z_0}^{z} \rho_y(x, \zeta) d\zeta \qquad (4.153a)$$

$$v(\mathbf{x}, z) = v_0(\mathbf{x}) - (g/f\rho_0) \int_{z_0}^{z} \rho_x(x, \zeta) d\zeta \qquad (4.153b)$$

where subscripts x, y refer to partial differentiation and $u_0(\mathbf{x})$, $v_0(\mathbf{x})$ are the velocity components at some reference depth. Equations (4.150)−(4.152) also give rise to the well-known Sverdrup interior vorticity balance

$$w_z = \beta v / f \qquad (4.154)$$

where β is the northward (y) gradient of the Coriolis parameter, and $f = f(y) = f_0 + \beta y$ is the beta-plane approximation.

These equations cannot be used alone to determine the full absolute velocity field ($\mathbf{u}$, w), even if the density field ρ were known. However, to resolve this indeterminacy, all we need is the velocity field at a particular depth where $\mathbf{u} = \mathbf{u}(\mathbf{x}, z_0)$ and $w = w(\mathbf{x}, z_0)$. Stommel and Schott (1977) demonstrated that these unknown reference values may be estimated by assuming the availability of measurements of some conservative tracer ϕ which satisfy the steady-state conservation law

$$\mathbf{u} \cdot \nabla \phi + w \phi_z = 0 \qquad (4.155)$$

This tracer might be salinity (S) or potential temperature (θ), or some function of both S and θ. Combining the vertical derivative of Eqn (4.155) with Eqns (4.153) and (4.154), yields

$$\left(\mathbf{u} \cdot \nabla + w \frac{\partial}{\partial z} \right) (f \phi_z) = (g/\rho_0) J \qquad (4.156)$$

where J is the Jacobian $J(\rho, \phi) = \rho_x \phi_y - \rho_y \phi_x$. In Eqn (4.156), $f \phi_z$ represents the potential vorticity, which would be conserved if density ρ were itself conserved. The tracer equation can be used again to eliminate the vertical velocity w

$$\mathbf{u} \cdot \mathbf{a} = (g/\rho_0) J(\rho, \phi) \qquad (4.157)$$

where the vector $\mathbf{a}$ is given by

$$a(\mathbf{x}, z) = \nabla(f \phi_z) - \frac{\nabla \phi}{\phi_z} f \phi_{zz} \qquad (4.158)$$

Using the integrated thermal wind Eqn (4.153a,b) yields

$$\mathbf{u}_0 \cdot \mathbf{a} = c \qquad (4.159)$$

where $\mathbf{u}_0$ is the horizontal velocity at depth z_0 and c is given by

$$c(\mathbf{x}, z) = -\mathbf{u}' \cdot \mathbf{a} + (g/\rho_0) J(\rho, \phi) \qquad (4.160)$$

In Eqn (4.160), the $\mathbf{u}'$ is that part of the horizontal velocity in the thermal wind relation that depends on the density field.

Since $\mathbf{a}$ and c depend on g, ρ, f, $\nabla \rho$, ∇f, ϕ_z and ϕ_{zz}, they can be determined using closely spaced hydrographic stations through measurements of $T(z)$ and $S(z)$. Thus, from Eqn (4.159), we can calculate $\mathbf{u}_0$ using the hydrographic data. Equations (4.159) holds at all levels so that two different levels can be used to specify u_0 and v_0. We can then calculate the vertical velocity w from Eqn (4.156). The full velocity solution should be independent of the levels chosen for these computations. In reality, Eqn (4.159) is

not an exact relation as it was derived from approximate dynamical laws and computed from data that contain measurement and sampling errors. As a consequence, our estimate of $\mathbf{u}_0$ from Eqn (4.159) should be done as a best fit to the data from the two levels chosen.

Suppose that N levels are chosen from the hydrographic data $(N \geq 2)$. Let $c_n = x(\mathbf{x}, z_n)$ and $\mathbf{a}_n = \mathbf{a}(\mathbf{x}, z_n)$ for $1 \leq n \leq N$. The simple least-squares best fit minimizes

$$R^2 = \sum_{n=1}^{N} R_n^2 = \sum_{n=1}^{N} (c_n - \mathbf{u}_0 \cdot \mathbf{a}_n)^2 \qquad (4.161)$$

where R_n is the residual at level n and R is the root-mean-square (RMS) total error. R^2 is a minimum if $\mathbf{u}_0$ satisfies a simple linear system

$$\mathbf{M} \mathbf{u}_0 = \mathbf{d} \qquad (4.162)$$

where the 2×2 systematic, nonnegative matrix $\mathbf{M}$ depends on the components of $\mathbf{a}_n$, while $\mathbf{d}$ depends on $\mathbf{a}_n$ and c. If $\mathbf{a}$ or c varies with depth, Eqn (4.157) implies that the total velocity vector $\mathbf{u}$ must also depend on depth. For the β-spiral problem, we find that the large-scale ocean currents constitute a spiral with depth at each station. The β-spiral in Figure 4.56 is from the study by Stommel and Schott (1977) who used hydrographic data from the North Atlantic to estimate $\mathbf{u}_0$ for a reference level of $z_0 = 1000$ m depth. In this application they found, $u_0 = 0.0034 \pm 0.00030$ m/s and $v_0 = 0.0060 \pm 0.00013$ m/s at $28°$ N, $36°$ W.

The β-spiral problem includes two of the basic concepts common to inverse methods. First, we deal with an incomplete set of physical laws (Eqns (4.150)–(4.152)), or their rearrangement, as in the case of the thermal wind Eqn (4.153a,b), which includes the unknown reference velocity. Second, we often resort to the indirect measurement of an additional quantity, which, in the case of the present example, is a conservative tracer. This application could have benefited from the inclusion of prior estimates of the errors in the dynamical equations and in the hydrographic data.

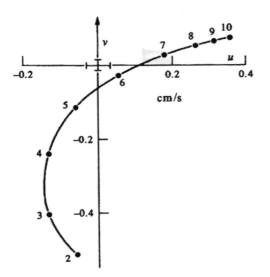

FIGURE 4.56 The β-spiral in horizontal velocity $\mathbf{u} = (u(z), v(z))$ at 28° N, 36° W, with depths in hundreds of meters. Error bars for the two components of velocity are given at the origin. *After Stommel and Schott (1977).*

4.12.2.2 Wunsch's Method

In a parallel development to the β-spiral technique, Wunsch (1977) used inverse methods to estimate reference velocities simultaneously around a closed path in the ocean (Bennett, 1992). As discussed by Davis (1978), Wunsch's method and the β-spiral method are closely related. Both approaches assume the vertically integrated thermal wind Eqn (4.153) and both provide estimates for the reference velocity $\mathbf{u}_0$. In Wunsch's method, the thermal wind velocity, $\mathbf{u}'$, is assumed to be zero at the reference level z_0, which in general may be a function of position $(z_0 = z_0(\mathbf{x}))$. Wunsch chose the reference level to be the ocean bottom at $z_0(\mathbf{x}) = H(\mathbf{x})$, with $\mathbf{u}_0(\mathbf{x})$ defined to be the bottom velocity. He then divided the water column into a number of layers defined by temperature ranges. This is consistent with the general water mass structure of the North Atlantic as defined by Worthington (1976). These layers need not be uniform in depth at each hydrographic station. Together with the coastline of the U.S., the hydrographic stations formed a closed path in the western North Atlantic (Figure 4.57).

FIGURE 4.57 The locations of hydrographic stations in the North Atlantic used by Wunsch to obtain absolute current estimates using inverse theory. *After Wunsch (1977).*

We now let v denote the outward component of velocity across the closed triangle formed by the lines of hydrographic stations in Figure 4.57. That is, $v = \mathbf{u} \cdot \mathbf{n}$ where $\mathbf{n}$ is the outward unit normal to the sections. We can further let $v' = \mathbf{u}' \cdot \mathbf{n}$ be the outward thermal wind velocity and $b = \mathbf{u}_o \cdot \mathbf{n}$ be the outward horizontal velocity at the seafloor. Let $v'_{\mathbf{n}}(z)$ and b_n denote the thermal wind velocity estimate and unknown bottom velocity midway between the nth station pair, where $1 \leq n \leq N$, and let v'_{mn} denote the average value of v'_n in the mth layer of the water column, where $1 \leq m \leq M$. Wunsch chose the Mth layer to be the total water column, thus the Mth tracer is the total mass of the water column. The assumption of tracer conservation within each layer can be written as

$$\sum_{i=1}^{N} \left(v'_{mn} + b_n \right) \Delta z_{mn} \Delta x_n = 0, \ 1 \leq m \leq M$$

(4.163)

where Δz_{mn} is the thickness of the mth layer at the nth station pair, and Δz_{mn} is the separation distance between the nth station pair. This system of M equations for N unknowns b_n, $1 \leq n \leq N$, may be written in matrix notation as

$$\mathbf{Ab} = \mathbf{c}$$

(4.164)

where $\mathbf{A}$ is an $M \times N$ matrix and $\mathbf{c}$ is a column vector of length M with elements

$$A_{mn} = \Delta z_{mn} \Delta x_n$$

(4.165a)

$$c_m = -\sum_{i=1}^{N} \overline{v'_{mn}} A_{mn}$$

(4.165b)

Wunsch used $M = 5$ layers as defined by the ranges 12–17 °C, 4–7 °C, 2.5–4 °C, and the entire water column (total mass). The hydrographic data were from $N = 43$ station pairs. For this problem, the matrix Eqn (4.165a,b) represents five equations for 43 unknown velocities, so that the system is underdetermined and has many different solutions.

As reported by Bennett (1992), Wunsch (1977) somewhat arbitrarily selected the vector $\mathbf{b}$ with

the shortest length. This was found by minimizing

$$t_1 = \mathbf{b}^T \mathbf{b} + 2\mathbf{I}^T (\mathbf{Ab} - \mathbf{c})$$

(4.166)

where the superscript T denotes the transpose of the matrix and $\mathbf{I}$ is an unknown Lagrange multiplier consisting of a column vector of length M. It can be shown that t_1 is a minimum when

$$\mathbf{b} + \mathbf{A}^T \mathbf{I} = \mathbf{0}$$

(4.167)

which gives the minimum solution

$$\mathbf{b} = \mathbf{A}^T \left(\mathbf{AA}^T \right)^{-1} \mathbf{c}$$

(4.168)

which satisfies Eqn (4.164). The symmetric matrix $\mathbf{AA}^T$ has dimensions $M \times M$ and is nonnegative (Bennett, 1992). However, $\mathbf{AA}^T$ may be singular. These singularities may be overcome by allowing errors in the hydrographic data and conservation laws; that is, by not seeking exact solutions of Eqn (4.164). We can instead write Eqn (4.164) in a quadratic form adding weights to each term. It can be shown that for positive weights, we are able to define an exact solution of the problem. This transfers the problem to the selection of these weights.

This cursory presentation of Wunsch's method for computing reference velocities demonstrates, once again, some of the basic elements of inverse methods: A system of incomplete physical laws and inexact measurements of related fields. It is necessary to admit errors into the equations and data values in order to stabilize the solution and to derive a unique solution. In his review, Davis (1978) concluded that both the underdetermined problem of Wunsch's method and the over determined problem of the β-spiral method are consequences of tacit assumptions made about noise levels and fundamental scales of motion. Davis suggested that a more orderly approach would be based on Gauss-Markov smoothing (Bennett, 1992) which should be an improvement, assuming explicit and quantitative estimates of the noise and its structure.

4.12.3 The IWEX Internal Wave Problem

Another oceanographic example of the inverse method is found in Olbers et al. (1976) and Willebrand et al. (1977). Here, inverse theory is used to determine the three-dimensional internal wave spectrum from an array of moored current meters (Figure 4.58). In this example, the Fredholm Eqn (4.143) is written in matrix form and becomes

$$y_i = A_{ij}x_j; \quad 1 \leq i \leq N; \ 1 \leq j \leq K \quad (4.169)$$

where y_i are N observed velocity cross-spectra (the data), A_{ij} are the kernel functions (for matrix **A**) representing the physical relations from internal wave theory and x_j are the K internal wave parameters to be determined by the inverse method. The inverse problem is to find the K parameters of the theoretical internal wave energy density cross-spectra using the N observed cross-spectra from the current meter array. We achieve this by using the least-squares method to minimize

$$\varepsilon^2(a) = [\widehat{y} - y(a)]\mathbf{W}[\widehat{y} - y(a)]^* \quad (4.170)$$

where a represents a set of trial values used to find the minimum and the asterisk (*) denotes the complex conjugate. In Eqn (4.170), **W** is a weighting matrix used to scale the problem and to produce statistical independence (Jackson, 1972).

It is common to expand the kernel function matrix **A** into eigenvectors (Jackson, 1972). Thus, we write

$$\mathbf{A}V_j = \lambda_j u_j, \quad \mathbf{A}^T u_j = \lambda_i V_i \quad (4.171)$$

FIGURE 4.58 Location of the IWEX study area showing the positions of the three current meter moorings on the Hatteras Plain in the western North Atlantic. *From Briscoe (1975).*

Following the SVD we conducted in the EOF analysis (Section 4.4.2), we can factor the matrix **A** as

$$\mathbf{A} = \mathbf{UBV}^T \qquad (4.172)$$

where **U** is an $N \times P$ matrix whose columns are the eigenvectors u_i, $i = 1, \ldots, P$; **V** is the $M \times P$ matrix whose columns are the eigenvectors v_i, $i = 1, \ldots, P$, and **B** is the diagonal matrix of eigenvalues. After **U** and **V** are formed from the eigenvectors corresponding to the P nonzero eigenvalues of **A**, there remain $(N - P)$ eigenvectors U_j and $(K - P)$ eigenvectors V_j, which correspond to zero eigenvalues. If we assemble these into columns of matrices, we have $\mathbf{U}_o$ (an $N \times (N - P)$ matrix) and $\mathbf{V}_o$ (a $K \times (K - P)$ matrix). This is called *annihilator space* and reveals that our model is composed of both real model space (which corresponds to the data) and annihilator space, which is linked to zeros in the data field. When we perform an inverse calculation, we usually recover a solution, which lies in real model space. We must remember, however, that any function in space a can be added to the solution and still produce a solution that fits the data. With the kernel functions transformed into an orthogonal framework (expanded into eigenvectors) we construct the "smallest" or minimum energy model-solution.

When $P = N$, there is a solution to Eqn (4.170) and $P = M$ guarantees that a solution, if it exists, is unique. For $P < N$, the system is said to be over constrained, while if $P < M$, the system is both over constrained and underdetermined. In the latter case, an exact solution may not exist but there will be an infinite number of solutions satisfying the least squares criterion. This is the case for the present internal wave example, which is both over constrained and underdetermined.

Returning to our internal wave problem, we find **W** in Eqn (4.170) using the least-squares method which produces the maximum likelihood estimator for a Gaussian distribution. This estimator is defined to be the inverse of the data covariance matrix. From the current meter array, 60 time series were divided into 25 overlapping segments. For each segment, cross-spectral estimates were computed for each of 600 equidistant frequencies. Averaging over segments and frequency bands to increase statistical significance, resulted in 3660 cross-spectra. The resultant 3660×3660 covariance matrix is difficult to invert. The diagonal of the weight matrix was selected to be

$$W = \operatorname{diag}\left[1/\operatorname{var}(y_i)\right] \qquad (4.173)$$

which reproduces the main features of the maximum likelihood weight matrix (Olbers et al., 1976). We note that, again for this problem, there are many more data points than parameters so that the system is over constrained.

The least-squares solution procedure for this internal wave example is as follows:

1. first find a parameter estimate $\hat{a}$ (the best guess);
2. linearize at the value $a = \hat{a}$, such that

$$\widehat{y}(a) = \widehat{y}(\widehat{a}) + \mathbf{D}(a - \widehat{a}) + \ldots \qquad (4.174)$$

where

$$\mathbf{D} = \left\{ \delta\widehat{y}_i \middle/ \delta a_j \right\}\Big|_{a=\widehat{a}} \qquad (4.175)$$

3. improve the parameter estimate by using

$$a - \widehat{a} = \mathbf{H}[\widehat{y}(a) - \widehat{y}(\widehat{a})] \qquad (4.176)$$

where the $N \times K$ matrix **H** is the generalized inverse of **D** derived from the linear terms of Eqn (4.174). If the matrix $\mathbf{D} < \mathbf{TWD}$ is nonsingular and well conditioned then

$$\mathbf{H} = \left(\mathbf{D}^T\mathbf{WD}\right)^{-1}\mathbf{D}^T\mathbf{W} \qquad (4.177)$$

and Eqn (4.172) becomes the least-squares solution of Eqn (4.171). Since $\mathbf{D}^T\mathbf{WD}$ is an $K \times K$ matrix, it can be easily inverted using standard diagonalization routines.

Having now arrived at a solution, $\mathbf{A} = \{A_{ij}\}$ of the problem in Eqn (4.169), we are left with two additional questions: (1) How well are the data reproduced by our solution? and (2) How accurately do we know our parameters $\alpha_{\min}$? Since our data are subject to random errors, we can treat y as a statistical quantity and test the hypothesis that y and the model estimate $\hat{y}(\alpha_{\min})$ are the same with a 95% probability (inverse estimate must be within the 95% confidence interval of our data point). Using the central limit theorem for our segment and frequency-averaged spectral values, we can approximate the 95% confidence interval on y as

$$\varepsilon^2_{95\%} = \overline{\delta y W \delta y}[1 + O(L^{-1})] = L \qquad (4.178)$$

where $\delta y = y - \hat{y}$, and $O(\cdot)$ indicates the order of magnitude. Now if

$$\varepsilon^2\left(a_{\min} \leq \varepsilon^2_{95\%}\right) \qquad (4.179)$$

the model is a statistically consistent representation of the data. The consistency of the IWEX model is provided by the results in Figure 4.59, where we have plotted the measured, $\varepsilon^2(a)$, and expected, ε^2, values of the parameter ε^2. In this case, all values have been normalized so that magnitudes provide some indication of the percentage to which the observed and estimated (modeled) values of the data, y, coincide. For the most part, the measured values of ε^2 are scattered about the expected values of this parameter. Except at the M_2 tidal frequency and for frequencies greater than one cph, the hybrid IWEX model gives a consistent description of the IWEX data set to the 95% level.

Our second question regarding the accuracy of the parameter solution $a_{\min}$, can be answered by calculating the covariance matrix of the parameters. Using Eqn (4.172), we obtain the $K \times K$ covariance matrix of the parameters,

$$\overline{\delta a \, \delta a} = \mathbf{H} \, \overline{\delta y \delta y} \, \mathbf{H}^T \qquad (4.180)$$

from the data covariance matrix $\overline{\delta y \delta y}$. As usual, there is a reciprocal relation between the

FIGURE 4.59 Consistency for the IWEX study. The error estimate ε^2 is the squared difference between the observed data and the modeled data obtained by inverse methods. Except for motions in the M_2 tidal band and at frequencies great than about 1 cph, the results are within the 95% confidence level. $N_{\max}$ and N_{dw} are the maximum Nyquist frequency and the Nyquist frequency for the deep water, respectively. *From Briscoe (1975).*

variance and the resolution of the parameters. Statistically uncorrected parameters can be found by diagonalizing the matrix in (4.180).

4.12.4 Summary of Inverse Methods

In this section we have presented the basic concepts of the general inverse problem and have set up the solution system for two different applications in physical oceanography. Our treatment is by no means comprehensive and is intended to serve only as a guide to understanding the process of forming linear inverse solutions to fit observed oceanographic data.

The first example we treated is the computation of absolute geostrophic velocity by specifying an unknown reference velocity. Both the β-spiral (Stommel and Schott, 1977) and Wunsch's (1977) method are discussed. The dynamics are restricted to geostrophy and the conservation of mass. The second example was the specification of parameters in theoretical internal wave cross-spectra to reproduce the velocity cross-spectra of an array of moored current meters. The statistical nature of both the data and the model are considered and the accuracy of the results are expressed in probabilistic terms. Readers interested in further discussion of these and other related applications of inverse methods are referred to Bennett (1992). This book contains a complete review of inverse methods along with discussion of most of the popular applications of inverse techniques in physical oceanography. We also direct the interested reader to the paper by Egbert et al. (1994) in which a generalized inverse method is used to determine the four principal tidal constituents (M_2, S_2, K_1, O_1) for open ocean tides. The tides are constrained (in a least squares sense) by the hydrodynamic equations and by observational data. In the first example, solutions are obtained using inversion of the harmonic constants from a set of 80 open ocean tide gauges. The second example uses cross-over data from TOPEX/POSEIDON satellite altimetry. According to the authors, "The inverse solution yields tidal fields which are simultaneously smoother, and in better agreement with altimetric and ground truth data, than previously proposed tidal models." In recent years, the acquisition of high resolution (better than 0.1 mm accuracy), rapidly sampled tsunami wave records have made it possible to augment the delineation of the seismic source regions for the tsunamigenic earthquakes. The inverse travel time estimates of tsunami source regions provided by seismologists provide critical information on the failure regions, which can then be input into high resolution tsunami wave propagation models. Once the tsunami data have been analyzed, these data can be input into the models and run in reverse to better more accurately define the boundaries of the seismic sources.

5

Time Series Analysis Methods

The advent of ocean observing satellites, long-term mooring capability, and cabled marine observatories, coupled with high-density storage devices, is enabling oceanographers to collect long time series of oceanic and meteorological data. Similarly, the use of rapid-response sensors on moving platforms such as Argo drifters and autonomous underwater vehicles (AUVs) has made it possible to generate snapshots of spatial structure over extensive distances. Time series data are collected from moored instrument arrays or by repeated measurements at the same location using ships, satellites, or other instrumented packages. Quasi-synoptic spatial data are obtained from ships, manned-submersibles, remotely operated vehicles, AUVs, satellites, Argo drifters, and satellite-tracked drifters. Satellite imaging also produces densely sampled spatial data whose two-dimensional coverage can repeat in time, yielding a three-dimensional data set that can be analyzed in one, two, or all three of these dimensions.

As discussed in Chapters 3 and 4, the first stage of analysis following data verification and editing usually involves estimates of arithmetic means, variances, correlation coefficients, and other sample-derived statistical quantities. These quantities tell the investigator how well the sensors are performing and help characterize the observed oceanographic variability. However, general statistical quantities provide little

insight into the different types of signals that are blended together to make the recorded data. The purpose of this chapter is to present methodologies that examine data series in terms of their frequency content. With the availability of modern high-speed computers, frequency-domain analysis has become much more central to our ability to decipher the cause and effect of oceanic change. The introduction of fast Fourier transform (FFT) techniques in the 1960s further aided the application of frequency-domain analysis methods in oceanography. Such analyses were not practical prior to the advent of modern digital processors. The pre-FFT computer algorithms used in time series analyses are now basically obsolete and no longer in use.

5.1 BASIC CONCEPTS

For historical reasons, the analysis of sequential data is known as *time series analysis*. As a form of data manipulation, it has been richly developed for a wide assortment of applications. While we present some of the latest techniques, the emphasis of this chapter will be on those "tried and proven" methods most widely accepted by the general oceanographic community. Even these established methods are commonly misunderstood and incorrectly applied. Where appropriate, references to other texts will be given for those interested in a more thorough description of analysis techniques. As with previous texts, the term "time series" will be applied to both temporal and spatial data series; methods that apply in the time domain also apply in the space domain. Similarly, the terms *frequency domain* and *wavenumber domain* (the formal transforms of time and spatial series, respectively) are used interchangeably. Wavenumber is the appropriate unit when applying these methods to spatial series and it is poor grammar to refer to wavenumber as "spatial frequency" since frequency only applies to time series.

A basic purpose of time series analysis methods is to define the variability of a data series in terms of dominant periodic functions. We also want to know the "shape" of the spectra. Of all oceanic phenomena, the barotropic astronomically forced tides most closely exhibit deterministic and stationary periodic behavior, making them the most readily predictable motions in the sea. In coastal waters, tidal observations over a period as short as one month can be used to predict local tidal elevations with a high degree of accuracy. Where accurate specification of the boundary conditions is possible, a reasonably good hydrodynamic numerical model that has been calibrated against observations can reproduce the regional tide heights to an accuracy of a few centimeters. Tidal currents are much less easily predicted because of the complexities introduced by stratification, seafloor topography, basin boundaries, and nonlinear interactions. For example, although baroclinic (internal) tides generated over abrupt topography in a stratified ocean contribute little to surface elevations, they can lead to strong baroclinic currents. These baroclinic currents typically have both deterministic and nondeterministic (i.e., stochastic) components, and hence are only fully predictable in a statistical sense.

Surface gravity waves are periodic and quasi-linear oceanic features but are generally only predictable in a stochastic sense due to inadequate knowledge of the surface wind fields, the air–sea momentum transfer, and oceanic boundary conditions. Refraction induced by wave-current interactions can be important but difficult to determine. Other oceanic phenomena such as coastal-trapped waves and near-inertial oscillations have marked periodic signatures but are intermittent because of the vagaries of the forcing mechanisms and changes in oceanic and topographic conditions along the direction of propagation. Other less obvious regular behavior can be found in observed time and space records. For instance, oceanic variability at the low-frequency end of the spectrum is dominated by fluctuations at the annual to decadal periods, consistent with baroclinic Rossby waves and short-term climate change, while that at the ultra-low frequencies is dominated by ice-age climate scale variations possibly associated with highly amplified feedback responses to weak Milankovitch-type forcing processes (changes in the caloric summer insolation at the top of the atmosphere arising from changes in the earth's orbital eccentricity, and tilt and precision of its rotation axis).

Common sense should always be a key element in any time series analysis. Attempts to use analytical techniques to find "hidden" signals in a time series often are not very convincing, especially if the expected signal is buried in the measurement noise. Because noise is always present in real data, it should be clear that, for accurate resolution of periodic behavior, data series should span at least a few repeat cycles of the timescale of interest, even for stationary processes. Thus, a daylong record of hourly values will not fully describe the diurnal cycle in the tide nor will a 12-month series of monthly values fully define the annual cycle of sea surface temperature (SST). For these short records, modern spectral analysis methods can help pinpoint the peak frequencies. As we noted in Chapter 1, a fundamental limitation to resolving time series fluctuations is given by the "sampling theorem", which states that the highest detectable frequency or wavenumber (the Nyquist frequency or wavenumber) is determined by the interval between the data points. For example, the highest frequency that we can hope to resolve by an hourly time series is one cycle per 2 h, or one cycle per $2\Delta t$, where Δt is the interval of time between points in the series.

For the most part, we fit series of well-known functions to the data in order to transform from the time domain to the frequency domain. As with the coefficients of the sine and cosine functions used in Fourier analysis, we generally assume that the functions have slowly varying amplitudes and phases, where "slowly" means that coefficients change little over the length of the record. Other linear combinations of orthogonal functions with similar limitations on the coefficients can be used to describe the series. However, the trigonometric functions are unique in that uniformly spaced samples covering an integer number of periods of the function form orthogonal sequences. Arbitrary orthogonal functions, with a similar sampling scheme, do not necessarily form orthogonal sequences. (Note that a basis function $\phi_k(t)$ is orthogonal if the integral over the data set $\langle \phi_k(t), \phi_l(t) \rangle = \int \phi_k(t)\phi_l(t)dt = 0, \quad k \neq l$). Another advantage of using common functions in any analysis is that the behavior of these functions is well understood and can be used to simplify the description of the data series in the frequency or wavenumber domain. In this chapter, we consider time series to consist of periodic and aperiodic components superimposed on a secular (long-term) trend and uncorrelated random noise. Fourier analysis and spectral analysis are among the tools used to characterize oceanic processes. Determination of the Fourier components of a time series can be used to determine a *periodogram*, which can then be used to define the spectral power density (*spectrum*) of the time series. However, the periodogram is not the only way to get at the spectral energy density. For example, prior to the introduction of the FFT, the common method for calculating

spectra was through the Fourier transform of the autocorrelation function. More modern spectral analysis methods involve autoregressive spectral analysis (including use of maximum entropy techniques), wavelet transforms, and fractal analysis.

5.2 STOCHASTIC PROCESSES AND STATIONARITY

A common goal of most time series analysis is to separate deterministic periodic oscillations in the data from random and aperiodic fluctuations associated with unresolved background noise (unwanted geophysical variability) or with instrument error. It is worth recalling that time series analyses are typically statistical procedures in which data series are regarded as subsets of a stochastic process. A simple example of a stochastic process is one generated by a linear operation on a purely random variable. For example, the function $x(t_i) = 0.5x(t_{i-1}) + \varepsilon(t_i)$, $i = 1, 2, \ldots,$ for which $x(t_0) = 0$, say, is a linear random process provided that the fluctuations $\varepsilon(t_i)$ are statistically independent. Stochastic processes are classified as either discrete or continuous. A continuous process is defined for all time steps while a discrete process is defined only at a finite number of points. The data series can be scalar (univariate series) or a series of vectors (multivariate series). While we will deal with discrete data, we assume that the underlying process is continuous.

If we regard each data series as a realization of a stochastic process, each series contains an infinite ensemble of data having the same basic physical properties. Since a particular data series is a sample of a stochastic process, we can apply the same kind of statistical arguments to our data series as we did to individual random variables. Thus, we will be making statistical probability statements about the results of frequency transformations of data series. This fact is important to remember since there is a great temptation to regard transformed values as inherently independent data points. Since many data collected in time or space are highly correlated because of the presence of low frequency, nearly deterministic components, such as long-period tides and the seasonal cycle, standard statistical methods do not really apply. Contrary to the requirements of stochastic theory, the values are not statistically independent. "What constitutes the ensemble of a possible time series in any given situation is dictated by good scientific judgment and not by purely statistical matters" (Jenkins and Watts, 1968). A good example of this problem is presented by Chelton (1982) who showed that the high correlation between the integrated transport through Drake Passage in the Southern Ocean and the circumpolar-averaged zonal wind stress "may largely be due to the presence of a strong semiannual signal in both time series". A strong statistical correlation does not necessarily mean there is a cause and effect relationship between the variables.

As implied by the previous section, the properties of a stochastic process generally are time dependent and the value $y(t)$ at any time, t, depends on the time elapsed since the process started. A simplifying assumption is that the series has reached a steady state or equilibrium in the sense that the statistical properties of the series are independent of absolute time. A minimum requirement for this condition is that the probability density function (PDF) is independent of time. Therefore, a stationary time series has constant mean, μ, and variance, σ^2. Another consequence of this equilibrium state is that the joint PDF depends only on the time difference $t_1 - t_2 = \tau$ and not on absolute times, t_1 and t_2. The term *ergodic* is commonly used in association with stochastic processes for which time averages can be used in place of ensemble averages (see Chapter 3). That is, we can average over "chunks" of a time series to get the mean, standard deviation, and other statistical quantities rather than having to produce repeated realizations of the time series. Any formalism involving

ensemble averaging is of little value as the analyst rarely has an ensemble at his or her disposal and typically must deal with a single realization. We need the ergodic theorem to enable us to use time averages in place of ensemble averages.

5.3 CORRELATION FUNCTIONS

Discrete or continuous random time series, $y(t)$, have a number of fundamental statistical properties that help characterize the variability of the series and make it easily possible to compare one time series against another. However, these statistical measures also contain less information than the original time series and, except in special cases, knowledge of these properties is insufficient to reconstruct the time series.

5.3.1 Mean and Variance

If y is a stochastic time series consisting of N values $y(t_i) = y_i$ measured at discrete times t_i {t_1, t_2,..., t_N}, the true mean value μ for the series can be estimated by

$$\mu \equiv E[y(t)] = \frac{1}{N} \sum_{i=1}^{N} y_i = \bar{y} \qquad (5.1)$$

where $E[y(t)]$ is the expected value, $E[|y(t)|] < \infty$ for all t and $\bar{y} = \frac{1}{N}\sum_{i=1}^{N} y_i$ is the sample mean. The estimated mean value is not necessarily constant in time; different segments of a time series can have different mean values if the series is nonstationary. If $E[y^2(t)] < \infty$ for all t, an estimate of the true variance function is given by

$$\sigma^2 \equiv E[\{y(t) - \mu\}^2] = \frac{1}{N} \sum_{i=1}^{N} [y_i - \bar{y}]^2. \quad (5.2)$$

The positive square root of the variance is the standard deviation, σ, or root-mean-square (RMS) value. See Chapter 3 for further discussion on the mean and variance.

5.3.2 Covariance and Correlation Functions

These terms are used to describe the covariability of given time series as functions of two different times, $t_1 = t$ and $t_2 = t + \tau$, where τ is the lag time. If the process is *stationary* (unchanging statistically with time) as we normally assume, then absolute time is irrelevant and the covariance functions depend only on τ.

Although the terms "covariance function" and "correlation function" are often used interchangeably in the literature, there is a fundamental difference between them. Specifically, covariance functions are derived from data series following removal of the true mean value, μ, which we typically approximate using the sample mean, $\overline{y(t)}$. Correlation functions use the "raw" data series before removal of the mean. The confusion arises because most analysts automatically remove the mean from any time series with which they are dealing. To further add to the confusion, many oceanographers define correlation as the covariance normalized by the variance, σ^2.

For a stationary process, the *autocovariance function*, C_{yy}, which is based on lagged correlation of a function with itself, is estimated by

$$C_{yy}(\tau) \equiv E[\{y(t) - \mu\}\{y(t + \tau) - \mu\}]$$
$$= \frac{1}{N-k} \sum_{i=1}^{N-k} [y_i - \overline{y}][y_{i+k} - \overline{y}] \quad (5.3)$$

where $\tau = \tau_k = k\Delta t \ (k = 0, \ldots, M)$ is the lag time for k sampling time increments, Δt, and $M \ll N$. The corresponding expression for the *autocorrelation function* R_{yy} is

$$R_{yy}(\tau) \equiv E[y(t)y(t + \tau)]$$
$$= \frac{1}{N-k} \sum_{i=1}^{N-k} (y_i y_{i+k}) \quad (5.4)$$

At zero lag ($\tau = 0$)

$$C_{yy}(0) = \sigma^2 = R_{yy}(0) - \mu^2 \quad (5.5)$$

where we must be careful to define σ^2 obtained from Eqn (5.2) in terms of the normalization factor $1/N$ rather than $1/(N-1)$ (see Chapter 3). From the above definitions, we find

$$C_{yy}(\tau) = C_{yy}(-\tau); \quad R_{yy}(\tau) = R_{yy}(-\tau) \quad (5.6)$$

showing that the autocovariance and autocorrelation functions are symmetric with respect to the time lag τ.

The autocovariance function can be normalized using the variance (Eqn (5.2)) to yield the normalized autocovariance function

$$\rho_{yy}(\tau) = \frac{C_{yy}(\tau)}{\sigma^2} \quad (5.7)$$

(Note: some oceanographers call Eqn (5.7) the autocorrelation function.)

The basic properties of the normalized autocovariance function are:

1. $\rho_{yy}(\tau) = 1$, for $\tau = 0$;
2. $\rho_{yy}(\tau) = \rho_{yy}(-\tau)$, for all τ;
3. $|\rho_{yy}(\tau)| \leq 1$, for all τ;
4. If the stochastic process is continuous, then $\rho_{yy}(\tau)$, must be a continuous function of τ.

If we now replace one of the $y(t)$ in the above relations with another function $x(t)$, we obtain the *cross-covariance function*

$$C_{xy}(\tau) \equiv E\left[\{y(t) - \mu_y\}\{x(t + \tau) - \mu_x\}\right]$$
$$= \frac{1}{N-k} \sum_{i=1}^{N-k} [y_i - \overline{y}][x_{i+k} - \overline{x}] \quad (5.8)$$

and the *cross-correlation function*

$$R_{xy}(\tau) \equiv E[y(t)x(t + \tau)]$$
$$= \frac{1}{N-k} \sum_{i=1}^{N-k} y_i x_{i+k} \quad (5.9)$$

The normalized cross-covariance function (or *correlation coefficient function*) for a stationary process is

$$\rho_{xy} \equiv \frac{C_{xy}(\tau)}{\sigma_x \sigma_y} \quad (5.10)$$

Here, $y(t)$ could be the alongshore component of daily mean wind stress and $x(t)$ the daily mean sea-level elevation at the coast. As a result of the Coriolis force, the alongshore current generated by the wind causes a sea-level setup along the coast. Typically, the sea level set up at mid to high latitudes lags the alongshore wind stress by about 1 day.

Care should be taken in interpreting covariance and correlation estimates for large lags. Problems arise if low-frequency components are present in the data since the averaging inherent in these functions becomes based on fewer and fewer samples and loses its statistical reliability as the lag increases. For example, at lag $\tau = 0.1T$ (i.e., 10% of the length of the time series) there are roughly 10 independent cycles of any variability on a timescale, $T_{0.1} = 0.1T$, while at lags of $0.5T$ there are only about two independent estimates of the timescale $T_{0.5}$. In many cases, low-frequency components in geophysical time series make it pointless to push the lag times much beyond 10–20% of the data series. Some authors argue that division by N rather than by $N - k$ reduces the bias at large lags. Although this is certainly true ($N \gg N - k$ at large lags), it does not mean that the results are a better representation of statistical reality. In essence, neither of these estimators is optimal. Ideally one should write down the likelihood function of the observed time series, if it exists. Differentiation of this likelihood function would then give a set of equations for the maximum likelihood estimates of the autocovariance function. Unfortunately, the derivatives are in general untraceable and one must work with estimators given above. Results for this section are summarized as follows:

1. Estimators with divisors $T = N\Delta t$ usually have smaller mean square errors (MSEs) than those based on $T - \tau$; also, those based on $1/T$ are positive definite while those based on $1/(T - \tau)$ may not be.

2. Some form of correction for low-frequency trends is required. In simple cases, one can simply remove a mean value while in others

the trend can be removed. Trend removal must be done carefully so that erroneous data are not introduced into the time series during the subtraction of the trend.

3. There will be strong correlations between values in the autocorrelation function if the correlation in the original series was moderately strong; the autocorrelation function, which can be regarded as a new time series derived from $y(t)$, will, in general, be more strongly correlated than the original series.

4. Due to the correlation in (3), the autocorrelation function may fail to dampen according to expectations; this will increase the basic length scale in the function.

5. Correlation is a relative measure only.

In addition to its direct application to time series analysis, the autocorrelation function was critical to the development of early spectral analysis techniques. Although modern methods typically calculate spectral density distributions directly from the Fourier transforms of the data series, earlier methods determined spectral estimates from the Fourier transform of the autocorrelation function. An important milestone in time series analysis was the proof by N. Wiener and A. Khinchin in the 1930s that the correlation function is related to the spectral density function through a Fourier transform relationship. According to the Wiener–Khinchin relations, the autospectrum of a time series is the Fourier transform of its autocorrelation function.

5.3.3 Analytical Correlation/Covariance Functions

The autocorrelation function of a zero-mean random process $\varepsilon(t)$ ("white noise") can be written as

$$R_{\varepsilon\varepsilon}(\tau) = \sigma_\varepsilon^2 \rho_{\varepsilon\varepsilon}(\tau) = \sigma_\varepsilon^2 \delta(\tau) \qquad (5.11)$$

where $\delta(\tau)$ is the Dirac delta function. In this example, σ_ε^2 is the variance of the data series. Another useful function is the cross-correlation between the time-lagged stationary signal

$y(t) = \alpha x(t - \tau) + \varepsilon$ and the original signal $x(t)$. For constant α

$$R_{xy}(\tau) = \alpha R_{xx}(\tau - \tau_0) + \sigma_\varepsilon^2 \qquad (5.12a)$$

which, for low noise, has a peak value

$$R_{xy}(\tau_0) = \alpha R_{xx}(0) = \alpha \sigma_x^2 \qquad (5.12b)$$

Functions of the type Eqn (5.12) have direct use in ocean acoustics where the time lag, τ_0, at the peak of the zero-mean autocorrelation function can be related to the phase speed c and distance of travel d of the transmitted signal $x(t)$ through the relation $\tau_0 = d/c$. It is through calculations of this type that modern acoustic Doppler current meters (ADCMs) and scintillation flow meters determine oceanic currents. In the case of ADCMs, knowing τ_0 and d gives the speed c and hence the change of the acoustic signal by the currents during the two-way travel time of the signal. Scintillation meters measure the delay τ_0 for acoustic signals sent between a transmitter–receiver pair along two parallel acoustic paths separated by a distance d. The relation $\tau_0 = d/v$ then gives the mean flow speed, v, normal to the direction of the acoustic path. Sending the signals both ways in the transmitter–receiver pairs gets around the problem of knowing the sound speed c in detail.

Although the calculation of autocorrelation and autocovariance functions is fairly straightforward, care is needed in interpreting the resulting values. For example, a stochastic process is said to be Gaussian (or normal) if the multivariate PDF is normal. Then the process is completely described by its mean, variance, and autocovariance function. However, there is a class of non-Gaussian processes that have the same normalized autocovariance function, ρ, as a given normal process. Consider the linear system

$$\tau_0 \frac{dy}{dt} + y(t) = z(t) \qquad (5.13)$$

where $z(t)$ is a white-noise input and $y(t)$ is the output. Here, $y(t)$ is called a "first-order autoregressive process", which has the normalized autocorrelation function

$$\rho_{yy}(\tau) = e^{-|\tau|/\tau_0} \qquad (5.14)$$

Thus, if the input to the first-order system has a normal distribution then by an extension of the central limit theorem it may be shown that the output is normal and is completely specified by the autocorrelation function.

Another process with an exponential autocorrelation function, which differs greatly from the normal process, is called the *random telegraph signal* (Figure 5.1). Alpha particles from a radioactive source are used to trigger a flip-flop between $+1$ and -1. Assuming the process was started at $t = -\infty$, we can derive the normalized autocorrelation function as

$$\rho_{yy}(\tau) = e^{-2\lambda|\tau|} \qquad (5.15)$$

If $\lambda = 1/(2\tau_0)$, then this is the same as the autocorrelation function of a normal process, which is characteristically different from the flip-flop

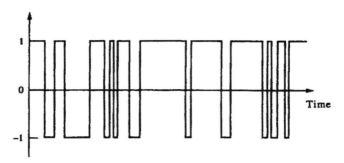

FIGURE 5.1 A realization of a random telegraph signal with digital amplitudes of ± 1 as a function of time.

time series. Again, one must be careful when interpreting autocorrelation functions. As with any correlation between two variables, the auto-correlation function only indicates how the time series vary together and says nothing about the magnitudes of their variations. A detailed examination of the data requires the analysis of the magnitudes of the values in the series and not just their correlation.

5.3.4 Observed Covariance Functions

To see what autocorrelation functions look like in practice, and to emphasize the fact that the methodology applies to spatial as well as temporal data, consider the acoustic profile data in Table 5.1. Here, we have tabulated the calibrated acoustic backscatter anomaly measured over 5-m depth increments in the upper ocean using a 150 kHz acoustic Doppler current profiler (ADCP) lowered from a ship. These spatial data are from the first bin of adjacent Beams 1 and 2 of a four-beam ADCP, and represent the back-scatter intensity anomaly (in decibels) from zooplankton ensonified at a distance of 5 m from the instrument transducers. Since each of the transducers is tilted at an angle of $30°$ to the vertical, the two 5-m increment profiles are separated horizontally by only 3.9 m and so the auto-correlations for the two series should be nearly identical at all lags. In this case, we use the normalized covariance Eqn (5.7) derived from Eqn (5.3) in which the sum is divided by the number of lag values, $N - k$, for lag $\tau = k\Delta z$, where $\Delta z = 5$ m. As indicated by the autocorrelation functions in Figure 5.2, the functions are similar at small lags where statistical reliability is large but diverge significantly at higher lags with the decrease in the number of independent covariance estimates.

5.3.5 Integral Timescales

The integral timescale, T^*, is defined as the sum of the normalized autocorrelation function (Eqn (5.7)) over the length $L = N\Delta\tau$ of the time series for N lag steps, $\Delta\tau$. Specifically, the estimate

$$
\begin{aligned}
T^* &= \frac{\Delta\tau}{2} \sum_{i=0}^{N'} [\rho(\tau_i) + \rho(\tau_{i+1})] \\
&= \frac{\Delta\tau}{2\sigma^2} \sum_{i=0}^{N'} [C(\tau_i) + C(\tau_{i+1})]
\end{aligned}
\tag{5.16}
$$

TABLE 5.1   Acoustic Backscatter Anomaly (Decibels) Measured in Bin#1 (Depth, m) from Two Adjacent Transducers (Beams 1 and 2) on a 4-Beam 150 kHz ADCP Lowered from a Ship in the Northeast Pacific

Beam	75 m	80	85	90	95	100	105	110	115	120
1	11.56	0.67	−8.33	−9.82	−13.91	−18.00	3.67	−2.00	−12.29	−13.71
2	14.67	3.00	−5.67	−9.64	−12.82	−16.00	−8.50	−11.00	−15.29	−16.71
125 m	**130**	**135**	**140**	**145**	**150**	**155**	**160**	**165**	**170**	**175**
−11.33	−8.00	24.14	38.13	40.00	35.00	29.63	24.00	26.50	28.75	30.63
−10.33	−2.00	23.71	36.63	41.00	33.14	24.38	15.00	20.63	26.25	31.88
180 m	**185**	**190**	**195**	**200**	**205**	**210**	**215**	**220**	**225**	**230**
30.50	31.00	36.00	31.63	21.00	12.25	3.00	−7.00	−4.43	−0.50	0.75
31.00	29.13	29.75	24.75	16.00	7.25	3.25	6.38	11.57	12.25	5.38

The data cover a depth range of 75–230 m at increments of 5 m (N = 32 Values). The two vertical profiles are separated horizontally by a distance of roughly 3.9 m. The first line in each set gives the depth of the observations, the next two lines the anomaly values for Beam 1 then Beam 2. The record means have not been removed from the data.

FIGURE 5.2 Autocorrelation functions of the acoustic backscatter data in Table 5.1. The thick line is for acoustic Beam 1, the dashed line for acoustic Beam 2.

for $N' \leq N - 1$ gives a measure of the dominant correlation timescale within a data series; for times longer than T^*, the data become decorrelated. There are roughly $\Delta\tau N/T^*$ actual degrees of freedom (DoF) within the time series. In reality, the summation typically is limited to $N' \ll N$ since low frequency components within the time series prevent the summation from converging to a constant value over the finite length of the record. In general, one should continue the summation until it reaches a near-constant value, which we take as the value for T^*. If no plateau is reached within a reasonable number of lags, no integral timescale exists. In that case, the integral timescale can be approximated by integrating only to the first zero crossing of the autocorrelation function (cf. Poulain and Niiler, 1989).

5.3.6 Correlation Analysis vs Linear Regression

Geophysical data are typically obtained from random temporal sequences or spatial fields that cannot be regarded as mutually independent. Because the data series depend on time and/or spatial coordinates, the use of linear regression to study relationships between data series may lead to incomplete or erroneous conclusions. As an example, consider two time series: A white-noise series, consisting of identically distributed and mutually independent random variables, and the same series but with a time shift. As the values of the time series are statistically independent, the cross-correlation coefficient will be zero at zero lag, even though the time series are strictly linearly related. Regression analysis would show no relationship between the two series. However, cross-correlation analysis would reveal the linear relationship (a coefficient of unity) for a lag equal to the time shift. Correlation analysis is often a better way to study relations among time series than traditional regression analysis.

5.4 SPECTRAL ANALYSIS

Spectral analysis is used to partition the variance of a time series as a function of frequency. For stochastic time series such as wind waves, contributions from the different frequency components are measured in terms of the *power spectral density* (PSD). For deterministic waveforms such as surface tides, either the PSD or the *energy spectral density* (ESD) can be used. Here, power is defined as energy per unit time. The need for two different spectral definitions lies in the boundedness of the integral of signal variance for increasing record length. In practice, the term *spectrum* is applied to all spectral functions including commonly used terms such as auto-spectrum and power spectrum. The term *cross-spectrum* is reserved for the "shared" power between two coincident time series. We also distinguish between *nonparametric* and *para-metric* spectral methods. Nonparametric methods, which are based on conventional Fourier transforms, are not data-specific while parametric techniques are data-specific and assign a predetermined model to the time series. In general, we use parametric methods for short time series (few cycles of the oscillations of interest) and nonparametric methods for long time series (many cycles of the oscillations of interest).

The word spectrum is a carryover from optics. The "colors" red, white, and blue of the electromagnetic spectrum are often used to describe the frequency distribution of oceanographic spectra. A spectrum whose spectral density decreases with increasing frequency is called a "red" spectrum, by analogy to visible light where red corresponds to longer wavelengths (lower frequencies). Similarly, a spectrum whose magnitude increases with frequency is called a "blue" spectrum. A "white" spectrum is one in which the spectral constituents have near-equal amplitude throughout the frequency range. In the ocean, long-period variability (periods greater than several days) tends to have red spectra while instrument noise tends to have white spectra. Blue spectra are confined to certain frequency bands such as the low-frequency portion of wind–wave spectra and within the "weather band" (2 < period < 10 days) for deep wind-generated currents. Spectra of wind-generated inertial currents in the deep ocean are often "blue-shifted" to frequencies a few percent higher than the local inertial frequency (Fu, 1981; Thomson et al., 1990).

In the days before modern computers, it was customary to compute the spectrum of discrete oceanic data from the Fourier transform of the autocorrelation function using a small number of lag intervals, or "lags". First formalized by Blackman and Tukey (1958), the autocorrelation method lacks the wide range of optional improvements to the computations and generalized "tinkering" permitted by more modern techniques. From a historical perspective, the autocorrelation approach has importance for the direct mathematical link it provides through the Wiener–Khinchin relations that link variance functions in the time domain to those in the frequency domain. Today, it is the spectral *periodogram* generated using the FFT or the Singleton Fourier transform that is most commonly used to estimate oceanic spectra. (We assume that the reader has a basic understanding of Fourier analysis and FFTs. Those unfamiliar with these concepts can proceed to Section 5.8 where the topics are discussed in considerable detail.)

Other methods have been developed over the years as a result of fundamental performance limitations with the periodogram method. These limitations are: (1) restricted frequency resolution when distinguishing between two or more signals, with frequency resolution dictated by the available record length independent of the characteristics of the data or its signal-to-noise ratio (SNR); (2) energy "leakage" between the main lobe of a spectral estimate and adjacent side-lobes, with a resulting distortion and smearing of the spectral estimates, suppression of weak signals, and the need to use smoothing windows; (3) an inability to adequately determine the spectral content of short time series; and (4) an inability to adjust to rapid changes in signal amplitude or phase. Other techniques such as the maximum entropy method (MEM, best suited to short time series) and the wavelet transform (best suited to event-like signals and signals whose frequency content changes over time) are addressed in this chapter.

Fundamental concepts: Several basic concepts are woven into the fabric of this chapter. First of all, the sample data we collect are subsets of either stochastic or deterministic processes. Deterministic processes are predictable, stochastic ones are not. Secondly, the very act of sampling to generate a time series of finite duration is analogous to viewing an infinitely long time series through a narrow "window" in the shape of a rectangular box-car function (Figure 5.3(a)). The characteristics of this window in the frequency domain can severely distort the frequency content of the original data series from which the sample has been drawn. As illustrated by Figure 5.3(b), the sampling process results in spectral energy being "rippled" away from one

frequency (the central lobe of the response function) to a wide number range of adjacent frequencies. The large side-lobes of the rectangular window are responsible for the leakage of spectral energy from the central frequency to nearby frequencies.

A third point is that the spectra of random processes are themselves random processes. Therefore, if we are to determine the frequency content of a data series with some degree of statistical reliability (i.e., to be able to put confidence intervals on spectral peaks), we need to precondition the time series and average the raw periodogram estimates. Averaging can be done in the time domain by using specially designed windows or in the frequency domain

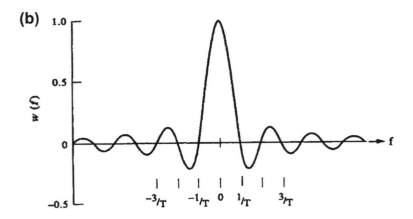

FIGURE 5.3 The box-car (rectangular) window, which creates a sample time series from a "long" time series. (a) The box-car window in the time (t) domain. Here, $w(t) = 1$, $-T/2 \leq t \leq T/2$, and $w = 0$ otherwise. (b) Frequency $W(f)$ response of the box-car window in (a). The central lobe straddles each spectral (frequency) component within the time series and has a width, $\Delta f = 2/T$. Zeros occur at $f = \pm m/T$, where $m = 1, 2,...$

by averaging together adjacent spectral estimates. Windows (which are discussed in detail in Section 5.4.6) suppress Gibbs' phenomenon associated with finite length data series and enable us to increase the number of *degrees of freedom* (DoF) used in each spectral estimate. (Here, the term "degrees of freedom" refers to the number of statistically independent variables or values used in a particular estimate. We note that the spectral values are chi-square variables and that the DoF now apply to that PDF.) We can also improve spectral estimates by partitioning a time series into a series of segments and then conducting spectral analysis on the separate pieces. Spectral values in each frequency band for each piece are then averaged as a block to improve statistical reliability. This is similar to averaging adjacent spectral values in the periodogram, which will give a similar increase in the DoF of the resulting spectral estimate. The penalty for doing this is a loss in frequency resolution. The alternative—calculating a single periodogram and then smoothing in the frequency domain—suffers the same loss of frequency resolution for a smoothing that gives the same DoF.

Regardless of which averaging approach we choose, the results will be tantamount to viewing the data through another window in the frequency domain. Any smoothing window used to improve the reliability of the spectral estimates will again distort the results and impose structure on the data, such as periodic behavior, when no such structure may exist in the original time series. In addition, conventional methods make the implicit assumption that the unobserved data or correlation lag values situated outside the measurement interval are zero, which is generally not the case. The smoothing window results in smeared spectral estimates. The more modern parametric methods allow us to make more realistic assumptions about the nature of the process outside the measurement interval, other than to assume it is zero or cyclic. This eliminates the need for window

functions. The improvement over conventional FFT spectral estimates can be quite dramatic, especially for short records. However, even then, there remain pitfalls, which have tended to detract from the usefulness of these methods to oceanography. Each new method has its own advantages and disadvantages that must be weighed in context of the particular data set and the way it has been collected. For time series with low SNR, most of the modern methods are no better than the conventional FFT approach.

Means and trends: Prior to spectral analysis, the record mean and trend are generally removed from any time series (Figure 5.4). Unless stated otherwise, we will assume that the time series $y(t)$ we wish to process has the form $y'(t) = y(t) - \overline{y(t)}$, where $\overline{y(t)} = y_0 + \alpha t$ is the mean value and αt is the linear trend (y_0 and α are constants). If the mean and trend are not removed prior to spectral analysis, they can distort the low-frequency components of the spectrum (see Section 5.4.12). Packaged spectral programs often include record mean and linear trend removal as part of the data preconditioning. Nonlinear trends are more difficult to remove, especially since a single function may not be appropriate for the entire data domain. The latter may apply also to linear trends.

The mean value removed from a record is not always the average for the entire record. For example, to examine interannual variability in the monthly time series of sea-level height, alongshore current velocity, or any other scalar, $\eta(t_m)$, we first calculate the mean monthly values $\overline{\eta(t_m)}$ for each month separately over the entire record (e.g., the individual means for all Januaries, all Februaries, etc.). These mean monthly values for $m = 1, 2, \ldots, 12$, rather than the simple average value for all values over the entire record, are then subtracted from the original data for the appropriate month to obtain monthly anomalies, $\eta'(t_m) = \eta(t_m) - \overline{\eta(t_m)}$. As with other averaging processes, the user will need to determine how many missing data values will be

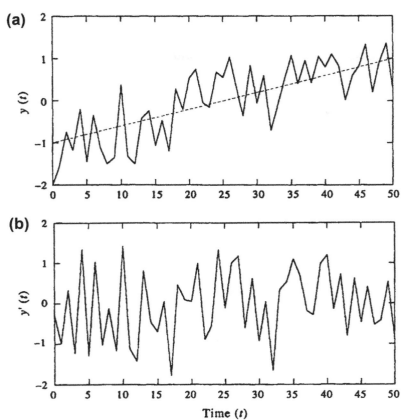

FIGURE 5.4 Mean and trend removal for an artificial time series $y(t)$. Here, $y_o = -1.0$, the trend, $\alpha = 0.025$, and the fluctuating component, $y'(t)$, was obtained using a uniformly distributed random number generator. (a) Original time series, showing the linear trend; (b) Time series with the mean and linear trend removed.

permitted for a given month before the monthly value is considered "missing" or not available. Trend removal can then be applied to the monthly anomalies to obtain the final anomaly record. As a final comment, we note that certain records, such as those from moored bottom pressure recorders and near-surface transmissometers or dissolved oxygen sensors, will contain long-term nonlinear trends that should be removed from the data record prior to spectral analysis. However, this must be done cautiously. Unless one has a justified physical model for a particular trend (including a linear trend), removal of the trend may itself add spurious frequency components to the detrended signal.

5.4.1 Spectra of Deterministic and Stochastic Processes

Time series data can originate with deterministic or stochastic processes, or a mixture of the two. Turbulence arising from eddy-like motions generated by strong tidal currents in a narrow coastal channel provides an example of mixed deterministic and stochastic processes. To see the difference between the two types of processes in terms of conventional spectral estimation, consider the case of a continuous *deterministic* signal, $y(t)$. If the total signal energy, E, is finite

$$E = \int_{-\infty}^{\infty} |y(t)|^2 dt < \infty \qquad (5.17)$$

then $y(t)$ is absolute-integrable over the entire domain and the Fourier transform $Y(f)$ of $y(t)$ exists. This leads to the standard transform pair

$$Y(f) = \int_{-\infty}^{\infty} y(t)e^{-i2\pi ft} dt \qquad (5.18a)$$

$$y(t) = \int_{-\infty}^{\infty} Y(f)e^{i2\pi ft} df = \frac{1}{2\pi} \int_{-\infty}^{\infty} Y(\omega)e^{i\omega t} d\omega$$

$$(5.18b)$$

where $e^{\pm i2\pi ft} = \cos(2\pi ft) \pm i\sin(2\pi ft)$, f is the frequency in cycles per unit time, and $\omega = 2\pi f$ is the angular frequency in radians per unit time. The square of the modulus of the Fourier transform for all frequencies

$$S_E(f) = Y(f)Y^*(f) = |Y(f)|^2 \qquad (5.19)$$

is then the Energy Spectral Density (ESD), $S_E(f)$, of $y(t)$. (As usual, the asterisk denotes the complex conjugate.) To show that Eqn (5.19) is an energy density, we use Parseval's theorem

$$\int_{-\infty}^{\infty} |y(t)|^2 dt = \int_{-\infty}^{\infty} |Y(f)|^2 df \qquad (5.20)$$

which states that the total energy, E, of the signal in the time domain is equal to the total energy, $\int S_E(f) df$, of the signal in the frequency domain. Thus, $S_E(f)$, is an energy density (energy per unit frequency) which, when multiplied by df, yields a measure of the total signal energy in the frequency band centered near frequency f. The "power" of a deterministic signal, E/T, is zero in the limit of very long time series ($T \to \infty$).

Now, suppose that $y(t)$ is a stationary *random* process rather than a deterministic waveform. Unlike the case for the finite energy deterministic

signal, the total energy in the stochastic process is unbounded (the characteristics of the process remain unchanged over time) and functions of the form (Eqn (5.18)) do not exist. In other words, the Fourier transform method introduced earlier fails in the sense that the total energy, as defined by Eqn (5.17), does not decrease as the length of the time series increases without bound. To get around this problem, we must deal with the frequency distribution of the signal *power* (the time average of energy or energy per unit time, E/T), which is a bounded function. The basis for spectral analysis of random processes is the autocorrelation function $R_{yy}(\tau) = E[y(t)y(t+\tau)]$. Using the Wiener–Khinchin relation, the PSD, $S(f)$, becomes

$$S(f) = \int_{-\infty}^{\infty} R_{yy}(\tau)e^{-i2\pi f\tau} d\tau \qquad (5.21a)$$

For an ergodic random process, for which ensemble averages can be replaced by time averages, R_{yy} has the from

$$R_{yy}(\tau) = \lim_{T \to \infty} \frac{1}{T} \int_{-T/2}^{T/2} [y(t)y^*(t+\tau)] dt \qquad (5.21b)$$

By definition, the energy and PSD functions quantify the signal variance per unit frequency. For example, in the case of a stationary random process, integration of $S(f)$ gives the relation

$$s^2 = \int_{f-\Delta f/2}^{f+\Delta f/2} S(f) df \qquad (5.22)$$

where s^2 is the integrated signal variance in the narrow frequency range $\Delta f = [f - \tfrac{1}{2}\Delta f, f + \tfrac{1}{2}\Delta f]$. If we assume that the spectrum is nearly uniform over this frequency range, we find

$$S(f) \approx \frac{s^2}{\Delta f} \qquad (5.23)$$

which defines the spectrum for a stochastic processes in terms of a power density, or variance

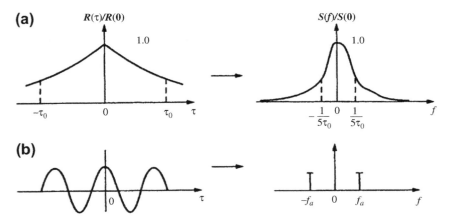

FIGURE 5.5 Examples of slowly decaying autocorrelation functions, $R(\tau)$, as a function of time lag, τ. Functions are normalized by their peak values. (a) The correlation function for a highly correlated signal leads to a relatively narrow power spectra density distribution, $S(f)$; (b) The case for autocorrelation, $R(\tau) \approx \cos(2\pi f_a \Delta t)$ for a single frequency component, f_a, and corresponding line spectra at frequencies $\pm f_a$. *(From Konyaev (1990).)*

per unit frequency. The product $S(f) \cdot \Delta f$ is the total signal variance within the frequency band Δf centered at frequency f.

At this point, there are several other basic concepts worth mentioning. First of all, a waveform whose autocorrelation function $R(\tau)$ attenuates slowly with time lag, τ, will have a narrow spectral distribution (Figure 5.5(a)) indicating that there are relatively few frequency components to destructively interfere with one another as τ increases from zero. In the limiting case of only one frequency component, f_a, we find $R(\tau) \approx \cos(2\pi f_a \tau)$ and Fourier *line spectra* appear at frequencies $\pm f_a$ (Figure 5.5(b)). Because they consist of near monotone signals, tidal motions are highly autocorrelated and produce sharp spectral lines. In contrast, a rapidly decaying autocorrelation function implies a broad spectral distribution (Figure 5.6(a)) and a large number of frequency components in the original waveform. In the limit $R(\tau) \to \delta(\tau)$ (Figure 5.6(b)), there is an infinite number of equal-amplitude frequency components in the waveform and the spectrum $S(f) \to$ constant (white spectrum).

Figure 5.7 provides an example of time series data generated by the relation $y(k) = A\cos(2\pi nk/N)$

$+ \varepsilon(k)$, where $k = 0,\ldots,N$ is time in units of $\Delta t = 1$, $n/(N\Delta t) = 0.25$ is the frequency in units of Δt^{-1}, and $\varepsilon(k)$ is a random number between -1 and $+1$. (We will often use this type of generic example rather than a specific example from the oceanographic literature. That way, readers can directly compare their computational results with ours.) In the present case, if we set $\Delta t = 1$ day, then the time series $y(k)$ could represent east–west current velocity oscillations of a synoptic (3- to 10-day) period associated with wind-forced motions (cf. Cannon and Thomson, 1996). Here, we set $A = 1$ and $\varepsilon(k) \neq 0$ for mostly deterministic data (Figure 5.7(a)) and $A = 0$ for random data (Figure 5.7(b)). In the analysis, the record has been padded with zeroes up to time $k = 2N$. For the mostly deterministic case, the noise causes partial decorrelation of the signal with lag, but the spectral peak remains prominent. For the purely random case, the spectrum resembles white noise but with isolated spectral peaks that one might mistake as originating with some physical process. The latter result is a good example of why we need to attach confidence limits to the peaks of spectral estimates (see Section 5.4.8). It is disconcerting to see the number

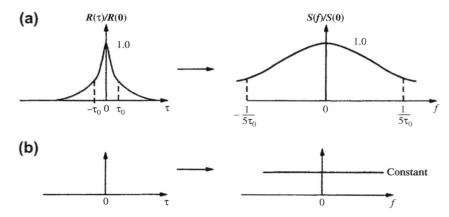

FIGURE 5.6 As for Figure 5.5 but for rapidly decaying autocorrelation functions, $R(\tau)$. (a) Correlation function for a weakly correlated signal leading to a broad power spectra density distribution. (b) The limiting case, $R(\tau) \approx (2\pi f_a \Delta t)$, and the related spectrum, $S(f) = $ constant (a white spectrum). *(From Konyaev (1990).)*

of papers that are published in reputable journals that present spectra without including confidence intervals. At the same time, one must be cognizant of the meaning of the confidence intervals being presented. Use of a very low significance level might suggest a high degree of confidence in the results. However, this confidence can be rendered meaningless if a low significance level is selected. Significance levels of 95 or 99% are commonly accepted as "meaningful".

5.4.2 Spectra of Discrete Series

Consider an infinitely long time series $y(t_n) = y_n$ sampled at equally spaced time increments $t_n = n\Delta t$, where Δt is the sampling interval and n is an integer, $-\infty < n < \infty$. From sampling theory, we know that a continuous representation of the discrete times series $y_s(t)$, can be represented as the product of the continuous time series $y(t)$ with an infinite set of delta functions, $\delta(t)$, such that

$$y_s(t) = y(t) \sum_{n=-\infty}^{\infty} \delta(t - n\Delta t)$$

$$= y(t) \frac{\Xi(t/\Delta t)}{\Delta t}$$

(5.24a)

where Ξ is the "sampling function", for which the Fourier transform is

$$Y(f) = \int_{-\infty}^{\infty} \left[\sum_{n=-\infty}^{\infty} y(t)\delta(t - n\Delta t)\Delta t \right] e^{-i2\pi ft} dt$$

$$= \Delta t \sum_{n=-\infty}^{\infty} y_n e^{-i2\pi ft}$$

(5.24b)

In effect, the original time series is multiplied by a "picket fence" of delta functions $\Xi(t/\Delta t) \approx \sum_{n=-\infty}^{\infty} \delta(t - n\Delta t)$, which are zero everywhere except for the infinitesimal rectangular region occupied by each delta function (Figure 5.8(a) and (b)). Comparison of the above expression with Eqn (5.18) shows that retention of the time step Δt ensures conservation of the rectangular area in the two expressions as $\Delta t \to 0$. Provided that the time series $y(t)$ has a limited number of frequencies (i.e., is band-limited), whereby all frequencies are contained in the Nyquist interval

$$-f_N \leq f_k \leq f_N \qquad (5.25)$$

in which $f_N \equiv f_{Nyquist} = 1/(2\Delta t)$ is the Nyquist frequency, the ESD

$$S_E(f) = |Y(f)|^2 \qquad (5.26)$$

FIGURE 5.7 Autocovariance function, $C(\tau)$, and corresponding spectrum, $S(f)$, for the time series, $y(k) = A \cos(2\pi nk/N) + \varepsilon(k)$; $k = 0,\ldots, N$, $\Delta t = 1$, $n/N = 0.25$ is the frequency, and $\varepsilon(k)$ is a random number between -1 and $+1$. (a) $C(\tau)$ and $S(f)$ for $A = 1$ and $\varepsilon \neq 0$ (mostly deterministic data); and (b) for $A = 0$ (purely random data). Records have been padded with zeros up to time $k = 2N = 32$.

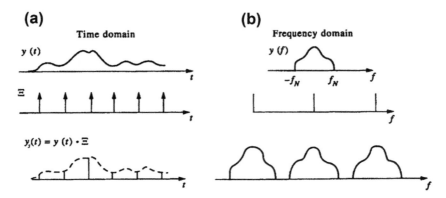

FIGURE 5.8 (a) A "picket fence" of delta functions, $\delta(t - n\Delta t)$, used to generate a discrete data series from a continuous time series. (b) The Fourier transform (schematic only) of the different functions. Here, $y(f)$ denotes $Y(f)$.

is identical to that for a continuous function. Conversely, if $Y(f) \neq 0$ for $|f| > f_N$ then the sampled and original times series do not have the same spectrum for $|f| < f_N$. The spectrum Eqn (5.26) obtained by Fourier analysis of discrete time series is called a *periodogram* spectral estimate, a term first coined by Schuster (1898) in a study of sunspot cycles. (Note that we always use f_N for the Nyquist frequency; the subscript N for Nyquist should never be confused with the subscript N used in summations or the N used for degrees of freedom (DoF).)

Real oceanographic time series data are discrete and have finite duration, $T = N\Delta t$. Returning to Eqn (5.24), this means that the summation is over a limited range $n = 1$ to N, and the spectral amplitude for the sample must be defined in terms of the discrete Fourier transform (DFT)

$$Y_k = \Delta t \sum_{n=1}^{N} y_n e^{-i2\pi f_k n \Delta t}$$
$$= \Delta t \sum_{n=1}^{N} y_n e^{-i2\pi kn/N};$$
$$f_k = k/N\Delta t, \quad k = 0, \dots, N \quad (5.27)$$

The frequencies f_k are confined to the Nyquist interval, with positive frequencies, $0 \leq f_k \leq f_N$, corresponding to the range $k = 0, \dots, N/2$ and

negative frequencies, $-f_N \leq f_k \leq 0$, to the range $k = N/2, \dots, N$. Since $f_{N-k} = f_k$, only the first $N/2$ Fourier transform values are unique. Specifically, $Y_k = Y_{N-k}$ so that we will generally confine our attention to the positive interval only.

The inverse Fourier transform (IFT) is defined as

$$y_n = \frac{1}{N\Delta t} \sum_{k=0}^{N-1} Y_k e^{i2\pi kn/N}, \quad n = 1, \dots, N \quad (5.28)$$

As indicated by Eqn (5.27), the Fourier transforms, Y_k, are specified for the discretized frequencies f_k, where $f_k = kf_1$ and $f_1 = 1/(N\Delta t) = 1/T$ characterizes both the fundamental frequency and the bandwidth, Δf, for the time series. The ESD for a discrete, finite-duration time series is then

$$S_E(f_k) = |Y_k|^2, \quad k = 0, \dots, N-1 \quad (5.29)$$

and Parseval's energy conservation theorem (5.20) becomes

$$\Delta t \sum_{n=1}^{N} |y_n|^2 = \Delta f \sum_{k=0}^{N-1} |Y_k|^2$$

where we have used $\Delta f = 1/(N\Delta t)$. A plot of $|Y_k|^2$ vs frequency, f_k, gives the discrete form of the periodogram spectral estimate.

Any geophysical data set we collect is subject to discrete sampling and windowing. As noted earlier, a time series of geophysical data, $y(t_n)$, sampled at time steps Δt can be considered the product of an infinitely long time series with a rectangular window that spans the duration ($T = N\Delta t$) of the measured data. The discrete spectrum $S(f_k)$ is the then the *convolution* of the true spectrum, $S(f)$, with the Fourier transform of the rectangular window (Figure 5.3(b)). Since the window allows us to see only a segment of the infinite time series, the spectrum $S(f_k)$ provides a distorted picture of the actual underlying spectrum. This distortion, created during the Fourier transform of the rectangular window, consists of a broadening of the central lobe and leakage of power from the central lobe into the side lobes. (The "ripples" on either side of the central lobe in Figure 5.3(b) are side lobes.) A further problem is that the function Y_k and its Fourier transform now become periodic with period N, although the original infinite time series $y(t)$, of which our sample data are a subset, may have been nonperiodic.

As noted in the previous section, the convergence of $|Y(f)|^2$ to $S(f)$ is smooth for deterministic functions in that the function $|Y'(f)|^2$, obtained by increasing the sample record length from T to T', would be a smoother version of $|Y(f)|^2$. For stochastic signals, the function $|Y'(f)|^2$ obtained from the longer time series (T') is just as erratic as the function for the shorter series. The sample spectra of a stochastic process do not converge in any statistical sense to a limiting value as T tends to infinity. Thus, the sample spectrum is not a consistent estimator in the sense that its PDF does not tend to cluster more closely about the true spectrum as the sample size increases. To show what we mean, consider the spectrum of a process consisting of $N = 400$ random, normally distributed deviates (Gaussian white noise) sampled at 1 s intervals. (True white noise is a mathematical construct and is as physically impossible as the spike of an impulse function.) The highest frequency we can hope to

measure with these data is the Nyquist frequency, $f_N = 0.5$ cps (cycles per second). The spectra computed from 50 and then from 100 values of the fully white-noise signal are presented in Figure 5.9(a). Also shown is the theoretical sample spectrum, corresponding to a uniform amplitude of 1.0. The shorter the sample used for the discrete spectral estimates, the greater the amplitude spikes in the power spectrum. This same tendency also is apparent in Table 5.2, which lists the means, variances, and MSEs (Mean Square Errors) computed from various subsamples of the white-noise signal. Here, MSE is defined as the variance plus bias of an estimator $\widehat{y}(t)$ of the true signal $y(t)$; that is

$$\text{MSE} = E\left[(\widehat{y} - y)^2\right] = V[\widehat{y}] + B^2 \qquad (5.30)$$

where $B = E[\widehat{y}] - y$ is the bias of the estimator. The mean is lower in both the $N = 50$ and $N = 400$ cases while it is greater in the case where $N = 100$ and is exactly 1.0 for $N = 200$. The variance increases as N increases, as does the MSE. However, if this were a purely random discrete process (discrete white noise), the sample spectral estimator of the variance would be independent of the number of observations.

Now consider the spectrum of a second-order autoregressive process for a sample of $N = 400$ measured at 1 s increments (Figure 5.9(b)). (An autoregressive process of order p is one in which the present value of y depends on a linear combination of the previous p values of y. See Section 5.5.2.) The Nyquist frequency is again 0.5 cps and the maximum bandwidth of the spectral resolution, $\Delta f = 1/(N\Delta t)$, is equal to 0.0025 cps. At higher frequencies, the sample spectrum appears to be a good estimator of the theoretical spectrum (the smooth solid line), while for the lower frequencies there are large spikes in the sample spectrum that are not characteristic of the true spectrum. This misleading appearance is largely a consequence of the fact that the theoretical spectrum has most of its energy at the lower frequencies. In reality, the computed raw spectrum

(a)

TABLE 5.2	Behavior of Sample Spectra of White Noise as the Record Length, N, is Increased			
Record Length (N)	**50**	**100**	**200**	**400**
Mean	0.85	1.07	1.00	0.95
Variance	0.630	0.777	0.886	0.826
Mean square error	0.652	0.782	0.886	0.828

Units are arbitrary. (After Jenkins and Watts (1968).)

(i.e., with no smoothing) can fluctuate by 100% about the mean spectrum. The fluctuations are much smaller at higher frequencies simply because the actual spectral level is correspondingly smaller.

The basic reason that Fourier analysis breaks down when applied to real time series is that it is based on the assumption of fixed (stationary) amplitudes, frequencies, and phases (see Section 5.8). Stochastic series are instead characterized by random changes in frequency, amplitude, and phase. Thus, our treatment must be a statistical approach that makes it possible to accommodate these types of changes in our computation of the power spectrum.

5.4.3 Conventional Spectral Methods

The two spectral estimation techniques founded on Fourier transform operations are the indirect autocorrelation approach popularized by Blackman and Tukey in the 1950s and the direct periodogram approach presently favored by the oceanographic community. The FFT is the most common algorithm for determining the periodogram. The autocorrelation approach is mainly discussed for completeness. These

(b)

FIGURE 5.9 Power spectra of discrete signals and their theoretical values. Frequency in cycles per second (cps); spectra are in units of amplitude-squared/cps. (a) Power spectrum for the first half ($N = 50$) and full ($N = 100$) realization of a discrete normal white-noise process measured at 1-s intervals. (b) Power spectrum for one realization of a second-order autoregressive process of $N = 400$ values measured at 1-s increments. $f_N = 0.5$ cps is the Nyquist frequency and the maximum bandwidth of the spectral resolution, $\Delta f = 1/N\Delta t = 0.0025/s$. *(From Jenkins and Watts (1968).)*

methods fall into the category of nonparametric techniques, which are defined independently of any specific time series. Parametric techniques, described later in this chapter, make assumptions about the variability of the time series and rely on the series for parameter determination.

The following sections first describe the two conventional spectral analysis methods without providing details on how to improve spectral estimates. We wish to first outline the procedures for calculating spectra before describing how to improve the statistical reliability of the spectral estimates. Once this is done, we give a thorough description of windowing, frequency-band averaging, and other spectral improvement techniques.

5.4.3.1 The Autocorrelation Method

In the Blackman–Tukey method, the autocovariance function, $C_{yy}(\tau)$ (which equals the autocorrelation function, $R_{yy}(\tau)$, if the record mean has been removed), is first computed as a function of lag, τ, and the Fourier transform of $C_{yy}(\tau)$ used to obtain the PSD as a function of frequency. An unbiased estimator for the autocovariance function for a data set consisting of N equally spaced values $\{y_1, y_2, ..., y_N\}$ is

$$C_{yy}(\tau_m; N-m) = \frac{1}{N-m} \sum_{n=1}^{N-m} y_n y_{n+m} \quad (5.31a)$$

where $m = 0, ...,$ M is the number of lags ($\tau_m = m\Delta t$) and $M < N$. In place of this estimator, some authors (cf. Kay and Marple, 1981) argue for the use of

$$C_{yy}(\tau_m; N) = \frac{1}{N} \sum_{n=1}^{N-m} y_n y_{n+m} \quad (5.31b)$$

which typically has a lower MSE than $C_{yy}(\tau_m; N-m)$ for most finite data sets. Because $E[C_{yy}(\tau_m; N)] = [(N-m)/N]C_{yy}(\tau_m; N-m)$, the function $C_{yy}(\tau; N)$ is a biased estimator for the autocovariance function. Despite this, we will often use the relation Eqn (5.31b) for the autocovariance function since it yields a PSD that is equivalent

to the PSD obtained from the direct application of the FFT, as discussed in the next section. Moreover, the weighting $(N-m)/N$ acts like a triangular (Bartlett) smoothing window to help reduce spectral leakage. We will use Eqn (5.31a) when we want a "stand-alone" unbiased estimator of the covariance function, keeping in mind that this formulation gives largest weight to the most poorly determined components in the Fourier analysis, which is often not desirable despite the reduction in bias.

The one-sided PSD, G_k, for an autocovariance function with a total of M lags is found from the Fourier transform of the autocovariance function

$$G_k = 2\Delta t \sum_{m=0}^{M} C_{yy}(\tau_m)e^{-i2\pi km/M}, \quad k = 0, ..., \frac{M}{2}$$

$$(5.32a)$$

where $\tau_m = m\Delta t$ and $2\Delta t = 1/f_N$. Since $C_{yy}(\tau_m)$ is an even function, the spectrum of $\{y_n\}$ can be calculated from the cosine transform

$$G_k = 2\Delta t \left[C_{yy}(0) + 2\sum_{m=1}^{M} C_{yy}(\tau_m)\cos\left(\frac{2\pi km}{N}\right) \right],$$

$$k = 0, ..., \frac{M}{2} \quad (5.32b)$$

where $G_k = 2S_k$ is centered at positive frequencies $f_k = k/N\Delta t$ and the Nyquist interval $0 \le f_k \le f_{Nyquist}$ is divided into $N/2$ segments (N is even). For the two-sided spectrum, S_k, the first $(N/2)+1$ frequencies are identical to those for the one-sided spectrum and correspond to positive frequencies in the range $0 \le f_k \le f_N$. The last $(N/2)-1$ spectral values for the two-sided spectral density, defined for $k = (N/2)+1, (N/2)+2, ..., N-1$, correspond to spectral density estimates for negative frequencies in the range $-f_N \le f_k \le 0$.

The solid line in Figure 5.10 shows the spectrum of monthly mean SSTs derived from the cosine transform using the Blackman–Tukey autocorrelation method for the version (Eqn (5.31b)) of the autocovariance function. The temperature data span the 36-month period from

FIGURE 5.10 Spectra $(°C)^2/$ cpm (cpm = cycles per month) vs frequency (per month) for monthly mean sea surface temperatures collected at a coastal station in the northeast Pacific for the period January 1982—December 1984 (cf. Table 5.3). The solid line is the un-smoothed spectrum from the Blackman—Tukey autocorrelation method (the cosine transform of the autocovariance function Eqn (5.31b)); dashed line is the un-smoothed spectrum from the fast Fourier transform (FFT) method based on the first 2^5 (=32) data values. Spectral peaks span the annual period ($f = 0.083$/month).

January 1982 to December 1984 for Amphitrite Point (Table 5.3). Since we wish to compare the Blackman—Tukey spectrum in Figure 5.10 with that derived from the data series using a pack-aged FFT routine (the dashed line in Figure 5.10), the lags used to generate the Blackman—Tukey were computed for the first 32 (2^5) points only, which is four fewer points than normally would be used in the Blackman—Tukey approach (see

Section 5.8 for a discussion of FFTs). In this case, artificially extending the lag correlation beyond 10—20% of the data, as recommended earlier, is a necessity if we are to obtain reason-able estimates of the spectra using the autocorre-lation method. As expected, results reveal a strong spectral peak centered near, but not at, the annual frequency ($f = 1.0$ cycles per year = 0.083 cycles per month). There are too

TABLE 5.3   Monthly Mean Sea Surface Temperatures SST (°C) at Amphitrite Point Lightstation (48°55.16′ N, 125°32.17′ W) on the West Coast of Canada for January 1982 through December 1984

YEAR 1982

n	1	2	3	4	5	6	7	8	9	10	11	12
SST	7.6	7.4	8.2	9.2	10.2	11.5	12.4	13.4	13.7	11.8	10.1	9.0

YEAR 1983

n	13	14	15	16	17	18	19	20	21	22	23	24
SST	8.9	9.5	10.6	11.4	12.9	12.7	13.9	14.2	13.5	11.4	10.9	8.1

YEAR 1984

n	25	26	27	28	29	30	31	32	33	34	35	36
SST	7.9	8.4	9.3	9.9	11.0	11.1	12.6	14.0	13.0	11.7	9.8	8.0

few data to enable us to accurately resolve the location of the frequency peak. In the present example, all spectral estimates are positive. However, the autocorrelation method can yield erroneous negative spectra for weak frequency components when there are gaps in the data record.

We emphasize that the spectra in Figure 5.10 have been constructed without any averaging or windowing. This means that each spectral estimate has the minimum possible two DoF (corresponding to the orthogonal sine and cosine components obtained from the Fourier transform) so that the error in each estimate is equal to the value of the estimate itself. Some form of averaging is needed if we are to place confidence limits on the spectra (see Sections 5.4.6 and 5.4.7). The two spectra are slightly different because the record used for the FFT method is shorter than that used for the autocovariance method.

5.4.3.2 The Periodogram Method

The preferred method for estimating the PSD of a discrete sample $\{y_1, y_2,..., y_N\}$ is the direct or periodogram method. Instead of first calculating the autocorrelation function, the data are transformed directly to obtain the Fourier components $Y(f)$ using Eqn (5.27). To help avoid end effects (Gibbs' phenomenon) and wrap-around problems, the original time series can be padded with $K \leq N$ zeroes after the mean has been removed from the time series. The padding will also increase the frequency resolution of the periodogram (see Section 5.4.9). Although use of $K = N$ zeroes is not recommended for computational reasons, it has one advantage: The N-lag covariance function obtained from the IFT of the $2N$-point PSD is identical to the N-lag covariance function (Eqn (5.31b)), as noted previously in Section 5.4.3.1. As with the autocorrelation method, improvements in the statistical reliability of the spectral estimates would be attained by "windowing" the time series prior

to spectral estimation or by averaging the raw periodogram estimates over several adjacent frequency bands (see Sections 5.4.6 and 5.4.7).

The two-sided PSD (or autospectral density) for frequency f in the Nyquist interval $-1/(2\Delta t) \leq f \leq 1/(2\Delta t)$ (i.e., $-f_N \leq f \leq f_N$) and a padding of K zeroes is

$$S_{yy}(f) = \frac{1}{(N+K)\Delta t}\left|\Delta t \sum_{n=0}^{N+K-1} y_n e^{-i2\pi f n\Delta t}\right|^2$$

$$= \frac{1}{(N+K)\Delta t}|Y(f)|^2$$

$$(5.33a)$$

while the one-sided PSD for the positive frequency interval only, $0 \leq f \leq 1/(2\Delta t)$, is

$$G_{yy}(f) = 2S_{yy}(f) = \frac{2}{(N+K)\Delta t}|Y(f)|^2$$

$$(5.33b)$$

Division by Δt transforms the ESD of Eqn (5.29) into a PSD, $S_{yy}(f)$.

Evaluation of Eqn (5.33a) using the FFT defines $Y(f)$ in terms of the DFT estimates, $Y(f_k) = Y_k$, where the f_k forms a discrete set of $(N+K)/2$ equally spaced frequencies $f_k = \pm k/[(N+K)\Delta t]$, $k = 0, 1,..., [(N+K)/2]-1$ in the Nyquist interval, $-1/2\Delta t \leq f_k \leq 1/2\Delta t$. The case $k = 0$ represents the mean component. The two-sided PSD is then

$$S_{yy}(0) = \frac{1}{(N+K)\Delta t}|Y_0|^2, \quad k = 0$$

$$S_{yy}(f_k) = \frac{1}{(N+K)\Delta t}\left[|Y_k|^2 + |Y_{N+K-k}|^2\right],$$

$$k = 1,...,\frac{(N+K)}{2}-1 \qquad (5.34a)$$

$$S_{yy}(f_N) = S_{yy}\left(f_{(N+K)/2-k}\right)$$

$$= \frac{1}{(N+K)\Delta t}\left|Y_{(N+K)/2}\right|^2, \quad k = \frac{(N+K)}{2}$$

and the one-sided PSD is

$$G_{yy}(0) = \frac{1}{(N+K)\Delta t}|Y_0|^2, \quad k = 0$$

$$G_{yy}(f_k) = \frac{2}{(N+K)\Delta t}|Y_k|^2,$$

$$k = 1,\ldots,\frac{(N+K)}{2} - 1 \qquad (5.34b)$$

$$G_{yy}(f_N) = G_{yy}\left(f_{(N+K)/2-k}\right)$$

$$= \frac{1}{(N+K)\Delta t}\left|Y_{(N+K)/2}\right|^2,$$

$$k = \frac{(N+K)}{2}$$

Multiplication of $S_{yy}(f) \equiv S_k$ (or G_k) by the bandwidth of the signal $\Delta f = 1/[(N+K)\Delta t]$ gives the estimated signal variance, σ_k^2, in the kth frequency band; i.e., $\sigma_k^2 = S_k' = S_k\Delta f$. The summation

$$\sum_{n=0}^{N+K-1} S_k' = \sum_{n=0}^{N+K-1} S_k\Delta f \qquad (5.35)$$

gives the variance and total power of the signal. The quantity

$$S_k' = \frac{1}{[(N+K)\Delta t]^2}\left[|Y_k|^2 + |Y_{N+K-k}|^2\right]$$

$$= \frac{1}{(N+K)^2}\sum_{n=0}^{N+K-1}\left|y_n e^{-i2\pi fn\Delta t}\right|^2 \qquad (5.36)$$

is often computed as the periodogram. However, this is not correctly scaled as a PSD but represents the "peak" in the spectral plot rather than the "area" under the plot of S_k vs Δf. The representation Eqn (5.36) is sometimes useful although most oceanographers are more familiar with the PSD form of the periodogram.

It bears repeating that the use of Fourier transforms assumes a periodic structure to the sampled data when no periodic structure may actually exist in the time series. That is, the FFT of a finite length data record is equivalent to

assuming that the record is periodic. We again note that autospectral functions are always real so that $S'_{yy}(f_k) = S'_{yy}(2f_N - f_k)$, and the one-sided autospectral periodogram estimate becomes

$$G'_{yy}(f_k) = 2S_k' = \frac{2}{[(N+K)\Delta t]^2}|Y(f_k)|^2 \quad (5.37)$$

Until the 1960s, the direct transform method first used by Schuster (1898) to study "hidden periodicities" in measured sunspot numbers was seldom used due to difficulties with statistical reliability and extensive computational time. The introduction of the first practical FFT algorithms for spectral analysis (Cooley and Tukey, 1965) greatly reduced the computational time by taking advantage of patterns in DFT functions (see Section 5.8). Problems with the statistical reliability of the spectral estimates are resolved through appropriate windowing and averaging techniques, which we discuss in Sections 5.4.6 and 5.4.7. Figure 5.10 compares the unsmoothed periodogram spectral estimate for the monthly mean SST data at Amphitrite Point (Table 5.3) with the corresponding spectrum obtained from the Blackman—Tukey method. As mentioned earlier, the FFT requires data lengths equal to powers of two so that we have shortened the series to $2^5 = 32$ months. As we found with the Blackman—Tukey autocorrelation method, the FFT spectrum of coastal temperatures has a strong peak near the annual period, albeit with a slightly different spectral amplitude.

5.4.3.3 The PSD for Periodic Data

For a strictly periodic digital time series $y(t)$ having an exact integer number of oscillations over the interval $[0, T]$, we can use the Fourier series expansion Eqn (5.256) and write

$$y(t) = \frac{1}{2}A_0 + \sum_{n=1}^{N}[A_n \cos(\omega_n t) + B_n \sin(\omega_n t)]$$

$$= \frac{1}{2}C_0 + \sum_{n=1}^{N}[C_n \cos(\omega_n t + \phi_n)]$$

$$(5.38)$$

in which the constants A_n, B_n are given by Eqn (5.258) and where

$$C_n = \left(A_n^2 + B_n^2\right)^{1/2}$$

$$\phi_n = \tan^{-1}(B_n/A_n)$$

(5.39)

are the amplitude and phase of the complex Fourier coefficient for the nth frequency component, $\omega_n = 2\pi f_n$. Since the data record contains periodic components only, a plot of $2|C_n|^2$ against n ($n = 0, \ldots, N - 1$) yields a series of distinct "spikes" or line spectra, S_n, with the variance divided equally between negative and positive frequencies

$$S_n = \frac{(\Delta t)^2}{T}\left[|C_n|^2 + |C_{N-n}|^2\right]$$

$$= \frac{2\Delta t}{N}|C_n|^2$$

(5.40)

where the record mean value C_0 has been subtracted from the record $y(t)$. Here we have assumed that $y(t)$ is a real function. The squared Fourier components $|C_n|^2$ give the contribution of the nth frequency component to the total variance and the various frequency components contribute additively to the total power of the time series. The contribution from each component is assumed to be independent of that from all other components.

5.4.3.4 Variance-Preserving Spectra

Because the PSD, $S_{yy}(f)$, and frequency, f, of a time series often range over several orders of magnitude, spectral distributions are usually plotted as the logarithm of $S_{yy}(f)$ vs the logarithm of frequency; i.e., $\log[S_{yy}(f)]$ vs $\log(f)$. This format allows the user to provide a compact presentation of the spectral distribution. The latter is also useful where a spectrum has a power law dependence of the form $S_{yy}(f) \sim f^{-p}$. In this case, the slope of the spectrum is given as $p = -\log[S_{yy}(f)]/\log(f)$. An example of a more narrowly focused format of $\log[S_{yy}(f)]$ vs f (a log–linear plot) is presented in Figure 5.11(a) where we have used time series data generated

by the relation $y(k) = A\cos(2\pi nk/N) + \varepsilon(k)$ from Section 5.4.1 (Figure 5.7). Spectral density has units of energy/frequency for the same units used for f. For example, the PSD of a current velocity record are typically in units of $(cm/s)^2/$ cph or $(cm/s)^2/$cpd plotted against log(frequency) or frequency in cph (cycles per hour) or cpd (cycles per day), respectively. (Sometimes, m/s are used in place of cm/s, and vice versa.)

In the log–linear format, the integration proceeds over frequency bands of width Δf centered at frequency f_c (where the "c", in this case, stands for center of the frequency band), so that the area under each small rectangular segment of the spectral curve is equal to a pseudo-variance

$$\sigma_*^2(f_c) = \int_{f_c - \Delta f/2}^{f_c + \Delta f/2} \log\left[S_{yy}(f)\right]df$$

(5.41)

Although log spectra plots have an appealing shape, the integral Eqn (5.41) is certainly not variance preserving. To preserve the signal variance, $\sigma^2(f_c)$, under the spectral curve, we need to plot $fS_{yy}(f)$ vs $\log(f)$, as in Figure 5.11(b). Replacing df in Eqn (5.41) with $d[\log(f)]$, the true *variance-preserving* form of the spectrum becomes

$$\sigma^2(f_c) = \int_{f_c - \Delta f/2}^{f_c + \Delta f/2} fS_{yy}(f)d[\log(f)]$$

$$= \int_{f_c - \Delta f/2}^{f_c + \Delta f/2} S_{yy}(f)df$$

(5.42)

where we have used the fact that $d[\log(f)] = df/f$. Equations (5.42) gives the true signal variance within the band Δf. In particular, if $S_{yy}(f) \approx S_c$ is nearly constant over the frequency increment Δf, then $\sigma^2(f_c) \approx S_c\Delta f$ is the signal variance in band Δf centered at frequency f_c. In this format,

FIGURE 5.11 Two common types of spectral plot derived for the time series $y(k) = A\cos(2\pi nk/N) + \varepsilon(k)$ (see Figure 5.7). (a) A plot of log power spectral density, log $[S_{yy}(f)]$, vs frequency, f; (b) A variance-preserving plot in which $f[S_{yy}(f)]$ is plotted against $\log(f)$.

there is a clear spectral peak at $f = 0.25$ cycles per unit time that is associated with the term $\cos(2\pi nk/N)$ in the original analytical expression.

5.4.3.5 The Chi-Squared Property of Spectral Estimators

Throughout this chapter, we have claimed that each spectral estimate for maximum frequency resolution, $1/T$, obtained from Fourier transforms of stochastic time series has two DoF. We now present a more formal justification for that claim for discrete spectral estimators by showing that each estimate is a stochastic chi-square (pronounced "ki-square") variable with two DoF (i.e., there are two independent squares entering the expression for the chi-square variable). Consider any stochastic white-noise process $\eta(t)$, for which $E[\eta(t)] = 0$. The Fourier components are

$$A(f) = \sum_{n=-N}^{N-1} \eta(n\Delta t)\cos(2\pi fn\Delta t)$$

$$B(f) = \sum_{n=-N}^{N-1} \eta(n\Delta t)\sin(2\pi fn\Delta t)$$

(5.43)

where as usual, $-1/(2\Delta t) \leq f \leq 1/(2\Delta t)$ is the Nyquist interval, and it follows that $E[A(f)] = 0 = E[B(f)]$. Thus, at the harmonic frequencies $f_k = k/N\Delta t$, the variance is

$$V[A(f_k)] = E[A^2(f_k)] = \sigma_\eta^2 \sum_{n=-N}^{N-1}\cos^2(2\pi f_k n\Delta t)$$

$$= \frac{1}{2}N\sigma_\eta^2, \quad k = \pm 1, \pm 2, \ldots, \pm(N-1)$$

$$= N\sigma_\eta^2, \quad k = 0, -N$$

(5.44a)

Similarly

$$V[B(f_k)] = \frac{1}{2}N\sigma_\eta^2, \quad k = \pm1, \pm2, ..., \pm(N-1)$$

$$= 0, \quad k = 0, -N \qquad (5.44b)$$

When $k \neq j$, the covariance is

$$C[A(f_k), A(f_j)] = \sigma_\eta^2 \sum_{n=-N}^{N-1} \cos(2\pi f_k n \Delta t)$$

$$\times \cos(2\pi f_j n \Delta t) = 0$$

$$(5.45a)$$

and

$$C[A(f_k), B(f_j)] = 0 \quad \text{(orthogonality condition)}$$

$$(5.45b)$$

Because $A(f_k)$ and $B(f_k)$ are linear functions of normal random variables, $A(f_k)$ and $B(f_k)$ are also distributed normally. Hence, the random variables

$$\frac{A^2(f_k)}{V[A(f_k)]} = \frac{2A^2(f_k)}{N\sigma_\eta^2}$$

$$(5.46)$$

$$\frac{B^2(f_k)}{V[B(f_k)]} = \frac{2B^2(f_k)}{N\sigma_\eta^2}$$

are each distributed as χ_1^2, which is a chi-square variable with one DoF.

Since the normal distributions $A(f_k)$ and $B(f_k)$ are independent random variables, the sum of their squares

$$\frac{2}{\sigma_\eta^2}[A^2(f_k) + B^2(f_k)] = \frac{2}{\Delta t \sigma_\eta^2} S_{yy}(f_k) \quad (5.47)$$

is distributed as χ_2^2, which is chi-square variable with two DoF. Here, $S_{yy}(f_k)$ is the sample spectrum. Thus

$$\frac{E[2S_{yy}(f_k)]}{\Delta t \sigma_\eta^2} = 2 \qquad (5.48)$$

and

$$E[S_{yy}(f_k)] = \sigma_\eta^2 \Delta t \qquad (5.49)$$

which is the spectrum. At the harmonic frequencies (set by the record length), the sample spectrum is an unbiased estimator of the white-noise spectrum of $\eta(t)$. Also, at these frequencies, the variance of the estimate is constant and independent of sample size. This explains the failure of the sample estimates of the variance to decrease with increasing sample size. We remark further that, even if $\eta(t)$ is not normally distributed, the random variables $A(f_k)$ and $B(f_k)$ are very nearly normally distributed by the central limit theorem. Hence, the distribution of the $S_{yy}(f)$ will be very nearly distributed as χ_2^2 regardless of the PDF of the $\eta(t)$ process.

5.4.4 Spectra of Vector Series

To calculate the spectra of vector time series such as current and wind, we first need to resolve the data into orthogonal components. Spectral analysis is then applied to the combined series of components and the results stored as a complex quantity in the computer. Raw data are recorded as speed and direction by rotor-type meters and as orthogonal components by acoustic and electromagnetic meters. The usual procedure is to convert recorded time series to an earth-referenced Cartesian coordinate system consisting of two orthogonal horizontal components and a vertical component (cf. Section 4.3.5). In the open ocean, horizontal velocities typically are resolved into components of eastward (zonal; u) and northward (meridional; v) time series, whereas in the coastal ocean it is preferable to resolve the vector components into cross-shore (u') and longshore (v') components through the rotation

$$\begin{pmatrix} u' \\ v' \end{pmatrix} = \begin{pmatrix} \cos\theta & \sin\theta \\ -\sin\theta & \cos\theta \end{pmatrix} \begin{pmatrix} u \\ v \end{pmatrix} \quad (5.50a)$$

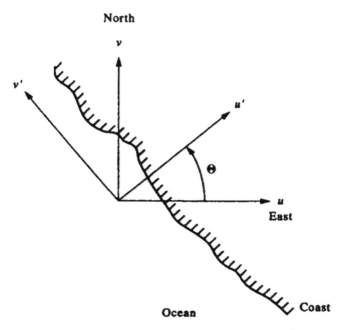

North

FIGURE 5.12 Cross-shore (u') and longshore (v') velocity components in a Cartesian coordinate system rotated through a positive (counterclockwise) angle from the eastward (u) and northward (v) directions.

$$u' = u \cos \theta + v \sin \theta$$
$$v' = -u \sin \theta + v \cos \theta \qquad (5.50b)$$

where the angle θ is the orientation of the coastline (or the local bottom contours) measured counterclockwise from the eastward direction (Figure 5.12). Thus, in the case where the coastline is rotated counterclockwise to lie along a parallel of latitude (i.e., $\theta = \pi/2$), we find $u' = v$ and $v' = -u$. Alternatively, one can let the current velocity observations define θ as the direction of the major axis obtained from principal component analysis; that is, the axis which maximizes the variance in a scatter plot of u vs v (see Figure 4.14).

In coastal regions, the principal axis is usually closely parallel to the coastline. For studies of highly circularly polarized motions, such as inertial waves and tidal currents, resolution into clockwise and counterclockwise rotary components is

often more useful. The choice of representation depends on the preference of the investigator and the type of process being investigated. More is said on this subject in Section 5.4.4.2.

5.4.4.1 Cartesian Component Rotary Spectra

The horizontal velocity vector can be represented in Cartesian coordinates as a complex function $w(t)$ whose real part, $u(t)$, is the projection of the vector on the zonal (or cross-shelf) axis and whose imaginary part, $v(t)$, is the projection of the vector on the meridional (or longshelf) axis (Figure 5.13).

$$w(t) = u(t) + iv(t) \qquad (5.51)$$

(The use of vector $w(t)$ follows the convention of Gonella (1972), Mooers (1973) and others in their discussion of rotary spectral analysis and is not to be confused with the weights $w(t)$, generally

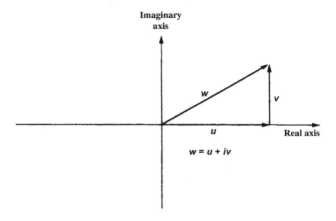

FIGURE 5.13 Horizontal velocity represented as a complex vector, $w = u + iv$, with components (u, v) along the real and imaginary axes, respectively.

written as $w(t_n)$, used in the sections on data windowing, or the vertical component of velocity, w. Gonella (1972) used u_1 and u_2 for the two horizontal velocity components.) A complete description of the time variability of a three-dimensional vector at a single point consists of six functions of frequency: three autospectra for the three velocity components, (u, v, w) and three cross-spectra. For the two-dimensional vectors considered in this section, there are two autospectra and one cross-spectrum. The DFT, $W(f_k) = U(f_k) + iV(f_k)$, $(f_k = k/N\Delta t$, $k = 1, \ldots, N$; $k = 0$ is the mean flow) is

$$W(f_k) = \Delta t \sum_{n=0}^{N-1} w(t)e^{-i2\pi kn/N}$$

$$= \Delta t \sum_{n=0}^{N-1} [u(t) + iv(t)]e^{-i2\pi kn/N} \quad (5.52)$$

where $U(f_k)$ and $V(f_k)$ are the Fourier transforms of $u(t)$ and $v(t)$, respectively. If the original record is separated into M blocks of length N', where $N = MN'$ is the total record length if no overlapping of segments is used, the spectral density function is given in terms of the number of segments used to form the block-averaged, one-sided autospectrum $(0 \leq f_k' < \infty)$

$$G_{ww}(f_k') = \frac{2}{N\Delta t} \sum_{m=1}^{M} |W_m(f_k')|^2$$

$$= \frac{2}{N\Delta t} \sum_{m=1}^{M} \left\{ [W_{Rm}(f_k')]^2 + [W_{Im}(f_k')]^2 \right\}$$

$$= \frac{2}{N\Delta t} \sum_{m=1}^{M} \left\{ [U_{Rm}(f_k') - V_{Im}(f_k')]^2 \right.$$
$$\left. + [U_{Im}(f_k') + V_{Rm}(f_k')]^2 \right\}$$

$$(5.53)$$

where $f_k' = k/N'\Delta t$, $k = 0, 1, \ldots, N'/2$ ($k = 0$ is the mean flow) and for FFT analysis, $N' = 2p$ (positive integer p), and where the subscripts R and I stand for the real and imaginary parts of the given Fourier components.

5.4.4.2 Rotary Component Spectra

Rotary analysis of currents involves the separation of the velocity vector for a specified frequency, ω, into clockwise and counterclockwise rotating circular components with amplitudes A^-, A^+, and relative phases θ^-, θ^+, respectively. Thus, instead of dealing with two Cartesian components (u, v) we deal with two circular components $(A^-, \theta^-; A^+, \theta^+)$. Several reasons

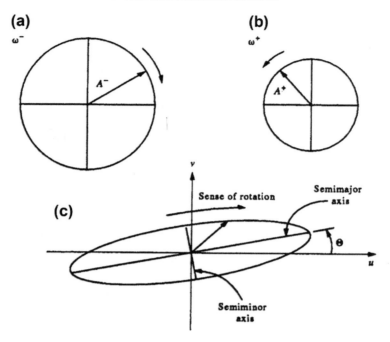

FIGURE 5.14 Current ellipses formed by the vector addition of two oppositely rotating vectors. (a) Clockwise component (ω^-) and (b) counterclockwise component (ω^+) with amplitudes, A^- and A^+, respectively. (c) General case of elliptical motion with major axis tilted at an angle θ counterclockwise from east. ε^- and ε^+ (not shown) are the angles of the two circular components at time $t = 0$.

can be given for using this approach: (1) The separation of a velocity vector into oppositely rotating components can reveal important aspects of the wave field at the specified frequencies. The method has proven especially useful for investigating currents over abrupt topography, wind-generated inertial motions, diurnal frequency continental shelf waves, and other forms of narrow-band oscillatory flow; (2) in many cases, one of the rotary components (typically, the clockwise component in the northern hemisphere and counterclockwise component in the southern hemisphere) dominates the currents so that we need to deal with one scalar quantity rather than two. Inertial motions, for example, are almost entirely clockwise (counterclockwise) rotary in the northern (southern) hemisphere so that the counterclockwise

(clockwise) component can be ignored for most applications; and (3) many of the rotary properties, such as spectral energy $S^-(\omega)$ and $S^+(\omega)$ and rotary coefficient, $r(\omega)$, are invariant under coordinate rotation so that local steering of the currents by bottom topography or the coastline are not factors in the analysis.

The vector addition of the two oppositely rotating circular vectors (Figure 5.14(a) and (b)) causes the tip of the combined vector (Figure 5.14(c)) to trace out an ellipse over one complete cycle. The eccentricity, e, of the ellipse is determined by the relative amplitudes of the two rotary components. Motions at frequency ω are circularly polarized if one of the two components is zero; motions are rectilinear (back-and-forth along the same line) if both circularly polarized components have the same magnitude. In rotary spectral format, the current vector $w(t)$ can be written as the Fourier series

$$w(t) = \overline{u(t)} + \sum_{k=1}^{N} U_k \cos(\omega_k t - \phi_k)$$

$$+ i\left[\overline{v(t)} + \sum_{k=1}^{N} V_k \cos(\omega_k t - \theta_k)\right]$$

$$= \left[\overline{u(t)} + i\overline{v(t)}\right] + \sum_{k=1}^{N}[U_k \cos(\omega_k t - \phi_k)$$

$$+ iV_k \cos(\omega_k t - \theta_k)]$$

$$(5.54)$$

in which $\overline{u(t)} + i\overline{v(t)}$ is the mean velocity, $\omega_k = 2\pi f_k = 2\pi k/N\Delta t$ is the angular frequency, t ($=n\Delta t$) is the time, and (U_k, V_k) and (ϕ_k, θ_k) are the amplitudes and phases, respectively, of the Fourier constituents for each frequency for the real and imaginary components. Subtracting the mean velocity and expanding the trigonometric functions, we find

$$w'(t) = w(t) - \left[\overline{u(t)} + i\overline{v(t)}\right]$$

$$= \sum_{k=1}^{N}\{U_{1k} \cos(\omega_k t) + U_{2k} \sin(\omega_k t)$$

$$+ i[V_{1k} \cos(\omega_k t) + V_{2k} \sin(\omega_k t)]\}$$

$$(5.55)$$

in which we have defined the even (U_{1k}, V_{1k}) and odd $(U_{2k}, V_{2k}.)$ functions as

$$U_{1k} = U_k \cos\phi_k, \quad U_{2k} = U_k \sin\phi_k \quad (5.56a)$$
$$V_{1k} = V_k \cos\phi_k, \quad V_{2k} = V_k \sin\theta_k \quad (5.56b)$$

Dropping the prime notation for $w'(t)$ and following some reorganization, we can write the kth frequency component of the series as the sum of counterclockwise $(+)$ and clockwise $(-)$ components

$$w_k(t) = w_k^+(t) + w_k^-(t)$$

$$= A_k^+ \exp(i\varepsilon_k^+)\exp(i\omega_k t) + A_k^- \exp(i\varepsilon_k^-)\exp(-i\omega_k t)$$

$$= \exp\left[\frac{i(\varepsilon_k^+ + \varepsilon_k^-)}{2}\right]\left\{[A_k^+ + A_k^-]\cos\left[\frac{\varepsilon_k^+ - \varepsilon_k^-}{2} + \omega_k t\right]\right.$$

$$\left. + i[A_k^+ - A_k^-]\sin\left[\frac{\varepsilon_k^+ - \varepsilon_k^-}{2} + \omega_k t\right]\right\}$$

$$(5.57)$$

where the counterclockwise and clockwise rotary component amplitudes are given by

$$A_k^+ = \frac{1}{2}\left\{[(U_{1k} + V_{2k})]^2 + [(U_{2k} - V_{1k})]^2\right\}^{1/2}$$

$$(5.58a)$$

$$A_k^- = \frac{1}{2}\left\{[(U_{1k} - V_{2k})]^2 + [(U_{2k} + V_{1k})]^2\right\}^{1/2}$$

$$(5.58b)$$

and the corresponding phase angles for time $t = 0$, by

$$\varepsilon_k^+ = \tan^{-1}[(V_{1k} - U_{2k})/(U_{1k} + V_{2k})] \quad (5.59a)$$

$$\varepsilon_k^- = \tan^{-1}[(U_{2k} + V_{1k})/(U_{1k} - V_{2k})] \quad (5.59b)$$

Each of the constituents contributing to Eqn (5.55) has the form of an ellipse with major semiaxis of length $L_M = (A_k^+ + A_k^-)$ and minor semiaxis of length $L_m = |A_k^+ - A_k^-|$ (Figure 5.14(c)). The ellipse is tilted at an angle of $\theta = \frac{1}{2}(\varepsilon_k^+ + \varepsilon_k^-)$ from the u-axis and the vector is along the major axis of the ellipse at time $t = (\varepsilon_k^+ - \varepsilon_k^-)/(4\pi f_k)$. The one-sided spectra $(G_k^+, G_k^-) = (S_k^+, S_k^-)$ for the two oppositely rotating components for frequencies $f_k = \omega_k/2\pi$ are

$$S\left(f_k^+\right) = S_k^+ = \frac{(A_k^+)^2}{N\Delta t}, \quad f_k = 0, \ldots, 1/(2\Delta t)$$

$$(5.60a)$$

$$S\left(f_k^-\right) = S_k^- = \frac{(A_k^-)^2}{N\Delta t}, \quad f_k = -1/(2\Delta t), \ldots, 0$$

$$(5.60b)$$

Plots of rotary spectra are generally presented in two ways. In Figure 5.15(a), both S^- and S^+ are plotted as functions of frequency magnitude, $|f| \geq 0$, with solid and dashed lines used for the clockwise and counterclockwise spectra, respectively. In Figure 5.15(b), we use the fact that clockwise spectra are defined for negative frequencies and counterclockwise spectra for positive frequencies. The spectra $S(f_k^+)$ and $S(f_k^-)$

FIGURE 5.15 Rotary current spectra for hourly currents measured at 40-m depth in the Beaufort Sea, Arctic Ocean (water depth = 170 m). Peaks are at the diurnal (D) and semidiurnal (SD) tidal frequencies. Frequency resolution is 0.0005 cph and there are 112 degrees of freedom (DoF) per spectral band. Vertical bar gives the 99% level of confidence, (a) One-sided rotary spectra, $S^{-}(f)$ and $S^{+}(f)$, vs f for positive frequency, f; (b) Two-sided rotary spectra, $S(f_{k}^{+}) = S^{+}$ and $S(f_{k}^{-}) = S^{-}$ vs log f for positive and negative frequencies, $f_{\pm k}$. *(Courtesy E. Carmack, A. Rabinovich, and E. Kulikov.)*

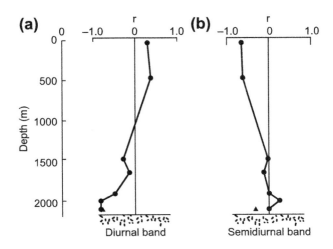

FIGURE 5.16 Rotary coefficient, $r(\omega)$, as a function of depth for current oscillations in (a) the diurnal frequency band ($\omega/2\pi \approx 0.04$ cph) and (b) the semidiurnal band ($\omega/2\pi \approx 0.08$ cph). *(From Allen and Thomson (1993).)*

used in Figure 5.15(a) are then plotted on opposite sides of zero frequency. In these spectra, peak energy occurs at the diurnal and semidiurnal periods. The predominantly clockwise rotary motions at semidiurnal periods suggest a combination of tidal and near-inertial motions (at this latitude the inertial period is close to the semidiurnal tidal period).

Another useful property is the rotary coefficient

$$r(\omega) = \frac{S_k^+ - S_k^-}{S_k^+ + S_k^-} \qquad (5.61)$$

which ranges from $r = -1$ for clockwise motion, to $r = 0$ for unidirectional flow, to $r = +1$ for counterclockwise motion. The rotary nature of the flow can change considerably with position, depth, and time. As indicated by Figure 5.16, the observed diurnal tidal currents over Endeavour Ridge in the northeast Pacific change from moderately positive to strongly negative rotation with depth. In contrast, the semidiurnal currents change from strongly negative near the surface to strongly rectilinear at depth. (Data, in this case, are from a string

of current meters moored for a period of 9 months.) We remark that the definition Eqn (5.61) differs in sign from that of Gonella (1972), who used $S_k^- - S_k^+$ rather than $S_k^+ - S_k^-$ in the numerator. Because many types of oceanic flow are predominantly clockwise rotary in the northern hemisphere, Gonella's definition has the advantage that clockwise rotating currents have positive rotary coefficients. However, we find Gonella's definition a bit awkward since clockwise motions, which are linked to *negative* frequencies, then have *positive* rotary coefficients.

5.4.4.3 Rotary Spectra (via Cartesian Components)

Gonella (1972) and Mooers (1973) present the rotary spectra in terms of their Cartesian counterparts and provide a number of rotational invariants for analyzing current and wind vectors at specified frequencies. Specifically, the one-side autospectra for the counterclockwise (CCW) and clockwise (CW) rotary components of the vector $w(t) = u(t) + iv(t)$ are, in terms of their Cartesian components

$$G\left(f_k^+\right) = \frac{1}{2}\left[G_{uu}\left(f_k\right) + G_{vv}\left(f_k\right) + Q_{uv}\left(f_k\right)\right],$$

$$f_k \geq 0 \ (CCW \ component) \qquad (5.62a)$$

$$G\left(f_k^-\right) = \frac{1}{2}\left[G_{uu}\left(f_k\right) + G_{vv}\left(f_k\right) - Q_{uv}\left(f_k\right)\right],$$

$$f_k \leq 0 \ (CW \ component) \qquad (5.62b)$$

where $G_{uu}(f_k)$ and $G_{vv}(f_k)$ are the one-sided auto-spectra of the u and v Cartesian components of velocity and $Q_{uv}(f_k)$ is the quadrature spectrum between the two components, where

$$Q_{uv}\left(f_k\right) = -Q_{uv}\left(-f_k\right) = \left(U_{1k}V_{2k} - V_{1k}U_{2k}\right)$$

$$(5.63)$$

As defined in Section 5.6, the spectrum can be written in terms of cospectrum (real part) and quadrature spectrum (imaginary part)

$$G_{uv}\left(f_k\right) = C_{uv}\left(f_k\right) - iQ_{uv}\left(f_k\right) \qquad (5.64)$$

5.4.5 Effect of Sampling on Spectral Estimates

Spectral estimates derived by conventional techniques are limited by two fundamental problems: (1) the finite length, T, of the time series; and (2) the discretization associated with the sampling interval, Δt. The first problem is inherent to all real data sets while the second is associated with finite instrument response times and/or the need to digitize the time series for the purposes of analysis.

Irrespective of the method used to calculate the power spectrum of a waveform, the record duration, $T = N\Delta t$, and sampling increment, Δt, impose severe limitations on the information that can be extracted. Ideally, we would like to have sensors that can sample rapidly enough (small Δt) that no significant frequency component goes unresolved. This also eliminates aliasing problems in which unresolved spectral energy at frequencies higher than the Nyquist frequency is folded back into lower frequencies.

At the same time, we wish to record for a sufficiently long period (large N) that we capture many cycles of the lowest frequency of interest. Long-term sampling also enables us to better resolve frequencies that are close together and to improve the statistics (confidence intervals) for spectral estimates. In reality, most data series are a compromise based on the frequencies of interest, the response limitations of the sensor, and cost. The choices of the sampling rate and the record duration are tailored to best meet the task at hand.

5.4.5.1 Effect of Finite Record Length

As noted earlier, we can think of a data sample $\{y(t)\}$ of duration $T = N\Delta t$ as the output from an infinite physical process $\{y'(t)\}$ viewed through a finite length window (Figure 5.3). The window has the shape of a "box-car" function, $w(t_n) = w_n = w(n\Delta t)$, which has unit amplitude and zero phase lag over the duration of the data sequence but is zero elsewhere. That is $y(t_n) = w(t_n) \cdot y'(t_n)$ where

$$\begin{aligned} w_n &= 1, \quad n = 0, ..., N-1 \\ w_n &= 0, \quad \text{for } n \geq N, \ n < 0 \end{aligned} \qquad (5.65)$$

Since it is truncated, the data set has endpoint discontinuities, which lead to Gibbs' phenomena ("ringing") and ripple effects in the frequency domain. The DFT $Y(f)$ of the truncated series $y_n = y(n\Delta t)$ is

$$Y(f) = \sum_{n=-\infty}^{\infty} w_n y'_n e^{-i2\pi f n \Delta t} \qquad (5.66)$$

In frequency space, $Y(f)$ is the convolution (written as *) of the Fourier transform of the infinite data set, $Y'(f)$, with the Fourier transform $W(f)$ of the function $w(t)$. That is

$$Y(f) = \int_{-\infty}^{\infty} Y'(f')W'(f-f')df' \qquad (5.67)$$

$$\equiv Y'(f) * W(f)$$

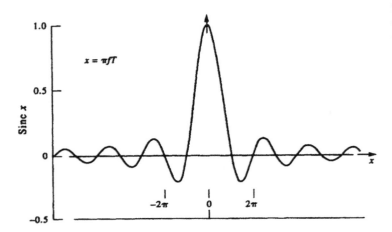

FIGURE 5.17 The function sinc$(x) = \sin(x)/x$ showing the large side-lobes, which are responsible for leakage of spectral power from a given frequency to adjacent frequencies.

where for a box-car function

$$W(f) = T \exp (i\pi fT) \frac{\sin (\pi fN\Delta t)}{(\pi fN\Delta t)}$$

$$\equiv T \exp (i\pi fT) \text{ sinc } (\pi fN\Delta t)$$

(5.68)

and sinc$(x) \equiv \sin(x)/x$. It is the large side-lobes or ripples of the sinc function (Figure 5.17), which are responsible for the leakage of spectral power from the main frequency components into neighboring frequency bands (and vice versa). In particular, $Y(f)$ for a specific frequency $f = f_0$ is spread to other frequencies according to the phase and amplitude weighting of the window function. Leakage has the effect of both reducing the spectral power in the central frequency component and contaminating it with spectral energy from adjacent frequency bands. Those familiar with the various mathematical forms for the Dirac delta function, $\delta(f)$, will recognize the formulation

$$\delta(f) = \lim_{f \to 0} \left[\frac{\sin(\pi f\Delta t)}{\pi f\Delta t} \right] = \lim_{f \to 0} [\text{sinc}(\pi f\Delta t)]$$

Thus, as the frequency resolution increases (i.e., $f \to 0$), $Y(f) \to Y'(f)$.

In addition to distorting the spectrum, the box-car window limits the frequency resolution of the periodogram, independently of the data. The convolution $Y'(f)*W(f)$ means that the narrowest spectral response of the resultant transform is confined to the main-lobe width of the window transform. For a given window, the main-lobe width (the width between the -3 dB $= 10 \log(\frac{1}{2})$ levels of the main lobe) determines the frequency resolution, Δf, of a particular window. For most windows, including the box-car window, this resolution is roughly the inverse of the observation time; $\Delta f \approx 1/T = 1/(N\Delta t)$.

5.4.5.2 Aliasing

Poor discretization of time series data due to limitations in the response time of the sensor, limitations in the recording and data storage rates, or through postprocessing methods may cause *aliasing* of certain frequency components in the original waveform (Figure 5.18(a)). An aliased frequency is one that masquerades as another frequency. In Figures 5.18(b), for example, the considerable tidal energy at diurnal and semidiurnal periods (1 and 2 cpd) that is well resolved by the hourly sampled record is folded back to lower frequencies of roughly 0.065, 0.10, and 0.15 cpd (periods of 14.8, 9.6, and 7.4 days, respectively) when the original sea-level record is subsampled at daily

FIGURE 5.18 The origin of aliasing. (a) The solid line is the tide height recorded at Victoria, British Columbia over a 60-day period from July 29 to September 27, 1975 (time in Julian days). The diamonds are the sea-level values one would obtain by only sampling once per day. (b) The power spectrum obtained from the two data series in (a). In this case, the high frequency energy (dashed curve) gets folded back into the spectrum at lower (aliased) frequencies (solid curve).

($\Delta t = 24$ h) intervals. The aliased signals are no-where near the original higher frequency tidal signals. If we knew nothing about the true spectrum, and were presented only with the aliased spectrum in Figure 5.18(b), we would be hard pressed to provide a physical explanation for the strong fortnightly and weather-band cycles in the sea-level time series.

As illustrated by Figure 5.18, it becomes impossible, for a specific sampling interval, to tell with certainty which frequency out of a large number of possible aliases is actually contributing to the signal variability. This leads to differences in the spectra between the continuous and discrete time series. Since we use the spectra of the discrete series to estimate the spectrum of the continuous series, the sampling interval must be properly selected to minimize the effect of the aliasing. If we know from previous analysis that there is little likelihood of significant energy at the disguised frequencies, then aliasing is not a problem. Otherwise, a degree of smoothing may be required to ensure that higher frequencies do not contaminate the lower frequencies. This smoothing must be performed prior to sampling or digitizing since aliased contributions cannot be recognized once they are present in the discrete data series.

The aliasing problem can be illustrated in a number of ways. To begin with, we note that for discrete data at equally spaced intervals, Δt,

we can measure only those frequency components lying within the principal frequency range,

$$-\omega_N \leq \omega \leq -\omega_0, \quad \omega_0 \leq \omega \leq \omega_N, \quad \omega_N \geq 0$$
(5.69a)

$$-f_N \leq f \leq -f_0, \quad f_0 \leq f \leq f_N, \quad f_N \geq 0 \quad (5.69b)$$

in which $\omega_N = \pi/\Delta t$ and $f_N = 1/(2\Delta t)$ are the usual Nyquist frequencies in radians and cycles per unit time, respectively, and $\omega_0 = 2\pi/T$ and $f_0 = 1/T$ are corresponding fundamental frequencies for a time series of duration T. The Nyquist frequency is the highest frequency that can be extracted from a time series having a sampling rate of $1/\Delta t$. Clearly, if the original time series has spectral power at frequencies for which $|f| \geq f_N$, these spectral contributions are unresolved and will contaminate power associated with frequencies within the principal range (Figure 5.19). The unresolved variance becomes lumped together with other frequency components. Familiar examples of aliasing are the slow reverse rotation of stage-coach wheels in classic western movies due to the undersampling by the frame rate of the movie camera. Even in modern TV commercials or movies, distinguishable features on moving automobile tires or wheel frames often can be seen to rotate rapidly backwards, slow to a stop, then turn forward at the correct rotation speed as the vehicle gradually comes to a stop. Automobile commercials can

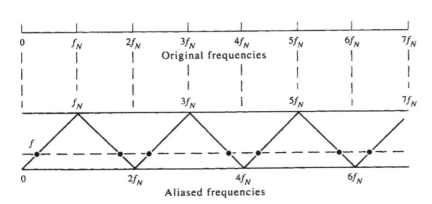

FIGURE 5.19 The spectral energies of all frequencies, $f = \omega/2\pi$, at the nodes (•) located along the dotted line are folded back to the left, accordion style, into the spectral estimate for the spectrum, $S(f)$ for the principal range, $0 \leq f \leq f_N$ ($0 \leq \omega \leq \omega_N$). *(Adapted from Bendat and Piersol (1986).)*

avoid this problem by equipping the wheels with featureless hubcaps, spokes, and tires– or by digitizing to a higher resolution.

If $\omega, f \geq 0$ are frequencies inside the principal intervals (Eqn (5.69)); the frequencies outside the interval, which form aliases with these frequencies are (in sequence)

$$2\omega_N \pm \omega, 4\omega_N \pm \omega, ..., 2p\omega_N \pm \omega \quad (5.70a)$$
$$2f_N \pm f, 4f_N \pm f, ..., 2pf_N \pm f \quad (5.70b)$$

where p is a positive integer. These results lead to the alternate term *folding* frequency for the Nyquist frequency since spectral power outside the principal range is folded back, accordion style, into the principal interval. As illustrated by Figure 5.19, folding the power spectrum about f_N produces aliasing of frequencies $2f_N - f$ with frequencies f; folding the spectrum at $2f_N$ produces aliasing of frequencies $2f_N + f$ with frequencies $2f_N - f$, which are then folded back about f_N into frequency f, and so forth. For example, if $f_N = 5$ rad/h, the observations at 2 rad/h are aliased with spectral contributions having frequencies of 8 and 12 rad/h, 18 and 22 rad/h, and so on.

We can verify that oscillations of frequency $2p\omega_N \pm \omega$ (or $2pf_N \pm f$) are indistinguishable from frequency ω (or f) by considering the data series $x_\omega(t)$ created by the single frequency component $x_\omega(t) = \cos(\omega t)$. Using the transformation $\omega \rightarrow (2p\omega_N \pm \omega)$, together with $t_n = n\Delta t$ and $\omega_N = \pi/\Delta t$, yields

$$\begin{aligned}
x_\omega(t_n) &= \cos\left[(2p\omega_N \pm \omega)t_n\right] \\
&= \text{Re}\left\{\exp[i(2p\omega_N \pm \omega)t_n]\right\} \\
&= \text{Re}\left\{\exp[i2p\omega_N t_n]\exp\left[\pm i\omega t_n\right]\right\} \quad (5.71) \\
&= (+1)^{pn} \text{Re}\left[\exp\left(\pm i\omega t_n\right)\right] \\
&= \cos\left(\omega t_n\right) = x_\omega(t_n)
\end{aligned}$$

In other words, the spectrum of $x(t)$ at frequency ω will be a superposition of spectral contributions from frequencies ω, $2p\omega_N \pm \omega$, $4p\omega_N \pm \omega$, and so forth. More specifically, it can

be shown that the aliased spectrum $S_a(\omega)$ for discrete data is given by

$$S_a(\omega) = \sum_{n=-\infty}^{\infty} S(\omega + 2n\omega_N) \quad (5.72a)$$
$$= S(\omega) + \sum_{n=1}^{\infty}[S(2n\omega_N - \omega) + S(2n\omega_N + \omega)] \quad (5.72b)$$

The true spectrum, S, gives the distorted spectrum, S_a, caused by the summation of overlapping copies of measured spectra in the principal interval. Only if the original record is devoid of spectral power at frequencies outside the principal frequency range will the spectrum of the observed record equal that of the actual oceanic variability. To avoid aliasing problems, one has no choice but to sample the data as frequently as justifiably possible (i.e., up to frequencies beyond which energy levels become small) or to filter the sampled data before they are recorded (as in the case of a stilling well used to eliminate gravity waves from a tidal record). A further example of spectral contamination by aliased frequencies is illustrated in Figure 5.20(a) and (b). In Figure 5.20(b), we have assumed that the wave recorder was inadvertently programed to record at 0.13 Hz, corresponding to a limiting wave period of 7.69 s. The energy from the shorter-period waves was not measured but contaminate the energy of the longer-period waves when folded back about the Nyquist frequency.

5.4.5.3 Nyquist Frequency Sampling

Sampling time series that has significant variability at the Nyquist frequency affords its own set of problems. Suppose we wish to represent $y(t)$ through the usual Fourier relation

$$y(t) = \int_{-\omega_N}^{\omega_N} Y(\omega)e^{i\omega t}d\omega \quad (5.73)$$

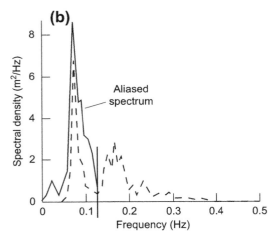

FIGURE 5.20 An aliased autospectrum. (a) The true spectrum, $S(f)$ (m²/cps), of wind-generated waves as a function of frequency (Hz = cycles per second); (b) Aliased spectrum, $S_a(f)$, that would arise from folding about a hypothetical Nyquist frequency, $f_N = 0.13$ Hz.

where we have assumed that $Y(\omega) = 0$ for $|\omega| > \omega_N$. In this case, there is no aliasing problem since there is no power at frequencies greater than w_N. The function $y(t)$ can be constructed from frequency components strictly in the interval $(-\omega_N, \omega_N)$. In discrete form for infinite length data

$$y(t) = \frac{1}{2\omega_N} \sum_{n=-\infty}^{\infty} \left[y_n \int_{-\omega_N}^{\omega_N} e^{i\omega(t-n\Delta t)} d\omega \right]$$

(5.74a)

where the integral has the form of a sinc function such that

$$y(t) = \sum_{n=-\infty}^{\infty} y_n \frac{\sin\left[\omega_N(t - n\Delta t)\right]}{\omega_N(t - n\Delta t)}$$

(5.74b)

Given the data $\{y_n\}$, we can construct $y(t)$. However, suppose that $y(t)$ fluctuates with the Nyquist frequency ω_N such that

$$y(t) = y_o \cos(\omega_N t + \theta)$$

(5.75)

where, for the sake of generality, the phase angle is arbitrary, $0 \leq \theta \leq 2\pi$. Then, using $\sin(n\pi) = 0$ for all n (an integer)

$$\begin{aligned} y_n = y(n\Delta t) &= y_o \cos(n\pi + \theta) \\ &= y_o[\cos(n\pi)\cos\theta] \\ &= y_o(-1)^n \cos\theta \end{aligned}$$

(5.76)

This leads to a component with amplitude $y_n = y_o(-1)^n \cos\theta$, which fluctuates in sign because of the term $(-1)^n$, $-\infty \leq n \leq \infty$. If θ is unknown, the function $y(t)$ cannot be constructed. If $\theta = k\pi/2$, so that $\cos(\omega_N t + \theta) = \sin(\omega_N t)$, the observer will find no signal at all. In general, $0 \leq |\cos\theta| \leq 1$ and the magnitude will always be less than y_0, resulting in biased data.

According to the above analysis, we should sample slightly more frequently than Δt if we are to fully resolve oscillations at the maximum frequency of interest (assumed to be the Nyquist frequency). A sampling rate of 2.5 samples per cycle of the frequency of interest appears to be acceptable whereby $\Delta t = 1/(2.5 f_N) = (2/5)(1/f_N) = (4/5)\pi/\omega_N$.

5.4.5.4 Frequency Resolution

The need to resolve spectral estimates in neighboring frequency bands is an important requirement of time series analysis. Without sufficient resolution, it is not possible to determine whether a given spectral peak is associated with a single frequency, or is a smeared response containing a number of separate spectral peaks. A good example of this for tides is presented by Munk and Cartwright (1966), who show that for long records, the main constituents in the diurnal and semidiurnal frequency bands can be resolved into a multitude of other tidal frequencies. How well the peaks can be resolved depends on the frequency differences, Δf, between the peaks and the length, T, of the data set used in the analysis. For an unsmoothed periodogram, the frequency resolution in hertz is roughly the reciprocal of the time duration in seconds of the data.

The distinction between well-resolved and poorly resolved spectral estimates is somewhat subjective and depends on how we wish to define "resolution." As illustrated by diffraction patterns in classical optics, we can follow the "Rayleigh criterion" for the separation of spectral peaks (Jenkins and White, 1957). Recall that the diffraction pattern for a given frequency, f, of light varies as $\mathrm{sinc}(\phi) = \sin [(\phi - \phi_f)]/(\phi - \phi_f)$, where ϕ is the angle of the incident light beam to the grating. This also is the functional form for the spectral peak of a truncated time series (see *windowing* in the next section). Two spectral lines are said to be "well resolved" if the separation between peaks exceeds the difference in frequency between the center frequency to the maximum at the first side-lobe and "just resolved" if the spectral peak of one pattern coincides with the first zero of the second pattern (Figure 5.21(a)–(c)). Here, the separation in frequency is equal to the difference in frequency between the peak of one spectrum and the first zero of the function $\sin(\phi)/\phi$ of the second (where $\phi = \omega T/2$). The spectral peaks are "not resolved" if this separation is less than that between the center frequency and the first zero of the $\sin(\phi)/\phi$ functions (Figure 5.21(d)).

Consider an oceanic record consisting of two sinusoidal components, both having amplitude y_o and constant phase lags such that

$$y(t) = y_o[\cos (\omega_1 t + \theta_1) + \cos (\omega_2 t + \theta_2)],$$
$$-T/2 \leq t \leq T/2 \tag{5.77}$$

where as usual $\omega = 2\pi f$. The one-sided, unsmoothed PSD, $S(\omega)$, for these data is then found from the Fourier transform

$$S(\omega) = \frac{1}{2} T y_o^2 \left\{ \frac{\sin\left[\frac{1}{2}T(\omega - \omega_1)\right]}{\left[\frac{1}{2}T(\omega - \omega_1)\right]} + \frac{\sin\left[\frac{1}{2}T(\omega - \omega_2)\right]}{\left[\frac{1}{2}T(\omega - \omega_2)\right]} \right\}$$

The power spectrum consists of two terms of the form $\sin(\phi)/\phi$ centered at frequencies ω_1 and ω_2. Using the Rayleigh criterion, we can just resolve the two peaks (i.e., determine if there is one or two sinusoids contributing to the spectrum) provided that the frequency separation $\Delta\omega = |\omega_1 - \omega_2|$ ($\Delta f = |f_1 - f_2|$) is equal to the frequency difference for the peak of one frequency and the first zero of $\sin(\phi)/\phi$ for the other frequency. Since zeroes of $\sin(\phi)/\phi$ occur at frequencies f equal to $\pm 1/T$, $\pm 2/T$, ..., $\pm p/T$, the frequencies are just resolved when

$$\Delta\omega = \frac{2\pi}{T}; \quad \Delta f = \frac{1}{T} \tag{5.78a}$$

and well resolved for

$$\Delta\omega > \frac{3\pi}{T}; \quad \Delta f > \frac{3}{2T} \tag{5.78b}$$

In summary, resolution of two frequencies f_k and $f_{k+1}(=f_k \pm \Delta f)$ using an unsmoothed periodogram or equivalently a rectangular window, requires a record of length T, where $\Delta f = 1/T$ frequency units. Note also that $1/T$ is equal to the fundamental frequency, f_1, which is the lowest frequency that we can calculate for the record. For some nonrectangular windows, the length of the data set must be increased to about $2T = 2/\Delta f$ to achieve the same frequency separation.

(a) $\Delta f = \frac{2}{T}$

(b) $\Delta f = \frac{3}{2T}$

(c) $\Delta f = \frac{1}{T}$

(d) $\Delta f = \frac{1}{2T}$

FIGURE 5.21 Resolution of spectral lines. (a, b) Well resolved; (c) just resolved; and (d) not resolved. *(From Jenkins and White (1957).)*

In a related study, Munk and Hasselman (1964) discuss the "super-resolution" of tidal frequency variability. The fact that time series of tidal heights vary at precise frequencies and have relatively large SNRs suggests that the traditional requirement (that a minimum record length T is required to separate tidal constituents separated by frequency difference $\Delta f = 1/T$) is "grossly incomplete." The modified resolvable frequency difference is

$$\Delta f = \frac{1}{rT}; \quad \Delta\omega = \frac{2\pi}{rT} \quad (5.79)$$

in which $r \equiv$ (signal level/noise level)$^{1/2}$. On this basis, the Rayleigh criterion must be considered a conservative measure of the resolution requirement for deterministic processes.

5.4.6 Smoothing Spectral Estimates (Windowing)

The need for statistical reliability of spectral estimates brings us to the topic of spectral averaging or smoothing. As we have seen, DFTs (Discrete Fourier Transforms) provide an elegant method for decomposing a data sequence into a set of discrete spectral estimates. For a data sequence of N values, the periodogram estimate of the spectrum can have a maximum of $N/2$ Fourier components. If we use all $N/2$ components to generate the periodogram, there are only two DoF per spectral estimate, corresponding to the coefficients A_n, B_n of the sine and cosine functions for each Fourier component (see Sections 5.4.3.1 and 5.4.3.5) or, alternatively, to the magnitude and phase of each Fourier component (see Section 5.4.3.3). Based on the assumption that data are drawn from a normally distributed random sample, we can define the confidence limits for the spectrum in terms of a chi-squared distribution, χ_n^2, where for n DoF

$$E[\chi_n^2] = \mu^2 = n, \quad E[(\chi_n^2 - \mu^2)] = \sigma^2 = 2n \quad (5.80)$$

Substituting $n = 2$ into these expressions, we find that the standard deviation, σ, is equal to the mean, μ, of the estimate, indicating that results based on two DoF are not statistically reliable. It is for this reason that some sort of ensemble averaging or smoothing of spectral estimates is required. The smoothing can be (1) applied directly to the time series through convolution with a sliding averaging function or by (2) averaging adjacent spectral estimates. A one-shot smoothing applied to the entire data record marginally increases the number of DoF per spectral estimate. In most practical applications, the full time series is broken into a series of short overlapping segments and smoothing is applied to each of the overlapping segments. The analyst then ensemble averages the smoothed spectra from each segment to increase the number of DoF per spectral estimate. The greater the smoothing, the greater the number of DoF per spectral band, the narrower the confidence limits, and the greater the reliability of any observed spectral peaks. The trade-off is a longer processing time and a loss of spectral resolution that can remove smaller peaks that may or may not be indicative of real processes (see Figure 5.22).

A window is a smoothing function applied to finite observations or their Fourier transforms to minimize "leakage" in the spectral domain. Convolution in the time domain and multiplication in the frequency domain are adjoint Fourier functions (see Appendix G regarding convolution). A practical window is one which allows little of the energy in the main spectral lobe to leak into the side-lobes, where it can obscure and distort other spectral estimates that are present. In fact, weak signal spectral responses can be masked by higher side-lobes from stronger spectral responses. Skillful selection of tapered data windows can reduce the side-lobe leakage, although always at the expense of reduced resolution. Thus, we want a window that minimizes the side-lobes and maximizes (concentrates) the energy near the frequency of interest in the main lobe. These two performance limitations are rather troublesome when analyzing short data records. Short data occur in practice because many measured processes are event-like (of short duration) or have slowly time-varying spectra that may be considered constant over only short record segments. The window is applied to data to reduce the order of the discontinuity at the boundary of the periodic extension since few harmonics will fit exactly into the length of the time series.

Signals with frequencies other than those of the basis set are not periodic in the observation window. The periodic extension of a signal, not commensurate with its natural period, exhibits discontinuities at the boundaries of the observational period. Such discontinuities are responsible for spectral contributions or leakage over the entire basis set. In the time domain, the windows are applied to the data as a multiplicative weighting (convolution) to reduce the order of the discontinuities at the boundary of the periodic extensions. The windowed data are brought to zero smoothly at the boundaries so that the periodic extensions of the data are continuous in many orders of the derivatives. The value of $Y(f)$ at a particular frequency f, say f_o, is the sum of all the spectral contributions at each f weighted by the window centered at f_o and measured at f

$$Y(f) = Y'(f) * W(f) \qquad (5.81)$$

There exist a multitude of data windows or tapers with different shapes and characteristics ranging from the rectangular (box-car) window discussed in the previous section, to the classic Hanning and Hamming windows, to more sophisticated windows such as the Dolph–Chebyshev window. The type of window used for a given application depends on the required degree of side-lobe suppression, the allowable widening of the central lobe, and the amount of computing one is willing to endure. We will briefly discuss several of the conventional windows plus the Kaiser–Bessel window recommended by Harris (1978). Additional details on the Kaiser–Bessel window and filter are provided in Section 6.9.

5.4.6.1 Desired Window Qualities

Windows affect the attributes of a given spectral analysis method, including its ability to detect and resolve periodic waveforms, its

FIGURE 5.22 Spectra of sea-level oscillations recorded by a bottom-pressure gauge in Malokurilsk Bay on the west coast of Shikotan Island, Russia. Time series length, $T = N\Delta t$, where $N = 20{,}160$ and $\Delta t = 1$ min. Segment lengths are $T_s = M\Delta t$, $M \ll N$. Each time series segment has been smoothed with a Kaiser–Bessel window with 50% overlap between segments. Block averaging has been used to smooth the spectral estimates. (a) Highly smoothed spectrum with $M = 128$ (2^7), degrees of freedom (DoF) = 628; (b) Moderately smoothed spectrum with $M = 512$ (2^9), DoF = 154; (c) Weakly smoothed spectrum with $M = 2048$ (2^{11}), DoF = 36; (d) Same as (c) except that DoF = 36 applies to the lowest frequency range only. For $f \geq 6 \times 10^{-2}$ cycles/min, the number of spectral estimates averaged together increases as 3×36, 5×36, and 7×36, for each of the next three frequency ranges. *(Courtesy of Alexander Rabinovich.)*

dynamic range, confidence intervals, and ease of implementation. Spectral estimates are affected not only by the broadband noise spectrum of the data but also by narrow-band signals that fall within the bandwidth of the window. Leakage of spectral power from a narrow-band spectral component, f_0, to another

frequency component, f_a, produces a bias in the amplitude and position of a spectral estimate. This bias is especially disruptive for the detection of weak signals in the presence of nearby strong signals. To reduce the bias, we need a "good" window. Although there are no universal standards for a good window, we would like it to possess the following characteristics in Fourier transform space:

1. The central main lobe of the window (which is centered on the frequency of interest) should be as narrow as possible to improve the frequency resolution of adjacent spectral peaks in the data set, and the first side-lobes should be greatly attenuated relative to the main lobe to avoid contamination from other frequency components. Here, the narrowness of the central lobe is measured by the positions of the -3 dB (half amplitude points, $10 \log \frac{1}{2}$) on either side of the lobe. Retention of a narrow central lobe, while suppressing the side-lobes, is not as easy as it sounds since suppression of the side-lobes invariably leads to a broadening of the central lobe;

2. The window should suppress the amplitudes of side-lobes at frequencies far removed from the central lobe. That is, the side-lobes should have a rapid asymptotic fall-off rate with frequency so that they leak relatively little energy into the spectral estimate at the central lobe (i.e., into the frequency of interest);

3. The coefficients of the window should be easy to generate for multiplication in the time domain and convolution in the Fourier transform domain.

A good performance indicator (PI) for the time domain window $w(t)$ can be defined as the difference between the equivalent noise bandwidth (ENBW) and the bandwidth (BW), located between the -3 dB levels of the central lobe (Harris, 1978)

$$PI = \frac{ENBW - BW}{BW} = \frac{\frac{1}{BW} \sum_n w^2(n\Delta t)}{\left[\sum_n w(n\Delta t) \right]^2} - 1 \qquad (5.82)$$

where we have normalized by the BW. The lower the value, the better the performance of

the filter; windows that perform well have values for this ratio ($\times 100\%$) of between 4.0 and 5.5%. A summary of the figures of merit for several well-known windows is presented in Table 5.4. PI values are obtained using columns four and five. For example, for the weakly performing box-car (rectangular) window, $PI = 0.124$ (12.4%), while for the strongly performing Kaiser window, $PI = 0.049$ (4.9%). The choice of window can be daunting; Harris lists more than 44 windows for smoothing spectral estimates.

5.4.6.2 Rectangular (Box-Car) and Triangular Windows

As discussed at the beginning of Section 5.4, a rectangular window has an amplitude of unity throughout the observation interval of duration $T = N\Delta t$, with the weighting given by

$$w(n\Delta t) = 1, \quad n = 0, 1, ...,$$
$$N - 1 \ (\text{or} \ -N/2 \le n \le N/2) \quad (5.83)$$
$$= 0, \quad \text{elsewhere}$$

(Figure 5.23(a)). Using the relation $\omega T = N\theta$, where $\theta = \omega \Delta t$ and $T = N\Delta t$, the spectral window from the DFT is

$$W(\theta) = Te^{-i(N-1)\theta/2}\frac{\sin (N\theta/2)}{N\theta/2} \qquad (5.84a)$$

$$|W(\theta)|^2 = T^2 \left[\frac{\sin(N\theta/2)}{N\theta/2} \right]^2 \qquad (5.84b)$$

(Figure 5.23(b)) where the exponential term in Eqn (5.84a) gives the phase shift of the window as a function of the frequency $\omega = \theta / \Delta t$. The function W, the Dirichlet kernal, has strong side-lobes, with the power of the first side-lobe down only -13 dB (factor of 0.22) from the main lobe. The remaining side-lobes fall off weakly at -6 dB per octave, which is the functional rate for a discontinuity (an "octave" corresponds to a factor of 2 change in frequency). Zeros of $W(\theta)$ occur at integer multiples of the frequency resolution,

TABLE 5.4   Windows, Figures of Merit and Performance Indicator (PI)

Window	Highest Side-Lobe Level (dB)	Side-Lobe Attenuation (dB/octave)	ENBW (Bins)	3 dB BW (Bins)	PI	Overlap Correlation 75%	Overlap Correlation 50%
Rectangle	−13	−6	1.00	0.89	0.124	0.750	0.500
Triangle	−27	−12	1.33	1.28	0.031	0.719	0.250
Hanning	−32	−18	1.50	1.44	0.042	0.659	0.167
Hamming	−43	−6	1.36	1.30	0.046	0.707	0.235
Parzen	−21	−12	1.20	1.16	0.035	0.765	0.344
Tukey $\alpha = 0.5$	−15	−18	1.22	1.15	0.061	0.727	0.364
Kaiser $\alpha = 2.0$	−46	−6	1.50	1.43	0.049	0.657	0.169
Bessel							
$\alpha = 2.5$	−57	−6	1.65	1.57	0.051	0.595	0.112
$\alpha = 3.0$	−69	−6	1.80	1.71	0.052	0.539	0.074
$\alpha = 3.5$	−82	−6	1.93	1.83	0.054	0.488	0.048

The last column gives the correlation between adjacent data segments for the specified percentage segment overlap. For completeness, we include the Tukey and Parzen Windows. (Adapted from Harris (1978).)

$f_1 = 1/T$, for which $N\theta/2 = \omega T/2 = \pm p\pi$. That is, where $f = \pm p/T (\pm 1/T, \pm 2/T, \ldots)$.

The triangular (Bartlett) window

$$w(n\Delta t) = \begin{cases} \dfrac{n}{(N/2)}, & n = 0, 1, \ldots, N/2 \\[2mm] \dfrac{N-n}{(N/2)}, & n = N/2, \ldots, N-1 \end{cases}$$

(5.85a)

$$= \frac{N/2 - |n|}{(N/2)}, \quad 0 \le |n| \le N/2 \qquad (5.85b)$$

(Figure 5.24(a)) has the DFT

$$W(\theta) = \frac{2T}{N} e^{-i(N-1)\theta/2} \left[\frac{\sin(N\theta/2)}{N\theta/2} \right]^2 \qquad (5.86a)$$

$$|W(\theta)|^2 = \frac{4T^2}{N^2} \left[\frac{\sin(N\theta/2)}{N\theta/2} \right]^4 \qquad (5.86b)$$

(Figure 5.24(b)) which we recognize as the square of the sinc function for the rectangular window. The main lobe between zero crossings has twice the width of the rectangular window but the level of the first side-lobe is down by −26 dB, twice that of the rectangular window. Despite the improvement over the box-car window, the side-lobes of the triangular window are still extensive and use of this window is not recommended if other windows are available.

The Parzen window

$$w(n\Delta t) = 1 - |n/(N/2)|^2, \quad 0 \le |n| \le N/2$$

(5.87)

is the squared counterpart to the Bartlett window. This is the simplest of the continuous polynomial windows and has first side-lobes down by −22 dB and falls off with frequency as $1/\omega^2$.

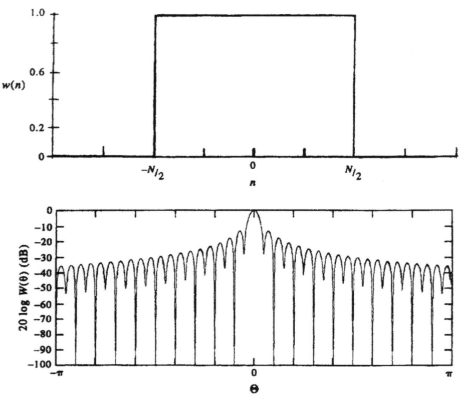

FIGURE 5.23 A box-car (rectangular) window for $N = 41$ weights. (a) Weights, $w(n) = 1.0$ in the time domain $(-20 \leq n \leq 20)$. (b) Fourier transform of the weights, $W(\theta)$, plotted as $20 \log|W(\theta)|$, where $\theta = \omega \Delta t / N = 40\pi / N$ is the frequency span of the window.

5.4.6.3 Hanning and Hamming Windows (50% Overlap)

The Hann window, or *Hanning window* as it is most commonly known, is named after the Austrian meteorologist Julius von Hann and is part of a family of trigonometric windows having the generic form $\cos^\alpha(n)$, where the exponent, α, is typically an integer from 1 through 4. The case $\alpha = 1$ leads to the *Tukey* (or *cosine-tapered*) *window* (Harris, 1978). As α becomes larger, the window becomes smoother, the side-lobes fall off faster, and the main lobe widens. The Hanning window ($\alpha = 2$), also known as the *raised cosine* and *sine-squared* window, is defined in the time domain as

$$w(n\Delta t) = \begin{cases} \sin^2(\pi n / N) = \dfrac{1}{2}[1 - \cos(2\pi n / N)], \\ \qquad n = 0, 1, \ldots, N - 1 \\ \sin^2[\pi(n + N/2)/N] = \\ \dfrac{1}{2}[1 - \cos[2\pi(n + N/2)/N]], \\ \qquad n = -N/2, \ldots, N/2 \end{cases}$$

(5.88)

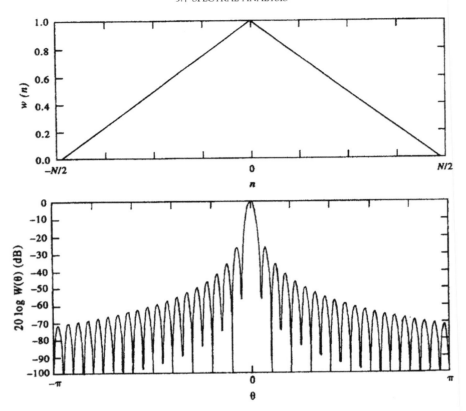

FIGURE 5.24 The triangular (Bartlett) window for $N = 51$ weights. (a) Weights, $w(n)$ in the time domain ($-20 \leq n \leq 20$). (b) Fourier transform of the weights, $W(\theta)$, plotted as $20 \log|W(\theta)|$ (cf. Figure 5.22).

(Figure 5.25(a)), which is a continuous function with a continuous first derivative. The DFT of this weighting function is

$$W(\theta) = \frac{1}{2}D(\theta) + \frac{1}{4}[D(\theta - \theta_1) + D(\theta + \theta_1)]$$

$$(5.89)$$

(Figure 5.25(b)), where $\theta_1 = 2\pi/N$ and

$$D(\theta) = Te^{i\theta/2}\frac{\sin(N\theta/2)}{N\theta/2} \qquad (5.90)$$

is the standard function (Dirichlet kernal) obtained for the rectangular and triangular

windows. Thus, the window consists of the summation of three sinc functions (Figure 5.25(c)), one centered at the origin, $\theta = 0$, and two other translated Dirichlet kernals having half the amplitude of the main kernal and offset by $\theta = \pm 2\pi/N$ from the central lobe. There are several important features of the window response $W(\theta)$. First of all, the functions D are discrete and defined only at points that are multiples of $2\pi/N$, which also correspond to the zero crossings of the central function, $D(\theta)$. Secondly, for all of zero crossings except those at $\theta_{\pm1} = \pm 2\pi/N$, the translated functions also have zero crossings at multiples of $2\pi/N$. As a result, only values at $-2\pi/N$, 0, and $+2\pi/N$

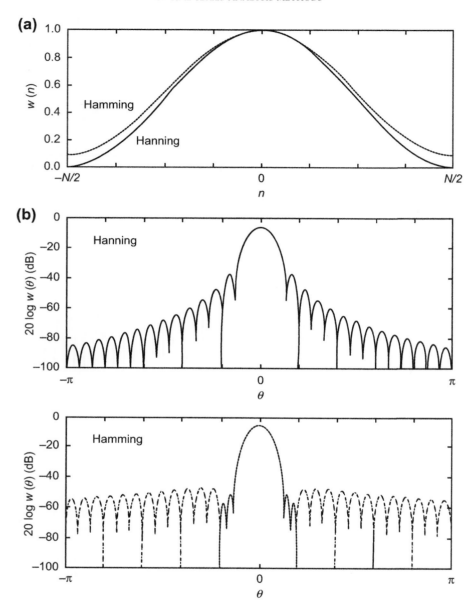

FIGURE 5.25 The Hanning and Hamming windows for $N = 41$ weights. (a) Weights, $w(n)$, $(-20 \leq n \leq 20)$. (b) Fourier transform of the weights, $W(\theta)$, plotted as $20 \log|W(\theta)|$ (cf. Figure 5.22). The response functions have not been re-scaled.

contribute to the window response. It is the widening of the main lobes of the translated functions that causes them to be nonzero at the first zero crossings of the central function. Lastly, because the translated functions are out of phase with the central function, they tend to cancel the side-lobe structure. The first side-lobe is down by -32 dB (factor of 0.025) from the main lobe. The remaining side-lobes diminish as $1/\omega^3$ or at about -18 dB per octave.

An attractive aspect of the Hanning window is that smoothing in the frequency domain can be accomplished using only three convolution terms corresponding to θ_o, $\theta_{\pm 1}$. The Hanning-windowed Fourier transform, $Y_H(f_k)$, representing the spectrum for the frequency, f_k, is then obtained from the raw spectra Y for the frequencies, f_k and the two adjoining frequencies, f_{k-1} and f_{k+1}; that is

$$Y_H(f_k) = \frac{1}{2}\left\{ Y(f_k) - \frac{1}{2}\left[Y(f_{k-1}) + Y(f_{k+1})\right] \right\}$$

(5.91)

The transform $Y(f_k)$ has already been rectangular-windowed by the very act of collecting the data but is "raw" in the sense that no additional smoothing has been applied. Other processing advantages of the Hanning window are discussed by Harris (1978). Since the squares of the weighting terms $(1/2)^2 + (1/4)^2 + (1/4)^2 = 3/8$, the total energy will be reduced following the application of the Hanning window. To compensate, the amplitudes of the Fourier transforms, $Y_H(f)$ should be multiplied by $\sqrt{8/3}$ prior to computation of the spectra. Specifically

$$Y_H(f_k) = \Delta t (8/3)^{1/2} \sum_{n=0}^{N-1} y_n$$

$$\times [1 - \cos(2\pi n/N)]e^{-i2\pi kn/N} \quad (5.92)$$

where $f_k = k/(N\Delta t)$.

The *Hamming window* is a variation on the Hanning window designed to cancel the first side-lobes. To accomplish this, the relative sizes of the three Dirichlet kernals are adjusted through a parameter, γ where

$$w(n\Delta t) = \gamma + (1-\gamma)\cos(2\pi n/N)],$$
$$n = -N/2, \dots, N/2 \quad (5.93a)$$

$$W(\theta) = \gamma D(\theta) + \frac{1}{2}(1-\gamma)[D(\theta - 2\pi/N)$$
$$+ D(\theta + 2\pi/N)]$$

(5.93b)

Perfect cancellation of the first side-lobes (located at $\theta_1 = 2.5\pi/N$) occurs when $\gamma = 25/46 \approx 0.543478$. Taking $\gamma = 0.54$ leads to near-perfect cancellation at $\theta_1 = 2.6\pi/N$ and a marked improvement in side-lobe level. The Hamming window is defined as

$$w(n\Delta t) = 0.54 + 0.46\cos(2\pi n/N),$$
$$n = -N/2, \dots, N/2 \quad (5.94)$$

and has a spectral distribution similar to that of the Hanning window with more "efficient" side-lobe attenuation. The highest side-lobe levels of the Hanning window occur at the first side-lobes and are down by -32 dB from the main lobe. For the Hamming window, the first side-lobe is highly attenuated and the highest side-lobe level (the third side-lobe) is down by -43 dB. To compensate for the filter, the amplitudes of the Fourier transforms $Y_{\text{Ham}}(f)$ should be multiplied by $\sqrt{5/2}$ prior to computation of the spectra. On a similar note, anyone using any of the windows in this section to calculate running mean time series should make sure each estimated value is divided by the sum of the weights used, $\sum_N w_n$.

5.4.6.4 *Kaiser–Bessel Window*

Harris (1978) identifies the Kaiser–Bessel window as the "top performer" among the many different types of windows he considered. Among other factors, the coefficients of the window are easy to generate and it has a high ENBW, one of the criteria used to separate good and bad windows. The trade-off is an increased main-lobe width for reduced side-lobe levels. In the time domain the filter is defined in terms of the zeroth-order modified Bessel functions of the first kind.

$$w(n\Delta t) = \frac{I_0(\pi\alpha\Omega)}{I_0(\pi\alpha)}, \quad 0 \le |n| \le N/2 \quad (5.95)$$

where the argument $\Omega = [1 - (2n/N)^2]^{1/2}$ and

$$I_0(x) = \sum_{k=0}^{\infty} \left[\frac{(-1)^k (x/2)^k}{\Gamma(k+1)k!} \right]^2 \quad (5.96)$$

The parameter $\pi\alpha$ is half of the time-bandwidth product, with α typically having values 2.0, 2.5, 3.0, and 3.5. The transform is approximated by

$$W(\theta) \approx [N/I_0(\pi\alpha)] \frac{\sinh\left\{ \left[\pi^2\alpha^2 - (N\theta/2)^2 \right]^{1/2} \right\}}{\left\{ \left[\pi^2\alpha^2 - (N\theta/2)^2 \right]^{1/2} \right\}} \quad (5.97)$$

Plots of the weighting function w and the DFT for W are presented in Figure 5.26 for two values of the parameter α (=2.0, 3.0). The modified Bessel function I_0 is defined as follows:
For $|x| \le 3.75$

$$I_0(x) = \{[\{[(4.5813 \times 10^{-3}Z + 3.60768 \times 10^{-2})Z$$
$$+2.659732 \times 10^{-1}]Z + 1.2067492\}Z$$
$$+3.0899424]Z + 3.5156229\}Z + 1.0 \quad (5.98a)$$

where for real x

$$Z = (x/3.75)^2 \quad (5.98b)$$

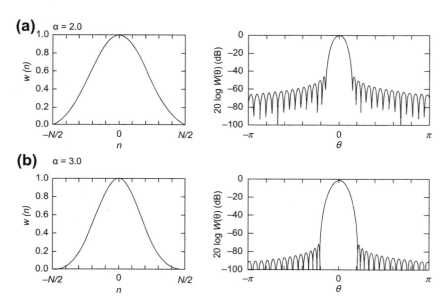

FIGURE 5.26 The Kaiser–Bessel window for $N = 51$ weights and $\alpha = 2.0$ and 3.0. (a) Weights, $w(n)$, ($-20 \le n \le 20$). (b) Fourier transform of the weights, $W(\theta)$, plotted as $20 \log|W(\theta)|$ (cf. Figure 5.22). *(From Harris (1978).)*

is distributed as a chi-square variable with ν DoF. It then follows that

$$P\left[\chi^2_{\alpha/2,\nu} < \frac{\nu \tilde{G}_{yy}(f)}{G_{yy}(f)} < \chi^2_{1-\alpha/2,\nu}\right] = 1 - \alpha$$

(5.105)

where

$$P\left[\chi^2_\nu \le \chi^2_{\alpha/2,\nu}\right] = \alpha/2 \qquad (5.106)$$

Thus, the true spectrum, $G_{yy}(f)$, is expected to fall into the interval

$$\frac{\nu \tilde{G}_{yy}(f)}{\chi^2_{1-\alpha/2,\nu}} < G_{yy}(f) < \frac{\nu \tilde{G}_{yy}(f)}{\chi^2_{\alpha/2,\nu}} \qquad (5.107)$$

with $(1 - \alpha)100\%$ confidence. In this form, the confidence limit applies only to the frequency f and not to other spectral estimates. We further point out that the DoF, ν, in the above expressions are different for windowed and nonwindowed time series. For windowed time series, we need to use the "equivalent" degrees of freedom (DoF), as presented in Table 5.5 for some of the more commonly used windows.

Another way to view these arguments is to equate $\tilde{G}_{yy}(f)$ with the measured standard

TABLE 5.5 Equivalent Degrees of Freedom for Spectra Calculated Using Different Windows

Type of Window	Equivalent Degrees of Freedom
Truncated periodogram	(N/M)
Bartlett window	$3(N/M)$
Daniell window	$2(N/M)$
Parzen window	$3.708614(N/M)$
Hanning window	$(8/3)(N/M)$
Hamming window	$2.5164(N/M)$

N is the number of data points in the time series and M is the half-width of the window in the time (or spatial) domain. (From Priestley (1981)). $N \ne M$ for the truncated periodogram.

deviation, $s^2(f)$, of the spectrum and $G_{yy}(f)$ with the true variance, $\sigma^2(f)$. Then

$$\frac{(\nu - 1)s^2(f)}{\chi^2_{1-\alpha/2,\nu}} < \sigma^2(f) < \frac{(\nu - 1)s^2(f)}{\chi^2_{\alpha/2,\nu}} \qquad (5.108)$$

If spectral peaks fall outside the range Eqn (5.108) then to the $(1 - \alpha)100\%$ confidence level they are unlikely to have occurred by chance. The confidence levels are found by looking up the values for $\chi^2_{1-\alpha/2,\nu}$ and $\chi^2_{\alpha/2,\nu}$ in a chi-square table, then calculating the intervals based on the observed standard deviation, s. (Confidence limits on spectral coherency functions are given in Section 5.6.6.1.)

5.4.8.1 Confidence Intervals on a Logarithmic Scale

The confidence intervals derived above apply only to individual frequencies, f. This results from the fact that the confidence interval is determined by the value $\tilde{G}_{yy}(f)$ of the spectral estimate and will be different for each spectral estimate. It would be convenient if we could have a single confidence interval that applies to all of the spectral values at all frequencies. To obtain such a confidence interval, we transform the spectrum using the $\log_{10}$ function. Transforming the above confidence limits we have

$$\log\left[\tilde{G}_{yy}(f)\right] + \log\left[\nu/\chi^2_{1-\alpha/2,\nu}\right] \le \log\left[G_{yy}(f)\right]$$

$$\le \log\left[\tilde{G}_{yy}(f)\right] + \log\left[\nu/\chi^2_{\alpha/2,\nu}\right] \qquad (5.109)$$

or

$$\log\left[\nu/\chi^2_{1-\alpha/2,\nu}\right] \le \log\left[G_{yy}(f)\right] - \log\left[\tilde{G}_{yy}(f)\right]$$

$$\le \log\left[\nu/\chi^2_{\alpha/2,\nu}\right] \qquad (5.110a)$$

$$\log\left[\nu/\chi^2_{1-\alpha/2,\nu}\right] \le \log\left[G_{yy}(f)/\tilde{G}_{yy}(f)\right]$$

$$\le \log\left[\nu/\chi^2_{\alpha/2,\nu}\right]$$

(5.110b)

FIGURE 5.29 Confidence intervals for current velocity spectra at 50-m depth for three locations (B, C, and 62) on the northeast Gulf of Alaska shelf (59.5° N, 142.2° W), March 15–April 15, 1976. (a) 95% interval for the low-pass filtered currents. The single vertical bar applies to all frequencies; (b) 95% interval for unfiltered records. Confidence interval narrows at higher frequencies with the increased number of degrees of freedom (4–36) used in selected frequency ranges. *(Adapted from Muench and Schumacher (1979).)*

where $\log[G_{yy}(f)/\tilde{G}_{yy}(f)] \to 0$ as the estimated spectrum approaches the real spectrum; i.e., $\tilde{G}_{yy}(f) \to G_{yy}(f)$. When the estimated spectrum is plotted on a log scale, a single vertical confidence interval is determined for all frequencies by the upper and lower bounds in the above expressions (Figure 5.29(a)). The spectral estimate $\tilde{G}_{yy}(f)$ itself is no longer a part of the confidence interval. This aspect, together with the fact that most spectral amplitudes span many orders of magnitude, is a principal reason for presenting spectra as log values. If larger numbers of spectral estimates are averaged together at higher frequencies (i.e., ν is increased), the confidence interval narrows with increasing frequency

(Figure 5.29(b)). Note that the length of the confidence interval is longer above the central point than below.

5.4.8.2 Fidelity and Stability

The general objective of all spectral analysis is to estimate the function $G_{yy}(f)$ as accurately as possible. This involves two basic requirements:

1. The mean smoothed spectrum, $\tilde{G}_{yy}(f)$, be as close as possible to the actual spectrum $G_{yy}(f)$. That is, the bias

$$B(f) = G_{yy}(f) - \tilde{G}_{yy}(f) \qquad (5.111)$$

should be small. If this is true for all frequencies, then $\tilde{G}_{yy}$ is said to reproduce $G_{yy}(f)$ with high *fidelity*.

2. For a time series of length T that has been segmented into M pieces for spectral estimation, the variance of the smoothed spectral estimator for bandwidth b_1 is

$$V\left[\tilde{G}_{yy}(f)\right] \approx \frac{(M/b_1)}{T}\left[G_{yy}(f)\right]^2 \qquad (5.112)$$

and should be small. If this is true, the spectral estimator is said to have high *stability*.

5.4.9 Zero-Padding and Prewhitening

For logistical reasons, many of the time series that oceanographers collect are too short for accurate definition of certain spectral peaks. The frequency resolution $\Delta f = 1/T$ for a record of length T may not be sufficient to resolve closely spaced spectral components. Also, discrete points in the computed spectrum may be too widely spaced to adequately delineate the actual frequency of the spectral peaks. Unfortunately, the first problem—that of trying to distinguish waveforms with nearly the same frequency—can only be solved by collecting a longer time series; i.e., by increasing T to sharpen up the frequency resolution f of the periodogram. However, the second problem—that of locating the frequency of a spectral peak more precisely—can be addressed by padding (extending) the time series with zeros prior to Fourier transforming. Transforming the data with zeros serves to refine the frequency scale through interpolation between PSD estimates within the Nyquist interval $-f_N \leq f \leq f_N$. That is, additional frequency components are added between those that would be obtained with a nonzero-padded transform. Adding zeros helps fill in the shape of the spectrum but in no case is there an improvement in the fundamental frequency resolution. *Zero-padding* is useful for: (1) smoothing the appearance of the periodogram estimates via

interpolation; (2) resolving potential ambiguities where the frequency difference between line spectra is greater than the fundamental frequency resolution; (3) helping define the exact frequency of spectral peaks by reducing the "quantization" accuracy error; and (4) extending the number of samples to an integer power of two for FFT analysis. An example of how zero-padding improves the spectral resolution of a simple digitized data set is provided in Figure 5.30. We again emphasize that increased zero-padding helps locate the frequency of discernible spectral peaks, in this case the peaks of the $\sin x/x$ function, but cannot help distinguish closely spaced frequency components that were unresolved by the original time series prior to padding.

Prewhitening is a filtering or smoothing technique used to improve the statistical reliability of spectral estimates by reducing the leakage from the most intense spectral components and low-frequency components of the time series that are poorly resolved. To reduce the biasing of these components, the data are smoothed by a window whose spectrum is inversely proportional to the unknown spectrum being considered. Within certain frequency bands, the spectrum becomes more uniformly distributed and approaches that of white noise. Information on the form of the window necessary to construct the white spectrum must be available prior to the application of the smoothing. In effect, the time series, $y(n\Delta t)$ is filtered with the weighting function, $w(n\Delta t)$ such that the output is

$$y'(n\Delta t) = w(n\Delta t) \cdot (n\Delta t) \qquad (5.113)$$

has a nearly white spectrum. Once the spectrum $S'_{yy}(\omega)$ is determined, the desired spectrum is derived directly as

$$S_{yy}(\omega) = \frac{S'_{yy}(\omega)}{|W(\omega)|^2} \qquad (5.114)$$

The best aspects of the parametric and nonparametric spectral techniques can be combined if a parametric model is used to prewhiten the time series prior to the application of a

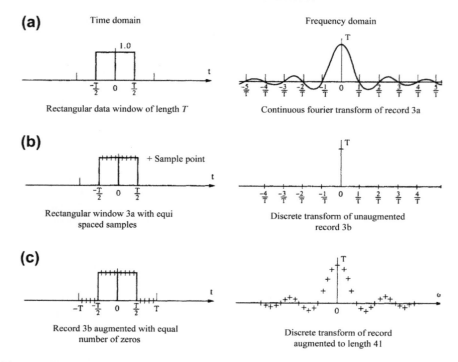

(a) Time domain   Frequency domain

Rectangular data window of length T Continuous fourier transform of record 3a

(b) + Sample point

Rectangular window 3a with equi   Discrete transform of unaugmented
spaced samples   record 3b

(c)

Record 3b augmented with equal   Discrete transform of record
number of zeros   augmented to length 41

FIGURE 5.30 Use of zero-padding to improve the delineation of spectral peaks. (a) A continuous box-car window of length, T and its continuous Fourier transform; (b) a discrete sample of (a) at equally spaced sampling intervals and its discrete Fourier transform; (c) same as (b) but with zero-padding of $2T$ data points. Note that the middle panel on the right is not a misprint. Transform values lie on the horizontal axis at the points $-4/T$, $-3/T$, and so on. *(From Henry and Graefe (1971).)*

smoothed periodogram analysis. In most pre-whitening situations, one is limited to using the first-difference filter in which the current data value has subtracted from it the next value multiplied by some weighting coefficient, $0 \leq \alpha \leq 1$. That is $y'(t) = y(t) - \alpha y(t + \Delta t)$. The weighting coefficient can be taken as equal to the correlation coefficient of the initial data series with a shift of one time step, Δt. The filter suppresses low frequencies and stresses high frequencies and has a frequency response

$$W(f) = \left[1 - \alpha e^{-i2\pi f \Delta t}\right]^2$$
$$= 1 - 2\alpha \cos(2\pi f \Delta t) + \alpha^2 \qquad (5.115)$$

Prewhitening reduces leakage and increases the effectiveness of frequency averaging of the spectral estimate (reduces the random error).

The reduced leakage gives rise to a greater dynamic range of the analysis and allows us to examine weak spectral components. Notice that, if $Y(f)$ is the Fourier transform of $y(t)$, then the Fourier transform of $y'(t)$ is

$$Y'(\omega) = \int_t y'(t)e^{-i\omega t}dt \approx \omega Y(\omega) \qquad (5.116)$$

so that *first differencing* is like a linear high-pass filter with amplitude $|W(\omega)| = |\omega|$. This effect shows up quite well in the processing of satellite-tracked drifter data. Spectra of the drifter positions (longitude, $x(t)$; latitude, $y(t)$ as functions of time, t, are generally "red" whereas the spectra of the corresponding drifter velocities (zonal, $u = \Delta x / \Delta t$; meridional, $v = \Delta y / \Delta t$) are considerably "whiter" (Figure 5.31).

FIGURE 5.31 Effect of a first-difference (high-pass) filter on resulting spectra. (a) Spectra of longitude (Δx) and latitude (Δy) displacements of a satellite-tracked drifter launched in the northeast Pacific in September 1990 ($\Delta t = 3$ h; duration, $T = 90$ days); (b) as with (a) but for the zonal ($u = \Delta x / \Delta t$) and meridional velocity ($v = \Delta y / \Delta t$). Mean position of the drifter was 49.6° N, 136.7° W; f denotes the mean inertial frequency; vertical line is the 95% confidence interval.

5.4.10 Spectral Analysis of Unevenly Spaced Time Series

Most discrete oceanographic time series data are recorded at equally spaced time increments. However, some situations arise where the recorded data are spaced unevenly in time or space. For example, positional data obtained from satellite-tracked drifters are sampled at irregular time intervals due to the eastward progression in the swaths of polar-orbiting satellites and to the advection of the drifters by surface currents. Repeated time series oceanic transects are typically spaced at irregular intervals due to the vagaries of ship scheduling and weather. In addition, instrumental problems and data dropouts generally lead to "gappy," irregularly spaced time series.

As noted in Section 3.17, a common technique for dealing with irregularly sampled or gappy data is to interpolate data values to a regular grid. This works well as long as there are not too many gaps and the gaps are of short duration relative to the signals of interest. Long data gaps can lead to the creation of erroneous low-frequency oscillations in the data at periods

comparable to the gap lengths. Only for the least-squares (LS) method for harmonic analysis described in Section 5.9 is unevenly sampled data perfectly acceptable. Vaníček (1971), Lomb (1976) and others have devised an LS spectral analysis method for unevenly spaced time series. The Lomb method described by Press et al. (1992) evaluates data, and associated sines and cosines, at the times, t_n, that the data are measured. For the N data values $x(t_n) = x_n$, $i = 1, ..., N$, the Lomb-normalized periodogram is defined as

$$
P(\omega) = \frac{1}{2\sigma^2} \left\{ \frac{\left[\sum_{n=1}^{N} (x_n - \overline{x}) \cos[\omega(t_n - \tau)] \right]^2}{\sum_{n=1}^{N} \cos^2[\omega(t_n - \tau)]} \right.
$$

$$
\left. + \frac{\left[\sum_{n=1}^{N} (x_n - \overline{x}) \sin[\omega(t_n - \tau)] \right]^2}{\sum_{n=1}^{N} \sin^2[\omega(t_n - \tau)]} \right\}
$$

(5.117)

where as usual

$$\overline{x} = \frac{1}{N} \sum_{n=1}^{N} x_n; \quad \sigma^2 = \frac{1}{N-1} \sum_{n=1}^{N} (x_n - \overline{x})^2 \tag{5.118}$$

are the mean and standard deviation of the time series, and the time offset, τ, is defined by

$$\tan(2\omega\tau) = \frac{\displaystyle\sum_{n=1}^{N} \sin(2\omega t_n)}{\displaystyle\sum_{n=1}^{N} \cos(2\omega t_n)} \tag{5.119}$$

The offset τ renders Eqn (5.117) identical to the equation we would derive if we attempted to estimate the harmonic content of a data set at frequency ω using the linear LS model

$$x(t) = A \cos(\omega t) + B \sin(\omega t) \tag{5.120}$$

In fact, Vaníček's founding paper on the technique refers to it as an LS spectral analysis method. The method, which gives superior results to FFT methods, weights the data on a per point basis rather than on a time-interval basis. By not using weights that span a constant time interval, the method reduces errors introduced by unevenly sampled data. For further details on the Lomb periodogram, including the introduction of significance testing of spectral peaks, the reader is referred to Press et al. (1992; pp. 569–577).

5.4.11 General Spectral Bandwidth and Q of the System

Once the PSD, $S(\omega)$, has been computed, the general spectral bandwidth BW may be determined from the three moments, m_k, of the spectra

$$m_k = \int_0^\infty \omega^k S(\omega)d\omega, \quad k = 0, 1, 2$$

$$\approx \sum_{i=0}^{N/2} \omega_i^k S(\omega_i)\Delta\omega \tag{5.121}$$

where $N/2$ is the number of spectral estimates and $\Delta\omega$ is the frequency resolution of the spectral estimates (cf. Masson, 1996). In particular

$$BW = \left[(m_2 m_0/m_1^2) - 1\right]^{1/2} \tag{5.122}$$

The bandwidth, $\Delta\omega_{BW}$, of a particular spectral peak within an oscillatory system can be used to estimate the dissipation of the system at the peak (resonant) frequency, ω_r. Specifically, the "Q" or *Quality factor* of the system measures the amount of energy, E, stored in a linear oscillator compared to the amount of energy lost per cycle through frictional dissipation, $\omega^{-1}dE/dt$ (Rabinovich, 2009). The Q-factor characterizes the sharpness of the resonant frequency and is commonly used as a direct measure of tidal dissipation in the ocean. Suppose that the energy of a simple linear system passes through a maximum at resonance frequency and that the energy of the system falls to 50% of its maximum value at frequencies $\omega \approx \omega_r \pm \Delta\omega_{BW}/2$. The Q of the system is then given by

$$Q = \frac{\omega E}{dE/dt} = \frac{\omega_r}{\Delta\omega_{BW}} = \alpha^{-1} \tag{5.123}$$

where $E = E_o e^{-\alpha\omega t}$ is the system energy as it decays from an initial value E_o with a dimensionless damping coefficient, α. For example, Wunsch (1972) finds $Q \approx 3.3$ for an apparent resonant period of 14.8 h for the North Atlantic Ocean while Garrett and Munk (1971) obtain a global-wide lower bound of 25 for normal modes near the semidiurnal frequency.

5.4.12 Summary of the Standard Spectral Analysis Approach

In summary, PSD estimates for time series $y(t)$ can be obtained as follows using the standard autocorrelation and periodogram approaches:

1. Remove the mean and trend from the time series. Failure to remove the trend can lead to spurious energy (power) at low

frequencies. Remove *obvious* "spikes" caused by errant sensor responses or other forms of recording glitch, and also try to adjust the data series for discontinuities caused by internal offsets in the instrument or to sudden changes in sensor position or depth (Figure 5.32(a)). Removing spikes and adjusting for offsets is not as easy as it sounds. However, if not taken into account in the original time series, spikes and offsets can lead to erroneous spectral distributions (Figure 5.32(b)).

2. If block averaging is to be used to improve the statistical reliability of the spectral estimates (i.e., to increase the number of DoF), divide the data series into M sequential blocks of N' data values each, where $N' = N/M$ (see Section 5.4.7). Depending on which type of window is to be applied, the sequential blocks can have up to 50% overlap.

3. To partially reduce end effects (Gibbs' phenomenon) or to increase the series length to a power of two for FFT analysis, pad the data with $K \leq N$ zeroes. Also pad the record with zeroes if you wish to increase the frequency resolution or center spectral estimates in specific frequency bands. To further reduce end effects and side-lobe leakage, taper the time series using a Hanning (raised cosine) window, Kaiser–Bessel window, or other appropriate window (see Section 5.4.6).

4. Compute the Fourier transforms, $Y(f_k)$, $k = 0$, 1, 2,..., $N - 1$, for the time series (for convenience, we have taken the number of padded values as $K = 0$). For block-segmented data, calculate the Fourier transforms, $Y_m(f_k)$, for each of the M blocks ($m = 1,..., M$) where $k = 0, 1,..., N' - 1$ and $N' < N$. To reduce the variance associated with the tapering in step 3, the transforms can be computed for overlapping segments.

5. Rescale the spectra to account for the loss of "energy" during application of the window.

That is, adjust the scale factor of $Y(f_k)$ (or $Y_m(f_k)$ in the case of smaller block size partitioning) to account for the reduction in spectral energy due to the tapering in step 3. For the Hanning window, multiply the amplitudes of the Fourier transforms by $\sqrt{8/3}$. The rescaling factors for other windows are listed in the right-hand column of Table 5.5.

6. Compute the raw PSD for the time series (or for each block) where for the two-sided spectral density estimates:

$$S_{yy}(f_k) = \frac{1}{N\Delta t} [Y^*(f_k)Y(f_k)],$$
$$k = 0, 1, 2, ..., N - 1$$

(no block averaging)

$$S_{yy}(f_k; m) = \frac{1}{N\Delta t} [Y_m^*(f_k)Y(f_k)],$$
$$k = 0, 1, 2, ..., N' - 1 \quad (5.124a)$$

(block averaging) and for the one-sided spectral density estimates

$$G_{yy}(f_k) = \frac{2}{N\Delta t} [Y^*(f_k)Y(f_k)],$$
$$k = 0, 1, 2, ..., N/2$$

(no block averaging)

$$G_{yy}(f_k; m) = \frac{2}{N\Delta t} [Y_m^*(f_k)Y(f_k)],$$
$$k = 0, 1, 2, ..., N'/2 \quad (5.124b)$$

(block averaging)

7. In the case of the block-segmented data, average the raw spectral density estimates from the M blocks of data, frequency-band by frequency-band, to obtain the smoothed periodogram for $S_{yy}(f_k)$ or $G_{yy}(f_k)$. Remember, the trade-off for increased smoothing (more DoF) is a decrease in frequency resolution.

FIGURE 5.32 Effects of data spikes and offsets on spectral estimates. (a) The top line shows a 1-min sampled sea-level times series for Victoria, British Columbia. The middle panel is the same series but in which five data values have been converted into "spikes" (single data points with anomalously high values). In the bottom panel, we have inserted a single negative offset of 0.5 m midway through the original time series; (b) spectra of the three time series in (a). There is considerable loss of high frequency information compared to the original time series and the addition of erroneous low frequency energy in the case of the offset time series. Numbers denote periods in minutes of selected spectral peaks. *(Courtesy of Alexander Rabinovich.)*

8. Incorporate 80, 90, and/or 95% confidence limits in spectral plots to indicate the statistical reliability of spectral peaks. Most authors use the 95% confidence intervals.

We can illustrate some additional points in the above summary using the log–log spectra of sea-level oscillations (Figure 5.22) recorded over 14 days (20,160 min) in 1991 at Malokurilsk Bay on the west coast of Shikotan Island in the western Pacific. The main spectral peak is centered at a period of 18.6 min and corresponds to a wind-generated seiche amplitude of about 25 cm (Rabinovich and Levyant, 1992). All spectra have been obtained using segmented versions of the 14-day time series. Each time series segment has been smoothed using a Kaiser–Bessel window with 50% overlap between segments, and each segment has been treated as an independent time series. An FFT algorithm was used to calculate the spectrum for each segment. The smoothest spectrum (Figure 5.22(a)) is based on block averaged spectral estimates from roughly 157 overlapping segments ($\sim$20,160 min/128 min), the moderately smooth spectrum (Figure 5.22(b)) from the average of 39 overlapping segments, and the noisiest spectrum (Figure 5.22(c)) from the average of 10 overlapping segments. Taking into account the 50% overlap between segments and the fact that there are two DoF per raw spectral estimate, there are 628 (=157 × 4), 154, and 36 DoF for the three spectra, respectively. The smoothed spectrum in Figure 5.22(d) is derived using a slightly different approach. Although the segment lengths are the same as those in Figure 5.22(c) (i.e., 2048 min), the number of DoF is increased with increasing frequency, ω. In this sliding scale, the lowest frequency range uses 36 DoF (as with Figure 5.22(c)), the next frequency band averages together the spectra for three adjacent frequencies to give 108 DoF, the next averages together the spectra for five adjacent frequencies to give 180 DoF, and so on.

As indicated by Figure 5.22, increasing the number of frequency bands averaged in each spectral estimate enhances the overall smoothness of the spectrum and improves the statistical reliability for specific spectral peaks. The number of degrees of freedom (DoF) increases and the confidence interval narrows. The penalty we pay for improved statistical confidence is reduced resolution of the spectral peaks. As in Figure 5.22(a), too much smoothing diminishes our ability to specify the frequency of spectral peaks and washes out peaks linked to some of the weaker seiches. Because each time series segment is so short, we also lose definition at the low-frequency end of the spectrum. As indicated by Figure 5.22(c), too little smoothing leads to a noisy spectrum for which few spectral peaks are associated with any physical processes. The sliding DoF scale in Figure 5.22(d) is a useful compromise.

One last point. Up until now, we have assumed that the sensors being used to collect the data have the sensitivity to record all of the variations of interest. If this is not the case, then no form of spectral analysis can extract information from the signal, regardless of the temporal resolution. Consider, for example, Figure 5.33(a), which shows a 9-day time series of bottom pressure (sea-level height) collected at 1-min intervals in Saanich Inlet on Vancouver Island. The top line shows the raw bottom pressure data sampled at 0.001 m (1 mm) equivalent vertical resolution. This is followed by time series generated by rounding off the 1-min data values to vertical resolutions that are factors of 10 and 100 lower than that of the original record. The impact of the lower vertical resolution is clearly displayed by the spectra in Figure 5.33(b). As would be the case for inadequate sensor resolution, the spectra of vertical displacements becoming increasingly degraded at higher frequencies. Although the sampling interval is the same for all time series, the spectral details

FIGURE 5.33 The importance of sensor resolution to the detection of physical signals using spectral analysis. (a) 1-min sea-level record collected by a modern pressure gauge at Patricia Bay, Saanich Inlet, British Columbia. The top panel shows with original time series at 0.001 m (1 mm) vertical resolution, followed by time series formed by degrading (using decimal runoff) the original series to 0.01 and 0.1 m vertical resolution; (b) Spectra for the three time series showing the loss of information with increased degradation in vertical resolution. Numbers refer to spectral peaks in periods of minutes. The sampling rate is the same in all cases. (*Courtesy, Alexander Rabinovich.*)

of the sea-level signal are lost, including the background roll-off as a function of frequency.

Covariance function: Since the covariance function, $C_{yy}(\tau)$, and the autospectrum are Fourier transform pairs, the above analysis can be used to obtain a smoothed or unsmoothed estimate of the covariance function. To do this, first calculate the Fourier transform, $Y(f)$, of the time series, and determine the product $S_{yy}(f) = N^{-1}\Delta t[Y^*(f)Y(f)]$. Then take the IFT of the autospectrum, $S_{yy}(f)$, to obtain the covariance function, $C_{yy}(\tau)$. If the spectrum is unsmoothed prior to the IFT (or inverse fast Fourier transform (IFFT) if the FFT was used), we obtain the raw covariance function. If, on the other hand, the autospectrum is smoothed prior to the above integral using one of the spectral windows, such as the Hanning window, the covariance function also will be a smoothed function.

A word of caution: Although everyone agrees on the basic formulation for the DFT and the inverse discrete Fourier transform (IDFT), there are several ways to normalize the relations using the number of records, N. In our definitions, Eqns (5.26) and (5.28), N appears in the denominator of the IDFT. Some authors normalize using $1/N$ in the DFT only, while others insist on symmetry by using $1/\sqrt{N}$ in both DFT and its inverse. When using "canned" programs to obtain DFTs and IDFTs, ensure that you know how the transforms are defined and adapt your analysis to fit the appropriate processing routines.

5.5 SPECTRAL ANALYSIS (PARAMETRIC METHODS)

If the analytical model for a time series was known exactly, a sensible spectral estimation method would be to fit the model spectrum to the observed spectrum and determine any unknown parameters. In general, however, oceanic variability is too complex to admit simple analytical models and parametric spectral estimates over the full frequency range of the data series. In addition, the imposition of an overly simplified spectral model could seriously degrade any estimation. On the other hand, it is reasonable that relatively simple spectral models might adequately reflect the system dynamics over limited frequency bands. Under some very general conditions, any stationary series can be represented in closed form by a statistical model in which the corresponding spectrum is a rational function of frequency (i.e., a ratio of two polynomials in ω).

If the time series under investigation is long relative to the timescales of interest, and if the spectrum is not overly complicated and does not have a too large dynamic range, the simple smoothed periodogram technique will probably yield adequate results. At a minimum, it will identify the major features in the spectrum. For shorter time series or in studies of fine spectral structure, other techniques may be more applicable. One such spectral analysis technique was developed by Burg (1967, 1972), who showed that it was possible to obtain the power spectrum by requiring the spectral estimate to be the most random (i.e., to have the maximum entropy) of any power spectrum, which is consistent with the measured data. This leads to a spectral estimate with a high frequency resolution since the method uses the available lags in the autocovariance function without modification and makes a nonzero estimate (prediction) of the autocorrelation function beyond those, which are routinely calculated from the data. Because the spectral values are computed using a maximum entropy condition, the resulting spectral estimates are not accurate in terms of spectral amplitude.

The most popular of the "modern" parametric techniques is the *autoregressive power spectral density* (AR PSD) model whose origins are in economic time series forecasting and statistcal estimation. Autoregressive estimation was introduced to the earth sciences in the 1960s where it was originally applied to geophysical time series data under the name MEM. The duality between AR and MEM estimation has been thoroughly explored by Ulrych and Bishop (1975). Autoregressive spectral estimation is attractive because it has superior frequency resolution compared to conventional FFT techniques. As an example of the frequency resolution capability, consider the 14-year time series of average monthly air temperature for New York City (Figure 5.34(a)). The unsmoothed periodogram and three smoothed periodograms reveal a broad spectral peak centered at a period of 1 year (Figure 5.34(b)). This compares to the much sharper annual peak obtained via AR estimation (Figure 5.34(c)). The results reveal another important difference between the two methods. With the nonparametric periodogram approach discussed in the previous section, we can determine confidence limits for the spectral peaks, while for the parametric method the significance levels for the peaks are unknown. For example, the MEM is good for finding the location of spectral peaks but is not reliable for computing the correct spectral energy at those peaks. (The periodogram smoothing in Figure 5.34(b) was performed using a Parzen window with truncation values $N = 16$, 32, and 64; the weights for these windows are $w(n) = 1 - |2n/N|^2$, with $0 \leq |n| \leq \frac{1}{2} N$.)

In general, autoregressive and maximum entropy PSD estimation are not as widely used in oceanography as traditional spectral analysis methods. The former finds its greatest application in analytical climate modeling and in wavenumber spectral estimation. Modern

FIGURE 5.34 (a) Time series of monthly average air temperature for New York city (1946–1959); (b) the unsmoothed (raw) periodogram and three smoothed periodograms for Parzen windows with truncation lengths of 16, 32, and 64 months; and (c) an autoregressive spectral estimate of (a) showing the sharp peak at 12-month period. *(From Pagano (1978).)*

parametric techniques are good as long as the model is good. On the other hand, if the model is false, the resulting spectrum estimate can be highly misleading. It follows that if one has no reason for believing a specific model, it is better to use a nonparametric model. For this reason, we limit our presentation to the essential elements of the two methods. The reader is directed to Marple (1987) for a thorough discussion of the topic, including an introduction to Fourier transform methods of spectral analysis.

5.5.1 Some Basic Concepts

Many deterministic and stochastic discrete-time series processes encountered in oceanography are closely approximated by a rational transfer model in which the input sequence $\{x_n\}$ and the output sequence $\{y_n\}$, which is meant to model the input data, are related by the linear difference relation

$$y_n = \sum_{k=0}^{q} b_k x_{n-k} - \sum_{m=1}^{p} a_m y_{n-m} \qquad (5.125)$$

Here, y_n is shorthand notation for $y(n\Delta t)$, also written as $y(n)$. In its most general form, the linear model Eqn (5.125) is termed an *autoregressive moving average* (ARMA) model. The PSD of the ARMA output process is

$$P_{ARMA}(f) = \sigma^2 \Delta t [A(f)/B(f)]^2 \qquad (5.126)$$

where σ^2 is the variance of the applied white-noise driving mechanism and $\sigma^2 \Delta t$ is the PSD of the noise for the Nyquist interval $-1/(2\Delta t) < f < 1/(2\Delta t)$. Here

$$\begin{aligned} A(f) &= \alpha[\exp{(i2\pi f \Delta t)}], \\ B(f) &= \beta[\exp{(i2\pi f \Delta t)}] \end{aligned} \qquad (5.127)$$

where the coefficients α, β are defined in terms of the z-transform, $X(z)$, of the variable $z = \exp(i2\pi f\Delta t)$

$[=\exp(i2\pi k/N)$ in discrete form] where k, $n = 0$, $1, \ldots, N-1$

$$X(z) = \sum_{n=0}^{N-1} x_n z^{-n} \qquad (5.128)$$

which maps a real-valued sequence into a complex plane. Note that Eqn (5.128) is defined through negative powers of z, the convention used in electrical engineering. Geophysicists expand in positive powers of z (z^{+n}) but define $z = \exp(-iz\pi f \cdot \Delta t)$ so the results are the same. The z-transform of the autoregressive branch is

$$\alpha(z) = \sum_{n} a_n z^{-n} \qquad (5.129a)$$

while that of the moving average branch is

$$\beta(z) = \sum_{n} b_n z^{-n} \qquad (5.129b)$$

Specification of the parameters $\{a_k\}$, termed the autoregressive coefficients, the parameters $\{b_k\}$, termed the moving-average coefficients, and the variance, σ^2, is equivalent to specifying the spectrum of the process $\{y_n\}$. Without loss of generality, one can assume $a_0 = 1$ and $b_0 = 1$ since any gain of the system (5.125) can be incorporated into σ^2. If all the $\{a_k\}$ terms except $a_0 = 1$ vanish then

$$y_n = \sum_{k=0}^{q} b_k x_{n-k} \qquad (5.130)$$

and the process is simply a moving average of order q, and

$$P_{MA}(f) = \sigma^2 \Delta t |A(f)|^2 \qquad (5.131)$$

This model is sometimes called an *all-zero model* since spectral peaks and valleys are formed through zeroes of the function $A(f)$. If all the $\{b_k\}$ terms except $b_0 = 1$ vanish, then

$$y_n = \sum_{m=1}^{p} a_m y_{n-m} + \varepsilon_n \qquad (5.132)$$

and the process is strictly an autoregressive model of order p. The process is called AR in the sense that the sequence y_n is a linear regression on itself with ε_n representing the error. With this model, the present value y_n is expressed as a weighted sum of past values plus a noise term. The PSD is

$$P_{AR}(f) = \frac{\sigma^2 \Delta t}{|B(f)|^2} \qquad (5.133)$$

In the engineering literature, this model is sometimes called an *all-pole model* since narrow spectral peaks can be sharply delineated through zeroes in the denominator.

5.5.2 Autoregressive Power Spectral Estimation

The discrete form of an autoregressive model $y(t)$ of order p is represented by the relationship

$$y(n) = a_1 y(n-1) + a_2 y(n-2) + \dots$$
$$+ a_p y(n-p) + \varepsilon(n) \qquad (5.134)$$

where time $t = n\Delta t$, the a_k $(k = 1, \dots, p)$ are constant coefficients, and $\varepsilon(t)$ is a white-noise series (usually called the "innovation" of the AR process) with zero mean and variance σ^2. Another interpretation of the AR process is one that links $y(t)$ with a value that is predicted from the previous $p-1$ values of the process with a prediction error equal to $\varepsilon(t)$. Thus, the a_k $(k = 1, \dots, p)$ represent a p-point prediction filter. If $Y(z)$ is the z-transform of $y(n)$ then

$$Y(z) = \sum_{n=0}^{p} y(n) z^n \qquad (5.135)$$

and

$$Y(z) - Y(z)\left(a_1 z + a_2 z^2 + \dots + a_p z^p\right) = D(z) \qquad (5.136)$$

so that

$$|Y(z)|^2 = \frac{|D(z)|^2}{\left|1 - a_1 z - a_2 z^2 \dots - a_p z^p\right|^2} \qquad (5.137)$$

Substituting $z = \exp(-i2\pi f\Delta t)$ we obtain half of the true power spectrum. If the autoregression is a reasonable model for the data, then the AR PSD estimate based on Eqn (5.133) is

$$P_{AR}(f) = \frac{\sigma^2 \Delta t}{\left|1 + \sum_{k=1}^{p} a_k \exp\left(-i2\pi fk\Delta t\right)\right|^2} \qquad (5.138)$$

To find the PSD we need to estimate only three things: (1) the autoregressive parameters $\{a_1, a_2, \dots, a_p\}$; (2) the variance, σ^2, of the white-noise process that is assumed to be driving the system; and (3) the order, p, of the process. The limitations of the AR model are the degrading effect of observational noise, spurious peaks, and some anomalous effects that occur when the data are dominated by sinusoidal components. Unlike conventional Fourier spectral estimates, the peak amplitudes in AR spectral estimates are not linearly proportional to the power when the input process consists of sinusoids in noise. For high SNRs, the peak is proportional to the square of the power with the area under the peak proportional to power.

5.5.2.1 Autoregressive Parameter Estimation

Yule–Walker (YW) equations: If the autocorrelation function, $R_{yy}(k)$, is known exactly, we can find the $\{a_k\}$ by the YW equations. This method relates the AR parameters to the known (or estimated) autocorrelation function of $y(n)$

$$R_{yy}(k) = \frac{1}{N} \sum_{n=1}^{N-k} \{[x(n-k) - \bar{x}][x(n) - \bar{x}]\};$$

$$\bar{x} = \frac{1}{n} \sum_{n=1}^{N} x(n) \qquad (5.139)$$

There are other methods of estimating R_{yy}, but this estimator has the attractive property that its mean-squared error is generally smaller than that of other estimators (Jenkins and Watts, 1968). Since it is generally assumed that the mean $\bar{x}$ has been removed from the data, the autocovariance and autocorrelation functions are equal. To obtain the AR parameters, one need only to choose p equations from the YW equations for $k > 0$, solve for $\{a_1, a_2, \ldots, a_p\}$, and then find σ^2 from Eqn (2.39) for $k = 0$. The matrix equation to derive the a_i values and σ^2 is

$$\begin{vmatrix} R_{yy}(0) & R_{yy}(-1) & \ldots & R_{yy}(-p) \\ R_{yy}(1) & R_{yy}(0) & \ldots & R_{yy}[-(p-1)] \\ \ldots & \ldots & \ldots & \ldots \\ \ldots & \ldots & \ldots & \ldots \\ R_{yy}(p) & R_{yy}(p-1) & \ldots & R_{yy}(0) \end{vmatrix}$$

$$\times \begin{vmatrix} 1 \\ a_1 \\ \ldots \\ \ldots \\ a_p \end{vmatrix} = \begin{vmatrix} \sigma^2 \\ 0 \\ \ldots \\ \ldots \\ 0 \end{vmatrix} \qquad (5.140)$$

Thus, to determine the AR parameters and the variance, σ^2, one must solve (5.140) using the $p+1$ autocorrelation lags, $R_{yy}(0)$, ..., $R_{yy}(p)$, where $R_{yy}(-k) = R^*_{yy}(k)$.

Solutions to the YW matrix equation can be found via the computationally efficient Levinson–Durbin algorithm which proceeds recursively to compute the parameter sets $\{a_{11}, \sigma^2_1\}$, $\{a_{21}, a_{22}, \sigma^2_2\}$,..., $\{a_{p1}, a_{p2}, \ldots, a_{pp}, \sigma^2_p\}$. The final set at order p (the first subscript) is the desired solution. The algorithm requires p^2 operations as opposed to the $O(p^3)$ operations of Gaussian elimination. More specifically, the recursion algorithm gives

$$a_{11} = \frac{-R_{yy}(1)}{R_{yy}(0)} \qquad (5.141a)$$

$$\sigma^2_1 = \left(1 - |a_{11}|^2\right) R_{yy}(0) \qquad (5.141b)$$

with the recursion for $k = 2, 3, \ldots, p$ given by

$$a_{kk} = \frac{-1}{\sigma^2_1} \left[R_{yy}(k) + \sum_{j=1}^{k-1} a_{k-1,j} R_{yy}^{(k-j)} \right] \qquad (5.142a)$$

$$a_{ki} = -a_{k-1,i} + a_{kk}\left(a_{k-1,k-i}\right)^* \qquad (5.142b)$$

$$\sigma^2_k = \left(1 - |a_{kk}|^2\right) \sigma^2_{k-1} \qquad (5.142c)$$

Burg algorithm: Box and Jenkins (1970) point out that the YW estimates of the AR coefficients are very sensitive to rounding errors, particularly when the AR process is close to becoming nonstationary. The assumption that $y(k) = 0$, for $|k| > p$ leads to a discontinuity in the autocorrelation function and a smearing of the estimated PSD. For this reason, the most popular method for determining the AR parameters (prediction error filter coefficients) is the Burg algorithm. This algorithm works directly on the data rather than on the autocorrelation function and is subject to the Levinson recursion Eqn (5.142b). As an illustration of the differences in the YW and the Burg estimates, the respective values of a_{11} for the series $y(t_k) = y(k)$ are

$$a_{11} = \frac{\sum_{k=2}^{p} y(k)y(k-1)}{\sum_{k=1}^{p} y(k)^2},$$

for the Yule – Walker estimate

$$a_{11} = \frac{\sum_{k=2}^{p} y(k)y(k-1)}{\frac{1}{2}x_1^2 + \sum_{k=1}^{p} y(k)^2 + \frac{1}{2}x_p^2},$$

for the Burg estimate   (5.143)

Detailed formulation of the Burg algorithm is provided by Kay and Marple (1981; p. 1392). Again, there are limitations to the Burg algorithm, including spectral line splitting and biases in the frequency estimate due to contamination by rounding errors. Spectral line splitting occurs when the spectral estimate exhibits two closely spaced peaks, falsely indicating a second sinusoid in the data.

LS estimators: Several LS estimation procedures exist that operate directly on the data to yield improved AR parameter estimates and spectra compared with the YW or Burg approaches. The two most common methods use forward linear prediction for the estimate, while a second employs a combination of forward and backward linear prediction. Ulrych and Bishop (1975) and Nuttall (1976) independently suggested this LS procedure for forward and backward prediction in which the Levison recursion constraint imposed by Burg is removed. The LS algorithm is almost as computationally efficient as the Burg algorithm requiring about 20 more computations. The improvement by the LS approach over the Burg algorithm is well worth the added computation time. Improvements include less bias in the frequency estimates, and absence of observed spectral line splitting for short sample sinusoidal data.

Barrodale and Erickson (1978) provide a FORTRAN program for an "optimal" LS solution to the linear prediction problem. The algorithm solves the underlying LS problem directly without forcing a Toeplitz structure on the model. Their algorithm can be used to determine the parameters of the AR model associated with the MEM and for estimating the order of the model to be used. As illustrated by the spectra in Figure 5.35, this approach leads to a more accurate frequency resolution for short sample harmonic processes. In this case, the test data were formed by summing 0.03 and 0.2 Hz sine waves generated in single precision and sampled 10 times per second. The reader is also referred to Kay and Marple (1981; p. 1393) for additional details.

5.5.2.2 *Order of the Autoregressive Process*

The order p of the autoregressive filter is generally not known a priori and is acknowledged as one of the most difficult tasks in time series modeling by parametric methods. The choice is to postulate several model orders then compute some error criterion that indicates which model order to pick. Too low a guess for the model order results in a highly smoothed spectral estimate. Too high an order introduces

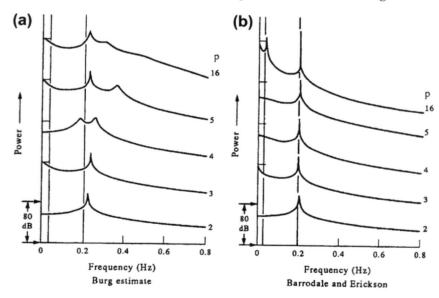

FIGURE 5.35 Maximum entropy method spectra obtained using (a) the Burg and (b) the Barrodale and Erickson algorithms. Signal consists of a combined 0.2 and 0.03 Hz (cps) sine wave. Spectra are plotted for increasing numbers of coefficients, *p*. *(From Barrodale and Erickson (1978).)*

spurious detail into the spectrum. One intuitive approach would be to construct AR models with increasing order until the computed prediction error power σ^2_k reaches a minimum. Thus, if a process is actually an AR process of order p, then $a_{p+1,k} = a_{pk}$ for $k = 1, 2, \ldots, p$. The point at which a_{pk} does not change would appear to be a good indicator of the correct model order. Unfortunately, both the YW equations and Burg algorithm involve prediction error powers

$$\sigma^2_k = \sigma^2_{k-1}\left[1 - |a_{kk}|^2\right] \qquad (5.144a)$$

that decrease monotonically with increasing order p, so that as long as $|a_{kk}|^2$ is nonzero (it must be ≤ 1) the prediction error power decreases. Thus, the prediction error power is not sufficient to indicate when to terminate the search. Alternative approaches (Kay and Marple, 1981) have been proposed by Akaike (termed the final prediction error, FPE, and the Akaike information criterion, AIC), and by Parzen (termed the criterion autoregressive transfer function). The AIC determines the model order by minimizing an information theoretic function. If the process has Gaussian statistics, the AIC is

$$\text{AIC}(p) = \ln\left(\sigma^2_p\right) + 2(p + 1)/N \qquad (5.144b)$$

where σ^2_p is the prediction error power and N is the number of data samples. The second term represents the penalty for the use of extra autoregressive coefficients that do not result in a substantial reduction in the prediction error power. The order p is the one that minimizes the AIC.

5.5.2.3 *Maximum Entropy Method*

The only constraint on the AR method is that the data yield the known autocorrelation function, $R_{yy}(k)$ for the interval $0 < k < p$. The assumption that $y(k) = 0$, for $|k| > p$ leads to a discontinuity in the autocorrelation function and a smearing of the estimated PSD. The MEM was designed, independently of autoregressive estimation, to eliminate the distortion of the spectrum caused by the truncated $R_{yy}(k)$. By adding a second constraint to improve the spectral estimation, the method gets away from the problems with the YW algorithm. In essence, the MEM is a way of extrapolating the known autocorrelation

SUMMARY OF ALGORITHMS

Method	Model Applied	Advantages	Disadvantages
Periodogram method using FFT or direct Fourier transform	Sum of harmonics (sines and cosines). No specific model needed.	1. Uses harmonic least-squares fit to the data; 2. Output $S(f)$ directly proportional to power; 3. Most computationally efficient; 4. Well-established methodology; 5. Confidence intervals easily computed; 6. Integral of $S(f)$ over frequency band Δf is equal to the variance of the signal in that band; 7. Easily generalized to cross-spectra and rotary spectra analyses.	1. Frequency resolution $\Delta f \approx 1/T$ dependent only on record length, T; 2. Poor performance for short data records; 3. Side-lobe leakage distorts spectra if appropriate windowing not done; windowing reduces frequency resolution, Δf; 4. Must average spectral estimates to improve statistical reliability.

(Continued)

SUMMARY OF ALGORITHMS (*cont'd*)

Method	Model Applied	Advantages	Disadvantages
Autoregressive, Yule–Walker algorithm.	Autoregressive (all-pole) process. Specific model.	1. Improved spectral resolution over Fourier transform methods; 2. Sharp spectral peaks; 3. No side-lobe leakage problems; 4. Minimum phase (stable) linear prediction filter guaranteed if biased lag estimates computed; 5. Related to linear prediction analysis and adaptive filtering.	1. AR model order, p, must be specified; 2. Spectral line splitting occurs; 3. Implied windowing distorts spectra; 4. Confidence intervals not readily computed.
Autoregressive, Burg algorithm.	Autoregressive (all-pole) process. Specific model.	1. Improved resolution over Fourier transform methods. Uses a constrained recursive least-squares approach; 2. No side-lobe leakage problems; 3. High resolution for low noise signals; 4. Good spectral fidelity for short data series; 5. No windowing implied; 6. Stable linear prediction filter guaranteed.	1. AR model order, p, must be specified; 2. Spectral line splitting can occur; 3. Confidence intervals not readily computed.
Autoregressive, least-squares method.	Autoregressive (all-pole) process. Specific model.	1. Sharper spectra than for other AR methods; 2. No side-lobes; 3. Good spectral fidelity for short data series; 4. No windowing; 5. No line splitting; 6. Uses exact recursive least-squares solution with no constraint.	1. AR model order must be specified; 2. Stable linear prediction filter not guaranteed, though stable filter results in most cases.

function to lags $k > p$, which are not known. In words, we assume that $\{R_{yy}(0), ..., R_{yy}(p)\}$ are known and find a logical way to extend to lags $\{R_{yy}(p+1),...\}$. As it turns out, the power spectral estimate for the MEM approach is equivalent to the power spectral estimate for the AR process.

In general, there exist an infinite number of possible extrapolations. Burg (1968) argued that preferred extrapolation should do two things: (1) yield the known R_{yy} for $0 \leq k \leq p$; and (2) generate an extrapolated R_{yy} for $k > p$ that causes the time series to have maximum entropy under the constraint (1). The time series that results is the most random one, which adheres to the known R_{yy} for the first $p+1$ lags. Alternatively, we can say that the PSD is the one with whitest noise (flattest spectrum) of all possible spectra for which $\{R_{yy}(0), ..., R_{yy}(p)\}$ is known. The reason for choosing the maximum entropy criterion is that it imposes the fewest constraints on the unknown time series by maximizing its randomness thereby causing minimum bias and operator intervention. For a Gaussian process, the entropy per sample is proportional to

$$\int_{-1/2\Delta}^{1/2\Delta t} \ln\left[P_y(f)\right] \mathrm{d}f \qquad (5.145)$$

where $P_y(f)$ is the PSD of y_n. The spectrum is found by maximizing Eqn (5.145) subject to the constraint that the $p+1$ known lags satisfy the Wiener–Khinchin relation

$$\int_{-1/2\Delta t}^{1/2\Delta t} P_y(f)e^{-i2\pi fn\Delta t}\mathrm{d}f = R_{yy}(n),$$

$$n = 0, 1, ..., p \qquad (5.146)$$

The solution is found using the Lagrange multiplier technique (see Ulrych and Bishop, 1975) as

$$P_y(f) = \frac{\sigma_p^2 \Delta t}{\left|1 + \sum_{k=1}^{p} a_{pk} \exp(-i2\pi fk\Delta t)\right|^2} \qquad (5.147)$$

where $\{a_{p1}, ..., a_{pp}\}$ and σ_p^2 are just the order-p predictor parameters and prediction error power, respectively. With knowledge of $\{R_{yy}(0), R_{yy}(1), ..., R_{yy}(p)\}$ the PSD of the MEM is equivalent to the PSD of the autoregressive method. That is, the MEM spectral analysis is equivalent to fitting an AR model to the random process. It is indeed interesting that the representation of a stochastic process by an AR model is that representation that exhibits maximum entropy. The duality of the AR model and MEM has enabled workers to apply the large body of literature on AR time series analysis to overcome shortcomings of the MEM.

The estimation of the MEM spectral density requires knowledge of the order of the AR process that we use to model the data. The importance of correctly estimating the order p is illustrated using the following AR process $y_n \equiv y(t_n)$ at times $t_n = n\Delta t$:

$$y_n = 0.75y_{n-1} - 0.5y_{n-2} + \varepsilon_n \qquad (5.148)$$

with noise variance $\sigma_\varepsilon^2 = 1$ (Figure 5.36(a)). Here $E[y(t)\varepsilon(t)] = \sigma_\varepsilon^2$, but $E[y(t)\varepsilon'(t)] = 0$ for any other additive noise, ε'. As indicated by Figure 5.36(b), which compares the theoretical power of a specified second-order AR process with the PSD computed from a realization of this process using $p = 2$ and $p = 11$ (Ulrych and Bishop, 1975), the correct choice of p is vital in obtaining a meaningful estimate of the power spectrum of the process. The peak value and the width of the spectral line of the MEM PSD estimate also may have considerable variance in the MEM estimates.

Although the MEM has numerous advantages over traditional nonparametric spectral techniques, especially for short data series, the usefulness of the approach is diminished by the lack of a straightforward criterion for choosing

FIGURE 5.36 Maximum entropy spectra. (a) Time series for the second-order AR process, $y_n = 0.75y_{n-1} - 0.5y_{n-2} + \varepsilon_n$ (Eqn (5.148)). (b) Spectral computation for the AR process. Solid line: the true power spectrum. Dot–dash line: maximum entropy method (MEM) estimate with 3-point ($p = 2$) prediction error filter. Dashed line: MEM estimate with 12-point ($p = 11$) error filter. *(From Ulrych and Bishop (1975).)*

the length (order) of the prediction model. Too short a length results in a highly smoothed spectrum obviating the resolution advantages of the MEM, whereas an excessive length introduces spurious detail into the spectrum.

Confidence intervals: A major shortcoming of MEM is the lack of a mathematically consistent variance estimator (confidence interval) for the spectral density. One approach is to approximate the confidence bounds in the same way that we compute the bounds in traditional multivariate spectral analysis (i.e., using a chi-square variable with ν DoF) under the assumption that the equivalent number of DoF is given by $\nu = N/p$, where N is the number of data points in the time series and p is the order of the model (Privalsky and Jensen, 1993, 1994). The order p should be chosen on the basis of objective criteria such as AIC, Parzen's criterion, and so on (see Lütkepohl, 1985).

5.5.2.4 An Autoregressive Model of Global Temperatures

One way to determine the effect of initial conditions and random noise on the global temperature predictions of computer-simulated general circulation models (GCMs) is to obtain a control realization, modify the initial conditions and

noise, obtain a second realization, and compare results. Since this could take several weeks to months of supercomputing time, a more practical approach is to employ a model of the global air temperature series, $T(t)$, derived by Jones (1988) (Figure 5.37). If we assume that the sensitivity of GCMs to changing conditions is similar to that of a stationary autoregressive model, then marked changes in the AR model that result from slight changes in the initial conditions or inherent noise are evidence that GCMs are too sensitive to these parameters to be reliable.

If $Z_n \equiv Z(t_n)$ represents the temperature deviation (departure from the long-term mean) at year t_n, then the maximum likelihood fourth order AR model for the temperature data in Figure 5.37 is

$$Z_n = 0.669Z_{n-1} - 0.095Z_{n-2} + 0.104Z_{n-3}$$
$$+ 0.247Z_{n-4} + \varepsilon_n$$

$$(5.149)$$

where $Z_n = T_n - \overline{T}$, and ε_n is an uncorrelated white-noise series with zero mean and variance equal to $0.0115\,^\circ C^2$ (Tsonis, 1991; Gray and Woodward, 1992). In general, we can state that for any AR process, the initial values will have little effect on forecasts if the sample size is large relative to the order of the process. For this

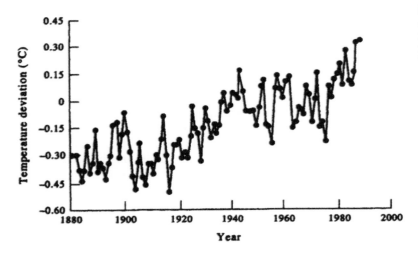

FIGURE 5.37 The annual global mean air temperatures from 1881 to 1988 as deviations (°C) from the 1951−1970 average. *(From Gray and Woodward (1992).)*

FIGURE 5.38 Two simulated realizations from the AR(4) model given by Eqn (5.149). (a) Same starting values but different and independently derived noise sequence; (b) different starting values but the same noise sequence. *(From Gray and Woodward (1992).)*

reason, AR processes are often known as "short memory" processes. In the above model, the correlation between $Z(t)$ and $Z(t + m\Delta t)$ is $0.9(0.96)^m$, for values of m greater than about five. For example, the correlation coefficient between $Z(t)$ and $Z(t + 30\Delta t)$ is 0.27, while that between $Z(t)$ and $Z(t + 50\Delta t)$ is 0.14. These correlations imply that, even if we started the model with the same initial values $Z_1, \ldots, Z_4$,

different realizations of the model would typically have low cross-correlation after 30 years and possess very little similarity beyond 50 years (Figure 5.38(a)). The dissimilarity is associated with the stochastic nature of the noise $\varepsilon(t)$, which quickly decorrelates the present value of the model from its past values. The fact that the two series take on similar levels near $t = 100$ years is not an indication that they are merging since extending these realizations to even longer times shows them departing from one another.

To show that initial conditions are much less important than noise, Gray and Woodward generated two samples with different starting values but with the same noise sequence. This was intended to mimic a specified set of random conditions driving the weather but having different starting values. As revealed by Figure 5.38(b), the realizations begin to merge by 30 years, demonstrating their insensitivity to the initial conditions. A further point is that for stationary AR processes, the forecast function is only a function of the sample mean and the last four observations. Because the starting values are independent of the last four observations, and small changes in the starting conditions have little effect on the sample mean for a long time series, the forecasts from such a model

will be insensitive to changes in initial conditions. In closing their article, Gray and Woodward note that conventional ARMA modeling methodology indicates that the temperature time series should first be differentiated. Application of a variety of techniques suggests an order 10 (AR(10)) model as the "optimum" model for the differentiated data, which gives rise to an AR(11) model for the original time series, not an AR(4) model used in the analysis. Lastly, Tsonis (1992) points out that it is not appropriate to change the noise of the signal without also changing the initial conditions.

5.5.3 Maximum Likelihood Spectral Estimation

As first demonstrated by Capon (1969), spectra can be defined using the maximum likelihood procedure. Instead of using a fixed

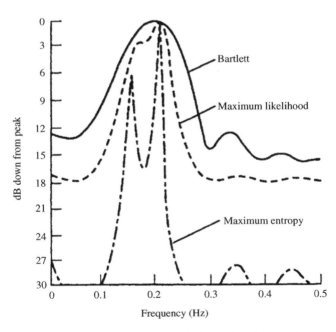

FIGURE 5.39 Power spectral estimates for a signal consisting of white noise plus two sine waves with frequencies 0.15 and 0.2 Hz (cps). Solid line: spectrum using the autocovariance method with a Bartlett smoothing window. Dashed line: maximum likelihood spectral estimate. Dash–dot line: maximum entropy spectrum. *(From Lacoss (1971).)*

FIGURE 5.40 Comparison of spectra from periodogram method and maximum entropy method (MEM). (a) A sinusoid with 10% white noise and truncated with a 1 s window; (b) the power spectrum of (a) computed as the square of the modulus of the Fourier transform; (c) the MEM power spectrum of (a). Frequency in Hz (cps). *(From Ulrych (1972).)*

window to operate on the autocorrelation function, the window shape is changed as a function of wavenumber or frequency. The window is designed to reject all frequency components in an optimal way, except for the one frequency component, which is desired.

Rather than go through the details of defining the procedure for the maximum likelihood spectrum, we offer here comparisons between the traditional method (in this case, represented by a spectrum computed using a Bartlett window), a maximum likelihood spectrum, and a spectrum computing using the maximum entropy procedure (Figure 5.39). As the figure illustrates, the maximum entropy spectrum has narrow

peaks while both the Bartlett window and maximum likelihood method yield much broader spectral peaks. Note also that, except for the maximum spectral values, the maximum entropy spectrum significantly underestimates the spectral estimates for the 0.15 Hz signal and white noise. The maximum entropy spectrum also has small side-lobe energy that is dramatically less than the off-peak energy in either of these two spectra. The maximum likelihood spectral values are also systematically lower than those using the standard method with a Bartlett window. A similar comparison is shown in Figure 5.40, which first shows a time series of a 1 Hz (1 cps) sinusoid with 10%

white noise added to it (Figure 5.40(a)). The power spectrum computed as the square of the Fourier coefficients is displayed in Figure 5.40(b). This can be compared with the narrow-peaked maximum entropy spectrum in Figure 5.40(c). The peaks are located at the same frequency representative of the 1 Hz, but the maximum entropy spectrum is extremely narrow while the Fourier and maximum likelihood power spectra have very wide peaks. It is easy to see that the MEM seriously underestimates the spectral values at frequencies other than the main peak.

5.6 CROSS-SPECTRAL ANALYSIS

Estimation of autospectral density functions deals only with the frequency characteristics of a single scalar or components of a vector time series, $x(t)$. Estimation of cross-spectral density functions performs a similar analysis but for two time series, $x_1(t)$ and $x_2(t)$, spanning concurrent times, $0 \leq t \leq T$. Although we often use time series from similar distributions, such as the velocity records from nearby moorings, cross-spectra may also be computed for two completely different quantities. In that sense, we *can* mix apples and oranges. For example, the cross-spectrum formed from the time-varying velocity fluctuations, $x_1(t) = u'(t)$, and the temperature fluctuations, $x_2(t) = T'(t)$, measured over the same time span at the same location gives an estimate of the local eddy heat flux, $q' = \rho C_\rho u'T'(t)$, as a function of frequency (ρ is the density and C_ρ is the specific heat of seawater). Because autospectra involve terms like $x_1 x_1^*$, where the asterisk denotes complex conjugate, the spectra are real-valued and all phase information in the original signal is lost. Cross-spectra, on the other hand, involve terms like $x_1 x_2^*$ and are generally complex quantities whose real and imaginary parts take into account the correlated portions of both the amplitudes and relative phases of the two signals.

There are two ways to quantify the real and imaginary parts of cross-spectra. One approach is to write the cross-spectrum as the product of an amplitude function, called the *cross-amplitude spectrum*, and a phase function called the *phase spectrum*. The sample cross-amplitude spectrum gives the distribution of coamplitudes with frequency while the sample phase spectrum indicates the angle (or time) by which one series leads or lags the other series as a function of frequency. Alternatively, the cross-spectrum can be decomposed into a *coincident spectral density function* (or *cospectrum*), which defines the degree of co-oscillation for those frequency constituents of the two time series that fluctuate in phase, and a *quadrature spectral density function* (or *quadspectrum*), which defines the degree of co-oscillation for frequency constituents of the two series that co-oscillate but are out of phase by $\pm 90°$. Statistical confidence intervals can be provided for normalized versions of the cross-spectral estimates.

5.6.1 Cross-Correlation Functions

In Section 5.4.3.1, we showed that the autocovariance function, $C_{xx}(\tau)$, and the autospectrum, $S_{xx}(f)$, are Fourier transform pairs. Similarly, for separate time series $x_1(t)$ and $x_2(t)$, the cross-covariance function, $C_{x1x2}(\tau)$, and the cross-spectrum, $S_{x1x2}(t)$, are transform pairs. Thus, we can take the Fourier transform of the lagged cross-covariance function to obtain the cross-spectrum or we can take the IFT of the cross-spectrum to obtain the cross-covariance function. As a prelude to cross-spectral analysis, it is worth presenting a brief summary of cross-correlation functions commonly used in oceanography for scalar and vector time series. The cross-correlation functions tell us how closely two records are "related" in the time domain, whereas the cross-spectrum tells us how oscillations within specific frequency bands are related in the frequency domain.

Using the abbreviation, $C_{12}(\tau)$ for the more awkward notation, $C_{x1x2}(\tau)$, the unbiased *cross-covariance function* is defined as

$$C_{12}(\tau) = \frac{1}{N-m} \sum_{N=0}^{N-m} x_1(n\Delta t) x_2(n\Delta t + \tau) \quad (5.150)$$

where $\tau = m\Delta t$ is the lag time for $m = 0, 1, ..., M$, $M \ll N$ (see also Section 5.4.3.1). Division of Eqn (5.150) by the product $C_{11}(0)C_{22}(0)$, corresponding to the autocovariance functions for each series at zero lag, gives the *cross-correlation coefficient function* for the data samples

$$\rho_{12}(\tau) = \frac{C_{12}(\tau)}{[C_{11}(0)C_{22}(0)]^{1/2}} \quad (5.151)$$

The time series $x_1(t)$ and $x_2(t)$ represent any two quantities we wish to compare. They also may represent quantities measured at different depths or locations for the same time period. For example, Kundu and Allen (1976) used the lagged covariance function

$$\rho(\mathbf{x}_1, \mathbf{x}_2, \tau) = \frac{\overline{v'(\mathbf{x}_1, t) v'(\mathbf{x}_2, t + \tau)}}{\left[\overline{(v'(\mathbf{x}_1, t))^2 (v'(\mathbf{x}_2, t))^2} \right]^{1/2}}$$

$$= \frac{\frac{1}{N-m} \sum_{n-1}^{N-m} v'(\mathbf{x}_1, n) v'(\mathbf{x}_2, n+m)}{\frac{1}{N} \left[\sum_{n=1}^{N} (v'(\mathbf{x}_1, n))^2 (v'(\mathbf{x}_2, n))^2 \right]^{1/2}},$$

$$m = 0, 1, ..., M \ll N \quad (5.152)$$

to examine the correlation between the alongshore (v) components of current for different coastal sites separated by a distance $d = |\mathbf{x}_1 - \mathbf{x}_2|$. Moreover, if τ_{max} is the lag that gives the maximum correlation, then the speed of propagation, c, of the coherent signal in the direction $\mathbf{d} = \mathbf{x}_1 - \mathbf{x}_2$ is $c = |\mathbf{d}|/\tau_{max}$, the direction of propagation determined from the sign of τ_{max} (Figure 5.41). In Figure 5.41, the lagged correlations between time series of low-pass filtered

FIGURE 5.41 The lag time of maximum correlation of the longshore component of current at 60-m depth vs the distance of separation for the Oregon coast for 1973. Results indicate a mean northward signal propagation of 120 km/day. *(From Kundu and Allen (1976).)*

alongshore currents, $v(\mathbf{x}, t)$, at different sites along the continental shelf are used to examine the poleward propagation of low-frequency coastal-trapped waves. Results in the figure are based on currents at 60-m depth. Letters refer to pairs of stations used; e.g., $C - P$ is the lag between the Carnation and Poinsettia stations.

A generalization of Eqn (5.152) is given by Kundu (1976). If $w = u + iv$ is the complex velocity for horizontal velocity components, (u, v), then the correlation between two rotating velocity vectors is given by the complex correlation coefficient

$$\rho(\mathbf{x}_1, \mathbf{x}_2, \tau) = \frac{\overline{w_1^*(t) w_2(t + \tau)}}{\left[\left(\overline{w_1^*(t) w_1(t)} \right)^{1/2} \left(\overline{w_2^*(t) w_2(t)} \right)^{1/2} \right]}$$

$$(5.153)$$

where subscripts denote locations 1 and 2, and the overbars denote the time or ensemble average. The correlation, ρ, which is independent of the choice of coordinate systems, is a complex quantity whose magnitude gives the overall measure of correlation and whose phase gives the average counterclockwise angle of the second vector with respect to the first.

5.6.2 Cross-Covariance Method

Following the Blackman–Tukey procedure for autospectral density estimation, the Fourier transform of the cross-covariance function, $C_{12}(\tau)$, can be used to find the cross-spectrum, $S_{12}(f)$. Although the cross-covariance method is straightforward to apply, the sample cross-covariance function, $C_{12}(\tau)$, suffers from the same disadvantage as the sample autocovariance function, $C_{11}(\tau)$, in that neighboring values tend to be highly correlated, thereby reducing the effective number of DoF. Moreover, the statistical significance falls off rapidly with increasing lag, τ, so that the number of lags, M, is much shorter than the record length ($M \ll N$). Calculation of cross-spectra is best performed using the direct Fourier transform method. In fact, it is common practice these days to use the IFT of the cross-spectrum to obtain the cross-covariance function.

5.6.3 Fourier Transform Method

As with autospectral analysis, estimates of cross-spectral density functions are most commonly derived using Fourier transforms. The steps in calculating the cross-spectrum using standard Fourier transforms or FFTs are similar to those discussed in Section 5.4.12 for spectra (see also Bendat and Piersol, 1986):

1. Ensure that the two time series, $x_1(t)$ and $x_2(t)$, span the same period of time, t_n, where $n = 0, 1, ..., N - 1$ and $T = N\Delta t$ is the length of each record. Remove their respective means and trends, and if needed, despike the data series. If block averaging is to be

used to improve the statistical reliability of the spectral estimates, divide the available data for each pair of time series into m sequential blocks of N' data values each, where $N' = N/m$.

2. To reduce side-lobe leakage, taper the time series, $x_1(t)$ and $x_2(t)$, using a Hanning (raised-cosine) window, Kaiser–Bessel window, or other appropriate taper. Rescale the spectra (step 4) to account for the loss of "energy" during application of the window (see Table 5.5).

3. Compute the Fourier transforms, $X_1(f_k)$, $X_2(f_k)$, $k = 0, 1, 2, ..., N - 1$, for the two time series, $x_1(t)$ and $x_2(t)$. For block-segmented data, calculate the Fourier transforms, $X_{1m}(f_k)$ and $X_{2m}(f_k)$, for each of the m blocks, where $k = 0, 1, ..., N' - 1$. To reduce the variance associated with the tapering in step 2, the transforms can be computed for overlapping segments.

4. Adjust the scale factor of $X_1(f_k)$ and $X_2(f_k)$ [or $X_{1m}(f_k)$, $X_{2m}(f_k)$] for the reduction in spectral energy due to the tapering in step 2. For the Hanning window, multiply the amplitudes of the Fourier transforms by $\sqrt{8/3}$.

5. Compute the raw cross-spectral power density estimates for each pair of time series (or each pair of blocks), where for the two-sided spectral density estimate for no block averaging

$$S_{12}(f_k) = \frac{1}{N\Delta t}[X_1^*(f_k)X_2(f_k)],$$
$$k = 0, 1, 2, ..., N - 1$$

or, for block averaging,

$$S_{12}(f_k; m) = \frac{1}{N\Delta t}[X_{1m}^*(f_k)X_{2m}(f_k)],$$
$$k = 0, 1, 2, ..., N' - 1 \qquad (5.154a)$$

and for the one-sided spectral density estimates for no block averaging

$$G_{12}(f_k) = \frac{2}{N\Delta t}\left[X_1^*(f_k)X_2(f_k)\right],$$

$$k = 0,1,2,\ldots,N/2$$

or, for block averaging

$$G_{12}(f_k;m) = \frac{2}{N\Delta t}\left[X_{1m}^*(f_k)X_{2m}(f_k)\right],$$

$$k = 0,1,2,\ldots,N'/2 \qquad (5.154b)$$

6. In the case of the block-segmented data, average the raw cross-spectral density estimates from the m blocks of data to obtain the smoothed periodogram for $S_{12}(f_k)$, the two-sided cross-spectrum, or $G_{12}(f_k)$, the one-sided cross-spectrum.

Cross-covariance function: Since the cross-covariance function, $C_{12}(\tau)$ $(=R_{12}(\tau)$, the cross-correlation function, if the mean is removed from the record), and the cross-spectrum are Fourier transform pairs, Eqn (5.154) can be used to obtain a smoothed or unsmoothed estimate of the cross-covariance function. To do this, we first calculate the Fourier transforms $X_1(f)$ and $X_2(f)$ of the individual time series, and then determine the product $S_{12}(f)=(N\Delta t)^{-1}[X_1^*(f)X_2(f)]$. We then take the IFT of the cross-spectrum, $S_{12}(f)$, to obtain the cross-covariance function

$$C_{12}(\tau) = \int\limits_{-\infty}^{\infty} S_{12}(f)e^{i2\pi f\tau}df \qquad (5.155)$$

If the spectrum is unsmoothed prior to the IFT (or IFFT if the number of spectral estimates is a power of two), we obtain the raw cross-covariance function. If, on the other hand, the cross-spectrum is smoothed prior to Eqn (5.155) using one of the spectral windows, such as the Hanning window, the cross-covariance function also will be a smoothed function.

We can use the acoustic backscatter data in Table 5.1 to illustrate the direct and indirect

TABLE 5.6 Unsmoothed, Normalized Cross-Covariance Function, $\rho_{12}(\tau)$, Given by Eqn (5.151), as a Function of Lag τ in Increments of 5 m for Bin 1 of Beams 1 and 2 of the Acoustic Backscatter Spatial Series (Profiles) Listed in Table 5.1

Lag τ (m)	0	5	10	15	20	25	30	35
	0.96	0.94	0.85	0.71	0.57	0.48	0.40	0.31
Lag τ (m)	40	45	50	55	60	65	70	75
	0.23	0.14	0.02	−0.19	−0.24	−0.37	−0.46	−0.48

The first acoustic bins of the two beams (bin#1in each case) are separated by a horizontal distance of roughly 3.9 m.

methods for calculating the cross-covariance function. In Table 5.6, we present the normalized, unsmoothed cross-covariance function, $\rho_{12}(\tau)=C_{12}(\tau)/[C_{11}(0)C_{22}(0)]^{1/2}$, obtained directly from the definition Eqn (5.150). In this case, the lag τ is in 5-m depth increments. The indirect approach is based on the Fourier estimates presented in Tables 5.7 and 5.8. Here, we first give the Fourier transforms, $X_1(f)$ and $X_2(f)$, of the two profile series as a function of wavenumber, f (Table 5.7). We next calculate the cross-spectrum, $S_{12}(f)=(N\Delta t)^{-1}[X_1^*(f)X_2(f)]$, and then take the inverse transform of $S_{12}(f)$ to obtain the cross-covariance function, $C_{12}(\tau)$, as a function of lag (Table 5.8). No smoothing was applied to either data set, and the results obtained from the IFT method are identical to those listed in Table 5.6, within roundoff error. The advantage of the transform approach is that it is straightforward to derive a smoothed cross-covariance function by windowing the cross-spectral estimate prior to Fourier inversion.

5.6.4 Phase and Cross-Amplitude Functions

Suppose that the constituents of the bivariate time series $\{x_1(t), x_2(t)\}$ have the same frequency, f_0, but different amplitudes (A_1, A_2) and different phases (ϕ_1, ϕ_2), respectively. In particular, let

$$x_k(t) = A_k \cos\left(2\pi f_0 t + \phi_k\right), \quad k = 1,2 \quad (5.156)$$

TABLE 5.7 Complex Fourier Transforms of $X_1(f_k)$ and $X_2(f_k)$ for the Profiles of Acoustic Backscatter Listed in Table 5.1

FFT	k = 0	1	2	3	4	5	6	7
$X_1(f_k)$	348.13	−289.32	71.17	15.52	55.16	97.59	−28.66	5.07
	0.00	214.96	−16.35	−117.25	105.57	−16.98	−21.37	4.28
$X_2(f_k)$	339.02	−226.53	119.54	55.84	−5.24	59.55	−36.39	4.22
	0.00	227.88	38.22	−93.12	122.33	−24.13	−6.57	−19.09
k = 8	**9**	**10**	**11**	**12**	**13**	**14**	**15**	**16**
1.13	−6.16	41.11	24.03	−1.79	4.63	3.74	4.09	27.13
6.87	21.29	−2.96	−36.43	−4.60	1.08	3.54	18.45	0.00
11.90	5.68	23.89	13.85	3.96	7.37	11.27	2.34	27.79
−5.35	−4.63	−5.13	−18.72	−1.67	−4.93	−4.47	9.00	0.00

For each wavenumber, f_k, the table lists the real part of the transform (top line of each X pair) followed by the imaginary part (bottom line of each X pair), where $X_j(f_k) = ReX_j(f_k) + iImX_j(f_k)$, $j = 1, 2$. The vertical wavenumber, $f_k = k_F$, $k = 0, 1,..., 16$, where the fundamental vertical wavenumber, $f' = 1/155\ m = 0.00645\ cpm$ (cycles per meter).

TABLE 5.8 The Inverse Fast Fourier Transform (IFFT) of the Cross-Spectrum $S_{12}(f_k) = (N\Delta t)^{-1}[X_1(f_k)^* X_2(f_k)^*]$ Using the Values in Table 5.7

	$\tau = 0$	1	2	3	4	5	6	7
$C_{12}(\tau)$	13483.7	12752.4	11151.5	9087.4	6992.3	5436.5	4589.9	3411.7
$\tau = 8$	**9**	**10**	**11**	**12**	**13**	**14**	**15**	**16**
2382.5	1393.8	160.6	−1103.6	−2096.0	−3103.5	−3610.5	−3623.8	−3222.1

The values represent the raw (unnormalized) estimates of the cross-covariance function, $C_{12}(\tau)$, as a function of lag τ ($0 \leq \tau \leq 16$) in increments of 5 m for bin 1 of Beams 1 and 2 of the acoustic backscatter spatial series (profiles) listed in Table 5.1.

The Fourier transform of $x_k(t)$, over $-T/2 \leq t \leq T/2$ is

$$X_k(f) = \frac{A_k}{2}\left\{ e^{i\phi_k}\frac{\{\sin[\pi(f - f_0)T]\}}{\pi(f - f_0)} + e^{-i\phi_k}\frac{\{\sin[\pi(f + f_0)T]\}}{\pi(f + f_0)} \right\},$$

$$i = 1, 2 \tag{5.157}$$

Hence, the sample cross-spectra of the two series is

$$S_{12}(f) = \frac{1}{T}[X_1^*(f)X_2(f)] \tag{5.158}$$

where X_1^* is the complex conjugate of X_1. From this expression, we obtain

$$S_{12}(f) = \frac{A_1 A_2}{4T}\left\{ e^{-i\phi_1}\frac{\sin[\pi(f - f_0)T]}{\pi(f - f_0)} + e^{i\phi_1}\frac{\sin[\pi(f + f_0)T]}{\pi(f + f_0)} \right\}$$
$$\times \left\{ e^{i\phi_2}\frac{\sin[\pi(f - f_0)T]}{\pi(f - f_0)} + e^{-i\phi_1}\frac{\sin[\pi(f + f_0)T]}{\pi(f + f_0)} \right\} \tag{5.159}$$

where

$$S_{12}(f) \xrightarrow[T \to \infty]{} \frac{A_1 A_2}{4} \left[e^{-i(\phi_2 - \phi_1)} \delta(f + f_0) \right.$$

$$\left. + e^{i(\phi_2 - \phi_1)} \delta(f - f_0) \right] \quad (5.160)$$

The phase difference, $(\phi_2 - \phi_1)$, in the above expressions determines the lead (or lag) of one cosine oscillation relative to the other for given frequency, f. The cross-amplitude, $A_1 A_2$, is the square of the geometric mean amplitude of the co-oscillation for frequency, f. From Eqn (5.151), the sample cross-spectrum is

$$S_{12}(f) = \frac{A_1(f) A_2(f)}{T} \left[e^{i[\phi_2(f) - \phi_1(f)]} \right] \quad (5.161)$$

or

$$S_{12}(f) = A_{12}(f) \left[e^{i\phi_{12}(f)} \right] \quad (5.162)$$

where the sample phase spectrum, $\phi_{12}(f) = \phi_2(f) - \phi_1(f)$, is an odd function of frequency, and the sample cross-amplitude spectrum, $A_{12}(f) = A_1(f) A_2(f)/T$, is a positive even function of f.

5.6.5 Coincident and Quadrature Spectra

An alternative description of the above information involves formulating the cross-spectra in terms of coincident (C) and quadrature (Q) spectra. In this case, we can write

$$S_{12}(f) = C_{12}(f) - iQ_{12}(f) \quad (5.163)$$

where

$$C_{12}(f) = A_{12}(f) \cos [\phi_{12}(f)];$$

$$Q_{12}(f) = -A_{12}(f) \sin [\phi_{12}(f)] \quad (5.164)$$

and

$$A_{12}^2(f) = C_{12}^2(f) + Q_{12}^2(f);$$

$$\phi_{12}(f) = \tan^{-1} \left[\frac{-Q_{12}(f)}{C_{12}(f)} \right] \quad (5.165)$$

Here $C_{12}(f)$ is an even function of frequency and $Q_{12}(f)$ is an odd function. The cospectral

density function, $C_{12}(\tau)$, for frequency, f is not to be confused with the covariance function, $C_{12}(\tau)$, at time lag τ. Where confusion may arise, we use the cross-correlation, $R_{12}(\tau)$, in place of $C_{12}(\tau)$.) If we consider the bivariate cosine example that we used in Eqn (5.156), we have

$$C_{12}(f) = \frac{A_1 A_2}{4} \cos (\phi_2 - \phi_1) [\delta(f + f_0) + \delta(f - f_0)]$$

$$= \left\{ \frac{A_1 \cos \phi_1 \cdot A_2 \cos \phi_2}{4} + \frac{A_1 \sin \phi_1 \cdot A_2 \sin \phi_2}{4} \right\}$$

$$\times [\delta(f + f_0) + \delta(f - f_0)] \quad (5.166)$$

The sample cospectrum, $C_{12}(f)$, measures the covariance between the two cosine components and the two sine components. That is, it measures the contributions to the cross-spectrum from those components of the two time series that are "in phase" (phase differences of 0 or 180°). The sample quadrature spectrum, $Q_{12}(f)$, determines the contributions from those components of the time series that are coherent but "out of phase" (phase difference ±90°).

5.6.5.1 Relationship of Co- and Quad-Spectra to Cross-Covariance

The inverse transform of the cross-spectrum gives the cross-covariance (cross-correlation)

$$R_{12}(\tau) = \int_{-\infty}^{\infty} [C_{12}(f) - iQ_{12}(f)] e^{i2\pi f \tau} df$$

$$= \int_{-\infty}^{\infty} C_{12}(f) \cos (2\pi f \tau) df$$

$$+ \int_{-\infty}^{\infty} Q_{12}(f) \sin (2\pi f \tau) df \quad (5.167)$$

Since $C_{12}(f)$ is an even function, $R_{12}(0) = \int_{-\infty}^{\infty} C_{12}(f) df$. If we define

$$C_{12}(f) = \int_{-T}^{T} R_{12}^{+}(\tau)\cos(2\pi f\tau)d\tau$$

$$(5.168)$$

$$Q_{12}(f) = \int_{-T}^{T} R_{12}^{-}(\tau)\sin(2\pi f\tau)d\tau$$

then

$$R_{12}^{+}(\tau) = \frac{1}{2}[R_{12}(\tau) + R_{12}(-\tau)] \quad \text{(the even part)}$$

$$R_{12}^{-}(\tau) = \frac{1}{2}[R_{12}(\tau) - R_{12}(-\tau)] \quad \text{(the odd part)}$$

$$(5.169)$$

5.6.6 Coherence Spectrum (Coherency)

The *squared coherency, coherence-squared function,* or *coherence spectrum* between two time series, $x_1(t)$ and $x_2(t)$, is defined for frequencies, f_k, $k = 0, 1, ..., N-1$, as

$$\gamma_{12}^2(f_k) = \frac{|G_{12}(f_k)|^2}{G_{11}(f_k)G_{22}(f_k)}$$

$$= \frac{|S_{12}(f_k)|^2}{S_{11}(f_k)S_{22}(f_k)} \qquad (5.170)$$

$$= \frac{[C_{12}^2(f_k) + Q_{12}^2(f_k)]^2}{S_{11}(f_k)S_{22}(f_k)}$$

where $G_{11}(f_k)$ is the one-sided spectrum (confined to $f_k \geq 0$), $S_{11}(f_k) = \frac{1}{2}G_{11}(f_k)$ is the two-sided spectrum defined for all frequencies and $G_{12}(f_k)$ is the one-sided cross-spectrum. Here

$$0 \leq |\gamma_{12}^2(f_k)| \leq 1 \qquad (5.171)$$

and

$$\gamma_{12}(f) = |\gamma_{12}^2(f_k)|^{1/2}e^{-i\phi_{12}f_k} \qquad (5.172)$$

where $|\gamma_{12}^2(f_k)|^{1/2}$ is the modulus of the coherence function and $\phi_{12}(f_k)$ the phase lag between the two signals at frequency f_k, (Figure 5.42). In the

literature, both the squared coherency, γ_{12}^2, and its square root are termed "the coherence" so that there is often a confusion in meaning (Julian, 1975). To avoid any ambiguity, it is best to use squared-coherency when conducting coherence analyses once the sign of the coherence function is determined. This has the added advantage that squared coherency represents the fraction of the variance in x_1 ascribable to x_2 through a linear relationship between x_1 and x_2. Two signals of frequency, f_k are considered highly coherent and in phase if $|\gamma_{12}^2(f_k)| \approx 1$ and $\phi_{12}(f_k) \approx 0$, respectively (Figure 5.42). The addition of random noise to the functions, x_1 and x_2, of a linear system decreases the coherence-squared estimate and increases the noisiness of the phase associated with the system parameters. Estimation of $\gamma_{12}^2(f_k)$ is one of the most difficult problems in time series analysis since it is so highly noise dependent. We also point out that phase estimates generally become unreliable where coherency amplitudes fall below the 90–95% confidence levels for a given frequency.

The real part of the coherence function, $\gamma_{12}(f_k)$, lies between -1 and $+1$ while the squared-coherency is between 0 and $+1$. If the noise spectrum, $S_{\varepsilon\varepsilon}(f_k)$, is equal to the output spectrum, then the coherence function is zero. This says that white noise is incoherent, as required. Also, when $S_{\varepsilon\varepsilon}(f_k) = 0$, we have $\gamma_{12}^2(f_k) = 1$; that is the coherence is perfect if there is no spectral noise in the input signal. We note that, if no spectral smoothing is applied, we are assuming that there is no spectral noise. In this case, the coherency spectrum will be unity for all frequencies, which is clearly not physically realistic. Noise can be introduced to the system by smoothing over adjacent frequencies. We also can overcome this problem by a prewhitening step that introduces some acceptable noise into the spectra.

5.6.6.1 Confidence Levels

The final step in any coherence analysis is to specify the confidence limits for the coherence-square estimates. If $1 - \alpha$ is the $(1-\alpha)100\%$

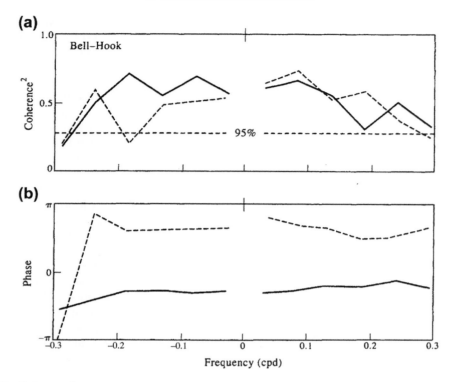

FIGURE 5.42   Coherence between current vector time series at sites Hook and Bell on the northeast coast of Australia (separation distance ≈ 300 km). (a) Coherence squared; (b) phase lag. Solid line: inner rotary coherence (rotary current components rotating in the same sense). Dashed line: outer rotary coherence (rotary current components rotating in the opposite sense). The increase in inner phase with frequency indicates equatorward phase propagation. Positive phase means that Hook leads Bell. *(From Middleton and Cunningham (1984).)*

confidence interval we wish to specify for a particular coherence function, then, for all frequencies, the limiting value for the coherence-square (i.e., the level up to which coherence-square values can occur by chance) is given by

$$\gamma_{1-\alpha}^2 = 1 - \alpha^{[1/(\text{EDoF}-1)]}$$
$$= 1 - \alpha^{[2/(\text{DoF}-2)]} \qquad (5.173)$$

where $\text{EDoF} = \text{DoF}/2$ is the number of independent cross-spectral realizations in each frequency band (Thompson, 1979). The commonly used confidence intervals of 90, 95, and 99% correspond to $\alpha = 0.10$, 0.05, and 0.01,

respectively. As an example, suppose that each of our coherence estimates is computed from an average over three adjacent cross-spectral Fourier components, then $\text{EDoF} = 3$ ($\text{DoF} = 6$). The 95% confidence level for the squared coherence would then be $\gamma_{95}^2 = 1 - (0.05)^{0.5} = 0.78$. Alternatively, if the cross-spectrum and spectra were first smoothed using a Hamming window spanning, the entire width of the data series, $\text{EDoF} = 2.5164$ (Table 5.5) and the 95% confidence interval $\gamma_{95}^2 = 1 - (0.05)^{0.6595} = 0.86$. For $\text{EDoF} = 2$, $\gamma_{1-\alpha}^2 = 1 - \alpha$ so that the confidence level for the 95% confidence interval is equal to itself.

A useful reference for coherence significance levels is Thompson (1979). In this paper, the author tests the reliability of significance levels, $\gamma^2_{1-\alpha}$, estimated from Eqn (5.173) with the coherence-square values obtained through the summations

$$\gamma^2(f) = \frac{\left| \sum_{k=1}^{K} X_{1k}(f) X_{2k}^*(f) \right|^2}{\sum_{k=1}^{K} |X_{1k}(f)|^2 \sum_{k=1}^{K} |X_{2k}(f)|^2}$$

$$(5.174)$$

In this expression, X_{1k} and X_{2k} are the Fourier transforms of the respective random time series, $x_{1k}(t)$ and $x_{2k}(t)$, generated by a Monte Carlo approach, and the asterisk denotes the complex conjugate. The upper limit K corresponds to the value of EDoF in Eqn (5.173a). Because $\gamma^2(f)$ is generated using random data, it should reflect the level of squared coherency that can occur by chance. For each value of K, $\gamma^2(f)$ was calculated 1000 times and the resultant values sorted as 90th, 95th, and 99th percentiles. The operation was repeated 10 times and the means and standard deviations were calculated. This amounts to a total of 20,000 Fourier transforms for each K (=EDoF). There is excellent agreement between the significance level derived from Eqn (5.173) and the coherence-square values for a white-noise Monte Carlo process (Table 5.9), lending considerable credibility to the use of Eqn (5.173) for computing coherence significance levels. The comparisons in Table 5.9 are limited to the 90 and 95% confidence intervals for $4 \leq K \leq 30$. Thompson (1979) includes the 99% interval and a wider range of K (EDoF) values.

Confidence intervals for coherence amplitudes, as well as for coherence phase, admittance, and other signal properties (see next section), can be derived using the data itself (Bendat and Piersol, 1986). Let $\widehat{\varphi}$ be an estimator for ϕ, a continuous, stationary random process, and define the standard error or random error of sample values as

TABLE 5.9   Monte Carlo Estimates, $\gamma^2(f)$, of the Significant Coherence-Squared and Prediction of This Value Using Eqn (5.173) for Significance Intervals $\alpha = 0.05$ and 0.10 for Equivalent Degrees of Freedom (EDoF) = 4, 5, 6, 8, 10, 20, and 30

	EDoF = 4	EDoF = 5	EDoF = 6	EDoF = 8	EDoF = 10	EDoF = 20	EDoF = 30
$\alpha = 0.10$							
$\gamma^2(f)$	0.539	0.437	0.371	0.288	0.230	0.114	0.076
$\gamma^2_{0.90}$	0.536	0.438	0.369	0.280	0.226	0.114	0.076
$\alpha = 0.05$							
$\gamma^2(f)$	0.629	0.531	0.452	0.354	0.288	0.144	0.099
$\gamma^2_{0.95}$	0.632	0.527	0.451	0.348	0.283	0.146	0.098

(After Thompson (1979).)

$$\text{random error} = \sigma[\widehat{\varphi}] = \left(E[\widehat{\varphi}^2] - E^2[\widehat{\varphi}] \right)^{1/2}$$

$$(5.175a)$$

and the RMS error as

$$\text{RSM error} = \left(E[(\widehat{\varphi} - \varphi)^2] \right)^{1/2}$$

$$= \left(\sigma^2[\widehat{\varphi}] + B^2[\widehat{\varphi}] \right)^{1/2} \quad (5.175b)$$

where B is the bias term $B[\widehat{\varphi}] = E[\widehat{\varphi}] - \phi$ and $E[x]$ is the expected value of x. If we now divide each error term by the quantity, φ being estimated, we obtain the normalized random error

$$\varepsilon_r = \frac{\sigma[\widehat{\varphi}]}{\varphi} = \frac{\left(E[\widehat{\varphi}^2] - E^2[\widehat{\varphi}] \right)^{1/2}}{\varphi} \quad (5.176a)$$

and the normalized RMS error

$$\varepsilon = \frac{\left(E[(\widehat{\varphi} - \varphi)^2] \right)^{1/2}}{\varphi} = \frac{\left(\sigma^2[\widehat{\varphi}] + B^2[\widehat{\varphi}] \right)^{1/2}}{\varphi}$$

$$(5.176b)$$

where it is assumed that $\varphi \neq 0$. Provided ε_r is small, the relation

$$\widehat{\varphi} = \varphi^2(1 \pm \varepsilon_r) \quad (5.177)$$

yields

$$\widehat{\varphi} = \varphi(1 \pm \varepsilon_r)^{1/2} \approx \varphi(1 \pm \varepsilon_r/2) \qquad (5.178)$$

so that

$$\varepsilon_r[\widehat{\varphi}^2] \approx 2\varepsilon_r[\widehat{\varphi}] \qquad (5.179)$$

Thus, for small ε_r the normalized error for squared estimates $\widehat{\varphi}^2$ is roughly twice the normalized error for unsquared estimates.

When the estimates $\widehat{\phi}$ have a small bias error, $B[\widehat{\varphi}] \approx 0$, and a small normalized error, e.g., $\varepsilon \leq 0.2$, the probability density for the estimates can be approximated by a Gaussian distribution. The confidence intervals for the unknown true parameter, φ, based on a single estimate, $\widehat{\varphi}$, are then

$$\widehat{\varphi}(1 - \varepsilon) \leq \varphi \leq \widehat{\varphi}(1 + \varepsilon) \quad \text{with 68\% confidence}$$
$$(5.180a)$$

$$\widehat{\varphi}(1 - 2\varepsilon) \leq \varphi \leq \widehat{\varphi}(1 + 2\varepsilon) \quad \text{with 95\% confidence}$$
$$(5.180b)$$

$$\widehat{\varphi}(1 - 3\varepsilon) \leq \varphi \leq \widehat{\varphi}(1 + 3\varepsilon) \quad \text{with 99\% confidence}$$
$$(5.180c)$$

5.6.7 Frequency Response of a Linear System

We define the admittance (or transfer) function of a linear system as

$$H_{12}(f_k) = \frac{S_{12}(f_k)}{S_{11}(f_k)} = \frac{G_{12}(f_k)}{G_{11}(f_k)},$$
$$f_k = k/T, \quad k = 1, \dots, N$$
$$= |H_{12}(f_k)| e^{-i\phi_{12}(f_k)} \qquad (5.181)$$

where $S_{11}(f_k)$ and $G_{11}(f_k)$ are, respectively, the two-sided and one-sided autospectrum estimates for the time series $x_1(t)$ selected here as the input time series. The gain (or admittance amplitude) function, $H(f_k)$, behaves like a

spectral regression coefficient at each frequency, f_k. Using the definition $G_{12}(f_k) = C_{12}(f_k) - iQ_{12}(f_k)$, where C is the co-spectrum and Q is the quadrature spectrum, we obtain

$$|H_{12}(f_k)| = \frac{G_{12}(f_k)}{G_{11}(f_k)}$$
$$(5.182)$$
$$= \frac{|C_{12}^2(f_k) + Q_{12}^2(f_k)|^{1/2}}{G_{11}(f_k)}$$

and where $\phi_{12}(f_k) = \tan^{-1}[-Q_{12}(f_k)/C_{12}(f_k)]$ by Eqn (5.165). Figure 5.43 shows the complex admittance for the observed alongshore component of oceanic wind velocity (time series 1) and the alongshore component of wind velocity derived from pressure-derived geostrophic winds (time series 2). As noted in the figure caption, the analysis is based on two separate methods for defining what is meant by the "alongshore component" of wind velocity. For both definitions of "alongshore" (as represented by the solid and dashed lines in the figure), the geostrophic winds closely approximate the amplitude and phase of the actual winds up to a frequency of about 0.05 cph (period = 20 h; $\log(0.05) = -1.3$ (cph)) after which the two signals no longer resemble one another. It is also at this frequency that the coherence consistently begins to fall below the 90% confidence level.

5.6.7.1 Multi-Input Systems Cross-Spectral Analysis

Many oceanographic time series are generated through the combined effects of several mutually coherent inputs. For example, low-frequency fluctuations in coastal sea level typically arise through the combined forcing of atmospheric pressure, along- and cross-shore wind stress, and surface buoyancy flux. Coherences between the forcing variables (e.g., pressure, alongshore wind stress, and runoff) are generally quite high. Because of this, it would be physically incorrect to use

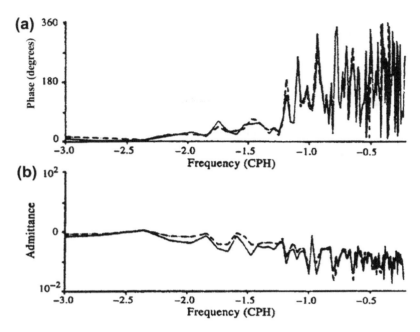

(a)

(b)

FIGURE 5.43 Complex admittance for observed (series 1) and computed (series 2) alongshore components of oceanic wind velocity for May—September 1980 off the coast of Vancouver Island. (a) Phase; (b) admittance amplitude. Positive phase means that series 1 leads series 2. Solid lines are for "alongshore" defined as parallel to the local shoreline; dashed lines are for "alongshore" derived using principal component analysis. The horizontal axes are log(frequency) where frequency is in CPH. *(From Thomson (1983).)*

ordinary cross-spectral analysis, which simply examines the correlation functions, $\gamma^2_{y:x}$, between the output, $y(t)$, and each of the inputs, $x(t)$, individually without taking into account the mutual correlation among all the inputs. If this is not done, the sum of the individual correlation functions can exceed unity. Provided that long-term sea-level fluctuations (the output time series) are linearly related to the individual forcing functions (the input time series), we can use *multi-input systems cross-spectral analysis* to calculate the relative contribution each of the input terms makes to the output. The effective correlation function for the total system will then be less than unity, as required. This concept was pioneered in oceanography by Cartwright (1968), Groves and Hannan (1968), and Wunsch (1972). All three studies were concerned with sea-level variations.

The purpose of this section is to provide a brief overview of multiple systems analysis.

For a thorough generalized presentation, the reader is directed to Bendat and Piersol (1986). Consider K constant-parameter linear systems associated with K stationary and ergodic input time series, $x_k(t)$, $k = 1, 2, \ldots, K$, a noise function, $\varepsilon(t)$, and a single output, $y(t)$, such that

$$y(t) = \sum_{k=1}^{K} y_k(t) + \varepsilon(t) \qquad (5.183)$$

where $y_k(t)$ are the outputs generated by each of the measured inputs, $x_k(t)$. We can only measure the accumulated response, $y(t)$, not the individual responses, $y_k(t)$. In the present context, $y(t)$ represents the measured time series of coastal sea level, $x_k(t)$ the corresponding weather variables, and $\varepsilon(t)$ the deviations from the ideal response due to instrument noise, remotely generated subinertial waves (waves with periods greater than the local inertial frequency), and other physical processes not correlated with the input functions. The Fourier transform of the output, $y(t)$ is

$$
\begin{aligned}
Y(f) &= \sum_{k=1}^{K} Y_k(f) + E(f) \\
&= \sum_{k=1}^{K} H_k(f)X_k(f) + E(f)
\end{aligned}
\tag{5.184}
$$

where

$$
H_k(f) = \frac{Y_k(f)}{X_k(f)}, \quad k = 1, 2, ..., K \tag{5.185}
$$

is the admittance (or transfer) function relating the kth input with the kth output at frequency, f. The frequency-domain spectral variables, $X_k(f)$ and $Y(f)$, can be computed from the measured time series, $x_k(t)$ and $y(t)$. Using these variables, we can then determine the functions, $H_k(f)$ and other properties of the system.

Multiplication of both sides of Eqn (5.184) by $X_j^*(f)$, the complex conjugate of $Xj(f)$, for any fixed $j = 1$, 2, ..., K, yields the power spectral relation

$$
\begin{aligned}
S_{jy}(f) &= \sum_{k=1}^{K} H_k(f)S_{jk}(f) + S_{je}(f), \\
&\quad j = 1, 2, ..., K
\end{aligned}
\tag{5.186}
$$

in which

$$
\begin{aligned}
S_{jy}(f) &= \overline{X_j^*(f)Y(f)}, \quad j = 1, 2, ..., K \\
S_{jk}(f) &= \overline{X_j^*(f)X_k(f)}, \quad j = 1, 2, ..., K
\end{aligned}
\tag{5.187}
$$

Here, the overbar denotes the average value, the $S_{jy}(f)$ are the cross-spectra between the K inputs and the single output, $S_{jk}(f)$ are the cross-spectra $(j \neq k)$ and spectra $(j = k)$ among the input variables, and $S_{j\varepsilon}(f)$ is the cross-spectrum between the input variables and the noise function. If the noise function, $\varepsilon(t)$, is uncorrelated with each input, x_k (as is normally

assumed), the cross-spectral terms, $S_{j\varepsilon}(f)$, will be zero and Eqn (5.186) becomes

$$
S_{jy}(f) = \sum_{k=1}^{K} H_k(f)S_{jk}(f), \quad j = 1, 2, ..., K \tag{5.188}
$$

This expression is a set of K equations in K unknowns—the $H_k(f)$ for $k = 1, 2, ..., K$—where all spectral terms can be computed from the measured records of $y(t)$ and $x_k(t)$. If the model is well defined, matrix techniques can be used to find the $H_k(f)$. Bendat and Piersol (1986) also define the problem in terms of the *multiple and partial coherence functions* for the system. The multiple coherence function is given by

$$
\gamma_{y:x}^2 = \frac{S_{vv}(f)}{S_{yy}(f)} = 1 - \frac{S_{\varepsilon\varepsilon}(f)}{S_{yy}(f)} \tag{5.189}
$$

where $S_{vv}(f)$ is the multiple coherent output spectrum, $S_{yy}(f)$ is the output spectrum, and $S_{\varepsilon\varepsilon}(f)$ is the noise spectrum. As with any squared coherence function, $0 \leq |\gamma_{y:x}^2| \leq 1$. For any problem with multiple inputs, $\gamma_{y:x}^2$ takes the form of a matrix whose off-diagonal elements take into account the coherent interactions among the different input terms. Expressions (5.188) and (5.189) simplify even further if the inputs themselves are mutually uncorrelated. In that case

$$
H_j(f) = \frac{S_{jy}(f)}{S_{jj}(f)}, \quad j = 1, 2, ..., K; \\
|H_j(f)|^2 S_{jj}(f) = \gamma_{jy}^2 S_{yy}(f) \tag{5.190}
$$

Hence, the contribution of the input variable, $x_j(t)$, to the output variable, $y(t)$, occurs only through the transfer (admittance) function, $H_j(f)$, of that particular input variable. No leakage of $x_j(t)$ takes place through any of the other transfer functions since $x_j(t)$ is uncorrelated with $x_k(t)$ for $k \neq j$.

In general, the output, $y(t)$, is forced not only by the mutually coherent parts of the various inputs but also by the noncoherent portions of the inputs that go directly to the output through their own transfer functions without being affected by other transfer functions. This leads to the need for *partial coherence functions*. If part of one record causes part or all of a second record, then turning off the first record will eliminate the correlated parts from the second record and leave only that part of the second record that is not due to the first record. Because we do not want to incorporate the coherent portions of given forcing terms in the partial coherence functions, the partial coherences are found by first subtracting out the coherent parts of the various input signals. Bendat and Piersol (1986) state that, if any correlation between $x_1(t)$ and $x_2(t)$ is due to $x_1(t)$, then the optimum linear effects of $x_1(t)$ to $x_2(t)$ should be found. Denoting this mutual effect as $x_{2 \cdot 1}(t)$, this should be subtracted from $x_2(t)$ to yield the conditioned (or residual) record, $x_{2 \cdot 1}(t)$ representing that part of $x_2(t)$ not due to $x_1(t)$.

Multi-input systems cross-spectral analysis takes into account the fact that any input record, $x_k(t)$, with nonzero correlations between other inputs will contribute to variations in the output, $y(t)$, by passage through any of the K linear systems, $H_k(f)$. The conditioned portion of $x_k(t)$ will contribute directly to the output through its own response function only. The problem is to determine what percentage contribution each input function makes to the total variance of $y(t)$ for a specified frequency band. The simplest case is a two-input system consisting of inputs $x_1(t)$ and $x_2(t)$ for which

$$Y(f) = H_1(f)X_1(f) + H_2(f)X_2(f) + E(f)$$
$$(5.191)$$

and, provided $\gamma_{12}^2 \neq 0$

$$H_1(f) = \frac{S_{1y}(f)\left[1 - \frac{S_{12}(f)S_{2y}(y)}{S_{22}(f)S_{1y}(y)}\right]}{S_{11}(f)\left[1 - \gamma_{12}^2(f)\right]} \qquad (5.192a)$$

$$H_2(f) = \frac{S_{2y}(f)\left[1 - \frac{S_{21}(f)S_{1y}(y)}{S_{11}(f)S_{2y}(y)}\right]}{S_{22}(f)\left[1 - \gamma_{12}^2(f)\right]} \qquad (5.192b)$$

What is important to note here is the nonzero coupling between the different input variables when the cross-coherence, $\gamma_{12}^2(f)$, is nonzero. The product, $H_1(f)S_{11}(f)$, in Eqn (5.192a) still represents the ordinary coherent spectrum between the input, x_1, and the output, y. However, when $|\gamma_{12}| \neq 0$, $x_1(t)$ influences $y(t)$ through the transfer function, $H_2(f)$, as well as through its own transfer function, $H_1(f)$. Similarly, $x_2(t)$ influences $y(t)$ through the transfer function $H_1(f)$ as well as through its transfer function $H_2(f)$ (Eqn (5.192b)). In general, the sum of $\gamma_{1y}^2(f)$ and $\gamma_{2y}^2(f)$ can be greater than unity when the outputs are correlated. The contributions from the conditioned records of $x_1(t)$ and $x_2(t)$ must also be taken into account when estimating the output response, $y(t)$. Once this is done, it becomes possible to construct reliable forecasting models for y.

Cartwright (1968) used the multiple input method to study tides and storm surges around eastern and northern Britain. He expanded the tide height, ζ, at each of the ports studied as a Taylor series of the atmospheric pressure, p, about the port location ($x = 0$, $y = 0$)

$$\zeta(x, y, t) = p_{00}(t) + xp_{10}(t) + yp_{01}(t) + x^2 p_{20}(t)$$
$$+ 2xy p_{11}(t) + y^2 p_{02}(t) + \dots$$
$$(5.193)$$

in which the pressure gradient terms (p_{10}, p_{01}) $= (\partial p / \partial x, \ \partial p / \partial y)$ are proportional to the geostrophic wind stress, the second derivatives (p_{20}, p_{02}) $= (\partial^2 p / \partial x^2, \ \partial^2 p / \partial y^2)$ are related to wind stress gradients, and so on. As indicated by Table 5.10, the variances in different frequency bands for the sea level at Aberdeen, Scotland are significantly reduced relative to the original values as the pressure, first derivatives, and second derivatives are successively included. Consequently, all of the mutually correlated weather variables are considered

TABLE 5.10 Residual Variances (cm^2) for Different F-requency Bands for Aberdeen, Scotland Sea-Level Oscillations

Variables included	0–0.5 cpd	0.5–0.8 cpd	1.1–1.8 cpd	2.1–2.8 cpd
Original variance	181	16	9.6	4.1
p_{00}	88	13	9.1	3.9
p_{00}, p_{10}, p_{01}	49	9	7.1	3.6
$p_{00}, p_{10}, p_{01}, \ldots, p_{02}$	38	6	5.3	3.3

The predictive model explains increasingly more of the variance as additional weather variables are incorporated in the analysis. Periods are in cycles per day (cpd). (Modified after Cartwright (1968).)

relevant to the predictability of sea level. In a more recent study, Sokolova et al. (1992) used the multiple spectral analysis technique to study sea-level oscillations measured from July to September, 1986 at different locations around the perimeter of the Sea of Japan (also, the East Sea). According to their analysis for both the multiple and partial coherences, 46–77% of the sea-level variance was coherent with atmospheric pressure and 5–37% was coherent with the wind stress.

5.6.8 Rotary Cross-Spectral Analysis

As outlined in Section 5.4.4, the decomposition of a complex horizontal velocity vector, $w(t) = u(t) + iv(t)$, into counter-rotating circularly polarized components can aid in the analysis and interpretation of oceanographic time series. (Here, u and v typically represent the eastward and northward or, alternatively, the alongshore and cross-shore, components of the current or wind velocity.) Many of the fundamentals of this approach can be found in Fofonoff (1969), Gonella (1972), Mooers (1973), Calman (1978), and Hayashi (1979). In rotary spectral analysis, the different frequency components of the vector, $w(t)$, are represented in terms of clockwise and counter-clockwise rotating vectors (Figure 5.14). The

counterclockwise component is considered to be rotating with positive angular frequency ($\omega \geq 0$) and the clockwise component with negative angular frequency ($\omega \leq 0$). Depending on which of the two components has the largest magnitude, the vector rotates clockwise or counterclockwise with time, with the tip of the vector tracing out an ellipse. If, for a given frequency, both components are of equal magnitude, the ellipse flattens to a line and the motions are *rectilinear* (back and forth along a straight line). Two one-sided autospectra and two one-sided cross-spectra can be computed for the rotary components. Mooers (1973) formulated these as two two-sided rotary autospectra called, respectively, the *inner* and *outer rotary autospectra*, the terminology originating from the resemblance of the inner and outer rotary autocovariance functions derived from the autospectra to the inner (dot) and outer (cross) products in mathematics. (A note on terminology: Mooers (1973) uses A and C for counterclockwise ($+$) and clockwise components ($-$) while Gonella (1972) uses $+/-$ subscripts for these components of the form u_+ and u_-. In this text, we use $+/-$ superscripts where, for example, the amplitude of the two vector components is written as A^+ and A^-.)

To simplify the mathematics, we assume that u and v are continuous, stationary processes with zero means and Fourier integral representations. The velocity vector, $w(t)$, can then be written in terms of its Fourier transform

$$w(t) = u(t) + iv(t) = \sum_p W_p e^{i\omega pt}$$
$$= \sum_p \{ [A_{1p} \cos(\omega_p t) + B_{1p} \sin(\omega_p t)]$$
$$+ i[A_{2p} \cos(\omega_p t) + B_{2p} \sin(\omega_p t)] \}$$
(5.194)

in which the Fourier transform component, W_p, is a complex quantity, A and B are constants, and ω_p is the frequency of the pth Fourier component. As outlined in Section 5.4.4, each Fourier component of frequency $\omega = \omega_p$ can be expressed as a combination of two circularly polarized components having counterclockwise ($\omega \geq 0$) and clockwise ($\omega \leq 0$) rotation. Each of two components has its own amplitude and phase, and the tip of the vector formed by the combination of the two oppositely rotating components traces out an ellipse over a period, $T = 2\pi/\omega$. The semimajor axis of the ellipse has length, $L_M = A^+(\omega) + A^-(\omega)$, and the semiminor axis has length, $L_m = |A^+(\omega) - A^-(\omega)|$. The angle, θ, of the major axis measured counterclockwise from the eastward direction gives the ellipse orientation.

If we specify $A_1(\omega)$ and $B_1(\omega)$ to be the amplitudes of the cosine and sine terms for the eastward (u) component in Eqn (5.194) and $A_2(\omega)$ and $B_2(\omega)$ to be the corresponding amplitudes for the northward (v) component, the amplitudes of the two counter-rotating vectors for a given frequency are

$$A^+(\omega) = \frac{1}{2}\left\{[B_2(\omega) + A_1(\omega)]^2 + [A_2(\omega) - B_1(\omega)]^2\right\}^{1/2}$$
(5.195a)

$$A^-(\omega) = \frac{1}{2}\left\{[B_2(\omega) - A_1(\omega)]^2 + [A_2(\omega) + B_1(\omega)]^2\right\}^{1/2}$$
(5.195b)

and their phases are

$$\tan(\theta^+) = [A_1(\omega) - B_1(\omega)]/[A_1(\omega) + B_2(\omega)]$$
(5.196a)

$$\tan(\theta^-) = [B_1(\omega) + A_2(\omega)]/[B_2(\omega) - A_1(\omega)]$$
(5.196b)

The eccentricity of the ellipse is

$$\varepsilon(\omega) = 2[A^+(\omega)A^-(\omega)]^{1/2}/[A^+(\omega) + A^-(\omega)]$$
(5.197)

where the ellipse traces out an area $\pi[(A^+)^2 - (A^-)^2]$ during one complete cycle of duration, $2\pi/\omega$. The use of rotary components leads to two-sided spectra; i.e., defined for both negative and positive frequencies. If $S^+(\omega)$ and $S^-(\omega)$ are the rotary spectra for the two components, then $A^\pm(\omega) \propto [S^\pm(\omega)]^{1/2}$ can be used to determine the ellipse eccentricity. The sense of rotation of the vector about the ellipse is given by the rotary coefficient (see Section 5.4.4.2)

$$r(\omega) = [S^+(\omega) - S^-(\omega)]/[S^+(\omega) + S^-(\omega)]$$
(5.198)

where $-1 \leq r \leq 1$. Values for which $r > 0$ indicate counterclockwise rotation while values of $r < 0$ indicate clockwise rotation; $r = 0$ is rectilinear motion.

Because u, v are orthogonal Cartesian components of the velocity vector, $w = (u, v)$, the rotary spectra can be expressed as

$$S^+(\omega) = [A^+(\omega)]^2, \quad \omega \geq 0$$
$$= \frac{1}{2}[S_{uu} + S_{vv} + 2Q_{uv}]$$
(5.199a)

$$S^-(\omega) = [A^-(\omega)]^2, \quad \omega \leq 0$$
$$= \frac{1}{2}[S_{uu} + S_{vv} - 2Q_{uv}]$$
(5.199b)

where S_{uu} and S_{vv} are the autospectra for the u and v components, and Q_{uv} is the quadrature spectrum between the two components. The stability of the ellipse is given by

$$\mu(\omega) = \frac{\left|\left\langle (A^-(\omega)A^+(\omega)\exp[i(\theta^+ - \theta^-)])\right\rangle\right|^2}{\left\langle (A^-)^2\right\rangle\left\langle (A^+)^2\right\rangle},$$

$$= \frac{|Y|}{[S^+(\omega)S^-(\omega)]^{1/2}} \quad \omega \geq 0$$
(5.200)

where

$$Y = \frac{1}{2}[S_{uu} - S_{vv} + i2S_{uv}]$$
(5.201)

and the ellipse has a mean orientation

$$\phi = \frac{1}{2}\tan^{-1}[2S_{uv}/(S_{uu} - S_{vv})]$$
(5.202)

where ϕ is measured counterclockwise from east (the function, ϕ is not coordinate invariant). The brackets $\langle \cdot \rangle$ denote an ensemble average or a band average in frequency space. The ellipse stability, $\mu(\omega)$, resembles the magnitude of a correlation function and is a measure of the confidence one might place in the estimate of the ellipse orientation (Gonella, 1972).

5.6.8.1 Rotary Analysis for a Pair of Time Series

Having summarized the rotary vector analysis for a single location, we now want to consider the coherence and cross-spectral properties for two time series measured simultaneously at two spatial locations. The objective of the rotary spectral analysis is to determine the "similarity" between the two time series in terms of their circularly polarized rotary components. For two vector time series, the inner and outer rotary cross-spectra can be computed. As the spectra are complex, they have both amplitude and phase. Hence, coherence and phase spectra can be computed, just as with the cross-spectra of two scalar time series. *Inner* functions describe corotating components (components rotating in the same direction) and *outer* functions describe counterrotating components (components rotating in opposite directions). We could, of course, use standard Cartesian components for this task. Unfortunately, Cartesian vectors and their derived relationships generally are dependent on the selected orientation of the coordinate system. The advantages of the rotary type of analysis are: (1) the coherence analysis is independent of the coordinate system (i.e., is coordinate invariant); and (2) the results encompass the coherence and phase of oppositely rotating, as well as like-rotating, components for motions that may be highly nonrectilinear. Because the counter-rotating components have circular symmetry, invariance under coordinate rotation follows for coherence.

We consider two vector time series defined by the relations

$$w_1(t) = (u_1, v_1); \quad w_2(t) = (u_2, v_2) \quad (5.203)$$

where, as before, $(u, v) = u + iv$ are complex quantities. If $W_1(\omega)$ and $W_2(\omega)$ are components of the

Fourier transforms of these time series, then the transforms can be expressed in the form

$$W(\omega) = \begin{cases} A^+ \exp\left(-i\theta^+\right), & \omega \geq 0 \\ A^- \exp\left(-i\theta^-\right), & \omega \leq 0 \end{cases} \quad (5.204)$$

with the same definitions for amplitudes and phases as in the previous subsection. These expressions equate the negative frequency components from the Fourier transform with the clockwise rotary components and the positive frequency components from the transform with the counterclockwise components.

Inner-cross spectrum: The inner cross-spectrum, $S_{w_j w_k}(\omega)$, provides an estimate of the joint energy content of two time series for rotary components rotating in the same direction (e.g., the clockwise component of series 1 with clockwise component of series 2; Figure 5.44). For all frequencies, $-\omega_N < \omega < \omega_N$

$$S_{w_j w_k}(\omega) = \left\langle W_j^*(\omega) W_k(\omega) \right\rangle, \quad j, k = 1, 2$$

$$= \begin{cases} A_j^+(\omega) A_k^+(\omega) \exp\left[-i\left(\theta_j^+ - \theta_k^+\right)\right], & \omega \geq 0 \\ A_j^-(\omega) A_k^-(\omega) \exp\left[i\left(\theta_j^- - \theta_k^-\right)\right], & \omega \leq 0 \end{cases}$$

$$(5.205)$$

where, as before, $\langle \cdot \rangle$ denotes an ensemble average or a band average in frequency space, and the asterisk denotes the complex conjugate. It follows that the inner-autospectrum for each time series is

$$S_{w_j w_j}(\omega) = \begin{cases} \left[A_j^+(\omega)\right]^2, & \omega \geq 0 \\ \left[A_j^-(\omega)\right]^2, & \omega \leq 0 \end{cases} \quad (5.206)$$

Thus, $S_{w_j w_j}(\omega)$ $(j = 1, 2)$ is the power spectrum of the counterclockwise component of the series j for $\omega \geq 0$, and the power spectrum for the clockwise component for $\omega \leq 0$. The area under the curve of $S_{w_j w_k}(\omega)$ vs frequency

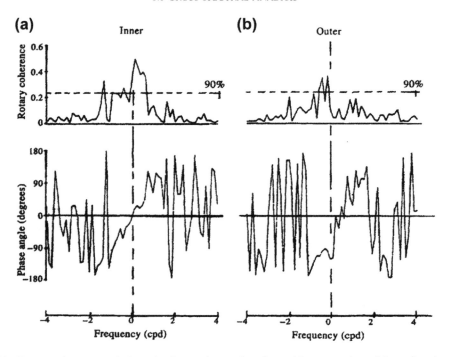

FIGURE 5.44 Rotary coherence and phase for 5-year time series of monthly mean winter (November through February) wind velocity from two sites off Alaska. (a) Corotating (inner) coherence and phase with 90% confidence level; (b) Counterrotating (outer) coherence and phase. *(From Livingstone and Royer (1980).)*

equals the sum of the variance of the u and v components. For $\omega \geq 0$, $S_{w_1 w_2}(\omega)$ is the cross-spectrum for the counterclockwise component of series 1 and 2, while for $\omega \leq 0$, $S_{w_j w_k}(\omega)$

two time series at frequency, ω, is defined in the usual manner. Specifically, using the previous definitions for the rotary components, we find

$$
\gamma_{12}^2(\omega) = \begin{cases} \left\{ \left\langle A_1^+ A_2^+ \cos(\theta_1^+ - \theta_2^+) \right\rangle^2 + \left\langle A_1^+ A_2^+ \sin(\theta_1^+ - \theta_2^+) \right\rangle^2 \right\} \Big/ \left\langle \left(A_1^+ \right)^2 \right\rangle \left\langle \left(A_2^+ \right)^2 \right\rangle, & \omega \geq 0 \\ \left\{ \left\langle A_1^- A_2^- \cos(\theta_1^- - \theta_2^-) \right\rangle^2 + \left\langle A_1^- A_2^- \sin(\theta_1^- - \theta_2^-) \right\rangle^2 \right\} \Big/ \left\langle \left(A_1^- \right)^2 \right\rangle \left\langle \left(A_2^- \right)^2 \right\rangle, & \omega \leq 0 \end{cases}
$$

$$(5.207)$$

represents the cross-spectrum for the clockwise rotary component.

Inner-coherence squared: The two-sided inner-coherence squared, $\gamma_{12}^2(\omega)$, between the

where $0 \leq \left| \gamma_{12}^2 \right| \leq 1$. A coherence of near zero indicates a negligible relationship between the two like-rotating series while a coherence near unity indicates a high degree of variability between the series. The inner-phase lag, ϕ_{12}, between the two vectors is

$$\phi_{12}(\omega) = \tan^{-1}[-\text{Im}\,(S_{w_1 w_2})\,/\,\text{Re}\,(S_{w_1 w_2})] \tag{5.208}$$

or, in terms of the clockwise and counterclockwise components

$$\tan(\phi_{12}) = \begin{cases} \langle A_1^+ A_2^+ \sin(\theta_1^+ - \theta_2^+)\rangle / \langle A_1^+ A_2^+ \cos(\theta_1^+ - \theta_2^+)\rangle, & \omega \geq 0 \\ \langle -A_1^- A_2^- \sin(\theta_1^- - \theta_2^-)\rangle / \langle A_1^- A_2^- \cos(\theta_1^- - \theta_2^-)\rangle, & \omega \leq 0 \end{cases} \tag{5.209}$$

The phase, which is the same for both the inner cross-spectrum and the inner coherence, is a measure of the phase lead of the rotary component of time series 1 with respect to that of time series 2. Figure 5.44(a) shows the inner rotary coherence and phase for 5 years of monthly winter (November–February) wind data measured off Alaska at Middleton Island (59.4° N, 146.3° W) and Environmental Weather Buoy EB03 (56.0° N, 148.0° W). Corotating wind vectors were generally coherent above the 90% confidence level for frequencies $-1 < f < 1$ cpd, with greater coherence at positive frequencies (Livingstone and Royer, 1980). The inner phase was nearly a straight line in the frequency range $-1 < f < 0$ cpd, increasing by 120° over this range.

Outer-cross spectrum: The outer cross-spectrum, $Y_{w_j w_k}(\omega)$, provides an estimate of the joint energy content between rotary components rotating in opposite directions (e.g., between the clockwise component of time series 1 and the counterclockwise component of time series 2). For frequencies in the Nyquist frequency range, $-\omega_N < \omega < \omega_N$

$$Y_{w_j w_k}(\omega) = \langle W_j(-\omega)W_k(\omega)\rangle, \quad j,k = 1,2$$
$$= \begin{cases} A_j^-(\omega)A_k^+(\omega)\exp\left[i\left(\theta_k^+ - \theta_j^-\right)\right], & \omega \geq 0 \\ A_j^+(\omega)A_k^-(\omega)\exp\left[-i\left(\theta_k^- - \theta_j^+\right)\right], & \omega \leq 0 \end{cases} \tag{5.210}$$

(Middleton, 1982). These relations resemble those for the inner-cross spectra but involve a combination of oppositely rotating vector amplitudes and phases. For the case of a single series, j, the outer rotary autospectrum is then

$$Y_{w_j w_j}(\omega) = A_j^-(\omega)A_j^+(\omega)\exp\left[i\left(\theta_j^+ - \theta_j^-\right)\right], \quad \omega \geq 0 \tag{5.211}$$

and is symmetric about $\omega = 0$, and so is defined for only $\omega \geq 0$. Hence, $Y_{w_j w_j}(\omega)$ is an even function of frequency; i.e., $Y_{w_j w_j}(\omega) = Y_{w_j w_j}(-\omega)$. As noted by Mooers, $Y_{w_j w_j}(\omega)$ is not a power spectrum in the ordinary physical sense because it is complex valued. Rather it is related to the spectrum of the uv-Reynolds stress.

Outer-coherence squared: After first performing the ensemble or band averages in the brackets $\langle \cdot \rangle$, the outer-rotary coherence squared between series j and k is expressed in terms of the Fourier coefficients as

$$\lambda_{jk}^2(\omega) = \begin{cases} \langle A_j^- A_k^+\rangle^2 \left[\langle\cos\left(\theta_k^+ - \theta_j^-\right)\rangle^2 + \langle\sin\left(\theta_k^+ - \theta_j^-\right)\rangle^2\right] / \langle (A_k^+)^2\rangle\langle (A_j^-)^2\rangle, & \omega \geq 0 \\ \langle A_j^+ A_k^-\rangle^2 \left[\langle\cos\left(\theta_j^+ - \theta_k^-\right)\rangle^2 + \langle\sin\left(\theta_j^+ - \theta_k^-\right)\rangle^2\right] / \langle (A_j^+)^2\rangle\langle (A_k^-)^2\rangle, & \omega \leq 0 \end{cases} \tag{5.212}$$

The phase lag, $\psi_{jk}(\omega)$, between the two oppositely rotating components of the two time series is then the same for the coherence and the cross-spectrum and is given by

If the values of

$$A_j^- A_k^+ \quad \text{and} \quad A_j^+ A_k^-$$

$$\tan(\psi_{12}) = \begin{cases} \left\langle A_j^- A_k^+ \sin\left(\theta_j^- - \theta_k^+\right)\right\rangle \Big/ \left\langle A_j^- A_k^+ \cos\left(\theta_j^- - \theta_k^+\right)\right\rangle, & \omega \geq 0 \\ \left\langle A_j^+ A_k^- \sin\left(\theta_k^- - \theta_j^+\right)\right\rangle \Big/ \left\langle A_j^+ A_k^- \cos\left(\theta_k^- - \theta_j^+\right)\right\rangle, & \omega \leq 0 \end{cases} \qquad (5.213)$$

change little over the averaging interval covered by the angular brackets, then

$$\psi_{jk}(\omega) = \begin{cases} \theta_j^- - \theta_k^+, & \omega \geq 0 \\ \theta_k^- - \theta_j^+, & \omega \leq 0 \end{cases} \qquad (5.214)$$

Figure 5.44(b) shows the outer rotary coherence and phase for 5-year records of winter winds off the coast of Alaska. Counter-rotating vectors were coherent at negative frequencies in the range $-1 < f < 0$ cpd and exhibited little coherence at positive frequencies. In this portion of the frequency band, the linear phase gradient was similar to that for the corotating vectors (Figure 5.44(a)).

Complex admittance function: If we think of the wind vector at location 1 as the source (or input) function and the current at location 2 as the response (or output) function, we can compute the complex inner admittance, Z_{12}, between two corotating vectors as

$$Z_{12}(\omega) = S_{w_1 w_2}(\omega)/S_{w_1 w_1}(\omega), \quad -\omega_N < \omega < \omega_N \qquad (5.215)$$

The amplitude and phase of this function are

$$|Z_{12}(\omega)| = |S_{w_1 w_2}(\omega)|/S_{w_1 w_1}(\omega) \qquad (5.216a)$$

$$\Phi_{12}(\omega) = \tan^{-1}\{\operatorname{Im}[S_{w_1 w_2}(\omega)]/\operatorname{Re}[S_{w_1 w_2}(\omega)]\} \qquad (5.216b)$$

For frequency ω, the absolute value of $Z_{12}(\omega)$ determines the amplitude of the clockwise (counterclockwise) rotating response one can expect at location 2 to a given clockwise (counterclockwise) rotating input at location 1. The phase, $\Phi_{12}(\omega)$, determines the lag of the response vector to the input vector.

The corresponding expressions for the complex outer admittance, Z_{12}, between two opposite-rotating vectors are

$$Z_{12}(\omega) = Y_{w_1 w_2}(\omega)/S_{w_1 w_1}(\omega), \quad -\omega_N < \omega < \omega_N \qquad (5.217)$$

with amplitude and phase

$$|Z_{12}(\omega)| = |Y_{w_1 w_2}(\omega)|/S_{w_1 w_1}(\omega) \qquad (5.218a)$$

$$\Phi_{12}(\omega) = \tan^{-1}\{\operatorname{Im}[Y_{w_1 w_2}(\omega)]/\operatorname{Re}[Y_{w_1 w_2}(\omega)]\} \qquad (5.218b)$$

For frequency ω, the absolute value of $Z_{12}(\omega)$ yields the amplitude of the clockwise (counterclockwise) rotating response one can expect at location 2 to a given counterclockwise (clockwise) rotating input at location 1. The phase, $\Phi_{12}(\omega)$, determines the lag of the response vector to the input vector.

5.7 WAVELET ANALYSIS

The terms "wavelet transform" and "wavelet analysis" are two recent additions to the lexicon of time series analysis. First introduced in the 1980s for processing seismic data (cf. Goupillaud et al., 1984), the technique has begun to attract attention in meteorology and oceanography, where it has been applied to time series measurements of turbulence (Farge, 1992; Shen and Mei, 1993), surface gravity waves (Shen et al., 1994), low-level cold fronts (Gamage and Blumen, 1993), and equatorial Yanai waves (Meyers et al., 1993).

As frequently noted in the literature, Fourier analysis does relatively poorly dealing with signals of the form $\varphi(t) = A(\tau)\cos(\omega t)$, where the amplitude, A, varies on the slow timescale, τ. Wavelet analysis has a number of advantages over Fourier analysis that are particularly attractive. Unlike the Fourier transform, which generates record-averaged values of amplitude and phase for each frequency component or harmonic, ω, the wavelet transform yields a localized, "instantaneous" estimate for the amplitude and phase of each spectral component in the data set. This gives wavelet analysis an advantage in the analysis of nonstationary data series in which the amplitudes and phases of the constituents may be changing rapidly in time or space. Where a Fourier transform of the nonstationary time series would smear out any detailed information on the changing processes, the wavelet analysis attempts to track the evolution of the signal characteristics through the data set. As with other transform techniques, problems can develop at the ends of the time series, and steps must be taken to mitigate these effects. Similar to other transform techniques involving finite length data, steps also must be taken to minimize the distortion of the transformed data caused by the nonperiodic behavior at the ends of the time series. Lastly, we note that increasing the temporal resolution, Δt, of the wavelet analysis decreases the frequency resolution, Δf, and vice versa, such that $\Delta t \Delta f < (1/4)\pi$, reminiscent of the Heisenberg uncertainty relation. The more accurately we want to resolve the frequency components of a time series, the less accurately we can resolve the changes in these frequency components with time.

5.7.1 The Wavelet Transform

As noted above, wavelet analysis enables us to study nonstationary signals, in which the amplitudes of the frequency components of the signal are changing with time. This contrasts with traditional spectral methods, which assume that the frequency components of the time series are stationary, thus allowing for a direct transfer into the frequency domain. Wavelet analysis provides a windowing technique with variable-sized windows, which permits customization of the frequency domain analysis by allowing for long time intervals, where more precise low-frequency information is wanted, and also for shorter time intervals, where high-frequency information is wanted. A wavelet is a "small wave" that grows and decays over a limited time span. The oldest and simplest wavelet is the Haar Wavelet, traditionally written as, $\psi(t)$, and introduced at the beginning of the twentieth century (Harr, 1910), where

$$\psi(t) \equiv \begin{cases} -\dfrac{1}{\sqrt{2}} & -1 < t \leq 0 \\ \dfrac{1}{\sqrt{2}} & 0 < t \leq 1 \\ 0 & \text{otherwise} \end{cases} \tag{5.219}$$

From Eqn (5.219), it can be shown that the translations of $\psi(t)$ are orthonormal to both its translated versions (shift in time) and its dilated versions (change in amplitude), i.e.,

$$\int\limits_{-\infty}^{\infty} \psi_{m,n}(t)\psi_{m,n'}^{*}(t)dt$$

$$= 2^{m} \int\limits_{-\infty}^{\infty} \psi(2^{m}t - n)\psi^{*}(2^{m}t - n')dt$$

$$= \int\limits_{-\infty}^{\infty} \psi(t - n)\psi^{*}(t - n')dt = \delta(n - n')$$

$$\tag{5.220}$$

where the asterisk (*) denotes the complex conjugate. As with the sines and cosines used in spectral analysis, the function, $\psi(t)$, therefore forms an orthonormal basis for the analysis of frequency-dependent variations.

Modern wavelet analysis involves the convolution of a real time-series, $x(t)$, with a set of functions, $g_{a\tau}(t) = g(t{:}\tau, a)$, that are derived from a "mother wavelet" or analyzing wavelet, $g(t)$, which is generally complex. In particular

$$g_{a\tau}(t) = \frac{1}{\sqrt{a}} g\left[a^{-1}(t - \tau)\right] \qquad (5.221)$$

where τ (real) is the *translation* parameter corresponding to the central point of the wavelet in the time series and a (real and positive) is the *scale dilation* parameter corresponding to the width of the wavelet. For the Gaussian-shaped Morlet wavelet (Figure 5.45) described in detail later in this section, the dilation parameter can be related to a corresponding Fourier frequency (or wavenumber).

The continuous wavelet transform, $X(t)$, of the time series with respect to the analyzing wavelet, $g(t)$, is defined through the convolution integral

$$X_g[\tau, a] = \frac{1}{\sqrt{a}} \int_{-\infty}^{\infty} g^*\left[a^{-1}(t - \tau)\right] x(t) dt \qquad (5.222)$$

in which g^* is the complex conjugate of g and variables τ, a are allowed to vary continuously through the domain $(-\infty, \infty)$. Wavelet analysis provides a two-dimensional unraveling of a one-dimensional time series into position, τ, and amplitude scale, a, as new independent variables. The wavelet transformation (Eqn (5.222)) is a sort of mathematical microscope, with magnification $1/a$, position τ, and optics given by the choice of the specific wavelet, $g(t)$ (Shen et al., 1994). Whereas Fourier analysis provides an average amplitude over the entire time series, wavelet analysis yields a measure of the localized amplitudes a as the wavelet moves through the time series with increasing values of τ. Although wavelets have a definite scale, they typically do not bear any resemblance to the sines and cosines of Fourier modes. Nevertheless, a correspondence between wavelength and scale, a, can sometimes be achieved.

To qualify for mother wavelet status, the function, $g(t)$, must satisfy several properties (Meyers et al., 1993):

1. Its amplitude $|g(t)|$ must decay rapidly to zero in the limit $|t| \rightarrow \infty$. It is this feature that produces the localized aspect of wavelet analysis since the transformed values, $X_g[\tau, a]$, are generated only by the signal in the cone of influence about $t = \tau$. In most instances, the wavelet, $g[(t - \tau)/a]$, is assumed to have an insignificant effect at some time $|t| = \tau_c$.
2. $g(t)$ must have zero mean. Known as the *admissibility condition*, this ensures the invertability of the wavelet transform. The original signal can then be obtained from the wavelet coefficients through the inverse transform

$$x(t) = \frac{1}{C} \int_{-\infty}^{\infty} \int_{-\infty}^{\infty} \left\{X_g[\tau, a] a^{-2} g_{a\tau}\right\} d\tau da$$

where

$$C^{-1} = \int_{-\infty}^{\infty} \left(\omega^{-1}|G(\omega)|^2\right) d\omega \qquad (5.223)$$

in which $G(\omega)$ is the Fourier transform of $g(t)$. For $1/C$ to remain finite, $G(0) = 0$.

3. Wavelets are often regular functions, such that $G(\omega < 0) = 0$. These are also called *progressive* wavelets. Elimination of negative frequencies means that wavelets need only be described in terms of positive frequencies.
4. Higher-order moments (such as variance and skewness) should vanish allowing the investigation of higher-order variations in the data. This requirement can be relaxed, depending on the application.

One of most extensively used wavelets is the standard (admissible and progressive) Morlet wavelet

$$g(t) = e^{-t^2/2} e^{+i\omega t} \qquad (5.224)$$

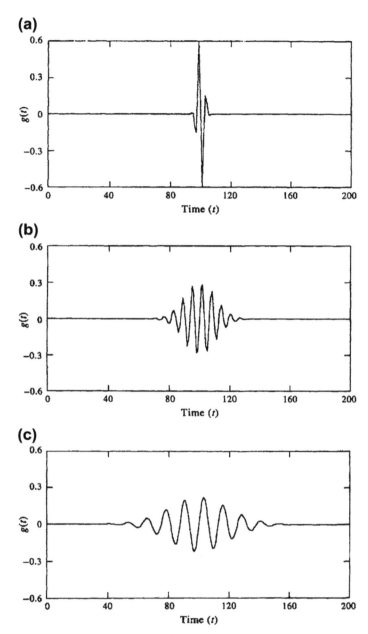

FIGURE 5.45 The Morlet wavelet, $g(t) = (1/\sqrt{a})e^{[(t-\tau)/a]^2}\{2sin[c(t-\tau)/a]\}$, where t is time in arbitrary units ($t = t_n$; $n = 1$, ..., 200). The example is for $c = 10$ and time lag $\tau = 100$ so that the wavelet is seen midway through the time series. (a) $a = 2$; (b) $a = 10$; (c) $a = 20$.

consisting of a plane wave of frequency ω (or wavenumber k in the spatial domain), which is modulated by a Gaussian envelope of unit width. Another possible wavelet, which is applicable to a signal with two frequencies, ω_1 and ω_2, is

$$g(t) = e^{-t^2/2}e^{i\omega_1 t}e^{i\omega_2 t} \quad (5.225)$$

while the wavelet

$$g(t) = e^{-t^2/2}e^{i\omega t}e^{ikt^2/2} \quad (5.226)$$

is applicable to short data segments with linearly increasing frequency ("chirps").

5.7.2 Wavelet Algorithms

The choice of $g(t)$ is dictated by the analytical requirements. More specifically, the wavelet should have the same pattern or signal characteristic as the pattern being sought in the time series. Large values of the transform $X_g(\tau, a)$ will then indicate where the time series $x(t)$ has the desired form. The simplest—and most time consuming—method for obtaining the wavelet transform is to compute the transform at arbitrary points in parameter (τ, a) space using the discrete form of Eqn (5.222) for known values of $x(t)$ and $g(t)$. If one integrates from $0 < a \le M$ and $0 < \tau \le N$, the integration time goes as MN^2. An alternate method is to use the convolution theorem and then obtain the wavelet transform in spectral space

$$X_g[\tau, a] = \frac{1}{\sqrt{a}} \int\limits_{-\infty}^{\infty} e^{i\tau\omega}G^*(a\omega)X(\omega)d\omega \quad (5.227)$$

where $G(\omega)$ and $X(\omega)$ are the Fourier transforms of $g(t)$ and $x(t)$, respectively. Since FFT transforms can now be exploited, the analysis time drops to $MN\log_2 N$. To use this method, $G(\omega)$ should be known analytically and the data must be preprocessed to avoid errors from the FFT algorithms. For example, if $x(t)$ is aperiodic, the discrete form of Eqn (5.226) will generate an artificial periodicity in the wavelet transform that greatly distorts the results for the end regions. Methods have been devised to work around this problem. Aliasing and bias in FFT routines must also be taken into account.

Meyers et al. (1993) used the standard Morlet wavelet Eqn (5.224), for which $g(t) = e^{-t^2/2} \cdot e^{i\omega t}$, to examine a signal that changes frequency halfway through the measurement. Here, we have broken with tradition and used ω instead of c for frequency. After considerable attempts (including use of raw data, cosine-weighted data, and other variations), the authors decided that the best approach was to taper or buffer the original time series with added data points that attenuate smoothly to zero past the ends of the time series. "The region of the transform corresponding to these points is then discarded after the transform. Without this buffering, a signal whose properties are different near its ends will result in a wavelet transform that has been forced to periodicity at all scales through a distortion (in some cases severe) of the end regions. The greater the aperiodicity of the signal, the greater the distortion."

For the Morlet wavelet, the dilation parameter, a, giving the maximum correlation between the wavelet and a plane Fourier component of frequency, ω_0 (i.e., a wave of the form $e^{i\omega_0 t}$) is

$$a_0 = \frac{\left[\omega + (2 + \omega^2)^{1/2}\right]}{4\pi}T_0 \quad (5.228)$$

where $T_0 = 2\pi/\omega_0$ is the Fourier period. (In wavenumber space, T_0 is replaced by wavelength λ_0 and ω_0 by k_0.) We note that any linear superposition of periodic components results in separate local maxima. Consequently, the wavelet transform of any function $x(t) = \sum A_j e^{ik_j t}$ will have modulus maxima at $a_j = [\omega + (2 + \omega^2)^{1/2}]/(2k_j)$.

5.7.3 Oceanographic Examples

In this section, we will consider two oceanographic wavelet examples (surface gravity wave heights and zonal velocity from a satellite-tracked drifter) using the standard Morlet wavelet

$$g(t) \rightarrow g[(t - \tau)/a] = \frac{1}{\sqrt{a}} e^{-\frac{1}{2}[(t-\tau)/a]^2}$$
$$\times \sin [\omega(t - \tau)/a] \quad (5.229)$$

In this real expression, the Gaussian function determines the envelope of the wavelet while the sine function determines the wavelengths that will be preferentially weighted by the wavelet. The wavelet function progresses through the time series with increasing τ, its cone of influence centered at times $t = \tau$. As a increases, the width of the Gaussian spreads in time from its center value (Figure 5.45(a)–(c)). Increasing ω increases the number of oscillations over the span of the function. The processing procedure is as follows: (1) read in the time series $x(n)$ ($n = 0$,

..., $N-1$) to be analyzed, where $N = 2^m$ (m is an integer). To reduce ringing, extend each end of the time series by adding a trigonometric taper, tap $= 1 - \sin\phi$, where tap $= 1.0$ at the end values $x(0)$ and $x(N - 1)$. The total length of the buffered time series must remain a power of two; (2) remove the mean of the new record and then take the FFT of the time series to obtain $X(\omega)$; (3) take the Fourier transform of the wavelet, $g(t)$, at given length scales, a, to obtain $G(a\omega)$; (4) calculate the integral Eqn (5.227) by convolving the product $G^*(a\omega)X(\omega)$ in Fourier space; (5) take the inverse FFT of the result to obtain $\sqrt{a}X_g(\tau,a)$ as a function of time dilation, τ, and amplitude, a.

In Figure 5.46(a) we have plotted a 300 s record of surface gravity wave heights measured off the west coast of Vancouver Island in the winter of 1993. Maximum wave amplitudes of

FIGURE 5.46 Morlet wavelet transform of surface gravity waves measured from a waverider buoy moored off the west coast of Vancouver Island. (a) Original 5-min time series of significant wave height for the winter of 1993; (b) wave amplitude (m) and (c) Phase (degrees) as function of time; (d) The value of a (wave period) at peak wave amplitude. *(Courtesy, Diane Masson.)*

around 3 m occurred midway through the time series. The Morlet wavelet transform of the record yields an estimate of the wave amplitude (Figure 5.46(b)) and phase (Figure 5.46(c)) as functions of the wave period (T) and time (t). Also plotted is the value of the wave period (T = scale a) at peak energy (Figure 5.46(d)). Comparison of Figure 5.46(b) and (d) reveals that the larger peaks near times of 75, 150, and 210 s all have about the same wavelet scale, a, corresponding to a peak wave period of around 8 s. Also, as one would expect, the 2π changes in phase between crests (Figure 5.46(c)) increase with increasing wave period (scale, a).

In our second example, we have applied a standard Morlet wavelet transform to a 90-day segment of 3-hourly sampled east—west (u) current velocity (Figure 5.47(a)) obtained from a satellite-tracked drifter launched in the northeast Pacific in August 1990 as part of the World Ocean Circulation Experiment. The drifter was drogued at 15-m depth and its motion was indicative of currents in the surface Ekman layer. The 90-day velocity record has been generated from positional data using a cubic spline interpolation algorithm. We focus our attention on the high-frequency end of the spectrum, $0 < a < 1.5$ days. As indicated by Figures 5.47(b) and (c), the first 30 days of the record, from Julian day (JD) 240—270, were dominated by weak semidiurnal tidal currents with periods of 0.5 days. Beginning on JD 270, strong wind-generated inertial motions with periods around 16 h ($f \approx 1.5$ cpd) dominated the spectrum. These energetic motions persisted through the record, except for a short hiatus near JD 295. A blow-up of the

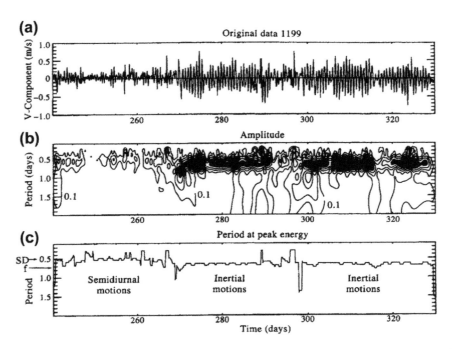

FIGURE 5.47 The Morlet wavelet transform of a 90-day record of the east—west velocity component from the trajectory of a satellite-tracked drifter in the northeast Pacific, September 1990. (a) Original 3-hourly time series; (b) amplitude (cm/s) vs time as a function of period, T, in the range, $0 < T < 2.0$ days; (c) period (days) of the current oscillations at peak amplitude. (*Courtesy, Jane Eert.*)

FIGURE 5.48 As for Figure 5.47 but for a larger range of periods. (a) Original 3-hourly velocity time series; (b) amplitude (cm/s) and (c) phase (degrees) vs time as a function of period, T, in the range, $2 < T \leq 20$ days; (d) period (days) of the current oscillations at peak amplitude.

segment from JD 240 to 270 shows a rapid change in signal phase associated with the shift from semidiurnal tidal currents to near-inertial motions. The contribution from the beat frequency between the M_2 tidal signal and the inertial oscillations, $fM_2 = 0.0805 + 0.0621$ cph $= 0.1426$ cph can also be seen in the transformed data at period $T \approx 0.29$ days. Examination of the longer period motions in Figure 5.48 (for $2 < a < 30$ days) suggests the presence of a long-period modulation of the high-frequency motions associated with the near-inertial wave events.

Our final example of wavelet examples is presented in Figure 5.49. The upper panel shows 8-month time series of the hourly alongshore components of current velocity, v, for two mooring sites located roughly 100 km apart along the 100 m depth isobath on the continental

shelf off the west coast of Canada. Brooks Peninsula is to the north of Estevan Point. The lower panel presents a Morlet wavelet analysis, showing how the coherence amplitude and phase lag between the two time series varies as a function of time and frequency. The plot covers motions with periods in the range of 6 h to about 2 months and is cut off at the bottom where the coherence function for the longest period motions is not resolvable (the longer the period of the motions, the shorter the time period over which the signal coherence can be determined). The coherence amplitudes are color-coded, with red and blue denoting strong and weak coherence amplitudes, respectively. The phase lag of the two signals for a given time and frequency is determined using the vectors and the circular scale at the bottom right of the figure. The phase scale goes from 0 to 180° in the

FIGURE 5.49 Morlet wavelet analysis of the coherence amplitude and phase of the alongshore component of current velocity, v, at 35 m depth for current meter mooring sites in 100 m of water off Brooks Peninsula and Estevan Point on the west coast of Vancouver Island, British Columbia. The top panel shows the hourly current velocity in m/s (see scale bar) for the two sites for September 28, 1997 to May 26, 1998. Instruments are separated alongshore by roughly 100 km. Coherence amplitude and phase values over the same time period are given by the scales on the lower left and right, respectively. See the text for a further explanation. *(Analysis and plots courtesy of Maxim Krassovski, Institute of Ocean Sciences.)*

counterclockwise and clockwise direction. If the phase vector in the figure has an upward component, the Brooks series *leads* the Estevan series, whereas if the phase vector has a downward component, the Brooks series *lags* the Estevan series (i.e., the Estevan series leads the Brooks series). Among other aspects of the flow, results show that there are persistent diurnal current motions associated with coastally trapped diurnal shelf waves (Crawford and Thomson, 1982, 1984; Cummins et al., 2000) and more intermittent semidiurnal currents associated with internal tides (Drakopoulos and Marsden, 1993; Cummins and Oey, 1997). The diurnal motions are highly coherent whereas the semidiurnal motions are much less coherent. Highly coherent coastally trapped waves with periods of around 10 days are also sometimes present in the record (Yao et al., 1984).

5.7.4 The S-Transformation

Wavelet transforms are not the only method for dealing with nonstationary oscillations with time-varying amplitudes and phases. The S-transformation (Stockwell et al., 1994) is an extension of the wavelet transform that has been used by Chu (1994) to examine the localized spectrum of sea level in the TOGA (Tropical Ocean Global Atmosphere) data sets. For this particular transform, the relationship between the S-transform, $S(\omega, \tau)$, and the data, $x(t)$, is given by

$$S(\omega, \tau) = \int_{-\infty}^{\infty} H(\omega + \alpha)e^{-(2\pi^2\alpha^2/\omega^2)}e^{i2\pi\alpha\tau}d\alpha$$

$$(5.230)$$

$$x(t) = \int_{-\infty}^{\infty}\int_{-\infty}^{\infty} S(\omega, \tau)e^{i2\pi\alpha\tau}d\omega d\tau \quad (5.231)$$

where

$$H(\omega + \alpha) = \int_{-\infty}^{\infty} x(t)e^{-i2\pi(\omega+\alpha)\tau}dt \quad (5.232a)$$

$$= \int_{-\infty}^{\infty} S(\omega + \alpha, \tau)d\tau \quad (5.232b)$$

is the standard Fourier transform of the input time series data. As indicated by Eqn (5.232b), the Fourier transform is the time average of the S-transform, such that $|H(\omega)|^2$ provides a record-averaged value of the localized spectra, $|S(\omega)|^2$, derived from the S-transform. Equation (5.231) can also be viewed as the decomposition of a time series, $x(t)$, into sinusoidal oscillations, which have time-varying amplitudes $S(\omega, \tau)$.

The discrete version of the S-transformation can be obtained as follows. As usual, let $x(t_n) = x(n\Delta t)$, $n = 0, 1, ..., N - 1$ be a discrete time series of total duration $T = N\Delta t$. The discrete version of Eqn (5.230) is then

$$S(0, \tau_q) = \frac{1}{N}\sum_{m=0}^{n-1} x(m/T), \quad p = 0 \quad (5.233a)$$

$$S(\omega_p, \tau_q) = \sum_{m=0}^{N-1}\left\{ H[(m+p)/T]e^{-(2\pi^2m^2/p^2)}e^{i2\pi mq/N}\right\},$$
$$p \neq 0$$

$$(5.233b)$$

where $S(0, \tau_q)$ is the mean value for the time series, $\omega_p = p/N\Delta t$ is the discrete frequency of the

signal, and $\tau_q = q\Delta t$ is the time lag. The DFT is given by

$$H(p/T) = \frac{1}{N}\sum_{k=0}^{N-1} x(k/T)e^{-i2\pi pk/N} \quad (5.234)$$

The S-transform is a complex function of frequency ω_p and time τ_q, with amplitude and phase defined by

$$A(\omega_p, \tau_q) = |S(\omega_p, \tau_q)| \quad (5.235a)$$

$$\Phi(\omega_p, \tau_q) = \tan^{-1}\{\text{Im}\,[S(\omega_p, \tau_q)]/\text{Re}\,[S(\omega_p, \tau_q)]\}$$
$$(5.235b)$$

For a sinusoidal function of the form

$$X(\omega_p, \tau) = A(\omega_p, \tau)\cos[2\pi\omega_p\tau + \Phi(\omega_p, \tau)]$$
$$(5.236)$$

the function X at frequency ω_p is called the "voice."

Chu (1994) applied the S-transform to the nondimensionalized sea-level records, $x(t)$, collected at Nauru (0°32′ S, 166°54′ W) in the western equatorial Pacific and La Libertad (2°12′ S, 80°55′ W) in the eastern equatorial Pacific. Here

$$x(t) = \frac{[\eta(t) - \overline{\eta}]}{\overline{\eta}} \quad (5.237)$$

and $\overline{\eta}(t)$ represents the mean value of the sea level, $\eta(t)$. A Fourier spectral analysis of the time series revealed a strong annual sea-level oscillation in the western Pacific and a weak annual oscillation in the eastern Pacific. Both stations had strong quasi-biennial oscillations with periods of 24–30 months. The S-transformation was then used to examine the temporal variability in these components throughout the 16- and 18-year time series. For example, the voices for the annual oscillation ($\omega_{16} = 16/T$; $T = 192$ months) were similar at the two locations with higher amplitudes in the late 1970s than in the late 1980s (Figure 5.50). At La Libertad, the annual cycle

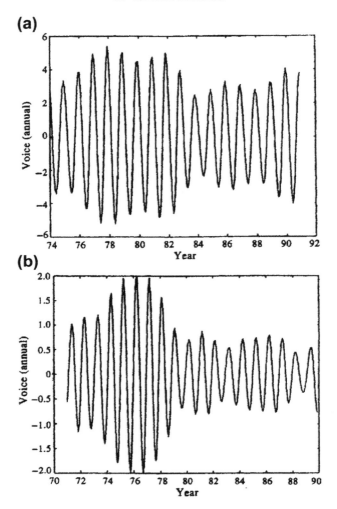

FIGURE 5.50 The "voices" for the annual oscillation ($\omega_{16} = 16/T$; $T = 192$ months) for (a) Nauru; and (b) La Libertad. Higher amplitudes were recorded in the late 1970s than in the late 1980s. *(Chu (1994).)*

became weak after 1979. The temporally varying quasi-biennial oscillations ($\omega_8 = 8/T$) were out of phase between the western and eastern Pacific (Figure 5.51).

5.7.5 The Multiple Filter Technique

The multiple filter technique is a form of signal demodulation that uses a set of narrow-band digital filters (windows) to examine variations in the amplitude and phase of dispersive signals as functions of time, t, and frequency, ω (or f). Originally designed to resolve complex transient seismic signals composed of several dominant frequencies (Dziewonski et al., 1969), the technique has recently been modified for the analysis of clockwise and counterclockwise rotary velocity components (Thomson et al., 1997) and in

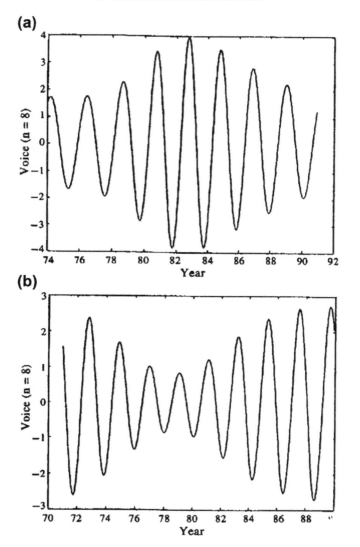

FIGURE 5.51 The "voices" for the quasi-biennial oscillations ($\omega_8 = 8/T$) for (a) Nauru; and (b) La Libertad. The oscillations were out of phase between the western and eastern Pacific. *(Chu (1994).)*

investigations of tsunami frequency content (Rabinovich et al., 2011a, b) and tsunami wave dispersion (Gonzalez and Kulikov, 1993).

The multiple filter technique relies on a series of band-pass filters centered on a range of narrow frequency bands to calculate the instantaneous signal amplitude or phase. Dziewonski

et al. (1969) filter in the frequency domain rather than the time domain, although the results are equivalent to within small processing errors. The filtering algorithm generates a matrix (grid) of amplitudes or phases with columns representing time and rows representing frequency (or period). The gridded values can then be

FIGURE 5.52 Multiple filter technique applied to tsunami sea-level heights measured at bottom pressure stations AK1 and AK2 in 5 km of water near 53° N, 156° W in the Gulf of Alaska. (a) November 30, 1987 and (b) 6 March 1988. Amplitude contours in the *f–t* diagram are normalized by the maximum value and drawn with a step of 1 dB. Solid curve denotes the theoretical arrival time for these highly dispersive waves. *(From Kulikov and González (1997); see also González and Kulikov (1993).)*

contoured to give a three-dimensional plot of the demodulated signal amplitude (or phase) as a function of time and frequency. González and Kulikov (1993) and Kulikov and González (1997) used the technique to examine the evolution of tsunami waves generated by an undersea earthquake in the Gulf of Alaska on March 6, 1987 (Figure 5.52). Sea-level heights measured by two bottom-pressure recorders deployed in the deep ocean to the south of Kodiak Island show that the tsunami waves were highly dispersive (low frequencies propagated faster than high frequencies) and that the arrival times of the waves closely followed the theoretical predictions for shallow-water wave motions. Peak spectral amplitudes were centered around a period of roughly 5 min, and the signal duration was about 40 min.

5.7.5.1 Theoretical Considerations

Since the technique is used to examine signal energy as a function of time and frequency, it is desirable that the filtering function has good resolution in the immediate vicinity of each center frequency and time value of the *f–t* diagram. The

Gaussian function was chosen to meet these requirements since the frequency–time resolution is greater for this function than any other type of nonband-limited function. A system of Gaussian filters with constant relative response leads to a constant resolution on a $\log(\omega)$ scale. If $\omega_n = 2\pi f_n$ denotes the center frequency of the nth row, the Gaussian window function can be written

$$H_n(\omega) = \exp\left\{-\alpha[(\omega - \omega_n)/\omega_n]^2\right\} \quad (5.238)$$

The Fourier transform of H_n, which bears a close resemblance to the Morlet wavelet Eqn (5.229), is

$$h_n(t) = \frac{\sqrt{\pi}}{2\alpha}\,\omega_n \exp\left[-(\omega_n^2 t^2/4\alpha)\right]\cos(\omega_n t)$$
$$(5.239)$$

The resolution is controlled by the parameter, α. The value of α that we choose depends on the dispersion characteristics in the original signal and, as the user of this method will soon discover, improved resolution in time means reduced resolution in frequency, and vice versa. We also need to truncate the filtering process.

Dziewonski et al. (1969) used a filter cut-off where the filter amplitude was down 30 dB from the maximum.

If we let BAND be the relative bandwidth, then the respective lower and upper limits of the symmetrical filter, denoted $\omega_{L,n}$ and $\omega_{U,n}$, are

$$\omega_{L,n} = (1 - BAND)\omega_n \qquad (5.240a)$$

$$\omega_{U,n} = (1 + BAND)\omega_n \qquad (5.240b)$$

The parameter α in Eqns (5.238) and (5.239) is expressed in terms of the bandwidth and the function β, where

$$\alpha = \beta/BAND^2 \qquad (5.241)$$

and

$$\beta = \ln\left[H_n(\omega_n)/H_n(\omega_{L,n})\right]$$
$$= \ln\left[H_n(\omega_n)/H_n(\omega_{U,n})\right] \qquad (5.242)$$

describes the decay of the window function, $H_n(\omega)$. The window function then takes the form

The flowchart for the analysis (Figure 5.53) is as follows:

1. Remove the mean and trend (linear or other obvious functional trend) from the digital time series, $y(t)$.
2. Fourier transform the time series. If an FFT algorithm is to be used for this purpose, augment the time series with zeroes to the nearest power of two.
3. Evaluate the center frequencies, $\omega_n = \omega_{n-1}/BAND$, for the array of narrow-band filters. The filters have a constant relative bandwidth, BAND, with the total width of each filter occupying the same number of rows in the log (frequency) scale. As noted on numerous occasions in the text, it is the length of the time series and the sampling rate, which determine the frequency of the Fourier components. Since it is often difficult to get the frequencies obtained from the Fourier analysis to line up exactly

$$H_n(\omega) = \begin{cases} 0 & \text{for } \omega(1 - BAND)\omega_n \\ \exp\left\{-\alpha[(\omega - \omega_n)/\omega_n]^2\right\} & \text{for } \left(1 - BAND\right)\omega_n \leq \omega \leq \left(1 + BAND\right)\omega_n \\ 0 & \text{for } \omega > (1 + BAND)\omega_n \end{cases} \qquad (5.243)$$

In their analysis of seismic waves, Dziewonski et al. (1969) used BAND $= 0.25$, $\beta = 3.15$, and $\alpha = \beta/BAND^2 = 50.3$.

The f–t diagram for the Alaska tsunamis (Figure 5.52) was obtained by windowing in the frequency domain with the truncated Gaussian function Eqn (5.243). In the time domain, the traces represent the convolution of the original data series with the Gaussian weighting function. The authors first set $\alpha = 25$ and chose $\beta = 1$, so that BAND $= 0.20$. The choice of β in Eqn (5.242) is arbitrary and can be set to unity, whereupon the bandwidth is determined by the e^{-1} values of the Gaussian function. For $\alpha = 25$ but $\beta = 2$, we have BAND $= 0.28$, and so on.

with the center frequencies of the filters, select those components of the Fourier analysis, which are closest to each member of the array and use these as the center frequencies.

4. Select equally spaced times (columns) for calculation of amplitude or phase, focusing mainly on the times following the arrival of the waves.
5. Filter the wave spectrum (sine and cosine functions of the Fourier transform) in the frequency domain with the Gaussian filter, $H_n(\omega)$. This filter is symmetric about the center frequencies, ω_n.
6. Take the IFT of the spectra using the same Fourier transform used in step 2. Since the IFT

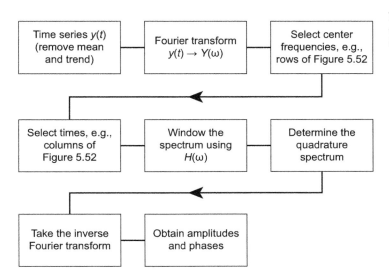

FIGURE 5.53 Flowchart for application of the multiple-filter technique. *(Adapted from Dziewonski et al. (1969).)*

for the wave spectrum as windowed by the function $H_n(\omega)$ yields only the in-phase component of the filtered signal for each ω_n, knowledge of the quadrature spectrum is also required for evaluation of the instantaneous spectral amplitudes and phases. The quadrature spectrum is found from the in-phase spectrum using

$$Q_n(\omega) = H_n(\omega)e^{i\pi/2} \qquad (5.244)$$

The amplitude and phase of the signal for each center frequency for each time are derived from the IFTs of the spectra and quadrature spectra.

7. Instantaneous spectral amplitudes and phases are computed for each time step. The procedure (5)–(7) is repeated for each center frequency.

The multiple filter technique can be used to examine rotary components of current velocity fields. In this case, the input is not a real variable, as it is for scalar time series, but a complex input, $w(t) = u(t) + iv(t)$. Figure 5.54 is obtained from the analysis of a 90-day time series of

surface currents measured by a 15 m drogued satellite-tracked drifter launched off the Kuril Islands in the western North Pacific on September 4, 1993 (Thomson et al., 1997). The 3-hourly sampling interval used for this time series was made possible by the roughly eight position fixes per day by the satellite-tracking system. Plots show the variation in spectral amplitude of the clockwise and counterclockwise rotary velocity components as functions of time and frequency. For illustrative purposes, we have focused separately on the high and low frequency ends of the spectrum (periods shorter and longer than 2 days). Several interesting features quickly emerge from these f–t diagrams. For example, the motions are entirely dominated by the clockwise rotary component except within the narrow channel (Friza Strait) between the southern Kuril Islands, where the motions become more rectilinear. The burst of clockwise rotary flow encountered by the drifter over the Kuril–Kamchatka Trench starting on day 28 was associated with wind-generated inertial waves, whereas the strong clockwise rotary diurnal currents first encountered on day 40 and again on day 55 were

FIGURE 5.54 Multiple-filter technique applied to the velocity of a near-surface (15 m drogued) satellite-tracked drifter launched off the Kuril Island in 1993. (a) S^- denotes the spectral amplitude (cm/s) of the clockwise rotary component vs frequency (cpd) and time (day); (b) S^+ denotes the spectral amplitude of the counterclockwise component. *(From Thomson et al. (1997).)*

associated with diurnal-period continental shelf waves propagating along the steep continental slope of the Kuril Islands.

5.8 FOURIER ANALYSIS

For many applications, including the spectral analysis discussed in Section 5.4, we can view time series as linear combinations of periodic or quasi-periodic components that are superimposed on a long-term trend and random noise. The periodic components are assumed to have fixed, or slowly varying, amplitudes and phases over the length of the record. The trends might include a slow drift in the sensor characteristics or a long-term

component of variability that cannot be resolved because of the limited duration of the data series. "Noise" includes random contributions from the instrument sensors and electronics, as well as frequency components that are outside the immediate range of interest (e.g., small-scale turbulence). A goal of time series analysis in the frequency domain is to reliably separate periodic oscillations from the random and aperiodic fluctuations. Fourier analysis is one of the most commonly used methods for identifying periodic components in near-stationary time series oceanographic data. If the time series are strongly nonstationary, more localized transforms such as the Hilbert and Wavelet transforms should be used.

The fundamentals of Fourier analysis were formalized in 1807 by the French mathematician Joseph Fourier (1768–1830) during his service as an administrator under Napoleon. Fourier developed his technique to solve the problem of heat conduction in a solid with specific application to heat dissipation in blocks of metal being turned into cannons. Fourier's basic premise was that any finite length, infinitely repeated time series, $y(t)$, defined over the principal interval $[0, T]$ can be reproduced using a linear summation of cosines and sines, or *Fourier series*, of the form

$$y(t) = \overline{y(t)} + \sum_p [A_p \cos(\omega_p t) + B_p \sin(\omega_p t)]$$

(5.245)

in which $\overline{y}$ is the mean value of the record, A_p, B_p are constants (the Fourier coefficients), and the specified angular frequencies, ω_p, are integer ($p = 1, 2, ...$) multiples of the fundamental frequency, $\omega_1 = 2\pi f_1 = 2\pi/T$, where T is the total length of the time series. Provided enough of these Fourier components are used, each value of the series can be accurately reconstructed over the principal interval. By the same token, the relative contribution a given component makes to the total variance of the time series is a measure of the importance of that particular frequency component in the observed signal. This concept is central to spectral analysis techniques. Specifically, the collection of Fourier coefficients having amplitudes A_p, B_p form a *periodogram*, which then defines the contribution that each oscillatory component, ω_p, makes to the total "energy" of the observed oceanic signal. Thus, we can use the Fourier components to estimate the power spectrum (energy per unit frequency bandwidth) of a time series, as described in Section 5.4. Since both A_p, B_p must be specified, there are two DoF per spectral estimate derived from the "raw" or unsmoothed periodogram.

5.8.1 Mathematical Formulation

Let $y(t)$ denote a continuous, finite-amplitude time series of finite duration. Examples include hourly sea-level records from a coastal tide gauge station or temperature records from a moored thermistor chain. If y is periodic, there is a period T^* such that $y(t) = y(t + T^*)$ for all t. In the language of Fourier analysis, the periodic functions are sines and cosines, which have the important properties that:

1. A finite number of Fourier coefficients provides the minimum MSE between the original data and a functional fit to the data series;
2. The functions are orthogonal so that coefficients for a given frequency can be determined independently.

Suppose that the time series is specified only at discrete times by subsampling the continuous series, $y(t)$, at a sample spacing of Δt (Figure 5.55). Since the series has a duration T, there are a total of $N = T/\Delta t$ sample intervals and $N + 1$ sample points located at times $y(t_n) = y(n\Delta t) = y_n$ ($n = 0, 1, ..., N$). Using Fourier analysis, it is possible to reproduce the original signal as a sum of sine or cosine waves of different amplitudes and phases. In Figure 5.55, we show a time series, $y(n\Delta t)$, of 41 data points followed by plots of the first, second, and sixth harmonics that were summed to create the time series. The frequencies of these harmonics are $f = 1/T$, $2/T$, and $6/T$, respectively, and each harmonic has the form $y_k(n\Delta t) = C_k \cos[2\pi k n/N + \phi_k]$, where (C_k, ϕ_k) are, respectively, the amplitudes and phases of the harmonics for $k = 1, 2,$ and 6. Here, $T = 40\Delta t$ and we have arbitrarily chosen $(C_1, \phi_1) = (2, \pi/4)$, $(C_2, \phi_2) = (0.75, \pi/2)$, and $(C_6, \phi_6) = (1.0, \pi/6)$ in order to generate the time series for this example. The $N/2$ harmonic, which is the highest frequency component that can be resolved by this sampling, has a frequency, $f_N = (N/2)/N\Delta t = 1/2\Delta t$ cycles per unit time and a period of $2\Delta t$. Called the *sampling* or *Nyquist* frequency, this represents the highest

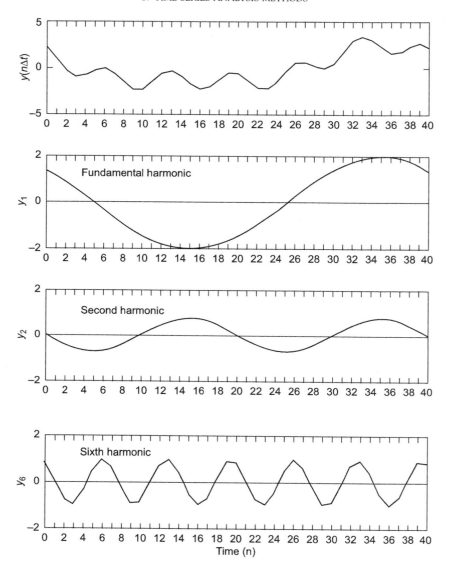

FIGURE 5.55 Discrete subsampling of a continuous signal, $y(t)$. The sampling interval is $\Delta t = 1$ time unit and the fundamental frequency is $f_1 = 1/T$, where $T = N\Delta t$ is the total record length and $N = 40$. The signal $y(t)$ is the sum of the first, second, and sixth harmonics which have the form $y_k(n\Delta t) = C_k \cos[(2\pi k n/N) + \phi_k]$; $k = 1, 2, 6$; $n = 0, 1, \ldots, 40$.

frequency resolved by the sample series in question. (As noted earlier in this chapter, we use the subscript N to denote the Nyquist frequency, which should not be confused with the integer N, as in $n = 1, 2, \ldots, N$, or the buoyancy frequency, $N(z)$.)

The fundamental frequency, $f_1 = 1/T$, is used to construct $y(t)$ through the infinite *Fourier series*

FIGURE 5.56 An example of a Fourier line spectrum with power at discrete frequencies, f, for a 24-h duration record with 1-h sampling increment.

length, detrended digital record having N (even) values at times $t_n = t_1, t_2, ..., t_N$, is

$$y(t_n) = \frac{1}{2}A_0 + \sum_{p=1}^{N/2}\left[A_p \cos\left(\omega_p t_n\right)\right. \qquad (5.256)$$
$$\left. + B_p \sin\left(\omega_p t_n\right)\right]$$

where the angular frequency, $\omega_p = 2\pi f_p = 2\pi p/T$. Using $t_n = n \cdot \Delta t$ together with (5.250) and (5.251), the final form for the discrete, finite Fourier series becomes

$$y(t_n) = \frac{1}{2}A_0 + \sum_{p=1}^{N/2}\left[A_p \cos(2\pi pn/N)\right.$$
$$\left. + B_p \sin(2\pi pn/N)\right]$$
$$= \frac{1}{2}C_0 + \sum_{p=1}^{N/2} C_p \cos\left[(2\pi pn/N) - \theta_p\right]$$

$$(5.257)$$

where the leading terms, $\frac{1}{2}A_0$ and $\frac{1}{2}C_0$, are the mean values of the record. The coefficients are

again determined using the orthogonality condition for the trigonometric functions. In fact, the main difference between the discrete case and the continuous case formulated in the last section (aside from the fact we can no longer have an infinite number of Fourier components) is that coefficients are now defined through the summations rather than through integrals

$$A_p = \frac{2}{N}\sum_{n=1}^{N} y_n \cos\left(2\pi pn/N\right),$$
$$p = 0, 1, 2, ..., N/2$$

$$A_0 = \frac{2}{N}\sum_{n=1}^{N} y_n, \quad B_0 = 0$$

$$A_{N/2} = \frac{1}{N}\sum_{n=1}^{N} y_n \cos\left(n\pi\right),$$

$$B_{N/2} = 0$$

$$B_p = \frac{2}{N}\sum_{n=1}^{N} y_n \sin\left(2\pi pn/N\right),$$
$$p = 1, 2, ..., (N/2) - 1 \qquad (5.258)$$

Notice that the summations in Eqn (5.258) consist of multiplying the data record by sine and cosine functions that "pick out" from the record those frequency components specific to their trigonometric arguments. Remember, the orthogonality condition requires that the arguments in the trigonometric functions be integer multiples of the total record length, $T = N\Delta t$, as they are in Eqn (5.258). If they are not, the sines and cosines do not form an orthonormal set of basis functions for the Fourier expansion and the original signal cannot be correctly replicated.

The arguments $2\pi pn/N$ in the above equations are based on a hierarchy of equally spaced frequencies, $\omega_p = 2\pi p/(N\Delta t)$, and time increment, "$n$". The summation goes to $N/2$, which is the limit of coefficients we can determine; for $p > N/2$ the trigonometric functions simply begin to cause repetition of coefficients already obtained for the interval, $p \le N/2$. Furthermore, it should be obvious that because there are as many coefficients as data points and because the trigonometric functions form an orthogonal basis set, the summation over the $2(N/2) = N$ discrete coefficients provides an exact replication of the time series, $y(t)$. Small differences between the original data and the Fourier series representation arise because of roundoff errors accumulated during the arithmetic calculations (see Chapter 3).

The steps in computing the Fourier coefficients are as follows. Step 1: Calculate the arguments, $\Phi_{pn} = 2\pi pn/N$, for each integer p and n. Step 2: For each $n = 1, 2, ..., N$, evaluate the corresponding values of $\cos\Phi_{pn}$ and $\sin\Phi_{pn}$, and collect sums of $y_n \cdot \cos \Phi_{pn}$ and $y_n \cdot \sin\Phi_{pn}$. Step 3: Increment p and repeat steps 1 and 2. The procedure requires roughly N^2 real multiply–add operations. For any real data sequence, roundoff errors plus errors associated with truncation of the total allowable number of desired Fourier components (maximum $f_p < f_{N/2}$) will give rise to a less than perfect fit to the data. The residual $\Delta y(t) = y(t) - y_{FS}(t)$ between the observations, $y(t)$ and the calculated Fourier series, $y_{FS}(t)$ will diminish with increased computational precision and increased numbers of allowable terms used in the series expansion. When computing the phases $\theta_p = \tan^{-1}[B_p/A_p]$ in the formulation (5.257), one must take care to examine in which quadrants A_p and B_p are situated. For example, $\tan^{-1}(0.2/0.7)$ differs from $\tan^{-1}(-0.2/-0.7)$ by 180°. The familiar ATAN2 function in FORTRAN is especially designed to take care of this problem.

5.8.3 A Computational Example

The best way to demonstrate the computational procedure for Fourier analysis is with an example. Consider the 2-year segment of monthly mean SSTs measured at the Amphitrite lightstation off the southwest coast of Vancouver Island (Table 5.11). Each monthly value is calculated from the average of daily surface thermometer observations collected around noon local time and tabulated to the nearest 0.1 °C. These data are known to contain a strong seasonal cycle of warming and cooling, which is modified by local effects of runoff, tidal stirring, and wind mixing.

The data in Table 5.11 are in the form $y(t_n)$, where $n = 1, 2, ..., N$ ($N = 24$). To calculate the coefficients A_p and B_p for these data, we use the summations Eqn (5.258) for each successive integer p, up to $p = N/2$. These coefficients are then used in Eqn (5.250) to calculate the magnitude $C_p = (A_p^2 + B_p^2)^{1/2}$ for each frequency component, $f_p = p/T$. Since C_p^2 is proportional to the variance at the specified frequency, the C_p enables us to rate the order of importance of each frequency component in the data series.

The mean value, $y(t) = \frac{1}{2}A_0$ and the 12 pairs of Fourier coefficients obtainable from the temperature record are listed in Table 5.12 together with the magnitude C_p. Values have been rounded to the nearest 0.01 °C. The Nyquist frequency, f_N, is 0.50 cycles per month (cpmo, $p = 12$) and the fundamental frequency, f_1, is 0.042 cpmo

TABLE 5.11   Monthly Mean Sea Surface Temperature (SST) (°C) at Amphitrite Point (48° 55.16′ N, 125° 32.17′ W) on the West Coast of Canada for January 1982 through December 1983

YEAR 1982

n	1	2	3	4	5	6	7	8	9	10	11	12
SST	7.6	7.4	8.2	9.2	10.2	11.5	12.4	13.4	13.7	11.8	10.1	9.0

YEAR 1983

n	13	14	15	16	17	18	19	20	21	22	23	24
SST	8.9	9.5	10.6	11.4	12.9	12.7	13.9	14.2	13.5	11.4	10.9	8.1

TABLE 5.12 Fourier Coefficients and Frequencies for the Amphitrite Point Monthly Mean Temperature Data

p	Frequency (cpmo)	Period (month)	Coefficient A_p (°C)	Coefficient B_p (°C)	Coefficient C_p (°C)	Phase θ_p (degrees)
0	0	—	21.89	0	21.89	0
1	0.042	24	−0.55	−0.90	1.05	−121.4
2	0.083	12	−1.77	−1.99	2.67	−131.7
3	0.125	8	0.22	−0.04	0.23	−10.3
4	0.167	6	−0.44	−0.06	0.45	−172.2
5	0.208	4.8	0.09	−0.07	0.11	−37.9
6	0.250	4	0.08	−0.04	0.09	−26.6
7	0.292	3.4	0.01	−0.16	0.16	−58.0
8	0.333	3	−0.03	−0.16	0.16	−100.6
9	0.375	2.7	−0.14	0.05	0.15	160.3
10	0.417	2.4	−0.09	−0.07	0.11	−142.1
11	0.458	2.2	−0.08	−0.12	0.14	−123.7
12	0.500	2	−0.15	0	0.15	0

Frequency is in cycles per month (ccpmo). $A_0/2$ is the mean temperature and θ_p is the phase lag for the pth component taken counterclockwise from the positive Ap axis.

($p = 1$). As we would anticipate from a visual inspection of the time series, the record is dominated by the annual cycle (period $= 12$ months) followed by weaker contributions from the biannual cycle (24 months) and semiannual cycle (6 months). For periods shorter than 6 months, the coefficients, C_p have similar amplitudes and likely represent the roundoff errors and background "noise" in the data series. This suggests that we can reconstruct the original time series to a high degree of accuracy using only the mean value ($p = 0$) and the first three Fourier coefficients ($p = 1, 2, 3$).

Figure 5.57 is a plot of the original SST time series and the reconstructed Fourier fit to this series using only the first three Fourier components from Table 5.12. Comparison of these two time series, shows that the reconstructed series does not adequately reproduce the skewed crest of the first year nor the high-frequency "ripples" in the

second year of the data record. There is also a slight mismatch in the maxima and minima between the series. Differences between the two curves are typically around a few tenths of a degree. In contrast, if we use all 12 components in Table 5.12, corresponding to 24 DoF, we get an exact replica of the original time series to within machine accuracy.

5.8.4 Fourier Analysis for Specified Frequencies

Analysis of time series for specific frequencies is a special case of Fourier analysis that involves adjustment of the record length to match the periods of the desired Fourier components. As we illustrate in the following sections, analysis for specific frequency components is best conducted using LS (Least Squares) fitting methods rather than Fourier analysis. LS analysis requires that

FIGURE 5.57 Monthly mean sea surface temperature (SST) record for Amphitrite Point on the west coast of Vancouver Island (see Table 5.11). The bold line is the original 24-month series; the dashed line is the SST time series generated using the first three Fourier components, f_p, $p = 0, 1, 2$, corresponding to the mean, 24-month, and 12-month cycles (Fourier components appear in Table 5.12). FFT, fast Fourier transform.

there be many fewer constituents than data values, which is usually the case for tidal analysis at the well-defined frequencies of the tide-generating potential. Problems arise if there are too few data values. For example, suppose that we have a few days of hourly water level measurements and we want to use Fourier analysis to determine the amplitudes and phases of the daily tidal constituents, f_k. To do this, we need to satisfy the orthogonality condition for the trigonometric basis functions for which terms like $\int \cos(2\pi f_j t) \cos(2\pi f_k t)dt = 0$ except where $f_j = f_k$ (the integral is over the entire length of the record, T). The approach is only acceptable when the length of the data set is an integer multiple of all the harmonic frequencies we are seeking. That is, the specified tidal frequencies, f_k, must be integer multiples of the fundamental frequency, $f_1 = 1/T$, such that $f_k \cdot T = 1, 2, \ldots, N$. If this holds, we can use Fourier analysis to find the constituent amplitudes and phases at the specified frequencies. In fact, this integer constraint on $f_k \cdot T$ is a principal reason why oceanographers prefer to use record lengths of 14, 29, 180, or 355 days when performing analyses of

tides. Since the periods of most of the major tidal constituents (K_1, M_2, etc.) are integer multiples of the fundamental tidal periods (1 lunar day, 1 lunar month ≈ 29 days, 1 year, 8.8 years, 18.6 years, etc.) of the above record lengths, the analysis is aided by the orthogonality of the trigonometric functions.

A note for those unfamiliar with tidal analysis terminology: Letters of tidal harmonics identify the different types ("species") of tide in each frequency band. Harmonic components of the tide-producing force that undergo one cycle per lunar day (≈ 25 h) have a subscript 1 (e.g., K_1), those with two cycles per lunar day have subscript 2 (e.g., M_2), and so on. Constituents having one cycle per day are called diurnal constituents, those with two cycles per day, semidiurnal constituents. The main daily tidal component, the K_1 constituent, has a frequency of 0.0418 cph (corresponding to an angular speed of 15.041° per mean solar hour) and is associated with the cyclic changes in the luni-solar declination. The main semidiurnal tidal constituent, the M_2 constituent, has a frequency of 0.0805 cph (corresponding to an angular speed of 28.984° per

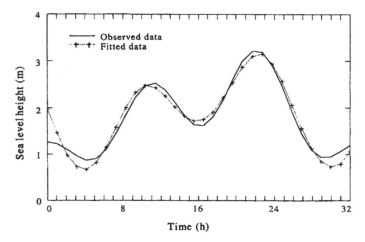

FIGURE 5.58 Hourly sea-level height (SLH) recorded at Tofino on the west coast of Vancouver Island (see Table 5.17). The bold line is the original 32-h series; the dotted line is the SLH series generated using the mean ($p = 0$) plus the next three Fourier components, f_p, $p = 1, 2, 3$ having nontidal periods, T_p, of 32, 16, and 8 h, respectively.

mean solar hour) and is associated with cyclic changes in the lunar position relative to the earth. Other major daily constituents are the O_1, P_1, S_2, N_2, and K_2 constituents. In terms of the tidal potential, the hierarchy of tidal constituents is M_2, K_1, S_2, O_1, P_1, N_2, K_2, and so on. Other important tidal harmonics are the lunar fortnightly constituent, M_f, the lunar monthly constituent, M_m, and the solar annual constituent, S_a. For further details the reader is referred to Thomson (1981), Foreman (1977, 1978), Pugh (1988), and Pawlowicz et al. (2002).

Returning to our discussion concerning Fourier analysis at specified frequencies, consider the 32-h tide gauge record for Tofino, British Columbia presented in Figure 5.58. As we show in Section 5.9, LS analysis can be used to reproduce this short record quite accurately using only the K_1 tidal constituent and the M_2 constituent. These are the dominant tidal constituents in all regions of the ocean except near amphidromic points. Because the record is 32-h long, the diurnal and semidiurnal frequencies are not integer multiples of the fundamental frequency, $f_1 = 1/T = 0.031$ cph, and are not among the sequence of 16 possible frequencies generated

from the Fourier analysis. In order to have frequency components centered more exactly at the K_1 and M_2 frequencies, we would need to shorten the record to 24 h or pad the existing record to 48 h using zeroes. In either case, the $f_k \cdot T$ for the tides would then be close to integers and a standard Fourier analysis would give an accurate fit to the observed time series. If we stick with the 32-h series, we find that the tidal energy in the diurnal and semidiurnal bands is partitioned among the first three Fourier components at frequencies, $f_1 = 0.031$, $f_2 = 0.062$, and $f_3 = 0.093$ cph. These frequencies are only vaguely close to those of the diurnal and semidiurnal constituents but do span the energy-containing frequency bands. As a result, the time series generated from the record mean combined with the first three Fourier components ($p = 1, 2, 3$) closely approximates the time series obtained using the true tidal frequencies (see Figure 5.59).

5.8.5 The Fast Fourier Transform

One of the main problems with both the autocovariance and the direct Fourier methods of spectral estimation is low computational

FIGURE 5.59 Hourly sea-level height (SLH) recorded at Tofino on the west coast of Vancouver Island (see Table 5.17; compare to Figure 5.58). The solid line is the original 32-h series; the dotted line is the SLH series obtained from a least-squares fit of the main diurnal (K_1, 0.042 cph) and main semi-diurnal (M_2, 0.081 cph) tidal frequencies to the mean-removed data (see Table 5.18).

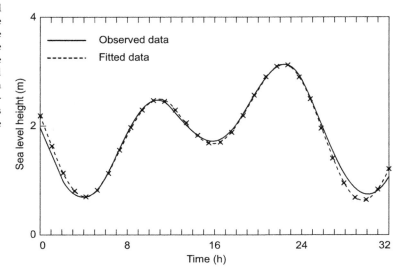

speed. The Fourier method requires the expansion into series of sine and cosine terms—a time-consuming procedure. The FFT is a way to speed up this computation while retaining the accuracy of the direct Fourier method. This makes the Fourier method computationally more attractive than the autocovariance approach.

To illustrate the improved efficiency of the FFT method, consider a series of N values for which $N = 2^p$ (p is a positive integer). The DFT of this series would require N^2 operations whereas the FFT method requires only $8N\log_2 N$ operations. The savings in computer time can be substantial. For example, if $N = 8192$, $N^2 = 67,108,864$ while $8N \log_2 N = 851,968$. Computers are much faster now than when the FFT method was introduced but the relative savings in computational efficiency remains the same. Bendat and Piersol (1986) define the speed ratio between the FFT and discrete Fourier method as $N/4p$. This becomes increasingly more important as the number of terms increases since the direct method computational time is $O(N^2)$, while for the FFT method it is $O(N)$. If one is seeking a smoothed power spectrum, it is often more efficient to compute the spectrum using the FFT technique and then smooth in spectral space by averaging over adjoining frequency bands rather than smoothing with an autocovariance lag window in the time domain.

To understand the FFT algorithm, we follow the derivation of Danielson and Lanczos (1942) who first helped pioneer the method. Consider a time series of x_t, where $t = 1, 2, ..., N$. We want to find the Fourier transform $X_m = X(m/N\Delta t)$, where $m = 0, 1, ..., N - 1$. To do this, we first partition x_t into two half-series, y_t and z_t, where $y_t = x_{2t-1}$, $z_t = x_{2t}$, $t = 1, 2, ..., N/2$. The series y_t contains values at the odd number times ($x_1, x_3, ...$) while the function z_t contains values at the even number times ($x_2, x_4, ...$). Both functions have $N/2$ values and their Fourier transforms are

$$Y_m^{(N/2)} = \frac{2}{N} \sum_{t=1}^{N/2} y_t \exp\left[\frac{(-i4\pi tm)}{N}\right] \quad (5.259a)$$

$$Z_m^{(N/2)} = \frac{2}{N} \sum_{t=1}^{N/2} z_t \exp\left[\frac{(-i4\pi tm)}{N}\right] \quad (5.259b)$$

where the superscript $(N/2)$ is used to denote the number of terms used in the expansion. But $X_m^{(N)}$, $Y_m^{(N/2)}$, and $Z_m^{(N/2)}$ are related since

$$
\begin{aligned}
X_m^{(N)} &= \frac{2}{N} \sum_{t=1}^{N/2} x_t \exp\left[\frac{-i4\pi tm}{N}\right] \\
&= \frac{1}{N} \sum_{t=1}^{N/2} \left\{ y_t \exp\left[\frac{-i4\pi tm}{N}(2t-1)\right] \right. \\
&\quad \left. + z_t \exp\left[\frac{-i4\pi tm}{N}(2t)\right] \right\} \\
&= \frac{1}{2}\exp\left[\frac{(i2\pi m)}{N}\right] Y_m^{(N/2)} + \frac{1}{2}Z_m^{(N/2)}, \\
&\quad 0 \le m \le (N/2) - 1 \qquad (5.260)
\end{aligned}
$$

Also

$$
Y_{m+N/2}^{(N/2)} = Y_m^{(N/2)}; \quad 0 \le m \le N/2 - 1
$$

$$
Z_{m+N/2}^{(N/2)} = Z_m^{(N/2)}; \quad 0 \le m \le N/2 - 1
$$

$$(5.261)$$

so that

$$
\begin{aligned}
X_{m+N/2}^{(N)} &= \frac{1}{2}\exp\left[i\left(\frac{2\pi}{N}\right)\left(m + \frac{N}{2}\right)\right] Y_m^{(N/2)} + \frac{1}{2}Z_m^{(N/2)} \\
&= -\frac{1}{2}\exp\left(\frac{i2\pi m}{N}\right) Y_m^{(N/2)} + \frac{1}{2}Z_m^{(N/2)}, \quad 0 \le m \le (N/2) - 1
\end{aligned}
$$

$$(5.262)$$

thus

$$
X_m^{(N)} = \frac{1}{2}\exp\left[i\frac{(2\pi m)}{N}\right] Y_m^{(N/2)} + \frac{1}{2}Z_m^{(N/2)},
$$

$$
0 \le m \le N/2 - 1 \qquad (5.263)
$$

and

$$
X_{m-N/2}^{(N)} = -\frac{1}{2}\exp\left[i\frac{(2\pi m)}{N}\right] Y_m^{(N/2)} + \frac{1}{2}Z_m^{(N/2)},
$$

$$
0 \le m \le N/2 - 1 \qquad (5.264)
$$

Thus, the Fourier transform for the series, x_t is found from the Fourier series of the half series, y_t and z_t. Since $N/2$ is even, this can be repeated. If the length of the data is not a power of two, it should be padded with zeros up to the next power of two. For a series of length $N = 2p$ (p a positive integer), the procedure is followed until partitions consist of only one term whose Fourier transform equals itself, or the procedure is followed until N becomes a prime number, i.e., $N = 3$. The Fourier transform is then found directly for the remaining short series.

5.9 HARMONIC ANALYSIS

Standard Fourier analysis involves the computation of Fourier amplitudes at equally spaced frequency intervals determined as integer multiples of the fundamental frequency, f_1. That is, for frequencies, $f_1, 2f_1, 3f_1, \ldots, f_N$ (f_N = Nyquist frequency). However, as we have shown in the previous section, standard Fourier analysis has major limitations when it comes to the analysis of data series in terms of predetermined frequencies. In the case of tidal motions, for example, it would be impractical to use any frequencies except those of the astronomical tidal forces. Equally importantly, we want to determine the amplitudes and phases of as many frequency components as possible by using as short a time series as possible. Since there are typically many more data values than prescribed frequencies, we have to deal with an overdetermined problem. This leads to a form of signal demodulation known as *harmonic analysis* in which the user specifies the frequencies to be examined and applies LS techniques to solve for the constituents. Harmonic analysis was originally designed for the analysis of tidal variability but applies equally to analysis at the annual and semiannual periods or any other well-defined cyclic oscillation. The familiar hierarchy of "harmonic" tidal constituents is dominated by diurnal and semidiurnal motions, followed by fortnightly, monthly, semiannual, and annual variability. In this section, we present a general discussion of harmonic analysis. The important subject of harmonic analysis of tides and tidal currents is treated separately in Section 5.9.3.

The harmonic analysis approach yields the required amplitudes and phase lags of the harmonic tidal coefficients or any other constituents we may wish to specify. For example, in studies of interannual variability, we may want to first define the *canonical seasonal cycle* as comprised of the amplitudes and phases of the 12-month (or 360 day) cycle plus the first and second subharmonics of 6 months (180 days) and 3 months (90 days). Once these coefficients have been determined, we can subtract the canonical cycle to then examine year-to-year variations in the original time series. In the case of tidal motions, subtraction of the reconstructed tidal signal from the original record yields a time series generally termed the *detided* or *residual* component of the time series. In many cases, it is the "detided" signal that is of primary interest. If we break the original time series into adjoining or overlapping segments, we can apply harmonic analysis to the segments to obtain a sequence of estimates for the amplitudes and phase lags of the various frequencies of interest. This leads to the notion of signal *demodulation*.

5.9.1 An LS Method

Suppose we wish to determine the harmonic constituents, A_q and B_q, for M-specified frequencies, which, in general, will differ from the Fourier frequencies defined by Eqn (5.247). In this case, $q = 0, 1, ..., M$ and $B_0 = 0$ so that there are a total of $2M + 1$ harmonic coefficients.

Assume that there are many more observations, N, than specified coefficients (i.e., that $2M + 1 \ll N$). The problem of fitting M harmonic curves to the digital time series is then overdetermined and must be solved using an optimization technique. Specifically, we estimate the amplitudes and phases of the various components by minimizing the squared difference (i.e., the LS) between the original data series and our fit to that series. The coefficients for each of the M resolvable constituents are found through solution of an $(M + 1) \times (M + 1)$ matrix equation.

For M possible harmonic constituents, the time series, $x(t_n)$, $n = 1, ..., N$ can be expanded as

$$x(t_n) = \bar{x} + \sum_{q=1}^{M} C_q \cos\left(2\pi f_q t_n - \phi_q\right) + x_r(t_n)$$

(5.265)

in which $\overline{x(t)}$ is the mean value of the record; x_r is the residual portion of the time series (which may contain other kinds of harmonic constituents); $t_n = n\Delta t$; and C_q, f_q, and ϕ_q are, respectively, the constant amplitude, frequency, and phase of the qth constituent that we have specified. In the present configuration, we assume that the specified frequencies have the form $f_q = q/(N\Delta t)$ so that the argument, $2\pi f_q t_n = 2\pi q n/N$. Reformulation of Eqn (5.265) as

$$x(t_n) = \bar{x} + \sum_{q=1}^{M} \left[A_q \cos\left(2\pi f_q t_n\right) \right.$$
$$\left. + B_q \sin\left(2\pi f_q t_n\right) \right] + x_r(t_n) \quad (5.266)$$

yields a representation in terms of the unknown coefficients, A_q, B_q, where

$$C_q = \left(A_q^2 + B_q^2 \right)^{1/2},$$

(frequency component amplitude)

$$\phi_q = \tan^{-1}\left(B_q/A_q \right),$$

(frequency component phase lag)

(5.267)

$1/\Delta f$. Obviously, the longer the record, the more constituents are permitted.

The choice of f_R is determined by the hierarchy of constituents within the tidal band of interest and level of noise in the observations. The hierarchy is in turn based on the contribution a particular constituent makes to the equilibrium tide, with the largest contribution usually coming from the M_2 tidal constituent (Cartwright and Edden, 1973). For the major contributors to the equilibrium tide, the magnitude ratios relative to M_2 in descending order are: $K_1/M_2 = 0.584$, $S_2/M_2 = 0.465$, and $O_1/M_2 = 0.415$. Depending on the level of noise in the observations, the principal semidiurnal constituent, M_2 (0.0805 cph), and the record mean, Z_0, can be determined for records longer than about 13-h duration, while the principal diurnal component, K_1 (0.0418 cph), can be determined for records longer than about 24 h. As a rough guide, separation of the next most significant semidiurnal constituent, S_2 (0.0833 cph), from the principal component M_2 requires a record length, $T > 1/|f(M_2) - f(S_2)| = 355$ h (14.7 days). Similarly, separation of the next most significant diurnal constituent, O_1 (0.0387 cph), from the principal component, K_1, requires an approximate record length, $T > 1/|f(K_1) - f(O_1)| = 328$ h (13.7 days). The frequencies, $f(K_1)$ and $f(O_1)$, then become the Rayleigh comparison frequencies for other neighboring tidal constituents in the diurnal band while the frequencies, $f(M_2)$ and $f(S_2)$, become the comparison frequencies for neighboring frequencies in the semidiurnal band. Extension of this procedure to longer and longer records eventually encompasses all the significant tidal constituents within the diurnal and semidiurnal bands. The first long-term constituent to be included in the analysis is the lunar—solar fortnightly cycle, M_{sf} (0.00282 cph), requiring an approximate record duration, $T > 14.8$ days, followed by the lunar monthly constituent, M_m (0.00151 cph), duration, $T > 31.8$ days, and the lunar fortnightly cycle, M_f (0.00305 cph), $T > 182.6$ days. These record length requirements are based on stochastic processes; shorter records can be used for deterministic processes such as tides provided that noise levels are low. Thus, in all cases, shorter record lengths can be used if the data are highly noise-free. By the same token, longer records are often needed to resolve the longer-period tides because of contamination from atmospheric effects.

A summary of the required record lengths for inclusion of the more important constituents is provided in Tables 5.14—5.16 together with a comparison of a given constituent's tidal potential magnitude relative to that of the principal component in the frequency band. Where possible, a candidate constituent is compared to the particular neighboring constituent, which has already been selected and is nearest in frequency.

TABLE 5.14 Record Lengths (in Hours) Needed to Resolve the Main Tidal Constituents in the Semidiurnal Tidal Band Assuming a Rayleigh Coefficient, $R = 1$

Tidal Constituent	Frequency (cph)	Comparison Constituent	Magnitude Ratio	Record Length (h)
M_2 (principal lunar)	0.0805	—	1	13
S_2 (principal solar)	0.0833	M_2	0.465	355
N_2 (larger lunar elliptic)	0.0790	M_2	0.192	662
K_2 (luni-solar)	0.0836	S_2	0.029	4383

Also listed are the comparison constituents and ratios of tidal potential to that of the principal semidiurnal constituent, M_2.

TABLE 5.15   Record Lengths (in Hours) Needed to Resolve the Main Tidal Constituents in the Diurnal Tidal Band Assuming a Rayleigh Coefficient, $R = 1$

Tidal Constituent	Frequency (cph)	Comparison Constituent	Magnitude Ratio	Record Length (h)
K_1 (luni-solar)	0.0418	—	0.584	24
O_1 (principal lunar)	0.0387	K_1	0.415	328
P_1 (principal solar)	0.0416	K_1	0.193	4383
Q_1	0.0372	O_1	0.079	662

Also listed are the comparison constituents and ratios of tidal potential to that of the principal semidiurnal constituent, M_2.

TABLE 5.16   Record Lengths (in Hours) Needed to Resolve the Main Tidal Constituents in the Long-Period Tidal Band Assuming a Rayleigh Coefficient, $R = 1$

Tidal Constituent	Frequency (cph)	Comparison Constituent	Magnitude Ratio	Record Length (h)
M_{sf} (mixed solar fortnightly)	0.002822	M_f	0.015	355
M_f (lunar fortnightly)	0.003050	—	0.172	4383
M_m (lunar monthly)	0.001512	M_{sm}	0.091	764
M_{sm} (solar monthly)	0.001310	—	0.017	4942
S_{sa} (solar semiannual)	0.000228	S_a	0.080	4383
S_a (solar annual)	0.000114	—	0.013	8766

Also listed are the comparison constituents and ratios of tidal potential to that of the principal semidiurnal constituent, M_2.

5.9.5 A Computational Example for Tides

As a simple example of the LS method of harmonic tidal analysis, consider the 32-hourly sea-level heights measured at Tofino, British Columbia during September 10–11, 1986 (Table 5.17). As indicated by Tables 5.14 and 5.15, we can at most resolve the K_1 and M_2 constituents. This problem is similar to that considered in Section 5.9.2, where we used the LS technique to fit the annual and semiannual components to a 24-month record of SST. Following the analysis in that section, the various matrices are written in terms of a mean component plus the contributions from the K_1 and M_2 frequencies, $f(K_1) = 0.0418$ cph and $f(M_2) = 0.0805$ cph, respectively. From Eqns (5.272) and (5.273), we find

TABLE 5.17 Hourly Values of Sea-Level Height (SLH) Measured at Tofino, British Columbia (49°09.0′°N, 125°54.0′°W) on the West Coast of Canada Starting September 10, 1986

n	1	2	3	4	2	6	7	8	9	10	11
SLH	1.97	1.46	0.98	0.73	0.67	0.82	1.15	1.58	2.00	2.33	2.48
n	12	13	14	15	16	17	18	19	20	21	22
SLH	2.43	2.25	2.02	1.82	1.72	1.75	1.91	2.22	2.54	2.87	3.10
n	23	24	25	26	27	28	29	30	31	32	
SLH	3.15	2.94	2.57	2.06	1.56	1.13	0.84	0.73	0.79	1.07	

Heights are in meters above the local datum.

$$\mathbf{D} = \begin{pmatrix} N & c_1 & c_2 & s & s_2 \\ c_1 & cc_{11} & cc_{12} & cs_{11} & cs_{12} \\ c_2 & cc_{21} & cc_{22} & cs_{21} & cs_{22} \\ s_1 & sc_{11} & sc_{12} & ss_{11} & ss_{12} \\ s_2 & sc_{21} & sc_{22} & ss_{21} & ss_{22} \end{pmatrix} \quad (5.281)$$

$$= \begin{pmatrix} 32 & 2.476 & -1.836 & 6.183 & 3.420 \\ 2.476 & 14.809 & 1.450 & 1.136 & 2.117 \\ -1.836 & 1.450 & 16.263 & -2.197 & 0.397 \\ 6.183 & 1.136 & -2.1 & 17.191 & 2.163 \\ 3.420 & 2.117 & 0.397 & 2.163 & 15.737 \end{pmatrix}$$
$$(5.282)$$

and from Eqns (5.273) and (5.275)

$$\mathbf{y} = \begin{pmatrix} yc_0 \\ yc_1 \\ yc_2 \\ ys_1 \\ ys_2 \end{pmatrix} = \begin{pmatrix} 57.640 \\ 6.514 \\ 6.138 \\ -0.199 \\ -3.335 \end{pmatrix} \quad (5.283)$$

where the elements of $\mathbf{D}$ and $\mathbf{y}$ have units of meters. The solution $\mathbf{z} = \mathbf{D}^{-1}\mathbf{y}$ is the vector

$$\mathbf{z} = \begin{pmatrix} A_0 \\ A_1 \\ A_2 \\ B_1 \\ B_2 \end{pmatrix} = \begin{pmatrix} 1.992\ m \\ 0.186\ m \\ 0.523\ m \\ -0.574\ m \\ -0.604\ m \end{pmatrix} \quad (5.284)$$

The results are summarized in Table 5.18. A plot of the original sea-level data and the fitted sea-level curve is presented in Figure 5.59. The standard deviation for the original record is 0.741 m while that for the fitted record is 0.736 m. For this short segment of the data record, the sum of the two tidal constituents accounts for over 99% of the total variance in the record. As a comparison, we have used the full analysis package without inference to analyze 29 days of the Tofino sea-level record beginning

TABLE 5.18 Least-Squares Estimates of the Amplitude (Fourier Coefficients) for the K_1 and M_2 Tidal Constituents for the 32-h Tofino Sea Level Starting at 2000, September 10, 1986

Q	Frequency (cph)	Period (h)	A_q (m)	B_q (m)	C_q (m)	C'_q (m)
0	—	—	3.984	0	3.984	4.100
1	0.042	24	0.186	−0.574	0.365	0.286
2	0.081	12	0.523	−0.604	0.638	0.986

The mean is $\frac{1}{2}A_0$ and q denotes the number of cycles per day (q = 0 is for the mean value). The last column, C'_q, gives the constituent amplitudes for a more extensive analysis that used a 29-day (685 h) data segment that had the same start time as the 32-h segment used to derive C_q.

at 20:00 hrs on September 10, 1986. The program finds a total of 30 constituents, including the mean, Z_0, with the sum of the tidal constituents accounting for 98% of the original variance in the signal. The record mean for the month is 2.05 m, and the K_1 and M_2 constituents have amplitudes of 0.286 and 0.986 m, respectively. As expected, these are quite different to the values derived on only 32 h of data (Table 5.18). Phases for the two constituents for the 29-day records are 122.0 and 12.5° compared with 107.9 and 130.9° for the same two constituents based on the 32-h records. Here, phase angles are, by convention, measured relative to a local (or, alternatively, the Greenwich) meridian of longitude. For example, off the west coast of Canada, it is common to use time in degrees relative to 120° W longitude as the reference time.

5.9.6 Complex Demodulation

In many applications, we seek to determine how the signal characteristics at a specific frequency, ω, change throughout the duration of a time series. For example, we might ask how the amplitude, phase, and orientation of the semidiurnal tidal current ellipses at different depths at a mooring location change with time. Wave packets associated with passing internal tides would be revealed through rapid changes in ellipse parameters at the M_2 and/or S_2 frequencies. The method for determining the temporal change of a particular frequency component for a velocity or scalar time series is called *complex demodulation*.

A common technique for finding the demodulated signal is to fit the desired parameters to sequential segments of the data series using LS algorithms. The analysis requires that there be many more data points than frequency components and each segment must span at least one cycle of the frequency of interest. As with any LS analysis, the observations do not have to be at regular time intervals. Inputs to complex demodulation algorithms require specification

of the start time of the first segment, the length of each segment, and the time between computation interval start times. Computation intervals may overlap, be end-to-end, or be interspersed with unused data. Following the LS analysis described under the section on harmonic analysis, the time increment between each estimate can be as short as one time step, Δt, thereby providing the maximum number of estimates for a given segment length, or as long as the entire record, thereby yielding a single estimate of the signal parameters.

For each segment of current velocity data, the fluctuating component of velocity at frequency, ω can be expressed as

$$\mathbf{u}(t) - \overline{\mathbf{u}(t)} = \left[u(t) - \overline{u(t)}\right] + i\left[v(t) - \overline{v(t)}\right]$$
$$= A^+ \exp\left[i(\omega t + \varepsilon^+)\right]$$
$$+ A^- \exp\left[-i(\omega t + \varepsilon^-)\right]$$

$$(5.285)$$

where $\overline{u(t)}, \overline{v(t)}$ are mean components of the velocity, and (A^+, A^-) are the amplitudes, and $(\varepsilon^+, \varepsilon^-)$ are the phases of the counterclockwise $(+)$ and clockwise $(-)$ rotating components, respectively. Data are at times, t_k, $(k = 1, ..., N)$ and solutions are found from the matrix equation

$$\mathbf{z} = \mathbf{D}^{-1}\mathbf{y} \qquad (5.286)$$

where

$$\mathbf{y} = \begin{pmatrix} u(t_1) \\ u(t_2) \\ \cdots \\ u(t_n) \\ v(t_1) \\ \cdots \\ v(t_n) \end{pmatrix};$$

$$\mathbf{z} = \begin{pmatrix} A^+ \cos(\varepsilon^+) \\ A^+ \sin(\varepsilon^+) \\ A^- \cos(\varepsilon^-) \\ A^- \sin(\varepsilon^-) \end{pmatrix} \equiv \begin{pmatrix} ACP \\ ASP \\ ACM \\ ASM \end{pmatrix} \qquad (5.286a)$$

and

$$
\mathbf{D} = \begin{pmatrix}
\cos(\omega t_1) & -\sin(\omega t_1) & \cos(\omega t_1) & \sin(\omega t_1) \\
\cos(\omega t_2) & -\sin(\omega t_2) & \cos(\omega t_2) & \sin(\omega t_2) \\
\cdots & \cdots & \cdots & \cdots \\
\cos(\omega t_n) & -\sin(\omega t_n) & \cos(\omega t_n) & \sin(\omega t_n) \\
\sin(\omega t_1) & \cos(\omega t_1) & -\sin(\omega t_1) & \cos(\omega t_1) \\
\cdots & \cdots & \cdots & \cdots \\
\sin(\omega t_n) & \cos(\omega t_n) & -\sin(\omega t_n) & \cos(\omega t_n)
\end{pmatrix}
$$

$$(5.286b)$$

Once the elements of $\mathbf{z}$ are found from the LS solution to the matrix equation (for example, using IMSL routine LLSQAR or specific MATLAB routines), we can find the various ellipse parameters from

$$A^+ = \left(ASP^2 + ACP^2\right)^{1/2};$$

$$A^- = \left(ASM^2 + ACM^2\right)^{1/2} \quad (5.287a)$$

$$\tan\left(\varepsilon^+\right) = \frac{ASP}{ACP}; \quad \tan\left(\varepsilon^-\right) = \frac{ASM}{ACM} \quad (5.287b)$$

For example, we could obtain the demodulated current amplitude and phase for near-inertial motions observed at a midlatitude mooring by setting $\omega = 2\Omega \sin\theta$ and obtaining LS solutions for a series of adjoining 24-h segments with no overlap (here, Ω is the angular earth rotation rate and θ is latitude). For the LS technique to be applicable, data would need to be sampled at roughly hourly intervals so that there were more data points per segment than parameters being estimated. Equatorward of $\theta = \pm 30°$, the period of inertial motions exceeds 24 h and the lengths of individual segments must be increased accordingly. Complex demodulation also can be used to examine inertial motions in Lagrangian-type data. In Figure 5.61(a), we have plotted the original and demodulated positions of a satellite-tracked drifter launched in the Canadian Arctic in the fall of 1988. The time series covers 60 days and was analyzed using overlapping 24-h subsections with the assumption that displacements occurred at the inertial period of 12.73 h for 70°N latitude.

Figure 5.61(b) presents a detailed analysis of the trajectory record for the 20 days ending October 11 when the buoy became trapped in growing sea ice. Note the intense inertial currents starting on September 30, the prevalence of the clockwise component of rotation, and the roughly $-6.4°$ per day drift in phase of the clockwise component of the current due to the changing latitude of the drifter relative to the reference latitude of 70° N.

5.10 REGIME SHIFT DETECTION

A regime can be broadly defined as an extended period of time over which a given variable or natural system assumes a statistically stable and stationary state. The rapid transition between states is termed a "regime shift." We can associate regimes and regime shifts with a wide variety of natural systems, including national governments, seasonal weather patterns, marine ecosystems, large-scale current patterns, and global climate. If we think of the seasons as distinct earth-ocean regimes, then the transition from winter to summer that occurs each spring (the "Spring Transition" (ST)) and the return to winter in fall (the "Fall Transition" (FT)) can be considered regime shifts. A square wave oscillator, in which the variable jumps abruptly from one stable "flat line" state to another, is a classic example of regimes and regime shifts. Note that "flat" in this context does not necessarily imply constant since there can be large fluctuations about the mean level. In marine and atmospheric sciences, regimes are generally data-determined and identified as periods of time having differing mean levels about which there can be pronounced fluctuations. Investigators might be interested in changes ranging from seasonal conditions during given years (cf., Agapitos and Gajewski, 2012) or in changes in average climate values over interannual timescales (cf., Rodionov and Overland, 2005). Although less common, regime shifts can also

(a)

FIGURE 5.61   (a). Complex demodulation at the inertial period of 12.73 h for the trajectory of a satellite-tracked drifter deployed in the Beaufort Sea in August 1988. (a) Original (solid line) and demodulated version (dashed line) of the drifter track. *(Courtesy of Humfrey Melling.)* (b). Complex demodulation at the inertial period of 12.73 h for the trajectory of a satellite-tracked drifter deployed in the Beaufort Sea in August 1988. (b) Parameters of the demodulation over a 20-day period of strong inertial motions. Top panel: phase of the clockwise (CW) rotary component (degrees). Remaining panels: amplitudes of the CW rotary, counterclockwise (CCW) rotary, and speed of the demodulated current. *(Courtesy of Humfrey Melling.)*

be identified through abrupt changes in the signal variance. For example, density-driven bottom water renewal events in coastal inlets can cause the bottom waters to transition from a quiescent state dominated by weak fluctuations in temperature, salinity, dissolved oxygen, and other oceanic variables, to a noisy, highly variable state characterized by pronounced fluctuations in these variables due to spatial gradients introduced by the event. Another example is the effect of a tsunami on water levels; the

mean level over days or longer does not change but shorter-term variations may be an order of magnitude greater than average, increasing the variance over a day or so. In ecosystems analysis, step changes in the variance of species abundance in adjacent trophic levels may signal cascading reorganization within the environment. Daskalov et al. (2007) found that a decrease or increase in species abundance variance in the Black Sea since the 1950s was related to a strengthening or weakening of top-down

(b)

Phase of CW-rotary current

Anticipated phase drift

Amplitude of CW-rotary current

Amplitude of CCW-rotary current

Amplitude of demodulated current

Beaufort Sea, 1988

FIGURE 5.61 (*continued*).

forcing of predatory fish linked through plankton to dissolved oxygen levels in the water column.

This section presents three nonparametric methods for defining regime shifts (statistical transitions) in the ocean. These are: "sequential *t*-test analysis of regime shifts" (STARS), "adaptive Kolmogorov–Zurbenko filters" (KZA), and "cumulative upwelling index" (CUI). A fourth approach—the use of subjective interpretation in which the investigator uses his or her expertise and knowledge of the data to partition specific data sets into regimes and regime shifts—should not be dismissed out of hand. As pointed out by Overland et al. (2006), "…the main difficulty in its (regime shifts) application is the inability to infer a unique underlying system structure from relatively short time series records." Moreover, all regime-shift algorithms require specification of certain parameters, thereby introducing a degree of subjectivity into the process. In our experience, it is impossible to design a purely objective method and that some degree of subjectivity (call it investigators "expertise" and "experience") is invariably required.

5.10.1 Sequential *t*-Test Analysis of Regime Shifts

STARS was originally designed for climatological time series (Rodionov, 2004, 2006) but has since found applications to other time series such as those for large-scale ecosystems in the Bering Sea (Rodionov and Overland, 2005). STARS was first written in FORTRAN but is now also available for Excel and MATLAB (available at http://www.climatelogic.com; earlier versions of the software are available at http://www.beringclimate.noaa.gov/regimes/). The program uses an algorithm for a sequential *t*-test (see Appendix Table D.3 for *t*-test values) in which deviations from the mean value are calculated and compared to a critical value. As each new observation is added to the analysis, a new t-test is performed. The test then determines the validity of the null hypothesis (H_0) that the mean values of two regimes (the original and a possible new regime, which is being considered based on newly added data values) are equal. In essence, the analysis is asking whether the new sequence of data values is part of a new regime or just a minor variation on the existing regime. The strength of a regime is quantified by the Regime Shift Index (RSI), which is dependent on a chosen cutoff length (L) and the probability level (p) of the t-test. Only regimes longer than the assigned cutoff length are detected, unless their RSI values are high enough to be detected for data segments shorter than L (specifically, at the beginning or end of a data series). As shown in Figure 5.62, the duration of regimes detected by STARS becomes shorter as the cutoff length, L, is reduced. The method also requires specification of the Huber weight parameter, H, which determines how outliers are to be weighted. All values within H standard deviations of the expected mean value of the new regime (default, $H = 1$) have a weight of one, and otherwise are weighted inversely proportional to their distance from the expected mean value of the new regime. Figure 5.63 shows that appropriate specification of the Huber weight parameter can be important. In this case, the identification of regimes (top panel) may be confounded by the presence of a single outlier (middle panel), unless H is adjusted appropriately (bottom panel).

Following Rodionov and Overland (2005), we let $x_1, x_2, \ldots, x_i, \ldots$ be a preexisting time series or, alternatively, a data series that has new data being added on a regular basis. The first step is to determine an initial critical regime by calculating the mean and standard deviation over the first L values, $x_1, x_2, \ldots, x_L$. Once this regime is calculated, the program then returns to the start of the time series and begins to recalculate a mean value starting with the mean of the first and second points, x_1 and x_2 (x_1 is assumed part of the

initial regime but x_2 may be part of a new regime). A check is performed to determine if the mean of these two values has a statistically significant deviation from the mean value of the current regime. If so, that value is marked as a potential change point, i_c, and subsequent observations are incorporated in the running mean to confirm or reject the null hypothesis. This approach allows detection of regime shifts near the beginning of time series for $i < L$. The hypothesis is tested throughout the data set using the RSI, which is calculated for each i_c as

$$RSI = \sum_{i=i_c}^{i_c+m} \frac{\widehat{x}_i}{L\sigma_L} \qquad (5.288)$$

where $m = 0, \ldots, L - 1$ is the number of values since the start of a new regime, σ_L is the average standard deviation for all the regime intervals in the time series to present, and $\widehat{x}_i = x_i - \overline{x}_{new}$ is the difference between the data value x_i, and the possible mean level for the new regime, $\overline{x}_{new}$ that is being tested. For $\overline{x}_{new}$ to be considered as the mean value for a new regime, the difference from the mean level for the current regime $\overline{x}_{cur}$ must be statistically significant according to a student's t-test specified by

$$difference = \overline{x}_{new} - \overline{x}_{cur} = t\sqrt{2\sigma_L^2/L} \quad (5.289)$$

where t is the value of the t-distribution with $2L - 1$ DoF at the given probability level, p. If, at any time from the start of the new regime, RSI becomes negative, the test fails and a zero value is assigned. If RSI remains positive throughout the range $(0, L - 1)$, then i_c is selected to be the time of a regime shift at the selected level $\leq p$. The search for the next regime shift starts at $i_c + L$ so as to ensure that its timing is detected correctly even if the actual duration of the new regime is short relative to the expected durations of actual regimes. Table 5.19 is a shortened version of that presented in Rodionov and Overland (2005) for pollock stock recruitment in the Bering Sea. Data are available from 1963 but

Pacific Decadal Oscillation Index Annual Mean

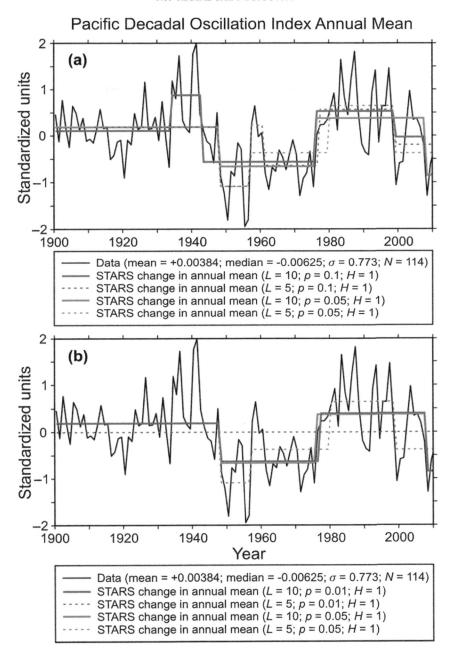

FIGURE 5.62 The effect of changing cutoff length, L, and probability level, p, on the statistically significant difference between regimes for the time series of annual means of the Pacific Decadal Oscillation (PDO) (the first spatial mode of sea surface temperature empirical orthogonal functions). Decreasing L and/or increasing p, increases the number of regimes found in a time series by decreasing the magnitude of the shifts to be detected. (a) Regimes detected in the annual mean of the PDO for $L = 5$ and 10, for $p = 0.1$ and 0.05 ($H = 1$); (b) same as (a) but for and $p = 0.05$ and 0.01. *(Plots and analysis courtesy of Roy Hourston, Institute of Ocean Sciences).*

FIGURE 5.63 The importance of considering the Huber weight parameter, H, when using sequential t-test analysis of regime shifts (STARS). The data in the three panels are identical except for the second data value, x_2, which has been changed from $x_2 \sim 0$ in the upper panel to $x_2 = 100$ in the middle and lower panels. The regimes identified by the STARS algorithm in the top panel can no longer be identified in the middle panel due to the addition of the outlier. When the Huber weight parameter is reduced from 1 to 0.1 in the bottom panel, the weighting of the outlier is reduced, and the original regimes are identified. *(Courtesy of Roy Hourston, Institute of Ocean Sciences).*

RSI values are only listed for the start of the first regime shift in 1978. Only for 3 years (1978, 1985, 1989) did RSI values remain positive up to $m = L - 1$.

The average value for the current regime, $\bar{x}_{cur}$, is calculated for the period $(i_c - L, i_c)$. If the transition from one regime to another is gradual, the program might not detect the change because $\bar{x}_{cur}$ is also changing as the window slides along the time axis. Specifically, the difference between the new arriving observations and $\bar{x}_{cur}$ may not be sufficiently statistically significant to become a change point and trigger the calculation of

RSI. In the version presented by Rodionov and Overland (2005), $\bar{x}_{cur}$ is calculated for the period that begins from the start of the previous regime shift to the point immediately before the current point in time. As a result, a stepwise function of regimes is produced in most cases. (In a previous version of the method presented by Rodionov (2004), the program could only detect abrupt changes in regimes.) To improve the performance at the beginning of the time series, testing for a new regime starts at x_2 rather than at x_{L+1} as the previous version. The average value, $\bar{x}_{cur}$, is still calculated for the entire initial period $(1, L)$

TABLE 5.19   Truncated Version of Regime Shift Index (RSI) Table for Pollock Recruitment ($L = 5$ years, $p = 0.1$; Huber Weight Parameter $= 1$)

Year ($i = i_c$)	$m = 0$	$m = 1$	$m = 2$	$m = 3$	$m = 4$
1978	0.51	0.45	0.44	0.21	0.57
1979	0	0	0	0	0
1980	0	0	0	0	0
1981	0.1	0	0	0	0
1982	0	0	0	0	0
1983	0.15	0	0	0	0
1984	0	0	0	0	0
1985	0.1	0.29	0.52	0.67	0.12
1986	0	0	0	0	0
1987	0	0	0	0	0
1988	0	0	0	0	0
1989	0.34	0.24	0.05	0.37	0.09
1990	0	0	0	0	0

The nonzero values of i_c sometimes triggered the calculation of RSI; however, only for the years highlighted in bold did the values remain positive and a regime shift declared.(From Rodionov and Overland (2005).)

but, if a regime shift occurred prior to $i = L$, it is now detected.

The STARS method may also be used to identify regime shifts based on changes in variance, in addition to changes in the mean. In this approach, shifts in the mean are identified first and then subtracted from the original data series. Changes in the variance of these residuals are then examined similarly to changes in the mean, but using an F-test of the ratio of variances rather than a t-test of the difference of means (Daskalov et al., 2007). Regime identification based on changes in variance may be viewed as supplementary and in support of regimes identified based on changes in the mean.

5.10.2 Adaptive Kolmogorov–Zurbenko Filters

The KZA algorithm is based on an iterative moving average Kolmogorov–Zurbenko (KZ) filter. In the KZ filter, a moving average over two times the half-window length plus one ($2q + 1$) is utilized over several iterations, in which the previously averaged values are used as input for the next successive iteration. The time rate of change of the moving averages is computed, and where large, the window length is shortened and moving averages are recomputed. This adaptive aspect of the filtering process defines the KZA filter. The significance of discontinuities may be estimated from the sample variances over the averaging windows. As an example, Figure 5.64 provides a comparison of the regime shift detection capabilities of STARS and KZA for the winter (DJF) time series of the Arctic Oscillation. Both are adjustable and necessitate specifying a window-width, or length of regime one wishes to detect. Both appear to perform reasonably well in detecting sharp discontinuities, or regime shifts, in the time series.

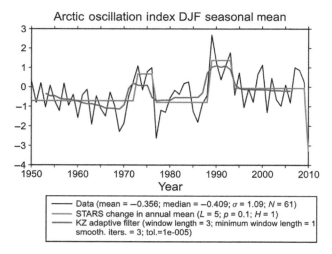

FIGURE 5.64 Comparison of regime shift detection using sequential *t*-test analysis of regime shifts (STARS) and adaptive Kolmogorov–Zurbenko filters (KZA) for the average winter (January–February–March) Arctic Oscillation Index. For KZA a window length of 3 appears to capture the regime changes better than a window length of 5 (not shown). *(Courtesy of Roy Hourston, Institute of Ocean Sciences).*

As mentioned in the introductory summary, the nonparametric KZA filter is a low-pass moving average KZ filter that dynamically adjusts the length of the filter according to the rate of change of the process being investigated. As the rate of change of the smoothed data set increases, the length of the filter decreases in order to better resolve the changes. Since the filter depends on an iterative moving average, it filters out high-frequency variations in the data set. Following the notation of Rodionov and Overland (2005), the simple moving average KZ filter is computed according to

$$y_i = \frac{1}{2q+1} \sum_{j=-q}^{q} x_{i+j} \qquad (5.290)$$

where x_i is the original data, y_i is the filtered data, and $2q+1$ is the length (number of data values) of the filter window. The filter is iterative, in that the filter Eqn (5.290) is applied, not once, but k-times. After the first pass ($k=1$), the low-pass filtered values, $y_i = y_{i,1}$, become the new x_i values and the filtering is repeated to produce

an even more smoothed version, $y_i = y_{i,2}$, of the original data and so on until $y_i = y_{i,k}$. The KZ filter is an efficient low-pass linear filter (Zurbenko, 1986) that can be defined as

$$Z(t) = KZ_{q,k}[X(t)] \qquad (5.291)$$

where $X(t)$ is the original time series, q is the half-length of the filter, and k is the number of iterations (successive applications) of the filter to generate an increasingly smooth data series, $y_{i,k}$.

The absolute value of the differentiated $Z(t)$ time series is defined by

$$D(t) = |Z(t+q) - Z(t-q)| \qquad (5.292)$$

while the localized time rate of change of $D(t)$ is given by

$$dD(t)/dt = D(t)' = D(t+1) - D(t) \qquad (5.293)$$

When a data point is located in a region of increasing $D(t)$, the half-length of the moving average in the tail region before the data point (q_{T-q}, q_T) is kept equal to the original half-length, q (as in Eqn (5.291)), while the half-length ahead of the data point

(q_H, q_{H+q}) is shortened as a function of $D(t)$. In the shortened region of $D(t)$, only the half-length behind the data point (the tail region) will be reduced. In the vicinity of the break point, the filter length is reduced, thus sharpening the regime shift resolution of the moving average. Modified after Eqn (5.290), the adaptive filter is defined by

$$Y_t = \frac{1}{q_H(t) + q_T(t)} \sum_{i=-q_T(t)}^{q_H(t)} X_{t+i} \qquad (5.294)$$

where

$$q_H(t) = \begin{cases} q, & \text{if } D'(t) < 0 \\ f(D(t))q, & \text{if } D'(t) \geq 0 \end{cases} \qquad (5.295a)$$

$$q_T(t) = \begin{cases} q, & \text{if } D'(t) > 0 \\ f(D(t))q, & \text{if } D'(t) \leq 0 \end{cases} \qquad (5.295b)$$

and q is the half-length of the filter in the initial $KZ_{q,k}$ filter. The function, $f(D(t))$ is defined as

$$f(D(t)) = 1 - \frac{D(t)}{\max[D(t)]} \qquad (5.296)$$

Note that $D(t)) = 0$ and $f = 1$ if the two ends of the filtered record $Z(t)$ are equal over the averaging interval, which remains as $(-q, q)$. However, $f = 0$ if the function, $D(t)$ reaches its maximum value $\max[D(t)]$ in the interval being considered. Because $\max[D(t)]$ is the largest change, the latter coincides with the break point.

Plots of the filtered data series, Y_t, reveal times of discontinuities (possible regime shifts) in the time series (Figure 5.65). This qualitative evidence for discontinuities can be placed on a more quantitative basis using the sample variances, σ_t^2, of Y_t defined by

$$\sigma_t^2 = \frac{\sum_{i=q_T}^{q_H} [Y_i - \overline{Y}_t]^2}{q_T + q_H} \qquad (5.297)$$

where $\overline{Y}_t$ is the record average. Zurbenko et al. (1996) compare the KZA with the parametric Schwarz criterion for identifying regime shifts in simulated and geophysical time series containing seasonal patterns and trends. Although both methods were successful in locating relatively large discontinuities, the KZA approach was rated more highly for two reasons: (1) it was more accurate in the case of simulated seasonal cycles; and (2) the Schwarz criterion depends on independent and trend-free data, conditions rarely satisfied in geophysical data.

5.10.3 Cumulative Upwelling Index

The CUI is a time-integrated version of the upwelling index (UI) devised by Bakun (1973) to define the transition (shift) to upwelling favorable winds within eastern boundary regions of the World Ocean. Examples of eastern boundary upwelling regions are the coasts of northwest and southwest Africa, Peru, Western Australia, and western North America. The timing of the transition from downwelling favorable winds in winter to upwelling favorable winds in summer (known as the Spring Transition, ST) is of particular importance to the functioning of large-scale marine ecosystems (Schwing et al., 1996; Bograd et al., 2009). Although developed for detecting the seasonal transition in the coastal ocean, the idea behind the CUI has possible applications to other time series experiencing seasonal to decadal or even longer-scale regime shifts, providing the parameter in question shifts between periods of positive and negative values.

Despite their name, "upwelling indices" are not based on measurements of upwelling (vertical displacements of isopycnals). In fact, most indices are not even based on oceanic measurements but rather on the alongshore (y-direction) component of coastal wind stress, τ_y, averaged over some time period (typically a day or a month) that is longer than the local inertial period $(2\pi/f)$, the time required for geostrophy to modify the circulation. Here, the wind stress serves as a proxy for the cross-shore (x-direction) surface Ekman layer transport, ME_x, generated along eastern boundary current regions. The

FIGURE 5.65 The results of the adaptive Kolmogorov–Zurbenko filters (KZA) filter for $q=100$, $k=3$ applied to 3000 synthetic standard normal random numbers with breaks of amplitude 0.5σ at times $t=1000$ and 2000, where σ is the standard deviation of the original time series. (a) Original data; (b) base line box-car function upon which the random data have been superimposed; (c) the KZA filtered data; and (d) the variance σ^2 of the filtered time series. (*Adapted from Zurbenko et al. (1996).*)

longest UI series begins in 1967 and is computed from the 6-hourly, $1 \times 1°$ resolution atmospheric sea-level pressure fields generated by the U.S. Navy Fleet Numerical Meteorology and Oceanography Center. In a steady state, the cross-shore transport in the wind-driven surface Ekman layer must be balanced by an oppositely directed cross-transport, MI_x, in the interior region below the Ekman layer of depth, z_E (i.e., $MI_x = -ME_x$). If we ignore the coastal boundaries and assume a constant vertical eddy viscosity, A_z, and a water depth much greater than the surface Ekman layer depth, $z_E = \pi(2A_z/|f|)^{1/2}$ (typically <100 m), the linearized, steady state,

alongshore momentum balance for uniform Coriolis parameter, f, is

$$-fu = \frac{1}{\rho} \frac{\partial \tau_y}{\partial z} \qquad (5.298)$$

where u is the cross-shore component of current velocity, ρ is the water density, and the depth, z, is measured vertically downward. Integrating Eqn (5.298) over the depth of the Ekman layer yields

$$ME_x = \int_0^{z_E} \rho u \, dz = \frac{\tau_y}{f} \qquad (5.299a)$$

$$MI_x = \int_{z_E}^{z_I} \rho u dz = -\frac{\tau_y}{f} \qquad (5.299b)$$

where τ_y is positive (negative) in the poleward (equatorward) direction in both hemispheres, and $f \geq 0$ (northern hemisphere) and $f \leq 0$ (southern hemisphere). Thus, ME_x is negative away from the coast in the Northern Hemisphere when the wind is blowing toward the equator, whereby the opposing interior flow, MI_x, is positive (toward the coast, to the right of the wind direction). This is classic wind-driven upwelling. In the Southern Hemisphere, an equatorward wind is also negative, but for a right-hand coordinate system, the x-direction is positive in the offshore direction. Once again, an equatorward wind leads to positive ME_x (away from the coast, to the left of the wind direction) and negative MI_x (toward the coast).

The CUI (Schwing et al., 2006; Pierce et al., 2006; Bograd et al., 2009) is generated by summing the daily mean upwelling indices derived using Eqn (5.299b) at coastal wind-grid locations starting on January 1 of a given year and continuing to the end of the year. As illustrated by Figure 5.66, the CUI can be used to define a variety of seasonal regime shift parameters: (1) the Julian date of the ST when the CUI reaches its minimum value at the end of the winter downwelling season; (2) the Julian date of the end of the upwelling season (which we can call the Fall Transition, FT) when the CUI reaches its seasonal maximum (termed END by Bograd et al., 2009); (3) the length of the upwelling season, equal to the total number of days between ST and FT; and (4) the intensity of the upwelling season, Total Upwelling Magnitude Index (TUMI), defined as $\text{TUMI} = \int_{ST}^{FT} CUI(t)$. Similarly, the Total Downwelling Magnitude Index (TDMI), a measure of the intensity of downwelling during the winter, is the total CUI integrated from the observed FT date to the date of the ST in the following year. Both TUMI and TDMI have units of m^3/s per unit length of coastline (e.g., m^3/s per 100 m). Results for the California Current System from 33 to 48° N (Bograd *et al.*, 2009) show that the upwelling season diminishes from roughly 357 days at 33° N to 151 days at 48° N, with the greatest

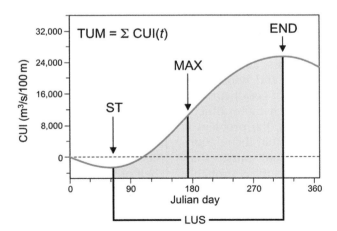

FIGURE 5.66 Schematic of the climatological annual cycle of cumulative upwelling index (CUI) ($m^3/s/100$ m) at 39° N in the California Current system. Times correspond to the following variables: ST (Spring Transition date); LUS (length of upwelling season); and TUM (total upwelling magnitude). END corresponds to the end of the upwelling season (start of the downwelling season, or Fall Transition). *(Adapted from Bograd et al. (2009).)*

intensity and variance off northern California (36−42° N). In the northern region of the California Current system, there has been a significant trend of 1 day/year ($r = 0.42$, $p = 0.0083$) toward a later ST, with an accompanying trend toward a shorter upwelling season at 48° N of −1.1 days/year. Bograd et al. (2009) find considerable inter-annual and decadal variability in the various upwelling indices, with El Niño events having a major impact on the upwelling phenology. The upwelling season began late and was of anomalously short duration in strong ENSO years (e.g., 1982−1983 and 1997−1998). Similar impacts are observed through changes in the currents in the California Current system arising from local and remotely wind-forced coastal trapped waves (Connolly et al., 2014).

5.11 VECTOR REGRESSION

Oceanographers often use time series from a more readily available variable as a surrogate for the time series of a less easily measured variable. For example, observations of tidal elevation can serve as a surrogate for the tidal currents while the alongshore component of wind stress from meteorological records is used as a surrogate for wind-induced upwelling (see previous section). In such instances, coastal stations, moored offshore platforms, and satellites provide oceanic sea-level data and wind fields whereas direct measurements of ocean currents and coastal upwelling are much more difficult to obtain. For vector processes, the problem is further complicated by the fact that the relationship can be direction-dependent rather than spatially isotropic. This complexity is probably best illustrated by the two-dimensional (vector) anisotropic response of ice-drift and ocean current velocity ("surface drift velocity") to wind forcing in the presence of coastal boundaries.

In this section, we outline regressional methods that can be used to examine possible dynamical relationships between vector time series, which, in turn, help determine if one variable can be used as a surrogate for the other variable. To summarize, we present a basic approach for using wind velocity observations to generate current or ice-drift velocities. The approach consists of estimating the "response matrixes" and corresponding "response ellipses," where the latter defines the drift or current velocity response to a unit wind velocity forcing. For each direction, φ, of the unit-magnitude wind velocity vector there is a corresponding oceanic response consisting of a "wind factor" $\alpha(\varphi)$ and "turning angle" $\theta(\varphi)$. These two factors represent (1) the speed of the current (or ice-drift) relative to the wind speed and (2) the angle that the drift velocity makes with the wind vector. The major ellipse axis corresponds to the direction of the "effective wind" ($\varphi = \varphi_{max}$) and the minor axis to the direction of the "noneffective" wind. The eigenvectors of the response matrix are along wind directions that are the same as the wind-induced drift velocity directions. Six analytical cases are possible, depending on the water depth, distance from the coast, and other factors. Results range from rectilinear response ellipses near the coast (where the orientation of the shoreline is prominent) to purely circular response ellipses in the open ocean, far from the influence of the coast. Responses derived from the 4-parameter vector-regression method are less constrained and therefore more representative of wind-induced surface motions than those derived using the traditional 2-parameter complex transfer function approach, also discussed in this section. Our presentation closely follows Rabinovich et al. (2007), who used the model to examine ice-drift along the western shelf of Sakhalin Island in the Sea of Okhotsk.

5.11.1 The 2-Parameter Complex Functional Approach

Estimates of ice-drift and currents for shelf and coastal regions are important for a wide variety of marine requirements including navigation, oil and gas exploration, and climate investigations. Seafloor topographic variations, wave-trapping effects, and the formation of land-fast ice (for freezing areas) are among the factors that make continental margins among the most challenging areas for coastal ocean prediction research (Wang et al., 2003). Wind stress is one of the major factors affecting oceanic motions along continental margins (Gill, 1982; Wadhams, 2000). Continental margins are also areas most likely to be instrumented with extensive, near real-time observing systems, thereby allowing for effective application of regression models for both diagnostic and forecasting purposes (cf. Thorndike and Colony, 1982; Fissel and Tang, 1991; Rabinovich et al., 2007).

Regressional analysis of two-vector time series is commonly based on a functional relationship between input and output vector series expressed as,

$$\mathbf{u} = \alpha \mathbf{V} \qquad (5.300)$$

where $\mathbf{V} = (U, V)$ is the input vector series (e.g., wind velocity), $\mathbf{u} = (u, v)$ is the output vector series (e.g., ice-drift or current velocity; herein, "drift velocity"), and $\alpha = a + ib$ is a complex coefficient determined using an LS regressional fit based on coincident wind and drift velocity observations. In the standard Cartesian coordinate system, u, U are positive to the east and v, V are positive to the north. The above formulation is widely used in studies of ice motion (Thorndike and Colony, 1982; Fissel and Tang, 1991; Greenan and Prinsenberg, 1998) and ocean currents (Cherniawsky et al., 2005). Equation (5.300) can also be written as

$$\mathbf{u} = \alpha_0 \exp\left(-i\theta_0\right)\mathbf{V} \qquad (5.301)$$

where $\alpha_0 = |\alpha| = \sqrt{a^2 + b^2}$ is the wind factor and where $\theta_0 = -\arctan(b/a)$, the turning angle of the drift velocity direction relative to the wind direction, is measured clockwise (counterclockwise) to the wind in the Northern (Southern) Hemisphere. The values, $\alpha_0 = 0.02$ (a wind of 1 m/s generates a 0.02 m/s ice-drift) and

$\theta_0 = 28°$, which describe the relationship between wind and free ice-drift in the open ocean, denote the "Nansen—Ekman ice-drift law" (Thorndike, 1986; Wadhams, 2000). Similar values ($\alpha_0 = 0.01 - 0.04$, $\theta_0 = 10 - 40°$) have been obtained for surface currents.

A major limitation of the above equations is that they describe an *isotropic* response of ocean to the wind, in which the parameters α_0 and θ_0 are invariant with respect to the wind direction, φ. This limitation contrasts with Overland and Pease (1988), who observed a markedly *anisotropic* response of ice-drift to the wind near coastal boundaries. Similarly, Fissel and Tang (1991) report that proximity to the coast and the direction of the wind relative to the coastline are major factors affecting the wind factor, $\alpha(\varphi)$, and turning angle, $\theta(\varphi)$. To account for these effects, we need to apply a two-dimensional (matrix) regression model based on regional winds.

5.11.2 The Vector Regressional Model

The influence of a nearby coast on wind-driven motions can be quantified using a two-dimensional vector model. This model relates the current or free ice-drift, $\mathbf{u}(t)$, to the wind, $\mathbf{V}(t)$, through the two-dimensional regression equation

$$\mathbf{u} = \mathbf{A}\mathbf{V} + \mathbf{E} \qquad (5.302a)$$

(Cooley and Lohnes, 1971; Maxwell, 1977) where

$$\mathbf{u} = \begin{pmatrix} u \\ v \end{pmatrix}, \quad \mathbf{V} = \begin{pmatrix} U \\ V \end{pmatrix},$$

$$\mathbf{A} = \begin{pmatrix} a_{11} & a_{12} \\ a_{21} & a_{22} \end{pmatrix}, \quad \text{and} \quad \mathbf{E} = \begin{pmatrix} \varepsilon_u \\ \varepsilon_v \end{pmatrix}.$$

$$(5.302b)$$

Here, a_{ij} are the regression (response) coefficients that link the cross-shore (u) and alongshore (v) components of drift velocity with the corresponding components (U, V) of the wind velocity, and coefficients (ε_u, ε_v) denote random noise.

Without loss of generality, we have assumed that the drift velocity and wind have zero mean speed. As a consequence, Eqn (5.302a) does not require a term representing the mean drift velocity.

The coefficients, a_{ij}, are obtained through an LS method (cf., Section 5.9.1) that minimizes the summations over the duration $(t = t_o, \ldots, T)$ of the applied data set,

$$\sum_{t=t_o}^{T} [a_{11} U(t) + a_{12} V(t) - u(t)]^2; \qquad (5.303a)$$

$$\sum_{t=t_o}^{T} [a_{21} U(t) + a_{22} V(t) - v(t)]^2 \qquad (5.303b)$$

From Eqn (5.303) we obtain the matrix relation (Cooley and Lohnes, 1971)

$$\mathbf{AD} = \mathbf{R}, \qquad (5.304a)$$

where $\mathbf{A}$ is defined by expression Eqn (5.302b), $\mathbf{D}$ is the auto-correlation matrix of the input variable (wind), and $\mathbf{R}$ is the cross-correlation matrix between input and output variables; specifically

$$\mathbf{D} = \begin{pmatrix} r_{UU}^2 & r_{UV}^2 \\ r_{UV}^2 & r_{VV}^2 \end{pmatrix} = \begin{pmatrix} \langle UU \rangle & \langle UV \rangle \\ \langle UV \rangle & \langle VV \rangle \end{pmatrix};$$

$$\mathbf{R} = \begin{pmatrix} r_{uU}^2 & r_{uV}^2 \\ r_{vU}^2 & r_{vV}^2 \end{pmatrix} = \begin{pmatrix} \langle uU \rangle & \langle uV \rangle \\ \langle vU \rangle & \langle vV \rangle \end{pmatrix},$$

$$(5.304b)$$

where $\langle \rangle$ denotes a time average. Terms involving random noise average to zero. From (5.304), it follows that

$$\mathbf{A} = \mathbf{RD}^{-1} \qquad (5.305)$$

which yields the four response coefficients, a_{11}, a_{12}, a_{21}, and a_{22}. Here, $\mathbf{D}^{-1}$ is the inverse of the symmetric matrix, $\mathbf{D}$. Because the matrix, $\mathbf{R}$ is generally nonsymmetric, the matrix, $\mathbf{A}$ is also generally nonsymmetric, leading to more complicated wind−ice and wind−current relationships than in Section 5.11.1.

5.11.2.1 Response Ellipses

The regression coefficients, a_{ij}, in Eqn (5.302) define a response ellipse, corresponding to the curve traced out by the tip of the output response vector $\boldsymbol{\alpha} = (\alpha_u, \alpha_v)$ of the drift velocity through one complete rotation of a unit amplitude input wind vector, $(U_0, V_0) = (\sin\varphi, \cos\varphi)$, where φ is the angle of the magnitude $= 1$ wind vector measured clockwise from north in the Northern Hemisphere. Specifically,

$$\alpha_u(\varphi) = a_{11} \sin\varphi + a_{12} \cos\varphi \qquad (5.306a)$$
$$\alpha_v(\varphi) = a_{21} \sin\varphi + a_{22} \cos\varphi \qquad (5.306b)$$

so that for each direction of the wind vector, φ, there is a corresponding "wind factor" $\boldsymbol{\alpha}$, having relative drift speed, $\alpha = |\boldsymbol{\alpha}(\varphi)|$, and direction, $\phi = \phi(\varphi)$. The angle between the drift velocity and wind vectors is the turning angle, $\theta = \phi - \varphi$. In the case of an isotropic response, $\alpha_u = \alpha_v =$ constant, $\theta =$ constant, so that the response ellipse is a circle.

In general, a response ellipse is described by four invariant parameters: (1, 2) the semimajor ($\alpha = A_{\max}$) and semiminor ($\alpha = A_{\min}$) axes corresponding to the maximum and minimum drift responses, respectively; (3) the orientation of the semimajor axis ($\phi = \phi_{\max}$); and (4) the direction of the "effective wind" ($\varphi = \varphi_{\max}$), corresponding to the wind direction that generates the maximum ice-drift or current response. Specifically, the *semimajor and semiminor axes* are

$$A_{\max} = \Big[(a_{11} \sin\varphi_{\max} + a_{12} \cos\varphi_{\max})^2 $$
$$+ (a_{21} \sin\varphi_{\max} + a_{22} \cos\varphi_{\max})^2 \Big]^{1/2}$$
$$(5.307a)$$

and

$$A_{\min} = \Big[(a_{11} \sin\varphi_{\min} + a_{12} \cos\varphi_{\min})^2 $$
$$+ (a_{21} \sin\varphi_{\min} + a_{22} \cos\varphi_{\min})^2 \Big]^{1/2}$$
$$(5.307b)$$

the *orientation of the semimajor axis* is

$$\phi_{max} = \arctan\left(\frac{a_{11}\sin\varphi_{max} + a_{12}\cos\varphi_{max}}{a_{21}\sin\varphi_{max} + a_{22}\cos\varphi_{max}}\right)$$

(5.307c)

(the orientation of the semiminor axis is then given by $\phi_{min} = \phi_{max} \pm 90°$), and the *direction of the effective wind*, φ_{max}, the wind direction angle producing the maximum drift response, is

$$\varphi_{max} = \frac{1}{2}\arctan\left[\frac{2(a_{11}a_{12} + a_{21}a_{22})}{\left(-a_{11}^2 + a_{12}^2 - a_{21}^2 + a_{22}^2\right)}\right]$$

(5.307d)

where φ_{max} is measured clockwise from north in the Northern Hemisphere. The direction, φ_{min}, of the *noneffective* wind (the direction giving the minimum wind response) is also measured clockwise from north and is related to φ_{max} by $\varphi_{min} = \varphi_{max} \pm 90°$.

5.11.2.2 Eigenvectors of Matrix A

If **A** is a square matrix and **V** is a column vector such that

$$\mathbf{AV} - \lambda\mathbf{V} = (\mathbf{A} - \lambda\mathbf{I})\mathbf{V} = 0 \qquad (5.308)$$

where

$$\mathbf{I} = \begin{pmatrix} 1 & 0 \\ 0 & 1 \end{pmatrix}$$

is the identity matrix, then **V** is said to be an *eigenvector* (*latent vector*) of the matrix **A** with scalar eigenvalues, λ (Maxwell, 1977). Each eigenvector is associated with a corresponding eigenvalue. The eigenvectors of the matrix **A** determine the transformation from the vector **V** to the vector **u**. Because the directions of the eigenvectors are unchanged by this transformation (i.e., $\varphi_\lambda = \phi_\lambda$), the eigenvectors give the direction along which the wind direction coincides with that of the wind-generated drift

velocity. The eigenvalues, λ, are found from the characteristic equation (Maxwell, 1977)

$$(\mathbf{A} - \lambda\mathbf{I}) = \begin{pmatrix} (a_{11} - \lambda) & a_{12} \\ a_{21} & (a_{22} - \lambda) \end{pmatrix} = 0$$

(5.309)

from which we obtain the quadratic equation

$$\lambda^2 - (a_{11} + a_{22})\lambda + (a_{11}a_{22} - a_{12}a_{21}) = 0.$$

(5.310)

Equations (5.310) is used to define properties of the matrix **A**.

There are three *invariants* of the matrix **A** determining the main properties of the transformation from wind to drift velocity (cf. Belyshev et al., 1983):

$$J_1 = a_{11} + a_{22} \quad \text{(the } trace\text{);} \qquad (5.311a)$$

$$J_2 = |a_{ij}| = a_{11}a_{22} - a_{12}a_{21} \text{(the } determinant\text{);}$$

(5.311b)

$$J_3 = a_{12} - a_{21} \quad \text{(the } index \ of \ asymmetry$$

or the *turn indicator*).

(5.311c)

The characteristic Eqn (5.310) has two roots (eigenvalues):

$$\lambda_{1,2} = \frac{1}{2}\left(J_1 \pm \sqrt{J_1^2 - 4J_2}\right) \qquad (5.312a)$$

whereby

$$J_1 = \lambda_1 + \lambda_2; \quad J_2 = \lambda_1\lambda_2. \qquad (5.312b)$$

The directions of the eigenvectors are found from (5.306) from which,

$$\lambda_j \sin\varphi_j = a_{11}\sin\varphi_j + a_{12}\cos\varphi_j, \quad j = 1,2.$$

(5.313a)

whereby,

$$\tan\varphi_j = a_{12}/(\lambda_j - a_{11}), \quad j = 1,2. \quad (5.313b)$$

When **A** is a symmetric matrix, $J_3 = 0$, whereby, for given eigenvalues λ_1 and λ_2, the eigenvectors V_1 and V_2 are orthogonal and correspond to the principal ellipse axes associated with the maximum and minimum response (amplification) of the output series (drift velocity) relative to the input series (wind). However, if $J_3 \neq 0$ then the eigenvectors of the response ellipses are nonorthogonal and are rotated relative to the principal ellipse axes. Depending on the sign of J_3, the turning angles, θ, have mainly positive or negative values. The angle between two eigenvectors, which is equal to $90°$ if $J_3 = 0$, becomes smaller with increasing $|J_3|$. In the case $\lambda_1 = \lambda_2$, there is only one eigenvector, and the turning angle always has the same sign except for the two zero-value points coincident with the eigenvector. The one-eigenvector condition ($\lambda_1 = \lambda_2$) may be presented as

$$D = J_1^2 - 4J_2 = (a_{11} - a_{22})^2 + 4a_{12}a_{21} = 0,$$
(5.314)

where D is the discriminant of Eqn (5.310). Case (5.314) is possible only if a_{12} and a_{21} have opposite signs. Moreover, if $J_1^2 < 4J_2$, then

$$(a_{11} - a_{22})^2 + 4a_{12}a_{21} < 0 \qquad (5.315)$$

so that Eqn (5.312a) does not have real roots and the turning angle always has the same sign.

5.11.2.3 Asymptotic Cases

There are two limiting cases for the ellipse structure: the flat ellipse (corresponding to rectilinear, one-dimensional motions) and the circle (corresponding to isotropic motions). The *flat ellipse* case occurs when the matrix $\mathbf{A}$ is *singular*, that is, when the determinant

$$J_2 = |a_{ij}| = a_{11}a_{22} - a_{12}a_{21} = 0. \qquad (5.316)$$

For a two-dimensional matrix, the determinant is zero if one row (or column) is proportional to the other row (or column), such that

$$\mathbf{A} = \begin{pmatrix} a_{11} & a_{12} \\ ka_{11} & ka_{12} \end{pmatrix} \qquad (5.317)$$

where k is a constant. This situation occurs when the wind-induced motions are rectilinear regardless of the wind direction, whereby the ice (or water) moves back and forth along one direction only. Motions of this type are possible in a narrow channel or near the coast, where the wind-generated drift is restricted by the presence of a boundary. The magnitude and sign of k determine the direction of these motions. For example, if $k = 0$ then the direction is west—east; if $k = 1$, then the direction is southwest—northeast (i.e., at $45°$ to the Cartesian coordinate system).

Using Eqn (5.307a) and (5.316), and several simple transformations, the directions of the most and least effective wind (those wind directions that produce the maximum and minimum drift response, respectively) are found from

$$(a_{11} \sin \varphi + a_{12} \cos \varphi)(a_{11} \cos \varphi - a_{12} \sin \varphi) = 0. \qquad (5.318)$$

From Eqn (5.306a) and (5.316), it is clear that when the first term in the brackets of Eqn (5.318) is equal to zero, there is no oceanic response to the wind. In this case,

$$\varphi_{min} = \varphi_0 = \arctan(-a_{12}/a_{11}), \qquad (5.319)$$

so that winds in the direction, $\varphi = \varphi_0 \pm 180°$ produce no oceanic response; for any other direction, there is always a nonzero response. The direction of the maximum response may be found by setting the second term of Eqn (5.318) to zero, yielding

$$\varphi_{max} = \arctan(a_{11}/a_{12}) = \varphi_0 \pm 90°. \qquad (5.320)$$

from which we obtain,

$$\sin \varphi_{max} = \frac{a_{11}}{\sqrt{a_{11}^2 + a_{12}^2}};$$

$$\cos \varphi_{max} = \frac{a_{12}}{\sqrt{a_{11}^2 + a_{12}^2}} \qquad (5.321)$$

From Eqn (5.307a) and (5.321) it follows that the magnitude of the maximum response has the simple form,

$$A_{\text{max}} = (1 + k^2)^{1/2}(a_{11}^2 + a_{12}^2)^{1/2} \qquad (5.322)$$

and the corresponding direction of the maximum drift velocity,

$$\phi_{\text{max}} = \arctan(1/k) \qquad (5.323)$$

depends only on the coefficient k. Consequently, any drift motions, including the maximum response, are in the direction, $\phi = \pm\arctan(1/k)$. According to Eqn (5.313), the matrix Eqn (5.317) for this case has two eigenvectors:

$$\lambda_1 = J_1, \quad \varphi_1 = \arctan(1/k); \qquad (5.324a)$$

$$\lambda_2 = 0, \quad \varphi_2 = \arctan(-a_{12}/a_{11}) = \varphi_0. \qquad (5.324b)$$

Note that, contrary to the more general case, the case of rectilinear motion is described by three, not four, independent parameters: a_{11}, a_{12}, and k (or φ_{max}, A_{max}, and ϕ_{max}).

The second limiting case, that of a *circular ellipse*, corresponds to an isotropic response for which the drift response ellipse is a circle. This occurs when the matrix $\mathbf{A}$ is antisymmetric

$(a_{21} = -a_{12})$ and the main diagonal coefficients are equal $(a_{22} = a_{11})$ so that,

$$\mathbf{A} = \begin{pmatrix} a_{11} & a_{12} \\ -a_{12} & a_{11} \end{pmatrix}. \qquad (5.325)$$

This case is equivalent to case Eqn (5.300) with $\mathbf{A} = \alpha$, whereby matrix Eqn (5.325) takes the form

$$\mathbf{A} = \alpha_0 \begin{pmatrix} \cos\theta & \sin\theta \\ -\sin\theta & \cos\theta \end{pmatrix}, \qquad (5.326)$$

where $\alpha_0 = \sqrt{a_{11}^2 + a_{12}^2}$ and $\theta = \tan^{-1}(a_{12}/a_{11})$. Matrix Eqn (5.326) describes a combination of stretching and rotation, so that the isotropic response is described by only two independent parameters, a_{11} and a_{12} (or α_0 and θ as in the 2-parameter outlined in Section 5.11.1). In general, matrix Eqn (5.325) has no eigenvectors. The one exception is the case of an isotropic response without any turning $(\theta = 0)$, which occurs when $a_{12} = 0$. In this case, wind vectors and response vectors always have the same directions and all vectors are the eigenvectors. The maximum turning angle, $\theta = 90°$ (which is the same for all wind directions) corresponds to the case when the main diagonal elements of the matrix Eqn (5.325) are equal to zero $(a_{11} = 0)$.

5.11.3 Wind vs Surface Drift: the Six Characteristic Cases

In this section, we consider matrices $\mathbf{A}$, whose parameters (the regression coefficients, a_{ij}) are representative of the six possible cases of ice or current response to wind forcing discussed in Section 5.11.2. These six test cases are:

$$(1): \quad \mathbf{A} = \begin{pmatrix} 1.0 & 1.25 \\ 1.6 & 2.0 \end{pmatrix}; \quad (2): \quad \mathbf{A} = \begin{pmatrix} 1.0 & 0.5 \\ 0.5 & 2.0 \end{pmatrix}; \quad (3): \quad \mathbf{A} = \begin{pmatrix} 1.0 & 1.0 \\ 0.5 & 2.0 \end{pmatrix};$$

$$(4): \quad \mathbf{A} = \begin{pmatrix} 1.0 & 0.5 \\ -0.5 & 2.0 \end{pmatrix}; \quad (5): \quad \mathbf{A} = \begin{pmatrix} 1.0 & 1.0 \\ -0.5 & 2.0 \end{pmatrix}; \quad (6): \quad \mathbf{A} = \begin{pmatrix} 2.0 & 1.0 \\ -1.0 & 2.0 \end{pmatrix}.$$

The coefficients, a_{ij} in the above matrices are expressed as the ratio of drift speed in CGS units to the wind speed in MKS units (i.e., cm/s and m/s, respectively) which also corresponds to a percentage. Note that the matrix elements in pairs (2) and (4) and (3) and (5) are identical except for the change in sign of a_{12}. Table 5.20 presents the derived matrix invariants, ellipse parameters, and eigenvector parameters for the six cases.

TABLE 5.20   Response Ellipse Parameters and Eigen Vectors for the Six Test Examples

	Matrix Invariants				Ellipse Parameters				Eigenvectors			
Case	J_1 (%)	J_2 (%)²	J_3 (%)	D (%)²	A_{max} (%)	A_{min} (%)	φ_{max} (°)	ϕ_{max} (°)	λ_1 (%)	φ_1 (°)	λ_2 (%)	φ_2 (°)
C1	3.00	0.00	−0.35	9.00	3.020	0.000	38.66	32.00	3.000	32.00	0.00	128.7
C2	3.00	1.75	0.00	2.00	2.207	0.793	22.50	22.50	2.207	22.50	0.793	112.5
C3	3.00	1.50	0.50	3.00	2.422	0.619	23.42	32.89	2.366	36.21	0.634	110.1
C4	3.00	2.25	1.00	0.00	2.081	1.081	−9.22	9.22	1.500	45.00	−	−
C5	3.00	2.50	1.50	−1.00	2.236	1.118	0.00	26.57	−	−	−	−
C6	4.00	5.00	2.00	−4.00	2.236	2.236	−	−	−	−	−	−

(From Rabinovich et al. (2007).)

Case 1: The determinant of matrix **A** is equal to zero ($J_2 = 0$), corresponding to a rectilinear (one-dimensional) response of the form Eqn (5.317) with $k = 1.6$. In near-shore regions, the orientation for this flat ellipse response ($\phi_{max} = 32°$ for the present case; Figure 5.67) would typically coincide with the orientation of the coastline, indicating that, regardless of which way the wind is blowing, only motions in the alongshore direction are possible. This means that the ocean response does not generally have a symmetric directional response to the wind. Moreover, because of Earth's rotation, the turning angle in the Northern (Southern) Hemisphere is expected to be mainly directed clockwise (counterclockwise) relative to the wind vector.

Case 2: The matrix **A** is symmetric ($J_3 = 0$), so that the eigenvectors, **V₁** and **V₂** for given eigenvalues, λ_1 and λ_2 are orthogonal and correspond to the principal ellipse axes, given here as $\lambda_1 = A_{max} = 2.21$, $\varphi_1 = \phi_{max} = 22.5°$; and $\lambda_2 = A_{min} = 0.79$, $\varphi_2 = \phi_{min} = 112.5°$ (Figure 5.68). The angle between the two eigenvectors is 90°. Because of Earth's rotation, this symmetric response case is possible only near the equator, where the Coriolis parameter, $f \approx 0$. Away from the equator, a solution requires that the response angle switch signs.

Case 3: Here, $J_3 \neq 0$ and the matrix **A** is nonsymmetric. This means that the eigenvectors are nonorthogonal and do not correspond to the principal ellipse axes (Figure 5.69). Because J_3 is positive in our example ($J_3 = 0.5$), the turning angles, θ, are mainly positive. The angle between the two eigenvectors is equal to 73.9°. For real oceanic conditions, the turning angle is expected to become increasingly smaller with increasing offshore distance.

Case 4: In this case, the discriminant Eqn (5.314) is equal to zero ($D = 0$), so there is only one eigenvector ($\lambda = \lambda_1 = \lambda_2 = 1.50$). The turning angle, θ, is always positive except for two zero-value points corresponding to the eigenvector (Figure 5.70). This case is observed in confined regions near a coastline.

Case 5: For this case, the discriminant is negative ($D = -1.0$), so that Eqn (5.312) does not have real roots and the turning angle is always positive (Figure 5.71). This is representative of wind-driven drift motions for offshore regions in the Northern Hemisphere; for the Southern Hemisphere, the turning angle will be negative.

Case 6: In this case, the matrix **A** is antisymmetric ($a_{21} = -a_{12} = 1.0$) and the diagonal coefficients are equal ($a_{22} = a_{11} = 2.0$). In this isotropic response example, both the wind factor, $\alpha_0 = 2.24$, and turning angle, $\theta = 26.6°$, are

FIGURE 5.67 Case 1. (a) Response ellipse describing rectilinear (one-dimensional) response of drift velocity (ice-drift or ocean current "drift") to the wind. Letters "W," "N," "E," and "S" give the direction of the drift velocity toward the west, north, east, and south, respectively. Numbers $0°$, $45°$,..., $315°$ indicate the direction of the wind; A_{max} and ϕ_{max} denote the magnitude of the maximum response and its direction, respectively; (b) Variations of the drift velocity response (wind factor) and turning angle as functions of the wind direction. λ_1 and λ_2 denote the eigenvalues; φ_1 and φ_2 are the corresponding wind directions. Shaded areas denote zones of positive turning angles. (*From Rabinovich et al. (2007).*)

uniform (Figure 5.72). This case corresponds to open-ocean regions where the influence of coasts is negligible and the current-ice-drift response to the wind has the same turning angle and response magnitude regardless of the wind direction.

The five cases 1, 3, 4, 5, and 6 characterize the way in which the surface drift currents respond to changes in the wind as a function of increasing distance from the shore, ranging from purely rectilinear (alongshore) wind-induced motions near the coast (case 1) to almost circular responses in the open ocean (case 6). The one remaining case (case 2) is only applicable to equatorial regions. In general, ice-drift response changes in a similar way to the ocean currents. However, ice-drift response is also dependent on ice concentration (Shevchenko et al., 2004). Higher ice concentration strengthens the internal ice stress, leading to marked attenuation in ice-motions, especially in the cross-shore direction. In contrast, reduced ice concentration leads to intensification of cross-shore motions, analogous to the effect of increased offshore distance.

5.11.3.1 Ice-Drift on the Sakhalin Island Shelf

A study of ice-drift and its vector regressional relationship to the wind as a function of offshore distance and ice concentration on the northeast coast of Sakhalin Island, Sea of Okhotsk, is presented in Rabinovich et al. (2007). To quantify the influence of ice concentration on the ice-drift response to wind, the ice-drift data for

FIGURE 5.68 Case 2. As in Figure 5.67, but describing the case of a symmetric response of the drift velocity to the wind. Eigenvectors, $\mathbf{V}_1$ and $\mathbf{V}_2$ are orthogonal with eigenvalues, λ_1 and λ_2. A_{max} and A_{min} denote the magnitude of the maximum and minimum responses; φ_{max} and φ_{min} indicate the corresponding directions of the wind. Shaded areas denote zones of positive turning angles. *(From Rabinovich et al. (2007).)*

1993 were divided into four sequential 18-day segments characterized by distinctly different ice types and concentrations: (1) the period March 12–30 consisted of ice concentrations of approximately 80–90%, with the ice field made up of large and small broken floes; (2) March 31–April 17 had the highest ice concentration (95–100%), consisting of large ice fields; (3) April 18–May 6 had reduced ice concentrations (60–80%) and diminished floe sizes; and (4) May 7–25 consisted of intensive melting with ice concentrations reduced by ∼40–50%. For each of the four time segments, two observational sites (S1 and S4) were examined and the matrix **A** computed for eight different cases.

Location S1 (located 4 km from shore) yielded the following matrices for the four time segments:

$$(1) \quad \mathbf{A} = \begin{pmatrix} 0.90 & -0.33 \\ -2.33 & 2.34 \end{pmatrix};$$

$$(2) \quad \mathbf{A} = \begin{pmatrix} 1.08 & -0.52 \\ -2.17 & 1.59 \end{pmatrix};$$

$$(3) \quad \mathbf{A} = \begin{pmatrix} 1.20 & -0.38 \\ -2.24 & 2.18 \end{pmatrix};$$

$$(4) \quad \mathbf{A} = \begin{pmatrix} 2.10 & 0.31 \\ -1.13 & 2.48 \end{pmatrix};$$

(a)

FIGURE 5.69 Case 3. As in Figure 5.67, but describing the case of a nonsymmetric response of the drift velocity to the wind. *(From Rabinovich et al. (2007).)*

(b)

and location S4 (16 km for shore) yielded

$$(1) \quad \mathbf{A} = \begin{pmatrix} 0.69 & -0.21 \\ -3.01 & 2.57 \end{pmatrix};$$

$$(2) \quad \mathbf{A} = \begin{pmatrix} 1.06 & -0.54 \\ -2.30 & 1.49 \end{pmatrix};$$

$$(3) \quad \mathbf{A} = \begin{pmatrix} 1.18 & -0.33 \\ -2.19 & 2.07 \end{pmatrix};$$

$$(4) \quad \mathbf{A} = \begin{pmatrix} 2.04 & 0.25 \\ -1.09 & 2.54 \end{pmatrix},$$

where the matrix coefficients, a_{ij}, are in units of cm/s or m/s, or percent. The pronounced flatness of the response ellipses for cases S4-2, S4-3, and S4-4 indicated that the wind-induced ice motions were strongly anisotropic, with the ice response in the alongshore direction much more pronounced than in the cross-shore direction (2.9–6.1% vs 0.2–1.9%, respectively). The alongshore values (2.6–5.4%) of the response coefficients (the wind factor) were similar to those obtained by Fissel and Tang (1991) for the Newfoundland shelf. Response coefficients for

FIGURE 5.70   Case 4. As in Figure 5.67, but for the case for which there is only one eigenvector, $\lambda_1 = \lambda_2 = 1.50$. In this case, the turning angle is always positive, except for two zero-value points of the eigenvector. *(From Rabinovich et al. (2007).)*

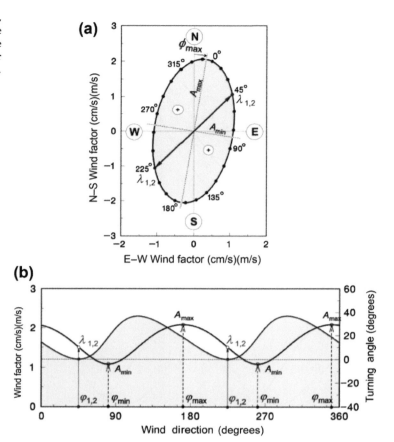

the more remote offshore observational area (S4) were greater than for the areas closest to *shore* (S1) by about 15–20%.

The results reveal marked temporal changes in the ice-drift response to the wind, apparently due to changes in ice properties. During the period of the highest ice concentration (period 2), the response ellipses are almost flat, indicating that the ice-drift response was rectilinear (along-shore). There are two eigenvalues, but the turning angles are mainly positive. In general, the ice-response ellipses resemble those for case 1 in Section 5.11.3. For the early spring (period 1), and especially during the late spring (period 3), the

response ellipses have larger magnitude and are more circular, indicating more intense cross-shore ice motions. Similarly, for the second period, the S1 and S4 matrixes have two eigen-values and a prevalence of positive turning angles. The response ellipses were of type 3. Finally, during the late spring (period 4), the response ellipses changed from flat to oval, similar to case 5. For period 4, the matrixes for both S1-4 and S4-4 had no eigenvalues and all turning angles were positive. According to this analysis, the last period was a time of free ice-drift, while the three other periods were times of high internal ice stress and influence of the coast.

(a)

(b)

FIGURE 5.71 Case 5. As in Figure 5.67, but for the case for which there are no eigenvalues and the turning angle is always positive. *(From Rabinovich et al. (2007).)*

As illustrated by the above study, the traditional 2-parameter approach for relating drift velocity to the wind is isotropic and, therefore, unrealistic for coastal regions where factors such as the orientation of the coastline and the regional bottom topography are important. The assumption of an isotropic response is likely invalid near the coast. When searching for dynamical relationships and associated surrogate variables, it is best to apply a two-dimensional (vector) regression model. In this model, the relationship between the wind and drift velocity (ice-drift or current velocity) is described by four independent regression (response) coefficients, a_{ij}, linking the cross-shore (u) and alongshore (v) components of the

drift to the corresponding components (U, V) of the wind velocity. For each direction of the wind vector, φ, the method prescribes a "wind factor," $\alpha(\varphi)$, (relative drift speed) and "turning angle," $\theta(\varphi)$, (the angle between the drift velocity and wind vector).

Because of its greater number of free coefficients, the 4-parameter vector model should yield a smaller residual variance than the traditional 2-parameter model. However, the number of DoF in the data set being analyzed decreases with an increase in the number of coefficients. As a consequence, calculation of the vector-regression coefficients to same level of confidence as the traditional model coefficients requires a longer time series. The stability of

FIGURE 5.72 Case 6. As in Figure 5.67, but for the case describing an isotropic response of the drift velocity to the wind. Both the wind factor, $\alpha_0 = 2.24$ and the turning angle, $\theta = 26.6°$ are spatially uniform. *(From Rabinovich et al. (2007).)*

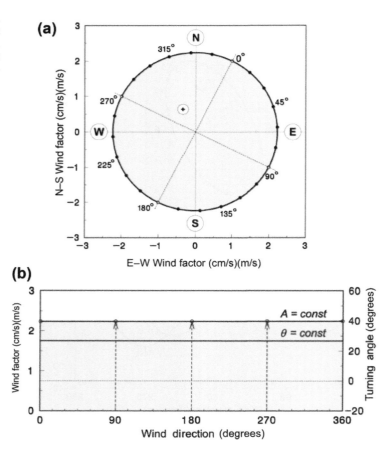

the response ellipse parameters (relative to small changes in the parameters of the input functions) is the main criterion for determining the reliability of the results. It is also important to note that the structure of these ellipses has physical meaning in the sense that it accounts for the significant difference in ice-drift response to alongshore and cross-shore winds. The results reveal that an anisotropic, vector-regression model is superior to an isotropic, 2-parameter model for examining wind-ice and wind-current processes in coastal zone regions. Moreover, the vector regression model is more likely to capture surface dynamical features of the wind response than the traditional model, and therefore is more reliable in the application of the wind velocity as a surrogate for surface drift velocity.

5.12 FRACTALS

The term "fractal" was coined by Mandelbrot (1967) to describe the bumpiness of geometrical curves and surfaces. Regardless of how closely we examine a fractal object, it fails to become smooth and its degree of jaggedness remains unchanged. Fractal objects are uneven at all scales and possess no characteristic length scales. Fractals are ubiquitous features whose presence has been reported in a wide variety

FIGURE 5.75 As in Figure 5.74 except for pseudo-drifter tracks generated using a random number generator. In this case, $H_y \approx 0.5$ and drifters perform a nonfractal random walk with dimension, $D \approx 2$.

FIGURE 5.76   Yardstick length, $L(\Delta)$, measured using a ruler with variable average yardstick length, Δ_{av} (in degrees of latitude), for three drifters launched in the Kuroshio Extension in 1977. (a) Linear coordinates; and (b) log—log coordinates. Note the divergence of the lengths for small Δ. *(From Osborne et al. (1989).)*

length scales, Δ (Figure 5.76). The case, $D_L = 1$ is the topological dimension for a smooth differential curve. For fractal dimensions, $D > 1$ and the length of the curve increases without bound for decreasing segment length.

A problem with applying Eqn (5.329) to irregularly sampled drifter records is that the data are unequally sampled both in time and space. Although it makes sense to use a spline-interpolation scheme to generate scalar coordinate data with equally spaced time increments, it is less meaningful to generate coordinate series with equally sampled positional increments. The reason is simple enough: Time is single-valued whereas location is not. Drifters often loop back on themselves. If the data are not equally spaced, we cannot define a sequence of fixed-length yardsticks but must measure the curve, $L(\Delta)$, as a function of the average yardstick

length, Δ_{av}. This averaging is valid provided the errors introduced by the averaging process are no worse than those arising from other sources (cf. Osborne et al., 1989). Another problem with the yardstick method is that it is based on the slope of Eqn (5.329) for small spatial scales. The measurement of these scales is often difficult in practice due to limitations in the response and/or positioning of the drifters, cyclone, or other Lagrangian particle.

5.12.3 Box-Counting Method

In this method, one counts the numbers, $N_m(L)$, of boxes of length, L in m-dimensional space that are needed to cover a "cloud" or set of points in the space. The Hausdorff—Besicovich dimension, D, of this set can be estimated by determining the number of cubes needed to cover the set in the limit as $L \to 0$. For a fractal curve, the number of boxes increases without bound as $L \to 0$. That is

$$N_m(L) \to L^{-D}, \quad L \to 0 \qquad (5.330)$$

If the original series is random, then $D = n$ for any dimension, n (a random process embedded in an n-dimensional space always fills that space). If, however, the value of D becomes independent of n (i.e., reaches a saturation value, D_0, say), it means that the system represented by the time series has some structure and should possess an attractor whose Hausdorff–Besicovitch dimension is equal to D_0. Once saturation is reached, extra dimensions are not needed to explain the dynamics of the system.

As an example, if we were to measure the area of surfaces embedded in three-dimensional space, we would count the number, $N_3(L)$, of cubic boxes of size, L required to cover the surface. The area, S, is then of order

$$S \approx N_3(L)L^2 \qquad (5.331)$$

For a nonfractal surface, the area asymptotes to a constant value independent of L, which is the true area of the surface. In general

$$N_3(L) \approx L^{-D}, \quad S \approx L^{2-D} \qquad (5.332)$$

5.12.4 Correlation Dimension

An important method for determining the self-similarity of monofractal curves has been proposed by Grassberger and Procaccia (1983). The technique also has found widespread use in studies of chaos and the dimensionality of strange attractors. Specifically, one determines the number of times that the computed distances, d_{ij}, between points in a time series, $x(t_i)$, (or pair of time series, $x_i(t)$ and $x_j(t)$) are less than a prescribed length scale, ε. That is, one finds what fraction of the total number of possible estimates of the distance, $d_{ij} = |x(t_i) - x(t_j)|$ that are less than

ε. For a single discrete vector time series, the Grassberger–Procaccia correlation function is defined as

$$C(\varepsilon) = \frac{1}{M(M-1)} \sum_{ij}^{M} H\left[\varepsilon - \left|x(t_i) - x\left(t_j\right)\right|\right],$$

$$M \to \infty \qquad (5.333)$$

where $H(\varepsilon, r_{ij})$ is the Heavyside step function ($=0$ for $\varepsilon < r$; $=1$ for $\varepsilon > r$) and M is the number of points in the time series. In (5.333), the vertical bars denote the norm of the vector, $d_{ij} = [(x(t_i) - x_j)^2 + (y(t_i) - y_j)^2]^{1/2}$. The fractal dimension for a self-affine curve is then obtained as the correlation dimension defined by

$$C(\varepsilon) \approx \varepsilon^v, \quad \varepsilon \to 0 \qquad (5.334)$$

The fractal dimension, v, is obtained from the log-transformed version of this equation (plots of C versus length scale are presented in Figure 5.77). According to Osborne et al. (1989), the correlation method gives the least uncertainty in the estimate of the fractal dimension, whereas largest errors are associated with the exponent scaling method.

5.12.5 Dimensions of Multifractal Functions

The various techniques discussed above will (within statistical error) give the same fractal dimension provided that the series being investigated exhibits self-similar monofractal behavior. However, because the techniques rely on different assumptions and measure different scaling properties of the series, the calculated dimensions will be different if the series has a multifractal structure. Multifractal properties are related to multiplicative random processes and are associated with different scaling properties at different scales.

A form of box counting can be used to study the multifractal properties of ocean drifters

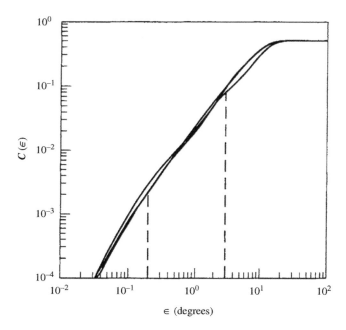

€ (degrees)

FIGURE 5.77 Correlation functions, $C(\varepsilon)$, for three drifters launched in the Kuroshio Extension in 1977. The slope of the function in log–log coordinates is a measure of the correlation dimension of the signal. The two vertical lines indicate the approximate limits of the scaling range. *(From Osborne et al. (1989).)*

(Osborne et al., 1989). Given a fractal curve on a plane, the plane is covered with adjacent square boxes of size, Δ and the probability, $p_i(\Delta)$, is computed that the ith box contains a piece of the fractal curve

$$p_i(\Delta) = \frac{n_i(\Delta)}{N} \qquad (5.335)$$

where n_i is the number of data points falling in the ith box and N is the total number of points in the time series. For fractal curves for small Δ

$$\sum_i \left[p_i(\Delta) \right]^q \approx \Delta^{(q-1)D} \qquad (5.336)$$

where the sum is extended over all nonempty boxes. The quantities, $D = D_q$, are the generalized fractal dimensions. A fundamental difference between monofractals and multifractals is

that for monofractals, D_q is the same for all q while for multifractals the different generalized dimensions are not equal. In general, $D_q < D_{q'}$ for $q > q'$.

5.12.6 Predictability

A box-counting method can be used to investigate the degree of chaotic behavior associated with the Lagrangian motions such as those of drifters and tropical cyclones. In this method, one counts the number, $N_n(\Delta)$, of boxes of dimension, Δ in n-dimensional space needed to cover a "cloud" or set of points in the space in the limit $\Delta \to 0$. In practice, the box-counting method is difficult to apply. Estimates of the predictability of drifter trajectories are more readily obtained using the correlation integral technique of Grassberger and Pocaccia (1983). In this case,

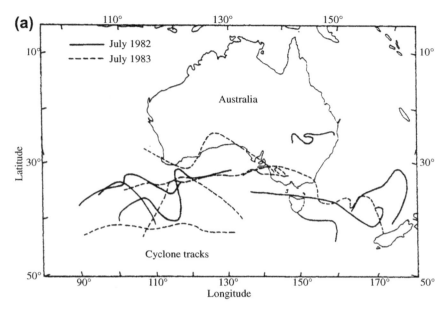

FIGURE 5.78 Use of fractals to study the predictability of cyclone tracks. (a) Cyclone tracks for Australia in July 1982, 1983. (b) Correlation integral or cumulative distance distributions, $C_m(\varepsilon)$ of pairs of independent cyclone trajectory pieces vs $\ln(\varepsilon)$. Each curve is for m times the data time step of 24 h ($m = 1–10$ from left to right in the figure). For increasing m, structure eventually becomes invariant at highest embedding dimensions, an indication that extra variables are not needed to account for the dynamics of the system. (c) Same as (b) but for a random-walk pseudo-cyclone generated using a random-number generator. The slopes approach $D_2 \rightarrow 2\,m$ for decreasing distance threshold, ε. *(From Fraedrich et al. (1990).)*

the degree of predictability is found from the dimension of the attractor derived from an embedded phase space created from all possible pairs of "drifters." The phase space serves, in turn, as a substitute for the state space needed to study the dynamics of a system (Tsonis and Elsner, 1990).

The analysis takes the following steps: (1) We first consider a pair of independent tracks of length, $m\Delta t$, where m is the embedding dimension and Δt is the sampling increment. Specifically, consider the cyclone tracks for Australia for July 1982 and 1983 (Figure 5.78(a)) examined by Fraedrich et al. (1990). For convenience, the start times and positions of the tracks are reinitialized so that they begin at the same time and location. Fraedrich and Leslie (1989) found that the errors introduced by reinitializing are less

than those from other sources; (2) We next examine the divergence of the paths by calculating the multiple track correlation function (or correlation integral), $C_m(\varepsilon)$, for the particular embedding dimension, m and path separation scale, ε. To this end, we count the number of tracks, $N_m(\varepsilon)$ of length, $m\Delta t$ for which the track length remains less than the great circle distance, ε, for all the segments in the track. For $m = 1$, each individual data point forms a unit-length segment of the drifter track. One then counts the number of times, $N_1(\varepsilon)$, that the distance between the drifter positions is less than ε for the $N = m$ possible drifter tracks. The distance between each drifter pair is considered; hence, for 10 drifters or cyclone tracks there would be $10 \times 10 = 100$ pairs. This process is repeated for all values of m to create a cloud of points in

(b)

(c)

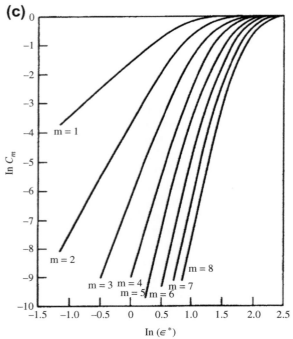

FIGURE 5.78 *(continued)*.

m-dimensional space, which then approximates the dynamics of the system from which the observations, $x(t)$, are drawn. The correlation integral is defined by

$$C_m(\varepsilon) = \frac{N_m(\varepsilon)}{[N_m - 1]^2} \qquad (5.337)$$

where $N_m(\varepsilon)$ is the number of pairs of trajectories of dimension m that remain less than a distance, ε from one another. Note that the numerator in the above expression is a squared quantity since it is based on the number of drifter pairs; (3) We then plot $\ln[C_m(\varepsilon)]$ vs $\ln(\varepsilon)$ to find the slope, D_2, of the curve

$$C_m(\varepsilon) \approx \varepsilon^{D_2}, \qquad \varepsilon \to 0 \qquad (5.338)$$

The subscript "2" indicates that pairs of points are used to create the phase space.

If both original time series are random, then $D_2 = 2m$. A random process embedded in a $2m$-dimensional space always fills that space. On the other hand, if D_2 becomes independent of m at some saturation value, D_0, it means that the system represented by the time series has some structure (i.e., predictability) and should possess an attractor whose Hausdorff–Besicovitch dimension is equal to D_0 (Figure 5.78(b)). The need to calculate D_0 from the observations arises because we do not know the value of m a priori. We, therefore, calculate D_0 for increasing m until we approach a structure that becomes invariant at higher embedding dimensions, an indication that extra variables are not needed to account for the dynamics of the system. The attractor can be a topological structure such as a point, limit cycle or torus, or a nontopological submanifold with fractal structure. For a random-walk regime, D_2 approaches $2m$ so that there is no corresponding limiting value, D_0.

The independent segments of the paired drifter trajectories of sufficiently long duration embed the attractor in a substitute phase space spanned by the time-lagged coordinates provided by the data. The correlation dimension,

D_2, measures the spatial correlation of the points that lie on the attractor. For a random time series there will be no such spatial correlation in any embedding dimension and thus no saturation will be observed in the exponent, D_2. We note that the dimensionality of an attractor, whether fractal or nonfractal, indicates the minimum number of variables present in the evolution of the corresponding dynamical system. In other words, the attractor must be embedded in a state space of at least its dimension. Therefore, the determination of the Hausdorff dimension of an attractor sets a number of constraints that should be satisfied by any numerical or analytical model used to predict the evolution of the system. The main concern is that we do not extend the interpretation when going from a densely populated low-dimensional space to a sparsely occupied high-dimensional space. We cannot go beyond the critical embedding dimension above which the scaling region cannot be accurately determined (Essex et al., 1987; Tsonis and Elsner, 1990).

6

Digital Filters

6.1 INTRODUCTION

Digital filtering is often an important step in the processing of oceanographic time series data. Applications involve the use of a series of specifically designed weights for smoothing and decimation of time series, removal of fluctuations in selected frequency bands, and the alteration of signal phase. The term "decimation" technically means the removal of every tenth point but is now commonly used for values other than 10. Digital filtering facilitates data processing by preconditioning the frequency content of the record. For example, filters can be used in studies of inertial waves to isolate current variability centered near the local Coriolis frequency, to remove background sea-level fluctuations in investigations of tsunamis, and to eliminate tidal frequency fluctuations in studies of low-frequency current oscillations (Figure 6.1). The terms "detided" or "residual" time series are commonly used to describe time series that have been filtered to remove tidal components. Filters also provide algorithms for data interpolation, for integration and differentiation of recorded signals, and for linear prediction models. Kalman filters used for estimation of the state of a data-controlled process are discussed separately in Section 4.10.

There is no single type of digital filter for general oceanographic use. Selection of an appropriate filter depends on a variety of factors, including the frequency content of the data and the kind of analysis to be performed on the filtered record. Personal preference and familiarity with one type of filter also can be deciding factors. Many oceanographers have their favorite filters and would not consider switching. However, in certain instances, one type of filter may be superior to another for a specific task, and proper filter selection involves some forethought. Often the type of filter must be tailored to the job at hand. For example, some of the so-called "tide-elimination" or "tide-killer" filters once used extensively in oceanography were designed for regions dominated by semidiurnal tides and are, therefore, inadequate for time series with marked diurnal period variability (Walters and Heston, 1982; Thompson, 1983). These filters permit leakage of unwanted diurnal tidal energy into the nontidal (residual) frequency bands of the filtered record. Elimination of this problem is possible through proper filter selection.

This chapter begins with a brief outline of basic filtering concepts and then proceeds to descriptions of some of the more useful digital filters presently used in marine research. We use the term "filter" to cover any linear operation on the data. In *optimal estimation* applications, the term applies specifically to an optimal estimate of the last measurement point. *Smoothing* is reserved for estimates spanned by the

FIGURE 6.1 Time series of hourly alongshore (top) and across-shore (bottom) components of current velocity at 53 m depth on the continental shelf of northern Vancouver Island during March 1980. Thin line, original hourly data; thick line, hourly data filtered with a low-pass Godin tide-elimination filter, $A_{25}^2 A_{24}/(25^2 24)$. *(From Huggett et al. (1987).)*

observations. Much of the emphasis in this chapter is on the design of low-pass digital filters that remove high-frequency oscillations from a given oceanographic time series. These filters can be used to construct other types of filters, including high-pass versions of a given type of low-pass filter. The running-mean filter, the Lanczos-window cosine (or Lanczos–cosine) filter, and the Butterworth filter are among those most commonly used in oceanography. The Kaiser–Bessel filter, which is one of our preferences, is discussed in Section 6.9. As with other time series analysis methods discussed in this book, all filters that we present in this chapter can also be applied to the spatial domain provided that the user just takes care to ensure that the filter algorithms are properly formulated.

6.2 BASIC CONCEPTS

From a practical standpoint, a good low-pass filter should have five essential qualities: (1) a sharp cutoff, so that unwanted high-frequency

components are effectively removed; (2) a comparatively flat pass-band that leaves the low-frequency components unchanged; (3) a clean transient response, so that rapid changes in the signal do not result in spurious oscillations or "ringing" within the filtered record; (4) zero phase shift; and (5) acceptable computation time. As a rule, many of these desirable features are mutually exclusive and there are severe limitations to achieving the desired filter. We are invariably faced with a trade-off between the ability of the filter to produce the required results and the amount of filter-induced data loss we can afford to tolerate. For example, improved statistical reliability (increased degrees of freedom) for specified frequency bands decreases the frequency resolution of a filter, while more sharply defined frequency cutoffs lead to greater ringing and associated data loss.

Consider a time series consisting of the sequence

$$x(t_n) = x_n, \quad n = 0, 1, ..., N - 1 \quad (6.1)$$

with observations at discrete times $t_n = t_o + n\Delta t$ in which t_o marks the start time of the record and Δt is the sampling increment. A digital filter is an algebraic process by which a sequential combination of the input $\{x_n\}$ is systematically converted into a sequential output $\{y_n\}$. In the case of linear filters, for which the output is linearly related to the input, the time-domain transformation is accomplished through convolution (or "blending") of the input with the weighting function of the filter. Filters having the general form

$$y_n = \sum_{k=-M}^{M} w_k x_{n-k} + \sum_{j=-L}^{L} g_j y_{n-j},$$

$$n = 0, 1, ..., N - 1 \qquad (6.2)$$

(in which M, L are integers and w_k, g_j are nonzero weighting functions or "weights") are classified as *recursive* filters since they generate the output by making use of a feedback loop specified by the second summation term. Such filters "remember" the past in the sense that all past output values contribute to all future output values. Filters based on the input data only (weights $g_j = 0$), are classified as *nonrecursive* filters. Any filter for which $-M \leq k \leq M$ is said to be physically unrealizable (in the sense of any real-time output) because both past and future data are needed to calculate the output. Filters of this type have widespread application in the analysis of prerecorded data for which all digital values are available beforehand. Filters which use only past and incoming data are said to be physically realizable or causal, and are used in real-time data acquisition and in forecasting procedures.

Impulse response: The output $\{y_n\}$ of a nonrecursive linear filter is obtained through the convolution

$$y_n = \sum_{k=-M}^{M} w_k x_{n-k} = \sum_{k=-M}^{M} w_{n-k} x_k,$$

$$n = 0, 1, ..., N - 1 \qquad (6.3)$$

where w_k are the time invariant weights and there are N data values $x_o, x_1, ..., x_{N-1}$. For a symmetric filter, the time-domain convolution becomes

$$y_n = \sum_{k=0}^{M} w_k(x_{n-k} + x_{n+k}), \quad n = 0, 1, ..., N - 1$$

$$(6.4)$$

in which $w_k = w_{-k}$. The set of weights $\{w_k\}$ is known as the *impulse response* function (IRF) and is the response of the filter to a spikelike impulse. To see this, we set $x_n = \delta_{0,n}$ where $\delta_{m,n}$ is the Kronecker delta function

$$\delta_{m,n} = 0, \quad m \neq n$$
$$= 1, \quad m = n \qquad (6.5)$$

Equations (6.3) then becomes

$$y_n = \sum_{k=-M}^{M} w_k \delta_{0,n-k} = w_n \qquad (6.6)$$

The summations in Eqns (6.3) and (6.4) are based on a total of $2M + 1$ specified weights with individual values of w_k labeled by subscripts $k = -M, -M + 1, ..., M$. To make practical sense, the number of weights is limited to $M \ll N/2$ where $(N - 1)\Delta t$ is the record length. In reality, it is not possible to use Eqn (6.3) to calculate an output value y_n for each time t_n. Because the response function spans a finite time, equal to $(2M - 1)\Delta t$, difficulties arise near the ends of the data record and we are forced to accept the fact that there are always fewer output data values than input values. There are three options: (1) we can make do with $2M$ fewer estimates of y_n (resulting from time losses of $M\Delta t$ at each end of the record), (2) we can create values of $x(t_n)$ for times outside the observed range $0 \leq t < (N - 1)\Delta t$ of the time series, or (3) we can progressively decrease the filter length, M, in accordance with the number of remaining input values. In the first approach, x_n is defined for $n = 0, 1, ..., N - 1$, whereas y_n is defined for the shortened range $n = M, M + 1, ..., N - (M + 1)$. In the second approach, the appended estimates of x_n should qualitatively resemble the data at either end of the record. For example, we could use the "mirror images" of the data reflected at the end points of the original time series. In the third approach, the values y_{M-1} and $y_{N-(M-1)}$ are based on $(M-1)$ weights, the values y_{M-2} and $y_{N-(M-2)}$ on $(M-2)$ weights, and so on.

Frequency response: The Fourier transform of $y(t_n)$ in Eqn (6.3) is

$$Y(\omega) = \sum_{n=-M}^{M} y_n e^{-i\omega n \Delta t}$$

$$= \sum_{k=-M}^{M} w_k e^{-i\omega k \Delta t} \sum_{n=-M}^{M} x_{n-k} e^{-i\omega(n-k)\Delta t}$$

$$= W(\omega)X(\omega)$$

$$(6.7)$$

so that convolution in the time domain corresponds to multiplication in the frequency domain. The function

$$W(\omega) = \frac{Y(\omega)}{X(\omega)} = \sum_{k=-M}^{M} w_k e^{-i\omega k \Delta t};$$

$$\omega \equiv \omega_n = 2\pi n/(N\Delta t)$$

$$(6.8)$$

$n = 0, \ldots, N/2$ is known as the *frequency response, or transfer function* (see Section 5.6.7) since it determines how a specific Fourier component $X(\omega)$ is modified as it is transformed from input to output. For the symmetric filter (Eqn (6.4)), the transfer function reduces to

$$W(\omega) = w_o + 2\sum_{k=1}^{M} w_k \cos(\omega k \Delta t) \qquad (6.9)$$

Once $W(\omega)$ is specified, the weights w_k are found through the inverse Fourier transform

$$w_k = \sum_{n=-N/2}^{N/2} W(\omega)e^{i\omega k \Delta t} \qquad (6.10)$$

In general, the frequency response (transfer function) $W(\omega)$ is a complex function that can written in the form

$$W(\omega) = |W(\omega)|e^{i\phi(\omega)} \qquad (6.11)$$

where the amplitude $|W(\omega)|$ is called the *gain* of the filter (a term originating with electrical circuitry) and $\phi(\omega)$ is the *phase lag* of the filter. The power $P(\omega)$ of the transfer function is given by

$$P(\omega) = W(\omega)W(-\omega) = W(\omega)W^*(\omega)$$

$$= |W(\omega)|^2 \qquad (6.12)$$

where, as usual, the asterisk denotes the complex conjugate.

6.3 IDEAL FILTERS

An ideal filter is one that has unity gain, $|W(\omega)| = 1$, at all frequencies within the specified *pass-band(s)* and zero gain at frequencies within the *stop-band(s)* (Figure 6.2). When processing recorded oceanographic data, it is generally advantageous to have $\phi(\omega) = 0$ for all ω so that the filter produces no alteration in the phase of the frequency components. As we discuss in conjunction with recursive filters, zero phase shift can be guaranteed by first passing the input forward then backward (after inversion) through the same set of weights. In the case of nonrecursive filters, zero phase is accomplished using symmetric filters (i.e., those with no imaginary components).

Digital filters commonly used in processing oceanographic data can be classified under the general headings of *low-pass, high-pass,* or *band-pass* filters. Although impossible to achieve, we would like the amplitudes of our ideal filters to satisfy the following relations (see Figure 6.2):

$$\text{Low-pass:} \ |W(\omega)| = 1 \ \text{for} \ |\omega| \leq \omega_c \qquad (6.13a)$$
$$= 0 \ \text{for} \ \omega_c \leq |\omega|$$

$$\text{High-pass:} \ |W(\omega)| = 0 \ \text{for} \ |\omega| \leq \omega_c$$
$$= 1 \ \text{for} \ \omega_c \leq |\omega|$$

$$(6.13b)$$

(a) Low-pass filter

(b) High-pass filter

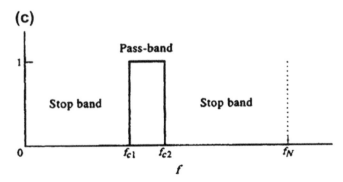

(c)

FIGURE 6.2 Frequency response (transfer) functions, $|W(f)|$, for ideal filters. (a) Low pass, (b) high pass, and (c) band-pass. The band-pass filter has been constructed from the combined low-pass and high-pass filters. f_N and f_c are the Nyquist and cutoff frequencies, respectively.

$$\text{Band} - \text{pass:} \ |W(\omega)| = 1 \ \text{for} \ \omega_{c1} \leq |\omega| \leq \omega_{c2}$$
$$= 0 \ \text{otherwise}$$

$$(6.13c)$$

The *cutoff frequency*, ω_c $(=2\pi f_c)$, marks the transition from the pass-band to the stop-band. For ideal filters, the transition is steplike, while for practical filters, the transition has a finite width. In the case of real filters, ω_c is defined as the frequency at which the mean filter amplitude in the pass-band is decreased by a factor of $\sqrt{2}$ and should roughly coincide with spectral minima in the time series being analyzed; the power of the filter is down by a factor of 2 (-3 dB) at the cutoff frequency. As its name implies, a low-pass filter lets through (or is "transparent" to) low-frequency signals but strongly attenuates high-frequency signals (cf. Figures 6.3(a) and (b)). High-pass filters let through the high-frequency components and strongly attenuate the low-frequency components (cf. Figures 6.3(a) and (c)). Band-pass filters permit only frequencies in a limited range (or band) to pass unattenuated.

Low-pass filters are the most common filters used in oceanographic data analysis. It is through these filters that low-frequency, long-term variability of oceanographic signals is determined. The running-mean filter, which involves a moving average over an odd number of values, is the simplest form of low-pass filter. More complex filters with better frequency responses, such as the low-pass Kaiser—Bessel window used in Figure 6.3(b), also are commonly used (see also Section 6.9). High-pass filtered data are readily obtained by subtracting the low-pass filtered data from the original record from which the low-pass data were derived. One does not need to create a separate high-pass filter. Similarly, band-pass filters can be formed by an appropriate combination of low-pass and high-pass filters. In the ocean, seawater acts as a form of natural low-pass filter, attenuating high-frequency wave or acoustic energy at a much more rapid rate than low-frequency energy. Acoustic waves of a few hertz (cycles per second) can propagate thousands of kilometers in the ocean, whereas acoustic waves of hundreds of kilohertz or more are strongly attenuated over a few hundred meters.

High-pass filters are less frequently used than low-pass filters. Applications include the delineation of high-frequency, high-wave number fluctuations in the internal wave band (roughly $2f < \omega < N$, where N is the Brunt—Väisälä frequency) and the isolation of seiche or tsunami motions in closed or semienclosed basins. Band-pass filters are used to isolate variability in relatively narrow frequency ranges such as the near-inertial frequency band or, in North America, the electronic-induced 60-cycle noise in high-frequency oceanic data caused by AC power supplies.

The maximum range of frequencies that can be covered by a digital filter is determined at the high-frequency end by the Nyquist frequency, $\omega_N = \pi/\Delta t$ (radians/unit time), and at the low-frequency end by the fundamental frequency, $\omega_1 = 2\pi/T$, where $T = N\Delta t$ is the length of the record. The corresponding range in cycles/unit time is determined by $f_N = 1/(2\Delta t)$ and $f_1 = 1/T$. Provided that the cutoff frequencies are sufficiently far removed from the ends of the intervals, digital filters can be applied throughout the range, $\omega_1 < |\omega| < \omega_N$ ($f_1 < |f| < f_N$).

As we discuss in Section 6.3.2, the filter response coefficients for the ideal filters described by Eqn (6.13a—c), where the desired filter amplitude is equal to 1 for all the pass-band frequencies and equal to 0 for all the stop-band frequencies, are obtained by taking the discrete Fourier transform (DFT) of the ideal frequency response. The problem is that this leads to infinitely long filter responses since the filters have to reproduce the infinitely steep discontinuities in the ideal frequency response at the band edges. To create a finite impulse response filter, the number of time-domain filter coefficients must be restricted by multiplying them by a finite width window function. The simplest window function is the rectangular window which corresponds to truncating the sequence after a certain number of terms. In order to suppress the side lobes and make the filter frequency response more closely approximate an ideal filter, the width of the window must be increased and the window function tapered down to zero at the ends. This will increase the width of the transition region between the pass- and stop-bands.

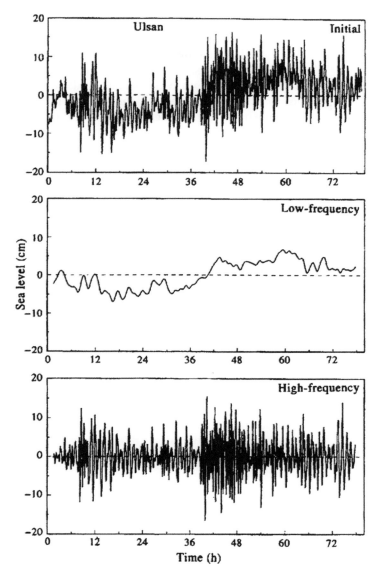

FIGURE 6.3 Filtering of a tide gauge record for Ulsan, Korea, using low- and high-pass Kaiser–Bessel filters (windows) with length $T/27 = 3$ h; $T = 81$ h is the record length and $\Delta t = 0.5$ min the sampling increment. (a) Original record; (b) low-pass filtered record; (c) high-pass filtered record. *(Courtesy, Alexander Rabinovich, Institute of Ocean Sciences.)*

6.3.1 Bandwidth

The difference in frequency between the two ends of a pass-band defines an important property known as the *bandwidth* of the filter. To illustrate the relevance of this property, we consider an ideal band-pass filter with constant gain, linear phase, and cutoff frequencies ω_{c1}, ω_{c2} such that

$$W(\omega) = W_o \exp(-i\omega t_o), \quad \omega_{c1} \le |\omega| < \omega_{c2}$$
$$= 0, \quad \text{otherwise}$$

$$(6.14)$$

From Eqn (6.13c), the impulse response is

$$
w_k = \frac{1}{2\pi} W_0 \left(\int_{\omega_{c1}}^{\omega_{c2}} e^{-i\omega t_o} e^{i\omega k \Delta y} \, d\omega \right.
$$

$$
\left. + \int_{\omega_{c1}}^{\omega_{c2}} e^{i\omega t_o} e^{-i\omega k \Delta y} \, d\omega \right)
$$

$$
= \frac{2W_0}{\pi} \Delta\omega \cos[\Omega(k\Delta t - t_o)] \frac{\sin[\Delta\omega(k\Delta t - t_o)]}{\Delta\omega(k\Delta t - t_o)}
$$

(6.15)

in which $\Omega = \frac{1}{2}(\omega_{c1} + \omega_{c2})$ is the center frequency and $\Delta\omega = \omega_{c2} - \omega_{c1}$ is the bandwidth. For high- or low-pass filters, the bandwidth is equal to the cutoff frequency.

Using the fact that $\sin p/p (\equiv \mathrm{sinc} p) \to 1$ as $p \to 0$, we find that the peak amplitude response of the filter (Eqn (6.15)) is directly proportional to the bandwidth $\Delta\omega$ as $\Delta\omega(k\Delta t - t_o) \to 0$. Note also that a narrow-band filter (one for which $\Delta\omega \to 0$) will oscillate longer (i.e., persist to higher values of k) than a broadband filter when subjected to a transient loading. Put another way, the persistence of the ringing that follows the application of the filter to a data set increases as the bandwidth decreases. From a practical point of view, this means that the ability of a filter to resolve sequential transient events is inversely proportional to the bandwidth. The narrower the bandwidth (i.e., the finer the resolution in frequency), the longer the time series needed to resolve individual events. For example, if we use a band-pass filter to isolate inertial frequency motions in the range 0.050–0.070 cph, the bandwidth $\Delta f = \Delta\omega/2\pi = 0.020$ cph and the filter could accurately resolve inertial events that occurred about $1/\Delta f = 50$ h apart. If we now reduce the bandwidth to 0.010 cph, the filter is only capable of resolving transient motions that occur more than 100 h apart. (The need to have long records to resolve closely spaced frequencies is exactly the problem

we faced in Section 5.4.5.4 regarding the Rayleigh criterion for tidal analysis.)

Another way of stating the above relationship is that the uncertainty in frequency, Δf (or $\Delta\omega$), is inversely proportional to the length of time T over which the signal oscillates (i.e., $\Delta f \approx 1/T$) so that $T\Delta f \approx 1$ for a given filter. If we wish to use a filter with a very narrow bandwidth, we need to analyze long time series records in which the signals of interest, such as the tides, have a high degree of persistence. In terms of observed data, the measured bandwidth of an oscillation in current speed, sea-level elevation, or other oceanic parameter is directly related to the persistence of the signal. For example, a wind-generated clockwise rotary inertial current having an observed bandwidth $\Delta f \approx 0.10$ cpd implies that the burst of inertial energy had a duration $T \approx 1/\Delta f = 10$ days.

6.3.2 Gibbs' Phenomenon

In practice, steplike transfer functions such as described by Eqn (6.13a–c) are not possible. Digital filters invariably possess finite slope transition zones between the stop- and pass-bands. To illustrate some of the fundamental impediments to creating ideal filters, consider the steplike transfer function

$$
\begin{aligned}
W(\omega) &= 1 & 0 < \omega \le \omega_N \\
&= 0 & -\omega_N \le \omega < 0
\end{aligned}
$$

(6.16)

(Figure 6.2(a)) where, for convenience, we specify a cutoff frequency $\omega_c = 0$. Assuming that $W(\omega)$ is repeated over multiples of the basic interval $(-\omega_N, \omega_N)$, the appropriate Fourier series expansion for Eqn (6.16) is given in the usual manner by

$$
W(\omega) = \frac{1}{2}a_o + \sum_{n=1}^{\infty} [a_n \cos(\omega n \Delta t)
$$

$$
+ b_n \sin(\omega n \Delta t)]
$$

(6.17)

with coefficients

$$a_n = \frac{1}{\omega_N} \int_{-\omega_N}^{\omega_N} W(\omega) \cos(\omega n \Delta t) d\omega \quad (6.18a)$$

$$b_n = \frac{1}{\omega_N} \int_{-\omega_N}^{\omega_N} W(\omega) \sin(\omega n \Delta t) d\omega \quad (6.18b)$$

The fact that $a_n = 1$, for all n, suggests reformulation of the problem in terms of the function

$$W_c(\omega) = W(\omega) - 1/2 \quad (6.19)$$

centered about $W(\omega) = 1/2$, the mean functional value at the discontinuity. Since $W(\omega)$ is then an odd function, cosine terms in Eqn (6.17) can be eliminated immediately. Moreover, W_c is symmetric about $\omega = \pm\frac{1}{2}\omega_N = \pm\pi/(2\Delta t)$ so that there are no even sine terms. For odd n, Eqn (6.18b) yields $b_n = 2/n\pi$ and Eqn (6.17) becomes

$$W(\omega) = \frac{1}{2} + \frac{2}{\pi} \left[\sin(\omega\Delta t) + \frac{\sin(3\omega\Delta t)}{3} + \frac{\sin(5\omega\Delta t)}{5} + \dots \right] \quad (6.20)$$

which must be truncated after a finite number of terms.

Successive approximations to the series (Eqn (6.20)), and hence to the function Eqn (6.16), are not convergent near discontinuities such as that for the steplike transition region of the ideal high-pass filter shown in Figure 6.5. In this example, the filter amplitude $|W(\omega)|$ is zero for $\omega < \omega_c$ (the stop-band) and unity for $\omega_c < \omega < \omega_N$ (the pass-band). The succession of overshoot ripples, or ringing, is known as *Gibbs' phenomenon*. The ripple period, $T_p = p\pi\Delta t$ (p is an integer), is fixed but increasing the number of terms in the Fourier series for $W(\omega)$ decreases the distortion due to the overshoot effects. However, even in the limit of infinitely many terms, Gibbs' phenomenon persists as the amplitude of the first overshoot diminishes asymptotically to about 0.18 or about 9% of the pass-band amplitude. The first minimum decreases asymptotically to about 5% of the

pass-band amplitude. In the limit of large $N \to \infty$, it can be shown (Godin, 1972; Hamming, 1977) that

$$W_\infty(0) \to \frac{1}{\pi} \int_0^\pi (\sin u/u) du \quad (6.21)$$

The values of $W_\infty(0)$ can be found in tables of the sine integral function. In the case of Figure 6.4, the value for the first maximum is 1.08949 (=1.0 + 0.08949), while that for the first undershoot is 0.9514 (=1.0 − 0.04858).

Gibbs' phenomenon has considerable importance in that it occurs whenever a function has a discontinuity. For example, suppose that we want to use Eqn (6.20) to remove spectral components near a cutoff frequency, ω_c. Unless the spectral components in the stop- and pass-bands are well separated relative to the width of the transition zone, the finite ripples will cause leakage of unwanted energy into the filtered record. Noise from the stop-band will not be completely removed and certain frequencies in the pass-band will be distorted. A critical aspect of filter design is the attenuation of the overshoot ripples using smoothing or tapering functions (windows). As discussed in Section 5.4.6, windows are important in reducing side-lobe leakage in spectral estimates.

Further difficulties arise when we apply the weights $\{w_k\}$ of an ideal filter in the time domain. Consider the nonrecursive, low-pass filter (positive frequency only)

$$W(\omega) = 1, \quad 0 \le \omega \le \omega_c \\ = 0 \quad \text{otherwise} \quad (6.22)$$

for which the impulse function is, for $k = -N, \dots, N$

$$w(t_k) = w_k = \frac{1}{\omega_N} \sum_{\omega=0}^{\omega_c} \cos(\omega k \Delta t) \Delta\omega$$

$$= \frac{\sin(\omega_c k \Delta t)}{\omega_N k \Delta t} \quad (6.23)$$

$$= \frac{f_c}{f_N} \frac{\sin(2\pi f_c k \Delta t)}{2\pi f_c k \Delta t}$$

FIGURE 6.4 Gibbs' phenomenon (overshoot ripples) arising from successive approximations to the steplike function $|W(\omega)| = 1$, $\omega_c < \omega \leq \omega_N$, and zero otherwise. $\omega_c = 2\pi f_c$ is the cutoff frequency. Curves are derived from Eqn (6.20) using $M = 3$, 7, and 11 terms.

in which $w_o = f_c/f_N$. The weights w_k attenuate slowly, as $1/k$, so that a large number of terms are needed if the filter frequency response $W(\omega)$ is to be effectively carried over to the time domain. In addition to being computationally inefficient, filters constructed from a large number of weights lead to considerable loss of information at the ends of the data sequence. Practical considerations force us to truncate the set of weights thereby enhancing the overshoot problem associated with Gibbs' phenomenon in the frequency domain. Moreover, if we truncate the length of the data set (Eqn (6.1)), we are unable to accurately replicate Eqn (6.22) in the frequency domain. This leads to a finite slope between the stop- and pass-bands of the filter.

The situation is similar for high-pass filters

$$W(\omega) = 0, \quad 0 \leq \omega \leq \omega_c$$
$$= 1, \quad \text{otherwise} \tag{6.24a}$$

In this case

$$w_k = \frac{1}{\omega_N} \sum_{\omega = \omega_c}^{\omega_N} \cos(\omega k \Delta t) \Delta \omega$$
$$= -\frac{f_c}{f_N} \frac{\sin(2\pi f_c k \Delta t)}{2\pi f_c k \Delta t}, \quad k = -N, ..., N \tag{6.24b}$$

where $w_o = 1 - f_c/f_o$. Notice that, except for the central term w_o, the weights w_k of the high-pass filter (Eqn (6.24b)) are equal to minus the weights w_k of the low-pass filter (Eqn (6.23)). The center value, w_o of the high-pass filter is found from w_o of the low-pass filter by: w_o (*high pass*) = $1 - w_o$ (*low pass*).

The difficulties that arise with Gibbs' phenomenon are somewhat alleviated by applying smoothing functions that attenuate the overshoot ripples. As usual, the price we pay for improved decay of the weighting terms is a broadening of the main lobe centered at the frequency being filtered. As we remarked earlier, the fact that the transition from the pass- to the stop-band takes place over a finite range of frequencies necessitates a working definition for the cutoff frequency, ω_c. Here, ω_c is defined as the frequency at which the power $|W(\omega)|^2$ of the filter is attenuated by a factor of 2 (-3 dB) from its mean pass-band value (power in decibels $= 20 \log(A/A_o)$ where A_o is a reference level for the signal amplitude, A, having power proportional to A^2). Alternatively, the cutoff frequency marks the frequency at which the amplitude $|W(\omega)|$ of the filter is reduced by a factor of $\sqrt{2}$ of the pass-band amplitude (amplitude in decibels $= 10 \log(A/A_o)$).

6.3.3 Recoloring

The transfer function amplitude $|W(\omega)|$ defines the effectiveness of a particular filter in transmitting or blocking power within specific frequency bands. Since no filter is perfect, in the sense that its transfer function is exactly unity throughout the pass-band(s) and zero in the stop-band(s), it is often necessary to "recolor" (rescale) the output $Y(\omega)$ so that the total variance in the pass-band spectral estimates equals the total variance of the input data for that frequency range. The need to recolor stems from practical considerations involving the choice of filter, cutoff frequency, and filter steepness through the transition band. For a pass-band of width $\Delta\omega$, multiplication of the filter output $|Y(\omega)|$ by a frequency-independent correction factor γ given by

$$\gamma(\Delta\omega) = \frac{\text{input variance within bandwidth}}{\text{output variance within bandwidth}}$$

ensures that the output power is adequately rescaled.

We can illustrate the recoloring process using the Hanning (von Hann) and Hamming windows. If $x(t)$ is any scalar time series of length N, and $y(t)$ is the filtered output of this series

following application of one of these windows, then the Fourier transform of the output, $Y(f_k)$, for discrete frequencies $f_k = (k/T)$, $k = 0, 1,...,$ $(N/2)$ is given by

$$Y(f_k) = 0.50X(f_k) - 0.25X(f_{k-1})$$
$$- 0.25X(f_{k+1}) \quad \text{(Hanning)} \quad (6.25a)$$
$$Y(f_k) = 0.54X(f_k) - 0.23X(f_{k-1})$$
$$- 0.23X(f_{k+1}) \quad \text{(Hamming)} \quad (6.25b)$$

where $X(f_k)$ is the Fourier transform of the original time series. The corresponding expected values for $|Y(f_k)|^2$ in Eqns (6.25a and 6.25b) are

$$E\left[\left|Y(f_k)\right|^2\right] = (0.50)^2 + (0.25)^2 + (0.25)^2$$
$$= 0.3750$$
$$(6.25c)$$

$$E\left[\left|Y(f_k)\right|^2\right] = (0.54)^2 + (0.23)^2 + (0.23)^2$$
$$= 0.3974$$
$$(6.25d)$$

so that the spectral density estimates $S(f_k) \approx |Y(f_k)|^2$ for each frequency component of a time series smoothed by a Hanning window should be rescaled by the exact factor $(0.375)^{-1} = 8/3$ to correct for the loss of power due to the filter. For the Hamming window, the factor is roughly $(0.397)^{-1} \approx 5/2$. Note that, according to Eqn (6.25a,b), we can easily obtain spectral estimates $S(f_k)$ for each windowed time series by summing up the squared amplitudes $|X(f)|^2$ of three adjacent Fourier components of the original time series

$$S(f_k) = C_o\left|X(f_k)\right|^2 + C_{-1}\left|X(f_{k-1})\right|^2$$
$$+ C_{+1}\left|X(f_{k+1})\right|^2 \quad (6.26)$$

where $C_o = 0.50$ and $C_{-1} = C_{+1} = -0.25$ for the Hanning window and $C_o = 0.54$ and $C_{-1} = C_{+1} = -0.23$ for the Hamming window.

6.4 DESIGN OF OCEANOGRAPHIC FILTERS

The isolation of signal variability within specific frequency bands requires filters with well-defined frequency characteristics. The design of application-specific filters can proceed in two basic ways. The first approach is to assemble a combination of simple filters, such as moving averages of variable length, and from them construct a filter with the required characteristics. This is referred to as *cascading* since the output from the lead-off filter is used as input to the second filter, output from the second filter is used as input to the third, and so on. Filter cascading is used in the design of Godin's (1972) tide-elimination filters and the squared Butterworth filters described later in this section. The second approach is to specify the desired characteristics of the filter precisely and then use poles and zeroes of mathematical functions to design a filter that meets these requirements as closely as possible. As an example, we might wish to eliminate the annual cycle from a long time series of upper ocean variability, such as sea surface temperature, so that weaker fluctuations are no longer overwhelmed by the dominant seasonal changes. The filter properties are then directly tailored to the processing requirements and to the data specific to the region of interest. (In this example, we could also use least squares analysis to determine the annual cycle and then subtract this cycle from the original data.)

Regardless of which approach is taken, it is important that the impulse and frequency response functions (FRFs) of the filter have a number of fundamental properties. (1) The FRF should have reasonably sharp transitions between adjacent stop- and pass-bands, especially if the data do not have wide "spectral gaps" between dominant frequencies within the two bands. At the same time, the transition should not be so steep as to introduce large side-lobe effects or cause the filter output to become unstable. (2) The transfer function should have nearly constant amplitude and zero phase (even symmetry) within the pass- and stop-bands so that corrections to amplitude and phase are easily applied. Linear phase change as a function of frequency is acceptable but requires corrective work at the end of the processing. (3) The impulse response should have as short a span as possible to both minimize the number of points lost (or that are need to be appended at the ends of the data) and reduce the amount of computation.

6.4.1 Frequency vs Time Domain Filtering

In most instances, filters are designed to precondition the frequency content of the data prior to further analysis. This immediately suggests that the design of a filter begins with specification of the transfer function, $W(\omega)$. Once $W(\omega)$ has been determined there are two ways to proceed. The standard time-domain approach (e.g., Hamming, 1977) is to inverse Fourier transform $W(\omega)$ to obtain the time domain filter weights, w_k, which are then used in the convolution Eqn (6.3) to determine the output $\{y_n\}$. The output is subsequently Fourier transformed to determine $Y(\omega)$. The frequency-domain approach (e.g., Walters and Heston, 1982; Middleton, 1983) makes use of the fact that $Y(\omega) = W(\omega)X(\omega)$, where $X(\omega)$ is the Fourier transform of the data $\{x(t)\}$. In this approach, the data are Fourier transformed to obtain $X(\omega_i)$, $i = 1, 2,...,N/2$, where $X(\omega)$ consists of a set of $N/2$ frequency-dependent amplitudes and phases $(A(\omega_i), \phi(\omega_i))$ at discrete frequencies. The filtered record is obtained by multiplying $X(\omega)$ by $W(\omega)$. The time domain series $\{y_n\}$ can be derived from the inverse Fourier transform of $Y(\omega)$.

There are pros and cons for both approaches. The time-domain approach uses the actual recorded data and filtering consists of simple sums and products. Moreover, the filtered series $\{y_n\}$ can be immediately plotted against the original input $\{x_n\}$ to see directly the effectiveness of the filter. Discontinuities in the time series, which lead to transient filter ringing effects, can be dealt with on the spot. However, if the calculation of $Y(\omega)$ and its associated spectral estimate $|Y(\omega)|^2$ are the ultimate goals, the time-domain approach requires application of two Fourier transforms: First, we use $W(\omega)$ to define the filter weights $\{w_k\}$ and then transform $y_n \rightarrow Y(\omega)$ to obtain the Fourier components. This can lead to roundoff and computational errors.

In the frequency-domain analysis, only one Fourier transform, $x_n \rightarrow X(\omega)$, is required. On this basis, it seems preferable to use the Fourier transform method and just set to zero all those frequency components outside the range of interest. The filtered data $\{y_n\}$ are then found through an inverse transform of the modified Fourier components, $Y(\omega) = W(\omega)X(\omega)$. One obvious difficulty with this procedure is that the discrete frequencies of the Fourier estimates may not be properly positioned relative to the required cutoff frequency of the filter, that is, the cutoff frequency may fall midway between two discrete Fourier components. Walters and Heston (1982) also pointed out that the sharp cutoff associated with this process causes ringing through the entire data set (Figure 6.5). For this reason, the Fourier coefficients must be reduced gradually to zero over a range of frequencies. For example, Nowlin et al. (1986) used a trapezoidal-shaped band-pass filter to study inertial oscillations in data collected in Drake Passage. In this particular instance, "Fourier coefficients within 0.03 cpd of the local inertial frequency were retained undiminished, and this central portion was flanked by two tapered sections 0.06 cpd wide in which the coefficients were reduced linearly to zero". The smooth filter transition results in a substantial reduction in ringing in the filtered data but is certainly reminiscent of data tapering required in the time-domain analysis. A more detailed discussion of frequency-domain filtering is presented in Section 6.10.

FIGURE 6.5 Frequency response functions for low-pass filters with different transition bands. Solid line, a steplike transition band; long-dashed line, a nine-point cosine-tapered transition band; short-dashed line, a three-point optimally designed transition band. The cutoff associated with each filter causes ringing through the entire data set. *(From Elgar, 1988.)*

6.4.2 Filter Cascades

In some instances, a desired filter $W(\omega)$ can be constructed from a series or *cascade* of basis filters $W_j(\omega)$ such that

$$W(\omega) = W_1(\omega) \times W_2(\omega) \times \cdots \times W_q(\omega) \quad (6.27)$$

where "$\times$" denotes successive applications of individual transfer functions, beginning with W_1. That is, the data are first processed with $W_1(\omega)$ and the output from this filter passed through $W_2(\omega)$; the output from $W_2(\omega)$ is then passed through $W_3(\omega)$, and so on until the last filter, $W_q(\omega)$. The final output from $W_q(\omega)$ corresponds to the sought-after output from $W(\omega)$. Although the technique is straightforward and helps to minimize roundoff error, it has a number of major drawbacks, including the need for extended computations and the possibility of repeated ringing as one filter after another is applied in succession.

A high-pass filter $W_H(\omega)$ is obtained from its low-pass counterpart $W_L(\omega)$ by the relation $W_H(\omega) = 1 - W_L(\omega)$ where, in theory, the combined output from the two filters simply recreates the original data, since $W_L(\omega) + W_H(\omega) = 1$. This has advantages in situations where $W_L(\omega)$ is easily derived or is already available. In the time domain, the high-pass filtered record $\{y'_n\}$ is obtained by subtracting the output $\{y_n\}$ from the low-pass filter from the input time series $\{x_n\}$. Care is needed to ensure that the times of y_n and x_n are properly aligned so that $y'_n = x_n - y_n$ and $n = M$, $M+1,\ldots, N - 2M$.

A band-pass filter can be constructed from an appropriate high- and low-pass filter using the method illustrated in Figure 6.2(c). Here, the cut-off frequency of the low-pass filter becomes the high-frequency cutoff of the band-pass filter; similarly, the cutoff frequency of the high-pass filter becomes the low-frequency cutoff of the band-pass filter. The cascade then has the form $W_B(\omega) = W_L(\omega) \times W_H(\omega)$.

Because nonrecursive filters are symmetric ($W(\omega)$ is a real function), there is no shift in phase between the input and output signals. This feature of the filters, as well as their general mathematical simplicity, has contributed to their popularity in oceanography. Recursive filters, on the other hand, are typically asymmetrical. This introduces a frequency-dependent phase shift between the input and output variables and adds to the complexity of these filters for oceanic applications. Despite these difficulties, recursive filters are useful additions to any processing repertoire. Note that, regardless of which type of filter is used, we can remove phase shifts introduced through the "forward" application of the filter by reversing the process and passing the data "backward" through the filter. In performing the latter step, we must be careful to invert the order of the record values between the forward and backward passes. Specifically, if the recursive filter introduces a phase shift $\phi(\omega)$ at frequency ω (or equivalently, a time shift $\phi/\omega = \phi/2\pi f$), it will introduce a compensating shift $-\phi(\omega)$ when the data are passed in the reverse order through the filter. To show this sequence, let $x_1, x_2,\ldots, x_n$ be the original data sequence used as input to a given filter with nonzero phase characteristics, and $y_1, y_2,\ldots, y_n$ the output from the filter (Figure 6.6). If we

FIGURE 6.6 The processing sequence for a nonsymmetrical recursive filter $W(\omega)$ which removes phase changes $\phi(\omega)$ introduced to the data sequence x_i ($i = 1,\ldots, n$) by the filter. This cascade produces a symmetric squared-filter response $|W(\omega)|^2$.

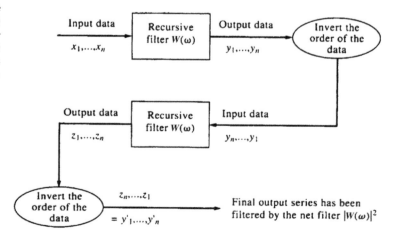

now invert the order of the output and pass the inverted signal through the filter again, we obtain a new output $z_1, z_2,..., z_n$. The order of the z-output is then inverted to form $z_n, z_{n-1},..., z_1$, which returns us to the proper time sequence. For simplicity we can rewrite this later sequence as $y'_1, y'_2, ...y'_n$. The act of applying the filter a second time cancels any phase change from the first pass through the filter. Note that this corresponds to squaring the transfer function so that the final transfer function for the recursive filter is $|W(\omega)|^2$.

As an example of a phase-dependent recursive filter, consider the high-pass *quasidifference filter*

$$y(n\Delta t) = x(n\Delta t) - \alpha x[(n-1)\Delta t] \qquad (6.28a)$$

where α is a parameter in the range $0 < \alpha \le 1$; $\alpha = 1$ corresponds to the simple difference filter (Koopmans, 1974). The frequency response (transfer function) for this filter is

$$W(\omega) = 1 - \alpha\, e^{-i\omega\Delta t} \qquad (6.28b)$$

and the phase function is

$$\phi(\omega) = \tan^{-1}[\alpha \sin(\omega\Delta t)/(1 - \alpha \cos(\omega\Delta t))] \qquad (6.28c)$$

Reversing the order of the output from the first pass of the data through the filter and then running the time-inverted record through the filter again is tantamount to passing the data through a second filter $W^*(\omega)$. This introduces a phase change $-\phi(\omega)$ which cancels the phase change $\phi(\omega)$ from the first filter (Figure 6.7). The symmetric filter obtained from this cascade is then

$$\begin{aligned}
|W(\omega)|^2 &= W(\omega) \times W^*(\omega) \\
&= \left(1 - \alpha\, e^{-i\omega\Delta t}\right)\left(1 - \alpha\, e^{+i\omega\Delta t}\right) \qquad (6.28d) \\
&= 1 - 2\alpha \cos(\omega\Delta t) + \alpha^2
\end{aligned}$$

6.5 RUNNING-MEAN FILTERS

The *running-mean* or *moving average filter* is the simplest and one of the most commonly used low-pass filters in physical oceanography. In a typical application, the filter (which is simply a moving rectangular window) consists of an odd number of $2M + 1$ equal weights, w_k, $k = 0, \pm 1,..., \pm M$, having constant values

FIGURE 6.7 The phase change $\phi(\omega)$ for a quasidifference filter (with $\alpha = 0.5$ as a function of frequency, ω).

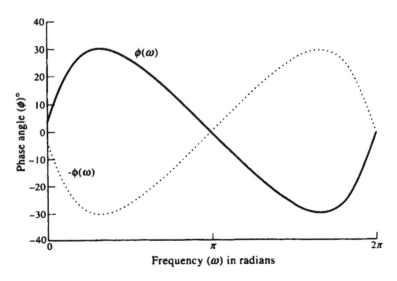

$$w_k = \frac{1}{2M+1} \quad (6.29a)$$

where w_k resembles a uniform probability density function in which all occurrences are equally likely. The running-mean filter produces zero phase alteration since it is symmetric about $k = 0$, it satisfies the normalization requirement

$$\sum_{k=-M}^{M} w_k = 1 \quad (6.29b)$$

and is straightforward to apply. To obtain the output sequence $\{y_m\}$ for input sequence $\{x_n\}$, the first $2M+1$ values of x_n (namely x_0, $x_1,..., x_{2M}$) are summed and then divided by $2M+1$, yielding the first filtered value $y_M = y(2M\Delta t/2) = y(M\Delta t)$. The subscript M reminds us that the filtered value replaces the original data record x_M at the appropriate location in the time series. The next value, y_{M+1}, is obtained by advancing the filter weights one time step Δt and repeating the process over the data sequence $x_1, x_2,..., x_{2M+1}$ and so on up to $N-2M$ output values. The $\{y_m\}$ consists of a "smoothed" data sequence with the degree of smoothing, and associated loss of information from the ends of the input, depending on the number of filter weights. Mathematically

$$y_{M+i} = \frac{1}{2M+1} \sum_{j=0}^{2M} x_{i+j}, \quad i = 0,..., N-2M \quad (6.30)$$

A high-pass running-mean filter can be generated by subtracting the output $\{y_m\}$ from the original data. The output $\{y'_m\}$ for the high-pass filter is

$$y'_m = x_m - y_m, \; m = M, M+1,..., N-2M \quad (6.31)$$

where we make certain that we subtract data values for the correct times. This technique of obtaining a high-pass filtered record from a low-pass filtered record will also be applied to other types of filters.

The frequency response $W(\omega)$ for the running-mean filter is given by Eqn (6.8). Using Eqn (6.29a) and the fact that $\Delta t = \pi/\omega_N$, we find that

$$W(\omega) = \frac{1}{2M+1}$$
$$\times \left\{ \frac{1 + 2\sin[(\pi/2M)(\omega/\omega_N)]\cos[(\pi/2(M+1))(\omega/\omega_N)]}{\sin[(\pi/2)(\omega/\omega_N)]} \right\} \quad (6.32a)$$

$$= \frac{1}{2M+1} \frac{\sin[(\pi/2(2M+1))(\omega/\omega_N)]}{\sin[(\pi/2)(\omega/\omega_N)]} \quad (6.32b)$$

where $W(\omega) \to 1$ as $\omega/\omega_N \to 0$. As M increases, the central lobe of the transfer function narrows (Figure 6.8) and the cutoff frequency (at which $|W(\omega)| = e^{-1}|W(0)|$ moves closer to zero frequency. The filter increasingly isolates the true mean of the signal. Unfortunately, the filter has considerable contamination in the stop-band due to the large, slowly attenuating side lobes. Reduction of these side-lobe effects requires a long filter which means severe loss of data at either end of the time series. The running-mean filter should therefore only be used with long data sets ("long" compared with the length of the filter). Accurate filtering requires use of more sophisticated filters.

For the three-point weighted average, $w_k = 1/3$ and Eqn (6.32b) yields

$$W(\omega; 3) = \frac{1}{3}[1 + 2\cos(\pi\omega/\omega_N)]$$
$$= \frac{1}{3} \frac{\sin[(3\pi/2)(\omega/\omega_N)]}{\sin[(\pi/2)(\omega/\omega_N)]} \quad (6.33)$$

while for five-point weighted average, $w_k = 1/5$ and

$$W(\omega; 5) = \frac{1}{5} \frac{\sin[(5\pi/2)(\omega/\omega_N)]}{\sin[(\pi/2)(\omega/\omega_N)]} \quad (6.34)$$

(Figure 6.8). Numerous examples of running-mean filters appear in the oceanographic

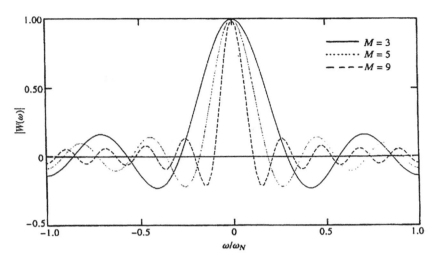

FIGURE 6.8 The frequency response functions, $|W(\omega)|$, for running-mean (weighted average, rectangular) filters for $M = 3$, 5, 9. $\omega_N = $ *Nyquist frequency.*

literature. A common use of running-mean filters is to convert data sampled at times t to an integer multiple of this time increment for use in standard analysis packages. Data collected at intervals Δt of 5, 10, 15, 20, or 30 min are usually converted to hourly data for use in tidal harmonic programs, although the least squares algorithms used in these programs also work with unequally spaced time series data (e.g., Foreman, 1977, 1978; revised 2004). Running-mean filters are also commonly used to create weekly, monthly, or annual time series (Figure 6.9).

6.6 GODIN-TYPE FILTERS

For the low-pass filtering of subhourly sampled tidal records prior to decimation to "standard" hourly values, Godin (1972) recommends the use of cascaded running-mean filters with response functions of the form

$$\frac{A_n^2 A_{n+1}}{n^2(n+1)}, \frac{A_n A_{n+1}^2}{n(n+1)^2} \qquad (6.35)$$

Here, A_n and A_{n+1} are the average values of n and $n + 1$ consecutive data points, respectively. Each filter smoothes the data three times. In the first version in Eqn (6.35), the smoothing is performed twice using the $\{n\}$-point average and once using the $\{n + 1\}$-point average. The alternative version uses the $\{n + 1\}$-point average twice and the n-point average once. Following the filter operation, the smoothed records can then be sub sampled at hourly intervals without concern for aliasing by higher frequency components. For the second version in Eqn (6.35), the response function is

$$W(\omega) = \frac{1}{n^2(n+1)}$$
$$\times \frac{\sin^2[(\pi/2n)(\omega/\omega_N)] \sin[(\pi/2(n+1))(\omega/\omega_N)]}{\sin^3[(\pi/2)(\omega/\omega_N)]}$$
$$(6.36)$$

Godin filters $(A_{12}^2 A_{14})/(12^2 14)$ are used routinely to smooth oceanographic time series sampled at multiples of 5-min increments prior to their use in tidal analysis programs. On the other hand, 30-min data would first be

FIGURE 6.9 Daily mean time series of across-shelf (top) and alongshelf (bottom) near-surface currents off Cape Romain in the South Atlantic Bight for the period January 10, 1979, to April 11, 1979. Thin line, daily average data; thick line, 30-day running-mean values. *(From McClain et al. (1988).)*

smoothed using the filter $(A_2^2 A_3)/(2^2 3)$ (Figure 6.10) and then decimated to hourly data. For example, the conversion of 30-min data collected by the early Aanderaa RCM4 mechanical current meters to hourly data requires such a three-stage running-average filter. The filter is needed to convert the instantaneous directions and average speeds from the current meter to quantities more closely resembling vector-averaged currents. Application of the

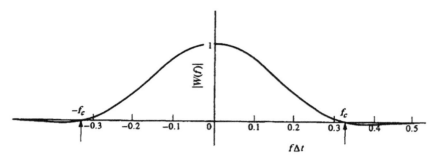

FIGURE 6.10 The frequency response function, $|W(f)|$, for the Godin-type filter $A_2^2 A_3/(2^2 3)$ used to smooth 30-min data to hourly values. The horizontal axis has units $f\Delta t$, with $f_N\Delta t = 0.5$; f_c is the cutoff frequency. *(From Godin (1972).)*

moving low-pass filter (Eqn (6.36)) removes high-frequency components and helps avoid the aliasing errors that would occur if the raw data were simply decimated to hourly values without any form of prior smoothing. Simply picking out a value each hour is, of course, akin to not having recorded the higher frequency variability in the first place. Some care is required in that the smoothing process reduces the amplitude of various Fourier components outside the tidal band. As a result, amplitudes of Fourier components derived after application of the filter must be corrected (recolored) in inverse proportion to the amplitude of the filter at the particular frequency. Phases of the Fourier components are unaltered by this symmetric filter.

The formulation (Eqn (6.35)) also can be used to generate low-pass filters to remove diurnal, semidiurnal, and shorter period fluctuations from the hourly records. Although these filters have been criticized in recent years because of their slow transition through the high-frequency end of the "weather band" (periods longer than two days), they are easy to apply, have good response in the daily tidal band, and consume relatively little data from the ends of the time series. The most commonly used version of the low-pass Godin filter is $(A^2{}_{24}A_{25})/(24^2 25)$ in which the hourly data are smoothed twice using the 24-point (24-h)

average and once using the 25-point average. The filter frequency response is

$$W(\omega) = \frac{1}{24^2 25}\sin^2[24(\pi/2)(\omega/\omega_N)]$$
$$\times \frac{\sin[25(\pi/2)(\omega/\omega_N)]}{\sin^3[(\pi/2)(\omega/\omega_N)]} \qquad (6.37)$$
$$= \frac{1}{24^2 25}\sin^2(24\pi f\Delta t)\frac{\sin(25\pi f\Delta t)}{\sin^3(\pi f\Delta t)}$$

whereas before $\omega = 2\pi f$ (f is in cycles per hour), $\omega_N = \pi/\Delta t$ and $\Delta t = 1$ h. Note that a total of 35 data points (i.e., 35 h) are lost from each end of the time series and that the filter has a half-amplitude point near 67 h (Figure 6.11). The weights of this symmetric 71-h-length filter are (Thompson, 1983)

$$w_k = \frac{1/2}{24^2 25}[1200 - (12-k)(13-k)$$
$$- (12+k)(13+k)], \quad 0 \le k \le 11$$
$$= \frac{1/2}{24^2 25}(36-k)(37-k), \quad 12 \le k \le 35$$
$$(6.38)$$

The Godin low-pass filter (Eqn (6.38)) effectively removes all daily tidal period energy except for slight leakage in the diurnal frequency band. More precisely, the filter eliminates variability due to the principal mixed diurnal

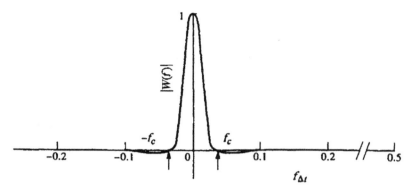

FIGURE 6.11   Same as Figure 6.10 but for the Godin-type low-pass filter $A_{25}^2 A_{24}/(25^2 24)$ used to eliminate tidal oscillations in hourly data. *(From Godin (1972).)*

constituent, K_1, for which the amplitude is down by 3.2×10^{-3}, and is only slightly less effective in removing variability due to the declinational diurnal constituent, O_1. The filter represents a marked improvement over the simple A_{24} and A_{25} running-mean filters and Doodson filter once commonly used earlier for tidal analysis (cf. Groves, 1955). The principal failing of the Godin filter is its relatively slow transition between the pass- and stop-bands which leads to significant attenuation of nontidal variability in the range of 2 to 3 days. This shortcoming of the filter has inspired a number of authors to investigate more efficient techniques for removing the high-frequency portion of oceanographic signals. The cosine-Lanczos filter, the transform filter, and the Butterworth filter are often preferred to the Godin filter, or earlier Doodson filter, because of their superior ability to remove tidal period variability from oceanic signals.

6.7 LANCZOS-WINDOW COSINE FILTERS

As mentioned in Section 6.3.2, transfer functions for ideal (rectangular) filters are formulated in terms of truncated Fourier series. This leads to overshoot ripples (Gibbs' phenomenon) near the cutoff frequency with subsequent leakage of

unwanted signal energy into the pass-band. *Lanczos-window cosine filters* are reformulated rectangular filters which incorporate a multiplicative factor (the *Lanczos window*) in rectangular filters to ensure more rapid attenuation of the overshoot ripples. A variety of other windows can also be used. The terms *Lanczos–cosine filter* and *cosine–Lanczos filter* are commonly used names for a family of filters using windows to reduce the side-lobe ripples. Owing to their simplicity and favorable characteristics, these filters have gained considerable popularity among physical oceanographers over the years (Mooers and Smith, 1967; Bryden, 1979; Freeland *et al.*, 1986).

6.7.1 Cosine Filters

We start with an ideal, low-pass filter with transfer function

$$W(\omega) = 1, 0 \leq |\omega| \leq \omega_c$$
$$= 0, \text{elsewhere} \tag{6.39}$$

and assume that the function $W(\omega)$ is periodic over multiples of the Nyquist frequency domain $(-\omega_N, \omega_N)$. Written as Fourier series, the response function is

$$W(\omega) = \frac{a_o}{2} + \sum_{k=1}^{M} [a_k \cos(\omega k \Delta t) + b_k \sin(\omega k \Delta t)]$$

$$\tag{6.40}$$

where we have truncated the series at $M \ll N$; as usual, N is the number of data points to be processed by the filter. To eliminate any frequency-dependent phase shift, we insist that $W(\omega) = W(-\omega)$, whereby $b_k = 0$. The resulting *cosine filter* has the frequency response

$$W(\omega) = w_0 + \sum_{k=1}^{M} w_k \cos(\pi k \omega / \omega_N) \quad (6.41)$$

where coefficients $w_k (= \frac{1}{2}a_k)$ are given by

$$w_k = \frac{1}{\omega_N} \int_0^{\omega_N} H(\omega) \cos(\pi k \omega / \omega_N) d\omega \quad (6.42)$$

with $k = 0, 1, \ldots, M$. The weighting terms w_k are those which determine the output series $\{y_n\}$ for given $\{x_n\}$. We assume that M is sufficiently large that $W(\omega)$ is close to unity in the pass-band and near zero in the stop-band.

For a low-pass cosine filter, $0 \leq |\omega| \leq \omega_c$ defines the bounds of the integral (Eqn (6.42)) and the weights are given by

$$w_k = \frac{\omega_c}{\omega_N} \frac{\sin(\pi k \omega_c / \omega_N)}{\pi k \omega_c / \omega_N}, \quad k = 0, \pm 1, \ldots, \pm M$$
$$(6.43)$$

for which $w_0 = \omega_c / \omega_N$. The corresponding weights for a high-pass filter, $|\omega| > \omega_c$, are

$$w_0 = 1 - \omega_c / \omega_N, \quad k = 0 \quad (6.44)$$

$$w_k = \frac{-\omega_c}{\omega_N} \frac{\sin(\pi k \omega_c / \omega_N)}{\pi k \omega_c / \omega_N}, \quad k = \pm 1, \ldots, \pm M$$
$$(6.45)$$

That is, w_0 (*high pass*) $= 1 - w_0$ (*low pass*), while for $k \neq 0$, the coefficients w_k are simply of opposite sign. The functions (Eqns (6.43) and (6.45)) are identical to those discussed in the context of Gibbs' phenomenon. Thus, the cosine filter is a poor choice for accurately modifying the frequency content of a given record based on preselected stop- and pass-bands. As an example of the response of

this filter, Figure 6.12 presents the transfer function

$$W(\omega) = 0.4 + 2 \sum_{k=1}^{9} [\sin(0.4k\pi)/k\pi] \cos(k\omega)$$

for a low-pass cosine filter with $\omega_c / \omega_N = 0.4$ and $M = 10$ terms. This filter response is compared to the ideal low-pass filter response and to the modified cosine filter using the Lanczos window (with sigma factors) discussed in the next section.

6.7.2 The Lanczos Window

Lanczos (1956) showed that the unwanted side-lobe oscillations of the form $\sin(p)/p$ in Eqns (6.43) and (6.45) could be made to attenuate more rapidly through use of a smoothing function or window. The window consists of a set of weights that successively average the (constant period) side-lobe fluctuations over one cycle, with the averaging period determined by the last term kept or the first term ignored in the Fourier expansion (Eqn (6.45)). In essence, the window acts as a low-pass filter of the weights of the cosine filter. The Lanczos window is defined in terms of the so-called *sigma factors* (cf. Hamming, 1977)

$$\sigma(M, k) = \frac{\sin(\pi k / M)}{\pi k / M} \quad (6.46)$$

in which M is the number of distinct filter coefficients, w_k, $k = 1, \ldots, M$, and $\omega_M = (M-1)/M$ is the frequency of the last term kept in the Fourier expansion. Multiplication of the weights of the cosine filter by the sigma factors yields the desired weights of the Lanczos-window cosine filter. Thus, the weights of the low-pass cosine–Lanczos filter become, using $\sigma(M, 0) = 1$

$$w_0 = \omega_c / \omega_M, \quad \text{for} \quad k = 0 \quad (6.47a)$$

$$w_k = \frac{\omega_c}{\omega_N} \frac{\sin(\pi k \omega_c / \omega_N)}{\pi k \omega_c / \omega_N} \sigma(M, k) \quad (6.47b)$$

FIGURE 6.12 Approximations to the frequency response of an ideal low-pass filter (dashed line). Solid curves give the frequency response for an unwindowed cosine filter, a Lanczos–cosine filter that uses sigma factors, and the response after double application of the Lanczos–cosine filter. Filters use $M = 10$ Fourier terms and $\omega_c = 0.4\omega_N$; $\omega_N =$ Nyquist frequency. Gibbs' effect is reduced by the sigma factors of the Lanczos window. *(From Hamming (1977).)*

for $k = \pm 1, \ldots, \pm M$ and $M \ll N$. The corresponding weights for the high-pass Lanczos–cosine filter are

$$w_0 = 1 - \omega_c/\omega_N, \quad \text{for} \quad k = 0 \qquad (6.48a)$$

$$w_k = -\frac{\omega_c}{\omega_N}\frac{\sin(\pi k\omega_c/\omega_N)}{\pi k\omega_c/\omega_N}\sigma(M,k) \qquad (6.48b)$$

The transfer function Eqn (6.47b) for a low-pass cosine–Lanczos filter is then

$$W_L(\omega)$$

$$= \frac{\omega_c}{\omega_N}\left[1 + 2\sum_{k=1}^{M-1}\sigma(M,k)\frac{\sin(\pi k\omega_c/\omega_N)}{\pi k\omega_c/\omega_N}\cos(\pi k\omega/\omega_N)\right]$$

$$(6.49)$$

while for the high-pass cosine–Lanczos filter

$$W_H(\omega) = 1 - W_L(\omega) \qquad (6.50)$$

Examination of the transfer functions in Figure 6.12 reveals that the side-lobe ripples are considerably reduced by the sigma factors of the Lanczos window. Again, the trade-off is a broadened central lobe, so that, although there is much less contamination from frequencies within the stop-band, the transition of the filter amplitude at the pass-band is less steep than that for the cosine filter. The effect of this smoothing, which represents a long period modulation of the weighting terms w_k in (6.43), can be illustrated numerically by taking a record length $N = 25$ and calculating the filter response $W(\omega/\omega_N)$ with and without the sigma factors. This exercise is instructive in other ways in that it emphasizes the effect of truncation errors during the calculations and indicates what happens if ω_c/ω_N is too near to the ends of the principal interval $0 \leq \omega/\omega_N \leq 1$. Consider the case $\omega_c/\omega_N = 0.022$, $N = 25$, and filter truncation at the fourth decimal place. For a high-pass cosine-type filter with no Lanczos window (which we want to have zero amplitude near zero frequency), we find $W(0) = 0.0740$, whereas use of the sigma factors (Lanczos window) yields $W(0) = 0.4015$. With the cutoff frequency so close to the end of the frequency range, the sigma factors clearly degrade the usefulness of the filter. Increasing the record length to $N = 50$ for the same cutoff frequency improves matters considerably; in this case, $W(0) = 0.0527$ and $W(1) = 0.9997$ using the sigma factors.

6.7.3 Practical Filter Design

Design of a low- or high-pass cosine–Lanczos filter begins with specification of: (1) the cutoff frequency and (2) the number M of weighting terms required to achieve the desired roll-off between the stop- and pass-bands. The cutoff frequency is then normalized by the Nyquist frequency, ω_N, obtained from the sampling interval Δt of the time series. As with other types of filters, it is advantageous to keep the normalized cutoff frequency away from the ends of the principal interval

$$0 \leq \omega/\omega_N \leq 1 \tag{6.51}$$

The weights w_k are then derived via Eqns (6.47a,b) and (6.48a,b).

Using Eqns (6.47a,b) and (6.49), and assuming an input $\{x_n\}$, $n = 0, 1, \ldots, N - 1$, the output for a low-pass cosine–Lanczos filter with $M + 1$ weights is

$$y_n = \frac{2\omega_c}{\omega_N}\left[x_n + \sum_{k=1}^{M} F(k)(x_{n-k} + x_{n+k})\right] \tag{6.52a}$$

in which

$$F(k) = \frac{1}{2}\frac{\sin(\pi k/M)}{\pi k/M}\frac{\sin(\pi k\omega_c/\omega_N)}{\pi k\omega_c/\omega_N} \tag{6.52b}$$

The output time series begins with $y_M = y(M\Delta t)$ corresponding to the first calculable value for the given filter length, M, and the assumption that the input data begin at $x_n = x_0$. That is

$$y_M = \frac{2\omega_c}{\omega_N}\left[x_M + \frac{1}{2}F(1)(x_{M-1} + x_{M+1})\right.$$
$$+ \frac{1}{2}F(2)(x_{M-2} + x_{M+2}) + \ldots \tag{6.53}$$
$$\left. + \frac{1}{2}F(M)(x_o + x_{2M})\right]$$

The chosen number of filter coefficients, M, is always a compromise between the desired roll-off of the filter at the cutoff frequency and the acceptable number of data points ($=2M$) that are lost from the two ends of the record. The greater the number M, the sharper the filter cut-off and the greater the data loss. Repeated (q times) processing of a given record by the same filter generates an increasingly sharper cascade filter response $(W(\omega/\omega_q))^q$ with an corresponding greater loss (qM) of data values from each end of the record. For a high-pass filter, M should be large enough that, in the time domain, the $2M$ weights for the corresponding low-pass filter span "many" periods of the higher frequency oscillations one is attempting to isolate using the filter.

The sum S of the weights w_k in Eqns (6.47a,b) and (6.48a,b)

$$S = \sum_{k=0}^{M} w_k \tag{6.54}$$

gives a qualitative measure of the filter performance. An ideal low-pass filter (i.e., one with no truncation or numerical roundoff effects) should give $S = 1$, while an ideal high-pass filter would have $S = 0$. Close proximity to these values indicates a numerically reliable filter.

6.7.4 The Hanning (von Hann) Window

A variety of cosine-type filters are presented in the recent oceanographic literature under the general term of Lanczos–cosine or cosine–Lanczos filters. A popular formulation having widespread application is the 5-day low-pass filter proposed by Mooers and Smith (1967) in a study of continental shelf waves off Oregon. In this study, a Hanning or raised cosine window defined by

$$w_k = \frac{1}{2}[1 + \cos(\pi k/M)], \quad |k| < M$$
$$= 0, \quad |k| > M \qquad (6.55)$$

replaces the sigma factors in Eqn (6.48b).

Let x_n, where $n = 1, 2, \ldots, N$, denote an hourly digital time series and $2M + 1 = 120$ be the total number of weights spanning a period of 120 h (5 days). The hourly output $\{y_n\}$ from the filter is then

$$y_n = \frac{1}{A}\left[x_n + \sum_{k=1}^{60} F(k)(x_{n-k} + x_{n+k}) \right] \qquad (6.56a)$$

where

$$F(k) = \frac{1}{2}[1 + \cos(\pi k/60)] \frac{\sin(p\pi k/12)}{p\pi k/12} \qquad (6.56b)$$

and

$$A = 1 + 2\sum_{k=1}^{60} F(k) \qquad (6.56c)$$

is the normalization factor. Once the number of filter weights k is specified (here, $k = 60$), the transfer function $W_L(\omega)$ is determined by the parameter p, the half-amplitude frequency of the filter in cycles per day (cpd). Specifically, we find

$$W_L(\omega) = \frac{1}{A}\left[1 + 2\sum_{k=1}^{60} F(k)\cos(\pi k\omega/\omega_N) \right] \qquad (6.57)$$

in which F and A are given by Eqns (6.56b) and (6.56c).

Comparison of Eqn (6.56b) with Eqn (6.52b) shows that

$$p = 12(\omega_c/\omega_N) = 24 f_c(\text{in cpd}) \qquad (6.58)$$

where $f_c = \omega_c/2\pi$ is the cutoff frequency in cycles per hour (cph) and where we have used the Nyquist frequency $f_N = 0.5$ cph for the hourly sampled data. The arguments of the angles in Eqns (6.52b) and (6.56b) are, therefore, identical. Where the filters differ is in the use of the sigma factors. Whereas the oscillations of $(1 + \cos(\pi k/M))$ are uniform with k, those of $\sin(\pi k/M)/(\pi k/M)$ decay with increasing k, similar to the way we have seen the term, $\sin(\pi k\omega_c/\omega_N)/(\pi k\omega_c/\omega_N)$, decay in amplitude (e.g., see the ripples in Figure 6.12). In this regard, the raised cosine window provides a more severe weighting of the truncated Fourier series than the sigma factors.

The value $p = 0.7$ cpd, corresponding to a cutoff period of 34.29 h, has been commonly used in the design of low-pass Lanczos–cosine filters (cf. Bryden, 1979). Although this produces an acceptable filter response for periods of two days and longer (where 2 days is generally the central period of the oceanic "spectral gap"), it has been shown to pass an unacceptable amount of high-frequency energy from the diurnal band, particularly from the O_1 and Q_1 tidal constituents (Walters and Heston, 1982). In an attempt to further reduce the leakage from the diurnal band, Mooers and Smith (1967) applied a separate filter to the low-pass filtered data from the $p = 0.7$ cpd filter or "Lancz7" filter (Thompson, 1983; Figure 6.13). Walters and Heston (1982) passed the data twice through the filter to produce the 10-day (Lancz7) filter. This not only results in a significantly improved filter amplitude throughout the diurnal band but also doubles the amount of data lost from the ends of the time series. Thompson (1983) suggested the use of a Lanczos–cosine filter with $p = 0.6$ cpd (the "Lancz6" filter) which equates to a cutoff period

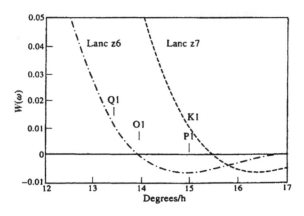

FIGURE 6.13 Expanded views of the filter responses for two tide-elimination filters for the diurnal frequency band. The Lancz6 and Lancz7 filters are low-pass Lanczos–cosine filters. $15°/h = 1.0$ cpd. *(Modified from Thompson (1983).)*

of 40 h. The Lancz6 filter essentially removes the leakage from the diurnal band but simultaneously shifts the low-pass portion of the filtered record to periods somewhat in excess of 2 days. The difference in the filters is quite subtle. For the Lanczos–cosine filter with $p = 0.7$ (Lancz7 filter), the first zero of the transfer function occurs at $15.4°/h$ (at 0.0428 cph), which is past the diurnal band (Figure 6.13); for the Lancz6 filter, the first zero is shifted to $14°/h$ (at 0.0389 cph) near the O_1 frequency of $13.9°/h$.

6.8 BUTTERWORTH FILTERS

The windowed cosine filters described in the previous section attempt to approximate an ideal rectangular transfer function using truncated Fourier cosine series. For nonrecursive filters, the output is a simple linear combination of the data and the role of the window is to attenuate the overshoot ripples created by truncation in the time domain (Gibbs' phenomenon). We now turn to a specific type of recursive filter for which the transfer function is created using a rational function in sines and cosines. Because this is a recursive filter, the output consists of both input data and past values of the output.

Let $\xi = \xi(\omega)$ be a monotonically increasing rational function of sines and cosines in the frequency, ω. The monotonic function

$$|W_L(\omega)|^2 = 1 \Big/ \left[1 + (\xi/\xi_c)^{2q}\right] \qquad (6.59)$$

(Figure 6.14) generates a particularly useful approximation to the squared gain of an ideal low-pass recursive filter with frequency cutoff ω_c. The filter design will eventually require $\xi(0) = 0$ so that the final version of $W_L(\omega)$ will closely resemble Eqn (6.59). (Note that we use the variable $\xi(\omega)$ instead of the usual variable notation, $w(\omega)$, to avoid any confusion with filter weights w.)

Butterworth filters of the form (Eqn (6.59)) have a number of desirable features (Roberts and Roberts, 1978). Unlike the transfer function of a linear nonrecursive filter constructed from a truncated Fourier series, the transfer function of a Butterworth filter is monotonically flat within the pass- and stop-bands, and has high tangency at both the origin ($\omega = 0$) and the Nyquist frequency, ω_N. The attenuation rate of $W_L(\omega)$ can be increased by increasing the *filter order*, q. However, once again we remark that too steep a transition from the stop-band to the pass-band can lead to ringing effects in the output due to Gibbs' phenomenon. Since it has a squared response, the Butterworth filter produces zero phase shift and its amplitude is attenuated by a factor of two at the cutoff frequency, for which $\xi/\xi_c = 1$ for all q. In contrast to nonrecursive filters, such as the Lanczos–cosine filter discussed in the previous section, there is no loss of output data from the ends of the record; N input values yield N output values. However, we do not expect to get something for nothing. The problem is that ringing distorts the data at the ends of the filtered output. As a consequence, we are forced to ignore output values near the ends of the filtered record, in analogy with the

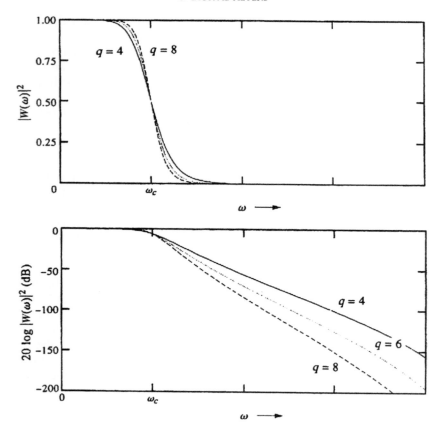

FIGURE 6.14   The frequency response functions $|W_L(\omega)|^2$ for an ideal squared, low-pass Butterworth filter for filter orders $q = 4$, 6, 8. Bottom panel gives response in decibels (dB). Power $= 0.5$ at the cutoff frequency, ω_c.

loss of data associated with nonrecursive filters. In effect, the loss is comparable to that from a nonrecursive filter of similar smoothing performance. A subjective decision is usually needed to determine where, at the two ends of the filtered record, the "bad" data end and the "good" data begin.

Butterworth filters fall into the category of physically realizable recursive filters having the time-domain formulation Eqn (6.2) with $k = 0, \ldots, M$. They may also be classified as infinite impulse response filters since the effects of a single impulse input can be predicted to an arbitrary time into the future. To see why we expect

$\xi(\omega)$ to be a rational function in sines and cosines, we use Eqn (6.2) and the fact that $W(\omega)$ is the ratio of the output to the input. We can then write

$$W(\omega) = \frac{\text{output}}{\text{input}} = \frac{\sum_{k=0}^{M} w_k e^{-i\omega k \Delta t}}{1 - \sum_{k=1}^{L} g_k e^{-i\omega k \Delta t}} \quad (6.60)$$

where the summations in the numerator and denominator involve polynomials in powers of the form $\exp(-i\omega k \Delta t)$ which can in turn be expressed through the variable ξ. The substitution $z = \exp(i\omega k \Delta t)$ leads to expression of the filter response $W(\omega)$ in terms of the z-transform and zeroes of poles.

6.8.1 High-Pass and Band-Pass Filters

High-pass and band-pass Butterworth filters can be constructed from the low-pass filter (Eqn (6.59)). For example, to construct a high-pass filter with cutoff frequency, ω_c, we use the transformation $\xi/\xi_c \rightarrow -(\xi/\xi_c)^{-1}$ in Eqn (6.59). The square transfer function of the high-pass filter is then

$$|W_H(\omega)|^2 = (\xi/\xi_c)^{2q} \Big/ \left[1 + (\xi/\xi_c)^{2q}\right] \qquad (6.61)$$

where, as required

$$|W_H(\omega)|^2 = 1 - |W_L(\omega)|^2 \qquad (6.62)$$

Band-pass Butterworth filters (and their counterparts, *stop-band* Butterworth filters) are constructed from a combination of low-pass and high-pass filters. For instance, the appropriate substitution in Eqn (6.59) for a band-pass filter is $\xi/\xi_c = \xi^*/\xi_c - (\xi^*/\xi_c)^{-1}$ which leads to the quadratic equation

$$(\xi^*/\xi_c)^2 - (\xi/\xi_c)(\xi^*/\xi_c) - 1 = 0 \qquad (6.63a)$$

with roots

$$\xi_{1,2}^*/\xi_c = (\xi/\xi_c)/2 \pm \left[(\xi/\xi_c)^2/4 + 1\right]^{1/2}. \quad (6.63b)$$

Substitution of $\xi/\xi_c = \pm 1$ (the cut-off points of the low-pass filter) yields the normalized cutoff functions $\xi_1^*/\xi_c = 0.618$ and $\xi_2^*/\xi_c = 1.618$ of the band-pass filter based on the cutoff frequency $\pm\omega_c$ of the associated low-pass filter. The corresponding band-pass cutoff functions for the cutoff frequency $-\omega_c$ of the low-pass filter are $\xi_1^*/\xi_c = -1.618$ and $\xi_2^*/\xi_c = -0.618$. Specification of the low-pass cutoff determines ξ_1^*/ξ_2^* of the band-pass filter. The bandwidth $\Delta\xi/\xi_c = -(\xi_1^* - \xi_2^*)/\xi_c = 1$ and the product $(\xi_1^*/\xi_c)(\xi_2^*/\xi_c) = 1$. Note that specification of ξ_1^* and ξ_2^* gives the associated function ξ_c of the low-pass filter

$$\xi_1^*\xi_2^* = \xi_c^2. \qquad (6.64)$$

6.8.2 Digital Formulation

The transfer functions (Eqns (6.59)–(6.62)) involve the continuous variable ξ whose structure is determined by sines and cosines of the frequency, ω. To determine a form for $\xi(\omega)$ applicable to digital data, we seek a rational expression with constant coefficients a to d such that the component $\exp(i\omega\Delta t)$ in Eqn (6.60) takes the form

$$\exp(i\omega\Delta t) = \frac{a\xi + b}{c\xi + d} \qquad (6.65)$$

(Here, we have replaced $-i\omega\Delta t$ with $+i\omega\Delta t$ without loss of generality.) As discussed by Hamming (1977), the constants are obtained by requiring that $\omega = 0$ corresponds to $\xi = 0$ and that $\omega \rightarrow \pi/\Delta t$ corresponds to $\xi \rightarrow \pm\infty$. Constants b and d (one of which is arbitrary) are set equal to unity. The final "scale" of the transformation is determined by setting $(\omega/2\pi)\Delta t = 1/4$ for $\xi = 1$. This yields

$$\exp(i\omega\Delta t) = \frac{1 + i\xi}{1 - i\xi} \qquad (6.66)$$

or, equating real and imaginary parts

$$\xi = \frac{2}{\Delta t}\left[\tan\left(\frac{1}{2}\omega\Delta t\right)\right]$$
$$= \frac{2}{\Delta t}[\tan(\pi\omega/\omega_s)], \quad -\omega_N < \omega < \omega_N \qquad (6.67)$$

where $\omega_s/(2\pi) = f_s$ is the sampling frequency ($f_s = 1/\Delta t$). We note that the derivation of Eqn (6.67) is equivalent to the conformal mapping

$$\xi = i\frac{2}{\Delta t}\frac{1 - z}{1 + z} \qquad (6.68a)$$

where

$$z = e^{2\pi i f \Delta t} = e^{i\omega\Delta t} \qquad (6.68b)$$

is the standard z-transform.

The transfer function of the (discrete) low-pass Butterworth filter is then (Rabiner and Gold, 1975)

$$|W_L(\omega)|^2 = \frac{1}{1 + [\tan(\pi\omega/\omega_s)/\tan(\pi\omega_c/\omega_s)]^{2q}}$$

(6.69a)

and that of the high-pass Butterworth filter is

$$|W_H(\omega)|^2 = \frac{[\tan(\pi\omega/\omega_s)/\tan(\pi\omega_c/\omega_s)]^{2q}}{1 + [\tan(\pi\omega/\omega_s)/\tan(\pi\omega_c/\omega_s)]^{2q}}$$

(6.69b)

The sampling and cutoff frequencies in these expressions are given by $\omega_s = 2\pi/\Delta t$ and $\omega_c = 2\pi/T_c$ in which $T_c = 1/f_c$ is the period of the cyclic cutoff frequency f_c. Plots of Eqn (6.69a) for various cutoff frequencies and filter order q are presented in Figure 6.15.

Use of the bilinear z-transform, $i(1-z)/(1+z)$, in Eqn (6.68a) eliminates aliasing errors that arise when the standard z-transform is used to derive the transfer function, these errors being large if the digitizing interval is large. Mathematically, the bilinear z-transform maps the inside of the unit circle ($|z| < 1$, for stability)

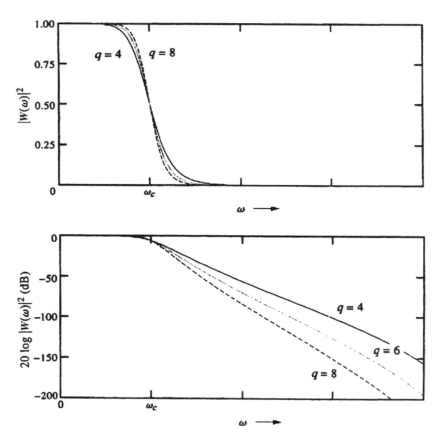

FIGURE 6.15 Same as Figure 6.14 but for discrete, low-pass squared Butterworth filters. *(After Rabiner and Gold (1975).)*

into the upper half plane. A thorough discussion of the derivation of pole and zeroes of Butterworth filters is presented in Kanasewich (1975) and Rabiner and Gold (1975).

We note that the above relationships define the square of the response of the filter $W(\omega)$ formed by multiplying the transfer function by its complex conjugate, $W^*(\omega) = W(-\omega)$. (In this instance, $W^*(\omega)$ and $W(-\omega)$ are equivalent since $i = \sqrt{-1}$) always occurs in conjunction with ω, as in $i\omega$. The product $W(\omega)W(-\omega)$ eliminates any frequency-dependent phase shift caused by the individual filters and produces a squared, and therefore sharper, frequency response than produced by $W(\omega)$ alone. The sharpness of the filter (as determined by the parameter q) is limited by filter ringing and stability problems. When q becomes too large, the filter begins to act like a step and Gibbs' phenomenon rapidly ensues.

Equations (6.68a,b) are used to design the filter in the frequency domain. In the time domain, we first determine the filter coefficients w_k and g_j for the low-pass filter (Eqn (6.2)) and then manipulate the output from the transfer function $W(\omega)$ to generate the output $|W(\omega)|^2$. To obtain the output for a high-pass Butterworth filter, $|W_H(\omega)|^2$, the output from the corresponding low-pass filter, $|W_L(\omega)|^2$, is first obtained and the resulting data values are subtracted from the original input values on a data-point-by-data-point basis.

6.8.3 Tangent vs Sine Filters

Equations (6.69a,b) define the transfer functions of *tangent* Butterworth low-pass filters. The corresponding transfer functions for *sine* Butterworth low-pass filters are given by

$$|W_H(\omega)|^2 = \frac{1}{1 + [\sin(\pi\omega/\omega_s)/\sin(\pi\omega_c/\omega_s)]^{2q}} \tag{6.70}$$

where we have simply replaced $\tan x$ with $\sin x$ in Eqn (6.69a,b). Although this book deals only

with the tangent version of the filter, there are situations where the sine version may be preferable (Otnes and Enochson, 1972). The tangent filter has "superior" attenuation within the stop-band but at a cost of doubled algebraic computation (the sine version has only recursive terms, while the tangent version has both recursive and nonrecursive terms).

6.8.4 Filter Design

The design of Butterworth filters is discussed in Hamming (1977). Our approach is slightly different but uses the same general concepts. We begin by specifying the sampling frequency $\omega_s = 2\pi f_s = 2\pi/\Delta t$ based on the sampling interval Δt for which

$$0 < \omega/\omega_s < 0.5 \tag{6.71}$$

and where the upper limit denotes the normalized Nyquist frequency, ω_N/ω_s. We next specify the desired cutoff frequency ω_c at the half-power point of the filter. For best results, the normalized cutoff frequency of the filter, ω_c/ω_s, should be such that the transition band of the filter does not overlap to any significant degree with the ends of the sampling domain (Eqn (6.71)). Once the normalized cutoff frequency (or frequencies) is (are) known, specification of the filter order q fully determines the characteristics of the filter response. Our experience suggests that the parameter q should be less than 10 and probably not larger than eight. Despite the use of double precision throughout the calculations, roundoff errors and ringing effects can distort the filter response for large q and render the filter impractical.

There are two approaches for Butterworth filter design once the cutoff frequency is specified. The first is to specify q so that the attenuation levels in the pass- and stop-bands are automatically determined. The second is to calculate q based on a required attenuation at a given frequency, taking advantage of the fact that we are working with strictly monotonic functions. Suppose we want an attenuation of $-D$ decibels at frequency ω_a in the stop-band of a low-pass filter having a cutoff frequency $\omega_c < \omega_a$. Using the definition for decibels and Eqn (6.59), we find that

$$q = 0.5 \frac{\log(10^{D/10} - 1)}{\log(\xi_a/\xi_c)}$$

$$\approx \frac{D/20}{\log(\xi_a/\xi_c)}, \quad \text{for } D > 10 \tag{6.72}$$

where D is a positive number measuring the decrease in filter amplitude in decibels and ξ is defined by Eqn (6.67). The nearest integer value can then be taken for the filter order provided that the various parameters (ω_a, D) have been correctly specified and q is less than 10. If the latter is not followed, the imposed constraints are too severe and new parameters need to be specified. The above calculations apply equally to specification of q based on the attenuation $-D$ at frequency $\omega_a < \omega_c$ in the stop-band of a high-pass filter, except that $\log(\xi_a/\xi_c)$ in Eqn (6.72) is replaced by $\log(\xi_c/\xi_a)$. Since $\log(x) = -\log(1/x)$, we can simply apply Eqn (6.72) to the high-pass filter, ignoring the minus sign in front of $\log(1/x)$.

6.8.5 Filter Coefficients

Once the characteristics of a transfer response have been specified, we need to derive the filter coefficients to be applied to the data in the time domain. We assume that the transfer function $W_L(\omega; q)$ of the low-pass filter can be constructed as a product, or cascade, of second-order ($q = 2$) Butterworth filters $W_L(\omega; 2)$ and, if necessary, one first-order ($q = 1$) Butterworth filter $W_L(\omega; 1)$. For example, suppose we required a filter of order $q = 5$. The transfer function would then be constructed via the cascade

$$W_L(\omega; 5) = W_L(\omega; 1) \times W_{L,1}(\omega; 2) \times W_{L,2}(\omega; 2) \tag{6.73}$$

in which the two second-order filters, $W_{L,1}$ and $W_{L,2}$, have different algebraic structure. Use of the cascade technique allows for variable order in the computer code for Butterworth filter programs without the necessity of computing a separate transfer function $W_L(\omega; q)$ each time. This eliminates a considerable amount of algebra and reduces the roundoff error that would arise in the "brute-force calculation" of W_L for each order.

The second-order transfer functions for a specified filter order q are given by

$$W_L(\omega; 2)$$

$$= \frac{[\xi_c^2(z^2 + 2z + 1)]}{a_k z^2 - 2z(\xi_c^2 - 1) + \{1 - 2\xi_c \sin[\pi(2k+1)/2q] + \xi_c^2\}} \tag{6.74a}$$

where ξ and z are defined by Eqns (6.67) and (6.68b)

$$a_k = 1 - 2\xi_c \sin[\pi(2k+1)/2q] + \xi_c^2 \tag{6.74b}$$

and k is an integer that takes on values in the range

$$0 \leq k < 0.5(q - 1) \tag{6.74c}$$

When q is an odd number, the first-order filter $W_L(\omega; 1)$ must also be used where

$$W_L(\omega; 1) = \left(\frac{\xi_c}{1 + \xi_c}\right) \frac{z + 1}{z - \left(\frac{1 - \xi_c}{1 + \xi_c}\right)} \tag{6.75}$$

Again, suppose that $q = 5$. The transfer function W_L is then composed of the lead filter $W_L(\omega; 1)$ given by Eqn (6.74a) and two second-order filters, for which k takes the values $k = 0$ and 1 in Eqn (6.74). Note that we have strictly adhered to the inequality in Eqn (6.74c). The first second-order filter is obtained by setting $k = 0$ in Eqn (6.74a); the second second-order filter is obtained by setting $k = 1$. For $q = 7$, a third second-order for $k = 2$ would be required, and so on.

The next step is to recognize that the first-order function (Eqn (6.75)) has the general form

$$W_L(\omega) = \frac{d_o z + d_1}{z - e_1} \tag{6.76}$$

and that the second-order function (Eqn (6.74a)) has the general form

$$W_L(\omega) = \frac{c_0 z^2 + c_1 z + c_2}{z^2 - b_1 z - b_2} \qquad (6.77)$$

where the sine terms in the coefficients of Eqn (6.74a) change with filter order q. The coefficients d, e in Eqn (6.76) are obtained by direct comparison with Eqn (6.75), while the coefficients b, c in Eqn (6.77) are obtained through comparison with Eqn (6.74a).

The recursive digital filters (Eqn (6.2)), whose time-domain algorithms have the transfer functions Eqn (6.76) and (6.77) are, respectively,

$$y_n = d_0 x_n + d_1 x_{n-1} + e_1 y_{n-1} \qquad (6.78)$$

and

$$y_n = c_0 x_n + c_1 x_{n-1} + c_2 x_{n-2} + b_1 y_{n-1} + b_2 y_{n-2} \qquad (6.79)$$

Direct comparison of Eqn (6.76) with Eqn (6.75) yields the time-domain coefficients for the first-order filter; comparison of Eqn (6.77) with Eqn (6.74a) yields the corresponding coefficients for the second-order filters for each value of k beginning with $k = 0$. In particular, we find, for the first-order filter

$$d_0 = d_1 = \frac{\xi_c}{1 + \xi_c}; e_1 = \frac{1 - \xi_c}{1 + \xi_c} \qquad (6.80)$$

and for the second-order filter

$$b_1 = 2(\xi_c^2 - 1)/a_k; \ b_2 = \left[(1 + \xi_c^2) - a_k\right]/a_k$$
$$c_o = \xi_c^2/a_k; \ c_1 = 2c_0; \ c_2 = c_o \qquad (6.81)$$

where the coefficients in Eqn (6.81) change with the parameter k according to the number of second-order filters needed to create the filter of order q.

To apply the $q = 5$ filter, we process the input data x_n ($n = 0, 1,..., N$) by the first-order filter (Eqn (6.78)). We then take the output from the first-order filter and process it by the first of the second-order filters (Eqn (6.79)) with $k = 0$. The resultant output is then processed by the next second-order filter (Eqn (6.79)) with $k = 1$. The sequence y'_n ($n = 0$, 1,...) derived from the three filter applications is the low-pass output for the fifth-order Butterworth filter $W_L(\omega; 5)$, as indicated by Eqn (6.73).

The task is only half complete since our ultimate goal is to remove any filter-induced phase shift by smoothing the data with the squared response of the filter $|W_L|^2$, given by Eqn (6.73). The sequence we require is: $\{x_n\}$ yields $\{y'_n\}$ as the output from $W_L(\omega)$ and $\{y'_n\}$ yields $\{y_n\}$ as the output from $|W_L(\omega)|^2$. To obtain the output $\{y_n\}$ for the square response of the filter, $|W_L(\omega)|^2$, we need to process the output $\{y'_n\}$, obtained from $W_L(\omega)$, with the filter $W_L(-\omega)$. There are three options. (1) We can separately design $W(-\omega)$, a relatively straightforward task involving some sign changes in Eqns (6.74a) and (6.75). (2) We can invert the order of the calculations such that the output $\{y'_n\}$ from $W_L(\omega)$ is passed through the inverted version of this filter. That is, the data from $W_L(\omega)$ are first run through the second-order filter ($k = 1$ for $q = 5$), with the output from this filter passed through the second-order filter ($k = 0$) and finally through the first-order filter. (3) We can simply invert the chronological order of the data $\{y'_n\}$ and pass the inverted sequence through the original filter $W_L(\omega)$. Since all the data are recorded beforehand, we recommend approach (3). The one caution is that the sequence of the final output must be inverted to regain the original chronological order of the data. In all cases, passing the inverted version of $\{y'_n\}$ through the filter cascade removes any phase shift associated with the first pass which produced $\{y'_n\}$ from $\{y_n\}$. A phase shift $\phi(\omega)$ caused by the first sequence of filters $W_1(\omega) \times W_2(\omega) \times ...$ is canceled by the phase shift $-\phi(\omega)$ due to the second sequence of filters, $W_L(-\omega)$.

Computer programs designed to carry out the Butterworth filter operations should assign the output $\{y'_n\}$ from each filter as the new input $\{x_n\}$ to the next filter in the cascade until the output corresponding to the filter $|W_L(\omega)|^2$ is achieved. The last set of output is then chronologically inverted and rerun through the same filter. Following the final set of calculations, the output sequence is inverted to ensure correct ordering in time.

To obtain the results for a *high-pass* Butterworth filter, one further operation is required. The final output $\{y_n\}$ ($n = 0, 1,...$) from the low-pass filter is subtracted point for point from the original input $\{x_n\}$ ($n = 0, 1,...$) to create the high-pass filtered data $y_n^* = x_n - y_n$. The procedure to obtain the low- and high-pass Butterworth filters is illustrated schematically in Figure 6.16.

6.9 KAISER–BESSEL FILTERS

As noted in Section 5.4.6, we consider the Kaiser–Bessel filter among the best filters for processing and analyzing digital oceanographic data. Designed by James Kaiser at Bell Laboratories (Kaiser, 1966), the Kaiser–Bessel filter requires specification of a single parameter, α, and has easy to generate coefficients with high equivalent noise bandwidth, a primary criterion for good digital filter design (note that the parameter $\beta = \pi\alpha$ is sometimes used in place of α to define the filter shape). In the time domain, the M values of the filter weights, $w(m)$, are defined as

$$w(m)$$
$$= \begin{cases} \dfrac{I_0(\pi\alpha\Omega)}{I_0(\pi\alpha)}, & -(M-1)/2 \le m \le +(M-1)/2 \\ 0 & \text{otherwise} \end{cases}$$

$$(6.82)$$

where I_0 is the zeroth-order modified Bessel function of the first kind, $\alpha > 0$ is an arbitrary real number that determines the shape of the filter, $(M-1)\Delta t$ is the width of the filter in the time domain for data sampling rate Δt, and

$$\Omega = \left[1 - \left(\frac{2m}{M-1} \right)^2 \right]^{1/2} \qquad (6.83)$$

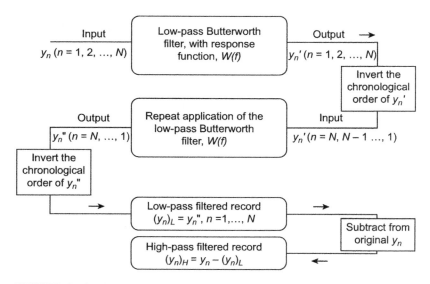

FIGURE 6.16 The procedure for obtaining low- and high-pass Butterworth filters.

The Bessel function

$$I_0(x) = \sum_{k=0}^{\infty}\left[\frac{(-1)^k}{\Gamma(k+1)}\frac{(x/2)^k}{k!}\right]^2 \qquad (6.84)$$

has a maximum value of unity at the origin ($I_0(0)=1$) and oscillates much like the cosine function with a decay rate proportional to $1/\sqrt{x}$, although the roots are not generally periodic, except asymptotically for large x (here, Γ is the Gamma function). Insertion of Eqns (6.83) and (6.84) in Eqn (6.82) shows that the filter impulse response peaks at $m=0$, where $w(0)=1$, and decays to either side of the central peak. The frequency response of the filter, $W(\omega)$, is obtained from the DFT of Eqn (6.82) and is approximated by

$$W(\omega) \approx \frac{(M-1)\Delta t}{I_0(\pi\alpha)}$$
$$\times \frac{\sinh\left\{\pi\left[(\alpha)^2 - ((M-1)\omega\Delta t/2\pi)^2\right]^{1/2}\right\}}{\pi\left[(\alpha)^2 - ((M-1)\omega\Delta t/2\pi)^2\right]^{1/2}}$$
$$(6.85)$$

where $\omega = 2\pi f$, is the angular frequency and the filter length $(M-1)\Delta t = 1/f_o$ is the inverse of the fundamental frequency, f_o, of the filter. Thus, the product $(M-1)\omega\Delta t/2\pi = f/f_o$ gives the normalized frequencies for the filter; here, $((M-1)\Delta t)f$ is referred to as the "DFT bin length". For the purposes of filter design, the modified Bessel function I_0 is defined in terms of a Taylor series expansion about $x=0$ as follows:

For $|x| \le 3.75$: $I_0(x)$
$$= \{[\{[(4.5813 \times 10^{-3}Z + 3.60768$$
$$\times 10^{-2})Z + 2.659732 \times 10^{-1}]Z$$
$$+ 1.2067492\}Z + 3.0899424]Z$$
$$+ 3.5156229\}Z + 1.0$$
$$(6.86a)$$

where for real x

$$Z = (x/3.75)^2. \qquad (6.86b)$$

For $|x| > 3.75 : I_0(x)$
$$= \exp(|x|)/|x|^{1/2}\{[(\{[\{[(3.92377$$
$$\times 10^{-3}Z - 1.647633 \times 10^{-2})Z$$
$$+ 2.635537 \times 10^{-2}]Z - 2.057706$$
$$\times 10^{-2}\}Z + 9.16281 \times 10^{-3}]Z$$
$$- 1.57565 \times 10^{-3}\}Z + 2.25319$$
$$\times 10^{-3})Z + 1.328592 \times 10^{-2}]Z$$
$$+ 3.39894228 \times 10^{-1}\}$$
$$(6.87a)$$

where

$$Z = 3.75/|x| \qquad (6.87b)$$

Examples of the filter weights and their corresponding DFT are presented in Figure 6.17 for two widely separated values of the parameter α.

Recall that the fundamental goal in filter design is to emulate as best as possible an ideal steplike filter whose frequency response is equal to 1 throughout the pass-band and equal to 0 throughout the stop-band. The ideal filter impulse response is then derived by taking the DFT of the ideal frequency response. Because the DFT of a step function leads to an infinite number of filter weights, truncation of the filter is needed, leading to a trade-off between filter width and side-lobe attenuation. As illustrated by Figure 6.17, varying the parameter α permits trade-offs between the width of the main lobe of the pass-band and the amplitudes of the side lobes within the stop-band of the filter response. As α increases, the main lobe increases in width, while the side lobes decrease in amplitude, with the filter taking on a Gaussian shape for large α in both the time and frequency domains. Other filters (such as the running-mean or rectangular

FIGURE 6.17 Frequency response function $W(f)$ and corresponding filter weights (impulse response) w_m for a Kaiser–Bessel filter defined by Eqn (6.82) for two values of α (=1.2 and 4.8) and $M = 31$.

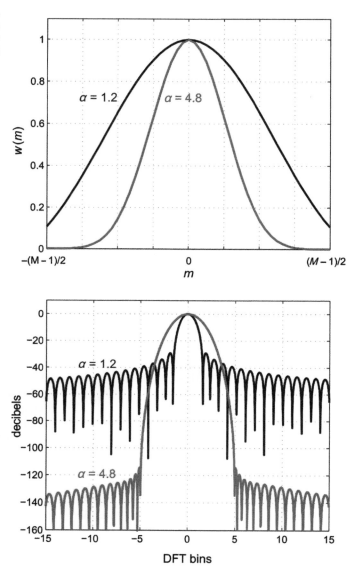

filter) have much steeper and confined central lobes but weakly attenuated side lobes compared to the Kaiser–Bessel filter. Because of the highly reduced side-lobe contamination of the Kaiser–Bessel filter, spectral estimates obtained using data segment with 50% overlaps retain near-statistical independence. The Kaiser–Bessel window can be made to approximate other windows by varying the α parameter (Table 6.1). A comparison between the filter weights, w_m, and the corresponding frequency response, $W(f)$, of the Kaiser–Bessel filter and several common filters (windows) is presented in Figure 6.18.

TABLE 6.1   Shapes of Kaiser–Bessel Filters Relative to Other Types of Filters for Different Values of the Parameter α and $\beta = \pi\alpha$

Alpha (α)	Beta (β)	Type of Window
0	0	Rectangular
1.6	5	Similar to Hamming
1.9	6	Similar to Hanning
2.7	8.6	Similar to Blackmann

6.9.1 A Low-Pass Kaiser–Bessel Filter

Many oceanic studies require daily mean time series from which the effects of diurnal and semidiurnal tides, inertial oscillations, internal waves, and other "high"-frequency motions have been removed. Except at latitudes less than 30° (the so-called "turning latitude"), where the inertial period exceeds the diurnal tidal period, removal of the above high-frequency constituents requires low-pass filters with cutoff periods capable of eliminating variations with periods less than roughly 25 h and specifically the diurnal K_1 and O_1 tidal motions with periods of 23.934 and 25.819 h, respectively (frequencies $f = \omega/2\pi = 1.00276$ and 0.92955 cpd, respectively). For this application, we recommend the use of a low-pass Kaiser–Bessel filter with cutoff periods of 25–50 h. Such low-pass filters ensure highly suppressed side lobes and, therefore, negligible contamination from higher frequency motions. We can illustrate how well low-pass Kaiser–Bessel filters reproduce an ideal low-pass filter whose stop-band is meant to remove daily tidal variations by examining the characteristics of the Kaiser–Bessel filter impulse amplitude and frequency response for commonly used values of α of 2.0, 2.5, 3.0, and 3.5 and filter lengths M of 25, 31, 37, and 49 hourly values for data series sampled at $\Delta t = 1$ h. Plots of the Kaiser–Bessel filter impulse response amplitudes, $w_m = w(m\Delta t)$, for $-(M-1)/2 \leq m \leq (M-1)/2$ and the corresponding frequency

response, $W(f)$, are presented in Figure 6.19. The filter characteristics are listed in Table 6.2 and the filter weights, in Table 6.3. The normalization coefficient, γ, is the value that multiplies each of the filter weights in order that the sum of the weights is equal to unity, viz.

$$\sum_{-(M-1)/2}^{(M-1)/2} \gamma w_m = 1. \tag{6.88}$$

Several general trade-off factors emerge from the above results: (1) all the filters achieve high side-lobe attenuation, with the first side lobe diminished by over −45 dB (a factor of $10^{-4.5} \cong 1/32,000$); (2) increasing α for a given filter length M increases the width of the filter pass-band (allowing greater leakage from signals in the ideal stop-band lying close to the cutoff period of 25 h, frequency $f = 0.96$ cpd), while at the same time enhancing attenuation of the side lobes (thereby greatly reducing leakage of higher frequency components located in the ideal stop-band); and (3) increasing the filter length, M, causes the filter cutoff frequency (defined as the 1/2 amplitude point or −3-dB level of the frequency response) to shift slightly into the pass-band away from the ideal cutoff period of 0.96 cpd (a negative effect), while greatly attenuating the diurnal K_1 and O_1 tidal signals within the stop-band (a positive effect). Semidiurnal motions with periods well into the stop-band are highly suppressed by all filters. Following low-pass filtering, the hourly record can be decimated to 24-h samples to obtain the daily mean time series.

6.10 FREQUENCY-DOMAIN (TRANSFORM) FILTERING

The type of digital filtering discussed in the previous sections involves convolution of the time series data with weighting functions called

FIGURE 6.18 Comparison of the filter weights (impulse response) w_m and corresponding frequency response function $W(f)$ for a Kaiser–Bessel filter (window) and several commonly used filters. All filters have a width of 50 units and are scaled to a maximum $w_0 = 1$. (Courtesy, Alexander Rabinovich, Institute of Ocean Sciences.)

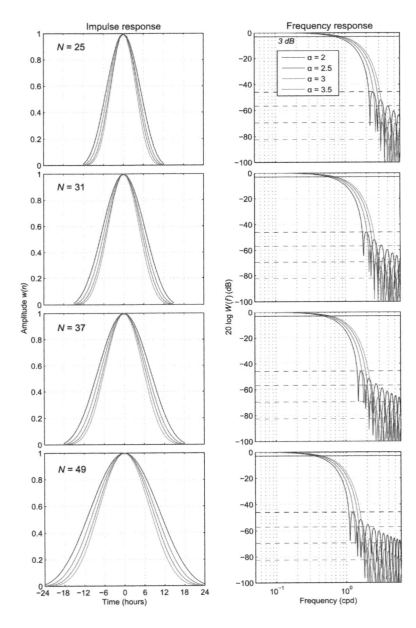

FIGURE 6.19 Frequency response function $W(f)$ and corresponding filter weights (impulse response) w_m for a low-pass Kaiser–Bessel filter designed for detiding hourly time series. Response functions are shown for four values of α (=2, 2.5, 3, and 3.5) and four values of the filter length M (=25, 31, 37, and 49 h). Because it is symmetrical about $f=0$, only the positive frequencies are shown for $W(f)$. *(Courtesy, Alexander Rabinovich, Institute of Ocean Sciences.)*

TABLE 6.2   Low-Pass Kaiser—Bessel Filter
Characteristics

Filter Length, M (h)	Alpha (α)	Cutoff (h)	Attenuation (dB)	Attenuation period (h)	Roll-off (h)
25	2	33.0	−46	11.0	16.5
	2.5	30.1	−57	9.1	13.1
	3	27.7	−69	7.7	10.7
	3.5	25.6	−82	6.6	9.0
31	2	41.0	−46	13.8	20.9
	2.5	37.9	−57	11.4	16.3
	3	34.1	−69	9.7	13.5
	3.5	32.0	−82	8.3	11.3
37	2	48.8	−46	16.5	25.0
	2.5	44.5	−57	13.7	19.7
	3	41.0	−69	11.5	16.0
	3.5	37.9	−82	9.9	13.5
49	2	64.0	−46	21.8	33.0
	2.5	60.2	−57	18.3	26.3
	3	53.9	−69	15.3	21.3
	3.5	51.2	−82	13.3	18.0

Columns are as follows: (1) filter length, M (number of hourly values used in the filter); (2) alpha (α) is the filter shape parameter; (3) cutoff is the period at which the frequency response W(ω) is attenuated by a factor of 2 (the -3-dB level of the maximum response amplitude); (4) attenuation is the amplitude decrease of the first side lobe in the stop-band; (5) attenuation period is the period (inverse frequency) at which the associated attenuation value is achieved; and (6) roll-off is the difference between the lowest frequency at which the "attenuation" is achieved and the cutoff frequency. Cutoff and roll-off are expressed in hours rather than frequency. (Courtesy, Maxim Krassovski, Institute of Ocean Sciences.)

impulse response functions that eliminate selected ranges of frequencies from the data. In the case of Fourier transform filtering, the weights are defined in terms of a Fourier transform window (frequency response function, *FRF*), $W(\omega)$, and filtering involves: (1) taking the FFT (Fast Fourier Transform) of the original data set, (2) multiplying the FFT output by the appropriate form of $W(\omega)$ that lets through the frequencies of interest and blocks all the others, and (3) taking the inverse FFT of the result to get back a filtered data set in the time domain. These steps are shown schematically in Figure 6.20. As an example, $W(\omega)$ might be a low-pass filter

designed to eliminate frequency components with periods $2\pi/\omega$ that are longer than 40 h. Alternatively, $W(\omega)$ could be a "notch" filter used to isolate oscillations centered near the local Coriolis frequency, or a two-notch filter designed to remove energy in the diurnal and semidiurnal tidal bands. Transform methods have been discussed from an oceanographic perspective by Walters and Heston (1982), Evans (1985), and Forbes (1988). As these papers indicate, the choice of an "appropriate" form for $W(\omega)$ is critical to the success of the method. Filtering in the frequency domain is attractive because of its simplicity compared to convolution in the time domain and because it is conceptually more in accord with our objective in filtering, namely, to remove specific periodicities in the data while retaining those of interest. Perhaps contrary to expectation, multiplication of the Fourier transform by a window is not always more computationally efficient than convolution of filter weights with the data (Evans, 1985).

We can outline use of the Fourier transform filtering as follows. Suppose we have a time series $\{x(t)\}$ with discrete values $x(n\Delta t) = x_n$, where n is an integer in the range $-N < n \leq N$. The Fourier transform of this time series is

$$X_k = \frac{1}{T} \sum_{n=-N+1}^{N} x_n \exp(-i\omega_k n\Delta t) \qquad (6.89)$$

where $T = 2N\Delta t$ is the record length and the Fourier frequencies are

$$\omega_k = 2\pi f_k = \frac{2\pi k}{T}, \quad -N < k \leq N. \qquad (6.90)$$

Let $w(r\Delta t) = w_r$, $-s \leq r \leq s$, represent a set of filter weights whose sum is unity to preserve the series mean and whose distribution is symmetric about $r = 0$ to preserve the phase information in the data. The number of weights, $S = 2s + 1$, is called the span of the filter. Since s points are lost from each end of the input data series, the filtered output series

TABLE 6.3 Filter Weights $w(m)$ for Different Filter Lengths (M, hrs) and Values of Alpha (α) for Low Pass Tide-Removing Kaiser–Bessel Filter

FILTER LENGTH, M (hours)

	25				31				37				49			
Alpha	2	2.5	3	3.5	2	2.5	3	3.5	2	2.5	3	3.5	2	2.5	3	3.5
NORMALIZATION COEFFICIENT																
	0.085	0.095	0.104	0.112	0.068	0.076	0.083	0.089	0.057	0.063	0.069	0.074	0.043	0.047	0.052	0.056
m																
−24													0.011	0.003	0.001	0.000
−23													0.023	0.007	0.002	0.001
−22													0.038	0.015	0.006	0.002
−21													0.058	0.026	0.011	0.005
−20													0.083	0.041	0.020	0.010
−19													0.113	0.061	0.033	0.018
−18									0.011	0.003	0.001	0.000	0.149	0.087	0.051	0.030
−17									0.027	0.009	0.003	0.001	0.190	0.119	0.075	0.047
−16									0.051	0.021	0.009	0.004	0.236	0.158	0.106	0.071
−15					0.011	0.003	0.001	0.000	0.083	0.041	0.020	0.010	0.287	0.203	0.144	0.102
−14					0.031	0.011	0.004	0.001	0.124	0.069	0.038	0.021	0.343	0.255	0.190	0.141
−13					0.063	0.028	0.013	0.006	0.175	0.108	0.066	0.041	0.403	0.313	0.243	0.189
−12	0.011	0.003	0.001	0.000	0.106	0.057	0.030	0.016	0.236	0.158	0.106	0.071	0.465	0.376	0.305	0.247
−11	0.038	0.015	0.006	0.002	0.164	0.099	0.060	0.036	0.305	0.220	0.158	0.114	0.529	0.444	0.373	0.313
−10	0.083	0.041	0.020	0.010	0.236	0.158	0.106	0.071	0.382	0.293	0.225	0.172	0.594	0.514	0.446	0.386
−9	0.149	0.087	0.051	0.030	0.320	0.234	0.170	0.124	0.465	0.376	0.305	0.247	0.658	0.586	0.523	0.466
−8	0.236	0.158	0.106	0.071	0.415	0.325	0.255	0.200	0.551	0.467	0.396	0.336	0.720	0.658	0.601	0.550
−7	0.343	0.255	0.190	0.141	0.516	0.430	0.359	0.299	0.637	0.562	0.497	0.439	0.779	0.727	0.679	0.634
−6	0.465	0.376	0.305	0.247	0.620	0.543	0.476	0.418	0.720	0.658	0.601	0.550	0.833	0.792	0.754	0.717
−5	0.594	0.514	0.446	0.386	0.720	0.658	0.601	0.550	0.798	0.750	0.705	0.662	0.881	0.851	0.823	0.795
−4	0.720	0.658	0.601	0.550	0.812	0.767	0.724	0.684	0.866	0.833	0.800	0.770	0.923	0.903	0.883	0.864
−3	0.833	0.792	0.754	0.717	0.890	0.862	0.835	0.809	0.923	0.903	0.883	0.864	0.956	0.944	0.933	0.921
−2	0.923	0.903	0.883	0.864	0.950	0.937	0.924	0.911	0.965	0.956	0.946	0.937	0.980	0.975	0.969	0.964
−1	0.980	0.975	0.969	0.964	0.987	0.984	0.980	0.977	0.991	0.989	0.986	0.984	0.995	0.994	0.992	0.991
0	1.000	1.000	1.000	1.000	1.000	1.000	1.000	1.000	1.000	1.000	1.000	1.000	1.000	1.000	1.000	1.000

(Continued)

TABLE 6.3   Filter Weights $w(m)$ for Different Filter Lengths (M, hrs) and Values of Alpha (α) for Low Pass Tide-Removing Kaiser—Bessel Filter (*cont'd*)

1	0.980	0.975	0.969	0.964	0.987	0.984	0.980	0.977	0.991	0.989	0.986	0.984	0.995	0.994	0.992	0.991
2	0.923	0.903	0.883	0.864	0.950	0.937	0.924	0.911	0.965	0.956	0.946	0.937	0.980	0.975	0.969	0.964
3	0.833	0.792	0.754	0.717	0.890	0.862	0.835	0.809	0.923	0.903	0.883	0.864	0.956	0.944	0.933	0.921
4	0.720	0.658	0.601	0.550	0.812	0.767	0.724	0.684	0.866	0.833	0.800	0.770	0.923	0.903	0.883	0.864
5	0.594	0.514	0.446	0.386	0.720	0.658	0.601	0.550	0.798	0.750	0.705	0.662	0.881	0.851	0.823	0.795
6	0.465	0.376	0.305	0.247	0.620	0.543	0.476	0.418	0.720	0.658	0.601	0.550	0.833	0.792	0.754	0.717
7	0.343	0.255	0.190	0.141	0.516	0.430	0.359	0.299	0.637	0.562	0.497	0.439	0.779	0.727	0.679	0.634
8	0.236	0.158	0.106	0.071	0.415	0.325	0.255	0.200	0.551	0.467	0.396	0.336	0.720	0.658	0.601	0.550
9	0.149	0.087	0.051	0.030	0.320	0.234	0.170	0.124	0.465	0.376	0.305	0.247	0.658	0.586	0.523	0.466
10	0.083	0.041	0.020	0.010	0.236	0.158	0.106	0.071	0.382	0.293	0.225	0.172	0.594	0.514	0.446	0.386
11	0.038	0.015	0.006	0.002	0.164	0.099	0.060	0.036	0.305	0.220	0.158	0.114	0.529	0.444	0.373	0.313
12	0.011	0.003	0.001	0.000	0.106	0.057	0.030	0.016	0.236	0.158	0.106	0.071	0.465	0.376	0.305	0.247
13					0.063	0.028	0.013	0.006	0.175	0.108	0.066	0.041	0.403	0.313	0.243	0.189
14					0.031	0.011	0.004	0.001	0.124	0.069	0.038	0.021	0.343	0.255	0.190	0.141
15					0.011	0.003	0.001	0.000	0.083	0.041	0.020	0.010	0.287	0.203	0.144	0.102
16									0.051	0.021	0.009	0.004	0.236	0.158	0.106	0.071
17									0.027	0.009	0.003	0.001	0.190	0.119	0.075	0.047
18									0.011	0.003	0.001	0.000	0.149	0.087	0.051	0.030
19													0.113	0.061	0.033	0.018
20													0.083	0.041	0.020	0.010
21													0.058	0.026	0.011	0.005
22													0.038	0.015	0.006	0.002
23													0.023	0.007	0.002	0.001
24													0.011	0.003	0.001	0.000

The Normalization Factor is the Value that Multiplies each Column of Filter Weights to Ensure that the sum of the Weights adds up to Unity (i.e., Sum Weights (w) = 1). M is the number of filter weights as measured in hours. Courtesy, Maxim Krassovski, Institute of Ocean Sciences.

$$y_n = \sum_{r=-s}^{s} w_r x_{n-r} = \sum_{r=-s}^{s} w_{n-r} x_r \qquad (6.91)$$

is shorter than the original series by $2s$ values. The effect of the convolution is to smear the signal $x(t)$ according to the weighting imposed by the IRF, $w(t)$. The FRF or transfer function

$$W(\omega) = \sum_{r=-s}^{s} w_r \exp(-i\omega r \Delta t)$$
$$= |W(\omega)| \exp(-i\phi(\omega)) \qquad (6.92)$$

gives the effect of the IRF on the transform of a sinusoid of unit amplitude and frequency

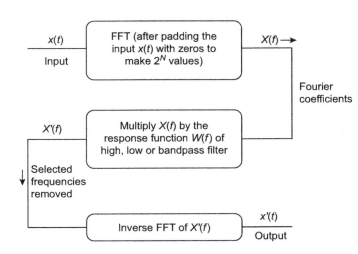

FIGURE 6.20 The procedure for obtaining DFT filters for application in the frequency domain.

ω ($=2\pi f$). As stated earlier, the absolute value $|W(\omega)|$ is the *gain factor* of the system and the associated phase angle, $\phi(\omega)$, the *phase factor* of the system. If a linear system is subjected to a sinusoidal input with a frequency ω and produces a sinusoidal output at the same frequency, then $|W(\omega)|$ is the ratio of the output amplitude to the input amplitude and $\phi(\omega)$ is the phase shift between the output and input. The frequency response function is viewed as a window or transfer function that lets through some frequencies and stops others. Note that W is defined at all frequencies such that $-\pi/\Delta t < \omega \leq \pi/\Delta t$, and not just at the Fourier frequencies, ω_k.

The key to Fourier transform filtering is that, for a constant-parameter linear system, the Fourier transform of the filtered data, $Y(\omega)$, is related to the Fourier transform of the input data, $X(\omega)$, through the product

$$Y(\omega) = W(\omega)X(\omega) \qquad (6.93)$$

In other words, convolution in the time domain, defined by Eqn (6.91), translates to multiplication in the frequency domain. The merits of a filter are judged by its FRF (frequency domain) and IRF (time domain). We would like the magnitude of the FRF to be near unity in the frequency bands to be passed by the filter and near zero in the bands to be stopped, i.e., $|W(\omega)| \approx 1$ and 0, respectively. The transition band between the stop- and pass-bands should be as narrow as possible since a broad transition band results in a filtered time series whose frequency content may be contaminated by unwanted frequencies. Similarly, the span of the IRF should be short so that the magnitude of weights decays to zero rapidly as r increases toward $\pm s$. If convolution is used, short filters are computationally more efficient and, moreover, result in less data loss. Unfortunately, the two criteria are at odds with one another. In general, the narrower the transition band in the frequency domain, the slower is the decay rate of the IRF in the time domain. Also, the steeper the maximum slope of the transition band, the larger are the side initial side lobes of the IRF that arise from the well-known Gibbs' phenomenon. In the limit of a step function-type FRF, in which the transition zone has zero width, the resulting IRF decays very slowly and has large side lobes (ringing). Thus, one must always compromise in specifying an FRF.

In all time-domain filtering (convolution), data are lost from each end of the original digital time series. For example, in the case of nonrecursive filters, in which the output is based on input time series alone, a known segment of the record of length $T/2$ is lost from either end of the time series ($T = (M - 1)\Delta t$ is the filter length). The same applies to recursive filters in which the present output from the filter is based on the original data series as well as previous values of the output. Here, the difficulty is that the amount of data we must discard from either end is not well defined because of ringing effects associated with the convolution and abrupt data discontinuities at the ends of the record. Transform windowing typically results in exactly the same amount of data loss as the equivalent time-domain filter (Walters and Heston, 1982). The Fourier transform treats the data outside the record as if it were zero, so that the ringing at the ends is introduced by the abrupt changes in the series from nonzero to zero and to the circular convolution of the window's IRF with the data. Ringing (Gibbs' phenomenon) occurs throughout the entire time series and becomes evident when the filtered FFT data are inverted to recover the desired filtered time series data. The effects of Gibbs' phenomenon are mitigated by tapering the frequency-domain filter using a linear or cosine function.

According to Thompson (1983), careful construction of weighting functions in the time domain can more effectively remove tidal components than Fourier transform filtering. This is because tidal frequencies do not generally coincide with Fourier frequencies of the record length. Design of IRF weights to minimize the squared deviation from some specified norm (least squares filter design) offers more control over the FRF at particular non-Fourier frequencies. On the other hand, broadband signals are best served by the FRF approach. Evans (1985) suggests that the ratio of convolution cost to windowing cost is $E = S/(2 \log_2(N))$, where S is the filter span. If $E > 1$, then windowing in the frequency domain is a more efficient method. Forbes (1988) addressed the problem

of removing tidal signals from the data while retaining the near-inertial signal and argues that Fourier transform filtering is effective provided that careful consideration is given to the filter bandwidth and the amount of tapering of the sides of the filter. Note that, in trying to remove strong tidal signals from a data series, it is sometimes beneficial to first calculate the tidal constituents and then subtract the harmonically predicted tidal signal from the data prior to filtering. This is time consuming and not an advantage if the filter is properly designed.

Figure 6.21(a) shows the energy-preserving power spectrum for a middepth current meter record from a Cape Howe mooring site ($37°35'$ S, $150°25'$ E) off the coast of New South Wales. To remove the strong tidal motions from this record, Forbes first used an untapered DFT with 12 and 17 adjacent Fourier coefficients set to zero in the diurnal and semidiurnal bands, respectively (Figure 6.21(b)). However, the greatest improvement in the Fourier transform filtering came from setting only three Fourier terms to zero and tapering the filter with a nine-point cosine taper in the frequency domain at the diurnal and semidiurnal frequencies (Figure 6.21(c)). Tapering the time series, not widening the filter by using more zero frequencies, was found to be a better way to improve filter characteristics. Perhaps, the most important conclusion from Forbes' work is that DFT filters are effective if the number of Fourier coefficients set to zero is sufficient to cover the unwanted frequency band and if the filter is cosine-tapered in the frequency domain to ensure a smooth transition to nonzero Fourier coefficients. In the nonintegral single-frequency case presented here (Forbes was looking at near-inertial motions) this amounted to a three-point filter with a nine-point cosine taper. The widths of the filter and taper must be determined for each application by a careful examination of the spectrum for leakage into adjacent frequencies, but once this is done, the technique is fast and simple to apply.

FIGURE 6.21 Energy-preserving spectra for a 4000-h current meter record at 720 m depth off Cape Howe, Australia. (a) Raw hourly data; (b) after applying a DFT filter with 12 and 17 adjacent Fourier coefficients set to zero in the diurnal and semidiurnal bands (no tapering); (c) after applying a DFT filter with three Fourier coefficients set to zero and nine Fourier coefficients cosine-tapered on each side of the zero coefficients. *(From Forbes, 1988.)*

To summarize the use of Fourier transform filtering:

1. Remove any linear trend (or nonlinear trend if it is well defined) from the data prior to filtering but do not be too concerned with cosine tapering the first and last 10% of the data. Fast Fourier transform the data.
2. Define the Fourier transform filter $W(\omega)$ for both positive and negative frequencies with the extreme frequencies given by $\pm 1/2\Delta t$.
3. If the measured data are real, and the filtered output is to be real, the filter should obey

$W(-\omega) = W^*(\omega)$, where the asterisk denotes complex conjugate. The easiest way to satisfy this condition is to pick $W(\omega)$ real and symmetric in frequency.

4. If $W(\omega)$ has sharp vertical edges then the impulse response of the filter (the response arising from a short impulse as input) will have damped ringing at frequencies corresponding to these edges. If this occurs, pick a smoother $W(\omega)$. Take the FFT inverse of $W(\omega)$ to see the impulse response of the filter. The more points used in the smoothing, the more rapid the falloff of the impulse response.

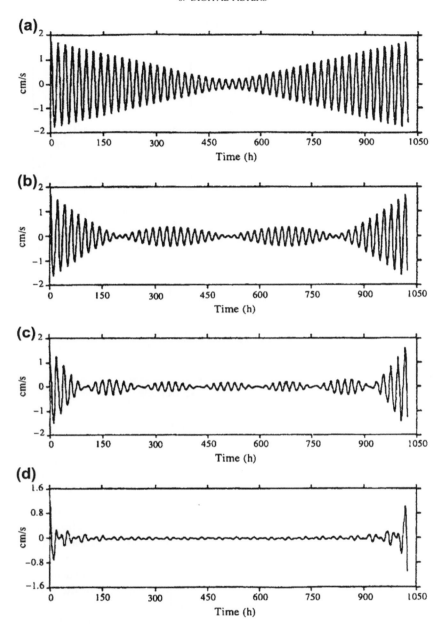

FIGURE 6.22 Ringing effects following application of different DFT filters to an artificial time series with frequency $f = 0.05$ cph and then inverting the transform. (a) Single Fourier coefficient at 0.05 cph set to zero, (b) three Fourier coefficients set to zero, (c) five Fourier coefficients set to zero, and (d) 21 coefficients set to zero. *(From Forbes (1988).)*

5. Multiply the transformed data series $X(\omega)$ by $W(\omega)$ and invert the resultant data series, $Y(\omega)$, to obtain the filtered data in the time domain. To eliminate ringing effects, discard $T/2$ data points from either end of the filtered time series, where T is the span of the IRF for the transform filter.

6.10.1 Truncation Effects

For all commonly used digital filters, a percentage of the end values from the filtered record must be omitted prior to further analysis. This loss of information from the ends of the output is linked to ringing effects associated with discontinuities at the ends of the input and to the nonexistence of integrable data prior to the start of the record. The ringing decays toward the interior of the data sequence after the end effects have been smoothed by a sufficient number of filter integrations (Figure 6.22). In the case of the squared Butterworth filter, both ends of the data are affected twice since the data are passed forward and backward through the filter. One approach is to assume that 10% of output data at each end of the filter output is contaminated and remove these points from the final output. However, each case is different and data elimination should be based on a trial and error approach using visual inspection to estimate the extent of the data removal. Padding the ends of the input with zeroes appears to serve no useful purpose. In some cases the ringing effect can be substantially reduced by using the zero cross-over points (for input centered about the mean record value) as the first record of the input.

References

Abdel-Hafez, M.F., Kim, D.J., Lee, E., Chun, S., Lee, Y.J., Kang, T., Sung, S., 2008. Performance improvement of the wald test for GPS RTK with the assistance of INS. Int. J. Cont. Automa. Syst. 6, 534–543.

Agapitos, K., Gajewski, K., 2012. Analysis of the seasonal cycle of the climate record of Ottawa, Ontario. Can. Meteor. Oceanogr. Soc. Bull. 40, 120–126.

Allen, S.E., Thomson, R.E., 1993. Bottom-trapped subinertial motions over mid-ocean ridges in a stratified rotating fluid. J. Phys. Oceanogr. 23, 566–581.

Anderson, D.L., 1993. ^{3}He from the mantle; primordial signal or cosmic dust? Science 261, 170–176.

Apel, J.R., Holbrook, J.R., Tsai, J., Liu, A.K., 1985. The Sulu Sea internal soliton experiment. J. Phys. Oceanogr. 15, 1625–1651.

Arnault, S., Menard, Y., Merle, J., 1990. Observing the tropical Atlantic Ocean in 1986–87 from altimetry. J. Geophys. Res. 95, 17,921–17,946.

Baker, E.T., Lavelle, J.W., 1984. The effect of particle size on the light attenuation coefficient of natural suspensions. J. Geophys. Res. 89, 8197–8203.

Baker, E.T., Lupton, J.E., 1990. Changes in submarine hydrothermal ^{3}He/heat ratios as an indicator of magmatic/tectonic activity. Nature 346, 556–558.

Baker, E.T., Massoth, G.J., 1986. Hydrothermal plume measurements: a regional perspective. Science 234, 980–982.

Baker, E.T., Massoth, G.J., 1987. Characteristics of hydrothermal plumes from two vent fields on the Juan de Fuca Ridge, northeast Pacific Ocean. Earth Planet. Sci. Lett. 85, 59–73.

Baker, E.T., Lavelle, J.W., Feely, R.A., Massoth, G.J., Walker, S.L., Lupton, J.E., 1989. Episodic venting of hydrothermal fluids from the Juan de Fuca Ridge. J. Geophys. Res. 94, 9237–9250.

Baker, E.T., German, C.R., Elderfield, H., 1995. Hydrothermal plumes over spreading-center axes: global distributions and geological inferences. In: Humphris, S.E., Zierenberg, R.A., Mullineaux, L.S., Thomson, R.E. (Eds.), Seafloor Hydrothermal Systems: Physical, Chemical, Biological and Geological Interactions, Geophysical Monograph, 91. AGU, pp. 47–71.

Baker Jr, D.J., 1981. Ocean instruments and experiment design. In: Warren, B.A., Wunsch, C. (Eds.), Evolution of Physical Oceanography. MIT Press, Cambridge, Mass, pp. 396–433.

Bakun, A., 1973. Coastal Upwelling Indices, West Coast of North America, 1946–71. NOAA Technical Report NMFS SSRF-671. US Department of Commerce.

Bard, E., Arnold, M., Östlund, H.G., Maurice, P., Monfray, P., Duplessy, J.-C., 1988. Penetration of bomb radiocarbon in the tropical Indian Ocean measured by means of accelerator mass spectrometry. Earth Planet. Sci. Lett. 87, 379–389.

Barkley, R.A., 1968. Oceanographic Atlas of Pacific Ocean. University of Hawaii Press, Honolulu.

Barnett, T.P., Davis, R.E., 1975. Eigenvector analysis and prediction of sea surface temperature fluctuation in the northern Pacific Ocean. In: Proc. WMO/IAMAP Symposium on Long Term Climatic Fluctuations. Norwich, England, pp. 439–450.

Barnett, T.P., Patzert, W.C., Webb, S.C., Brown, B.R., 1979. Climatological usefulness of satellite determined sea-surface temperatures in the tropical Pacific. Bull. Am. Met. Soc. 60, 197–205.

Barnett, T.P., 1983. Recent changes in sea level and their possible causes. Clim. Change 5, 15–38.

Barrodale, I., Erickson, R.E., 1978. Algorithms for Least-squares Linear Prediction and Maximum Entropy Spectral Analysis. MS report, DM-142-IR. University of Victoria.

Barth, J.A., Menge, B.A., Lubchenco, J., Chan, F., Bane, J.M., Kirincich, A.R., McManus, M.A., Nielsen, K.J., Pierce, S.D., Washburn, L., 2008. Delayed upwelling alters nearshore coastal ocean ecosystems in the northern California Current. Proc. Natl. Acad. Sci. U.S.A. 104, 3719–3724.

Bartz, R., Zaneveld, J., Pak, H., 1978. A transmissometer for profiling and moored observations in water. Soc. Photo-Opt. Instrum. Eng. J. 160 (V), 102–108.

Batchelor, G.K., 1967. An Introduction to Fluid Dynamics. Cambridge University Press, Cambridge.

Bates, J.J., Diaz, H.F., 1991. Evaluation of multichannel sea surface temperature product quality for climate monitoring. J. Geophys. Res. 96, 20,613–20,622.

Bates, J.J., Smith, W.L., 1985. Sea surface temperatures from geostationary satellites. J. Geophys. Res. 90, 11,609–11,618.

Bauer, S., Swenson, M., Griffa, A., 2002. Eddy-mean flow decomposition and eddy diffusivity estimates in the tropical Pacific Ocean. 2. Results. J. Geophys. Res. 107, 3154–3171.

Beardsley, R.C., Boicourt, W.C., 1981. On estuarine and continental-shelf circulation in the Middle Atlantic Bight. In: Warren, B.A., Wunsch, C. (Eds.), Evolution of Physical Oceanography. MIT Press, Cambridge, Mass, pp. 198–233.

Beardsley, R.C., Boicourt, W.C., Huff, L.C., McCullough, J.R., Scott, J., 1981. CMICE: a near-surface current meter inter-comparison experiment. Deep-Sea Res. 28A, 1577—1603.

Beardsley, R.C., 1987. A comparison of the vector-averaging current meter and new Edgerton, Germeshausen, and Gier, Inc., vector-measuring current meter on a surface mooring in Coastal Ocean Dynamics Experiment 1. J. Geophys. Res. 92, 1845—1859.

Belkin, I.M., O'Reilly, J.E., 2009. An algorithm for oceanic front detection in chlorophyll and sea surface tempera-ture satellite imagery. J. Mar. Syst. 78, 317—326.

Belkin, I.M., Cornillon, P.C., Sherman, K., 2009. Fronts in large marine ecosystems. Prog. Oceanogr. 81 (1—4), 223—236.

Belyshev, A.P., Klevantsov, YuP., Rozhkov, V.A., 1983. Probability Analysis of the Sea Currents (In Russian). Gidrometeoizdat, Leningrad, 264 pp.

Bendat, J.S., Piersol, A.G., 1986. Random Data: Analysis and Measurement Procedures. John Wiley, New York.

Bennett, A.F., 1976. Poleward heat fluxes in southern hemi-sphere oceans. J. Phys. Oceanogr. 4, 785—798.

Bennett, A.F., 1992. Inverse Methods in Physical Oceanog-raphy. Cambridge University Press, Cambridge.

Bernard, E.N., González, F.I., Meinig, C., Milburn, H.B., August 2001. Early detection and real-time reporting of deep-ocean tsunamis. In: Proceedings of the International Tsunami Symposium 2001 (ITS 2001), NTHMP Review Session, R-6, Seattle, WA, pp. 7—10 (on CD-ROM), 97—108.

Bernstein, R.L., Chelton, D.B., 1985. Large-scale sea surface temperature variability from satellite and shipboard measurements. J. Geophys. Res. 90, 11,619—11,630.

Bernstein, R.L., 1982. Sea surface temperature estimation using the NOAA-6 satellite advanced very high resolu-tion radiometer. J. Geophys. Res. 87, 9455—9466.

Berteaux, H.O., 1990. Program SSMOOR: Users' Instruc-tions. Cable Dynamics and Mooring Systems (CDMS), Woods Hole, Mass.

Berteaux, H.O., 1991. Coastal and Oceanic Buoy Engi-neering. H.O. Berceaux, Woods Hole, MA., USA.

Blackman, R.B., Tukey, J.W., 1958. The Measurement of Power Spectra. Dover, New York.

Bograd, S.J., Rabinovich, A.B., Thomson, R.E., Eert, A.J., 1999. On sampling strategies and interpolation schemes for satellite-tracked drifters. J. Atmos. Oceanic Technol. 16, 893—904.

Bograd, S.J., Schroeder, I., Sarkar, N., Qiu, X., Sydeman, W.J., Schwing, F.B., 2009. Phenology of coastal upwelling in the California Current. Geophys. Res. Lett. 36, L01602 http://dx.doi.org/10.1029/2008GL035933.

Bohling, G., 2005. Introduction to Geostatistics and Vario-gram Analysis. http://people.ku.edu/~gbohling/cpe940.

Boicourt, W.C., 1982. The recent history of ocean current meter measurement. Proc. IEEE Second Workshop Conf. Curr. Meas., 9—13.

Born, G.H., Richards, M.A., Rosborough, G.W., 1982. An empirical determination of the effects of sea-state bias on the SEASAT altimeter. J. Geophys. Res. 87, 3221—3226.

Born, G.H., Mitchell, J.L., Heyler, G.A., 1987. GEOSAT-ERM mission design. J. Astron. Sci. 35, 119—134.

Bowditch, N., 1977. American Practical Navigator. Defense Mapping Agency Hydrographic Center. DMA No. NVPUB9V1.

Box, G.E.P., Jenkins, G.M., 1970. Time Series Analysis: Fore-casting and Control. Holden-Day, San Francisco, Calif.

Boyd, P.W., Jickells, T., Law, C.S., Blain, S., Boyle, E.A., Buesseler, K.O., Coale, K.H., Cullen, J.J., et al., 2007. Mesoscale iron enrichment experiments 1993-2005: synthesis and future directions. Science 315 (5812), 612—617.

Brainerd, K.E., Gregg, M.C., 1995. Surface mixed and mixing layer depths. Deep Sea Res. 42, 1521—1543.

Bretherton, F.P., McWilliams, J.C., 1980. Estimations from irregular arrays. Rev. Geophys. Space Phys. 18, 789—812.

Bretherton, F.P., Davis, R.E., Fandry, C.B., 1976. A technique for objective analysis and design of oceanographic experiments applied to MODE-73. Deep-Sea Res. 23, 559—582.

Brink, K.H., Chapman, D.C., 1987. Programs for Computing Properties of Coastal-trapped Waves and Wind-driven Motions over the Continental Shelf and Slope, second ed. Woods Hole Oceanographic Institution. Technical Report No. WHOI-87—24.

Briscoe, M.G., 1975. Internal waves in the ocean. Rev. Geo-phys. Space Res. 13, 591—598.

Broecker, W.S., Peng, T.-H., 1982. Tracers in the Sea. Lamont-Doherty Geological Observatory. Eldigio Press, New York.

Broecker, W.S., Peng, T.H., Östlund, G., Stuiver, M., 1985. The distribution of bomb radiocarbon in the ocean. J. Geophys. Res. 90, 6953—6970.

Broecker, W.S., Peng, T.H., Ostlund, G., 1986. The distribu-tion of bomb tritium in the ocean. J. Geophys. Res. 91, 14,331—14,344.

Broecker, W.S., Virgilio, A., Peng, T.-H., 1991. Radiocarbon age of waters in the deep Atlantic revisited. Geophys. Res. Lett. 18, 1—3.

Brooks, C.F., 1926. Observing water-surface temperatures at sea. Mon. Wea. Rev. 54, 241—254.

Brower, R.L., Gohrband, G.S., Pichel, W.G., Signore, T.L., Walton, C.C., 1976. Satellite Derived Sea-surface Temperatures from NOAA Spacecraft. NOAA Technical Memo, Washington, DC. NESS No. 79.

Brown, N.L., Morrison, G.K., 1978. WHOI/Brown Conduc-tivity, Temperature and Depth Profiler. WHOI Report No. 78—23.

Brown, O.B., Brown, J.W., Evans, R.H., 1985. Calibration of advanced very high resolution radiometer infrared observations. J. Geophys. Res. 90, 11,667–11,678.

Brown, N.L., 1974. A precision CTD microprofiler. IEEE Publication 74 CHO873–0 OEC. In: Ocean 74 Record, 1974 IEEE Conference on Engineering in the Ocean Environment, vol. 2. Institute of Electrical and Electronics Engineers, New York, pp. 270–278.

Brown, R.A., 1983. On a satellite scatterometer as an anemometer. J. Geophys. Res. 88, 1663–1673.

Bruland, K.W., Donat, J.R., Hutchins, D.A., 1991. Interactive influences of bioactive trace metals on biological production in oceanic waters. Limnol. Oceanogr. 36, 1555–1577.

Bryden, H., 1979. Poleward heat flux and conversion of available potential energy in Drake Passage. J. Mar. Res. 37, 1–22.

Bullinaria, J.A., 2004. Introduction to Neural Networks. http://www.cs.bham.ac.uk/~jxb/inn.html.

Bullister, J.L., Weiss, R.F., 1988. Determination of $CC1_3F$ and $CC1_2F_2$ in seawater and air. Deep-Sea Res. 35, 839–853.

Bullister, J.L., 1989. Chlorofluorocarbons as time-dependent tracers in the ocean. Oceanography 2 (2), 12–17.

Burd, B.J., Thomson, R.E., 1993. Flow volume calculations based on three-dimensional current and net orientation data. Deep-Sea Res. 40, 1141–1153.

Burd, B.J., Thomson, R.E., 1994. Hydrothermal venting at Endeavour Ridge: effect on zooplankton biomass throughout the water column. Deep-Sea Res. 41, 1407–1423.

Burd, B.J., Thomson, R.E., 2012. Estimating zooplankton biomass distribution in the water column near Endeavour Ridge using acoustic backscatter and concurrently towed nets. Oceanography 25 (1), 269–276. http://dx.doi.org/10.5670/oceanog.2012.25.

Burg, J.P., 1967. In: Maximum Entropy Spectral Analysis, Paper Presented at the 37th Annual International Meeting, Soc. of Explor. Geophys., Oklahoma City, Okla., Oct. 31, 1967.

Burg, J.P., August 12–23, 1968. A new analysis technique for time series data. In: NATO Advanced Study Institute on Signal Processing Emphasis on Underwater Acoustics. Enschede, The Netherlands.

Burg, J.P., 1972. The relationship between maximum entropy spectra and maximum likelihood spectra. Geophysics 37, 375–376.

Busalacchi, A.J., O'Brien, J.J., 1981. Interannual variability of the equatorial Pacific in the 1960's. J. Geophys. Res. 86, 10,901–10,907.

Cahill, M.L., Middleton, J.H., Stanton, B.R., 1991. Coastal-trapped waves on the west coast of the South Island, New Zealand. J. Phys. Oceanogr. 21, 541–557.

Caiman, J., 1978. On the interpretation of ocean current spectra. J. Phys. Oceanogr. 8, 627–652.

Calder, M., 1975. Calibration of echo sounders for offshore sounding using temperature and depth. Int. Hydrographic Rev. LII (2), 13–17.

Cannon, G.A., Thomson, R.E., 1996. Characteristics of 4-day oscillations trapped by the Juan de Fuca Ridge. Geophys. Res. Lett. 23, 1613–1616.

Capon, J., 1969. High resolution frequency-wavenumber spectral analysis. Proc. IEEE 57, 1408–1418.

Cartwright, D.E., Edden, A.C., 1973. Corrected tables of tidal harmonics. Geophys. J. R. Astron. Soc. 33, 253–264.

Cartwright, D.E., 1968. A unified analysis of tides and surges round north and east Britain. Phil. Trans. Roy. Soc. London A263, 1–55.

Chapman, D.C., 1983. On the influence of stratification and continental shelf and slope topography on the dispersion of subinertial coastally trapped waves. J. Phys. Oceanogr. 13, 1641–1652.

Charnock, H., Lovelock, J.E., Liss, P.S., Whitfield, M., 1988. Tracers in the ocean. In: Proceedings of a Royal Society Meeting Held on 21 and 22 May, 1987. The Royal Society, London.

Chave, A.D., Luther, D.S., 1990. Low-frequency, motionally induced electromagnetic fields in the ocean. J. Geophys. Res. 95, 7185–7200.

Chave, A.D., Luther, D.S., Filloux, J.H., 1992. The barotropic electromagnetic and pressure experiment. 1. Barotropic current response to atmospheric forcing. J. Geophys. Res. 97, 9565–9593.

Chelton, D.B., Davis, R.E., 1982. Monthly mean sea level variability along the west coast of North America. J. Phys. Oceanogr. 6, 757–784.

Chelton, D.B., Schlax, M.G., 1996. Global observations of oceanic Rossby waves. Science 272, 234–238.

Chelton, D.B., Bernal, P.A., McGowan, J.A., 1982. Large-scale interannual physical and biological interaction in the California Current. J. Mar. Res. 40, 1095–1125.

Chelton, D.B., 1982. Statistical reliability and the seasonal cycle: comments on "Bottom pressure measurements across the Antarctic Circumpolar Current and their relation to the wind". Deep-Sea Res. 29, 1381–1388.

Chelton, D.B., 1983. Effects of sampling errors in statistical estimation. Deep-Sea Res. 30, 1083–1101.

Chelton, D.B., 1988. WOCE/NASA Altimeter Algorithm Workshop. US WOCE Technical Report, No. 2.

Chen, J.L., Wilson, C.R., Tapley, B.D., 2006. Satellite gravity measurements confirm accelerated melting of Greenland ice sheet. Science 313 (5795), 1958–1960.

Cheney, R.E., Marsh, J.G., Becley, B.D., 1983. Global meso-scale variability from collinear tracks of SEASAT altimeter data. J. Geophys. Res. 88, 4343–4354.

Cheng, W.C., Shen, C.-W., Chen, C.-Y., Cheng, M.-J., 2009. Stability analysis of an oceanic structure using the Lyapunov method. Eng. Comput. 27 (2), 186−204.

Cherniawsky, J.Y., Crawford, W.R., 1996. Comparison between weather buoy and comprehensive ocean-atmosphere data set wind data for the west coast of Canada. J. Geophys. Res. 101, 18377−18389.

Cherniawsky, J.Y., Foreman, M.G.G., Crawford, W.R., Henry, R.F., 2001. Ocean tides from TOPEX/Poseidon sea level data. J. Atmos. Oceanic Technol. 18, 649−664. http://dx.doi.org/10.1175/1520-0426(2001)018<0649: OTFTPS>2.0.CO;2.

Cherniawsky, J.Y., Crawford, W.R., Nikitin, O., Carmack, E.C., 2005. Bering Strait transports from satellite altimetry. J. Mar. Res. 63, 887−900.

Chisholm, S.W., Morel, F.M.M. (Eds.), 1991. What Controls Phytoplankton Production in Nutrient-rich Areas of the Open Sea?. Limnol. Oceanogr. 36, 1507−1970.

Chiswell, S.M., Wimbush, M., Luckas, R., 1988. Comparison of dynamic height measurements from an inverted echo sounder and an island tide gauge in the central Pacific. J. Geophys. Res. 93, 2277−2283.

Chiswell, S., October 1992. Inverted Echo Sounders at the WOCE Deep-water Station. WOCE Notes 4(4), 1−6. US Office, Department of Oceanography. Texas A&M University, College Station.

Christensen, E.J., Haines, B.J., Keihm, S.J., Morris, C.S., Norman, R.A., Purcell, G.H., Williams, B.G., Wilson, B.D., Born, G.H., Parke, M.E., Gill, S.K., Shum, C.K., Tapley, B.D., Kolenkiewicz, R., Nerem, R.S., 1994. Calibration of TOPEX/POSEIDON at platform Harvest. J. Geophys. Res. 99, 24,465−24,487.

Chu, P.C., 1994. Localized TOGA sea level spectra obtained from the S-transformation. TOGA Notes 17, 5−8.

Church, J.A., Freeland, H.J., Smith, R.L., 1986. Coastal-trapped waves on the east Australian continental shelf. Part I: propagation of modes. J. Phys. Oceanogr. 16, 1929−1943.

Clayson, C.A., Emery, W.J., Savage, R., 1993. Wind speed comparisons between GEOSAT, SSM/I, buoy and ECMWF wind speeds. J. Geophys. Res. 19, 558−563.

Coale, K.H., Johnson, K.S., Fitzwater, S.E., Blain, S.P.G., Stanton, T.P., Coley, T.L., 1998. IronEx-I, an in situ iron-enrichment experiment: experimental design, implementation and results. In: Coale, K.H. (Ed.), The Galapagos Iron Experiments: A tribute to John Martin. Deep Sea Research Part II: Topical Studies in Oceanography, 45(6), pp. 919−945.

Collins, C.A., Giovando, L.F., Abbott-Smith, K.A., 1975. Comparison of Canadian and Japanese merchant ship observations of sea surface temperature in the vicinity of the present ocean station P, 1927−53. J. Fish. Res. Bd. Can. 32, 253−258.

Connolly, S.R., D., M., Bellwood, D.R., Hughes, T.P., 2009. Testing species abundance models: a new bootstrap approach applied to Indo-Pacific coral reefs. Ecology 90 (11), 3138−3149.

Connolly, T.P., Hickey, B.M., Shulman, T., Thomson, R.E., 2014. Coastally trapped waves, alongshore pressure gradients, and the California Undercurrent. J. Phys. Oceanogr. 44, 319−342. http://dx.doi.org/10.1175/JPO-D-13-095.1.

Conway, K.W., Barrie, J.V., Thomson, R.E., 2012. Submarine slope failure and tsunami hazard in coastal British Columbia: Douglas Channel and Kitimat Arm. Geol. Surv. Can. Curr. Res. 2012−10, 13. http://dx.doi.org/10.4095/291732.

Cooley, W.W., Lohnes, P.R., 1971. Multivariate Data Analysis. J. Wiley, New York, 364 pp.

Cooley, J.W., Tukey, J.W., 1965. An algorithm for machine calculation of complex fourier series. Math. Comput. 19, 297−301.

Cox, R.A., Culkin, F., Riley, J.P., 1967. The electrical conductivity/chlorinity relationship in natural sea water. Deep-Sea Res. 14, 203−220.

Cox, R.A., 1963. The salinity problem. Prog. Oceanogr. 1, 243−261.

Craig, H., 1994. Retention of helium in subducted interplanetary dust particles. Science 265, 1892−1893.

Crawford, W.R., Thomson, R.E., 1982. Diurnal-period continental shelf waves along Vancouver Island: a comparison of observations with theoretical models. J. Phys. Oceanogr. 14, 1629−1646.

Crawford, W.R., Thomson, R.E., 1984. Diurnal-period continental shelf waves along Vancouver Island: a comparison of observations with theoretical models. J. Phys. Oceanogr. 14, 1629−1646.

Crawford, A.B., 1969. Sea Surface Temperatures, Some Instruments, Methods and Comparison. Tech. Note No. 103. WMO, 117−129.

Cresswell, G.R., 1976. A drifting buoy tracked by satellite in the Tasman Sea. Aust. J. Mar. Freshwat. Res. 27, 251−262.

Cummins, P.F., Masson, D., Foreman, M.G.G., 2000. Stratification and mean flow effects on diurnal tidal currents off Vancouver Island. J. Phys. Oceanogr. 30, 15−30. http://dx.doi.org/10.1175/1520-0485(2000)030<0015: SAMFEO>2.0.CO;2.

Cummins, P.F., Freeland, H.J., Thomson, R.E., 2011. Transport of Japan Tsunami Marine Debris to the Coast of British Columbia: An Updated Review. Science Response 2012-006. Canadian Science Advisory Secretariat (CSAS). http://www.dfo-mpo.gc.ca/csas-sccs/Publications/ScR-RS/2012/2012_006-eng.pdf.

Curry, P., Roy, C., 1989. Optimal environmental window and pelagic fish recruitment success in upwelling areas. Can. J. Aquat. Sci. 46, 670−680.

Dahlen, J.M., Chhabra, N.K., 1983. slippage errors and dynamic response of four drogued buoys measured at sea. Report P-1729, C.C. Draper Laboratory. In: Presented at Marine Technology Society/NOAA Data Buoy Center, 1983 Symposium on Buoy Technology, New Orleans, April 1983.

Dallimore, A., Thomson, R.E., Betrum, M.A., 2005. Late Holocene laminated sediments of Effingham Inlet, Vancouver Island, British Columbia: implications for regional geologic, paleoseismic, oceanographic and climatic history. Marine Geol. 219 (1), 47−69.

Danielson, G.C., Lanczos, C., 1942. Some improvements in practical fourier analysis and their application to X-ray scattering from liquids. J. Franklin Inst. 233, 365−380 and 435−452.

Dantzler, H.L., 1974. Dynamic salinity calibrations of continuous salinity/temperature/depth data. Deep-Sea Res. 21, 675−682.

Daskalov, G.M., Grishin, A.N., Rodionov, S., Mihneva, V., 2007. Trophic cascades triggered by overfishing reveal possible mechanisms of ecosystem regime shifts. Proc. Natl. Acad. Sci. 104 (25), 10518−10523.

Datawell bv, 1992. Directional Waverider. Datawell bv, Haarlem, The Netherlands.

Davis, R.E., Regier, L.A., 1977. Methods for estimating directional wave spectra from multielement arrays. J. Mar. Res. 35, 453−477.

Davis, R.E., Webb, D.C., Regier, L.A., Dufour, J., 1992. The Autonomous Lagrangian Circulation Explorer (ALACE). J. Atmos. Oceanic Technol. 9, 264−285.

Davis, R.E., 1976. Predictability of sea surface temperature and sea level pressure anomalies over the North Pacific Ocean. J. Phys. Oceanogr. 6, 249−266.

Davis, R.E., 1977. Techniques for statistical analysis and prediction of geophysical fluid systems. Geophys. Astrophys. Fluid Dyn. 8, 245−277.

Davis, R.E., 1978. Predictability of sea level pressure anomalies over the North Pacific Ocean. J. Phys. Oceanogr. 8, 233−246.

Davis, R.E., 1991. Observing the general-circulation with floats. Deep Sea Res. 38, 531−571.

Defant, A., 1936. Schichtung und Zirkulation des Atlantischen Ozeans. Die Troposphare. In: Wissenschaftliche Ergebnisse der Deutschen Altantishcen Expedition auf dem Forschungs- und Vermessungsschiff "Meteor" 1925−1927.

Defant, A., 1937. Stratification and circulation of the Atlantic Ocean. English Translation. In: Emery, W.J. (Ed.), The Troposphere, Scientific Results of the German Atlantic Expedition of the Research Vessel, Meteor, 1925−27. Amerind, New Delhi, 1981.

Defant, A., 1961. Physical Oceanography, VI. Pergamon Press, New York.

Denbo, D.W., Allen, J.S., 1984. Rotary empirical orthogonal function analysis of currents near the Oregon coast. J. Phys. Oceanogr. 14, 35−46.

Denbo, D.W., Allen, J.S., 1986. Reply to "Comments on: rotary empirical orthogonal function analysis of currents near the Oregon Coast". J. Phys. Oceanogr. 16, 793−794.

Denman, K.L., Freeland, H.J., 1985. Correlation scales, objective mapping and a statistical test of geostrophy over the continental shelf. J. Mar. Res. 43, 517−539.

Di Iorio, D., Lavelle, J.W., Rona, P., Bemis, K., Xu, G., Germanovich, L., Lowell, R., Genc, G., 2012. Measurements and models of meat flux and plumes from hydrothermal discharges near the deep sea floor. Oceanography 25 (1), 108−119. http://dx.doi.org/10.5670/oceanog.2012.14.

Di Lorenzo, E., Foreman, M.G.G., Crawford, W.R., 2005. Modelling the generation of Haida Eddies. Deep-Sea Res. Pt. II-Top. Stud. Oceanogr. 52 (7−8), 853−873. http://dx.doi.org/10.1016/j.dsr2.2005.02.007.

Diaconis, P., Efron, B., 1983. Computer-intensive methods in statistics. Sci. Am. 248, 116−130.

Dickson, A.G., Sabine, C.L., Christian, J.R. (Eds.), 2007. Guide to Best Practices for Ocean CO_2 Measurements. PICES Special Publication 3, p. 191. http://cdiac.ornl.gov/ftp/oceans/Handbook_2007/Guide_all_in_one.pdf.

Dillon, T.D., 1982. Vertical overturns: a comparison of thorpe and ozmidov length scales. J. Geophys. Res. 87, 9601−9613.

Dodimead, A.J., Favorite, F., Hirano, T., 1963. Salmon of the North Pacific. II: review of oceanography of the subarctic Pacific region. Int. North Pac. Fish. Commun. Bull. 13, 195.

Dodson, S.I., 1990. Predicting diel vertical migration of zooplankton. Limnol. Oceanogr. 35, 1195−1200.

Doney, S.C., Fabry, V.J., Feely, R.A., Kleypas, J.A., 2009. Ocean acidification: the other CO_2 problem. Annu. Rev. Mar. Sci. 1, 169−192. http://dx.doi.org/10.1146/annurev.marine.010908.163834.

Donlon, C.J., Minett, P.J., Gentemann, C., Nightingale, T.J., Barton, I.J., Ward, B., Murray, M.J., 2002. Toward improved validation of satellite sea surface skin temperature measurements for climate research. J. Clim. 15, 353−369.

Doodson, A.T., Warburg, H.D., 1941. Admiralty Manual of Tides. Hydrographic Department, Admiralty, London.

Drakopoulos, P.G., Marsden, R.F., 1993. The internal tide off the west coast of Vancouver Island. J. Phys. Oceanogr. 23, 758−775. http://dx.doi.org/10.1175/1520-0485(1993)023<0758:TITOTW>2.0.CO;2.

Druffel, E.R.M., Williams, P.M., 1991. Radiocarbon in seawater and organisms from the Pacific coast of Baja California. Radiocarbon 33, 291−296.

Druffel, E.R.M., 1989. Decade time scale variability of ventilation in the North Atlantic: high-resolution measurements of bomb radiocarbon in banded corals. J. Geophys. Res. 94, 3271−3285.

Dziewonski, A., Bloch, S., Landisman, M., 1969. A technique for the analysis of transient seismic signals. Bull. Seismolo. Soc. Am. 59, 421–444.

Edmond, J.M., Measures, C.M., McDuff, R.E., Chan, L.H., Collier, R., Grant, B., Gordon, L.I., Corliss, J.B., 1979. Ridge crest hydrothermal activity and the balance of the major and minor elements in the ocean: the Galapagos data. Earth Planet. Sci. Lett. 46, 1–18.

Efron, B., Gong, G., 1983. A leisurely look at the bootstrap, the jackknife, and cross-validation. Am. Stat. 37, 36–48.

Efron, B., 1977. Discussion on maximum likelihood from incomplete data via the EM algorithm (by A. Dempster, N. Laird, and D. Rubin). J. Royal Stat. Soc., 1–38.

Egbert, G.D., Bennett, A.F., Foreman, M.G.G., 1994. TOPEX/POSEIDON tides estimated using a global inverse model. J. Geophys. Res. 99, 24,821–24,852.

Eisner, J.B., Tsonis, A.A., 1991a. Do bidecadal oscillations exist in the global temperature record. Nature 353, 551–553.

Eisner, J.B., Tsonis, A.A., 1991b. Comparison of observed northern hemisphere surface air temperature records. Geophys. Res. Lett. 18, 1229–1232.

Ekman, V.W., 1905. On the influence of the earth's rotation on ocean currents. Arkiv für mathematik astronomi, cch fysik. 2.

Ekman, V.W., 1932. On an improved type of current meter. J. Cons. Int. Explor. Mer. 7, 3–10.

Elgar, S., 1988. Comment on "Fourier transform filtering: a cautionary note" by A.M. Forbes. J. Geophys. Res. 93, 15,755–15,756.

Emery, W.J., Meincke, J., 1986. Global water masses: summary and review. Oceanoglogica Acta 9, 383–391.

Emery, W.J., Thomson, R.E., 2001. Data Analysis Methods in Physical Oceanography, Second and revised ed. Elsevier, New York. 638 pp.

Emery, W.J., Royer, T.C., Reynolds, R.W., 1985. The anomalous tracks of North Pacific drifting buoys 1981–83. Deep-Sea Res. 32, 315–347.

Emery, W.J., Thomas, A.C., Collins, M.J., Crawford, W.R., Mackas, D.L., 1986. An objective procedure to compute surface advective velocities from sequential infrared satellite images. J. Geophys. Res. 91, 12,865–12,879.

Emery, W.J., Born, G.H., Baldwin, D.G., Norris, C.L., 1989a. Satellite derived water vapor corrections for GEOSAT altimetry. J. Geophys. Res. 95, 2953–2964.

Emery, W.J., Brown, J., Novak, V.P., 1989b. AVHRR image navigation; summary and review. Photog. Eng. Rem. Sens. 8, 1175–1183.

Emery, W.J., Fowler, C.W., Clayson, C.A., 1992. Satellite image derived Gulf Stream currents. J. Oceanic Atmos. Sci. Tech. 9, 285–304.

Emery, W.J., Yu, Y., Wick, G., Schluessel, P., Reynolds, R.W., 1994a. Improving satellite infrared sea surface temperature estimates by including independent water vapor observations. J. Geophys. Res. 99, 5219–5236.

Emery, W.J., Fowler, C.W., Maslanik, J., 1994b. Arctic sea ice concentrations from special sensor microwave imager and advanced very high resolution radiometer satellite data. J. Geophys. Res. 99, 18,329–18,342.

Emery, W.J., 1983. On the geographical variability of the upper level mean and eddy fields in the North Atlantic and North Pacific. J. Phys. Oceanogr. 12, 269–291.

Enfield, D.B., Cid, L., 1990. Statistical analysis of El Niño/Southern Oscillation over the last 500 years. TOGA Notes 1, 1–4.

Esfahani, S.N., 2014. Temporal and spatial evolution of the mixed layer in the southern Beaufort Sea and the Amundsen Gulf, Ph.D. Thesis. Université du Québec, Institut National de la Recherche Scientique, Centre Eau Terre Environnement.

Espinosa, A.L., 2012. Determination of acidification state of the Canadian Pacific coastal waters using empirical relationships with hydrographic data (M.Sc. thesis). School of Earth and Ocean Sciences, University of Victoria, Canada.

Essex, C., Lookman, T., Nerenberg, M.A.H., 1987. The climate attractor over short timescales. Nature 326, 64–66.

Evans, J.C., 1985. Selection of a numerical filtering method: convolution or transform windowing? J. Geophys. Res. 90, 4991–4994.

Fabry, V.J., Seibel, B.A., Feely, R.A., Orr, J.C., 2008. Impacts of ocean acidification on marine fauna and ecosystem processes. ICES J. Marine Sci. 65, 414–432. http://dx.doi.org/10.1093/icesjms/fsn048.

Farge, M., 1992. Wavelet transforms and their applications to turbulence. Ann. Rev. Fluid Mech. 24, 395–457.

Farmer, D.M., Armi, L., 1999. The generation and trapping of solitary waves over topography. Science 283, 188–190. http://dx.doi.org/10.1126/science.283.5399.188.

Farmer, D., Smith, J.D., 1980. Nonlinear internal waves in a fjord. In: Freeland, H.J., Farmer, D.M., Levings, C.D. (Eds.), Fjord Oceanography. Plenum, New York, pp. 465–493.

Farmer, D.M., Clifford, S.F., Verral, J.A., 1987. Scintillation structure of a turbulent tidal flow. J. Geophys. Res. 92, 5369–5382.

Favorite, F., Dodimead, A.J., Nasu, K., 1976. Oceanography of the subarctic Pacific region. Int. North Pac. Fish. Comm. Bull. 33, 187.

Feely, R.A., Gendron, J.F., Baker, E.T., Lebon, G.T., 1994. Hydrothermal plumes along the East Pacific Rise, 8°40′ to 11°50′N: particle distribution and composition. Earth Planet. Sci. Lett. 128, 19–36.

Feely, R.A., Sabine, C.L., Lee, K., Berelson, W., Kleypas, J., Fabry, V.J., Millero, F.J., 2004. Impact of anthropogenic CO_2 on the $CaCO_3$ system in the oceans. Science 305, 362—366.

Feely, R.A., Sabine, C.L., Byrne, R.H., Millero, F.J., Dickson, A.G., Wanninkhof, R., Murata, A., Miller, L.A., Greeley, D., 2012. Decadal changes in the aragonite and calcite saturation state of the Pacific Ocean. Glob. Biogeochem. Cycles 26, 15. http://dx.doi.org/10.1029/2011GB004157.

Fine, I.V., Rabinovich, A.B., Bornhold, B.D., Thomson, R.E., Kulikov, E.A., 2005. The Grand Banks landslide-generated tsunami of November 18, 1929: preliminary analysis and numerical modeling. Marine Geol. 215, 45—57.

Fine, R.A., 1985. Direct evidence using tritium data for throughflow from the Pacific into the Indian ocean. Nature 315, 478—480.

Fissel, D.B., Tang, C.L., 1991. Response of sea ice drift to wind forcing on the northeastern Newfoundland shelf. J. Geophys. Res. 96 (C10), 18,397—18,409.

Flagg, C.N., Smith, S.L., 1989. On the use of the acoustic Doppler current profiler to measure zooplankton abundance. Deep-sea Res. 36, 455—479.

Flierl, G., Robinson, A.R., 1977. XBT measurements of thermal gradient in the MODE eddy. J. Phys. Oceanogr. 7, 300—302.

Fofonoff, N.P., Froese, C., 1960. Programs for Oceanographic Computations and Data Processing on the Electronic Digital Computer ALWAC III-e. M-l Miscellaneous Programs. In: Fish. Res. Board Canada, Manuscript Report Oceanogr. And Limnol, 72.

Fofonoff, N.P., Tabata, S., 1966. Variability of oceanographic conditions between ocean station P and Swiftsure Bank off the Pacific coast of Canada. J. Fish. Res. Bd. Can. 23, 825—868.

Fofonoff, N.P., Hayes, S.P., Millard, R.C., 1974. WHOI/Brown CTD Microprofiler: Methods of Calibration and Data Handling. WHOI, 74—89.

Fofonoff, N.P., 1960. Transport Computations for the North Pacific Ocean 1955—1958. Fish. Res. Board Canada. Manuscript Report Oceanogr. and Limnol., No. 77—80.

Fofonoff, N.P., 1969. Spectral characteristics of internal waves in the ocean. Frederick C. Fuglister sixtieth anniversary volume. Deep-Sea Res. 76 (suppl.), 58—71.

Forbes, A.M.G., 1988. Fourier transform filtering: a cautionary note. J. Geophys. Res. 93, 6958—6962.

Forch, C., Knudsen, M., Sorensen, S.P.L., 1902. Berichte ueber die Konstantenbestimmungen zur Aufstellunt der hydrographischen Tabellen. Kgl. Danske Vedenskab. Selskab Skrifter, 7 Taekke, Naaturvidensk, og Mex.

Foreman, M.G.G., Czajko, P., Stucchi, D.J., Guo, M., 2009. A finite volume model simulation for the Broughton Archipelago, Canada. Ocean Model. 30, 29—47. http://dx.doi.org/10.1016/j.ocemod.2009.05.009.

Foreman, M.G.G., 1977. Manual for tidal height analysis and prediction. Pac. Mar. Sci. Rep. 77 (10) (Institute of Ocean Sciences, Sidney, BC).

Foreman, M.G.G., 1978. Manual for tidal currents analysis and prediction. Pac. Mar. Sci. Rep. 78 (6) (Institute of Ocean Sciences, Sidney, BC).

Fornari, D.J., Beaulieu, S.E., Holden, J.F., Mullineaux, L.S., Tolstoy, M. (Eds.), 2012. Introduction to the Special Issue: In RIDGE to Ridge 2000. Oceanography, 25 (1), pp. 12—17. http://dx.doi.org/10.5670/oceanog.2012.01.

Fraedrich, K., Leslie, L.M., 1989. Estimates of cyclone track predictability. I: tropical cyclones in the Australian region. Q.J. R. Meteorol. Soc. 115, 79—92.

Fraedrich, K., Grotjahn, R., Leslie, L.M., 1990. Estimates of cyclone track predictability. II: fractal analysis of mid-latitude cyclones. Q.J. R. Meteorol. Soc. 116, 317—335.

Fraedrich, K., McBride, J.L., Frank, W.M., Wang, R., 1997. Extended EOF-analysis of tropical disturbances: TOGA COARE. J. Sci. 54, 2363—2372.

Frank, W.M., 1979. Individual time period analyses over the GATE ship array. Mon. Wea. Rev. 107, 1600—1616.

Fratantoni, D.M., 2001. North Atlantic surface circulation during the 1990's observed with satellite-tracked drifters. J. Geophys. Res. 106, 22067—22093.

Freeland, H.J., Gould, W.J., 1976. Objective analysis of mesoscale ocean circulation features. Deep-sea Res. 23, 915—923.

Freeland, H.J., Rhines, P.B., Rossby, T., 1975. Statistical observations of the trajectories of neutrally buoyant floats in the North Atlantic. J. Marine Res. 33, 383—404.

Freeland, H.J., Church, J.A., Smith, R.L., Boland, F.M., 1985. Current Meter Data from the Australian Coastal Experiment: A Data Report. CSIRO Marine Laboratories, Hobart, Australia. Report 169.

Freeland, H.J., Boland, F.M., Church, J.A., Clarke, A.J., Forbes, A.M.G., Huyer, A., Smith, R.L., Thompson, R.O.R.Y., White, N.J., 1986. The Australian Coastal Experiment: a search for coastal-trapped waves. J. Phys. Oceanogr. 16, 1230—1249.

Freeland, H., Denman, K., Wong, C.S., Whitney, F., Jaques, R., 1997. Evidence of change in the mixed layer in the northeast Pacific Ocean. Deep Sea Res. 44, 2117—2129.

Freeland, H.J., 2013. Evidence of change in the winter mixed layer in the northeast Pacific Ocean: a problem revisited. Atmos. Ocean 50, 126—133.

Friehe, C.A., Pazan, S.E., 1978. Performance of an air-sea interaction buoy. J. Appl. Meteor. 17, 1488—1497.

Fu, L.-L., 1981. Observations and models of inertial waves in the deep ocean. Rev. Geophys. 19, 141—170.

Fuglister, F.C., 1960. Atlantic Ocean atlas of temperature and salinity profiles and data from the International Geophysical Year of 1957—1958. Woods Hole Oceanogr. Inst. Atlas Ser. 1.

Fukuoka, A., 1951. A study of 10-day forecast (A synthetic report). Geophys. Mag. 22, 177—208.

Furrer, R., Bengtsson, T., 2007. Estimation of high-dimensional prior and posterior covariance matrices in kalman filter variants. J. Multivar. Anal. 98, 227—255.

Gamage, N., Blumen, W., 1993. Comparative analysis of low-level cold fronts: wavelet, fourier, and empirical orthogonal function decompositions. Month. Weather Rev. 121, 2867—2878.

Gammon, R.H., Cline, J., Wisegarver, D., 1982. Chlorofluoromethanes in the northeast Pacific Ocean: measured vertical distributions and application as transient tracers of upper ocean mixing. J. Geophys. Res. 87, 9441—9454.

Ganachaud, A., Wunsch, C., 2000. Improved estimates of global ocean circulation, heat transport and mixing from hydrographic data. Nature 408, 453—456.

Gandin, L.S., 1965. Objective Analysis of Meteorological Fields. Israel Program for Scientific Translations, Jerusalem.

Garcia-Berdeal, I., Hautala, S.L., Thomas, L.N., Johnson, H.P., 2006. Vertical structure of time dependent currents in a mid-ocean ridge axial valley. Deep-sea Res. 53, 367—386.

Gargett, A.E., Östlund, G., Wong, C.S., 1986. Tritium time series from ocean station P. J. Phys. Oceanogr. 16, 1720—1726.

Gargett, A.E., 1994. Observing turbulence with a modified acoustic Doppler current profiler. J. Atmos. Oceanic Technol. 11, 1592—1610.

Garrett, C.J.R., Munk, W.H., 1971. The age of the tide and the "Q" of the oceans. Deep-Sea Res. 18, 493—503.

Garrett, C.J.R., Munk, W.H., 1979. Internal waves in the ocean. Ann. Rev. Fluid Mech. 11, 339—369.

Garrett, C.J.R., Petrie, B., 1981. Dynamical aspects of the flow through the Strait of Belle Isle. J. Phys. Oceanogr. 11, 376—393.

Gaspar, P., 1988. Modelling the seasonal cycle of the upper ocean. J. Phys. Oceanogr. 18, 161—180.

Gendron, J.F., Cowen, J.P., Feely, R.A., Baker, E.T., 1993. Age estimate for the 1987 megaplume on the southern Juan de Fuca Ridge using excess radon and manganese partitioning. Deep-Sea Res. 40, 1559—1567.

Gentemann, C.L., Minnett, P.J., 2008. Radiometric measurements of ocean surface thermal variability. J. Geophys. Res. 113, C08017. http://dx.doi.org/10.1029/2007JC004353.

Georgi, D.T., Dean, J.P., Chase, J.A., 1980. Temperature calibration of expendable bathythermographs. Ocean Eng. 7, 491—499.

German, C.R., Lin, Jian, Parson, L.M. (Eds.), 2004. Mid-Ocean Ridges: Hydrothermal Interactions Between the Lithosphere and Oceans. Geophysical Monograph Series, 148. American Geophysical Union.

Gilbert, D., 2012. A Milestone for Ocean Observation: International Argo Program Reaches One-Millionth Profile. Argo, Canada. http://www.dfo-mpo.gc.ca/science/publications/article/2012/12-13-12-eng.html.

Gill, A.E., 1982. Atmosphere — Ocean Dynamics. Academic Press, New York, 662 pp.

Godfrey, J.S., Golding, T.J., 1981. The Sverdrup relation in the Indian Ocean and the effect of Pacific-Indian Ocean throughflow on the Indian Ocean circulation and on the East Australia Current. J. Phys. Oceanogr. 11, 771—779.

Godin, G., 1972. The Analysis of Tides. University of Toronto Press.

Goldstein, R.M., Barnett, T.P., Zebker, H.A., 1989. Remote sensing of ocean currents. Science 246, 1282—1285.

Gonella, J., 1972. A rotary component method for analyzing meteorological and oceanographic vector time series. Deep-Sea Res. 19, 833—846.

González, F.I., Kulikov, Ye A., 1993. Tsunami dispersion observed in the deep ocean. In: Tsunamis in the World. Kluwer, pp. 7—16.

González, F.I., Milburn, H.M., Bernard, E.N., Newman, J., 19—22 January 1998. Deep-ocean assessment and reporting of tsunamis (DART): brief overview and status report. In: Proceedings of the International Workshop on Tsunami Disaster Mitigation, Tokyo, Japan, pp. 118—129.

González, F.I., Bernard, E.N., Meinig, C., Eble, M., Mofjeld, H.O., Stalin, S., 2005. The NTHMP tsunameter network. Nat. Hazards 35 (1), 25—39. http://dx.doi.org/10.1007/s11069-004-2402-4. Special Issue, U.S. National Tsunami Hazard Mitigation Program.

Gooberlet, M.A., Swift, C.T., Wilkerson, J.C., 1990. Ocean surface wind speed measurements of the special sensor microwave imager (SSM/I). IEEE Trans. Geosci. Remote Sens. 28, 823—828.

Gordon, A.L., 1986. Interocean exchange of thermocline water. J. Geophys. Res. 91, 5037—5046.

Gordon, R.L., 1996. Acoustic Doppler Current Profilers. Principles of Operation: A Practical Primer, second ed. RD Instruments, San Diego, Calif.

Gould, W.J., Sambuco, E., 1975. The effect of mooring type on measured values of ocean currents. Deep-Sea Res. 22, 55—62.

Gould, W.J., 1973. Effects of non-linearities of current meter compasses. Deep-Sea Res. 20, 423—427.

Goupillaud, P., Grossmann, A., Morlet, J., 1984. Cycle-octave and related transforms in seismic signal analysis. Geoexploration 23, 85—105.

Grassberger, P., Procaccia, I., 1983. Measuring the strangeness of strange attractors. Physica 9D, 189—208.

Grasshoff, K., 1983. Methods of Seawater Analysis. Verlag Chemie, Weinheim.

Grassl, H., 1976. The dependence of the measured cool skin of the ocean on wind stress and total heat flux. Bound. Layer Meteorol. 10, 465—474.

Gray, H.L., Woodward, W.A., 1992. Autoregressive models not sensitive to initial conditions. EOS Trans. AGU 73 (25), 267–268.

Green, A., 1984. Bulk dynamics of the expendable bathythermograph (XBT). Deep-Sea Res. 31, 415–426.

Green, D., 18–21 September 2006. Transitioning NOAA moored buoy systems from research to operations. In: Proceedings of OCEANS'06 MTS/IEEE Conference. Boston.

Greenan, B.J.W., Prinsenberg, S.J., 1998. Wind forcing of ice cover in the Labrador shelf marginal ice zone. Atmos. Ocean 36 (2), 71–93.

Grinsted, A., Moore, J.C., Jevrejeva, S., 2009. Reconstructing sea level from paleo and projected temperatures 200 to 2100 AD. Clim. Dyn., 10. http://dx.doi.org/10.1007/s00382-008-0507-2.

Groves, G.W., Hannan, E.J., 1968. Time series regression of sea level on weather. Rev. Geophys. 6, 129–174.

Groves, G.W., 1955. Numerical filters for discrimination against tidal periodicities. Trans. Am. Geophys. Union 36, 1073–1084.

Gruza, G.V., Ran'kova, E. Ya, Rocheva, E.V., 1988. Analysis of global data variations in surface air temperature during instrument observation period. Meteor. Gridr., 16–24.

Gutzler, D.S., Kiladis, G.N., Meehl, O.A., Weickmann, K.M., Wheeler, M., 1994. The global climate of December 1992–Frebruary 1993. Part II: large-scale variability across the tropical western Pacific during TOGA_COARE. J. Clim. 7, 1606–1622.

Guymer, T.H., Businger, J.A., Jones, W.L., Stewart, R.H., 1981. Anomalous wind estimates from SEASAT scatterometer. Nature 294, 735–737.

Haar, A., 1910. Zur Theorie der orthogonalen functionen Systeme. Math. Ann. 69, 331–371.

Haidvogel, D.B., Wilkin, J.L., Young, R., 1991. A semispectral primitive equation ocean circulation model using vertical sigma and orthogonal curvilinear horizontal coordinates. J. Comput. Phys. 94, 151–185.

Halpern, D., Weller, R.A., Briscoe, M.G., Davis, R.E., McCullough, J.R., 1981. Intercomparison tests of moored current measurements in the upper ocean. J. Geophys. Res. 86, 419–428.

Halpern, D., Knauss, W., Brown, O., Wentz, F., 1993. An Atlas of Monthly Mean Distributions of SSMI Surface Wind Speed, ARGOS Buoy Drift, AVHRR/2 Sea Surface Temperature, and ECMWF Surface Wind Components during 1991. Jet Propulsion Laboratory, Pasadena. JPL Publications 93–10.

Halpern, D., 1978. Mooring motion influences on current measurements. In: Proceedings of a Workshop Conference on Current Measurement. College of Marine Studies of Delaware, Newark. Technical Report DEL-SG-3–78.

Halverson, M.J., Pawlowicz, R., 2008. Estuarine forcing of a river plume by river flow and tides. J. Geophys. Res. 113, c09033. http://dx.doi.org/10.1029/2008JC004844.

Hamlington, B.D., Leben, R., Nerem, S., et al., 2011. Reconstructing sea level using cyclostationary empirical orthogonal functions. J. Geophys. Res. 116, C12015. http://dx.doi.org/10.1029/2011JC007529.

Hamme, R.C., Webley, P.W., Crawford, W.R., Whitney, F.A., DeGrandpre, M.D., Emerson, S.R., Eriksen, C.C., Giesbrecht, K.E., Gower, J.F.R., Kavanaugh, M.T., Peña, M.A., Sabine, C.L., Batten, S.D., Coogan, L.A., Grundle, D.S., Lockwood, D., 2010. Volcanic ash fuels anomalous plankton bloom in subarctic northeast Pacific. Geophys. Res. Lett. 37. http://dx.doi.org/10.1029/2010GL044629.

Hamming, R.W., 1977. Digital Filters. Prentice-Hall, Englewood Cliffs, N.J.

Hamon, B.V., Brown, N.L., 1958. A temperature-chlorinity-depth recorder for use at sea. J. Scientific Instru. 35, 452–458.

Hamon, B.V., 1955. A temperature-salinity-depth recorder. Conseil permanent international pour le exploration de la Mer. J. du Conseil 21, 22–73.

Hanafin, J.A., Minnett, P.J., 2001. Profiling temperature in the sea surface skin layer using FTIR measurements. In: Donelan, M.A., Drennan, W.M., Saltzmann, E.S., Wanninkhof, R. (Eds.), Gas Transfer at Water Surfaces. American Geophysical Union Monograph, pp. 161–166.

Hanafin, J.A., 2002. On Sea Surface Properties and Characteristics in the Infrared (Ph.D.). Meteorology and Physical Oceanography, University of Miami, 111 pp.

Hanawa, K., Yasuda, T., 1992. New detection method for XBT depth error and relationship between the depth error and coefficients in the depth-time equation. J. Oceanogr. 48, 221–230.

Hanawa, K., Yoritaka, H., 1987. Detection of systematic errors in XBT data and their correction. J. Oceanogr. Soc. Jpn. 32, 68–76.

Hanawa, K., Yoshikawa, Y., 1991. Re-examination of depth error in XBT data. J. Atmos. Oceanic Technol. 8, 422–429.

Hannachi, A., Jolliffe, I.T., Stephenson, D.B., Trendafilov, N., 2006. In search of simple structures in climate: simplifying EOFs. Int. J. Climatol. 26, 7–28.

Hansen, J., Lebedeff, S., 1987. Global trends of measured surface air temperature. J. Geophys. Res. 92, 13,345–13,372.

Hansson, D., Stigebrandt, A., Liljebladh, B., 2013. Modelling the Orust Fjord system on the Swedish west coast. J. Marine Syst. 113, 29–41.

Harada, K., Tsunogai, S., 1986. Ra-226 in the Japan Sea and the residence time of the Japan Sea water. Earth Planet. Sci. Lett. 77, 236–244.

Harnett, D.L., Murphy, J.L., 1975. Introductory Statistical Analysis. Addison-Wesley, Reading, Mass.

Harris, F.J., 1978. On the use of windows for harmonic analysis with the discrete fourier transform. Proc. IEEE 66, 51–83.

Haxby, W.F., 1985. Gravity Field of the World's Oceans, Chart Scale 1:51,400,000. Office of Naval Research, Washington, D.C.

Hayashi, Y., Tsushima, H., Hirata, K., Kimura, K., Maeda, K., 2011. Tsunami source area of the 2011 off the Pacific coast of Tohoku earthquake determined from tsunami arrival times at offshore observation stations. Earth Planets Space 63, 809–813.

Hayashi, Y., 1979. Space-time spectral analysis of rotary vector series. J. Atmos. Sci. 36, 757–766.

Hayne, G.S., Hancock, D.W., 1982. Sea state-related altitude errors in the Seasat altimeter. J. Geophys. Res. 87, 3227–3231.

Heinmiller, R.H., Ebbesmeyer, C.C., Taft, B.A., Olson, D.B., Nitkin, G.P., 1983. Systematic errors in expendable bathythermograph (XBT) profiles. Deep-Sea Res. 30, 1185–1197.

Heinmiller, R.H., 1968. Acoustic Release Systems. WHOI Technical Report 68–48. Woods Hole, Mass.

Helland-Hansen, B., 1918. Nogen hydrografiske metoder. Forh. Skand. Naturforskeres 16 (Kristiania).

Hellerman, S., Rosenstein, M., 1983. Normal monthly wind stress over the world ocean with error estimates. J. Phys. Oceanogr. 13, 1093–1104.

Hendry, R., 1993. Canadian Technical Report of Hydrography and Ocean Sciences. Bedford Institute of Oceanography CTD Trials. BIO, Dartmouth, Nova Scotia. B2Y 4A2.

Henry, R.F., Graefe, P.W.U., 1971. Zero Padding as a Means of Improving Definition of Computed Spectra. Canadian Department of Energy, Mines and Resources, Ottawa. Manuscript Series Report Series No. 20.

Hewitson, B.C., Crane, R.G., 1994. Neural Nets: Applications in Geography. The GeoJournal Library. Kluwer Academic Publishers.

Hichman, M.L., 1978. Measurement of Dissolved Oxygen. John Wiley, New York.

Hickey, B.M., Dobbins, E., Allen, S.E., 2003. Local and remote forcing of currents and temperature in the central southern California Bight. J. Geophys. Res. 108 (C3), 3081. http://dx.doi.org/10.1029/2000JC00313.

Hill, M.N. (Ed.), 1962. The Sea, Volume 1, Physical Oceanography. Interscience, New York.

Hilland, J.E., Chelton, D.B., Njoku, E.G., 1985. Production of global sea surface temperature fields for the Jet Propulsion Laboratory workshop. J. Geophys. Res. 90, 11,642–11,650.

Hiller, W., Käse, R.H., 1983. Objective analysis of hydrographic data sets from mesoscale surveys. Ber. Inst. Meereskd. Univ. Kiel 116.

Hiyagon, H., 1994. Retention of solar helium and neon in IDPs in deep sea sediment. Science 263, 1257–1259.

Hogg, N., 1977. Topographic waves along 70°W on the continental rise. J. Mar. Res. 39, 627–649.

Holgate, S.J., Matthews, A., Woodworth, P.L., Rickards, L.J., Tamisiea, M.E., Bradshaw, E., Foden, P.R., Gordon, K.M., Jevrejeva, S., Pugh, J., 2013. New data systems and products at the permanent service for mean sea level. J. Coastal Res. 29 (3), 493–504. http://dx.doi.org/10.2112/JCOASTRES-D-12-00175.1.

Holl, M.M., Mendenhall, B.R., 1972. Fields by Information Blending, Sea-level Pressure Version. Technical Note 72–2. Fleet Numerical Weather Central, Monterey, Calif.

Hollinger, J., 1989. Final Report. DMSP Special Sensor Microwave/Imager Calibration/validation, vol. 1. Navy Research Laboratory, Washington, DC.

Holte, J., Talley, L., 2009. A new algorithm for finding mixed layer depths with applications to Argo data and subantarctic mode water formation. J. Atmos. Ocean. Technol. 26, 1920–1939.

Horel, J.D., 1984. Complex principal component analysis: theory and examples. J. Clim. Appl. Meteorol. 23, 1660–1673.

Horne, E.P.W., Toole, J.M., 1980. Sensor response mismatches and lag correction techniques for temperature-salinity profilers. J. Phys. Oceanogr. 10, 1122–1130.

Hsieh, W.W., 1986. Comments on: "Rotary empirical orthogonal function analysis of currents near the Oregon Coast". J. Phys. Oceanogr. 16, 791–792.

Hsu, K., Gupta, H.V., Gao, X., Sorooshian, S., Iman, B., 2002. Self-organizing linear output (SONO): an artificial neural network suitable for hydrologic modeling and analysis. Water Resour. Res. 38, 1302–1318.

Huang, N.E., Leitao, C.D., Parra, C.G., 1978. Large-scale Gulf Stream frontal study using GEOS-3 radar altimeter data. J. Geophys. Res. 83, 4673–4682.

Huggett, W.S., Crawford, W.R., Thomson, R.E., Woodward, M.V., 1987. Data Record of Current Observations. In: Coastal Ocean Dynamics Experiment (CODE), volume XIX. Institute of Ocean Sciences. Part 1.

Hydes, D.J., Hartman, M.C., Kaiser, J., Campbell, J.M., 2009. Measurement of dissolved oxygen using optodes in a ferrybox system. Estuar. Coast. Shelf Sci. 83, 485–490.

Iguchi, T., Kozu, T., Meneghini, R., Awaka, J., Okamoto, K., 2000. Rain-profiling algorithm for the TRMM precipitation radar. J. Appl. Meteorol. 39, 2038–2052. http://dx.doi.org/10.1175/1520-0450(2001)040<2038: RPAFTT>2.0.CO;2.

Iler, R.K., 1979. The Chemistry of Silica. John Wiley, New York.

IPCC, Climate Change, 2013. The physical science basis. In: Stocker, T.F., Qin, D., Plattner, G.-K., Tignor, M., Allen, S.K., Boschung, J., Nauels, A., Xia, Y., Bex, V., Midgley, P.M. (Eds.), Contribution of Working Group I to the Fifth Assessment Report of the Intergovernmental Panel on Climate Change. Cambridge University Press, Cambridge, United Kingdom and New York, NY, USA, 1535 pp.

IPCC, Climate Change, 2007. The physical science basis. In: Solomon, S., et al. (Eds.), Contribution of Working Group I to the Fourth Assessment Report of the Intergovernmental Panel on Climate Change. Cambridge University Press, Cambridge, UK, and New York,.

Jackson, D.R., Jones, C.D., Rona, P.A., Bemis, K.G., 2003. A method for Doppler acoustic measurement of black smoker flow fields. G-Cubed 4, 1095. http://dx.doi.org/10.1029/2003G000509, 1–12.

Jackson, D.D., 1972. Interpretation of inaccurate, insufficient and inconsistent data. Geophys. J. R. Astron. Soc. 28, 97–109.

Jacobsen, A.W., 1948. An instrument for recording continuously the salinity, temperature and depth of sea water. Trans. Am. Meteorol. Soc., 1057–1070.

James, R.W., Fox, P.T., 1972. Comparative Sea-surface Temperature Measurements. Report No. 5, Report on Marine Science Affairs. WMO, Geneva, pp. 117–129.

James, T.S., Simon, K.M., Forbes, D.L., Dyke, A.S., Mate, D.J., 2011. Sea-level Projections for Five Pilot Communities of the Nunavut Climate Change Partnership. In: Geological Survey of Canada, Open File, 6715, 23.

Jamous, D., Mémery, L., Andrié, C., Jean-Baptiste, P., Merlivat, L., 1992. The distribution of helium 3 in the deep western and southern Indian Ocean. J. Geophys. Res. 97, 2243–2250.

Jannasch, H.W., Coletti, L.J., Johnson, K.S., Fitzwater, S.E., Needoba, J.A., Plant, J.N., 2008. The land/ocean biogeochemical observatory: a robust networked mooring system for continuously monitoring complex biogeochemical cycles in estuaries. Limnol. Oceanogr. Methods 6, 263–273.

Jansen, E., Overpeck, J., Briffa, K.R., Duplessy, J.-C., Joos, F., Masson-Delmotte, V., Olago, D., Otto-Bliesner, B., Peltier, W.R., Rahmstorf, S., Ramesh, R., Raynaud, D., Rind, D., Solomina, O., Villalba, R., Zhang, D., 2007. Palaeoclimate. In: Solomon, S., Qin, D., Manning, M., Chen, Z., Marquis, M., Averyt, K.B., Tignor, M., Miller, H.L. (Eds.), Climate Change 2007: The Physical Science Basis. Contribution of Working Group I to the Fourth Assessment Report of the Intergovernmental Panel on Climate Change. Cambridge University Press, Cambridge, United Kingdom and New York, NY, USA.

Jenkins, G.M., Watts, D.G., 1968. Spectral Analysis and its Applications. Holden-Day, San Francisco, Calif.

Jenkins, F.A., White, H.E., 1957. Fundamentals of Optics. McGraw-Hill, New York.

Jenkins, W.J., Edmond, J.M., Corliss, J.B., 1978. Excess ^{3}He and ^{4}He in Galapagos submarine hydrothermal waters. Nature 272, 156–158.

Jenkins, W.J., 1988. The use of anthropogenic tritium and helium-3 to study subtropical gyre ventilation and circulation. Phil. Trans. R. Soc. London A325, 43–61.

Jerlov, N.G., 1976. Marine Optics. Elsevier, New York.

Johnson, K.S., Needoba, J.A., Riser, S.C., Showers, W.J., 2007. Chemical sensor networks for the aquatic environment. Chem. Rev. 107, 623–640.

Johnson, K.S., Riser, S.C., Karl, D.M., 2010. Nitrate supply from deep to near surface waters of the North Pacific subtropical gyre. Nat. Lett. 465 (24), 1062–1065.

Jones, P.D., Wigley, T.M.L., Wright, P.B., 1986. Global temperature variations between 1861 and 1984. Nature 322, 430–434.

Jones, P.D., 1988. Hemispheric surface air temperature variations: recent trends and an update to 1987. J. Clim. 1, 654.

Joos, F., Plattner, G.-K., Stocker, T.F., Körtzinger, A., Wallace, D.W.R., 2003. Trends in marine dissolved oxygen: Implications for ocean circulation changes and the carbon budget. EOS 84 (21), 197–201, 27.

Joyce, R., Arkin, P.A., 1997. Improved estimates of tropical and subtropical precipitation using the GOES precipitation index. J. Atmos. Ocean. Tech. 10, 997–1011.

Julian, P.R., 1975. Comments on the determination of significance levels of the coherence statistic. J. Atmos. Sci. 32, 836–837.

Kadko, D., Rosenburg, N.D., Lupton, J.E., Collier, R., Lilley, M., 1990. Chemical reaction rates and entrainment within the Endeavour Ridge hydrothermal plume. Earth Planet. Sci. Lett. 99, 315–335.

Kaiser, J.F., 1966. Digital filters. In: Kuo, F.F., Kaiser, J.F. (Eds.), System Analysis by Digital Computer. Wiley, New York. Chap. 7.

Kalnay, et al., 1996. The NCEP/NCAR 40-year reanalysis project. Bull. Amer. Meteor. Soc. 77, 437–470.

Kanasewich, E.R., 1975. Time Series Analysis in Geophysics. University of Alberta Press.

Kara, A.B., Metzger, J., Bourassa, M.A., 2007. Ocean current and wave effects on wind stress drag coefficient and fluxes over the global ocean. Geophys. Res. Lett. 34, L01604. http://dx.doi.org/10.1029/2006GL027849.

Katsaros, K.B., Liu, W.T., Businger, J.A., Tillman, J.E., 1977. Heat transport and thermal structure in the interfacial boundary layer measured in an open tank of water in turbulent free convection. J. Fluid Mech. 83, 311–335.

Kautsky, H., 1939. Quenching of luminescence by oxygen. Trans. Faraday Soc. 35, 216–219.

Kay, S.M., Marple Jr, S.L., 1981. Spectrum analysis-a modern perspective. Proc. IEEE 69, 1380–1417.

Kelly, K.A., Strub, P.T., 1992. Comparison of velocity estimates from advanced very high resolution radiometer.
J. Geophys. Res. 97, 9653—9668.

Kelly, K.A., 1988. Comment on "Empirical orthogonal
function analysis of advanced very high resolution
radiometer surface temperature patterns in Santa Barbara
Channel" by G.S.E. Lagerloef and R.L. Berstein. J. Geophys. Res. 93, 15,753—15,754.

Keyte, F.K., 1965. On the formulae for correcting reversing
thermometers. Deep-Sea Res. 12, 163—172.

Kim, K.-Y., Chung, C., 2001. On the evolution of the annual
cycle in the tropical Pacific. J. Clim. 14, 991—994. http://
dx.doi.org/10.1175/1520-0442(2001)014<0991:OTEOTA>
2.0.CO;2.

Kim, K.-Y., north, G.R., Huang, J., 1996. EOFs of one-
dimensional cyclostationary time series: computations,
examples, and stochastic modelling. J. Atmos. Sci. 53,
1007—1017.

Kim, S.Y., Terrill, J., Cornuelle, B.D., Jones, B., Washburn, L.,
Moline, M.A., Paduan, J.D., Garfield, N., Largier, J.L.,
Crawford, G., Kosro, P.M., 2011. Mapping the
U.S. west coast surface circulation: a multiyear analysis of
high-frequency radar observations. J. Geophys. Res. 116.
http://dx.doi.org/10.1029/2010JC006669, 2011.

Kimoto, M., Ghil, M., Mo, K.C., 1991. Spatial structure of the
extratropical 40-day oscillation. In: Proc. Eighth Conf. On
Atmospheric and Oceanic Waves and Stability. Amer.
Meteor. Soc., Denver, CO, pp. 115—116.

Kipphut, G.W., 1990. Glacial meltwater input to the Alaska
Coastal Current: evidence from oxygen isotope
measurements. J. Geophys. Res. 95, 5177—5181.

Kirwan Jr, A.D., Chang, M.S., 1976. On the micropolar
Ekman problem. Int. J. Eng. Sci. 14, 685—692.

Kirwan Jr, A.D., McNally, G., Chang, M.-S., Molinari, R.,
1975. The effect of wind and surface currents on drifters.
J. Phys. Oceanogr. 5, 361—368.

Kirwan Jr, A.D., McNally, G.J., Reyna, E., Merrell Jr, W.J.,
1978. The near-surface circulation of the eastern North
Pacific. J. Phys. Oceanogr. 8, 937—945.

Kirwan Jr, A.D., McNally, G., Pazan, S., Wert, R., 1979.
Analysis of surface current response to wind. J. Phys.
Oceanogr. 9, 401—412.

Kistler, R., et al., 2001. The NCEP-NCAR 50-year reanalysis:
monthly means CD-ROM and documentation. Bull.
Amer. Meteor. Soc. 82, 247—268.

Knudsen, M. (Ed.), 1901. Hydrographical Tables. GEC,
Copenhagen.

Kohonen, T., 1982. Self-organized formation of topologically
correct feature maps. Biol. Cybern. 43, 59—69.

Kohonen, T., 2001. Self-organizing Maps, Third, extended
ed. Springer.

Konyaev, K.V., 1990. Spectral Analysis of Physical
Oceanographic Data. National Science Foundation,
Washington, D.C.

Koopmans, L.H., 1974. The Spectral Analysis of Time Series.
Academic Press, New York.

Koračin, D., Dorman, C., 2001. Marine atmospheric
boundary layer divergence and clouds along California
in June 1996. Mon. Wea. Rev. 129, 2040—2055.

Krauss, W., Kasë, R.H., 1984. Mean circulation and eddy
kinetic energy in the eastern North Atlantic. J. Geophys.
Res. 84, 3407—3415.

Krauss, W., 1993. Ekman drift in homogenous water.
J. Geophys. Res. 98, 20,187—20,209.

Kremling, K., 1972. Comparison of specific gravity in
natural sea-water from hydrographical tables and
measurements by a new density instrument. Deep-Sea
Res. 19, 377—383.

Kuhn, H., Quadfasel, D., Schott, F., Zenk, W., 1980. On
simultaneous measurements with rotor, wing and
acoustic current meters, moored in shallow water. Dtsch.
Hydrogr. Z. 33, 1—18.

Kulikov, E.A., González, F.I., 1995. On Reconstruction of the
Initial Tsunami Signal from Distant Bottom Pressure
Records. Doklady Akademii Nauk.

Kulikov, E.A., González, F.I., 1996. Recovery of the shape of
a tsunami signal at the source from measurements of
oscillations in the ocean level by a remote hydrostatic
pressure sensor, Trans. (Doklady) Russian Acad. Sci.,
Earth Sci. Sect. 345A, 585.

Kummerow, C., Hong, Y., Olson, W.S., Yang, S., Adler, R.F.,
McCollum, J., Ferraro, R., Petty, G., Shin, D.-B.,
Wilheit, T.T., 2001. The evolution of the Goddard
profiling algorithm (GPROF) for rainfall estimation from
passive microwave sensors,. J. Appl. Meteorol. 40,
1801—1820.

Kundu, P.K., Allen, J.S., 1976. Some three-dimensional
characteristics of low-frequency current fluctuations
near the Oregon coast. J. Phys. Oceanogr. 6, 181—199.

Kundu, P.K., Allen, J.S., Smith, R.L., 1975. Modal decomposition of the velocity field near the Oregon coast.
J. Phys. Oceanogr. 5, 683—704.

Kundu, P.K., 1976. An analysis of inertial oscillations
observed near the Oregon coast. J. Phys. Oceanogr. 6,
879—893.

Kundu, P.K., 1990. Fluid Mechanics. Academic Press, San
Diego, Calif.

Labrecque, A.M., Thomson, R.E., Stacey, M.W., Buckley, J.R.,
1994. Residual currents in Juan de Fuca Strait. Atmos.
Ocean 32, 375—394.

Lacoss, R.T., 1971. Data adaptive spectral analysis methods.
Geophysical 36, 661—675.

Ladd, C., Bond, N.A., 2002. Evaluation of the NCEP/NCAR
reanalysis in the NE Pacific and the Bering Sea. J. Geophys. Res. 107 (C10), 3158. http://dx.doi.org/10.1029/
2001JC001157.

LaFond, E.C., 1951. Processing Oceanographic Data. HO
Publication No. 614. US Hydrographic Office.

Lanczos, C., 1956. Applied Analysis. Prentice-Hall, Englewood Cliffs, NJ (Reprinted in 1988, Dover, New York.).

Lane, R.W.P.M., Manuels, M.W., Staal, W., 1984. A procedure for enriching and cleaning up rhodamine B and rhodamine WT in natural waters, using a sep-pak C18 cartridge. Water Res. 18, 163.

Langdon, C., 1984. Dissolved oxygen monitoring system using a pulsed electrode: design, performance and evaluation. Deep-Sea Res. 31, 1357–1367.

Large, W.G., McWilliams, J.C., Doney, S.C., 1994. Ocean vertical mixing: a review and a model with a nonlocal boundary layer parameterization. Rev. Geophys. 32, 363–403.

Laxon, S., McAdoo, D., 1994. Arctic Ocean gravity field derived from ERS-1 satellite altimetry. Science 265, 621–625.

LeBlond, P.H., Mysak, L.A., 1979. Waves in the Ocean. Elsevier, Amsterdam.

Lemon, D.D., Farmer, D.M., April 3–5, 1990. Experience with a multi-depth scintillation flowmeter in the Fraser estuary. In: Proceedings IEEE Fourth Working Conference on Current Measurement, Clinton, MD, pp. 290–298.

Lemon, D.D., Thomson, R.E., Delaney, J.R., Farmer, D.M., Rowe, F., Chave, R.A.J., 1996. Acoustic Scintillation Velocity Measurement of a Buoyant Hydrothermal Plume. Preliminary Report. ASL Environmental Science.

Lenn, Y.D., Chereskin, T.K., 2009. Observations of Ekman currents in the Southern Ocean. J. Phys. Oceanogr. 39, 768–779.

Lerner, R., Hollinger, J., 1977. Analysis of 1.4 GHz radiometric measurements from Skylab. Remote Sens. Environ. 6, 251–269.

Levitus, S., 1982. Climatological Atlas of the World Ocean. NOAA Professional Paper 13. US Department of Commerce, Rockville, Md.

Lewis, E.L., Perkin, R.G., 1981. The practical salinity scale 1978: conversion of existing data. Deep-Sea Res. 28A, 307–328.

Lewis, E.L., 1980. The practical salinity scale 1978 and its antecedents. J. Oceanic Eng. 5, 3–8.

Li, Q., Farmer, D.M., Duda, T.F., Ramp, S., 2009. Acoustical measurement of nonlinear internal waves using the Inverted Echo Sounder. J. Atmos. Oceanic Tech. 26, 2228–2242.

Liitkepohl, H., 1985. Comparison criteria for estimating the order of a vector autoregressive process. J. Time Ser. Anal. 6, 35–52.

Lilley, M.D., Feely, R.A., Trefry, J.H., 1995. Chemical and biochemical transformations in hydrothermal plumes. In: Humphris, S.E., Zierenberg, R.A., Mullineaux, L.S., Thomson, R.E. (Eds.), Seafloor Hydrothermal Systems: Physical, Chemical, Biological and Geological Interactions, Geophysical Monograph, 91. American Geophysical Union, pp. 369–391.

Lipps, F.B., Hemler, R.S., 1992. On the downward transfer of tritium to the ocean by a cloud model. J. Geophys. Res. 97, 12,889–12,900.

Liu, Y., Weisberg, R.H., 2005. Patterns of ocean current variability on the west Florida shelf using the self-organizing map. J. Geophys. Res. 110, C06003. http://dx.doi.org/10.1029/2004JC002786.

Liu, Y., Weisberg, R.H., Mooers, C.N.K., 2006. Performance evaluation of the self-organizing map for feature extraction. J. Geophys. Res. 111. C05018. http://dx.doi.org/10.1029/2005JC003117.

Liu, C., Zipser, E.,J., Nesbitt, S.W., 2007. Global distribution of tropical deep convection: differences using infrared and radar as the primary data source. J. Clim. 20, 489–503. http://dx.doi.org/10.1175/JCLI4023.1.

Livingstone, D., Royer, T.C., 1980. Eddy propagation determined from rotary spectra. Deep-Sea Res. 27A, 883–835.

Llewellyn-Jones, D.T., Minett, P.J., Saunders, R.W., Zavody, A.M., 1984. Satellite multichannel infrared measurements of sea surface temperature of the northeast Atlantic Ocean using AVHRR/2. Q.J. R. Meteorol. Soc. 110, 613–631.

Loeve, M., 1978. Probability Theory II (46). Graduate texts in mathematics, 0–387.

Lohrmann, A., 2001. Monitoring sediment concentration with acoustic backscattering instruments. Nortek Technical Notes, No. N4000-712, 5 pp.

Lomb, N.R., 1976. Least-squares frequency-analysis of unequally spaced data. Astrophys. Space Sci. 39, 447–462.

Lorenz, E., 1956. Empirical Orthogonal Functions and Statistical Weather Prediction. Air Force Cambridge Research Center, Air Research and Development Command, Cambridge Mass. Scientific Report No. 1.

Lueck, R.G., Picklo, J.J., 1990. Thermal inertia of conductivity cells: observations with a Sea-Bird cell. J. Atmos. Oceanic Tech. 7, 756–768.

Lueck, R.G., Hertzman, O., Osborn, T.R., 1977. The spectral response of thermistors. Deep-Sea Res. 24, 951–970.

Lueck, R.G., 1990. Thermal inertia of conductivity cells: theory. J. Atmos. Oceanic Technol. 7, 741–755.

Lukas, R., 1994. HOT results show interannual variability of Pacific deep and bottom waters. WOCE Notes 6 (2), 4.

Lumpkin, R., Garraffo, Z., 2005. Evaluating the decomposition of tropical Atlantic drifter observations. J. Atmos. Oceanic Technol. 22, 1403–1415.

Lumpkin, R., Treguier, A.-M., Speer, K., 2002. Lagrangian eddy scales in the northern Atlantic Ocean. J. Phys. Oceanogr. 32, 2425–2440.

Lumpkin, R., 2003. Decomposition of surface drifter observations in the Atlantic Ocean. Geophys. Res. Lett. 30, 1753. http://dx.doi.org/10.1029/2003GL017519.

Lupton, J.E., Craig, H., 1981. A major helium-3 source at 15°S on the East Pacific Rise. Science 214, 13−18.

Lupton, J.E., Delaney, J.R., Johnson, H.P., Tivey, M.K., 1985. Entrainment and vertical transport of deep-ocean water by buoyant hydrothermal plumes. Nature 316, 621−623.

Lupton, J.E., Baker, E.T., Massoth, G.J., 1989. Variable ^{3}He/heat ratios in submarine hydrothermal systems: evidence from two plumes over the Juan de Fuca Ridge. Nature 337, 161−164.

Lupton, J.E., Baker, E.T., Mottl, M.J., Sansone, F.J., Wheat, C.G., Resing, J.A., Massoth, G.J., Measures, C.I., Feely, R.A., 1993. Chemical and physical diversity of hydrothermal plumes along the East Pacific Rise, 8°45′N to 11°50′N. Geophys. Res. Lett. 20, 2913−2916.

Lynn, R.J., Reid, J.L., 1968. Characteristics and circulation of deep and abyssal waters. Deep-Sea Res. 15, 577−598.

Lynn, R.J., Svejkovsky, J., 1984. Remotely sensed sea surface temperature variability off California during a "Santa Ana" clearing. J. Geophys. Res. 89, 8151−8162.

Macdonald, R.W., McLaughlin, F.A., Wong, C.S., 1986. The storage of reactive silicate samples by freezing. Limnol. Oceanogr. 31, 1139−1142.

Mackas, D.L., Denman, K.L., Bennett, A.F., 1987. Least squares multiple tracer analysis of water mass composition. J. Geophys. Res. 92, 2907−2918.

Mackenzie, K.V., 1981. Nine term equation for sound speed in the oceans. J. Acoust. Soc. Am. 70, 807−812.

Macklin, S.A., Stabeno, P.J., Schumacher, J.D., 1993. A comparison of gradient and observed over-the-water winds along a mountainous coast. J. Geophys. Res. 98, 16,555−16,569.

MacPhee, S.B., 1976. Acoustics and echo sounding instrumentation. Can. Hydrographic Serv. Technical Report 76−1.

Mandelbrot, B.B., 1967. How long is the coast of Britain? Statistical self-similarity and fractional dimension. Science 155, 636−638.

Mann, M.E., Bradley, R.S., Hughes, M.K., 1998. Global-scale temperature patterns and climate forcing over the past six centuries. Nature 392, 779−787.

Mantyla, A.W., Reid, J.L., 1983. Abyssal characteristics of the world ocean waters. Deep-Sea Res. 30 (8A), 805−833.

Marks, K.M., McAdoo, D.C., Smith, W.H.F., 1993. Mapping the Southeast Indian Ridge with GEOSTAT. EOS 74 (8), 81−86.

Marple Jr, S.L., 1987. Digital Spectral Analysis. Prentice-Hall, Englewood Cliffs, N.J.

Marsden, R.F., 1987. A comparison between geostrophic and directly measured surface winds over the northeast Pacific Ocean. Atmos. Ocean 25, 387−401.

Martin, J.H., Knauer, G.A., 1973. The elemental composition of plankton. Geochem. Cosmochim. Acta 37, 1639−1653.

Martin, M., Talley, L.D., de Szoeke, R.A., 1987. Physical, chemical and CTD data from Marathon Expedition. R/V Thomas Washington 261, 4 May−4 June 1984. Oregon State University, College of Oceanography. Data Report 131, Ref. 87−15, May 1987.

Martini, M., Butman, B., 2007. Long-term Performance of Aanderaa Optodes and Sea-Bird SBE-43 Dissolved-oxygen Sensors Bottom Mounted at 32 m in Massachusetts Bay. U.S. Geological Survey, Woods Hole Science Center, Woods Hole, Massachusetts.

Masson, D., 1996. A case study of wave−current interaction in a strong tidal current. J. Phys. Oceanogr. 26, 359−372.

Maul, G., Bravo, N.J., 1983. Fitting of satellite and in-situ ocean surface temperatures: results for POLYMODE during the winter of 1977−1978. J. Geophys. Res. 88, 9605−9616.

Maxwell, A.E., 1977. Multivariate Analysis in Behavioral Research. Chapman and Hall, London, 164 pp.

McBride, J.L., Davidson, D.E., Puri, K., Tyrell, G.C., 1995. The flow during TOGA COARE as diagnosed by the BMRC tropical analysis and prediction scheme. Mon. Wea. Rev. 123, 717−736.

McClain, E.P., Pichel, W.G., Walton, C.C., Ahmad, Z., Sutton, J., 1983. Multi-channel improvements to satellite-derived global sea surface temperatures. Adv. Space Res. 2, 43−47.

McClain, E.P., Pichel, W.G., Walton, C.C., 1985. Comparative performance of AVHRR-based multichannel sea surface temperatures. J. Geophys. Res. Oceans 90, 11587−11601.

McClain, C.R., Yoder, J.A., Atkinson, L.P., Blanton, J.O., Lee, T.N., Singer, J.J., Müller-Karger, F., 1988. Variability of surface pigment concentration in the South Atlantic Bight. J. Geophys. Res. 93, 10,675−10,697.

McClain, E.P., 1981. Multiple atmosphere-window techniques for satellite sea surface temperatures. In: Gower, J.F.R. (Ed.), Oceanography from Space. Plenum, New York, pp. 73−85.

McDougall, R.J., 1985a. Double-diffusive interleaving/part 1: linear stability analysis. J. Phys. Oceanogr. 15, 1532−1541.

McDougall, R.J., 1985b. Double-diffusive interleaving/part 2: finite amplitude, steady state interleaving. J. Phys. Oceanogr. 15, 1542−1556.

McDuff, R.E., 1988. Effects of vent fluid properties on the hydrography of hydrothermal plumes. EOS Trans. AGU 69, 1497.

McDuff, R.E., 1995. Physical dynamics of deep-sea hydrothermal plumes. In: Humphris, S.E., Zierenberg, R.A., Mullineaux, L.S., Thomson, R.E. (Eds.), Seafloor Hydrothermal Systems: Physical, Chemical, Biological, and Geological Interactions, Geophysical Monograph, 91. American Geophysical Union, Washington, D.C, pp. 357–368.

McManus, J., Collier, R.W., Chen, C.-T.A., Dymond, J., 1992. Physical properties of Crater Lake, Oregon: a method for the determination of conductivity- and temperature-dependent expression of salinity. Limnol. Oceanogr. 37, 41–53.

McNally, G.J., White, W.B., 1985. Wind driven flow in the mixed layer observed by drifting buoys during autumn-winter in midlatitude North Pacific. J. Phys. Oceanogr. 15, 684–694.

McNally, G.J., Patzert, W.C., Kirwan Jr, A.D., Vastano, A.C., 1983. The near-surface circulation of the North Pacific using satellite tracked drifting buoys. J. Geophys. Res. 88, 7507–7518.

McNally, G.J., 1981. Satellite-tracked drift buoy observations of the near-surface flow in the eastern mid-latitude North Pacific. J. Geophys. Res. 86, 8022–8030.

McTaggart, K.E., Johnson, G.C., Johnson, M.C., Delahoyde, F.M., Swift, J.H., 2010. The GO-SHIP repeat hydrography manual: a collection of expert reports and Guidelines. IOCCP Report 14, ICPO Publication Series No. 134, Version 1.

McWilliams, J.C., 1976. Maps from the mid-ocean dynamics Experiment: part I. Geostrophic streamfunction. J. Phys. Oceanogr. 6, 810–827.

Meckler, A.N., Sigman, D.M., Gibson, K.A., François, R., Martínez-García, A., Jaccard, S.L., Röhl, U., Peterson, L.C., Tiedemann, R., Haug, G.H., 2013. Deglacial pulses of deep-ocean silicate into the subtropical North Atlantic Ocean. Nature 495, 495–498.

Meinen, C.S., Watts, D.R., 2000. Vertical structure and transport on a transect across the North Atlantic Current near 42°N: time series and mean. J. Geophys. Res. 105, 21869–21891.

Meinen, C.S., Garzoli, S.L., Johns, W.E., Baringer, M.O., 2004. Transport variability of the deep western boundary current and the Antilles Current off Abaco Island, Bahamas. Deep Sea Res. 51 (11), 1397–1415.

Meinig, C., Stalin, S.E., Nakamura, A.I., Milburn, H.B., 2005. Real-Time deep-ocean tsunami measuring, monitoring, and reporting system: the NOAA DART II description and Disclosure.

Meisel, D.D., 1978. Fourier transforms of data sampled at unequaled observational intervals. Astron. J. 83, 538–545.

Meisel, D.D., 1979. Fourier transforms of data sampled in unequally spaced segments. Astron. J. 84, 116–126.

Memery, L., Wunsch, C., 1990. Constraining the North Atlantic circulation with tritium data. J. Geophys. Res. 95, 5239–5256.

Mero, T.M., 1982. Performance results for the EG7G vector-measuring current meter (VMCM). In: Dursi, M., Woodward, W. (Eds.), Proceedings of IEEE Second Working Conference on Current Measurement. Institute of Electrical and Electronics Engineers, New York, pp. 159–164.

Merrifield, M.A., Guza, R.T., 1990. Detecting propagating signals with complex empirical orthogonal functions: a cautionary note. J. Phys. Oceanogr. 20, 1628–1633.

Meyers, S.D., Kelly, B.G., O'Brien, J.J., 1993. An introduction to wavelet analysis in oceanography and meteorology: with application to the dispersion of Yanai waves. Mon. Weather Rev. 121, 2858–2866.

Middleton, J.H., Cunningham, A., 1984. Wind-forced continental shelf waves from geographical origin. Cont. Shelf Res. 3, 215–232.

Middleton, J.H., 1982. Outer rotary cross spectra, coherences and phases. Deep-Sea Res. 29 (10A), 1267–1269.

Middleton, J.H., 1983. Low-frequency trapped waves on a wide, reef-fringed continental shelf. J. Phys. Oceanography 13, 1371–1382.

Miller, L., Cheney, R., 1990. Large-scale meridional transport in the tropical Pacific Ocean during the 1986–1987 El Niño from Geosat. J. Geophys. Res. 95, 17,905–17,919.

Minnett, P., and Kaiser-Weiss, A., 2012. Near-surface oceanic temperature gradients, GHRSST, discussion document, Version 12, 7 pp.

Minnett, P.J., Smith, M., Ward, B., 2011. Measurements of the oceanic thermal skin effect. Deep Sea Res. Pt. II Top. Stud. Oceanogr. 58, 861–868.

Minnett, P.J., 2003. Radiometric measurements of the sea-surface skin temperature - the competing roles of the diurnal thermocline and the cool skin. Int. J. Remote Sens. 24 (24), 5033–5047.

Mitrovica, J.X., Tamisiea, M.E., Davis, J.L., Milne, G.A., 2001. Recent mass balance of polar ice sheets inferred from patterns of global sea-level change. Nature 409, 1026–1029. http://dx.doi.org/10.1038/35059054.

Miyake, Y., Saruhashi, K., 1967. A study on the dissolved oxygen in the ocean. In: Geochemistry Study of the Ocean and the Atmosphere, Yasuo Miyake Seventieth Anniversary. Geochemical Laboratory, Meteorological Research Institute, Tokyo, pp. 91–98, 1978.

Miyakoda, K., Rosati, A., 1982. The variation of sea surface temperature in 1976 and 1977, I: the data analysis. J. Geophys. Res. 87, 5667–5680.

Mofjeld, H.O., Whitmore, P.M., Eble, M.C., González, F.I., Newman, J.C., 2001. Seismic-wave contributions to bottom pressure fluctuations in the North Pacific — implications for the DART tsunami array. In: Proc. Int. Tsunami Symp. 2001, Seattle, WA, CD, pp. 97—108.

Mofjeld, H.O., 2009. Tsunami measurements (Chapter 7). In: The Sea. Tsunamis, volume 15. Harvard University Press, Cambridge, MA and London, England, pp. 201—235.

Montgomery, R.B., 1938. Circulation in the upper layer of the southern North Atlantic deduced with the aid of isentropic analysis. Pap. Phys. Oceanogr. Meteorol. 6 (2), 55.

Montgomery, R.B., 1958. Water characteristics of Atlantic Ocean and of world ocean. Deep-Sea Res. 5, 134—148.

Mooers, C.N.K., Smith, R.L., 1967. Continental shelf waves off Oregon. J. Geophys. Res. 73, 549—557.

Mooers, C.N.K., 1973. A technique for the cross spectrum analysis of pairs of complex-valued time series, with emphasis on properties of polarized components and rotational invariants. Deep-Sea Res. 20, 1129—1141.

Muench, R.D., Schumacher, J.D., 1979. Some Observations of Physical Oceanographic Conditions on the Northeast Gulf of Alaska Continental Shelf. NOAA Technical Memorandum ERL PMEL-17, Seattle, Wash.

Mungov, G., Eblé, M., Bouchard, R., 2012. DART® tsunameter retrospective and real-time data: a reflection on 10 years of processing in support of tsunami research and operations. Pure Appl. Geophys. http://dx.doi.org/10.1007/s00024-012-0477-5.

Munk, W.H., Cartwright, D.E., 1966. Tidal spectroscopy and prediction. Phil. Trans. R. Soc. Lond. A259, 533—581.

Munk, W.H., Hasselman, C., 1964. Upper resolution of tides. In: Studies in Oceanography. Tokyo Geophysical Institute, University of Tokyo, pp. 339—344.

Murty, T.S., 1984. Storm surges: meteorological ocean tides. Can. Bull. Fish. Aquatic Sci. 212, 897.

Mysak, L.A., LeBlond, P.H., Emery, W.J., 1979. Trench waves. J. Phys. Oceanogr. 9, 1001—1013.

Nemac, A.F.L., Brinkhurst, R.O., 1988. Using the bootstrap to assess statistical significance in the cluster analysis of species abundance data. Can. J. Fish. Aquat. Sci. 45, 965—970.

Nerem, R.S., 2005. The Record of Sea Level Change from Satellite Measurements: What Have We Learned? American Geophysical Union Bowie Lecture.

Nicholson, D., Emerson, S., Eriksen, C.C., 2008. Net community production in the deep euphotic zone of the subtropical North Pacific gyre from glider surveys. Limnol. Oceanogr. 53, 2226—2236.

Niiler, P.P., Davis, R.E., White, H.J., 1987. Water-following characteristics of a mixed layer drifter. Deep-Sea Res. 34, 1867—1882.

Niiler, P.P., Maximenko, N.A., McWilliams, J.C., 2004. Dynamically balanced absolute sea level of the global ocean derived from near-surface velocity observations. Geophys. Res. Lett. 30, 2164. http://dx.doi.org/10.1029/2003GL018628.

Niiler, P.P., 2003. A Brief History of Drifter Technology. Autonomous and Lagrangian Platforms and Sensors Workshop. Scripps Institution of Oceanography, La Jolla, California.

Ninnis, R.N., Emery, W.J., Collins, M.J., 1986. Automated extraction of sea ice motion from AVHRR imagery. J. Geophys. Res. 91, 10,725—10,734.

Nowlin, W.D., Bottero, J.S., Pillsbury, R.D., 1986. Observations of internal and near-inertial oscillations at Drake Passage. J. Phys. Oceanogr. 16, 87—108.

Nuttall, A.H., Carter, G.C., 1980. A generalized framework for power spectral estimation. IEEE Trans. Acoust. Speech Signal Process ASSP-28, 334—335.

Nuttall, A.H., 1976. Spectral Analysis of a Univariate Process with Bad Data Points, via Maximum Entropy, and Linear Predictive Techniques. Technical Document 5419. Naval Underwater systems Center, New London, CT.

Odelson, B.J., Rajamani, M.R., Rawlings, J.B., 2005. A New Autocovariance Least-squares Method for Estimating Noise Covariances. Texas-Wisconsin Modeling and Control Consortium (TWMCC). Tech. Rpt. 2003—04.

Ohring, G., Wielicki, B., Spencer, R., Emery, B., Datla, R., 2005. Satellite instrument calibration for measuring global climate change: report of a workshop. Bull. Am. Meteor. Soc. 86, 1303—1313.

Olbers, D.J., Müller, P., Willebrand, J., 1976. Inverse technique analysis of large data sets. Phys. Earth Planet. Int. 12, 248—252.

Olbers, D.J., Wenzel, M., Willebrand, J., 1985. The inference of North Atlantic circulation patterns from climatological hydrographic data. Rev. Geophys. 23, 313—356.

Oldenburg, D.W., 1984. An introduction to linear inverse theory. IEEE Geosci. Rem. Sens. GW-22, 665—674.

Osborne, A.R., Kirwan, A.D., Provenzale, A., Bergamasco, L., 1989. Fractal drifter trajectories in the Kuroshio Extension. Tellus 41A, 416—435.

Östlund, H.G., Rooth, C.H., 1990. The North Atlantic tritium and radiocarbon transients 1972—1983. J. Geophys. Res. 95, 20,147—20,165.

Otnes, R.K., Enochson, L., 1972. Digital Time Series Analysis. John Wiley, New York.

Overland, J.E., Pease, C.H., 1988. Modeling ice dynamics of coastal seas. J. Geophys. Res. 93 (C12), 15,619—15,637.

Overland, J.E., Percival, D.B., Mofjeld, H.O., 2006. Regime shifts and red noise in the North Pacific. Deep Sea Res. Pt. I Oceanogr. Res. Pap. 53 (4), 582—588.

Pagano, M., 1978. Some recent advances in autoregressive processes. In: Brillinger, D.R., Tiao, G.C. (Eds.), Directions in Time Series. Institute for Mechanical Statistics.

Papadakis, J.E., 1981. Determination of the wind mixed layer by an extension of Newton's method. Pac. Marine Sci. Rep., 81–89. Inst. Ocean Sciences, Sidney, BC. Canada, 32 pp.

Papadakis, J.E., 1985. On a class of form oscillators. Speculations Sci. Technol. 8 (5), 291–304.

Paros, J.M., 1976. Digital pressure transducers. Measurements and Data 10 (2), 74–79.

Parsons, T.R., Whitney, F.A., 2012. Did volcanic ash from Mt. Kasatoshi in 2008 contribute to a phenomenal increase in Fraser River sockeye salmon (*Oncorhynchus nerka*) in 2010? Fish. Oceanogr. 21 (5), 374–377.

Parsons, T.R., Maita, Y., Lalli, C.M., 1984. A Manual of Chemical and Biological Methods for Seawater Analysis. Pergamon Press, Oxford.

Patterson, R.T., Prokoph, A., Reinhardt, A., Roe, H.M., 2007. Climate cyclicity in late Holocene anoxic marine sediments from the Seymour-Belize Inlet complex, British Columbia. Marine Geol. 242, 123–140.

Patterson, R.T., Chang, A.S., Prokoph, A., Roe, H.M., Swindles, G.T., 2013. Influence of the Pacific Decadal Oscillation, El Niño-Southern Oscillation and solar forcing on climate and primary productivity changes in the northeast Pacific. Quatern. Int. 310, 124–139.

Paulson, C.A., Simpson, J.J., 1981. The temperature difference across the cool skin of the ocean. J. Geophys. Res. 86, 11,044–11,054.

Pavlidis, T., Horowitz, S.L., 1974. Segmentation of plane curves. IEEE Trans. Comput. 23, 860–870.

Pawlowicz, R., Beardesley, B., Beardsley, R., Lentz, S., 2002. Classical tidal harmonic analysis including error estimates in MATLAB using T_TIDE. Comput. Geosci. 28, 929–937.

Pawlowicz, R., Riche, O., Halverson, M., 2007. The circulation and residence time of the Strait of Georgia using a simple mixing-box approach. Atmos. Ocean 45 (2), 173–193.

Pearson, C.A., Schumacher, J.D., Muench, R.D., 1981. Effects of wave-induced mooring noise on tidal and low-frequency current observations. Deep-Sea Res. 28A, 1223–1229.

Pearson, K., 1901. Principal components analysis. London, Edinburgh, Dublin Philosophical Magazine J. Sci. 6 (2), 559.

Peltier, W.R., 1990. Glacial isostatic adjustment and relative sea-level change. In: Sea-level Change. National Academy Press, Washington, D.C, pp. 73–87.

Peltier, W.R., 2009. Closure of the budget of global sea level rise over the GRACE era: the importance and magnitudes of the required corrections for global glacial isostatic adjustment. Quat. Sci. Rev. 28, 1658–1674. http://dx.doi.org/10.1016/j.quascirev.2009.04.004.

Peng, T.-H., Broecker, W.S., Mathieu, G.G., Li, Y.-H., 1979. Radon invasion rates in the Atlantic and Pacific oceans as determined during the GEOSECS program. J. Geophys. Res. 84, 2471–2486.

Peters, H., Gregg, M.C., Toole, J.M., 1988. On the parameterization of equatorial turbulence. J. Geophys. Res. 93, 1199–1218.

Peterson, J.I., Fitzgerald, R.V., Buckhold, D.K., 1984. Fiberoptic probe for in vivo measurement of oxygen partial pressure. Anal. Chem. 56, 62–67.

Pettigrew, N.R., Irish, J.D., September 1983. An evaluation of a bottom mounted Doppler acoustic profiling current meter. In: Proceedings Oceans '83.

Pettigrew, N.R., Beardsley, R.C., Irish, J.D., 1986. Field evaluations of a bottom mounted acoustic Doppler current profiler and conventional current meter moorings. In: Proceedings of the IEEE Third Working Conference on Current Measurement, January 22–24, 1986, Airlie, Virginia, pp. 153–162.

Pfeffer, W.T., Harper, J.T., O'Neel, S., 2008. Kinematic constraints on glacier contributions to 21st-Century sea-level rise. Science 321, 1340–1343. http://dx.doi.org/10.1126/science.1159099.

Pham, D.T., Verron, J., Roubaud, M.C., 1998. A singular evolutive Kalman filter for data assimilation in oceanography. J. Marine Syst. 16, 323–340.

Phillips, O.M., Gu, D., Donelan, M., 1993. Expected structure of extreme waves in a Gaussian sea. Part I: theory and SWADE buoy measurements. J. Phys. Oceanogr. 23, 992–1000.

Pickard, G.L., Emery, W.J., 1992. Descriptive Physical Oceanography: An Introduction, fifth ed. Pergamon Press, New York.

Pierce, S.D., Barth, J.A., Thomas, R.E., Fleischer, G.W., 2006. Anomalously warm July 2005 in the northern California Current: historical context and the significance of cumulative wind stress. Geophys. Res. Lett. 33 (22).

Pierson, W.J., 1981. The variability of winds over the ocean. In: Beal, R., DeLeonibus, P.S., Katz, I. (Eds.), Space-borne Synthetic Aperture Radar for Oceanography, Johns Hopkins Oceanographic Studies, vol. 7. Johns Hopkins University Press, Baltimore, Md.

Pillsbury, R.D., Bottero, J.S., Still, R.E., Gilbert, W.E., 1974. A Compilation of Observations from Moored Current Meters, vols. VI and VII. School of Oceanography, Oregon State University, Corvallis, Oregon. Refs 74–2 and 74–7.

Piola, A.R., Gordon, A.L., 1984. Pacific and Indian ocean upper-layer salinity budget. J. Phys. Oceanogr. 14, 747–753.

Pitcher, G.C., Figueiras, F.G., Hickey, B.M., Moita, M.T., April–May, 2010. The physical oceanography of upwelling systems and the development of harmful algal blooms. Prog. Oceanogr. 85 (1–2), 5–32.

Pizarro, O., Shaffer, G., 1998. Wind-driven, coastal-trapped waves off the island of Gotland, Baltic Sea. J. Phys. Oceanogr. 28, 2117–2129.

Plaut, G., Vautard, R., 1994. Spells of low-frequency oscillations and weather regimes in the northern hemisphere. J. Atmos. Sci. 51, 210–236.

Poulain, P.-M., Niiler, P.P., 1989. Statistical analysis of the surface circulation in the California Current System using satellite-tracked drifters. J. Phys. Oceanogr. 19, 1588–1603.

Pratt, J.H., 1859: see Vogt and Jung, 1991.

Pratt, J.H., 1871: see Vogt and Jung, 1991.

Preisendorfer, R.W., 1988. Principal Component Analysis in Meteorology and Oceanography. Developments in Atmospheric Science, 17. Elsevier, Amsterdam.

Press, W.H., Teukolsky, S.A., Vetterling, W.T., Flannery, B.P., 1992. Numerical Recipes in Fortran, second ed. Cambridge University Press, Cambridge.

Price, J.F., Weller, R.A., Pinkel, R., 1986. Diurnal cycling: observations and model of the upper ocean response to diurnal heating, cooling and wind mixing. J. Geophys. Res. Oceans 91, 8411–8427.

Priestley, M.B., 1981. Spectral Analysis and Time Series. Academic Press, London.

Privalsky, V.E., Jensen, D.T., 1993. Time Series Analysis Package. Utah Climate Center, Logan.

Privalsky, V.E., Jensen, D.T., 1994. Assessment of the influence of ENSO on annual global air temperatures. Dyn. Atmos. Oceans 22, 161–178.

Pugh, D.T., Spence, N.E., Woodworth, P.L., 1987. Data Holdings of the Permanent Service for Mean Sea Level. Permanent Service for Mean Sea Level, Bidston, Birkenhead, 156 pp.

Qazi, W.A., Emery, W.J., Fox-Kemper, B., 2013. Computing ocean surface currents over the coastal California Current System using 30-minute lag sequential SAR images. TGRS-2013-00144 IEEE Trans. Geosci. Remote Sens., 1–35.

Quadfasel, D., Schott, F., 1979. Comparison of different methods of current measurements. Dt. Hydrogr. Z. 32, 27–38.

Quay, P.D., Stuiver, M., Broecker, W.S., 1983. Upwelling rates for the equatorial Pacific Ocean derived from the bomb ^{14}C distribution. J. Marine Res. 41, 769–792.

Quinn, W.H., Neal, V.T., Antunez de Mayolo, S., 1987. El Niño occurrences over the past four and a half centuries. J. Geophys. Res. 92, 14,449–14,461.

Rabiner, L., Gold, B., 1975. Theory and Application of Digital Signal Processing. Prentice-Hall, Englewood Cliffs, N.J.

Rabinovich, A.B., Levyant, A.S., 1992. Influence of seiche oscillations on the formation of the long-wave spectrum near the coast of the southern Kuriles. Oceanology 32, 17–23.

Rabinovich, A.B., Thomson, R.E., 2001. Evidence of diurnal shelf waves in satellite-tracked drifter trajectories off the Kuril Islands. J. Phys. Oceanogr. 31 (9), 2650–2668.

Rabinovich, A.B., Thomson, R.E., Bograd, S.J., 2002. Drifter observations of anticyclonic eddies off Bussol' Strait, Kuril Islands. J. Oceanogr. 58, 661–671.

Rabinovich, A.B., Shevchenko, G.V., Thomson, R.E., 2007. Sea ice and current response to the wind: a vector regressional analysis approach,. J. Atmos. Oceanic Technol. 24, 1086–1101. http://dx.doi.org/10.1175/JTECH2015.1.

Rabinovich, A.B., Candella, R., Thomson, R.E., 2011a. Energy decay of the 2004 Sumatra tsunami in the world ocean. Pure Appl. Geophys., 32. http://dx.doi.org/10.1007/s00024-011-0279-1.

Rabinovich, A.B., Stroker, K., Thomson, R.E., Davis, E.E., 2011b. DARTs and CORK: high-resolution observations of the 2004 Sumatra tsunami in the northeast Pacific. Geophys. Res. Lett. 38, L08607. http://dx.doi.org/10.1029/2011GL047027.

Rabinovich, A.B., Thomson, R.E., Fine, I.V., 2012. The 2010 Chilean tsunami off the west coast of Canada and the Pacific northwest coast of the United States. Pure Appl. Geophys. http://dx.doi.org/10.1007/s00024-012-0541-1.

Rabinovich, A.B., Candella, R.N., Thomson, R.E., 2013. The open ocean energy decay of three recent trans-Pacific tsunamis. Geophys. Res. Lett. 40, 1–5. http://dx.doi.org/10.1002/grl.50639.

Rabinovich, A.B., 2009. Seiches and harbor oscillations. In: Kim, Y.C. (Ed.), Handbook of Coastal and Ocean Engineering. World Scientific Publ., Singapore, pp. 193–236.

Rao, P.K., Smith, W.L., Koffler, R., 1972. Global sea surface temperature distribution determined from an environmental satellite. Mon. Wea. Rev. 100, 10–14.

Ray, R.D., 1998. Ocean self-attraction and loading in numerical tidal models. Marine Geodesy 21, 181–192,.

RD Instruments, (see also Gordon, R.L.), 1989. Acoustic Doppler Current Profilers. Principles of Operation: A Practical Primer. RD Instruments, San Diego, CA.

Redfield, A.C., Ketchum, B.H., Richards, F.A., 1963. The influence of organisms on the composition of sea-water. In: Hill, M.N. (Ed.), The Sea, vol. 2. Interscience, New York, pp. 26–77.

Redfield, A.C., 1958. The biological control of chemical factors in the environment. Am. Scientist 46, 205–221.

Reid, J.L., Lynn, R.J., 1971. On the influence of the Norwegian-Greenland and Weddell seas upon the bottom waters of the Indian and Pacific oceans. Deep-Sea Res. 18, 1063–1088.

Reid, J.L., Mantyla, A.W., 1978. On the mid-depth circulation of the North Pacific Ocean. J. Phys. Oceanogr. 8, 946–951.

Reid, J.L., 1965. Intermediate waters of the Pacific Ocean. Johns Hopkins Oceanogr. Stud. 2.

Reid, J.L., 1982. Evidence of an effect of heat flux from the East Pacific Rise upon the characteristics of the mid-depth waters. Geophys. Res. Lett. 9, 381–384.

Reul, N., Saux-Picart, S., Chapron, B., Vandemark, D., Tournadre, J., Salisbury, J., 2009. Demonstration of ocean surface salinity microwave measurements from space using AMSR-E data over the Amazon plume. Geophys. Res. Lett. 36, L13607. http://dx.doi.org/10.1029/2009GL038860.

Reusch, D.G., Alley, R.B., Hewitson, B.C., 2007. North Atlantic climate variability from a self-organizing map perspective. J. Geophys. Res. 112, D02104. http://dx.doi.org/10.1029/2006JD007460.

Reynolds, R.W., Smith, T.M., 1994. Improved global sea surface temperature analyses using optimum interpolation. J. Clim. 7, 929–948.

Reynolds, R.W., Rayner, N.A., Smith, T.M., Stokes, D.C., Wang, W., 2002. An improved in situ and satellite SST analysis for climate. J. Clim. 15, 1609–1625.

Reynolds, R.W., 1982. A Monthly Averaged Climatology of Sea Surface Temperature. NOAA Technical Report NWS-31 Nat. Oceanic Atmos. Admin. (Silver Springs, Md).

Reynolds, R.W., 1983. A comparison of sea surface temperature climatologies. J. Clim. App. Meteorol. 22, 447–459.

Richardon, P.L., 1993. A census of eddies observed in North Atlantic SOFAR float data. Prog. Oceanogr. 31, 1–50.

Richardson, W.S., Stimson, P.B., Wilkins, C.H., 1963. Current measurements from moored buoys. Deep-Sea Res. 10, 369–388.

Richardson, P.L., Price, J.F., Owens, W.B., Schmitz Jr, W.J., 1981. North Atlantic subtropical gyre: SOFAR floats tracked by moored listening stations. Science 213, 435–437.

Richardson, A.J., Pfaff, M.C., Field, J.G., Silulwane, N.F., Shillington, F.A., 2002. Identifying characteristic chlorophyll a profiles in the coastal domain using an artificial neural network. J. Plankton Res. 24, 1289–1303.

Riche, O., 2011. Time-dependent Inverse Box-model for the Estuarine Circulation and Primary Productivity in the Strait of Georgia. University of British Columbia, 1–228. https://circle.ubc.ca/handle/2429/37738.

Richman, M.B., 1986. Rotation of principal components. J. Climatol. 6, 293–335.

Riser, S.C., 1982. The quasi-Lagrangian nature of SOFAR floats. Deep-Sea Res. 29, 1587–1602.

Risien, C.M., Reason, C.J.C., Shillington, F.A., Chelton, D.B., 2004. Variability in satellite winds over the Benguela upwelling system during 1999–2000. J. Geophys. Res. 109 (1978–2012).

Riva, R.E.M., Bamber, J.L., Lavellée, D.A., Wouters, B., 2010. Sea-level fingerprint of continental water and ice mass change from GRACE. Geophys. Res. Lett. 37, L19605.

Roache, P.J., 1972. Computational Fluid Dynamics. Hermosa, Albuquerque.

Roberts, J., Roberts, T.D., 1978. Use of the Butterworth low-pass filter for oceanographic data. J. Geophys. Res. 83, 5510–5514.

Robinson, A., McGillicuddy, D.J., Calman, J., Ducklow, H.W., Fasham, M.J.R., Hogc, F.E., Leslie, W.G., McCarthy, J.J., Podewski, S., Porter, D.L., Saure, G., Yoder, J.A., 1993. Mesoscale and upper ocean variabilities during the 1989 JGOFS bloom study. Deep-Sea Res. 40, 9–35.

Robinson, I.S., 1985. Satellite Oceanography. Ellis Horwood, Chichester.

Rodionov, S., Overland, J.E., 2005. Application of a sequential regime shift detection method to the Bering Sea ecosystem. ICES J. Marine Sci. 62, 328–332. http://www.beringclimate.noaa.gov/regimes/JMSPublArticle.pdf.

Rodionov, S.N., 2004. A sequential algorithm for testing climate regime shifts. Geophys. Res. Lett. 31, L09204. http://dx.doi.org/10.1029/2004GL019448. https://docs.google.com/file/d/0B8eNwWtdAAJbWGg4dFhRM1RqT28/edit.

Rodionov, S.N., 2006. Use of prewhitening in climate regime shift detection. Geophys. Res. Lett. 33, L12707. http://dx.doi.org/10.1029/2006GL025904.

Roemmich, D., Cornuelle, B., 1987. Digitization and calibration of the expendable bathythermograph. Deep-Sea Res. 34, 299–307.

Roesler, C.J., Emery, W.J., Kim, S.Y., 2013. Evaluating the use of high-frequency radar coastal current to correct satellite altimetry. J. Geophys. Res. 118 http://dx.doi.org/10.1002/jgrc.20220.

Roll, H.U., 1951. Wassertemperaturemessungen an deck und in Maschinenraum. Ann. Meteor. 4, 439–443.

Rørbæk, K., 1994. Comparison of Aanderaa Instruments DCM 12 Doppler Current Meter with RD Instruments Broadband Direct Reading 600 Khz ADCP. Danish Hydraulic Institute, Copenhagen.

Rosenberg, N.D., Lupton, J.E., Kadko, D., Collier, R., Lilley, M.D., Pak, H., 1988. Estimation of heat and chemical fluxes from a seafloor hydrothermal vent field using radon measurements. Nature 334, 604–607.

Rossby, H.T., Webb, D., 1970. Observing abyssal motions by tracking swallow floats in the SOFAR channel. Deep-Sea Res. 17, 359–365.

Rossby, H.T., Dorson, D., Fontaine, J., 1986. The RAFOS systems. J. Atmos. Oceanic Tech. 3, 672–679.

Rossby, H.T., 1969. On monitoring depth variations of the main thermocline acoustically. J. Geophys. Res. 74, 5542–5546.

Royer, T.C., 1981. Baroclinic transport in the Gulf of Alaska. Part II. A fresh water driven coastal current. J. Mar. Res. 39, 251–266.

Rual, P., 1991. XBT depth correction. In: Addendum to the Summary Report of the Ad Hoc Meeting of the IGOSS Task Team on Quality Control for Automated Systems, Marion, Mass., USA, June 1991, pp. 131–144. IOC/INF-888 Add.

Sanderson, B.G., Okubo, A., Goulding, A., 1990. The fractal dimension of relative Lagrangian motion. Tellus 42A, 550–556.

Sandford, T.B., 1971. Motionally induced electric and magnetic fields in the sea. J. Geophys. Res. 76, 3476–3492.

Sarmiento, J.L., Feely, H.W., Moore, W.S., Bainbridge, A.E., Broecker, W.S., 1976. The relationship between vertical eddy diffusion and buoyancy gradient in the deep sea. Earth Planet. Sci. Lett. 32, 357–370.

Sarmiento, J.L., Toggweiller, J.R., Najjar, R., 1988. Ocean carbon cycle dynamics and atmospheric pCO_2. Phil. Trans. R. Soc. A325, 3–21.

Satake, K., 1993. Depth distribution of coseismic slip along the Nankai Trough, Japan, from joint inversion of geodetic and tsunami data. J. Geophys. Res. 98, 4553–4565.

Saunders, P.M., 1976. Near-surface current measurements. Deep-Sea Res. 23, 249–258.

Saunders, P.M., 1980. Overspeeding of a Savonious rotor. Deep-Sea Res. 27A, 755–759.

Saur, T., 1963. A study of the quality of sea water temperatures reported in logs of ships' weather observations. J. Appl. Meteorol. 2, 417–425.

Sayles, M.A., Aagaard, K., Coachman, L.K., 1979. Oceanographic Atlas of the Bering Sea Basin. University of Washington Press, Seattle, Wash.

Scarborough, J.B., 1966. Numerical Mathematical Analysis. Johns Hopkins Press.

Scarlet, R.I., 1975. A data processing method for salinity, temperature, depth profiles. Deep-Sea Res. 27, 509–515.

Schaad, T., March 10, 2009. Oceanic, Atmospheric and Seismic Sensors with Parts-per Billion resolution. G8218 Rev E.. Paroscientific, Inc. (Technical Note).

Schlosser, P., Bönisch, G., Rhein, M., Bayer, R., 1991. Reduction of deepwater formation in the Greenland Sea during the 1980s: evidence from tracer data. Science 251, 1054–1056.

Schluessel, P., Shin, H.Y., Emery, W.J., Grassl, H., 1987. Comparison of satellite derived sea surface temperature with in situ skin measurements. J. Geophys. Res. 92, 2859–2874.

Schneider, N., Müller, P., 1990. The meridional and seasonal structure of the mixed-layer depth and its diurnal amplitude observed during the Hawaii-to-Tahiti shuttle experiment. J. Phys. Oceanogr. 20, 1395–1404.

Schott, F., Leaman, K.D., 1991. Observations with moored acoustic Doppler current profilers in the convection regime in the Golfe du Lion. J. Phys. Oceanogr. 21, 558–574.

Schott, F., 1986. Medium-range vertical acoustic Doppler current profiling from submerged buoys. Deep-Sea Res. 33, 1279–1292.

Schrama, E.J.O., Ray, R.D., 1994. A preliminary tidal analysis of Topex/Poseidon altimetry. J. Geophys. Res. 99, 24799–24808.

Schumacher, J.D., Reed, R.K., 1986. On the Alaska Coastal Current in the western Gulf of Alaska. J. Geophys. Res. 91, 9655–9661.

Schuster, A., 1898. On the investigation of hidden periodicities with application to a supposed 26 day period of meteorological phenomena. Terrestrial Magnetism 3, 13–41.

Schwing, F.B., O'Farrell, M., Steger, J.M., Baltz, K., 1996. Coastal upwelling indices, west coast of North America, 1946–1995. NOAA Tech. Memo. NOAA-TM-NMFS-SWFSC-231, 144 pp.

Schwing, F.B., Bond, N.A., Bograd, S.J., Mitchell, T., Alexander, M.A., Mantua, N., 2006. Delayed coastal upwelling along the US west coast in 2005: a historical perspective. Geophys. Res. Lett. 33 (22).

Seaver, G.A., Kuleshov, S., 1982. Experimental and analytical error of the expendable bathythermograph. J. Phys. Oceanogr. 12, 592–600.

Shen, Z., Mei, L., 1993. Equilibrium spectra of water waves forced by intermittent wind turbulence. J. Phys. Oceanogr. 23, 505–531.

Shen, Z., Wang, W., Mei, L., 1994. Finestructure of wind waves analyzed with wavelet transform. J. Phys. Oceanogr. 24, 1085–1094.

Shevchenko, G.V., Rabinovich, A.B., Thomson, R.E., 2004. Sea-ice drift on the northeastern shelf of Sakhalin Island. J. Phys. Oceanogr. 34 (11), 2470–2491.

Shum, C.K., Werner, R.A., Sandwell, D.T., Zhang, B.H., Nerem, R.S., Tapley, B.D., 1990. Variations of global mesoscale eddy energy observed from Geosat. J. Geophys. Res. 95, 17,865–17,876.

Siemens, C.W., 1876. On determining the depth of the sea without the use of a sounding line. Phil. Trans. R. Soc. London 166, 671–692.

Skyllingstad, E.D., Smyth, W.D., Moum, J.N., Wijesekera, H., 1999. Upper-ocean turbulence during a westerly wind burst: a comparison of large-eddy simulation results and microstructure measurements. J. Phys. Oceanogr. 29, 5–28.

Smith, W.L., Rao, P.K., Koffler, R., Curtis, W.R., 1970. The determination of sea-surface temperature from satellite high resolution infrared window radiation measurements. Mon. Wea. Rev. 95, 604–611.

Smyth, P., Burl, M., Fayyad, U., Perona, P., 1996. Modeling subjective uncertainty in image annotation. In: AKDDM. AAAI/MIT Press, pp. 517–540.

Smyth, W.D., Hebert, D., Moum, J.N., 1996a. Local ocean response to a multiphase westerly windburst. Part 1: the dynamic response. J. Geophys. Res. 101, 22,495–22,512.

Smyth, W.D., Hebert, D., Moum, J.N., 1996b. Local ocean response to a multiphase westerly windburst. Part 2: thermal and freshwater responses. J. Geophys. Res. 101, 22,513−22,533.

Snedecor, G.W., Cochran, W.G., 1967. Statistical Methods. Iowa State University Press, Ames, Iowa.

Snodgrass, F.E., 1968. Deep sea instrument capsule. Science 162, 78−87.

Sokolova, S.E., Rabinovich, A.B., Chu, K.S., 1992. On the atmosphere-induced sea level variations along the western coast of the Sea of Japan. La Mer 30, 191−212.

Soloviev, A.V., Schlüssel, P., 1994. Parameterization of the cool skin of the ocean and of the air-ocean gas transfer on the basis of modeling surface renewal. J. Phys. Oceanogr. 24, 1339−1346.

Spear, D.J., Thomson, R.E., 2012. Thermohaline staircases in a British Columbia fjord. Atmos. Ocean 50, 127−133. http://dx.doi.org/10.1080/07055900.2011.649034. First Article, 1−7.

Spencer, R.W., Hood, H.M., Hood, R.E., 1989a. Precipitation retrieval over land and ocean with the SSM/I: identification and characteristics of the scattering signal. J. Atmos. Oceanic Tech. 6, 254−273.

Spencer, R.W., Hinton, B.B., Olson, W.S., 1989b. Nimbus-7 37 GHz radiances correlated with radar rain rates over the Gulf of Mexico. J. Clim. Appl. Meteor. 22, 2095−2099.

Spiess, F., 1928. The Meteor Expedition-research and Experiences during the German Atlantic Expedition 1925−27. Amerind, New Delhi, 1985.

Sprent, P., Dolby, G.A., 1980. The geometric mean functional relationship. Biometrics 36, 547−550.

Sreenivasan, K.R., Ramshankar, R., Meneveau, C., 1989. Mixing, entrainment and fractal dimensions of surfaces in turbulent fluids. Proc. R. Soc. London A421, 79−109.

Stacey, M.W., Pond, S., LeBlond, P.H., 1988. An objective analysis of the low-frequency currents in the Strait of Georgia. Atmos. Ocean 26, 1−15.

Stegen, G.R., Delisi, D.P., Von Collins, R.C., 1975. A portable, digital recording, expendable bathythermograph (XBT) system. Deep-Sea Res. 22, 447−453.

Steinhart, J.C., Hart, S.R., 1968. Calibration curves for thermistors. Deep-Sea Res. 15, 497−503.

Stockwell, R.G., Mansinha, L., Lowe, R.P., 1994. Localization of the complex spectrum: the S transformation. AGU Trans. 55.

Stommel, H., Schott, F., 1977. The beta spiral and the determination of the absolute velocity field from hydrographic station data. Deep-Sea Res. 24, 325−329.

Strickland, J.D.H., Parsons, T.R., 1968. A Practical Handbook of Seawater Analysis. Bull. Fish. Res. Board Canada.

Strickland, J.D.H., Parsons, T.R., 1972. A Practical Handbook of Seawater Analysis, second ed. Bull. Fish. Res. Board Canada.

Strong, A.E., Pritchard, J.A., 1980. Regular monthly mean temperatures of the Earth's oceans from satellites. Bull. Am. Meteorol. Soc. 61, 553−559.

Stuiver, M.P., Quay, P.D., Östlund, N.D., 1982. Abyssal water carbon-14 distribution and the age of the world oceans. Science 219, 849−851.

Sturges, W., 1983. On interpolating gappy records for time-series analysis. J. Geophys. Res. 88, 9736−9740.

Suijlon, C.E.C., Buyse, J., 1994. Potentials of photolytic rhodamine WT as a large-scale water tracer assessed in a long-term experiment in the Loosdrecht lakes. Limnol. Oceanogr. 39, 1411−1423.

Sverdrup, H.U., Johnson, M.W., Fleming, R.H., 1942. The Oceans: Their Physics, Chemistry, and General Biology. Prentice-Hall, Engelwood Cliffs, N.J.

Sverdrup, H.U., 1947. Wind driven currents in a baroclinic ocean with applications to the equatorial currents in the eastern Pacific. Proc. Natl. Acad. Sci. U.S.A. 33, 318−336.

Swallow, J.C., 1955. A neutrally-buoyant float for measuring deep current. Deep-Sea Res. 3, 74−81.

Swift, C.T., McIntosh, R.E., 1983. Considerations for microwave remote sensing of ocean-surface salinity. IEEE Trans. Geosci. Remote Sens. 21, 480−491.

Sybrandy, A.L., Niiler, P.P., 1990. The WOCE/TOGA SVP Lagrangian Drifter Construction Manual. Scripps Institution of Oceanography, University of California, San Diego. SIO Reference 90−248.

Sylvester, J.J., 1889. On the reduction of a bilinear quantic of the nth order to the form of a sum of n products by a double orthogonal substitution. Messenger Math. 19, 42−46.

Tabata, S., Stickland, J.A., 1972. Summary of Oceanographic Records Obtained from Moored Instruments in the Strait of Georgia, 1969−70. Current Velocity and Seawater Temperature from Station H-06. Pacific Marine Science Report 72−7. Environment Canada.

Tabata, S., 1978a. On the accuracy of sea-surface temperatures and salinities observed in the northeast Pacific. Atmos. Ocean 16, 237−247.

Tabata, S., 1978b. Comparison of observations of sea-surface temperatures at ocean station "P" and N.O.A.A. buoy stations and those made by merchant ships travelling in their vicinities, in the northeast Pacific Ocean. J. Appl. Meteorol. 17, 374−385.

Talley, L.D., Joyce, T.M., 1992. Double silica maximum in the North Pacific. J. Geophys. Res. 97, 5465−5480.

Talley, L.D., Martin, M., Salameth, P., 1988. Trans Pacific Section in the Subpolar Gyre (TPS47): Physical, Chemical, and CTD Data. R/V Thomas Thompson TT190, 4 August 1985−7 September 1985. SIO Ref. 88−9. Scripps Institute of Oceanography, La Jolla, Calif.

Talley, L.D., Joyce, T.M., deSzoeke, R.A., 1991. Trans-Pacific sections at 47°N and 152°W: distribution of properties. Deep-Sea Res. 38, 563−582.

Talley, L.D., Pickard, G.L., Emery, W.J., Swift, J.H., 2011. Descriptive Physical Oceanography: An Introduction. Elsevier Ltd.

Tapley, B.D., Born, G.H., Park, M.E., 1982. The Seasat altimeter data and its accuracy assessment. J. Geophys. Res. 87, 3179−3188.

Tauber, G.M., 1969. The Comparative Measurements of Sea Surface Temperature in the USSR. Technical Note 103, Sea Surface Temperature. WMO, pp. 141−151.

Taylor, K.E., 2005. Taylor Diagram Primer. Wikipedia, 4pp.; (Taylor, K.E., 2001: Summarizing multiple aspect of model performance in a single diagram. J. Geophys. Res., 106, 7183−7192.).

Tchernia, P., 1980. Descriptive Regional Oceanography. In: Pergamon Marine Series, vol. 3. Pergamon Press, Oxford.

Tengberg, A., Hovdenes, J., Andersson, J.H., Brocandel, O., Diaz, R., Hebert, D., Arnerich, T., Huber, C., Körtzinger, A., Khripounoff, A., Rey, F., Rönning, C., Sommer, S., Stangelmayer, A., 2006. Evaluation of a life time based optode to measure oxygen in aquatic systems. Limnol. Oceanogr. Methods 4, 7−17.

Tennant, W., 2004. Considerations when using pre-1979 NCEP/NCAR reanalysis in the southern hemisphere. Geophys. Res. Lett. 31. http://dx.doi.org/10.1029/2004GL019751.

Thompson, T.W., Weissman, D.E., González, F.I., 1983. L-band radar backscatter dependence upon surface wind stress: a summary of new Seasat-I and aircraft observations. J. Geophys. Res. 88, 1727−1735.

Thompson, R., 1971. Spectral estimation from irregularly spaced data. IEEE Trans. Geosci. Electron. GE-9, 107−119.

Thompson, R.O.R.Y., 1979. Coherence significance levels. J. Atmos. Sci. 36, 2020−2021.

Thompson, R.O.R.Y., 1983. Low-pass filters to suppress inertial and tidal frequencies. J. Phys. Oceanogr. 13, 1077−1083.

Thomson, R.E., Fine, I.V., 2009. A diagnostic model for mixed layer depth estimation with application to ocean station "P" in the northeast Pacific. J. Phys. Oceanogr. 39, 1399−1415. http://dx.doi.org/10.1175/2008JPO3984.1.

Thomson, R.E., Freeland, H.J., 1999. Lagrangian measurement of mid-depth currents in the eastern tropical Pacific. Geophys. Res. Lett. 26, 3125−3128.

Thomson, R.E., Huggett, W.S., 1980. M2 baroclinic tides in Johnstone Strait, British Columbia. J. Phys. Oceanogr. 10, 1509−1539.

Thomson, R.E., Krassovski, M.V., 2010. The poleward reach of the California Undercurrent extension. J. Geophys. Res. Oceans 115, C09027. http://dx.doi.org/10.1029/2010JC006280.

Thomson, R.E., Ware, D.M., 1996. A current velocity index of ocean variability. J. Geophys. Res. 101, 14,297−14,310.

Thomson, R.E., Crawford, W.R., Huggett, W.S., 1985. Low-pass filtered current records for the west coast of Vancouver Island: Coastal Oceanic Dynamics Experiment, 1979-81. Can. Data Rep. Hydrography Ocean Sci. 40, 102.

Thomson, R.E., Curran, T.A., Hamilton, M.C., McFarlane, R., 1988. Time series measurements from a moored fluorescence-based dissolved oxygen sensor. J. Atmos. Oceanic Technol. 5, 614−624.

Thomson, R.E., LeBlond, P.H., Emery, W.J., 1990. Analysis of deep-drogued satellite-tracked drifter measurements in the northeast Pacific. Atmos. Ocean 28, 409−443.

Thomson, R.E., Gordon, R.L., Dolling, A.G., 1991. An intense acoustic back-scattering layer at the top of a mid-ocean ridge hydrothermal plume. J. Geophys. Res. 96, 4839−4844.

Thomson, R.E., Burd, B.J., Dolling, A.G., Gordon, R.L., Jamieson, G.S., 1992. The deep scattering layer associated with the Endeavour Ridge hydrothermal plume. Deep-Sea Res. 39, 55−73.

Thomson, R.E., Delaney, J.R., McDuff, R.E., Janecky, D.R., McLain, J.S., 1992. Physical characteristics of the Endeavour Ridge hydrothermal plume during July 1988. Earth Planetary Sci. Lett. 111, 141−154.

Thomson, R.E., LeBlond, P.H., Rabinovich, A.B., 1997. Oceanic odyssey of a satellite-tracked drifter: North Pacific variability delineated by a single drifter trajectory. J. Oceanogr. 53, 81−87.

Thomson, R.E., LeBlond, P.H., Rabinovich, A.B., 1998. Satellite-tracked drifter measurements of inertial and semidiurnal currents in the northeast Pacific. J. Geophys. Res. 103 (1), 1039−1071.

Thomson, R.E., Mihály, S.F., Kulikov, E.A., 2007. Estuarine versus transient flow regimes in Juan de Fuca Strait. J. Geophys. Res. Oceans 112, C09022. http://dx.doi.org/10.1029/2006JC003925.

Thomson, R.E., Davis, E.E., Heesemann, M., Villinger, H., 2010. Observations of long-duration episodic bottom currents in the Middle America Trench: evidence for tidally initiated turbidity flows. J. Geophys. Res. Oceans 115, C10020. http://dx.doi.org/10.1029/2010JC006166.

Thomson, R.E., Fine, I.V., Rabinovich, A.B., Mihály, S.F., Davis, E.E., Heesemann, M., Krassovski, M.V., 2011. Observations of the 2009 Samoa tsunami by the NEPTUNE-Canada cabled observatory: test data for an operational regional tsunami model. Geophys. Res. Lett. 38, L11701. http://dx.doi.org/10.1029/2011GL046728.

Thomson, R.E., Fine, I.V., Krassovski, M.V., Cherniawsky, J.Y., Conway, K.W., Wills, P., 2012. Numerical Simulation of Tsunamis Generated by Submarine Slope Failures in Douglas Channel, British Columbia. DFO Canadian Science Advisory Secretariat (CSAS). Research Document 2012/155. Vi+38pp.

Thomson, R.E., 1977. Currents in Johnstone Strait, British Columbia: supplementary data on the Vancouver island side. J. Fish. Res. Bd. Can. 34, 697—703.

Thomson, R.E., 1981. Oceanography of the British Columbia coast. Can. Spec. Pub. Fish. Aquat. Sci. 56 (Ottawa).

Thomson, R.E., 1983. A comparison between computed and measured oceanic winds near the British Columbia coast. J. Geophys. Res. 88, 2675—2683.

Thorndike, A.S., Colony, R., 1982. Sea ice motion response to geostrophic winds. J. Geophys. Res. 87 (C8), 5845—5852.

Thorndike, A.S., 1986. Kinematics of the sea ice. In: Untersteiner, N. (Ed.), The Geophysics of Sea Ice. Plenum, New York, pp. 489—549.

Thorpe, S.A., 1977. Turbulence and mixing in a Scottish loch. Phil. Trans. R. Soc. London A286, 125—181.

Tichelaar, B.W., Ruff, L.J., 1989. How good are our best models? Jackknifing, bootstrapping, and earthquake depth. EOS 70 (20), 593—605.

Tinis, S.W., Thomson, R.E., Mass, C.F., Hickey, B.M., 2006. Comparison of MM5 and meteorological buoy winds for the west coast of North America. Atmos. Ocean 44 (1), 65—81.

Titov, Rabinovich, V.A.B., Mofjeld, H.O., Thomson, R.E., González, F.I., 2005. The global reach of the 26 December 2004 Indian Ocean tsunami. Science 309, 2045—2048.

Toggweiler, J.R., Trumbore, S., 1985. Bomb-test ^{90}Sr in Pacific and Indian Ocean surface water as recorded by banded corals. Earth Planet. Sci. Lett. 74, 306—314.

Tokamamkian, R., Strub, P.T., McClean-Padman, J., 1990. Evaluation of the maximum cross-correlation method of estimating sea surface velocities from sequential satellite images. J. Atmos. Oceanic Technol. 7, 852—865.

Topham, D.R., Perkins, R.G., 1988. CTD sensor characteristics and their matching for salinity calculations. IEEE J. Oceanic Eng. 13, 107—117.

Trenberth, K.E., Olson, J.G., 1988. ECMWF Global Analysis 1979—1986: Circulation Statistics and Data Evaluation. National Center for Atmospheric Research. NCAR Technical Report Note NCAR/TN-300+STR.

Trenberth, K.E., 1975. A quasi-biennial standing wave in the southern hemisphere and interrelations with sea surface temperature. Quart. J. Roy. Meteor. Soc. 101, 55—74.

Trumbore, S.E., Jacobs, S.S., Smethie Jr, W.M., 1991. Chlorofluorocarbon evidence for rapid ventilation of the Ross Sea. Deep-Sea Res. 38, 845—870.

Trump, W., 1983. Effect of ship's roll on the quality of precision CTD data. Deep-Sea Res. 30 (11 A), 1173—1183.

Tsonis, A.A., Eisner, J.B., 1990. Comments on "Dimension analysis of climatic data". J. Clim. 3, 1502—1505.

Tsonis, A.A., 1991. Sensitivity of the global climate system to initial conditions. EOS Trans. AGU 72, 313—328.

Tsonis, A.A., 1992. Autoregressive models not sensitive to initial conditions. Reply. EOS Trans. AGU 25, 268.

Turner, J.S., Kraus, E.B., 1967. A one-dimensional model of the seasonal thermocline. I. A laboratory experiment and its interpretation. Tellus 19, 88—97.

Tushingham, A.M., Peltier, W.R., 1991. ICE-3-G: a new global model of late Pleistocene deglaciation based upon geophysical predictions of post glacial relative sea level change. J. Geophys. Res. 96, 4497—4523.

Ulrych, T.J., Bishop, T.N., 1975. Maximum entropy spectral analysis and autoregressive decomposition. Rev. Geophys. Space Phys. 13, 183—200.

Ulrych, T.J., 1972. Maximum entropy spectrum of truncated sinusoids. J. Geophys. Res. 77, 1396—1400.

UNESCO, 1966. International Oceanographic Tables. Unesco, Place de Fontenoy. National UNESCO Office of Oceanography, Institute of Oceanography, Wormley, Paris.

Urick, R.J., 1967. Principles of Underwater Sound. McGraw-Hill, New York.

Vachon, W.A., 1973. Scale Model Testing of Drogues for Free Drifting Buoys. Technical Report. The Charles Stark Draper Laboratory, Inc., Cambridge, Mass.

van Beers, W.C.M., Kleijnen, J.P.C., 2003. Kriging for interpolation in random simulation. J. Oper. Res. Soc. 54, 255—262.

van Leer, J., Düing, W., Erath, R., Kennelly, E., Speidel, A., 1974. The cyclesonde: an unattended vertical profiler for scalar and vector quantities in the upper ocean. Deep-Sea Res. 21, 385—400.

van Scoy, K.A., Fine, R.A., Östlund, H.G., 1991. Two decades of missing tritium into the North Pacific Ocean. Deep-Sea Res. 38, S191—S219.

Vanicek, P., 1971. Further development and properties of the spectral analysis by least-squares. Astrophys. Space Sci. 12, 10—73.

Vazquez, J., Zlotnicki, V., Fu, L.-L., 1990. Sea level variability in the Gulf Stream between Cape Hateras and 50°N, a GEOSAT study. J. Geophys. Res. 95, 17,957—17,964.

Vesanto, J., 2000. Using SOM in Data Mining. Interactive Data Visualization Using Focusing and Linking. Helsinki University of Technology (thesis) for the Degree of Licentiate of Science in Technology, Espoo, Finland, 49 pp.

Vogt, P.R., Jung, W.-Y., 1991. Satellite radar altimetry aids seafloor mapping. EOS 72 (43), 465, 468—469.

Volkov, D.L., Fu, L-L., Lee, T., 2010. Mechanisms of the meridional heat transport in the Southern Ocean. Ocean Dyn. 60, 791—801.

von Arx, W.S., 1950. An electromagnetic method for measuring the velocities of ocean currents from a ship under way. Pap. Phys. Oceanogr. Meteor. 11, 1—61.

von Storch, H., Zwiers, F.W., 1999. Statistical Analysis in Climate Research. Cambridge University Press, 484 pp.

Wadhams, P., 2000. Ice in the Ocean. Gordon and Breach Sci. Publ., Amsterdam, 351 pp.

Wadsworth, G.P., Bryan, J.G., Gordon, C.H., 1948. Short Range and Extended Forecasting by Statistical Methods. Air Weather Service Tech. Report No. 105-38. U.S. Air Force, Washington, D.C, 186 pp.

Walden, H., 1966. Zur messung der Wassertemperatur auf Handelsschiffen. Dtsch Hydro. Z. 19, 21—28.

Waliser, D.E., Gautier, C., 1993. Comparison of buoy and SSM/I-derived wind speeds in the tropical Pacific. TOGA Notes 12, 1—7.

Walker, E.R., Chapman, K.D., 1973. Salinity-conductivity Formulae Compared. Pacific Marine Science Report 73—5. Institute of Ocean Sciences, Sidney, British Columbia, Canada.

Wallace, J.M., Dickinson, R.E., 1972. Empirical orthogonal representation of time series in the frequency domain. Part I: theoretical considerations. J. Appl. Meteor. 11, 887—892.

Wallace, D.W.R., Lazier, J.R.N., 1988. Anthropogenic chlorofluoromethanes in newly formed Labrador Sea water. Nature 332, 61—63.

Wallace, J.M., 1972. Empirical orthogonal representation of time series in the frequency domain. Part II: application to the study of tropical wave disturbances. J. Appl. Meteor. 11, 893—900.

Walters, R.A., Heston, C., 1982. Removing tidal-period variations from time-series data using low-pass digital filters. J. Phys. Oceanogr. 12, 112—115.

Wang, D.-P., Mooers, C.N.K., 1977. Long coast trapped waves off the west coast of the United States, summer 1973. J. Phys. Oceanogr. 7, 856—864.

Wang, D.-P., Oey, L.-Y., Ezer, T., Hamilton, P., 2003. Near-surface currents in DeSoto Canyon (1997-99): comparison of current meters, satellite observation, and model simulation. J. Phys. Oceanogr. 33 (1), 313—326.

Ward, B., Minnett, P.J., 2001. An autonomous profiler for near surface temperature measurements. Gas Transfer Water Surfaces, Geophys. Monogr. 127. Amer. Geophys. Union, 167—172.

Warren, B.A., 1970. General circulation of the South Pacific. In: Wooster, W.S. (Ed.), Scientific Exploration of the South Pacific. National Academy of Sciences, Washington, D.C, pp. 33—49.

Warren, B.A., 1983. Why is no deep water formed in the North Pacific? J. Mar. Res. 41, 327—347.

Washburn, L., Emery, B.M., Jones, B.H., Ondercin, D.G., 1998. Eddy stirring and phytoplankton patchiness in the subarctic North Atlantic in late summer. Deep-Sea Res. 45, 1411—1439.

Watanabe, Y.W., Watanabe, S., Tsunogai, S., 1991. Tritium in the Japan Sea and the renewal time of Japan Sea deep water. Mar. Chem. 34, 97—108.

Watson, A.J., Ledwell, J.R., 1988. Purposefully released tracers. Phil. Trans. R. Soc. London A325, 189—200.

Watson, A.J., Liddicoat, M.I., 1985. Recent history of atmospheric trace gas concentrations deduced from measurements in the deep sea: application to sulpher hexafluoride and carbon tetrachloride. Atmos. Environ. 19, 1477—1484.

Watts, D.R., Rossby, H.T., 1977. Measuring dynamic heights with inverted echo sounders: results from MODE. J. Phys. Oceanogr. 7, 345—358.

Weare, B.C., Nasstrom, J.S., 1982. Examples of extended empirical orthogonal function analyses. Mon. Weath. Rev. 110, 481—485.

Weare, B.C., Navato, A.R., Newell, R.E., 1976. Empirical orthogonal analysis of Pacific sea surface temperatures. J. Phys. Oceanogr. 6, 671—678.

Wearn, R.B., Baker Jr, D.J., 1980. Bottom pressure measurements across the Antarctic Circumpolar Current and their relation to the wind. Deep-Sea Res. 21 A, 875—888.

Webb, A.J., Pond, S., 1986. A modal decomposition of the internal tide in a deep, strongly stratified inlet: Knight Inlet, British Columbia. J. Geophys. Res. 91, 9721—9738.

Weinreb, M.P., Hamilton, G., Brown, S., Koczor, R.J., 1990. Nonlinearity corrections in calibration of advanced very high resolution radiometer infrared channels. J. Geophys. Res. 95, 7381—7388.

Weiss, W., Roether, W., 1980. The rates of tritium input to the world oceans. Earth Planet Sci. Lett. 49, 453—446.

Welch, G., Bishop, G., 2006. An Introduction to the Kalman Filter. Department of Computer Science, University of North Carolina at Chapel Hill. TR 95—041.

Weller, R.A., Davis, R.E., 1980. A vector measuring current meter. Deep-Sea Res. 27, 565—582.

Weller, R.A., Plueddemann, A.J., 1996. Observations of the vertical structure of the oceanic boundary layer. J. Geophys.L Res. Oceans 101, 8789—8806.

Wenner, F., Smith, E.H., Soule, F.M., 1930. Apparatus for the determination aboard ship of the salinity of sea water by the electrical conductivity method. Bur. Stand. J. Res. 5, 711—732.

Wentz, F.J., Mattox, L.A., Peteherych, W., 1986. New algorithms for microwave measurements of ocean winds: applications to SEASAT and the special sensor microwave imager. J. Geophys. Res. 91, 2289—2307.

Wüst, G., 1957. Stromgeschwindigkeiten und Strommengen in den Tiefen des Atlantischen Ozeans. In: Wissenschaftliche Ergebnisse der Deutschen Atlantischen Expedition Meteor 1925—1927, 6, 261—420.

Wijesekera, H.W., Gregg, M.C., 1996. Surface layer response to weak winds, westerly bursts, and rain squalls in the western Pacific warm pool. J. Geophys. Res. Oceans 101, 977—997.

Wijffels, S.E., Firing, E., Bryden, H.L., 1994. Direct observations of the Ekman balance at 10°N in the Pacific. J. Phys. Oceanogr. 24, 1666—1679.

(cont'd)

Geopotential	1 dynamic centimeter (dyne cm) $= 10^3$ cm^2/s$^2 \approx 1.02$ cm
	1 joule/kg (1 J/kg) $= 1$ m^2/s$^2 = 10$ dyne cm
	Specific volume anomaly: $\delta = \alpha_{S,T,P} - \alpha_{35,0,P}$ where $\alpha = \rho^{-1}$ m^3/kg
	Geopotential anomaly (or dynamic height anomaly in the older literature):
	$\Delta\Phi(\text{or}\Delta D) = \Phi_{P_2} - \Phi_{P_1} = -\int_{P_1}^{P_2} \delta \mathrm{d}P$ joules/kg
	Potential energy:
	$PE = g^{-1} \int_{P_1}^{P_2} \delta(P)P\mathrm{d}P$ joules/m^2
Transport	1 sverdrup (Sv) $= 10^6$ m^3/s
Specific heat	Liquid water: $c_p = 4.1855$ J/(g·K) $= 1$ cal/g°C (at 15 °C, 101.325 kPa)
Capacity	Air: $c_p = 1.006$ J/(g·K) (at -50 to 40 °C; 101.325 kPa pressure)

Glossary of Statistical Terminology

Alternative hypothesis: Value of a parameter of a population other than the value hypothesized or believed to be true by the investigator.

Asymptotically normal distribution: A distribution of values that is not truly normal, but which approaches a normal distribution as the number of samples becomes very large.

Autocorrelation: In a time series, $x(t)$, with zero mean, the statistical relationship between values of a variable taken at certain times in the series and values of a variable taken at other times (a function of the time lag, τ, between the two series) written as,

$$R_{xx}(\tau) = E[x(t)x(t + \tau)]$$

where E is the expected value. Autocovariance is similar except that the mean of the record is not subtracted prior to the analysis.

Biased estimator: An estimator $\hat{x}$ for which the expected value $E[\hat{x}]$ of a sample has a systematic error with respect to the true expected value, μ_x; i.e., $E[\hat{x}] \neq \mu_x$.

Bin interval: A specified arbitrary interval, which partitions a quantity whose number of occurrences are being measured; used for constructing a histogram (frequency of occurrence distribution) of the data set.

Central limit theorem: States that the distribution of sample means taken from a large population approaches a normal (Gaussian) distribution.

Chi-squared distribution: The distribution function generated by the random variable,

$$\chi_n^2 = X_1^2 + X_2^2 + X_3^2 + \ldots + X_n^2$$

where $X_1, X_2, \ldots, X_n$ are n independent random variables drawn from a normal population with zero mean. The chi-squared variable χ_n^2 has an expected value (mean) of n and a variance of $2n$.

Confidence interval: An interval that has a specified probability of containing a given value or characteristic. If the values (e.g., measurements) are normally distributed, the commonly used 95% confidence interval implies that 95% of the values will lie within ± 2 standard deviations of the mean.

Continuous random variable: A random variable that may be characterized as a continuous function.

Correlation (covariance): A quantitative measure of the interdependence or association between two variables.

Countable: Either finite or denumerable.

Cross correlation: The correlation between corresponding members of two or more different series of the same duration: if $x(t) = (x_1, x_2, \ldots, x_n)$ and $y(t) = (y_1, y_2, \ldots, y_n)$ are two series, the cross correlation is the correlation between $x(t)$ and $y(t)$, or between $x(t)$ and $y(t+\tau)$ for lag τ such that,

$$R_{xy}(\tau) = E[x(t)y(t + \tau)]$$

(*see Autocorrelation*).

Cumulative distribution function: The integral of the probability density function, $p(x)$, over some specified interval, (x_1, x_2). The integral or sum of $p(x)$ from $-\infty$ to x gives the cumulative total of all values whose value is less than or equal to x.

Degrees of freedom: The number of truly independent samples used to estimate a parameter; when each sample of a series of length N is independent of all other values, the degrees of freedom, $\nu = N - 1$.

Discrete random variable: A random variable that may be characterized as a discrete function.

Ensemble average: An average over several realizations of a random variable taken over times of equal duration.

Ergodic hypothesis: The hypothesis that replaces statistical replications (ensemble averages) with averages in space or time. Allows us to compute averages in space or time as ensemble averages.

Estimator bias: The difference between an estimate and the true value of the parameter being estimated.

Expected value: For a random variable x with a probability function $f(x)$, this is the integral from $-\infty$ to ∞ of $xf(x)$; also known as the expectation and written as,

$$E(x) = \int_{-\infty}^{\infty} xf(x)\mathrm{d}x = \mu$$

Gamma distribution: A normal distribution whose probability distribution function involves the gamma function as,

$$f(x) = \frac{x^{\alpha-1}e^{-x/\beta}}{\beta^{\alpha}\Gamma(\alpha)}, \alpha, \beta > 0; \ 0 \le x \le \infty$$

$$= 0 \ \text{elsewhere}$$

where α and β are parameters of the distribution and $\Gamma(\alpha)$ is the gamma function.

$$\Gamma(\alpha) = \int_{0}^{\infty} x^{\alpha-1}e^{-x}\mathrm{d}x.$$

Gauss–Markov theorem: An unbiased linear estimator of a parameter has minimum variance, that is, is the best estimator, when it is determined by the least squares method.

Hypothesis testing: The branch of statistics that considers the problem of choosing between two actions on the basis of the observed value of a random variable whose distribution depends on a parameter, the value of which would indicate the correct choice to make.

Independent random variables: The discrete random variables, $X_1, X_2, ..., X_n$ are independent if for arbitrary values $x_1, x_2, ..., x_n$ of the variables, the probability that $X_1 = x_1, X_2 = x_2, ...$ is equal to the product of the probabilities that $X_i = x_i$, for $i = 1, 2, ..., n$; in this case, the random variables are unrelated.

Inference: The act of passing from statistical sample data to generalizations (as when inferring the values of true population parameters) with calculated degrees of certainty (confidence intervals at selected significance levels).

Joint probability density function: The distribution which gives the probability that $X_i = x_i$, for $i = 1, 2, ..., n$ for all values x_i of the random variable X_i.

Jointly sufficient statistics: Let $X_1, X_2, ..., X_n$ be a random sample from a probability density function $p(X; \theta)$, where θ is an unknown statistical parameter such as the mean or variance; the statistics $S_1, ..., S_r$ are defined to be jointly sufficient if and only if the conditional distribution of $X_1, X_2, ..., X_n$ given $S_1 = s_1, ..., S_r = s_r$ does not depend on θ.

Least squares method: A technique for fitting a line, polynomial, or other curve to a given distribution of points, which minimizes the sum of the squares of the deviations of the given points from the fitted curve.

Likelihood L(x): The likelihood of occurrence of a sample of independent values of $x_1, x_2, \ldots, x_n$, with a probability function $f(x)$, is the product $f(x_1) \cdot f(x_2) \cdot \ldots f(x_n)$.

Likelihood ratio: The probability of a random drawing of a specified sample from a population, assuming a given hypothesis about the parameter of the population, divided by the probability of a random drawing of the same sample assuming that the parameters of the population are such that this probability is maximized; i.e., $L(x_1)/L(x_2)$.

Linear regression: The straight line running through the points of a scatter diagram about which the amount of scatter is a minimum in the least squares sense.

Lognormal distributions: Because many parameters (especially biological quantities such as zooplankton biomass and surface chlorophyll-*a* concentration) change in an exponential manner in response to growth and mortality rates, they can be considered to have a lognormal distribution. The use of log-transformed data (using the natural logarithm, $\ln(x)$ for data values, x) permits use of standard statistical analysis assumptions and associated methodologies.

Maximum likelihood estimation: A method whereby the likelihood distribution is maximized to produce an estimate of the random variable.

Mean square error: A measure of the extent to which a collection of numbers, $x_1, x_2, \ldots, x_n$, is unequal and defined by the expression,

$$\begin{aligned}
\text{mean square error} &= E\left[(\hat{x} - x)^2\right] \\
&= E\left[(\hat{x} - E[\hat{x}])^2\right] \\
&\quad + E\left[(E[\hat{x}] - x)^2\right]
\end{aligned}$$

where $E[x]$ is the expected (mean) value and $\hat{x}$ is the estimator for x.

Method of moments: A procedure for estimating the parameters (such as the mean and variance) of a distribution.

Minimal sufficient statistic: A set of jointly sufficient statistics is defined to be minimal sufficient if and only if it is a function of every other set of sufficient statistics.

Moments: The nth moment of a distribution $f(x)$ about a point x_0 is the expected value $E[f^n(x - x_0)]$; the first moment is the mean of the distribution, while the variance is a function of the first and second moments (this definition only applies to moments about the mean and not the origin).

Moment generating function: Let x be a random variable with probability density function $p(x)$; the expected value of some function $f(rx)$, $E[f(rx)]$, is defined to be the moment generating function $m(r)$ of f if the expected value exists for every value of r in some interval ($\eta < r < \eta$; $\eta > 0$) whereby,

$$m(r) = E[f(rx)] = \int_{-\eta}^{\eta} f(rx)p(x)\mathrm{d}x.$$

Moment of a power spectrum, $S(f)$: Defined as $m_n = \int_0^\infty S(f)f^n df$ where f is the frequency. In the case of surface gravity waves, the *significant wave height* $H_s = (16m_o)^{1/2}$ and the *mean zero crossing period* is $T = (m_0/m_2)^{1/2}$.

Multivariate analysis: The study of random variables, which are multidimensional.

Normal (or Gaussian) probability distribution function: A normally distributed frequency distribution of a random variable x with a mean μ and standard deviation σ given by,

$$p(x) = \frac{1}{\sigma\sqrt{2\pi}}\exp\left[\frac{(x - \mu)^2}{2\sigma^2}\right].$$

Null hypothesis: The hypothesis that there is no validity to the specific claim that two variations (treatments) of the same thing can be distinguished by a specific procedure.

Pivotal statistic. The statistic that allows one to compute a confidence interval for a specific estimate.

Population: Any finite or infinite collection of individuals or elements that can be specified or labeled.

Population distribution: The distribution that characterizes a population; may be displayed using a histogram or frequency of occurrence diagram.

Population mean and variance: The population mean, μ, is the arithmetic average of values x_i ($i = 1, \ldots, N$) obtained from all members in a population of size N based on the measurement of some quantity, x, associated with each member; population variance is the arithmetic average of the numbers $(x_i - \mu)^2$.

Population moment: The rth moment associated with a particular population.

Probability: The probability of an event is the ratio of the number of times the event occurs relative to the total number of trials that take place.

Probability density function (PDF): A real-valued function whose integral over any set gives the probability that a random variable has values in this set.

Random variable: A well-defined function that allows us to assign a real number to any outcome of an experiment. Specifically, the random outcome of a particular experiment, indexed by k, can be represented by a real number x_k called the random variable.

Relative frequency distribution: A frequency distribution in which the individual class frequencies are expressed as a fraction of the total frequency range.

Sample: A selection of values from a larger collection of values.

Sample mean and variance: The mean value $\overline{x}$ and variance s of a sample taken from a given data set, X. In general these differ from the true mean, μ, and variance, σ, of the population.

Sample moment: The moment of a sample taken from a given set of samples.

Significance level: The probability of a false rejection of the null hypothesis.

Standard error: A measure of the variability any statistical constant would be expected to show in taking repeated random samples of a given size from the same universe of observations.

Standard normal variable: A normal (Gaussian) distributed random variable with specified mean and standard deviation, which has been transformed to a random variable with a mean of zero ($\overline{X} = 0$) and a standard deviation of unity ($s = 1$).

Stationarity: The property by which the statistics of a random variable do not change with time. For a stationary time series, quantities such as the mean and variance are nearly identical for different segments of the record.

Student's t-distribution: A probability distribution used to test the hypothesis that a random sample of n observations comes from a normal population with a given mean.

Sufficiency: Condition of an estimator that uses all the information about the population parameter contained in the sample observations.

Sufficient statistics: Let $X = (x_1, x_2, \ldots, x_n)$ be a random sample from the probability density function $p(x; \theta)$; a statistic $S = s(x_1, x_2, \ldots, x_n)$ is defined to be a sufficient statistic if and only if the conditional distribution of X given S, does not depend on θ for any statistic $R = r(x_1, x_2, \ldots, x_n)$.

Tschebysheff's theorem: Given a nonnegative random variable $f(x)$, and $k > 0$, the probability that $f(x) \geq k\sigma$ is less than or equal to the expected value of f divided by k^2.

Unbiased estimator: An estimate $\widehat{\theta}$ for a parameter θ whose expected value is $E[\widehat{\theta}] = \theta$.

Uniform probability density function: The distribution of a random variable in which each value has the same probability of occurrence.

Means, Variances and Moment-Generating Functions for Some Common Continuous Variables

Distribution	Probability Function	Mean	Variance	Moment-Generating Function
Uniform	$f(x) = \dfrac{1}{\theta_2 - \theta_1};\ \theta_1 \leq \infty \leq \theta_2$	$\dfrac{\theta_2 + \theta_1}{2}$	$\dfrac{(\theta_2 - \theta_1)^2}{12}$	$\dfrac{e^{t\theta_2} - e^{t\theta_1}}{t(\theta_2 - \theta_1)}$
Normal	$f(x) = \dfrac{1}{\sigma\sqrt{2\pi}}\exp\left[-\left(\dfrac{1}{2\sigma^2}\right)(x - \mu)^2\right]$	μ	σ^2	$\exp\left(\mu t + \dfrac{t^2\alpha^2}{2}\right)$
Gamma	$f(x) = \left[\dfrac{1}{\Gamma(\alpha)\beta^\alpha}\right]x^{\alpha-1}e^{-x/\beta}$	$\alpha\beta$	$\alpha\beta^2$	$(1 - \beta t)^{-\alpha}$
Chi-squared	$f(\chi^2) = \dfrac{(\chi^2)^{(\nu/2)-1}e^{-\chi^2/2}}{2^{\nu/2}\Gamma(\nu/2)}$	ν	2ν	$(1 - 2t)^{-\nu/2}$
Beta	$f(x) = \left[\dfrac{\Gamma(\alpha + \beta)}{\Gamma(\alpha)\Gamma(\beta)}\right]x^{\alpha-1}(1 - x)^{\beta-1}$	$\dfrac{\alpha}{\alpha + \beta}$	$\dfrac{\alpha\beta}{(\alpha + \beta)^2(\alpha + \beta + 1)}$	Does not exist in closed form

D

Statistical Tables

TABLE A4.1 Cumulative Normal Distribution. The Area or Cumulative Distribution, $F(z)$, Under the Standardized Normal Distribution Curve for $z \leq z_F$ Such that the Probability $P(z < z_F) = F(z)$. For Example, $P(z < z_F = 1.21) = 0.8869$, and $P(z > z_F = 1.21) = 1 - 0.8869 = 0.1131$

$$F(z) = \int_{-\infty}^{z} \frac{1}{\sqrt{2\pi}} e^{-z^2/2} dz$$

z	0.00	0.01	0.02	0.03	0.04	0.05	0.06	0.07	0.08	0.09
0.0	0.5000	0.5040	0.5080	0.5120	0.5160	0.5199	0.5239	0.5279	0.5319	0.5359
0.1	0.5398	0.5438	0.5478	0.5517	0.5557	0.5596	0.5636	0.5675	0.5714	0.5753
0.2	0.5793	0.5832	0.5871	0.5910	0.5948	0.5987	0.6026	0.6064	0.6103	0.6141
0.3	0.6179	0.6217	0.6255	0.6293	0.6331	0.6368	0.6406	0.6443	0.6480	0.6517
0.4	0.6554	0.6591	0.6628	0.6664	0.6700	0.6736	0.6772	0.6808	0.6844	0.6879
0.5	0.6915	0.6950	0.6985	0.7019	0.7054	0.7088	0.7123	0.7157	0.7190	0.7224
0.6	0.7257	0.7291	0.7324	0.7357	0.7389	0.7422	0.7454	0.7486	0.7517	0.7549
0.7	0.7580	0.7611	0.7642	0.7673	0.7704	0.7734	0.7764	0.7794	0.7823	0.7852
0.8	0.7881	0.7910	0.7939	0.7967	0.7995	0.8023	0.8051	0.8078	0.8106	0.8133
0.9	0.8159	0.8186	0.8212	0.8238	0.8264	0.8289	0.8315	0.8340	0.8365	0.8389
1.0	0.8413	0.8438	0.8461	0.8485	0.8508	0.8531	0.8554	0.8577	0.8599	0.8621
1.1	0.8643	0.8665	0.8686	0.8708	0.8729	0.8749	0.8770	0.8790	0.8810	0.8830
1.2	0.8849	0.8869	0.8888	0.8907	0.8925	0.8944	0.8962	0.8980	0.8997	0.9015
1.3	0.9032	0.9049	0.9066	0.9082	0.9099	0.9115	0.9131	0.9147	0.9162	0.9177
1.4	0.9192	0.9207	0.9222	0.9236	0.9251	0.9265	0.9279	0.9292	0.9306	0.9319
1.5	0.9332	0.9345	0.9357	0.9370	0.9382	0.9394	0.9406	0.9418	0.9429	0.9441

(Continued)

TABLE A4.1 Cumulative Normal Distribution. The Area or Cumulative Distribution, $F(z)$, Under the Standardized Normal Distribution Curve for $z \leq z_F$ Such that the Probability $P(z < z_F) = F(z)$. For Example, $P(z < z_F = 1.21) = 0.8869$, and $P(z > z_F = 1.21) = 1 - 0.8869 = 0.1131$

$$F(z) = \int\limits_{-\infty}^{z} \frac{1}{\sqrt{2\pi}} e^{-z^2/2} dz$$

(cont'd)

z	0.00	0.01	0.02	0.03	0.04	0.05	0.06	0.07	0.08	0.09
1.6	0.9452	0.9463	0.9474	0.9484	0.9495	0.9505	0.9515	0.9525	0.9535	0.9545
1.7	0.9554	0.9564	0.9573	0.9582	0.9591	0.9599	0.9608	0.9616	0.9625	0.9633
1.8	0.9641	0.9649	0.9656	0.9664	0.9671	0.9678	0.9686	0.9693	0.9699	0.9706
1.9	0.9713	0.9719	0.9726	0.9732	0.9738	0.9744	0.9750	0.9756	0.9761	0.9767
2.0	0.9772	0.9778	0.9783	0.9788	0.9793	0.9798	0.9803	0.9808	0.9812	0.9817
2.1	0.9821	0.9826	0.9830	0.9834	0.9838	0.9842	0.9846	0.9850	0.9854	0.9857
2.2	0.9861	0.9864	0.9868	0.9871	0.9875	0.9878	0.9881	0.9884	0.9887	0.9890
2.3	0.9893	0.9896	0.9898	0.9901	0.9904	0.9906	0.9909	0.9911	0.9913	0.9916
2.4	0.9918	0.9920	0.9922	0.9925	0.9927	0.9929	0.9931	0.9932	0.9934	0.9936
2.5	0.9938	0.9940	0.9941	0.9943	0.9945	0.9946	0.9948	0.9949	0.9951	0.9952
2.6	0.9953	0.9955	0.9956	0.9957	0.9959	0.9960	0.9961	0.9962	0.9963	0.9964
2.7	0.9965	0.9966	0.9967	0.9968	0.9969	0.9970	0.9971	0.9972	0.9973	0.9974
2.8	0.9974	0.9975	0.9976	0.9977	0.9977	0.9978	0.9979	0.9979	0.9980	0.9981
2.9	0.9981	0.9982	0.9982	0.9983	0.9984	0.9984	0.9985	0.9985	0.9986	0.9986
3.0	0.9987	0.9987	0.9987	0.9988	0.9988	0.9989	0.9989	0.9989	0.9990	0.9990
3.1	0.9990	0.9991	0.9991	0.9991	0.9992	0.9992	0.9992	0.9992	0.9993	0.9993
3.2	0.9993	0.9993	0.9994	0.9994	0.9994	0.9994	0.9994	0.9995	0.9995	0.9995
3.3	0.9995	0.9995	0.9995	0.9996	0.9996	0.9996	0.9996	0.9996	0.9996	0.9997
3.4	0.9997	0.9997	0.9997	0.9997	0.9997	0.9997	0.9997	0.9997	0.9997	0.9998

Adapted from Introductory Statistical Analysis by D. L. Harnett and J. L. Murphy, Addison-Wesley, 1976.

TABLE A4.2   Cumulative Chi-square Distribution. The Area or Cumulative Distribution, $F(\chi^2)$, Under the χ^2 Distribution Curve for Different Degrees of Freedom, ν, Such that the Probability $P(\chi^2_\nu < \chi^2_{\nu;F}) = F(\chi^2)$. For Example, for $\nu = 16$, the Probability $P(\chi^2_{16} < \chi^2_{16;F} = 26.3) = F(26.3) = 0.950$. Consequently, $P(\chi^2_{16} > \chi^2_{16;F} = 26.3) = 1 - F(26.3) = 0.050$

$$F(\chi^2) = \int_0^{\chi^2} \frac{x^{(\nu-2)/2}e^{-x/2}dx}{2^{\nu/2}[(\nu-2)/2]!}$$

						F							
ν	0.005	0.010	0.025	0.050	0.100	0.250	0.500	0.750	0.900	0.950	0.975	0.990	0.995
1	0.0^4393	0.0^3157	0.0^3982	0.0^2393	0.0158	0.102	0.455	1.32	2.71	3.84	5.02	6.63	7.88
2	0.0100	0.0201	0.0506	0.103	0.211	0.575	1.39	2.77	4.61	5.99	7.38	9.21	10.6
3	0.0717	0.115	0.216	0.352	0.584	1.21	2.37	4.11	6.25	7.81	9.35	11.3	12.8
4	0.207	0.297	0.484	0.711	1.06	1.92	3.36	5.39	7.78	9.49	11.1	13.3	14.9
5	0.412	0.554	0.831	1.15	1.61	2.67	4.35	6.63	9.24	11.1	12.8	15.1	16.7
6	0.676	0.872	1.24	1.64	2.20	3.45	5.35	7.84	10.6	12.6	14.4	16.8	18.5
7	0.989	1.24	1.69	2.17	2.83	4.25	6.35	9.04	12.0	14.1	16.0	18.5	20.3
8	1.34	1.65	2.18	2.73	3.49	5.07	7.34	10.2	13.4	15.5	17.5	20.1	22.0
9	1.73	2.09	2.70	3.33	4.17	5.90	8.34	11.4	14.7	16.9	19.0	21.7	23.6
10	2.16	2.56	3.25	3.94	4.87	6.74	9.34	12.5	16.0	18.3	20.5	23.2	25.2
11	2.60	3.05	3.82	4.57	5.58	7.58	10.3	13.7	17.3	19.7	21.9	24.7	26.8
12	3.07	3.57	4.40	5.23	6.30	8.44	11.3	14.8	18.5	21.0	23.3	26.2	28.3
13	3.57	4.11	5.01	5.89	7.04	9.30	12.3	16.0	19.8	22.4	24.7	27.7	29.8
14	4.07	4.66	5.63	6.57	7.79	10.2	13.3	17.1	21.1	23.7	26.1	29.1	31.3
15	4.60	5.23	6.26	7.26	8.55	11.0	14.3	18.2	22.3	25.0	27.5	30.6	32.8
16	5.14	5.81	6.91	7.96	9.31	11.9	15.3	19.4	23.5	26.3	28.8	32.0	34.3
17	5.70	6.41	7.56	8.67	10.1	12.8	16.3	20.5	24.8	27.6	30.2	33.4	35.7
18	6.26	7.01	8.23	9.39	10.9	13.7	17.3	21.6	26.0	28.9	31.5	34.8	37.2
19	6.84	7.63	8.91	10.1	11.7	14.6	18.3	22.7	27.2	30.1	32.9	36.2	38.6
20	7.43	8.26	9.59	10.9	12.4	15.5	19.3	23.8	28.4	31.4	34.2	37.6	40.0
21	8.03	8.90	10.3	11.6	13.2	16.3	20.3	24.9	29.6	32.7	35.5	38.9	41.4
22	8.64	9.54	11.0	12.3	14.0	17.2	21.3	26.0	30.8	33.9	36.8	40.3	42.8
23	9.26	10.2	11.7	13.1	14.8	18.1	22.3	27.1	32.0	35.2	38.1	41.6	44.2
24	9.89	10.9	12.4	13.8	15.7	19.0	23.3	28.2	33.2	36.4	39.4	43.0	45.6

(Continued)

TABLE A4.2 Cumulative Chi-square Distribution. The Area or Cumulative Distribution, $F(\chi^2)$, Under the χ^2 Distribution Curve for Different Degrees of Freedom, ν, Such that the Probability $P(\chi^2_\nu < \chi^2_{\nu;F}) = F(\chi^2)$. For Example, for $\nu = 16$, the Probability $P(\chi^2_{16} < \chi^2_{16;F} = 26.3) = F(26.3) = 0.950$. Consequently, $P(\chi^2_{16} > \chi^2_{16;F} = 26.3) = 1 - F(26.3) = 0.050$

$$F(\chi^2) = \int_0^{\chi^2} \frac{x^{(\nu-2)/2}e^{-x/2}dx}{2^{\nu/2}[(\nu-2)/2]!}$$

(cont'd)

| | | | | | | | F | | | | | | |
ν	0.005	0.010	0.025	0.050	0.100	0.250	0.500	0.750	0.900	0.950	0.975	0.990	0.995
25	10.5	11.5	13.1	14.6	16.5	19.9	24.3	29.3	34.4	37.7	40.6	44.3	46.9
26	11.2	12.2	13.8	15.4	17.3	20.8	25.3	30.4	35.6	38.9	41.9	45.6	48.3
27	11.8	12.9	14.6	16.2	18.1	21.7	26.3	31.5	36.7	40.1	43.2	47.0	49.6
28	12.5	13.6	15.3	16.9	18.9	22.7	27.3	32.6	37.9	41.3	44.5	48.3	51.0
29	13.1	14.3	16.0	17.7	19.8	23.6	28.3	33.7	39.1	42.6	45.7	49.6	52.3
30	13.8	15.0	16.8	18.5	20.6	24.5	29.3	34.8	40.3	43.8	47.0	50.9	53.7

Adapted from Introductory Statistical Analysis by D.L. Harnett and J.L. Murphy, Addison-Wesley, 1976; abridged from Tables of percentage points of the incomplete beta function and of the chi-square distribution C.M. Thompson, Biometrika, Vol. 32 (1941).

TABLE A4.3A Cumulative t-distribution. The Area or Cumulative Distribution, $F(t)$, Under the t-distribution Curve for Different Degrees of Freedom, ν, Such that the Probability $P(t_\nu < t_{\nu;F}) = F(t)$. The example here is for $n = 20$. For Example, for $\nu = 9$, the Probability $P(t_9 < t_{9;F} = 2.262) = F(2.262) = 0.975$ and $P(t_9 > t_{9;F} = 2.262) = 1 - F(2.262) = 0.025$, Corresponding to the 95% Confidence Interval $(F = F_{0.025})$. Note that $F_{0.100}$, $F_{0.50}$, and $F_{0.005}$ Correspond to the 80, 90, and 99% Levels, Respectively

$$F(t) = \int_{-\infty}^{t} \frac{\left(\frac{\nu-1}{2}\right)!}{\left(\frac{\nu-2}{2}\right)! \sqrt{\pi n}\left(1 + \frac{t^2}{\nu}\right)^{(\nu+1)/2}} dt$$

				F			
ν	0.75	0.90	0.95	0.975	0.99	0.995	0.9995
1	1.000	3.078	6.314	12.706	31.821	63.657	636.615
2	0.816	1.886	2.920	4.303	6.965	9.925	31.598
3	0.765	1.638	2.353	3.182	4.541	5.841	12.941
4	0.741	1.533	2.132	2.776	3.747	4.604	8.610
5	0.727	1.476	2.015	2.571	3.365	4.032	6.859
6	0.718	1.440	1.943	2.447	3.143	3.707	5.959
7	0.711	1.415	1.895	2.365	2.998	3.499	5.405
8	0.706	1.397	1.860	2.306	2.896	3.355	5.041
9	0.703	1.383	1.833	2.262	2.821	3.250	4.781
10	0.700	1.372	1.812	2.228	2.764	3.169	4.587
11	0.697	1.363	1.796	2.201	2.718	3.106	4.437
12	0.695	1.356	1.782	2.179	2.681	3.055	4.318
13	0.694	1.350	1.771	2.160	2.650	3.012	4.221
14	0.692	1.345	1.761	2.145	2.624	2.977	4.140
15	0.691	1.341	1.753	2.131	2.602	2.947	4.073
16	0.690	1.337	1.746	2.120	2.583	2.921	4.015
17	0.689	1.333	1.740	2.110	2.567	2.898	3.965
18	0.688	1.330	1.734	2.101	2.552	2.878	3.922
19	0.688	1.328	1.729	2.093	2.539	2.861	3.883
20	0.687	1.325	1.725	2.086	2.528	2.845	3.850
21	0.686	1.323	1.721	2.080	2.518	2.831	3.819
22	0.686	1.321	1.717	2.074	2.508	2.819	3.792
23	0.685	1.319	1.714	2.069	2.500	2.807	3.767
24	0.685	1.318	1.711	2.064	2.492	2.797	3.745

(Continued)

TABLE A4.3A Cumulative t-distribution. The Area or Cumulative Distribution, $F(t)$, Under the t-distribution Curve for Different Degrees of Freedom, ν, Such that the Probability $P(t_\nu < t_{\nu;F}) = F(t)$. The example here is for $n = 20$. For Example, for $\nu = 9$, the Probability $P(t_9 < t_{9;F} = 2.262) = F(2.262) = 0.975$ and $P(t_9 > t_{9;F} = 2.262) = 1 - F(2.262) = 0.025$, Corresponding to the 95% Confidence Interval $(F = F_{0.025})$. Note that $F_{0.100}$, $F_{0.50}$, and $F_{0.005}$ Correspond to the 80, 90, and 99% Levels, Respectively

$$F(t) = \int_{-\infty}^{t} \frac{\left(\frac{\nu-1}{2}\right)!}{\left(\frac{\nu-2}{2}\right)!\sqrt{\pi n}\left(1 + \frac{t^2}{\nu}\right)^{(\nu+1)/2}} dt$$

$(cont'd)$

| ν | \multicolumn{7}{c}{F} |
	0.75	0.90	0.95	0.975	0.99	0.995	0.9995
25	0.684	1.316	1.708	2.060	2.485	2.787	3.725
26	0.684	1.315	1.706	2.056	2.479	2.779	3.707
27	0.684	1.314	1.703	2.052	2.473	2.771	3.690
28	0.683	1.313	1.701	2.048	2.467	2.763	3.674
29	0.683	1.311	1.699	2.045	2.462	2.756	3.659
30	0.683	1.310	1.697	2.042	2.457	2.750	3.646
40	0.681	1.303	1.684	2.021	2.423	2.704	3.551
60	0.679	1.296	1.671	2.000	2.390	2.660	3.460
120	0.677	1.289	1.658	1.980	2.358	2.617	3.373
∞	0.674	1.282	1.645	1.960	2.326	2.576	3.291

Adapted from Introductory Statistical Analysis by D. L. Harnett and J. L. Murphy, Addison-Wesley, 1976; abridged from the Statistical tables of R.A. Fisher and Frank Yates, Oliver & Boyd, Edinburgh and London, 1938.

TABLE A4.3B Cumulative t-distribution (Two-tailed Tests). Similar to Table A4.3A Except that Values Give Cumulative Distribution, $F(t)$, Under the t-distribution Curve for Different Degrees of Freedom, ν, Regardless of Sign, Such that the Probability $P(|t_\nu| > |t_{\nu,F}|) = F(t)$. The example here is for $n = 20$. For Example, for $\nu = 9$, the Probability $P(|t_9| > |t_{9,F}| = 2.262) = F(2.262) = 0.05$ and $P(|t_9| < |t_{9,F}| = 2.262) = 1 - F(2.262) = 0.95$, Corresponding to the 95% Confidence Interval. Note that $F_{0.200}$, $F_{0.100}$, and $F_{0.010}$ Correspond to the 80, 90, and 99% Levels, Respectively

	F Probability of a Larger Value, Sign Ignored								
ν	0.500	0.400	0.200	0.100	0.050	0.025	0.010	0.005	0.001
1	1.000	1.376	3.078	6.314	12.706	25.452	63.657		
2	0.816	1.061	1.886	2.920	4.303	6.205	9.925	14.089	31.598
3	0.765	0.978	1.638	2.353	3.182	4.176	5.841	7.453	12.941
4	0.741	0.941	1.533	2.132	2.776	3.495	4.604	5.598	8.610
5	0.727	0.920	1.476	2.015	2.571	3.163	4.032	4.773	6.859
6	0.718	0.906	1.440	1.943	2.447	2.969	3.707	4.317	5.959
7	0.711	0.896	1.415	1.895	2.365	2.841	3.499	4.029	5.405
8	0.706	0.889	1.397	1.860	2.306	2.732	3.355	3.832	5.041
9	0.703	0.883	1.383	1.833	2.262	2.685	3.250	3.690	4.781
10	0.700	0.879	1.372	1.812	2.228	2.634	3.169	3.581	4.587
11	0.697	0.876	1.363	1.796	2.201	2.593	3.106	3.497	4.437
12	0.695	0.873	1.356	1.782	2.179	2.560	3.055	3.428	4.318
13	0.694	0.870	1.350	1.771	2.160	2.533	3.012	3.372	4.221
14	0.692	0.868	1.345	1.761	2.145	2.510	2.977	3.326	4.140
15	0.691	0.866	1.341	1.753	2.131	2.490	2.947	3.286	4.073
16	0.690	0.865	1.337	1.746	2.120	2.473	2.921	3.252	4.015
17	0.689	0.863	1.333	1.740	2.110	2.458	2.898	3.222	3.965
18	0.688	0.862	1.330	1.734	2.101	2.445	2.878	3.197	3.922
19	0.688	0.861	1.328	1.729	2.093	2.433	2.861	3.174	3.883
20	0.687	0.860	1.325	1.725	2.086	2.423	2.845	3.153	3.850
21	0.686	0.859	1.323	1.721	2.080	2.414	2.831	3.135	3.819
22	0.686	0.858	1.321	1.717	2.074	2.406	2.819	3.119	3.792
23	0.685	0.858	1.319	1.714	2.069	2.398	2.807	3.104	3.767
24	0.685	0.857	1.318	1.711	2.064	2.391	2.797	3.090	3.745
25	0.684	0.856	1.316	1.708	2.060	2.385	2.787	3.078	3.725
26	0.684	0.856	1.315	1.706	2.056	2.379	2.779	3.067	3.707

(Continued)

TABLE A4.3B Cumulative t-distribution (Two-tailed Tests). Similar to Table A4.3A Except that Values Give Cumulative Distribution, $F(t)$, Under the t-distribution Curve for Different Degrees of Freedom, ν, Regardless of Sign, Such that the Probability $P(|t_\nu| > |t_{\nu,F}|) = F(t)$. The example here is for $n = 20$. For Example, for $\nu = 9$, the Probability $P(|t_9| > |t_{9,F}| = 2.262) = F(2.262) = 0.05$ and $P(|t_9| < |t_{9,F}| = 2.262) = 1 - F(2.262) = 0.95$, Corresponding to the 95% Confidence Interval. Note that $F_{0.200}$, $F_{0.100}$, and $F_{0.010}$ Correspond to the 80, 90, and 99% Levels, Respectively (*cont'd*)

					F Probability of a Larger Value, Sign Ignored				
ν	0.500	0.400	0.200	0.100	0.050	0.025	0.010	0.005	0.001
27	0.684	0.855	1.314	1.703	2.052	2.373	2.771	3.056	3.690
28	0.683	0.855	1.313	1.701	2.048	2.368	2.763	3.047	3.674
29	0.683	0.854	1.311	1.699	2.045	2.364	2.756	3.038	3.659
30	0.683	0.854	1.310	1.697	2.042	2.360	2.750	3.030	3.646
35	0.682	0.852	1.306	1.690	2.030	2.342	2.724	2.996	3.591
40	0.681	0.851	1.303	1.684	2.021	2.329	2.704	2.971	3.551
45	0.680	0.850	1.301	1.680	2.014	2.319	2.690	2.952	3.520
50	0.680	0.849	1.299	1.676	2.008	2.310	2.678	2.937	3.496
55	0.679	0.849	1.297	1.673	2.004	2.304	2.669	2.925	3.476
60	0.679	0.848	1.296	1.671	2.000	2.299	2.660	2.915	3.460
70	0.678	0.847	1.294	1.667	1.994	2.290	2.648	2.899	3.435
80	0.678	0.847	1.293	1.665	1.989	2.284	2.638	2.887	3.416
90	0.678	0.846	1.291	1.662	1.986	2.279	2.631	2.878	3.402
100	0.677	0.846	1.290	1.661	1.982	2.276	2.625	2.871	3.390
120	0.677	0.845	1.289	1.658	1.980	2.270	2.617	2.860	3.373
∞	0.6745	0.8416	1.2816	1.6448	1.9600	2.214	2.5758	2.8070	3.2905

TABLE A4.4A Critical Values of the F-distribution for $\alpha = 0.05$. The Distributions Represent the Area Exceeding the Value of $F_{0.05, \nu_1, \nu_2}$ and $F_{0.01, \nu_1, \nu_2}$ as Shown by the Shaded Area in the Figure for Different Degrees of Freedom, ν. For Example, if $\nu_1 = 15$ and $\nu_2 = 20$, then the Critical Value for $\alpha = 0.05$ is 2.20

Values of $F_{0.05, \nu_1, \nu_2}$

ν_1 = Degrees of freedom for numerator

ν_2/ν_1	1	2	3	4	5	6	7	8	9	10	12	15	20	24	30	40	60	120	∞
1	161	200	216	225	230	234	237	239	241	242	244	246	248	249	250	251	252	253	254
2	18.5	19.0	19.2	19.2	19.3	19.3	19.4	19.4	19.4	19.4	19.4	19.4	19.4	19.5	19.5	19.5	19.5	19.5	19.5
3	10.1	9.55	9.28	9.12	9.01	8.94	8.89	8.85	8.81	8.79	8.74	8.70	8.66	8.64	8.62	8.59	8.57	8.55	8.53
4	7.71	6.94	6.59	6.39	6.26	6.16	6.09	6.04	6.00	5.96	5.91	5.86	5.80	5.77	5.75	5.72	5.69	5.66	5.63
5	6.61	5.79	5.41	5.19	5.05	4.95	4.88	4.82	4.77	4.74	4.68	4.62	4.56	4.53	4.50	4.46	4.43	4.40	4.37
6	5.99	5.14	4.76	4.53	4.39	4.28	4.21	4.15	4.10	4.06	4.00	3.94	3.87	3.84	3.81	3.77	3.74	3.70	3.67
7	5.59	4.74	4.35	4.12	3.97	3.87	3.79	3.73	3.68	3.64	3.57	3.51	3.44	3.41	3.38	3.34	3.30	3.27	3.23
8	5.32	4.46	4.07	3.84	3.69	3.58	3.50	3.44	3.39	3.35	3.28	3.22	3.15	3.12	3.08	3.04	3.01	2.97	2.93
9	5.12	4.26	3.86	3.63	3.48	3.37	3.29	3.23	3.18	3.14	3.07	3.01	2.94	2.90	2.86	2.83	2.79	2.75	2.71
10	4.96	4.10	3.71	3.48	3.33	3.22	3.14	3.07	3.02	2.98	2.91	2.85	2.77	2.74	2.70	2.66	2.62	2.58	2.54
11	4.84	3.98	3.59	3.36	3.20	3.09	3.01	2.95	2.90	2.85	2.79	2.72	2.65	2.61	2.57	2.53	2.49	2.45	2.40
12	4.75	3.89	3.49	3.26	3.11	3.00	2.91	2.85	2.80	2.75	2.69	2.62	2.54	2.51	2.47	2.43	2.38	2.34	2.30
13	4.67	3.81	3.41	3.18	3.03	2.92	2.83	2.77	2.71	2.67	2.60	2.53	2.46	2.42	2.38	2.34	2.30	2.25	2.21
14	4.60	3.74	3.34	3.11	2.96	2.85	2.76	2.70	2.65	2.60	2.53	2.46	2.39	2.35	2.31	2.27	2.22	2.18	2.13
15	4.54	3.68	3.29	3.06	2.90	2.79	2.71	2.64	2.59	2.54	2.48	2.40	2.33	2.29	2.25	2.20	2.16	2.11	2.07

ν_2 = Degrees of freedom for denominator

(Continued)

TABLE A4.4A Critical Values of the F-distribution for $\alpha = 0.05$. The Distributions Represent the Area Exceeding the Value of $F_{0.05,\,\nu_1,\nu_2}$, and $F_{0.01,\,\nu_1,\nu_2}$ as Shown by the Shaded Area in the Figure for Different Degrees of Freedom, ν. For Example, if $\nu_1 = 15$ and $\nu_2 = 20$, then the Critical Value for $\alpha = 0.05$ is 2.20 (cont'd)

Values of $F_{0.05,\,\nu_1,\,\nu_2}$

ν_1 = Degrees of freedom for numerator

ν_2/ν_1	1	2	3	4	5	6	7	8	9	10	12	15	20	24	30	40	60	120	∞
16	4.49	3.63	3.24	3.01	2.85	2.74	2.66	2.59	2.54	2.49	2.42	2.35	2.28	2.24	2.19	2.15	2.11	2.06	2.01
17	4.45	3.59	3.20	2.96	2.81	2.70	2.61	2.55	2.49	2.45	2.38	2.31	2.23	2.19	2.15	2.10	2.06	2.01	1.96
18	4.41	3.55	3.16	2.93	2.77	2.66	2.58	2.51	2.46	2.41	2.34	2.27	2.19	2.15	2.11	2.06	2.02	1.97	1.92
19	4.38	3.52	3.13	2.90	2.74	2.63	2.54	2.48	2.42	2.38	2.31	2.23	2.16	2.11	2.07	2.03	1.98	1.93	1.88
20	4.35	3.49	3.10	2.87	2.71	2.60	2.51	2.45	2.39	2.35	2.28	2.20	2.12	2.08	2.04	1.99	1.95	1.90	1.84
21	4.32	3.47	3.07	2.84	2.68	2.57	2.49	2.42	2.37	2.32	2.25	2.18	2.10	2.05	2.01	1.96	1.92	1.87	1.81
22	4.30	3.44	3.05	2.82	2.66	2.55	2.46	2.40	2.34	2.30	2.23	2.15	2.07	2.03	1.98	1.94	1.89	1.84	1.78
23	4.28	3.42	3.03	2.80	2.64	2.53	2.44	2.37	2.32	2.27	2.20	2.13	2.05	2.01	1.96	1.91	1.86	1.81	1.76
24	4.26	3.40	3.01	2.78	2.62	2.51	2.42	2.36	2.30	2.25	2.18	2.11	2.03	1.98	1.94	1.89	1.84	1.79	1.73
25	4.24	3.39	2.99	2.76	2.60	2.49	2.40	2.34	2.28	2.24	2.16	2.09	2.01	1.96	1.92	1.87	1.82	1.77	1.71
30	4.17	3.32	2.92	2.69	2.53	2.42	2.33	2.27	2.21	2.16	2.09	2.01	1.93	1.89	1.84	1.79	1.74	1.68	1.62
40	4.08	3.23	2.84	2.61	2.45	2.34	2.25	2.18	2.12	2.08	2.00	1.92	1.84	1.79	1.74	1.69	1.64	1.58	1.51
60	4.00	3.15	2.76	2.53	2.37	2.25	2.17	2.10	2.04	1.99	1.92	1.84	1.75	1.70	1.65	1.59	1.53	1.47	1.39
120	3.92	3.07	2.68	2.45	2.29	2.18	2.09	2.02	1.96	1.91	1.83	1.75	1.66	1.61	1.55	1.50	1.43	1.35	1.25
∞	3.84	3.00	2.60	2.37	2.21	2.10	2.01	1.94	1.88	1.83	1.75	1.67	1.57	1.52	1.46	1.39	1.32	1.22	1.00

ν_2 = Degrees of freedom for denominator

ν_1 = Degrees of freedom for numerator.
ν_2 = Degrees of freedom for denominator.
$P(F > 2.20) = 0.05$.
$P(F < 2.20) = 0.95$.

TABLE A4.4B Critical Values of the F-distribution for $\alpha = 0.01$ The Distributions Represent the Area Exceeding the Value of $F_{0.05, \nu_1, \nu_2}$, and $F_{0.01, \nu_1, \nu_2}$ as Shown by the Shaded Area in the Figure for Different Degrees of Freedom, ν. For Example, if $\nu_1 = 15$ and $\nu_2 = 20$, then the Critical Value for $\alpha = 0.01$ is 3.09

Values of $F_{0.01, \nu_1, \nu_2}$

ν_1 = Degrees of freedom for numerator

ν_1/ν_2	1	2	3	4	5	6	7	8	9	10	12	15	20	24	30	40	60	120	∞
1	4052	5000	5403	5625	5764	5859	5928	5982	6023	6056	6106	6157	6209	6235	6261	6287	6313	6339	6366
2	98.5	99.0	99.2	99.2	99.3	99.3	99.4	99.4	99.4	99.4	99.4	99.4	99.4	99.5	99.5	99.5	99.5	99.5	99.5
3	34.1	30.8	29.5	28.7	28.2	27.9	27.7	27.5	27.3	27.2	27.1	26.9	26.7	26.6	26.5	26.4	26.3	26.2	26.1
4	21.2	18.0	16.7	16.0	15.5	15.2	15.0	14.8	14.7	14.5	14.4	14.2	14.0	13.9	13.8	13.7	13.7	13.6	13.5
5	16.3	13.3	12.1	11.4	11.0	10.7	10.5	10.3	10.2	10.1	9.89	9.72	9.55	9.47	9.38	9.29	9.20	9.11	9.02
6	13.7	10.9	9.78	9.15	8.75	8.47	8.26	8.10	7.98	7.87	7.72	7.56	7.40	7.31	7.23	7.14	7.06	6.97	6.88
7	12.2	9.55	8.45	7.85	7.46	7.19	6.99	6.84	6.72	6.62	6.47	6.31	6.16	6.07	5.99	5.91	5.82	5.74	5.65
8	11.3	8.65	7.59	7.01	6.63	6.37	6.18	6.03	5.91	5.81	5.67	5.52	5.36	5.28	5.20	5.12	5.03	4.95	4.86
9	10.6	8.02	6.99	6.42	6.06	5.80	5.61	5.47	5.35	5.26	5.11	4.96	4.81	4.73	4.65	4.57	4.48	4.40	4.31
10	10.0	7.56	6.55	5.99	5.64	5.39	5.20	5.06	4.94	4.85	4.71	4.56	4.41	4.33	4.25	4.17	4.08	4.00	3.91
11	9.65	7.21	6.22	5.67	5.32	5.07	4.89	4.74	4.63	4.54	4.40	4.25	4.10	4.02	3.94	3.86	3.78	3.69	3.60
12	9.33	6.93	5.95	5.41	5.06	4.82	4.64	4.50	4.39	4.30	4.16	4.01	3.86	3.78	3.70	3.62	3.54	3.45	3.36
13	9.07	6.70	5.74	5.21	4.86	4.62	4.44	4.30	4.19	4.10	3.96	3.82	3.66	3.59	3.51	3.43	3.34	3.25	3.17
14	8.86	6.51	5.56	5.04	4.70	4.46	4.28	4.14	4.03	3.94	3.80	3.66	3.51	3.43	3.35	3.27	3.18	3.09	3.00
15	8.68	6.36	5.42	4.89	4.56	4.32	4.14	4.00	3.89	3.80	3.67	3.52	3.37	3.29	3.21	3.13	3.05	2.96	2.87

(Continued)

ν_2 = Degrees of freedom for denominator

TABLE A4.4B   Critical Values of the F-distribution for $\alpha = 0.01$ The Distributions Represent the Area Exceeding the Value of $F_{0.05, \nu_1, \nu_2}$ and $F_{0.01, \nu_1, \nu_2}$ as Shown by the Shaded Area in the Figure for Different Degrees of Freedom, ν. For Example, if $\nu_1 = 15$ and $\nu_2 = 20$, then the Critical Value for $\alpha = 0.01$ is 3.09 (cont'd)

Values of $F_{0.01, \nu_1, \nu_2}$

ν_1 = Degrees of freedom for numerator

ν_1/ν_2	1	2	3	4	5	6	7	8	9	10	12	15	20	24	30	40	60	120	∞
16	8.53	6.23	5.29	4.77	4.44	4.20	4.03	3.89	3.78	3.69	3.55	3.41	3.26	3.18	3.10	3.02	2.93	2.84	2.75
17	8.40	6.11	5.19	4.67	4.34	4.10	3.93	3.79	3.68	3.59	3.46	3.31	3.16	3.08	3.00	2.92	2.83	2.75	2.65
18	8.29	6.01	5.09	4.58	4.25	4.01	3.84	3.71	3.60	3.51	3.37	3.23	3.08	3.00	2.92	2.84	2.75	2.66	2.57
19	8.19	5.93	5.01	4.50	4.17	3.94	3.77	3.63	3.52	3.43	3.30	3.15	3.00	2.92	2.84	2.76	2.67	2.58	2.49
20	8.10	5.85	4.94	4.43	4.10	3.87	3.70	3.56	3.46	3.37	3.23	3.09	2.94	2.86	2.78	2.69	2.61	2.52	2.42
21	8.02	5.78	4.87	4.37	4.04	3.81	3.64	3.51	3.40	3.31	3.17	3.03	2.88	2.80	2.72	2.64	2.55	2.46	2.36
22	7.95	5.72	4.82	4.31	3.99	3.76	3.59	3.45	3.35	3.26	3.12	2.98	2.83	2.75	2.67	2.58	2.50	2.40	2.31
23	7.88	5.66	4.76	4.26	3.94	3.71	3.54	3.41	3.30	3.21	3.07	2.93	2.78	2.70	2.62	2.54	2.45	2.35	2.26
24	7.82	5.61	4.72	4.22	3.90	3.67	3.50	3.36	3.26	3.17	3.03	2.89	2.74	2.66	2.58	2.49	2.40	2.31	2.21
25	7.77	5.57	4.68	4.18	3.86	3.63	3.46	3.32	3.22	3.13	2.99	2.85	2.70	2.62	2.53	2.45	2.36	2.27	2.17
30	7.56	5.39	4.51	4.02	3.70	3.47	3.30	3.17	3.07	2.98	2.84	2.70	2.55	2.47	2.39	2.30	2.21	2.11	2.01
40	7.31	5.18	4.31	3.83	3.51	3.29	3.12	2.99	2.89	2.80	2.66	2.52	2.37	2.29	2.20	2.11	2.02	1.92	1.80
60	7.08	4.98	4.13	3.65	3.34	3.12	2.95	2.82	2.72	2.63	2.50	2.35	2.20	2.12	2.03	1.94	1.84	1.73	1.60
120	6.85	4.79	3.95	3.48	3.17	2.96	2.79	2.66	2.56	2.47	2.34	2.19	2.03	1.95	1.86	1.76	1.66	1.53	1.38
∞	6.63	4.61	3.78	3.32	3.02	2.80	2.64	2.51	2.41	2.32	2.18	2.04	1.88	1.79	1.70	1.59	1.47	1.32	1.00

ν_2 = Degrees of freedom for denominator

ν_1 = Degrees of freedom for numerator.
ν_2 = Degrees of freedom for denominator.
$P(F > 3.09) = 0.01$.
$P(F < 3.09) = 0.99$.

E

Correlation Coefficients at the 5% and 1% Levels of Significance for Various Degrees of Freedom ν

Degrees of Freedom	5%	1%
1	0.997	1.000
2	0.950	0.990
3	0.878	0.959
4	0.811	0.917
5	0.754	0.874
6	0.707	0.834
7	0.666	0.798
8	0.632	0.765
9	0.602	0.735
10	0.576	0.708
11	0.553	0.684
12	0.532	0.661
13	0.514	0.641
14	0.497	0.623
15	0.482	0.606
16	0.468	0.590
17	0.456	0.576
18	0.444	0.561

(cont'd)

Degrees of Freedom	5%	1%
19	0.433	0.549
20	0.423	0.537
21	0.413	0.526
22	0.404	0.515
23	0.396	0.505
24	0.388	0.496
25	0.381	0.487
26	0.374	0.478
27	0.367	0.470
28	0.361	0.463
29	0.355	0.456
30	0.349	0.449
35	0.325	0.418
40	0.304	0.393
45	0.288	0.372
50	0.273	0.354
60	0.250	0.325

(Continued)

(cont'd)

Degrees of Freedom	5%	1%
70	0.232	0.302
80	0.217	0.283
90	0.205	0.267
100	0.195	0.254
125	0.174	0.228
150	0.159	0.208
200	0.138	0.181
300	0.113	0.148
400	0.098	0.128
500	0.088	0.115
1000	0.062	0.081

Approximations and Nondimensional Numbers in Physical Oceanography

Beta parameter, β*: A nondimensionalized form of β (the beta parameter) defined as the ratio of the horizontal gradient in relative vorticity, $\nabla_h \zeta$, to the horizontal gradient in planetary vorticity, $\nabla_h f$

$$\beta^* \equiv \frac{\beta L^2}{U}$$

Here, $\zeta = \partial v / \partial x - \partial u / \partial y$ is the relative vorticity (for velocity components u, v in the x, y directions, respectively), f is the local Coriolis parameter, U is a horizontal velocity scale, L is a horizontal length scale (*see also Rhines length*) and β is defined as

$$\beta \equiv \nabla_h f \approx 10^{-11} / \text{m s}$$

In the *beta-plane approximation*, the curved surface of the earth is approximated by a flat plane tangent to the earth for which $f = f_o + \beta y$, where f_o is a reference value for f and y is the latitude. For this case, $\beta = df/dy$ is a constant.

Boussinesq approximation: Assumes that density changes in the fluid can be neglected except where density, $\rho = \rho_o + \rho'$, is multiplied by the acceleration of gravity, g. That is, the effects of density fluctuations, ρ', can be neglected in terms of the form $\rho F = (\rho_o + \rho')F$ for any variable F except for those involving, g (i.e., $\rho'g$).

Here $\rho_o = \rho_o(z)$ is the mean density and $\rho'/\rho_o \approx 10^{-3}$. At large *Mach* numbers ($U/c > 1$), the compressibility of the fluid becomes important and large density changes can occur. Since the speed of sound in water, $c \approx 1500$ m/s, is almost always large compared to the flow speed, U, the approximation is good for normal oceanic conditions.

Brunt—Väisälä frequency, N(z) (also Väisälä or Buoyancy frequency): The natural frequency of oscillation of a parcel of water displaced vertically (z-direction upward) from its level of equilibrium:

$$N(z) = \sqrt{-\left(\frac{g}{\rho_o}\frac{d\rho}{dz} + \frac{g^2}{c^2}\right)} \cong \sqrt{-\frac{g}{\rho_o}\frac{d\rho}{dz}}$$

where $g = 9.81$ m/s is the acceleration due to gravity, c is the speed of sound, $d\rho/dz \leq 0$ is the vertical in situ density gradient, and ρ_o is a reference density. N marks the maximum intrinsic frequency of oscillation obtainable by internal gravity waves. The compressibility term, $g^2/c^2 \approx 5 \times 10^{-5}/$s, associated with adiabatic displacement of the fluid can generally be ignored in the upper few thousand meters of the ocean where $10^{-4} < N < 10^{-2}/$s (periods of 17.5 h to 10.5 min). However, this is not the case in the deep ocean where $d\rho/dz$ is small and N can be of order 10^{-5}/s. Derivation of N using conductivity-temperature-depth data usually requires considerable low-pass filtering to eliminate erroneously high or negative values of the density gradient.

Burger number, B (also Stratification parameter, S): The squared ratio of the internal (Rossby) radius of deformation, r_i, to a longwave scale, L (such as the wavelength of a Rossby or coastal-trapped wave)

$$B \equiv \frac{N^2 H^2}{f^2 L^2}$$

where N is a characteristic Brunt-Väisälä frequency for the water depth H. For continental shelf-slope regions influenced by coastal-trapped waves, motions are baroclinic when $B \gg 1$ and barotropic when $B \ll 1$. For internal waves over a sloping bottom tilted at an angle θ to the horizontal, $H = L \sin \theta$ and

$$B \equiv \frac{N^2 \sin^2 \theta}{f^2}$$

Coriolis parameter (also Inertial frequency, Coriolis frequency, Planetary vorticity): The local vertical component of the earth's rate of rotation given by $f = 2\Omega \sin \theta$, where θ is the latitude and $\Omega = 0.72921 \times 10^{-4}$/s is the angular rate of rotation based on a *sidereal day* of 23 h 56 min 4 s. A sidereal day is the time for the earth to complete one rotation relative to an absolute reference point in space. Because of the movement of the earth about the sun, a sidereal day differs slightly from the solar day of 24 h. Latitude θ is positive for the northern hemisphere and negative for the southern hemisphere. At 50° latitude, $|f| = 1.117 \times 10^{-4}$/s (rad/s), corresponding to a cyclic frequency of 0.0640 cph and a period $T_f = 2\pi/|f|$ of 15.6 h; at 10° latitude, $|f| = 0.253 \times 10^{-4}$/s, corresponding to a period of 68.92 h = 2.87 days.

Cox number, C_θ: A relative measure of high vertical wavenumber temperature structure (temperature "noisiness") defined as the ratio of the mean vertical gradient squared to the mean-square vertical gradient for temperature, $T(z)$;

$$C_\theta \equiv \frac{\langle dT/dz \rangle^2}{\langle (dT/dz)^2 \rangle}$$

Ekman number, E: A nondimensional number giving the relative importance of frictional forces at a boundary to the Coriolis force

$$E \equiv \frac{\text{frictional force}}{\text{Coriolis force}} = \frac{\nu}{fD^2}$$

where ν is the turbulent eddy viscosity, f is the Coriolis parameter, and D is the depth of the fluid. The characteristic thickness, δ, of the Ekman layer is given by

$$\delta = \sqrt{\frac{2\nu_v}{f}}$$

where ν_v is the vertical component of eddy viscosity. For ν_v of order 10^{-2} m^2/s, $\delta \approx 20$ m at mid-latitudes ($f \approx 1 \times 10^{-4}$/s).

Froude number, F_r: The square root of the ratio of the inertial force to the gravitational force for barotropic motions with a free surface

$$F_r \approx \left[\frac{\text{inertial force}}{\text{gravitational force}} \right]^{1/2} = \frac{U}{\sqrt{gH}}$$

where U is the flow velocity and $c = \sqrt{gH}$ is the phase speed of a surface wave in a fluid of depth H. The flow is *supercritical* if $F_r > 1$ and *subcritical* if $F_r < 1$. The Froude number is analogous to the *Mach number* used for compressible fluids, such as air. Hydraulic jumps occur where the fluid speed transitions from supercritical to subcritical flow.

Froude number (internal), F'_r: The square root of the ratio of the inertial force to the buoyancy force for baroclinic motions in a stratified fluid

$$F'_r \equiv \left[\frac{\text{inertial force}}{\text{buoyancy force}} \right]^{1/2} = \frac{U}{\sqrt{g'H_n}}$$

where $g' = g(\rho_2 - \rho_1)/\rho_2$ is the reduced gravity and $c'_n = \sqrt{g'H_n}$ is the phase speed of a mode n internal wave in a fluid with an effective depth

H_n (see *Internal Rossby radius*). The flow is *supercritical* if $F'_r > 1$ and *subcritical* if $F'_r < 1$. The internal Froude number is used in studies of density-driven turbidity currents.

Geostrophic approximation: Assumes that the Rossby number is small ($Ro \ll 1$) so that horizontal motions are mainly a balance between the Coriolis force and the horizontal pressure gradient. It takes just over one inertial period, $T_f = 2\pi / |f|$, for a perturbed geostrophic flow to return to near geostrophic balance.

Hydrostatic approximation: Assumes that the vertical velocity, w, can be ignored in the vertical component of the momentum balance and that the vertical pressure gradient $\partial p / \partial z$ is proportional to the density, ρ:

$$\frac{\partial p}{\partial z} = -g\rho$$

Integration from depth z to the ocean surface $z = \eta$ gives, for near-uniform density $\rho \approx \rho_0$

$$p = p_0 + g\rho_0(\eta - z)$$

where p_0 is the atmospheric pressure at the ocean surface. The approximation cannot be used to study high-frequency internal wave dynamics.

Inertial period, T_f: The period of oscillation for the Coriolis frequency, f (see *Coriolis frequency*)

$$T_f = \frac{2\pi}{|f|} = \frac{\pi}{\Omega |\sin \theta|}$$

$T_f \approx 68.92$, 15.62, and $12.74\,\text{h}$ for inertial motions at latitudes θ of $10°$, $50°$, and $70°$, respectively.

Intrinsic frequency: If ω_O is the frequency of a wave measured at a fixed point and $\mathbf{k}$ the wavenumber vector of the wave, the intrinsic frequency, ω, of the wave as seen by an observer in a coordinate system moving with the mean flow, $\mathbf{U}$, is given by

$$\omega = \omega_o - \mathbf{k} \cdot \mathbf{U}$$

Thus, the frequency of the wave measured at fixed point is Doppler shifted by the amount $+\mathbf{k} \cdot \mathbf{U}$ relative to the intrinsic frequency. For most oceanic motions, the Doppler shift measured at fixed point is within a few percent of the intrinsic frequency, ω.

Kolmogorov microscale, η: The length scale at which turbulent motions begin to be damped out by small-scale molecular viscosity, ν

$$\eta \equiv 2\pi \left(\frac{\nu^3}{\varepsilon}\right)^{1/4}$$

in which $\varepsilon = 2\nu \langle (\partial u_i / \partial x_j)^2 \rangle$ is the mean rate of dissipation of turbulent kinetic energy (see *Ozmidov scale*). In the upper ocean, η is a few centimeters.

Mach number, M: The relative importance of fluid compressibility defined by the relation

$$M \equiv \left[\frac{\text{inertial force}}{\text{compressibility force}}\right]^{1/2} = \frac{U}{c}$$

where c is the speed of sound ($\approx 1500\,\text{m/s}$ in water) and U is the velocity of the fluid. Flows are subsonic if $M < 1$ and supersonic if $M > 1$. Compressibility effects can be ignored if $M < 0.3$.

Monin—Obukhov length, L_M: The height above a heated boundary at which mechanical (shear) production of turbulent kinetic energy equals the buoyant (convective) destruction of turbulent kinetic energy

$$L_M \equiv \frac{\text{shear production}}{\text{buoyant destruction}} = \frac{u_*^3}{k\alpha g \overline{wT'}}$$

where u_*, k, g, and α are, respectively, the friction velocity, the von Karman constant, the acceleration of gravity, and the coefficient of thermal expansion, and $\overline{wT'}$ is the mean heat flux for vertical velocity fluctuations w and temperature fluctuations T'.

Ozmidov (buoyancy) scale, η_b (or L_R): The ratio of nonlinear to buoyancy scales in a turbulent fluid; the scale above which eddy-like motions are damped by stratification

$$\eta_b \equiv 2\pi \left(\frac{\varepsilon}{N^3}\right)^{1/2}$$

Here, $\varepsilon = 2\nu \langle (\partial u_i / \partial x_j)^2 \rangle$ is the rate of dissipation of turbulent kinetic energy, and u_i is the ith component of velocity in the jth direction, x_j ($i, j = 1, 2, 3$ corresponding to the x, y, z directions, respectively). In the upper ocean, η_b can be up to a few meters.

Péclet Number, Pe: The diffusivity analog to the Reynolds number (Re):

$$Pe \equiv \frac{UL}{K} = Pr \cdot Re$$

where K is the diffusivity of heat or salt. In geophysical fluid dynamics, K corresponds to the turbulent eddy diffusivity (see *Prandtl number, Pr* and *Reynolds number, Re*).

Prandtl number, Pr: The ratio of momentum to heat (or salt) diffusivity:

$$Pr \equiv \frac{\text{momentum diffusivity}}{\text{heat diffusivity}} = \frac{\nu}{K_T}$$

For typical values of molecular viscosity $\nu \approx 10^{-2}\,\text{cm}^2/\text{s}$ ($10^{-6}\,\text{m}^2/\text{s}$) and molecular heat diffusivity $K_T \approx 10^{-3}\,\text{cm}^2/\text{s}$ ($10^{-7}\,\text{m}^2/\text{s}$), $Pr \approx 10$. For salt, $K_S \approx 10^{-5}\,\text{cm}^2/\text{s}$ ($10^{-9}\,\text{m}^2/\text{s}$) and $Pr \approx 1000$. A turbulent *Prandtl number* can be defined in terms of the turbulent eddy viscosity and turbulent diffusivities of heat and salt. The *Schmidt number* is similar to the Prandtl number with momentum diffusivity replaced by mass diffusivity.

Rayleigh number, Ra: The ratio of the destabilizing effect of the buoyancy force to the stabilizing effect of the viscous force:

$$Ra \equiv \frac{g\alpha\Gamma d^4}{K_T\nu}$$

where α is the coefficient of thermal expansion, $\Gamma = -d\langle T\rangle/dz$ is the vertical gradient of the background temperature $\langle T\rangle$ (the adiabatic temperature gradient, also known as the "lapse rate" by meteorologists), d is the depth of the layer, K_T is the thermal diffusivity, and ν is the kinematic viscosity. The "lapse rate" is the fastest rate at which the temperature can decrease with height without causing instability.

Reynolds number, Re: The ratio of the inertial (nonlinear) force to the viscous force

$$Re \equiv \frac{\text{inertial force}}{\text{viscous force}} = \frac{UL}{\nu}$$

where U is the flow velocity, L is a characteristic length scale, and ν is the kinetic viscosity; $\nu \approx 0.01\,\text{cm}^2/\text{s}$ ($0.01 \times 10^{-4}\,\text{m}^2/\text{s}$) for molecular processes. Viscous effects become important at small Reynolds numbers, $Re \ll 1$. In geophysical fluid dynamics, such as in the formation of mesoscale vortex streets, ν appears to correspond to the turbulent eddy viscosity.

Rhines length, ℓ: The scale at which mesoscale eddies transform from individual features to Rossby wave packets (the scale at which the planetary β-effect becomes comparable to nonlinear effects). Rossby wave propagation causes anisotropic elongation of the eddies in the zonal (east—west) direction and the eddy size in the meridional (north—south, y) direction stops growing at the scale

$$\ell = \sqrt{\frac{u}{\beta}}$$

where u is the root-mean-square velocity and β is the north—south gradient of the Coriolis parameter, $f = f_0 + \beta y$ (see *Beta parameter*).

Richardson number, Ri: A measure of the dynamic stability of the water column. In a two-layer fluid with reduced gravity g', mean flow U, and horizontal length scale, L, the *local Richardson number* is defined as

$$Ri \equiv \frac{g'L}{U^2}$$

while for a continuously stratified fluid with buoyancy frequency $N(z)$

$$Ri \equiv \frac{N^2L^2}{U^2}$$

The above expressions also are known as the *bulk Richardson number* since they define the overall stability characteristics of the water

Convolution

CONVOLUTION AND FOURIER TRANSFORMS

Consider the time-dependent functions $g(t)$ and $h(t)$ and their respective frequency-dependent Fourier transforms $G(f)$ and $H(f)$. The convolution of the two original functions (written $g * h$) is defined as

$$g * h \equiv \int_{-\infty}^{\infty} g(t)h(t - \tau)\mathrm{d}t \qquad (A7.1)$$

where $g * h$ is a function of the time lag, τ, and $g * h = h * g$. There is a one-to-one relationship between the function $g * h$ and the product of the Fourier transforms of the two functions such that

$$g * h \leftrightarrow G(f) \cdot H(f) \qquad (A7.2)$$

Known as the convolution theorem (A7.2), states that the Fourier transform of the convolution term on the left is the product of the Fourier transforms of the individual functions on the right side. In other words, convolution in one domain equates to the multiplication in the other domain. We further note that the correlation of g and h [corr(g, h); see Section 5.3] is written as

$$\mathrm{corr}(g, h) \equiv \int_{-\infty}^{\infty} g(t + \tau)h(t)\mathrm{d}t \qquad (A7.3)$$

which is also a function of the lag τ. As with convolution, we can form the transform pair

$$\mathrm{corr}(g, h) \leftrightarrow G(f) \cdot H(f)^{*} \qquad (A7.4)$$

called the correlation theorem, where $H(f)^{*}$ is the complex conjugate of $H(f)$ and $H(f)^{*} = H(-f)$, since we are restricting discussion to the usual case in which g and h are real functions. As this relationship indicates, multiplying the Fourier transform of one function by the complex conjugate of the Fourier transform of the other function yields the Fourier transform of their correlation. The correlation of a function with itself is called its autocorrelation (Section 5.3).

CONVOLUTION OF DISCRETE DATA

The analysis of geophysical data commonly involves the convolution of specially designed "data windows" (convolution functions or filters) with time series records in order to smooth the spectral estimates obtained from these data and to improve the statistical reliability of spectral peaks. Good filters are those that minimize unwanted spectral leakage associated with the filter's side lobes in the frequency domain. Consider a filter $h(t_k)$ applied to a discrete data series $g(t_j)$, where the t_j and t_k (j, $k = 0$, ...) are discrete times in the data series. The filter will have nonzero values over a short segment of the data to which it is being applied and will be zero elsewhere, yielding a single value for the central time of the filter for that specific piece of the data. The filter $h(t_k)$ typically has a central peak and falls off to zero on either side of the maximum.

The convolution theorem can be extended to discrete time series as follows. Assume that the time series, $g(t_j)$, has duration N and is completely determined by the N values $g(t_0)$, ..., $g(t_{N-1})$. The convolution of this function with the window, $h(t_k)$, is a member of the discrete Fourier transform pair

$$\sum_{k=-N/2+1}^{N/2} g\left(t_{j-k}\right)h(t_k) \leftrightarrow G_n H_n \qquad \text{(A7.5)}$$

where G_n ($n = 0, ..., N-1$) is the discrete Fourier transform of the time series $g(t_j)$ ($j = 0, ..., N-1$), and H_n ($n = 0, ..., N-1$) is the discrete Fourier transform of the function $h(t_k)$, ($k = 0, ..., N-1$). The values of $h(t_k)$ typically span a small fraction of the full data range $k = -N/2 + 1, ..., N/2$.

In Figure A7.1, the original time series, $g(t_j)$ (we have chosen normalized monthly values of the Southern Oscillation Index, SOI) is shown at the top and the convolution function, $h(t_k)$, used to filter the time series is presented in the middle panel. Here, we have used a simple five-year long Hamming window [see Chapter 6]. The window (filter) is symmetrical, uses 61 monthly weights (with nonzero first and last weights), and begins with the first month of the time series.

The bottom panel in Figure A7.1 shows the convolution of $h(t_k)$ with $g(t_j)$. As the filtered result clearly demonstrates, $h(t_k)$ acts as a smoothing function that flattens out the "bumpiness" of $g(t_j)$, reducing sharp year-to-year changes in the normalized SOI. This smoothing depends on the duration and the shape of the window, $h(t_k)$. A more sharply peaked $h(t_k)$ would produce less time series smoothing, leaving more of the large year-to-year variability. The function, $h(t_k)$, has exactly the same purpose as a moving average, except that the weights of the filter (the filter coefficients) are specially designed to reduce side lobe spectral leakage problems. For the moving average, all weights are of equal value. The convolved data (bottom panel of Figure A7.1) consists of variations longer than five years. Note the extended period of El Niño events (negative SIO) in the 1980s and 1990s.

CONVOLUTION AS TRUNCATION OF AN INFINITE TIME SERIES

An observed time series, $x(t)$, can be considered a subset of an unlimited duration time series $g(t)$, obtained by convolving $g(t)$ with a rectangular window $h(t)$ of the form

$$h(t) = \begin{cases} 1 & 0 \leq t \leq T \\ 0 & \text{otherwise} \end{cases} \qquad \text{(A7.6)}$$

As illustrated in Figure A7.2, the series $x(t)$ can be defined as

$$x(t) = h(t)g(t) \qquad \text{(A7.7)}$$

It follows that the Fourier transform of $x(t)$ is the convolution of the Fourier transforms of $h(t)$ and $g(t)$, namely

$$X(f) = \int_{-\infty}^{\infty} H(\zeta)G(f - \zeta)d\zeta \qquad \text{(A7.8)}$$

In this case, multiplication in the time corresponds to convolution in the frequency domain, whereas in the previous case we examined convolution in the time domain (as with a running average) and multiplication in the frequency domain. These concepts are essential for the application of data windows in both the time and frequency domains.

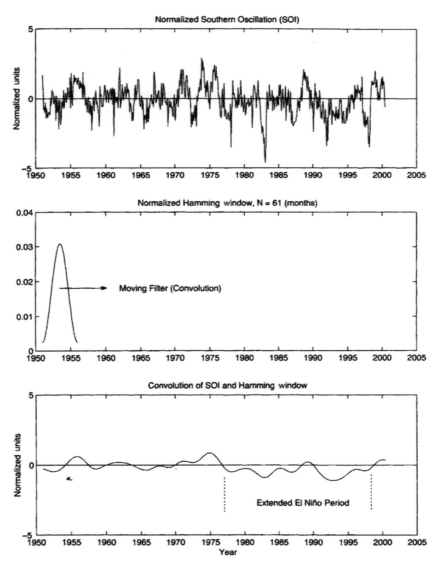

FIGURE A7.1 Convolution of the monthly time series of normalized Southern Oscillation Index (see U.S. government web site ftp.necp.noaa.gov/pub/cpc/wd52dg/data/indices/...) using a 61-month Hamming window (filter). Negative (positive) values of the index are associated with El Niño (La Niña) events. The convolution emphasizes the low-frequency variability of the El Niño-La Niña phenomenon in the equatorial Pacific.

FIGURE A7.2 Sampling a time series segment of duration T. The measurement is analogous to application of a rectangular window, $h(t)$, of amplitude 1.0 and duration T to an extensive time series $g(t)$.

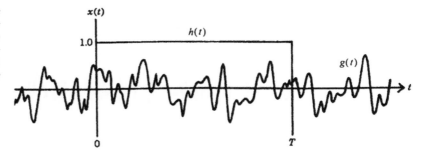

DECONVOLUTION

Deconvolution is the process of reversing (undoing) the smoothing that took place during application of the "data window", either in the time or frequency domains. It is assumed in this case that the response function is known and the process of deconvolution requires only a reverse of the process described above. Thus, the equation for deconvolution follows from that for convolution presented in Eqn (A7.1).

Index